BASIC ELECTRICITY FOR ELECTRONICS

BASIC ELECTRICITY FOR ELECTRONICS

Richard Blitzer

RCA Institutes, now called Technical Career Institutes, New York, New York
and
Rockland Community College, Suffern, New York

JOHN WILEY & SONS

New York • Chichester • Brisbane • Toronto

Library of Congress Cataloging in Publication Data:

Blitzer, Richard
Basic electricity for electronics.

1. Electronics. 2. Electric engineering.
I. Title.

TK7816.B55 621.3 73-20102
ISBN 0-471-08160-4

Printed in the United States of America

10 9

To my wife Connie, for everything, and to our granddaughter Lana who brings us great joy.

Preface

This book is intended for the beginning student in first- and second-semester courses for electronic technicians. Each chapter begins with a simplified approach and proceeds to a more thorough discussion. My purpose is to make the various topics in electricity and electronics *completely* understandable to the technical institute student, the junior college student, and the employee in company-training programs.

The mathematics used consists of algebra and right-triangle trigonometry. Where employed, the math is explained and illustrated with the same detail that I have found successful at the blackboard with thousands of students.

The various electronic topics are presented in a simple, logical manner, starting with the electron theory of matter and progressing, step by step, through electrical *circuits*. Each new topic builds upon the previous ones, slowly increasing the student's knowledge and understanding. The sequence and presentation of subjects are similar to that taught in the elementary terms of the electronic technician course at RCA Institutes.

Since I have a background of many years of teaching basic and advanced electronics, and years of electronic servicing, radar research laboratory engineering, and computer and guided-missile technical writing, I feel that I can "reach" the reader through this kind of presentation and stress the aspects of electrical theory that will do him the most good in his career as an electronic technician-engineer.

Fully worked-out examples follow *each group* of information within the chapter. Reference is then made to specific problems at the end of the chapter. In this way, the reader can immediately apply his knowledge of the information just learned. A cross-reference is also given within each group of end-of-chapter problems, referring to the chapter section number where this information is discussed. Examples and problems have been "class-tested" by use in laboratory experiments, on quizzes, and in exams.

I thank many of my fellow faculty members for their helpful suggestions. I particularly thank my son, Robert Blitzer, mathematics instructor at Miami-Dade Community College in Florida, for writing the "Math for Electronics" section. For excellent manuscript typing, my friend, Flora Dering, is hereby crowned "Queen of Secretaries." Also I am grateful to the many electronic companies who granted permission to use information and data. Credit is given where applicable.

Finally, I express heartfelt thanks and appreciation to my wife, Connie, for her work on the manuscript and for her patience, understanding, tolerance, and encouragement.

Richard Blitzer

Contents

Chapter 1 *Basic Theory of Electricity* 1

1-1. Matter 2
1-2. Elements and Compounds 2
1-3. The Atom 7
1-4. Valence Ring 11
1-5. Ions, Conductors, Insulators and Semiconductors 11
1-6. The Atom, Revisited 16
1-7. The Coulomb 17
1-8. Subshells 18
1-9. Electron Spin 18
1-10. Atomic Bonding 20
1-11. Ionic Bonding 20
1-12. Metallic Bonding 21
1-13. Covalent Bonding 23
1-14. The Electron and Energy 26

Chapter 2 *Electrical Units* 32

2-1. Voltage 32
2-2. Current 33
2-3. Resistance and Conductance 37
2-4. Prefixes 38
2-5. Ohm's Law, Relationship of Current to Voltage and Resistance 41
2-6. Ohm's Law, Relationship of Voltage to Current and Resistance 44
2-7. Ohm's Law, Relationship of Resistance to Voltage and Current 46
2-8. MKS System 48
2-9. Energy, Work, and Power 51

Chapter 3 *Series Resistor Circuits* 61

3-1. Series Current 61
3-2. Series Resistance 62
3-3. Series Voltages 63
3-4. Power Dissipation in Series Circuits 64
3-5. Voltages Across Series Resistors 68
3-6. Ground, Negative and Positive Voltages 70
3-7. Resistor Voltage Dividers, No Load 74
3-8. Series Filaments 79
3-9. Open Resistors in Series Circuits 81
3-10. More Than One Applied Voltage 84

Chapter 4 *Parallel Resistor Circuits* 96

4-1. Two Resistors in Parallel 96
4-2. Current Distribution in Two Parallel Resistor Circuits 103
4-3. More than Two Resistors in Parallel 105
4-4. Equal Resistors in Parallel 112
4-5. Defective Resistors in Parallel Circuits 113
4-6. Current Distribution in Two, Three, or More Parallel Resistors 116
4-7. Adding a Parallel Resistor To Produce the Desired $R_{\text{equivalent}}$ 120

Chapter 5 *Series-Parallel Combination Resistor Circuits* 126

5-1. Parallel-Series Circuits 126
5-2. Series-Parallel Circuits 131
5-3. More Complex Series-Parallel Circuits 141
5-4. Bridge Circuit 146
5-5. Resistor Voltage Divider with Loads 148
5-6. Effects of Open Resistors in Parallel-Series Circuits 153
5-7. Effects of Open Resistors in Series-Parallel Circuits 154

Chapter 6 *Resistor Network Circuits* 161

6-1. Resistance Between Two Points 161
6-2. Thévenin's Theorem 163
6-3. Norton's Theorem 166
6-4. Maxwell's Cyclic Currents, or Loop Currents 168
6-5. Nodal Analysis 173
6-6. Series Resistors, Two Voltage Sources 179
6-7. Superposition Theorem 180
6-8. Thévenin Theorem, Two- and Three-Voltage Sources 188
6-9. Norton Theorem, Two- and Three-Voltage Sources 193

6-10. Maxwell's Cyclic Currents, Two- and Three-Voltage Sources 201
6-11. Nodal Analysis, Two- and Three-Voltage Sources 204
6-12. Millman's Theorem, Two- and Three-Voltage Sources 217
6-13. Delta-to-Wye Transformation 222
6-14. Wye-to-Delta Transformation 229

Chapter **7** *Batteries* 243

7-1. The Primary Cell 243
7-2. Secondary or Rechargeable Cells 245
7-3. Batteries: Cells Connected in Series and Parallel 246
7-4. Internal Resistance 249
7-5. Testing a Battery 252

Chapter **8** *Resistors and Conductors* 254

8-1. Resistors 254
8-2. Conductors 260

Chapter **9** *Magnetism and Electromagnetism* 269

9-1. Natural Magnets 269
9-2. Magnetic Field 270
9-3. Magnetic Theory 273
9-4. Magnetic Materials 273
9-5. Electromagnetism in a Wire 278
9-6. Electromagnetism in a Coil 280
9-7. Strength of an Electromagnet 283
9-8. Practical Uses of Magnets and Electromagnets 284
9-8A. The Relay 285
9-8B. The Meter 286
9-8C. The Loudspeaker 287
9-8D. The Electric Motor Principle 288
9-8E. Electromagnetic Deflection of Electron Beam in CRT 289
9-9. The Earth's Magnetism 291
9-10. A Magnet 293
9-11. Magnetic Units 294
9-12. Flux 294
9-13. Magnetomotive Force 294
9-14. Flux Density 298
9-15. Magnetizing Force 300
9-16. Permeability, Permeance, and Reluctance 302
9-17. *B-H* Magnetization Curve and Hysteresis 310

Chapter **10** *DC Meters and Motor Principle* 325

10-1. The Permanent-Magnet Moving-Coil Meter 325
10-2. Ammeter, Multi-Range Conversions 327
10-3. Using the Multi-Range DC Ammeter 330
10-4. Voltmeter, Multi-Range Conversions 334
10-5. Using the Multi-Range DC Voltmeter 340
10-6. The Series Ohmmeter 345
10-7. Using the Ohmmeter 349
10-8. The DC Motor 351
10-9. Electrodynamometer 354
10-10. Ayrton Shunt 356
10-11. Wheatstone Bridge 361
10-12. Using the Wheatstone Bridge 364
10-13. Shunt Ohmmeter 366

Chapter **11** *Induced Voltage, Inductance, and Alternating Current* 380

11-1. Induced Voltage 380
11-2. Simple AC Generator 383
11-3. Inductance 387
11-4. Self-Inductance 387
11-5. Mutual Inductance and Coefficient of Coupling 388
11-6. The Simple Transformer 389
11-7. Transformer Currents 393
11-8. AC Voltage Sine Wave 397
11-9. Inductance, More Advanced 401
11-10. Self-Induced Voltage 405
11-11. Mutual Inductance and Coefficient of Coupling (Advanced) 407
11-12. Transformer (Advanced) 411
11-13. Transformer Impedance Matching 416
11-14. Simple Rotating Generators 419
11-15. AC Sine Waves 421
11-16. Average Value of Sine Wave 428
11-17. AC Voltmeters 432

Chapter **12** *Inductors and Resistors in DC and AC Circuits* 441

12-1. Inductor and Resistor in Series DC Circuit 441
12-2. Inductor in AC Circuit 445
12-3. Inductor and Resistor in AC Series Circuit 447
12-4. Inductor and Resistor in AC Parallel Circuit 458
12-5. L and R in DC Series Circuit (More Advanced) 462
12-6. Inductor and Resistor in AC Series Circuit (More Advanced) 469

12-7. Inductors and Resistors in AC Parallel Circuits (More Advanced) 477
12-8. Energy in a Coil 484
12-9. Voltage Square Waves and *LR* Circuits 487
12-10. Inductance and Internal Resistance 490

Chapter **13** *Capacitors and Resistors in DC and AC Circuits* 500

13-1. The Capacitor 501
13-2. Capacitors, Series and Parallel 510
13-3. Quantity (Q) of Charge on a Capacitor 515
13-4. RC Time Constant, Capacitor Charging 521
13-5. Capacitor with AC Applied Voltage 528
13-6. Capacitor and Resistor in Series with AC Voltage Applied 532
13-7. Capacitor and Resistor in Parallel, Phase Relationship 547
13-8. Uncharged Series Capacitors with DC Applied Voltage (More Advanced) 556
13-9. RC Time Constant (More Advanced) 571
13-10. Capacitive Reactance and Resistance in Series AC Circuits (More Advanced) 580
13-11. Capacitive Reactance and Resistance in Parallel AC Circuits (More Advanced) 583
13-12. Power and Power Factor in AC R and X_C Circuits 586
13-13. Capacitor Value 590
13-14. Capacitor Current 592
13-15. Voltage Square Waves and *RC* Circuits 596

Chapter **14** *Inductance, Capacitance, and Resistance in AC Circuits* 616

14-1. *L*, *C*, and *R* in Series Circuits 616
14-2. *L*, *C*, and *R* in Parallel Circuits 619
14-3. *L*, *C*, and *R* in Parallel-Series Circuits 622
14-4. Conductance, Susceptance, and Admittance 628

Chapter **15** *Resonance and Filters* 631

15-1. Series Resonant Circuits 631
15-2. Parallel Resonant Circuits 643
15-3. Filters 649

Chapter **16** *Introduction to Vacuum Tubes* 658

16-1. The Diode Vacuum Tube 658
16-2. Diode Circuit 662
16-3. The Triode Vacuum Tube 662

16-4. Triode Amplifier Circuit 664
16-5. Tetrode and Pentode Vacuum Tube 668
16-6. Tube Characteristics or Parameters 670

Chapter **17** *Introduction to Semiconductors* 674

17-1. Semiconductors 674
17-2. *P-N* Junction 677
17-3. The Transistor 681

Appendix **I** Mathematics for Electronics 687
Appendix **II** More Complete List of Prefixes 699
Appendix **III** Ohm's Law and Power Equation Chart for Resistor Circuits; also for *X* and *R* 700
Appendix **IV** Trigonometric Functions 702
Appendix **V** Exponential Functions—Values, and Common Logarithms 704
Appendix **VI** Color Codes, Resistors and Capacitors 710

Answers to Odd-numbered Problems 717

Index 723

BASIC ELECTRICITY FOR ELECTRONICS

chapter

1

Basic Theory of Electricity

In this part of the first chapter we present a simple introduction to electrical theory and electronics. Later, a more detailed discussion is given. The explanations are often only *theories.* As new discoveries in a field are made, theories often undergo changes. The electrical theories presented here are the accepted ones of today.

The term *electronics* comes from the word *electron,* which is the name of that unbelievably tiny particle found in all matter such as air, wood, metal, water, and the like. *Electron* is derived from the Greek *elektron* meaning amber. In ancient times, about 600 B.C., Thales of Miletus, a Greek, noticed that when a yellow resin called *amber* (elektron) was rubbed, it had the strange property of attracting, small light objects. *Electronics* today includes the fields of radio, television, computers, radar, guided missiles, space vehicles the automated factory, and the nuclear power plant.

William Gilbert an English physician and experimenter in the year 1600 named other substances that, when rubbed, exhibited the same attraction characteristics as the resin amber, or *elektron.* He called these electrics, from which the expressions *electricity* and *electrical* are derived. Electricity is simply a part of the larger field of electronics. Usually, the term *electricity* is used to describe the more fundamental or basic components and circuits, whereas *electronics* covers the more advanced and complex things in this field.

1-1. Matter

To learn electronics, an understanding of the actual construction of *matter* is first required. *Matter* may be defined as anything that occupies space and has mass or weight. This covers every known substance. The pages of this book, the ink on these pages, the clothes you are wearing, the chair on which you sit, even your yourself, are all *matter*. These examples of matter are solids. Other forms of matter are liquids such as water, and gases such as hydrogen or air. Matter often consists of combinations of several substances.

Refer to End-of-Chapter Problems 1-1 to 1-3

1-2. Elements and Compounds

When matter is made up of only one substance alone, such as iron, copper, or oxygen it is said to be an *element*. Tables 1-1 and 1-2 list the 103 elements known at the time of this writing, with additional ones being discovered from time to time. Table 1-1 shows an alphabetical listing of the elements with the chemical symbol and the *atomic number* of each. These terms are discussed shortly.

TABLE 1-1 The Elements Listed Alphabetically

Name of Element	Symbol	Atomic Number	Name of Element	Symbol	Atomic Number
Actinium	Ac	89	Calcium	Ca	20
Aluminum	Al	13	Calfornium	Cf	98
Americium	Am	95	Carbon	C	6
Antimony	Sb	51	Cerium	Ce	58
Argon	A	18	Cesium	Cs	55
Arsenic	As	33	Chlorine	Cl	17
Astatine	At	85	Chromium	Cr	24
Barium	Ba	56	Cobalt	Co	27
Berkelium	Bk	97	Copper	Cu	29
Beryllium	Be	4	Curium	Cm	96
Bismuth	Bi	83	Dysprosium	Dy	66
Boron	B	5	Einsteinium	Es	99
Bromine	Br	35	Erbium	Er	68
Cadmium	Cd	48	Europium	Eu	63

(*Continued*)

TABLE 1-1 (Cont.)

Name of Element	Symbol	Atomic Number	Name of Element	Symbol	Atomic Number
Fermium	Fm	100	Plutonium	Pu	94
Fluorine	F	9	Polonium	Po	84
Francium	Fr	87	Potassium	K	19
Gadolinium	Gd	64	Praseodymium	Pr	59
Gallium	Ga	31	Promethium	Pm	61
Germanium	Ge	32	Protactinium	Pa	91
Gold	Au	79	Radium	Ra	88
Hafnium	Hf	72	Radon	Rn	86
Helium	He	2	Rhenium	Re	75
Holmium	Ho	67	Rhodium	Rh	45
Hydrogen	H	1	Rubidium	Rb	37
Indium	In	49	Ruthenium	Ru	44
Iodine	I	53	Samarium	Sm	62
Iridium	Ir	77	Scandium	Sc	21
Iron	Fe	26	Selenium	Se	34
Krypton	Kr	36	Silicon	Si	14
Lanthanum	La	57	Silver	Ag	47
Lawrencium	Lw	103	Sodium	Na	11
Lead	Pb	82	Strontium	Sr	38
Lithium	Li	3	Sulphur	S	16
Lutetium	Lu	71	Tantalum	Ta	73
Mangesium	Mg	12	Technetium	Tc	43
Manganese	Mn	25	Tellurium	Te	52
Mendelevium	Md	101	Terbium	Tb	65
Mercury	Hg	80	Thallium	Tl	81
Molybdenum	Mo	42	Thorium	Th	90
Neodymium	Nd	60	Thulium	Tm	69
Neon	Ne	10	Tin	Sn	50
Neptunium	Np	93	Titanium	Ti	22
Nickel	Ni	28	Tungsten	W	74
Niobium	Nb	41	Uranium	U	92
Nitrogen	N	7	Vanadium	V	23
Nobelium	No	102	Xenon	Xe	54
Osmium	Os	76	Ytterbium	Yb	70
Oxygen	O	8	Yttrium	Y	39
Palladium	Pd	46	Zinc	Zn	30
Phosphorous	P	15	Zirconium	Zr	40
Platinum	Pt	78			

TABLE 1-2 Table of the Elements (in Order of Atomic Number)

Name of Element	Symbol	Atomic Number	Number of Electrons on Each Shell						
			K	L	M	N	O	P	Q
Hydrogen	H	1	1						
Helium	He	2	2						
Lithium	Li	3	2	1					
Beryllium	Be	4	2	2					
Boron	B	5	2	3					
Carbon	C	6	2	4					
Nitrogen	N	7	2	5					
Oxygen	O	8	2	6					
Fluorine	F	9	2	7					
Neon	Ne	10	2	8					
Sodium	Na	11	2	8	1				
Magnesium	Mg	12	2	8	2				
Aluminum	Al	13	2	8	3				
Silicon	Si	14	2	8	4				
Phosphorous	P	15	2	8	5				
Sulphur	S	16	2	8	6				
Chlorine	Cl	17	2	8	7				
Argon	Ar	18	2	8	8				
Potassium	K	19	2	8	8	1			
Calcium	Ca	20	2	8	8	2			
Scandium	Sc	21	2	8	9	2			
Titanium	Ti	22	2	8	10	2			
Vanadium	V	23	2	8	11	2			
Chromium	Cr	24	2	8	13	1			
Manganese	Mn	25	2	8	13	2			
Iron	Fe	26	2	8	14	2			
Cobalt	Co	27	2	8	15	2			
Nickel	Ni	28	2	8	16	2			
Copper	Cu	29	2	8	18	1			
Zinc	Zn	30	2	8	18	2			
Gallium	Ga	31	2	8	18	3			
Germanium	Ge	32	2	8	18	4			
Arsenic	As	33	2	8	18	5			
Selenium	Se	34	2	8	18	6			
Bromine	Br	35	2	8	18	7			
Krypton	Kr	36	2	8	18	8			
Rubidium	Rb	37	2	8	18	8	1		

(Continued)

TABLE 1-2 (Cont.)

Name of Element	Symbol	Atomic Number	Number of Electrons on Each Shell						
			K	L	M	N	O	P	Q
Strontium	Sr	38	2	8	18	8	2		
Yttrium	Y	39	2	8	18	9	2		
Zinconium	Zr	40	2	8	18	10	2		
Niobium	Nb	41	2	8	18	12	1		
Molybdenum	Mo	42	2	8	18	13	1		
Technetium	Tc	43	2	8	18	13	2		
Ruthenium	Ru	44	2	8	18	15	1		
Rhodium	Rh	45	2	8	18	16	1		
Palladium	Pd	46	2	8	18	18			
Silver	Ag	47	2	8	18	18	1		
Cadmium	Cd	48	2	8	18	18	2		
Indium	In	49	2	8	18	18	3		
Tin	Sn	50	2	8	18	18	4		
Antimony	Sb	51	2	8	18	18	5		
Tellurium	Te	52	2	8	18	18	6		
Iodine	I	53	2	8	18	18	7		
Zenon	Xe	54	2	8	18	18	8		
Cesium	Cs	55	2	8	18	18	8	1	
Barium	Ba	56	2	8	18	18	8	2	
Lanthanum	La	57	2	8	18	18	9	2	
Cerium	Ce	58	2	8	18	20	8	2	
Praseodymium	Pr	59	2	8	18	21	8	2	
Neodymium	Nd	60	2	8	18	22	8	2	
Promethium	Pm	61	2	8	18	23	8	2	
Samarium	Sm	62	2	8	18	24	8	2	
Europium	Eu	63	2	8	18	25	8	2	
Gadolinium	Gd	64	2	8	18	25	9	2	
Terbium	Tb	65	2	8	18	27	8	2	
Dysprosium	Dy	66	2	8	18	28	8	2	
Holmium	Ho	67	2	8	18	29	8	2	
Erbium	Er	68	2	8	18	30	8	2	
Thulium	Tm	69	2	8	18	31	8	2	
Ytterbium	Yb	70	2	8	18	32	8	2	
Lutetium	Lu	71	2	8	18	32	9	2	
Hafnium	Hf	72	2	8	18	32	10	2	
Tantalum	Ta	73	2	8	18	32	11	2	
Tungsten	W	74	2	8	18	32	12	2	

(Continued)

TABLE 1-2 (Cont.)

Name of Element	Symbol	Atomic Number	Number of Electrons on Each Shell						
			K	L	M	N	O	P	Q
Rhenium	Re	75	2	8	18	32	13	2	
Osmium	Os	76	2	8	18	32	14	2	
Iridium	Ir	77	2	8	18	32	15	2	
Platinum	Pt	78	2	8	18	32	17	1	
Gold	Au	79	2	8	18	32	18	1	
Mercury	Hg	80	2	8	18	32	18	2	
Thallium	Tl	81	2	8	18	32	18	3	
Lead	Pb	82	2	8	18	32	18	4	
Bismuth	Bi	83	2	8	18	32	18	5	
Polonium	Po	84	2	8	18	32	18	6	
Astatine	At	85	2	8	18	32	18	7	
Radon	Rn	86	2	8	18	32	18	8	
Francium	Fr	87	2	8	18	32	18	8	1
Radium	Ra	88	2	8	18	32	18	8	2
Actinium	Ac	89	2	8	18	32	18	9	2
Thorium	Th	90	2	8	18	32	18	10	2
Protactinium	Pa	91	2	8	18	32	20	9	2
Uranium	U	92	2	8	18	32	21	9	2
Neptunium	Np	93	2	8	18	32	22	9	2
Plutonium	Pu	94	2	8	18	32	23	9	2
Americium	Am	95	2	9	18	32	25	8	2
Curium	Cm	96	2	8	18	32	25	9	2
Berkelium	Bk	97	2	8	18	32	26	9	2
Californium	Cf	98	2	8	18	32	27	9	2
Einsteinium	Es	99	2	8	18	32	28	9	2
Fermium	Fm	100	2	8	18	32	29	9	2
Mendelevium	Md	101	2	8	18	32	30	9	2
Nobelium	No	102	2	8	18	32	31	9	2
Lawrencium	Lw	103	2	8	18	32	32	9	2

Matter that consists of two or more elements is called a *compound.* A very common compound with which we are familiar is plain table salt, made up of the elements sodium (Na) and chlorine (Cl), or written chemically NaCl. Water, another compound, consists of the elements hydrogen (H) and oxygen (O). The well-known chemical symbol with water, H_2O, simply means that it is made up of two parts of hydrogen and one part of oxygen.

If half the water in a glass is poured off, the remainder, of course, is half a glass of water. If, in turn, half of this is poured away, what is left is now a quarter of a glass of water. If this process is continuously repeated, the remaining substance is still water, except that less of it remains. Eventually a time occurs when the quantity of water is extremely tiny, barely visible, but it is still water. Now imagine that this tiny speck of water could still be cut in half, and although invisible to the naked eye, could be further subdivided again and again. Eventually, a point is reached where any further subsivision results in a substance that no longer has the properties of water. Just before this last subdivision takes place, a very small, invisible, particle of water remains that still retains the physical and chemical characteristics of water. This small particle is called a *molecule*. Now, if one more division occurs, that which is left is no longer water, but is now either hydrogen (H) or oxygen (O), depending on which was discarded. A *molecule* then is the smallest particle of any substance (such as the compound water) which retains its physical and chemical characteristics.

In a compound, a molecule consists of at least two atoms, one or more atoms of each element making up that compound. For example, a molecule of water consists of two hydrogen atoms and one oxygen atom. Hydrogen and oxygen are each elements. A molecule of common table salt (a compound) is made up of one atom of sodium and one atom of chlorine. Sodium and chlorine each are elements. A molecule of an element such as copper consists of only one atom of that element, but in another element such as hydrogen, a molecule consists of two atoms of this element.

Refer to End-of-Chapter Problems 1-4 to 1-9

1-3. The Atom

As discussed in the previous paragraph, an *atom* is the smallest part of an element that retains its characteristics, (see Table 1-1). The atom itself, strangely enough, consists principally of a tiny solar system, similar to our sun and its planets. This entire tiny solar system is less than a hundred millionth of an inch in size. Just as the planets Mercury, Venus, Earth, Mars, and the others in our system revolve around our sun, and countless other planets revolve around their suns in their systems, each atom acts as a complete solar system in itself. This atomic description was given by Bohr, the Danish scientist, in 1913.

As shown in Fig. 1-1*a*, the simplest atom is that of the element hydrogen. This atom consists of a center core called the *nucleus*, inside of which is a particle called the *proton*. Outside, some distance from the nucleus, is another particle called the *electron*, or *planetary electron*. The electron revolves around the nucleus just as the earth travels around the sun. The *path* or *orbit* that the electron takes in its revolutions around the nucleus is also called the *ring* or *shell*.

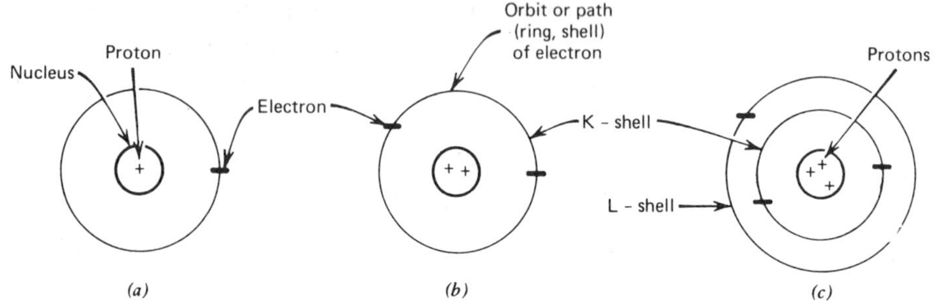

Figure 1-1. Simple atoms. (*a*) Hydrogen atom (atomic number 1). (*b*) Helium atom (atomic number 2). (c) Lithium atom (atomic number 3).

It has been found that the proton inside the nucleus, and the electron outside, have opposite charges. A positive sign has been assigned to the proton, while a negative sign has been given to the electron. These are shown in Fig. 1-1. There are numerous other particles in the nucleus, but these do not concern us in the study of electricity and electronics, and are not shown here.

The *atomic number* of each element shown in Tables 1-1 and 1-2 actually describes the number of planetary electrons revolving around the nucleus. There are normally the same number of protons as there are electrons. As shown in Fig. 1-1*a*, the atomic number of hydrogen is one. This means that an atom of hydrogen has one electron and also one proton.

Helium, with an atomic number of 2, has two electrons and also two protons. This is depicted in Fig. 1-1*b*. The element Lithium, with an atomic number of three, has three electrons and three protons, as illustrared in Fig. 1-1*c*.

Similar to our solar system where the planets move around the sun at different distances from the sun, some electrons revolve around the nucleus at different distances. Fig. 1-1*a* and *b* show an atom with only one orbit or shell, while Fig. 1-1*c* illustrates an atom with two shells. These shells, starting with the one nearest the nucleus, are called the K, L, M, N, O, P, and Q shells. Fig 1-1*c* depicts an atom of Lithium with three electrons. Two of these are revolving around the nucleus at the same distance, placing them on the *K shell*; the third electron, which is at a greater distance from the nucleus, is on the second ring or the *L shell*.

The number of shells or orbits depends on the atomic number (the number of electrons). Table 1-2 lists the elements in order of these numbers, and also shows the number of electrons on each shell. The following pattern may be seen: the maximum number of electrons that can be present on any shell is given by this relationship:

> *Maximum* number of electrons on any ring equals *twice* the *square* of the *number of that shell* (in its nearness to the nucleus).

This is usually written as the equation:

$$\text{maximum number} = 2\,(n^2) \tag{1-1}$$

where n is the number of the shell in its nearness to the nucleus. The K shell is closest to the nucleus and n would be 1. This means that the *maximum* number of electrons that could be "held" on the K shell is

$$\begin{aligned}\text{maximum number} &= 2\,(n^2)\\ &= 2\,(1^2)\\ &= 2 \text{ electrons}\end{aligned}$$

Notice that this agrees with the maximum number of electrons shown in Table 1-2 for the K shell, where no more than two are listed in this column.

Similarly, the L shell, the *second* ring, where n would be 2. can "hold" a maximum number of:

$$\begin{aligned}\text{maximum number} &= 2\,(n^2) \qquad (1\text{-}1)\\ &= 2\,(2^2)\\ &= 2\,(4)\\ &= 8 \text{ electrons}\end{aligned}$$

This, too, agrees with the maximum number of electrons shown in Table 1-2 for the L shell, where no more than eight are listed in this column. Figure 1-2 depicts an atom with its eight shells, and also shows the maximum number of electrons that could be on each.

Example 1-1

Calculate the maximum number of electrons that could appear on

(a) the M shell
(b) the N shell

Solution

(a) Since the M shell is *third* ring from the nucleus, n is 3 here, and

$$\begin{aligned}\text{maximum number} &= (n^2) \qquad (1\text{-}1)\\ &= 2\,(3^2)\\ &= 2\,(9)\\ &= 18 \text{ electrons}\end{aligned}$$

Notice that this is the *maximum* number of electrons listed in Table 1-2 in the M shell column, and also shown in Fig. 1-2.

Figure 1-2. A theoretical atom, showing the maximum number of electrons on each shell.

(b) Since the N shell is the *fourth* ring from the nucleus, n (in the equation) is 4 here, and

$$\begin{aligned} \text{maximum number} &= 2\,(n^2) \qquad (1\text{-}1) \\ &= 2\,(4^2) \\ &= 2\,(16) \\ &= 32 \text{ electrons} \end{aligned}$$

Notice that this is the *maximum* number of electrons listed in Table 1-2 in the N shell column, and also shown in Fig. 1-2.

For a Similar Problem Refer to End-of-Chapter Problems 1-10 to 1-14

1-4. Valence Ring

Just as the *maximum* number of electrons on any shell or ring of an atom is known (equation 1-1), it can also be seen from Table 1-2 that the *outermost* ring almost never "holds" more than 8 electrons. This outermost shell is called the *valence* ring. Very often most rules have exceptions, and there is one here too. The element *palladium* (Pd), atomic number 46, has 18 electrons on its outermost ring which is its N shell. This is the only atom with more than 8 electrons on its outermost ring.

Refer to End-of-Chapter Problems 1-15 and 1-16

1-5. Ions, Conductors, Insulators, and Semiconductors

When the outermost shell (valence ring) of an atom has eight electrons, it is said to be complete. As a result, this atom will not easily permit an electron to leave the atom, nor will it easily allow an electron to join those already part of the atom. Elements such as neon (atomic number 10), argon (18), krypton (36), xenon (54), and radon (86) each have eight electrons on their valence rings, as is shown in Table 1-2. As a result, these atoms are called *inert* because they will not readily give up or accept electrons.

When the number of valence electrons (those on the outermost ring) is small, that is, far less than the maximum number of eight, it seems that these valence electrons are not held too tightly by the nucleus. As a result, it is then possible to force the atom to lose one or more electrons, or even to accept some additional electrons. An atom of any element is normally neutral. That is, it has neither a positive nor a negative charge, since the number of protons (positive particles) is normally equal to the number of planetary electrons (negative particles).

If an atom is forced to lose one or more of its electrons, then this atom now has more protons in its nucleus than it has electrons rotating around the nucleus. The positive charges (protons) now outnumber the negative charges (electrons), and the atom is no longer neutral but has a positive charge. It is called a positive *ion*. Figure 1-3*b* depicts a simple positive ion. Similarly, if one or more additional electrons were added to a neutral atom, then there would

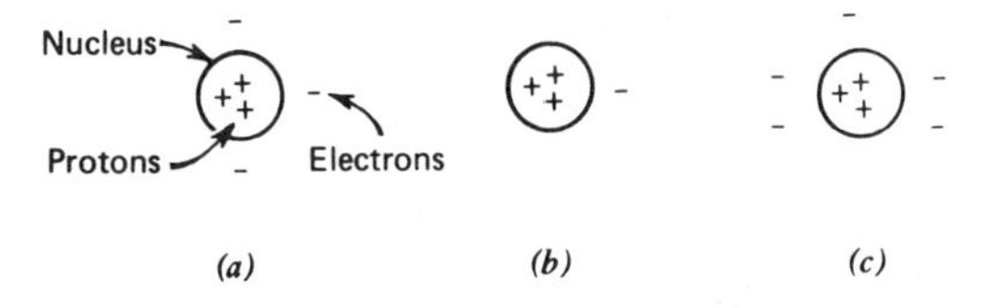

Figure 1-3. Simple atom and ions. (*a*) Neutral atom. (*b*) Positive ion (deficiency of electrons). (c) Negative ion (surplus of electrons).

be a greater number of negative charges (electrons) than positive particles (protons). The atom is no longer neutral, but it now negatively charged and is called a negative *ion*. Figure 1-3*c* shows a simple negative ion.

Electrons may be removed from an atom or added onto an atom in several ways, and they all involve expenditure of energy. One simple way that this may be achieved is by friction (mechanical and heat energy), and the result is called a *static* electric charge. When a person walks across a rug (friction) and then touches a metal object or another person, a tiny spark (and a slight electric shock) may occur. The rubbing or friction of the person's shoes against the rug may either remove electrons from the shoes, adding them to the rug, or do just the reverse. In either case, the person and his shoes become charged, having either gained or lost some electrons. Rubbing many dissimilar materials likewise produces charges on each. Rubbing a glass rod with a piece of silk cloth, or a hard rubber bar with fur, or a plastic comb rubbed briskly in one's hair, results in charging each of the materials, one negatively and the other positively. For example, glass rubbed with a silk cloth causes some electrons of the glass atoms to be removed. As a result, the glass rod becomes positively charged, and the silk gains electrons from the glass which adhere to the silk, so that the silk becomes negatively charged. Charges acquired in this way are called *static* charges.

Ions may also be formed as a result of a process that occurs in an electron tube (discussed in a later chapter) called *electron emission*. Here, because of heat energy, valence electrons are hurled or literally boiled off a heated metallic part of the tube. Some electrons strike atoms of gas inside the electron tube. Depending on the speed of the emitted electrons and the angle at which they collide with gas atoms, some electrons may become attached to gas atoms, forming negative ions. Other emitted electrons may not stick to the gas atoms but may knock off electrons that were originally part of the gas atoms. These atoms, having lost electrons, become positive ions.

Ions are also produced by chemical means (as we show later in this chapter) when atoms of one element combine with atoms of another element to form a compound

Conductors. When the valence ring (outermost ring) of an atom only contains a few electrons (far less than the complete number of 8), such as one, two, or three, the nucleus does not seem to hold these valence electrons too tightly, and they may easily be removed as discussed previously. These valence electrons are called *free* electrons since, with the application of very little persuasion in the form of energy, they are free to leave the atom. The energy may be the friction of rubbing, as described previously, or it could be electrical as a result of the forces of *attraction* and *repulsion*.

A study of electrostatics shows that a repelling effect or repulsion takes place between particles of similar polarity. That is, since all electrons are defined as negative particles, then two electrons placed near each other will attempt to repel one another. It is therefore said that *like charges repel*.

Particles having opposite charges, on the other hand, *attract* one another. A proton, for example, and an electron attract each other and would tend to move toward each other. It is said that *opposite charges attract.* The planetary electrons revolving around the nucleus are attracted to the protons within the nucleus. Because of the great speed at which the electrons are traveling in their orbits around the nucleus, they have high kinetic energies. Their centrifugal force would tend to have them fly off their orbits, leaving the atom. However, the electrostatic attraction between the orbital electrons and the nuclear protons prevents this.

When the number of valence electrons of an atom are few, say one, two, or three, they seem to be loosely held by the nucleus. These *free* electrons may easily be made to leave the atom. If another electron, or a negatively charged ion, is brought near an atom with free electrons, the electrostatic repulsion causes one or more of these free electrons to leave the atom. The same result may also be achieved if a positive ion were brought near an atom having free electrons. One or more of these free electrons would be attracted by the positively charged ion, and would leave the atom.

When an atom of some element (usually metallic) has only a few valence electrons, these free electrons may be made to easily move from one atom to an adjacent atom. If the movement of these free electrons between many atoms is in the *same* direction, it is called an *electric current* (to be discussed in the next chapter), and the (metallic) element which has these free electrons is called a *conductor.* Some elements have a greater number of free electrons per unit volume (such as a cubic centimeter) than others, making some conductors better than others. The approximate number of free electrons may be calculated knowing the weight of a cubic centimeter of some metallic element, its atomic weight (not given in Table 1-2), and the number of valence electrons per atom.

From Table 1-2, it may be seen that many metals only have one, two, or three valence electrons (which is far from the maximum number of eight), and these are therefore conductors. Silver (atomic number 47) with one valence electron, also has the greatest number of free electrons per unit volume, and is the best conductor. Copper (atomic number 29), also with one valence electron, has slightly fewer free electrons than silver, and is the next best conductor. Gold (atomic number 79) has one electron in its valence ring, and is next in order of conducting ability.

Table 1-3 lists several elements in the order of conducting ability, showing also the number of valence electrons on each atom, and the approximate value of each element's *resistance* or opposition to the movement of electrons from atom to atom (electric current). This resistance is measured in *ohms.* Note that the *resistance* of each conducting element is a very small decimal number. ($1.59 \times 10^{-6}\ \Omega$ is a method of writing $0.00000159\ \Omega$.)

Insulators. The opposite of a conductor is the *insulator* or *nonconductor.* When an atom has its outer ring (valence ring) complete, having eight elec-

TABLE 1-3 Metallic Elements in the Order of Conducting Ability

Element	Atomic Number	Valence Electrons	Resistance (Approximate) in Ohms per Cubic Centimeter at 20°C
Silver	47	1	[a]1.59×10^{-6}
Copper	29	1	1.72×10^{-6}
Gold	79	1	2.44×10^{-6}
Aluminum	13	3	2.82×10^{-6}
Magnesium	12	2	4.6×10^{-6}
Tungsten	74	2	5.6×10^{-6}
Molybdenum	42	1	5.69×10^{-6}
Zinc	30	2	5.79×10^{-6}
Cadmium	48	2	7.6×10^{-6}
Nickel	28	2	7.8×10^{-6}
Platinum	78	1	$10. \times 10^{-6}$
Iron	26	2	$10. \times 10^{-6}$

[a] 1.59×10^{-6} is 0.00000159.
1.72×10^{-6} is 0.00000172.
Etc.

trons on it, these valence electrons act as if they were tightly held by the nucleus. As a result, it is difficult to make an electron leave this atom. Some of the elements, as discussed previously, such as neon and argon have their valence rings complete with eight electrons. These elements are described as chemically *inert*, and resist having electrons removed from their atoms. Such elements are extremely poor conductors and are, therefore, good insulators except when sufficient voltage (difference of charge) causes a breakdown or ionization.

The insulators used in electrical circuit are usually compounds made up of two or more elements that have combined naturally, such as rubber (a combination of several elements including carbon and sulfur), or mica (a rock-forming mineral.) Even air is an insulator; with dry air being much better than moist or humid air. Many excellent insulators do not exist naturally, but are man-made. Among them are porcelain (a type of baked earth clay), bakelite (a baked combination of carbon, hydrogen, and oxygen), glass (sand heated to a liquid state, and then cooled), and numerous plastics called the *polymers* (such as polystyrene, vinyl, nylon, polyethylene, synthetic rubber, and many others).

Any *insulator* may develop into a *conductor* if sufficient force is applied. It is possible that an electron may be separated from an atom despite its valence ring being complete with eight electrons if a sufficiently large *positive charge* (a large number of + ions) is brought near the 8-valence-ring atom. The *difference* in *charge* between the neutral atom and the large positive group of ions is called a *difference of potential* or a *voltage*. When this *voltage* is large enough, it produces a *breakdown* of the insulator, and electrons can then leave the atoms of the material. When this occurs, the insulator is no longer doing its job. The opposition or resistance to the movement of electrons is measured in *ohms*. A theoretically perfect insulator would have an infinitely high resistance. In actual practice, any insulator offers a high resistance (not infinity) to the movement of electrons; the higher, the better is its insulating quality.

Table 1-4 lists some commonly used electrical insulating materials, the approximate *voltage breakdown rating* of each, and the approximate *resistance* in *ohms* when operating properly. Note that these *resistance* values are very large (5×10^8 means: 500,000,000). When a voltage is equal to, or exceeds the breakdown rating of the insulator, a spark occurs through the material. This damages the insulator and destroys or reduces any further insulation properties. The material has to be replaced.

TABLE 1-4 Common Electrical Insulators and Their Characteristics

Insulator	Voltage Breakdown (Approximate) per Mil (0.001 in.) Thickness	Resistance (Approximate) Ohms per Cubic Centimeter
Air, dry	19 to 23	
Porcelain	40 to 150	[a]5×10^8
Bakelite	300 to 550	[b]1×10^{12} to 1×10^{13}
Nylon	305	1×10^{13}
Glass	335 to 2000	1×10^{14} to 8×10^{14}
Vinyls	400 to 500	1×10^{14}
Rubber, hard	450	1×10^{12} to 1×10^{15}
Varnished cloth	450 to 550	
Polystyrene	508 to 760	1×10^{17}
Mica	600 to 1500	5×10^{13} to 2×10^{17}
Shellac	900	1×10^{16}
Polyethylene	1000	1×10^{17}
Paper, waxed, high quality	1250	

[a] 5×10^8 is 5 followed by 8 zeroes, or 500,000,000.
[b] 1×10^{12} is 1 followed by 12 zeroes, or 1,000,000,000,000.
Etc.

Semiconductors. Some elements called *conductors*, such as silver and copper, have an extremely *small* opposition to the movement of electrons from atom to atom. Some of these conductors are listed in Table 1-3. Copper, for example, has a resistance of $1.72 \times 10^{-6}\Omega$ per cubic centimeter, or 0.00000172 Ω. *Insulators*, on the other hand, usually compounds such as porcelain, bakelite, hard rubber, and the like (see Table 1-4) have tremendously *high* resistance to electric current. Bakelite, for example, has a resistance of about $1 \times 10^{12}\Omega$ per cubic centimeter, which is 1,000,000,000,000 Ω, or one trillion.

Certain elements called *semiconductors* offer resistance to electric current, but this is very much *smaller* than the *insulators* (which run into the millions of ohms, or billions, or larger), and much *larger* than the *conductors* (which run to a tiny fraction of an ohm, such as 0.00000159 Ω, which is 159 *trillionths* of an ohm.) Elements such as silicon (atomic number 14) and germanium (32), as shown in Table 1-2, have four electrons in the outermost ring (valence ring.) This is exactly one half of the maximum number of eight valence electrons. A good conductor has only a few valence electrons, while a good insulator has its valence ring either complete or almost complete. Silicon and germanium atoms, each with four valence-ring electrons, have a resistance in between that of the conductors and the insulators. For example, germanium has a resistance of approximately 60 Ω per cubic centimeter, while silicon is about a thousand times as much, or 60,000 Ω. In the following part of this chapter, and in a later chapter on transistors, a more thorough discussion is given on how a semiconductor element (silicon or germanium) plus another element added in tiny amounts are combined to make the semiconductor much more of a conductor in order to produce the amplification qualities of a transistor.

Refer to End-of-Chapter Problems 1-17 to 1-35

MORE ADVANCED THEORY OF ELECTRICITY

1-6. The Atom, Revisited

In Section 1-3, an elementary description of the atom is given, as shown in Fig. 1-1. The central core of the atom, or the nucleus, for most atoms, contains the proton and other particles too.The presently accepted theory and knowledge of the atom can account for more than 30 different particles. Among them are electrons (not the planetary ones), neutrons, neutrinos, several kinds of mesons, hyperons, and numerous others.

The mass of some of the atomic particles is thought to be known: a proton has a mass of 1.672×10^{-27} kilograms (1 kg = 35.27 oz): a neutron has a mass approximately the same as the proton, or about 1.675×10^{-27} kg; an electron is much lighter, having a mass of about 9.11×10^{-31} kg. (*Note:* 1.672×10^{-27} kg means that the decimal point is actually 27 places to the *left* of its position between the 1 and the 6, or that there are 26 zeroes before the 1672, or 0.000 000 000 000 000 000 000 000 00 1672 kg.)

Despite the fact that the electron has so much less mass than the proton, the electron is approximately the same size. The diameter of an electron has been variously estimated at somewhere between 1.4 and 2.5 $\times$ 10^{-13} centimeters (1 cm = 0.3937 in.).

Refer to End-of-Chapter Problems 1-36 to 1-38

1-7. The Coulomb

The charge on an electron is assumed to be negative and is measured as 1,601 $\times$ 10^{-19} *coulomb*, or

$$1 \text{ electron} = 1.601 \times 10^{-19} \text{ coulomb} \quad (1\text{-}2)$$

From the above, a coulomb may be found, using simple algebra:

$$\frac{1 \text{ electron}}{1.601 \times 10^{-19}} = 1 \text{ coulomb}$$

$$0.624 \times 10^{19} = \text{electrons} = 1 \text{ coulomb}$$

or

$$6.24 \times 10^{18} \text{ electrons} = 1 \text{ coulomb} \quad (1\text{-}3)$$

A *coulomb*, then, is this large number of electrons, and has been named to honor Charles A Coulomb who, in 1785, proved the relationship of magnetic and electric attraction with distance.

The charge on a proton is opposite in polarity to that of an electron, and is expressed as a positive charge. This charge is equal to that of the electron, and the *coulomb* is also employed to express the proton's charge:

$$1 \text{ proton} = 1.601 \times 10^{-19} \text{ coulomb} \quad (1\text{-}4)$$

and

$$1 \text{ coulomb} = 6.24 \times 10^{18} \text{ protons} \quad (1\text{-}5)$$

The letter Q represents the *quantity* of electrons or protons in *coulombs*, or

$$1Q \text{ (in coulombs)} = 6.24 \times 10^{18} \text{ electrons (or protons)} \quad (1\text{-}6)$$

Example 1-2

How many *surplus* electrons are on a body which is negative and has a charge of three coulombs?

Solution

$$Q = 6.24 \times 10^{18} \text{ electrons} \quad (1\text{-}3)$$

and

$$3Q = 3(6.24 \times 10^{18})$$

$$3Q = 18.72 \times 10^{18} \text{ electrons}$$

Refer to End-of-Chapter Problems 1-39 to 1-42

1-8. Subshells

In the previous description of the physical makeup of the atom (Section 1-3), it is shown that the planetary electrons revolve around the nucleus at various distances from the nucleus. Some electrons travel an orbit close to the nucleus; others on different orbits or shells travel further from the nucleus. The number of electrons on each shell of the atoms of the various elements are given in Table 1-2, with the theoretical *maximum* number on each shell shown in Fig. 1-2.

Actually, delving a little deeper into the Bohr atom, the theory is that the shells really consist of *subshells*. Each shell, except the one nearest to the nucleus, consists of two or more paths or subshells. The *first* shell (K) nearest the nucleus, is made up of only *one* path or orbit. The *second* shell (L) is comprised of *two* orbits or subshells, one just slightly further from the nucleus, as shown in Fig. 1-4. Similarly, the *third* shell (M) is made up of *three* orbits or subshells. This same regularity continues for all the remaining shells; the *fourth* shell (N) is really *four* subshells; the *fifth* (O), *five* subshells; the *sixth* (P), *six* subshells; and finally the *seventh* (Q) with *seven* subshells. This is depicted in Fig. 1-4.

Note that in this more detailed drawing of the atom, *all* the subshells of *one* particular group are referred to as *that* particular shell or ring. For example, in Fig. 1-4 the *four subshells* of the fourth ring (the N shell) are usually simply referred to as the *N shell*. Note also that the first or K shell has only one subshell, and has a maximum of two electrons. The first or *inner* subshell of *each* of the *groups* similarly can hold a maximum of two electrons. The next outer subshell of each group can hold a maximum of *six* electrons (*four* more than the previous subshell). Each of the next following subshells can hold four more than its predecessor. As an example, in Fig. 1-4 note that the *five subshells* of the fifth ring, or *O shell*, starting with the innermost subshell, hold maximum numbers of electrons of 2, 6, 10, 14 ,and 18, respectively. The total *theoretical* maximum then for this O shell is 50 electrons, although no element has been discovered yet having more than 32 electrons in its O shell (as shown in Table 1-2).

Refer to End-of-Chapter Problems 1-43 to 1-45

1-9. Electron Spin

The electron revolving around the nucleus is called a *planetary* electron, named, of course, because of its similarity to a planet traveling around the sun. Theory also has been suggested that the tiny electron is similar to the planet in even another respect. Like the planet, the electron also spins or rotates on its own axis while revolving around the nucleus. This *spin* of the electron on its own axis helps explain the theory of magnetism (discussed in a later chapter).

Refer to End-of-Chapter Problem 1-46

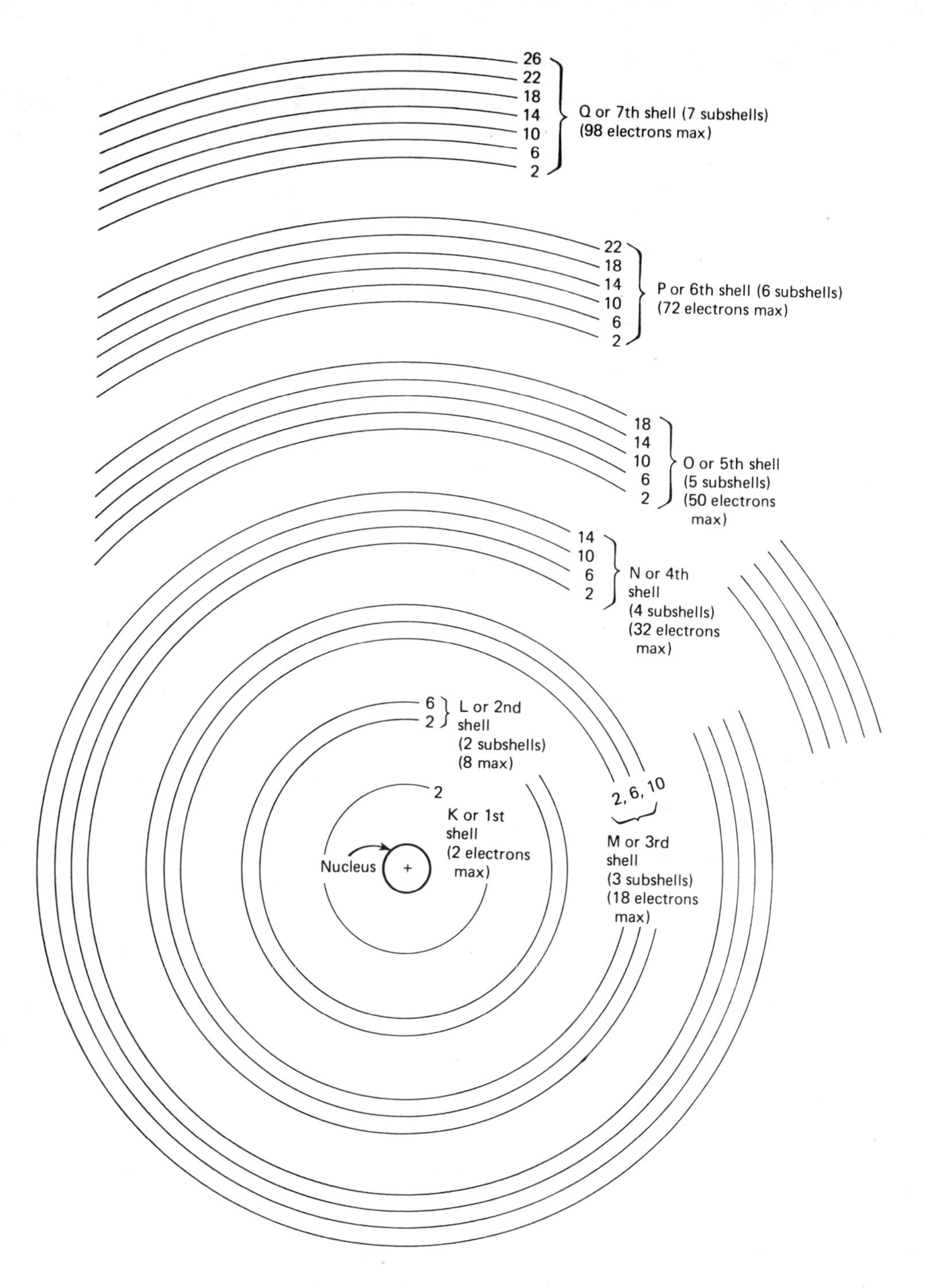

Figure 1-4. The atom, showing shells and subshells.

1-10. Atomic Bonding

In Sections 1-4 and 1-5 it is shown that the outermost ring or shell (the *valence* ring) acts as if it were complete when eight electrons are present on it. Such an atom does not easily combine with other atoms. When the valence ring holds only one or two electrons, these are easily "lost" by the atom, leaving a positively charged atom (+ ion). If a valence ring is not quite complete, it will allow an electron or two to join the shell, making it complete. This atom is now a negatively charged atom (− ion). When atoms of one or more elements combine, it is called *atomic bonding*. Three types are: *ionic bonding, metallic bonding, and covalent bonding.*

Refer to End-of-Chapter Problems 1-47 to 1-48

1-11. Ionic Bonding

Ionic bonding explains how atoms of *different* elements cling together to form a compound. In Section 1-5 it is shown that when an atom gains one or more electrons, the atom is no longer uncharged or neutral, but now becomes a *negative ion*. Similarly, if an atom has been forced to give up one or more electrons, this atom now becomes a *positive ion*.

The compound *sodium chloride* (NaCl), which is the chemical name for ordinary table salt, is made up of the elements *sodium* (Na) and *chlorine* (Cl). As shown in Fig. 1-5*a* and Table 1-2, sodium has atomic number of 11. With 11 protons in its nucleus, it has 2 electrons on its K shell, 8 (2 + 6) on its L subshells, and only one on its M shell. Since the number of electrons (11) is equal to the number of protons, the sodium atom is neutral.

In Fig. 1-5*b* and Table 1-2, chlorine is shown having an atomic number of 17, with 17 protons in the nucleus and 17 electrons revolving around the nucleus. These planetary electrons, as depicted in Fig. 1-5*b*, are arranged 2 on the K shell, 8 (2 + 6) on the L subshells, and 7 (2 + 5) on the M subshells. With equal numbers of protons (17) and planetary electrons (17), the chlorine atom is neutral.

The one electron on the outermost shell of the sodium atom (Fig. 1-5*a*) is not held very tightly by the nucleus, and can easily leave the atom (Fig. 1-5*c*) to join the outer shell of the chlorine atom. The chlorine atom normally has seven valence electrons (2 + 5) on its outermost shell (Fig. 1-5*b*), and by accepting an electron from the sodium atom, the chlorine now has its outer ring complete with eight electrons (Fig. 1-5*d*).

Since the sodium atom has lost an electron (Fig. 1-5*c*), this atom is now a +ion. The chlorine atom has gained an electron (Fig. 1-5*d*) and is now a −ion. The attraction between the −ion (Chlorine) and the +ion (sodium) keeps the two atoms joined together as a molecule of the new combination *sodium chloride*, and the process is called ionic bonding.

Refer to End-of-Chapter Problems 1-49 to 1-51

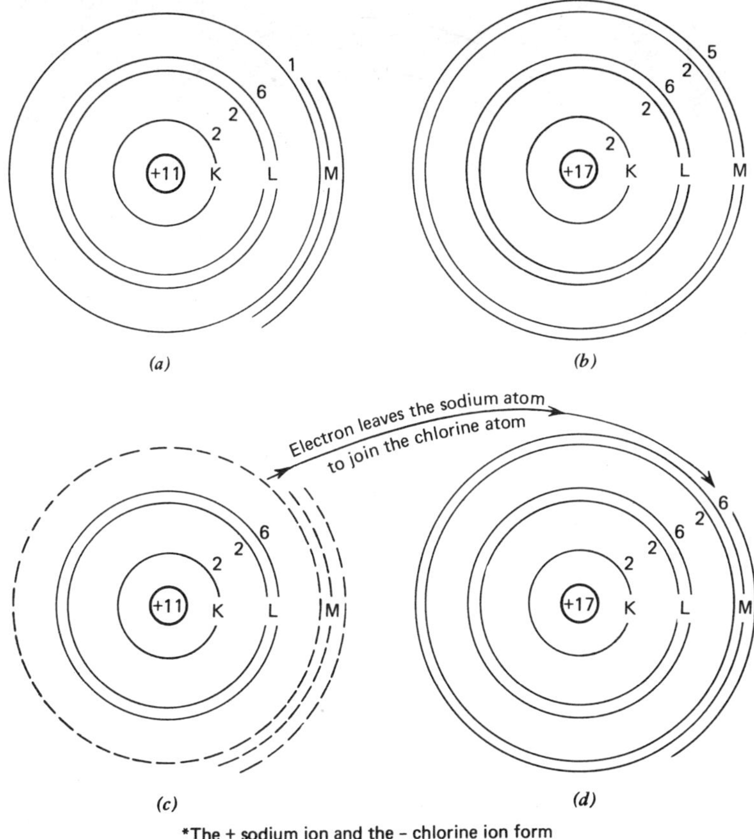

*The + sodium ion and the - chlorine ion form a molecule of sodium chloride.

Figure 1-5. Ionic bonding. (*a*) Sodium atom, neutral. (*b*) Chlorine atom, neutral. (*c*) Sodium atom has lost an electron from its M shell, and is now a + ion.* (*d*) Chlorine atom has gained an electron on its center M subshell, and is now a − ion.* (*The + sodium ion and the − chlorine ion form a molecule of sodium chloride.)

1-12. Metallic Bonding

Another form of cohesion between atoms is called *metallic bonding*, which accounts for the atoms of a metal clinging together. Taking silver as an example, the atomic number (Table 1-2) is 47, meaning, of course, that an atom of silver has 47 protons in the nucleus and 47 planetary electrons revolving around outside the nucleus. The electron distribution listed in Table 1-2 and shown in Fig. 1-6 is 2, 8, 18, 18 and 1 in the K to O shells, respectively. Note that there is only one electron in the outermost shell (valence ring). This electron is not held very tightly by the nucleus and, as a result, the valence electron from one atom can easily move to an adjacent atom. This is especially

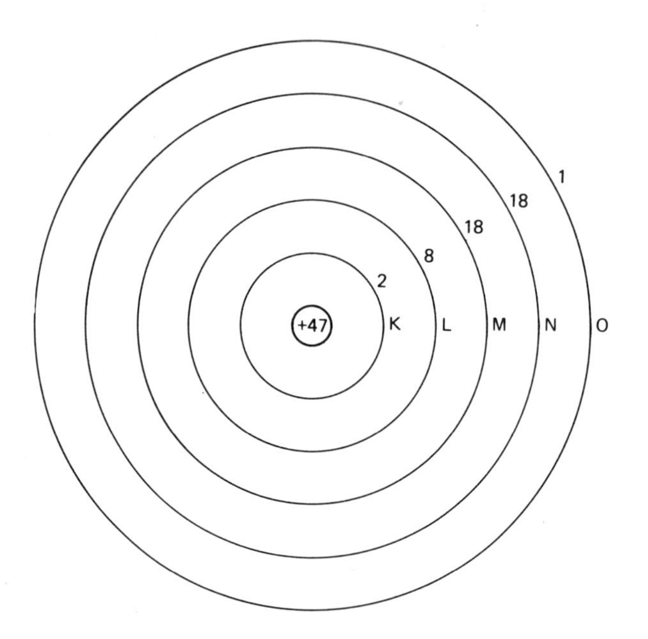

Figure 1-6. An atom of silver.

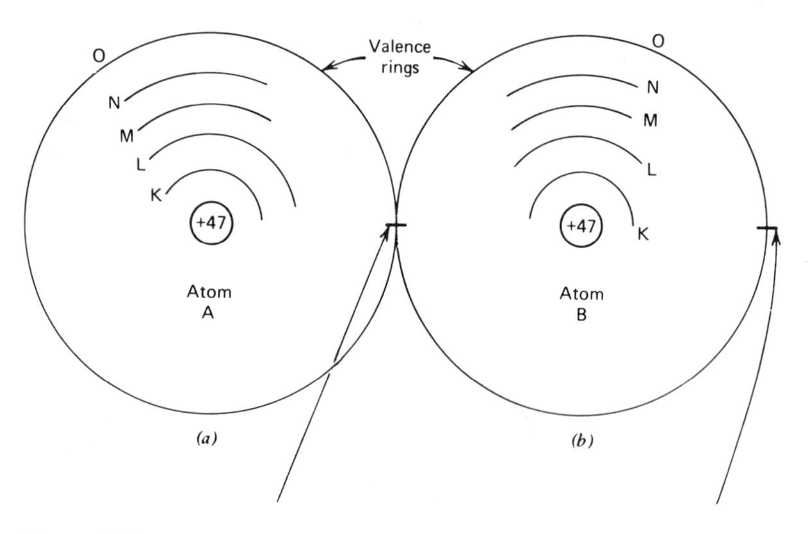

Figure 1-7. Metallic bonding of silver atoms. (*a*) This valence electron, originally part of atom *A*, enters the valance ring of atom *B*. This makes atom *A* positively charged (+ ion). (*b*) This valence electron, originally part of atom *B*, is forced off atom B by the electron coming from atom *A*. The *B* electron leaves atom *B* and is a ''free'' electron, becoming part of a cloud of drifting electrons.

true when a valence electron is midway between the nuclei of two adjacent atoms, as depicted in Fig. 1-7. When an electron leaves silver atom A, joining atom B, A becomes a +ion. The effect of the new electron entering atom B is to cause a valence electron to leave atom B. This "free" electron and others from numerous atoms form an electron cloud, or a negatively charged body. The *attraction* between the +ions (atoms which have lost valence electrons) and the electron clouds, and the *repulsion* between each +ion, results in the atoms becoming arranged in a fixed pattern producing the seemingly solid silver metal.

Refer to End-of-Chapter Problems 1-52 to 1-54

1-13. Covalent Bonding

Atoms of some elements which have four valence electrons adhere to each other as a result of a phenomenon called *covalent bonding*. This is the sharing of valence electrons (those on the outermost shell) by adjacent atoms. Certain elements such as carbon, silicon, and germanium can exist in a crystal-lattice form. A very small part of the crystal-lattice is shown in Fig. 1-8. Four atoms,

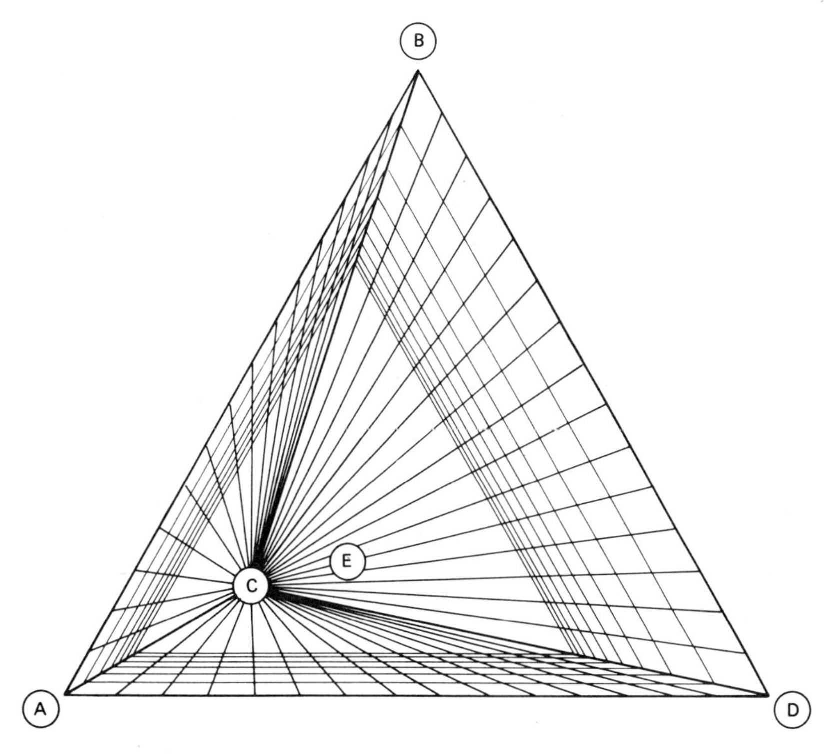

Figure 1-8. Atom *E* surrounded by four other atoms (*A*, *B*, *C*, and *D*) in a tetrahedral-shaped three-dimensional configuration.

A, B, C, and D, are arranged so that each is located at a corner of a three-dimensional tetrahedral-shaped form. A tetrahedron is a four-sided figure, sort of a pup tent where the two sides slope to a common point at the ground. In the drawing, one side, at the left, is the flat plane made up of the spaces between atoms A, B, and C. A second side, nearest the viewer, is between atoms A. B. and D. A third side, at the rear, is between B, C, and D. Finally, the fourth side, the bottom, is between atoms A, C, and D. Using the pup-tent analogy, the viewer is looking into the tent from its open end, in Fig. 1-8.

At the center, between the four corner atoms, is another atom, E. In Fig. 1-9, each of these atoms, A, B, C, D, and E is shown in a flat two-dimensional drawing. From Table 1-2, carbon (atomic number 6), silicon (14), and germanium (32) each are listed as having four valence-ring electrons. Notice that in Fig. 1-9, the center atom, E, is surrounded by atoms A, B, C, and D. Only the valence rings are shown. Atom E has its own four valence electrons, plus one more from *each* of the surrounding atoms, A, B, C, and D. Therefore, atom E *seems* to have a total of eight valence electrons, which is the maximum number on an outermost shell. As a result, this valence ring of atom E acts as if it were complete.

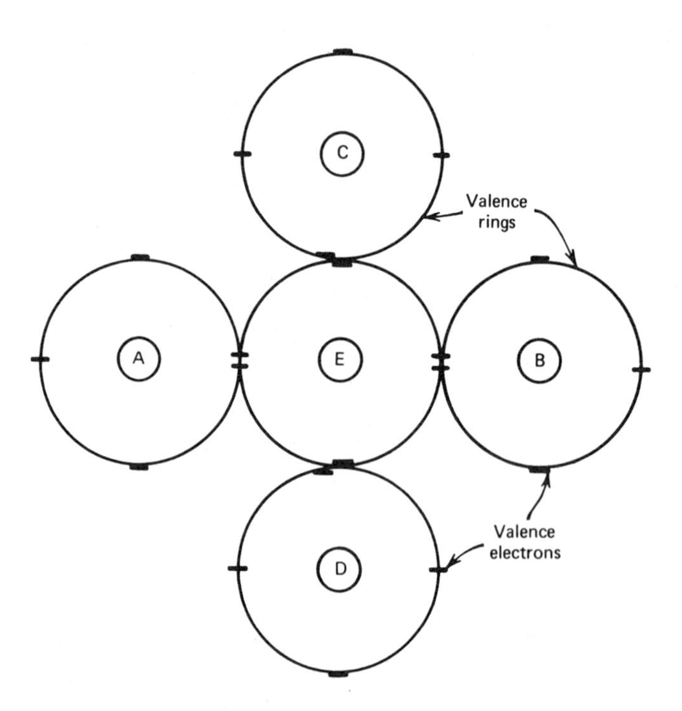

Figure 1-9. Covalent bonding. Atom *E* has eight valence-electrons; its own four, plus one each from atoms *A*, *B*, *C*, and *D*.

The group of four atoms, A, B, C, and D surrounding atom E of Figs. 1-8 and 1-9 constitute only a tiny portion of the crystal-lattice configuration. Actually, there are countless other tetrahedral formations, with every atom A, B, C, D, etc., *each* being the center atom of its own tetrahedron grouping. As a result, each atom seemingly has its valence ring complete with *eight* electrons (its own four; plus one from each of the four surrounding atoms). This is again depicted in the two-dimensional drawing of Fig. 1-10. Here, the original atoms A, B, C, and D of Figs. 1-9 and 1-10 are again shown surrounding atom E, but atoms A, B, C, and D are likewise *each* surrounded by others.

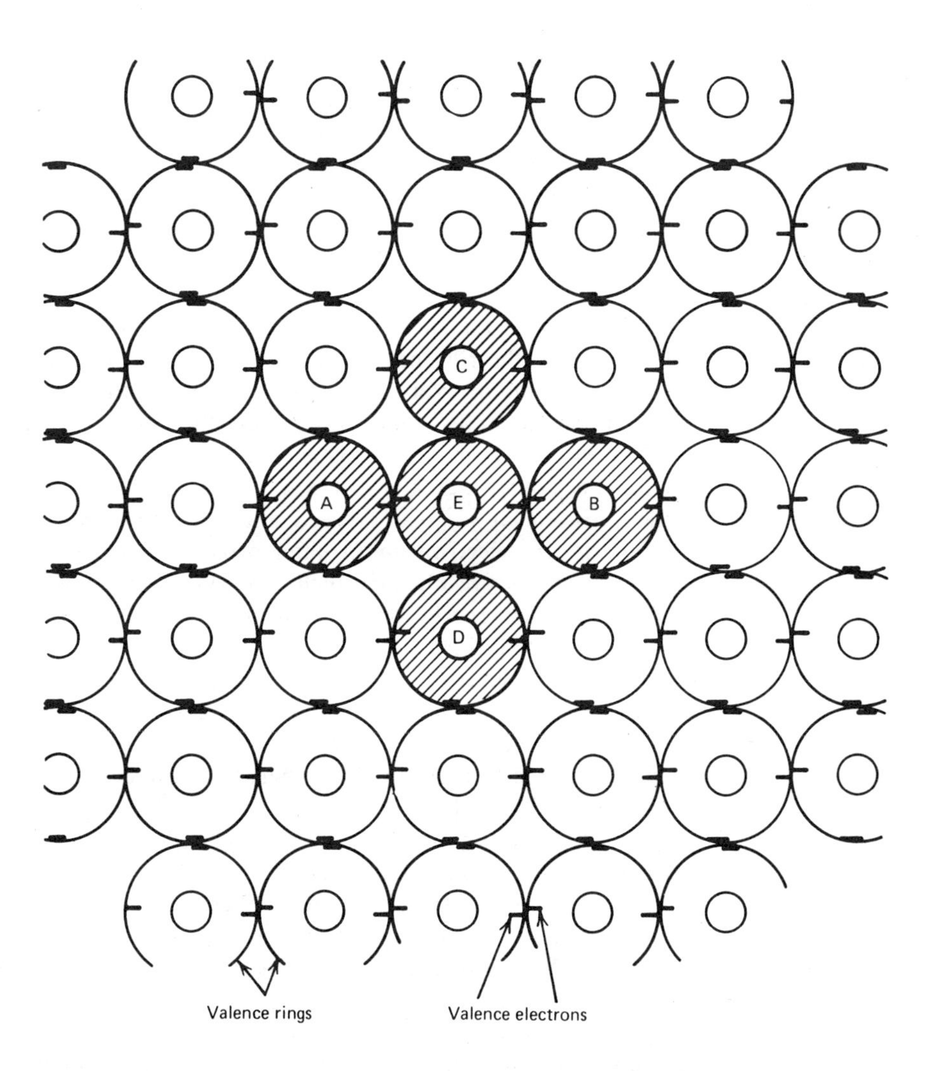

Figure 1-10. Crystal-lattice configuration of covalent bonding.

With each atom having its valence ring complete, the atoms of the element become stable and remain locked in this condition.

Refer to End-of-Chapter Problems 1-55 to 1-58

1-14. The Electron and Energy

The electron, as has been shown previously, is a particle of matter, having dimensions and mass. However, it leads a double life, sort of a Jekyl and Hyde existence. The electron is also said to act as a *wave of electromagnetic energy*. The term *energy* means having the *ability* to do work. The movement of the electron traveling in its orbit around the nucleus produces a *kinetic energy* in the electron. This is the energy due to motion. In Chapter 2 a discussion of energy is given. It is shown that the amount of energy is measured in units called *joules*, and in much smaller units called *electron volts*. The amount of kinetic energy present in an electron is directly proportional to the speed or velocity of the electron in its orbit around the nucleus.

Electrons on the shell furthest from the nucleus travel at a faster speed than electrons on the ring nearest the nucleus. This may be likened somewhat to two pennies on a revolving record turntable, with one penny placed near the center spindle, and the second one near the outer edge of the turntable. Both coins, of course, make the same number of revolutions per minute, but the outer one covers a greater distance in making a complete trip. The outer coin, then, travels at a greater speed and, therefore, has more kinetic energy. Similarly, the electron on an outer orbit has greater energy than one on an inner orbit. As a result, each shell may be thought of as different energy levels, with minimum energy on the first shell, and progressively higher levels on each succeeding ring.

The energy level of each ring is in a definite amount of electron volts. An electron will orbit only at definite distances from the nucleus, but not *in between* a pair of rings. The area between rings is called a *forbidden energy gap*, while the area of a group of subshells is referred to as an *energy band*. This is depicted in Fig. 1-11.

An electron may have its energy increased in one of several ways. Heat or *thermal energy* consists of particles called phonons. The energy of these phonons is proportional to the temperature. Light rays or *radiant energy* consists of electromagnetic energy particles called *photons*, and the energy of these photons depends on the *frequency* or the *wavelength*.

When an electron collides with a *phonon* of heat energy, or with a *photon* of light, additional energy is acquired by the electron. If the new energy of the electron is now equal to that of the next higher energy band, the electron will

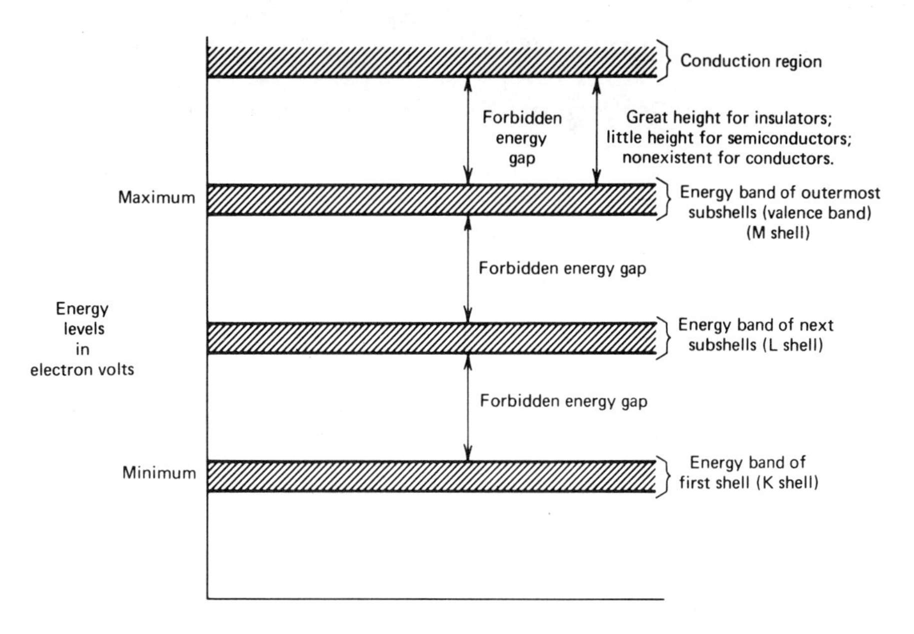

Figure 1-11. An example of the energy bonds of an atom.

momentarily leave its orbit and join the next outer shell. After a brief instant, the electron falls back into its normal orbit, giving up its additional energy either in the form of heat, or in the form of light. The radiation given off by the electron as it loses energy is an electromagnetic wave. As a result of this action, the electron is said to act as a particle and also as a wave of electromagnetic energy.

If, as shown in Fig. 1-11, an electron in the valence band acquires sufficient energy to cause it to jump across the forbidden energy gap into the *conduction region* (also called the *conduction band*), the electron may now enter another atom.

The width or *height* of the *forbidden energy gap* between the *valence energy band* and the *conduction region* of Fig. 1-11 is large for insulators, smaller for semiconductors, and practically nonexistent for conductors. In conductors, the upper edge of the valence energy band, and the lower edge of the conduction region actually overlap. The result is that valence electrons at ordinary room temperature acquire sufficient energy to jump from the valence shell to the conduction region.

Refer to End-of-Chapter Problems 1-59 to 1-69

PROBLEMS

See section 1-1 for discussion covered by the following.

1-1. Define *matter*.
1-2. Give three examples of matter which is visible.
1-3. Give an example of matter which is invisible.

See section 1-2 for discussion covered by the following.

1-4. Give three example of *elements*.
1-5. Define a *compound*
1-6. Give two examples of compounds, and state the elements that form these compounds.
1-7. Define a *molecule*.
1-8. Define an *atom*.
1-9. How many elements are presently known?

See section 1-3 for discussion covered by the following.

1-10. Draw a simple atom and identify each part.
1-11. What polarity has been assigned to (a) the electron, (b) the proton?
1-12. What does the *atomic number* of an element signify?
1-13. What is the equation predicting the maximum number of electrons on any ring or shell of an atom?
1-14. What is the theoretical maximum number of electrons on the fifth or "O" ring?

See section 1-4 for discussion covered by the following.

1-15. What is the outermost ring of an atom called?
1-16. What is the maximum number of electrons that could appear on the valence ring?

See section 1-5 for discussion covered by the following.

1-17. When an atom has lost one or more electrons, what is the polarity of the charge remaining on the atom?
1-18. What is the atom of the previous question called?
1-19. When one or more electrons are added to an atom, what is the polarity of the charge on this atom?

1-20. Name three methods whereby electrons may be removed from, or added to, an atom?
1-21. When the valence ring of an atom contains only one or two electrons, are they held tightly or loosely by the nucleus?
1-22. What are free electrons?
1-23. Do similar or *like* charges attract or repel each other?
1-24. Do opposite or *unlike* charges attract or repel each other?
1-25. When the atoms of some element have few free electrons, is this element a conductor or an insulator?
1-26. When the atoms of an element have many free electrons, is this element a conductor or an insulator?
1-27. Name three elements which are good conductors.
1-28. What makes up an electric current?
1-29. When the valence ring of an atom is complete with eight electrons, or almost complete with six or seven electrons, are these electrons loosely or tightly held by the atom?
1-30. Name three good insulator materials.
1-31. How might an insulator become a conductor?
1-32. What is a *voltage*?
1-33. What does *voltage breakdown* of an insulator mean?
1-34. When the atoms of some element have four electrons (half of the maximum number of eight) on the valence ring, is this element a conductor, semiconductor, or insulator?
1-35. Name two elements that are commonly used in the type of material described in the previous question.

More Challenging Questions: See section 1-6 for discussion covered by the following.

1-36. Name another particle, other than the electron and the proton, that is present in an atom.
1-37. Which is heavier, a proton or an electron?
1-38. Which is larger, a proton or an electron?

See section 1-7 for discussion covered by the following.

1-39. What is a *coulomb* in terms of electrons?
1-40. What is a *coulomb* in terms of protons?
1-41. What does the letter Q represent?
1-42. How many surplus electrons are present on a body which has a negative charge and is represented as -2.5 coulombs?

See section 1-8 for discussion covered by the following.

1-43. What are subshells of atoms?
1-44. How many subshells comprise the fifth ring or orbit?
1-45. What are the maximum number of electrons on each subshell?

See section 1-9 for discussion covered by the following.

1-46. Describe two ways in which an electron behaves as the planet Earth.

See section 1-10 for discussion covered by the following.

1-47. What is atomic bonding?
1-48. Name three forms of atomic bonding.

See section 1-11 for discussion covered by the following.

1-49. Describe briefly what occurs in *ionic bonding.*
1-50. Give an example of two elements united by ionic bonding.
1-51. In ionic bonding, why do the atoms of the two different elements adhere to each other?

See section 1-12 for discussion covered by the following.

1-52. What polarity is the cloud of free electrons produced in metallic bonding?
1-53. What polarity is the atom that has lost an electron to an adjacent atom in metallic bonding?
1-54. Briefly describe what occurs in metallic bonding.

See section 1-13 for discussion covered by the following.

1-55. In covalent bonding, each atom is surrounded by four others in a *tetrahedral* formation. Describe or draw this shape.
1-56. How many valence electrons must each atom have to produce covalent bonding?
1-57. In covalent bonding, how many valence electrons does each atom *act* as if it actually had?
1-58. Briefly describe the sharing of valence electrons by the atoms in covalent bonding.

See section 1-14 for discussion covered by the following.

1-59. What does the term *energy* mean?

1-60. What is *kinetic* energy?

1-61. What produces kinetic energy in an electron?

1-62. Which electrons have greater energy, those further from the nucleus, or those nearer the nucleus?

1-63. Why is the correct answer to the previous question true?

1-64. What is a *forbidden-energy gap*?

1-65. If an orbiting electron collides with a particle of heat energy called a *phonon*, or with a particle of light energy called a *photon*, what may happen to the energy of the electron?

1-66. If an orbiting electron acquires sufficient additional energy, what may happen to the electron?

1-67. What occurs when an electron falls from an outer orbit to an inner shell?

1-68. Does a conductor, an insulator, or a semiconductor have a large difference in energy levels (forbidden energy gap) between the valence energy band and the conduction band?

1-69. Which, the conductor, the insulator, or the semiconductor, has no difference in energy levels (forbidden energy gap) between the valence band and the conduction band?

chapter

2

Electrical Units

In order to explain thoroughly electric current and what affects it, we discuss in this chapter the parts or components that make up circuits, the symbols for these parts, and the units of measurement used in electrical and electronic theory.

2-1. Voltage

When some electrons have been removed from an uncharged or neutral body, those atoms that have lost electrons are now positively charged and are called +*ions* (as discussed previously in Section 1-5). Similarly, when some electrons have been added to an uncharged or neutral body, those atoms that have gained electrons are now negatively charged and are called −*ions* The *difference in charge* between the positively, and the negatively charged bodies is called a *voltage* or a *difference in potential.* The term *potential* as used here means the ability to do work. Another term used to describe *voltage* is *electromotive force* or *emf.* This describes the force or pressure required to move electrons.

Voltage is measured in units called *volts*, in honor of Alessandra Volta, the Italian physicist who in 1799 constructed the first voltage source using chemical action, the forerunner of today's batteries. A *volt* has been defined as the electrical pressure required to move one *coulomb* (6.24×10^{18} electrons) per second through an opposition or resistance with a value of one *ohm.* The volt has also been defined with respect to *foot-pounds of work.* If a 4-lb weight

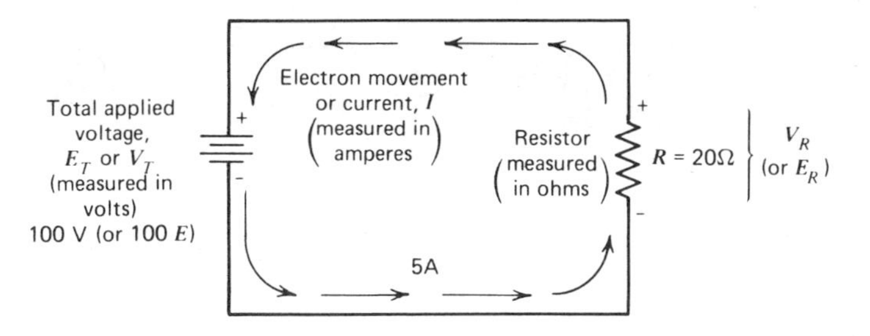

Figure 2-1. A simple electrical circuit.

is lifted a distance of 3 ft, then 12 ft-lb of work have been performed. A one volt difference of potential is present between two points when 0.738 ft-lb of work (called a *joule*) has been performed in order to move 6.24×10^{18} electrons (1 coulomb) between these two points.

The letter E or the letter V is usually used to denote the term *voltage* or *difference of potential.* The symbol for an applied voltage source which does not reverse its polarity (d-c source) is shown in Fig. 2-1 as a pair, or several pairs, of parallel lines, one longer than the other. This is also the symbol for a battery. Note that the *shorter* line denotes the *negative* end, while the *longer* one is the *positive* polarity. The letters E_T or V_T stand for the term total *voltage.* Similarly, if the applied voltage had a value of 100 volts, either the letter V or E again is used after the number 100. This is shown in Fig. 2-1 as 100 V or 100 E. As is discussed in a later section of this chapter, when electric current flows through an opposition called a resistor (shown in Fig. 2-1), a *difference of potential* or a *voltage* is present from one end to the other end of this resistor. This voltage *across* the resistor is called V_R or E_R. Note that a voltage is always the *difference between two points.*

Refer to End-of-Chapter Problems 2-1 to 2-5

2-2. Current

Often, to understand something new, it helps to compare it to a more familiar item. The idea of water flow is usually used then it is desired to explain electric current and pressure (voltage) for the first time. In Fig. 2-2*a* tank X is full of water, while another equal size one, tank Y is only partly filled. The two tanks are joined at their lower ends by a pipe, but a valve is kept closed so that water cannot move from either tank. The weight of the water in the full tank X is greater than that in the partially filled tank Y. There is therefore a *difference* in the weight or pressure of the water on the bottoms of the two tanks.

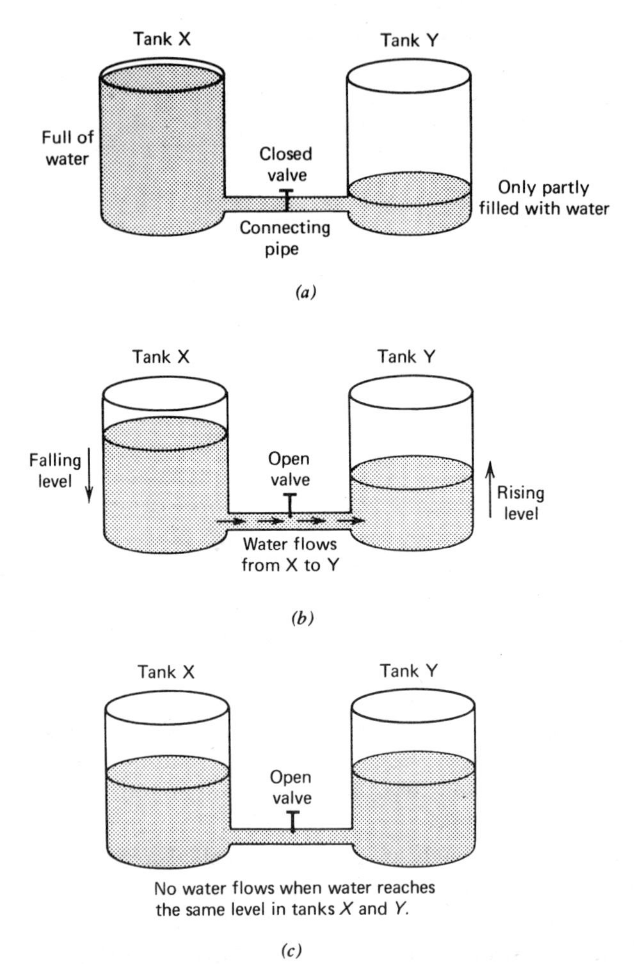

Figure 2-2. Water current and pressure. (c) No water flows when water reaches the same level in tanks *X* and *Y*.

When the valve is opened, as shown in Fig. 2-2*b*, water flows from tank X through the pipe into tank Y. The water level in tank X decreases, while that in tank Y increases. The *difference* in the water levels in the two tanks becomes less. Water continues to flow until the levels in the two tanks are equal. At this time, there is no difference in the pressures, and current ceases. This is depicted in Fig. 2-2*c*. Water is said to "seek its own level."

When electrons have been removed from one body (producing +ions) and placed onto a second (resulting in −ions), as shown in Fig. 2-3*a*, the charged ions, both negative and positive, will attempt to return to an uncharged or neutral state. If a connection is made between the two charged

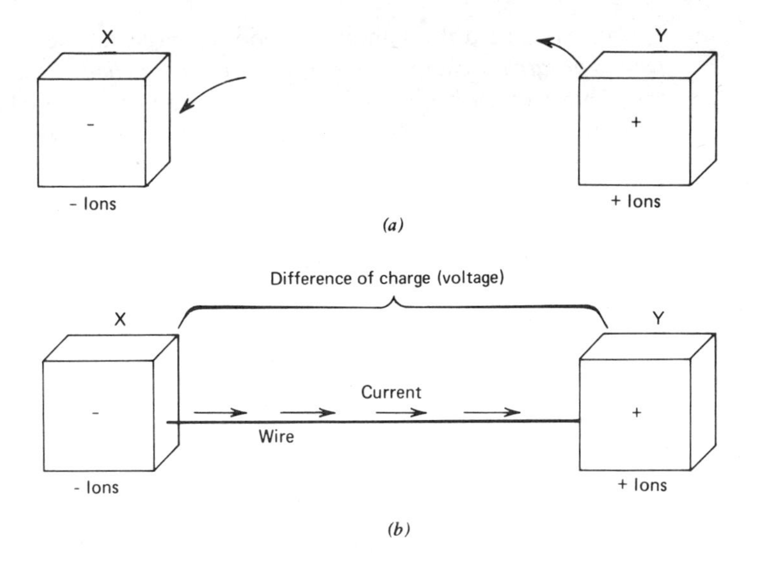

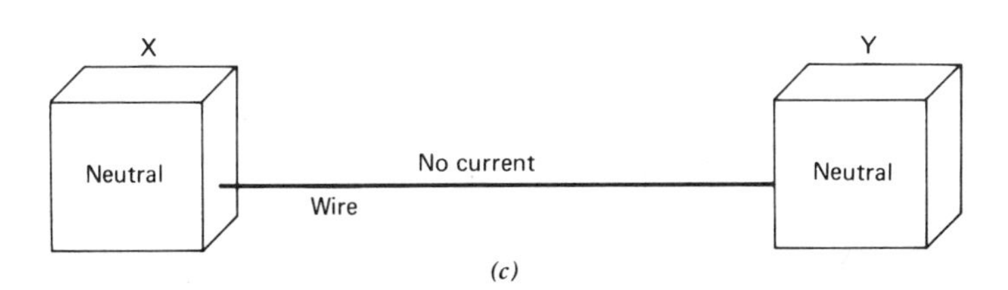

Figure 2-3. Electric current and difference of charge. (*a*) Removing electrons from *Y* and adding them to *X* makes *Y* positive and *X* negative. (*b*) Connecting a wire between *X* and *Y* permits a movement of electrons from the negative *X* to the positive *Y*. (c) No current flows when *X* and *Y* are uncharged or neutral.

bodies using a conductor, as shown in Fig. 2-3*b*, the surplus electrons from the negative body move through the conductor to the positive body. This movement of electrons *from negative to positive* is called electric *current*, or sometimes *current flow*, although the word *flow* is repetitious or redundant. This current continues as long as there is a difference of charge or a voltage between the two bodies. As shown in Fig. 2-3*c*, when the two bodies X and Y have been restored to a neutral state, there is no difference (zero volts) between them, and no more current flows.

Current as used throughout the book, consists of the *movement of electrons from a negatively charged, to a positively charged body.* Many years ago, before the theory of electrons and atoms was known, the makeup of an electric current was unknown. It was guessed, then, to be an invisible "liquid" which was often compared to water flow. Since water flows from a hilltop down to the lower ground below, and the positive end of a voltage source was thought of as the top of the "electric hill," then it was said that electric current flows from positive to negative, or from a "high point" to a "low point." This is just the reverse of the electron movement (current) direction. This older description of "current" going from "*plus to minus*" is today usually referred to as "*conventional current*" to distinguish it from the "electron movement" current. The technician-level textbooks usually use the *electron current* explanation (*negative to positive*), while engineering textbooks employ the *conventional current* theory (positive to negative). The use of either in a circuit discussion produces the same net conclusions. As a result, the technician-engineer should be able to follow both explanations. However, in any particular circuit, he must be consistent, sticking to one direction for current and not changing from one to the other. As mentioned before, *current* described throughout this text is the *movement of electrons*, which is from *negative to positive.* This is shown in Figs. 2-1 and 2-3*b*.

Current is measured in *amperes*, where one ampere represents a *coulomb* (6.24×10^{18} electrons) moving past a point in *one second.* The *ampere* was named to honor the French physicist André M. Ampère, who in 1820 demonstrated electromagnetism. The term *amperage* is often used for the word *current*. The letter "a" or "A" is used for *ampere*.

The movement of electrons, when a voltage is applied, from the negative end of the voltage source to the positive terminal, as shown in Fig. 2-1 constitutes an electric current. The electrons move along the wire (the conductor) from one atom to the next. Each electron entering the wire from the negative terminal of the applied voltage causes an electron to be repelled off the nearest atom of the wire. The repelled electron moves to the next adjacent atom, causing another electron to be pushed off this second atom. This action continues along the circuit of Fig. 2-1, and eventually an electron leaves the last atom of the wire, entering the positive terminal of the applied voltage source. This movement of electrons is relatively slow, but its *effect* is very fast. That is, even though each electron may only move the tiniest fraction of an inch going from an atom to the next adjacent one, and no further, the *effect* of an electron entering one end of the wire (the negative end) of Fig. 2-1 is to cause a different electron to leave the other end of the wire (the positive end). If the *path* through the wire from the negative terminal to the positive terminal of the applied voltage source were 186,000 miles long, then it would take one second (approximately) from the instant an electron entered the wire, until another electron left the other end of the wire. Therefore, the *effect* of the electron drift through the circuit is that some invisible wave motion travels around the circuit at almost the speed of a light wave in free space, 186,000 miles per second, or 300,000,000 meters (300×10^6) per second. This "invisible wave motion" that travels so rapidly through the circuit is not, of course, the actual electron movement (current). It can be likened somewhat to a long row of standing dominoes (small, flat, rectangularly shaped objects), each spaced a distance apart of about half their respective heights. When the first object is pushed over so that it falls against the adjacent one, the second one falls against the third, the third against the fourth, and so on, each domino has only moved an inch or so. However a motion travels down the long line and reaches the far end.

The letter I is used to represent the word *current*, since the letter C is used for another item called *capacitance* (to be discussed in a later chapter).

Refer to End-of-Chapter Problems 2-6 to 2-11

2-3. Resistance and Conductance

The *opposition* to an electric current offered by some components is called *resistance*. This resistance is measured in units called *ohms*, honoring Georg S. Ohm, who in 1827 first showed the relationship between voltage, current, and resistance. The symbol Ω (Greek letter omega) is used to denote the word *ohm*, while the current diagram symbol for *resistance* is a wavy series of saw-tooth lines. There are both shown in Fig. 2-1. Practically everything in an electrical circuit has some resistance to current. A good conductor such as copper wire has very little resistance (Chapter 8), while certain components called *resistors* have oppositions that may be as small as a fraction of an ohm, or may be as high as millions of ohms (Chapter 8). An insulator should have a theoretical resistance of *infinity*, or millions upon millions of ohms.

One ohm has been defined as the resistance of a one square millimeter tube of mercury, 106.3 centimeters long, at zero degrees centigrade. It is also the amount of opposition that results in one ampere of current when one volt is applied to a circuit. One ohm of resistance is also defined as the amount of opposition that generates 0.24 calories of heat when one ampere of current flows for one second (where one calorie raises the temperature of one gram of water one degree centigrade). Resistance may also be defined as the ratio between voltage and current.

Although every part in an electrical circuit has some opposition as its characteristic, certain components are specifically placed in a circuit to add resistance. These are called *resistors*, and are discussed in Chapter 8. The letter R is used to represent the term *resistance*

Refer to End-of-Chapter Problems 2-12 to 2-15

Conductance. This is simply the reciprocal of resistance, or conductance = 1/resistance. *Conductance* is measured in a unit called *mho*. Note that this is the resistance unit *ohm* spelled in reverse. The symbol for the term *conductance* is the letter G. Therefore, $G = 1/R$. The symbol for the unit *mho* is ℧, which is the ohm symbol, Ω, turned upside down. Resistance is also the reciprocal of conductance, or $R = 1/G$. *Conductance* as will be shown in later discussions, may be used in preference to *resistance* when solving certain types of circuits.

Refer to End-of-Chapter Problems 2-16 to 2-20

2-4. Prefixes

Current, voltage, and resistance units are often very small or very large. A current of 0.005 A or even 0.000002 A in a vacuum tube or in a transistor is not unusual. Similarly, a resistor may be 25,000 Ω or even 15,000,000 Ω. To avoid having to use extremely small or large numbers in discussing values of current, voltage, or resistance, *prefixes* are added in front of the actual basic unit. For example, the prefix *milli* means a *thousandth* part of (something). Therefore, instead of saying that a current is *0.005A* or *5/1000 A*, it is preferable to say *5 milliamperes*, or 5 mA.

Another example is the prefix *kilo*, meaning a *thousand* (something). Again, instead of saying that a resistor has a value of 25,000 Ω, it is usually preferred to say *25 kilohms*, or 25 kΩ.

As shown by the two examples using *milli* (a thousandth part of) and *kilo* (a thousand), it may be seen that the prefixes use the decimal system. Table 2-1 lists the most commonly used prefixes.

When one solves an electrical circuit involving current, voltage, and resistance, it is necessary to use the *basic units* of *amperes*, *volts*, and *ohms*, respectively. This is true even though the units may have been used with prefixes. As shown in Table 2-1 it is necessary therefore to convert from a *prefixed* unit to its basic form. If a current is given in *milliamperes*, it must be changed into *amperes* when solving problems. Similarly, a resistor value in *kilohms* or *megohms* must be converted into *ohms*, while a voltage stated as *kilovolts* or *millivolts* or *microvolts* must be switched to *volts*.

The first *prefix* listed in Table 2-1 is *meg* which means *million*, or 1,000,000. This number may be written in a more convenient form without the string of zeroes. This form is called *scientific or engineering notation*. Briefly, it involves changing a very large or a very small number with a string of zeroes by *relocating the decimal point* and then *multiplying* the new number *by a power of ten* (which tells you how many places the decimal point must be moved to get *back* to where it *really belongs*). For example, the number 1,000,000 may be rewritten as $1. \times 10^6$, since relocating the decimal point just to the right of the digit *1* means that the decimal point really belongs six places to the right, hence the $\times 10^6$.

TABLE 2-1 List of Prefixes Most Commonly Used[b]

Prefix	Symbol	Value	Conversion to Basic Unit (Example)
Meg	M	Million	$\times 10^6$ (2 megohm = 2×10^6 ohms) or 2 MΩ
Kilo	K	Thousand	$\times 10^3$ (5 kilohm = 5×10^3 ohms) or 5 KΩ
Milli	m	Thousandth	$\times 10^{-3}$ (25 milliamperes = 25×10^{-3} ampere) or 25 mA
Micro	μ	Millionth	$\times 10^{-6}$ (150 microvolts = 150×10^{-6} volts) or 150 μV
Micro micro, (or pico)	μμ (or p)	Millionth of a millionth	$\times 10^{-12}$ (250 micro micro [a]farads = 250×10^{-12} farads) or 250 μμF also 250 picofarads or 250 pF

[a] Unit used to measure a capacitor.
[b] See also Appendix II, for a more complete prefix list.

A number such as *5000.* could therefore be written 5×10^3; and *20,000.* could be written 20×10^3 or even 2×10^4.

Note that in the previous couple of paragraphs it was necessary in every case to move the decimal point to the *right* to bring it *back* to where it originally belonged. This involved *positive* powers of ten, with the + sign omitted (no sign means it is positive). However, when the decimal point must be moved to the *left* to bring it *back* to where it belongs, a *negative* power of ten is used.

Some examples are the following. A number such as *.005* could be rewritten as $5. \times 10^{-3}$ since the decimal point has been relocated just to the right of the digit 5, and the point must be moved three places to the *left* to bring it *back* to its original position; hence $\times 10^{-3}$.

The number *0.025* then becomes 25×10^{-3} or even 2.5×10^{-2}. The number *0.00015* could be rewritten $15. \times 10^{-5}$ or 1.5×10^{-4} or even $150. \times 10^{-6}$.

A discussion of each of the prefixes and the examples shown in Table 2-1 follows:

The prefix *meg* (M) has a value of a million. A *2-megohm* resistor is therefore 2,000,000. ohms (or 2×10^6 Ω).

The prefix *kilo* (K) has a value of a thousand. A *5-kilohm* resistor is therefore 5000. ohms (or $5 \times 10^3\ \Omega$).

The prefix *milli* (m) has a value of a *thousandth*. Therefore, 25 milliamperes (25 mA) is 0.025 ampere (or 25×10^{-3}, or 2.5×10^{-2} A).

The prefix *micro* (μ) has a value of a *millionth*. Therefore 150 microvolts (150 μF) is 0.000150 volts (or 150×10^{-6}, or 1.5×10^{-4} V).

The double prefix *micro micro* ($\mu\mu$) also caled *pico* (p) is worth a value of a *millionth of a millionth*. A device called a capacitor could have a value of 250 micro microfarads (250 $\mu\mu$f) or 250 picofarads (250 pF), which is the same as 0.000 000 000 250 farads (or 250×10^{-12}, or 2.5×10^{-10} F).

The following example further illustrates the previous discussion.

Example 2-1

Perform the following conversion. (Note that when converting into a *larger* unit, the number becomes *smaller*, such as 12 in. = 1 ft. Similarly, when converting into a *smaller* unit, the number does the *opposite* and becomes *larger* such as 3 ft = 36 in.)

(a) 300,000 meters	= ?	kilometers (km)	= ? $\times 10^?$ meters
(b) 0.25 ampere	= ?	milliamperes (mA)	= ? $\times 10^?$ amperes
(c) 0.065 seconds	= ?	microseconds (μsec)	= ? $\times 10^?$ seconds
(d) 0.25 megohms	= ?	ohms (Ω)	= ? $\times 10^?$ ohms
(e) 500 picofarads	= ?	farads (F)	= ? $\times 10^?$ farads

Solution

(a) Here, *meters* are to be converted into the *larger* unit *kilometers*, which means that the number must get *smaller*. Since *kilo* is a thousand, then 300,000. meters is 300 kilometers, and is also 300×10^3 meters (or 3×10^5 m).

(b) Changing *amperes* into the *smaller* unit *milliamperes* should result in a *larger* number, which means that the decimal point of 0.25 amperes should be moved to the *right*. Since the prefix *milli* means a *thousandth*, it involves *three* decimal places. Therefore, moving the decimal point of 0.25 ampere three places to the right, yields *250. milliamperes* (note that the number became larger). Rewriting 0.25 ampere in *scientific notation* yields *2.5×10^{-1} ampere.*

(c) Changing *0.065 seconds* into the *smaller* unit *microseconds* should produce a *larger* number, which means moving the decimal point to the *right*. Since the prefix *micro* means *millionth*, it involves six decimal places. Therefore *0.065 seconds* becomes *65000. microseconds*. Rewriting *0.065 seconds* in scientific notation, it becomes *6.5×10^{-2} seconds.*

(d) Converting 0.25 *megohms* into the *smaller* unit *ohms* means that the number should become *larger*, requiring the decimal point to be moved to the *right*. Since the prefix *meg* means million, it involves six decimal places.

Therefore, *0.25 megohms* becomes *250,000. ohms*. Rewriting 250,000 ohms in scientific form, it becomes 2.5×10^5 *ohms*.

(e) Changing 500 *picofarads* into the *larger* unit *farads* should result in a *smaller* number, produced by moving the decimal point to the *left*. The prefix *pico* means a *millionth of a millionth*, involving *12* decimal places. Therefore, moving the decimal point in the *500. picofarads 12* places to the *left* produces *0.000 000 000 500 farads*. Rewriting this unwieldy number in scientific notation form gives $5. \times 10^{-10}$ *farads*.

Appendix II has a more complete listing of prefixes and their values.

For Similar Problems Refer to End-of-Chapter Problems 2-21 to 2-22

2-5. Ohm's Law, Relationship of Current to Voltage and Resistance

A basic electrical circuit consists of an applied source of voltage connected to a *load*. The load, usually depicted as a resistor as in Fig. 2-4, may be a light bulb, a motor for a fan or drill, a heating coil such as a toaster or an iron, or many other things. In Sections 2-1, 2-2, and 2-3, voltage, current, and resistance were discussed independently. Here, the relationship of each to the other is delved into.

The amount of current in a circuit depends on the applied voltage and the size of the resistor (the load). Current I is measured in units called *amperes*, and is *directly* proportional to the amount of the applied voltage (E). The letter "V" is also often used to designate *voltage*. The applied voltage acts in a manner similar to a pressure pump that forces water to flow through a pipe. If the pressure of the pump is increased, the water current is increased, and more gallons per second flow. In the electrical circuit, if the applied voltage (the difference of charge between the two terminals of the source) is increased, *more* electrons move past a point than before. Hence, current has increased.

Current is *inversely* proportional to the size of the resistor (R) in the circuit. This means that if R is made larger, current *decreases*. Again returning to the water comparison, if the inner diameter of the water pipe (the size of its passage) were made smaller, it would offer greater opposition to the flow of

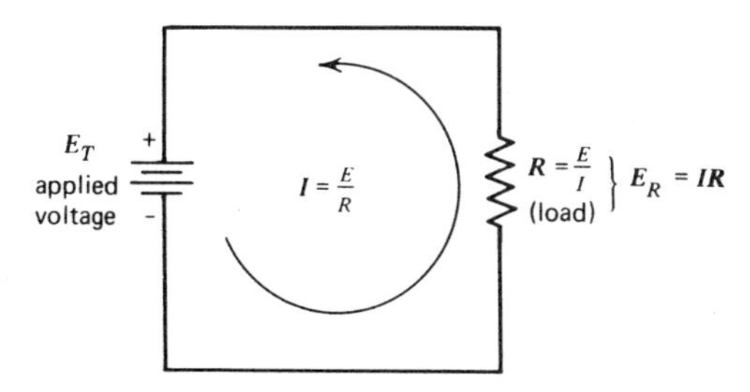

Figure 2-4. Ohm's law, the basis electrical circuit.

water, resulting in fewer gallons pet second. In the electric circuit a similar thing occurs. When the opposition or resistance *increases*, fewer electrons move past a point than previously.

This relationship of current (I), voltage (E), and Resistance (R), is called *Ohm's law*, and appears as

$$\text{current} = \frac{\text{voltage}}{\text{resistance}} \tag{2-1}$$

or

$$I = \frac{E}{R} \tag{2-1}$$

or

$$I = \frac{V}{R}$$

Note that the above equation repeats the fact that the *current* (I) is *directly proportional* to the *voltage* (E), and *inversely proportional* to the *resistance* (R). In equation 2-1, current is in amperes; voltage is in volts; and resistance is in ohms.

In a circuit such as is shown in Fig. 2-4, if the applied voltage is 120 V and the resistor is 30 Ω, then the current could be found.

$$I = \frac{E}{R} \tag{2-1}$$

$$I = \frac{120\ \text{V}}{30\ \Omega}$$

$$I = 4\ \text{A}$$

If the applied *voltage* were *increased* to 150 V, then the *current* would also *increase* to a value found in the following:

$$I = \frac{V}{R} \tag{2-1}$$

$$I = \frac{150\ \text{V}}{30\ \Omega}$$

$$I = 5\ \text{A}$$

Returning to the original circuit consisting of 120 V and 30 Ω, with a current of 4 A, if the *resistance* were now *increased* to 60 Ω, the *current* would *decrease* (inversely proportional) to the value found in the following:

$$I = \frac{E}{R} \tag{2-1}$$

$$I = \frac{120}{60}$$

$$I = 2\ \text{A}$$

The following examples illustrates Ohm's law, solving for the current.

Example 2-2

In Fig. 2-4, if E is 6 V and $R = 15\ \text{k}\Omega$, find the current.

Solution

In Ohm's law, equation 2-1, R must be in *ohms*. Therefore, since R is 15 kΩ, it is converted to 15,000 Ω or $15 \times 10^3\ \Omega$. Solving for current, I,

$$I = \frac{E}{R} \tag{2-1}$$

$$I = \frac{6}{15 \times 10^3}$$

Dividing 6 by 15 gives 0.4, and moving the 10^3 from the *denominator* to the *numerator* requires reversing the *sign* of the exponent or power 3 to a −3. (Refer to Appendix I "Mathematics for Electronics," for further discussion.) Current I, is then

$$I = 0.4 \times 10^{-3}\ \text{A}$$

or

$$0.0004\ \text{A}$$

or

$$0.4\ \text{mA}$$

Example 2-3

How much current flows through a 20 kΩ resistor with 300 V applied across it?

Solution

$$I = \frac{V}{R} \tag{2-1}$$

$$I = \frac{300}{20 \times 10^3}$$

$$I = 15 \times 10^{-3}\ \text{A}$$

or

$$0.015\ \text{A}$$

or

$$15\ \text{mA}$$

Example 2-4

A 3-MΩ resistor has 15 kV applied across it. What current flows in this circuit?

Solution

Both the 3-MΩ resistor and the voltage of 15 kV must be converted into basic units of *ohms*, and *volts* to be used in the Ohm's law equation. Converting, 3 MΩ is 3×10^6, and 15 kV is 15×10^3 V. Solving for the current I,

$$I = \frac{E}{R} \tag{2-1}$$

$$I = \frac{15 \times 10^3}{3 \times 10^6}$$

Dividing 15 by 3 gives *5*. Dividing 10^3 by 10^6 involves an algebraic *subtraction* of the powers or exponents $+3$ and $+6$, or $+3$ *minus* $+6$. The *subtrahend* $+6$ must first be reversed in sign to a -6, and then the $+3$ and the -6 are combined in algebraic *addition*. With opposite signs ($+3$, -6), take the *smaller* ($+3$) from the *larger* (-6), and give the answer the *sign* of the *larger*, giving an answer of -3. (Refer to Appendix I, "Mathematics for Electronics," for further detailed discussion.) Therefore, 10^3 divided by 10^6 is 10^{-3}, and the current is

$$I = 5 \times 10^{-3} \text{ A}$$

or

$$0.005 \text{ A}$$

or

$$5 \text{ mA}$$

For Similar Problems Refer to End-of-Chapter Problems 2-23 and 2-24

2-6. Ohm's Law, Relationship of Voltage to Current and Resistance

In the previous section it was shown that current (I) is directly proportional to voltage (E), but inversely proportional to resistance (R), and this is shown in the Ohm's law equation.

$$I = \frac{E}{R} \tag{2-1}$$

If it were desired to solve the above equation in terms of the *voltage* E, it would be necessary to remove R from the right side of the equation, thus leaving E by itself. To do this, we multiply both sides of the equation by R, as shown:

$$I(R) = \frac{E}{R}(R)$$

Canceling the R's at the right side yields,

$$IR = E \qquad (2\text{-}2)$$

or, rewriting for convenience,

$$E = IR \qquad (2\text{-}2)$$

or

$$V = IR$$

This form of Ohm's law states that the *voltage* (E) is equal to the *product* of the *current* (I) times the *resistance* (R). In an electrical circuit such as is shown in Fig. 2-4, when the *current* (I) and the *resistor* (R) are known, it is simple to find the *voltage* (E) by using this second form of Ohm's law, equation 2-2. The following examples illustrate this.

Example 2-5

If 3 A flow through a resistance of 70 Ω, what is the *voltage* applied across the resistor?

Solution

$$\text{voltage} = (\text{current})\,(\text{resistance}) \qquad (2\text{-}2)$$

$$E\ (\text{or } V) = IR$$

$$E = (3)\,(70)$$

$$E = 210\text{ V}$$

Example 2-6

A 0.25 MΩ resistor has 120 mA flowing through it. What is the voltage across the resistor?

Solution

The 0.25 MΩ must be changed to *ohms*, and the 120 mA must be converted to *amperes* in equation 2-2. Converting, 0.25 MΩ becomes 250,000 Ω or 2.5×10^5 Ω, and 120 mA become 0.12 A or 1.2×10^{-1} A. Therefore, the voltage is

$$V\ (\text{or } E) = IR \qquad (2\text{-}2)$$

$$V = (120\text{ mA})\,(0.25\text{ M}\Omega)$$

$$V = (1.2 \times 10^{-1})\,(2.5 \times 10^5)$$

$$V = 3 \times 10^4\text{ V}$$

or

$$30{,}000.\text{ V}$$

or

$$30\text{ kV}$$

Example 2-7

What is the voltage applied to a 3.5 kΩ resistor, causing 150 μA of current?

Solution

The 3.5 kΩ must be converted into *ohms*, and the 150 μA must be changed to *amperes* to be used in the Ohms law equation. Converting, 3.5 kΩ become $3.5 \times 10^3\ \Omega$, and 150 μA become 150×10^{-6} A or 1.5×10^{-4}A. Therefore, the voltage is

$$V \text{ (or } E) = IR \qquad (2\text{-}2)$$

$$V = (150\ \mu\text{A})\ (3.5\ \text{k}\Omega)$$

$$V = (1.5 \times 10^{-4})\ (3.5 \times 10^{3})$$

$$V = 5.25 \times 10^{-1}\ \text{V} \qquad \text{or} \qquad 0.525\ \text{V}$$

For Similar Problems Refer to End-of-Chapter Problems 2-25 to 2-27

2-7. Ohm's Law, Relationship of Resistance to Voltage and Current

In the preceding two sections these relationships were discussed:

$$\text{current} = \frac{\text{voltage}}{\text{resistance}} \qquad (2\text{-}1)$$

$$I = \frac{E}{R} \qquad (2\text{-}1)$$

and

$$\text{voltage} = (\text{current})\ (\text{resistance}) \qquad (2\text{-}2)$$

or

$$E = (I)\ (R) \qquad (2\text{-}2)$$

If we wish to solve equation 2-2 in terms of R, the letter I should be removed from the right side of the equation, thus leaving R by itself. To accomplish this, we divide both sides of the equation by I, producing

$$\frac{E}{I} = \frac{IR}{I}$$

canceling the I's on the right side of the equation gives

$$\frac{E}{I} = R \qquad (2\text{-}3)$$

Rewriting for convenience,

$$R = \frac{E \text{ (or } V)}{I} \qquad (2\text{-}3)$$

or, in words,

$$\text{resistance} = \frac{\text{voltage}}{\text{current}} \tag{2-3}$$

From equation 2-3 it may be seen that if the *voltage* across a resistor is known, and the *current* through it is also known, then the value of this *resistor* may be found by using this third variation of the Ohm's law equation. The following examples illustrate this.

Example 2-8

What is the value of a resistor if there is 60 V across it and 2 A flowing through it?

Solution

Using Ohm's law, equation 2-3:

$$R = \frac{E}{I} \tag{2-3}$$

$$R = \frac{60}{2}$$

$$R = 30\ \Omega$$

Example 2-9

A 120-V motor has 1.5 A flowing through it. What is the resistance of this motor?

Solution

$$R = \frac{E}{I} \tag{2-3}$$

$$R = \frac{120}{1.5}$$

$$R = 80\ \Omega$$

Example 2-10

A resistor is placed in a circuit where 120 μA are flowing. If 60 mV should be across this resistor, what size should it be?

Solution

In the Ohm's law equation, the 120 μA and the 60 mV must be changed to *amperes* and *volts*.

Converting, 120 μA is 120×10^{-6} A, and 60 mV is 60×10^{-3} V. Therefore, the resistor is

$$R = \frac{E}{I} \qquad (2\text{-}3)$$

$$R = \frac{60 \text{ mV}}{120 \ \mu\text{A}}$$

$$R = \frac{60 \times 10^{-3}}{120 \times 10^{-6}}$$

Dividing 10^{-3} by 10^{-6} means algebraically subtracting the exponents, the -3 and the -6. This becomes -3 *minus* -6, or -3 and $+6$, yielding an answer of $+3$ for the power of ten.

$$R = 0.5 \times 10^{3}$$

$$\text{or } 500 \ \Omega$$

For Similar Problems Refer to End-of-Chapter Problems 2-28 to 2-31

Figure 2-5 is a helpful aid that students find useful in learning the three forms of Ohm's law (equation 2-1, $I = E/R$; equation 2-2, $E = IR$; and equation 2-3, $R = E/I$).

By placing your finger so that it covers any *one* letter in Fig. 2-5 (E, I, or R), the relationship between it and the other two is seen. Covering the I, shows E/R (equation 2-1). Similarly, covering the E, shows IR (equation 2-2). Covering the R, shows E/I (equation 2-3).

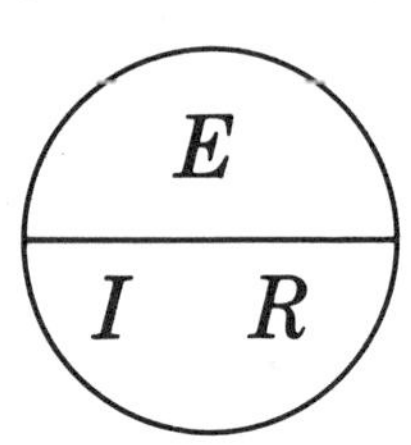

Figure 2-5. Ohm's law aid.

2-8. MKS System

The system of units most often used in electronics is called the mks system. The letters denote the basic units for length, measured in *meters* (1 m = 39.37 in.); for mass, measured in *kilograms* (1 kg = 2.205 lb); and for time, measured in *seconds*. The term *mass* is often mistakenly described as *weight*. This is understandable, since the difference is difficult to visualize. An object has the *same mass* whether it's on the earth or on the moon, but the object would *weigh* less on the moon. Weight, then, is related to gravity, but mass is not, being related to acceleration and force. Mass has been defined as the force required, divided by the resultant acceleration.

The meter-kilogram-second (mks) system is preferred over another metric system employing smaller units called the centimeter-gram-second (cgs) system and also over the English unit system called foot-pound-second (fps). Table 2-2 lists some of the items or terms and the units used for each in the three systems, along with some conversions from one system to another.

TABLE 2-2 Units in the Three Systems

Item	mks Units	cgs Units	fps Units
Length	Meter (1 m = 100 cm) (1 m = 39.37 in.)	Centimeter	Foot
Mass	Kilogram (1 kg = 1000 g) (1 kg = 2.205 lb)	Gram	Slug (1 Slug = 14.59 kg)
Time	Second	Second	Second
Acceleration	Meters per second per second	centimeters per second per second	Feet per second per second
Energy (work)	Joule (Newton-meter) (1 joule is the work done when 1 volt causes 1 coulomb of movement) 1 J = 0.738 ft lb) (1 J = 1×10^7 ergs) (1 J = 6.25×10^{18} elV)	Erg (dyne-cm)	Foot-pound
Force (or weight)	Newton (1 N = 1×10^5 dynes) [1 N accelerates 1 kg 1 m/(sec)/(sec)]	Dyne [1 dyne accelerates 1 g 1 cm/(sec)/(sec)]	Pound (1 lb = 4.45 N)
Heat	Kilocalorie (1 kcal raises the temperature of 1 kg of water 1°C) (1 J = 2.39×10^{-4} kcal)	Calorie (1 cal raises the temperature of 1 g of water 1°C)	British Thermal Unit (Btu) 1 (Btu = 252 cal)
Power	Joule per second (1 J/sec = 1 W)	Erg per second (10^7 ergs/see = 1 W)	Foot-pound per second (550 ft-lb/sec = 1hp) (1 hp = 746 W)
Velocity	Meter per second	Centimeter per second	Feet per second

The following examples refer to Table 2-2.

Example 2-11

Convert 20 lb of force in the fps system into newtons in the mks system.

Solution

Since 1 lb = 4.45 N, then 20 lb converted into newtons is 20 *multiplied* by 4.45 or 89 N. (When converting into the *smaller* unit, newtons, the resulting number must become *larger*; therefore, multiplication is necessary.)

Example 2-12

Convert 255 ft-lb of work in the fps system into joules in the mks system.

Solution

Since 1 J = 0.738 ft-lb, then 255 ft-lb converted into joules is 255 *divided* by 0.738, or 346 J.

Example 2-13

A force of 25,000 dynes in the cgs system is equivalent to how many newtons in the mks system?

Solution

One newton = 1×10^5 dynes. Therefore, converting *dynes* into the *larger* unit *newtons* should result in a *smaller* number. As a result, *division* must be employed. 25,000 dynes *divided* by 1×10^5 results in

$$N = \frac{25{,}000}{10^5}$$

$$N = 25{,}000 \times 10^{-5}$$

$$N = 0.25$$

For Similar Problems Refer to End-of-Chapter Problems 2-32 to 2-39

2-9. Energy, Work, and Power

Energy is usually defined as the ability or capacity to do *work*. *Work* may be defined as a physical movement against some opposition. *Power* is the *rate* of performing work.

Energy may result from the movement of a body; this is called *kinetic* energy. The energy present when a *voltage* or *difference of potential* is present is called *potential* energy. In Section 2-1 we stated that one volt is the difference of potential that exists when one joule of *work* (0.738 ft-lb) has been performed moving one coulomb (6.24×10^{18} electrons). This can be expressed by the equation:

$$\text{volt} = \frac{\text{joule}}{\text{coulomb}} \tag{2-4}$$

or

$$\text{volt} = \text{joule per coulomb}$$

(The unit of work, the *joule*, was named honoring James P. Joule an English scientist of the 1800s.)

Current is defined in Section 2-2 as the drift or movement of electrons in the same direction. One *ampere* of *current* is a *coulomb* (6.24×10^{18}) passing a point in a circuit in *one second*. Therefore, expressed as an equation, *current* is

$$\text{current} = \frac{\text{coulomb}}{\text{second}} \tag{2-5}$$

or

$$\text{current (in amperes)} = \text{coulombs per second}$$

Power, the *rate* of performing *work*, is actually then *joules* per *second*, or

$$\text{power} = \frac{\text{joule}}{\text{second}} \tag{2-6}$$

This equation 2-6 is actually derived from multiplying equations 2-4 and 2-5, as shown in thc following:

$$\text{volt} = \frac{\text{joule}}{\text{coulomb}} \tag{2-4}$$

$$\text{current} = \frac{\text{coulomb}}{\text{second}} \tag{2-5}$$

Multiplying equations 2-4 and 2-5 gives

$$(\text{volt})\,(\text{current}) = \left(\frac{\text{joule}}{\text{coulomb}}\right)\left(\frac{\text{coulomb}}{\text{second}}\right)$$

$$(\text{volt})\,(\text{current}) = \frac{\text{joule}}{\text{second}} \tag{2-7}$$

The right halves of equations 2-6 and 2-7 are identical; therefore, the left halves must be equal to each other, yielding

$$\text{power} = (\text{volt})\,(\text{current}) \tag{2-8}$$

Power is measured in *watts*, honoring the Scottish scientist James Watt. Rewriting equation 2-8, it becomes

$$\text{power} = EI \tag{2-8}$$

where *power* is in *watts*, if E is in *volts*, and I is in *amperes*. The letter W (for watts) is often used in equation 2-8 in place of P (for power) so that the equation may be either: $P = EI$, or $W = EI$.

By substituting from Ohm's law in equation 2-8, two other useful equations for power may be derived. In

$$P = E(I) \tag{2-8}$$

substitute E from Ohm's law equation 2-2, $E = IR$, giving

$$P = E(I)$$

$$P = IR(I)$$

$$P = I^2R \tag{2-9}$$

where P is in watts, if I is in *amperes*, and R is in *ohms*. This is the second power formula.

A third useful power equation may also be derived from

$$P = E(I) \tag{2-8}$$

by substituting for I from Ohm's law equation 2-1, $I = E/R$, giving

$$P = E(I) \tag{2-8}$$

$$P = E\left(\frac{E}{R}\right)$$

$$P = \frac{E^2}{R} \tag{2-10}$$

The three power equations, sometimes referred to as *Watt's laws*, then are

$$P = EI \tag{2-8}$$

and

$$P = I^2R \tag{2-9}$$

and

$$P = \frac{E^2}{R} \tag{2-10}$$

The chart in Appendix III lists the three forms of the power equation and Ohm's law in a useful reference form.

Power, Voltage, and Current. When *voltage*, *current*, and *power* are being discussed, equation 2-8, $P = EI$, should be used. The following examples illustrate this. Often *power* is referred to as power *dissipated*, or power *consumed*, or even as *energy dissipated*. But these terms are not really technically correct, since neither is being *dissipated* nor *consumed*. *Power* is simply the *rate* at which the work is being done, while *energy* is simply being changed from one form into another. A battery, for example, transforms chemical energy into electrical energy, and is said to generate power. A resistor, through which current is flowing, transforms electrical energy into heat. In the case of a toaster or a light bulb, the heat produced is necessary and useful. Heat produced in a resistor is wasted and is a loss of power.

Example 2-14

A circuit has 200 V applied to it, with 25 mA flowing. Find the power.

Solution

$$P = E(I) \qquad (2\text{-}8)$$

$$P = (200)\,(25\text{ mA})$$

$$P = (200)\,(25 \times 10^{-3}\text{A})$$

$$P = 5000 \times 10^{-3}\text{ W}$$

$$P = 5\text{ W}$$

Example 2-15

A resistor has 1500 mA flowing through it, and 500 mV across it. How much power is being dissipated in the resistor?

Solution

$$P = E(I) \qquad (2\text{-}8)$$

$$P = (500\text{ mV})\,(1500\text{ mA})$$

$$P = (500 \times 10^{-3})\,(1500 \times 10^{-3})$$

$$P = 750{,}000 \times 10^{-6}\text{ W}$$

$$P = 0.75\text{ W}$$

Example 2-16

How much current is flowing through the filament of a 120 V, 300-W light bulb which is being operated properly?

Solution

$$P = EI \qquad (2\text{-}8)$$

$$300 = 120I$$

$$\frac{300}{120} = I$$

$$2.5\ \text{A} = I$$

The chart of Appendix III shows that $I = P/E$.

For Similar Problems Refer to End-of-Chapter Problems 2-40 to 2-43

Power, Current, and Resistance. When *power*, *current*, and *resistance* are discussed, equation 2-9, $P = I^2R$, should be used. The following examples illustrate this.

Example 2-17

A 10 kΩ resistor has 20 mA flowing through it. What is the power being consumed?

Solution

$$P = I^2(R) \qquad (2\text{-}9)$$

$$P = (20\ \text{mA})^2\ (10\ \text{k}\Omega)$$

$$P = (20 \times 10^{-3})^2\ (10 \times 10^3)$$

$$P = (400 \times 10^{-6})\ (10 \times 10^3)$$

$$P = 4000 \times 10^{-3}\ \text{W}$$

$$P = 4\ \text{W}$$

Example 2-18

A resistor has 400 mA flowing through it, and is dissipating 80 W. What is the value of the resistor?

Solution

$$P = I^2R \qquad (2\text{-}9)$$

$$80 = (400\ \text{mA})^2\ R$$

$$80 = (0.4)^2\ R$$

$$80 = 0.16\ R$$

$$\frac{80}{0.16} = R$$

$$500\ \Omega = R$$

Example 2-19

A 25-Ω resistor is dissipating 100 W of power. What is the current through the resistor?

Solution

$$P = I^2R \tag{2-9}$$

$$100 = I^2\,(25)$$

$$\frac{100}{25} = I^2$$

Taking the square root of both sides gives

$$\sqrt{\frac{100}{25}} = \sqrt{I^2}$$

$$\sqrt{4} = I$$

$$2\text{ A} = I$$

The chart of Appendix III shows that $I = \sqrt{P/R}$.

Example 2-20

What current flows through a 150-Ω resistor if the power dissipation is 3.75 mW?

Solution

$$P = I^2R \tag{2-9}$$

$$3.75\text{ mW} = I^2\,(150)$$

$$3.75 \times 10^{-3} = I^2\,(1.5 \times 10^2)$$

$$\frac{3.75 \times 10^{-3}}{1.5 \times 10^2} = I^2$$

Taking the square root of both sides produces

$$\sqrt{\frac{3.75 \times 10^{-3}}{1.5 \times 10^2}} = \sqrt{I^2}$$

$$\sqrt{2.5 \times 10^{-5}} = I$$

Rewrite the number under the radical (square root) sign so that it has an *even* power of 10 (such as $25. \times 10^{-6}$), since it is almost meaningless to take the square root of an *odd* power of 10. Taking the square root of an an *even* power of ten simply involves *dividing* the *power* by 2, as shown in the following:

Repeating

$$\sqrt{2.5 \times 10^{-5}} = I$$

Rewriting

$$\sqrt{25. \times 10^{-6}} = I$$

$$5. \times 10^{-3} = I$$

$$0.005 \text{ A} = I$$

or

$$5\text{mA} = I$$

The chart of Appendix III shows that $I = \sqrt{P/R}$.

For Similar Problems Refer to End-of-Chapter Problems 2-44 to 2-46

Power, Voltage, and Resistance. The third power formula, equation 2-10, $P = E^2/R$, should be employed whenever *power*, *voltage*, and *resistance* are the variables being used. The following examples illustrate this:

Example 2-21

How much power is being transformed in a 5-kΩ resistor which has 30 V across it?

Solution

$$P = \frac{E^2}{R} \tag{2-10}$$

$$P = \frac{30^2}{5 \times 10^3}$$

$$P = \frac{900}{5 \times 10^3}$$

$$P = 180 \times 10^{-3} \text{ W}$$

or

$$0.18 \text{ W}$$

or

$$180 \text{ mW}$$

Example 2-22

What is the resistance of the heating element of a 1000-W electric iron operating correctly with 120 V applied to it?

Solution

$$P = \frac{E^2}{R} \tag{2-10}$$

Multiplying both sides of equation by R gives

$$PR = E^2$$

Dividing both sides of P gives

$$R = \frac{E^2}{P}$$

The above is also listed in the chart of Appendix III.

$$R = \frac{120^2}{1000}$$

$$R = \frac{14400}{1000}$$

$$R = 14.4\ \Omega$$

Example 2-23

What is the voltage across a 200-kΩ resistor that is dissipating 80 W?

Solution

$$P = \frac{E^2}{R} \tag{2-10}$$

Multiplying both sides of the equation by R gives

$$PR = E^2$$

Taking the square root of both sides yields

$$\sqrt{PR} = \sqrt{E^2}$$

$$\sqrt{PR} - E$$

This is also shown in Appendix III.

$$\sqrt{(80)\,(200\text{K})} = E$$

$$\sqrt{(8 \times 10)\,(2 \times 10^5)} = E$$

$$\sqrt{16 \times 10^6} = E$$

$$4 \times 10^3\ \text{V} = E$$

or

$$4000\ \text{V} = E$$

For Similar Problems Refer to End-of-Chapter Problems 2-47 to 2-50

PROBLEMS

See section 2-1 for discussion covered by the following.

2-1. What are some terms used to describe the difference between two bodies that have opposite polarity charges?

2-2. What does *potential* mean in electricity?

2-3. What is the basic unit of measurement of voltage?

2-4. Define a volt.

2-5. (a) Write, using simple letters, that which denotes the *total applied voltage*.

(b) Using simple letters, write that which means *the voltage across a resistor*.

See section 2-2 for discussion covered by the following.

2-6. What is electric current?

2-7. What two things are required to have an electric current?

2-8. What is conventional current?

2-9. What is the unit of measurement of electric current?

2-10. What constitutes an ampere?

2-11. Which letter is used to denote current?

See section 2-3 for discussion covered by the following.

2-12. What is the term used for opposition to current?

2-13. What is the unit of measurement for current opposition?

2-14. What is the symbol used for the unit in the previous question?

2-15. Which has a low resistance, a conductor or an insulator?

2-16. What is the term used for the reciprocal of resistance?

2-17. What is the unit of measurement for reciprocal of resistance?

2-18. What letter denotes conductance?

2-19. What symbol is used for mho?

2-20. Write an equation for resistance in terms of conductance?

See section 2-4 for discussion covered by the following.

2-21. Make the following conversions (use "scientific notation" if more convenient).

(a) 0.05 μF into farads.

(b) 0.05 MΩ into ohms.

(c) 75 mA into ampere.

(d) 350 μV into volts.

2-22. Make the following conversions (use "scientific motation" if more convenient).
(a) 0.035 A into mA.
(b) 25 $\mu\mu$F into farads.
(c) 27,500 Ω into kΩ.
(d) 450 μA into mA.
(e) 0.75 MΩ into kΩ.

See section 2-5 for discussion covered by the following.

2-23. A 250-kΩ resistor has 150 V applied across it. What is the current through the resistor?
2-24. If 75 μV is applied across a 0.3 MΩ resistor, what is the current through the resistor?

See section 2-6 for discussion covered by the following.

2-25. What is the voltage across a 40-Ω resistor when 3 A flow through it?
2-26. A 50-kΩ resistor has 300 μA flowing through it. What is the voltage across this resistor?
2-27. 30 mA flow through a 5.6 MΩ resistor. What is the voltage across it?

See section 2-7 for discussion covered by the following.

2-28. What is the resistance of a 120-V light bulb if 2.5 A flow through it?
2-29. A 6-V electric bell requires 30 mA to operate it. What is its resistance?
2-30. What value resistor is required if the voltage across it is 300 mV with a current of 150 μA?
2-31. A device requires 1000 μV and 2.5 mA. What is its resistance?

See section 2-8 for discussion covered by the following.

2-32. What basic units are used in the mks system for *length*, *mass*, and *time*?
2-33. What is the unit for *force* or weight in the mks system?
2-34. What is the unit for *energy* in the mks system?
2-35. Convert a force of 27 lb in the fps system into newtons in the mks system.

2-36. Convert a force of 3000 dynes in the cgs system into newtons in the mks system.
2-37. Convert 44.4 ft-lb of work in the fps system into joules in the mks system.
2-38. How many joules in the mks system are equivalent to 50,000 ergs of energy in the cgs system?
2-39. One horsepower (746 W) in the fps system is equivalent to how many joules per second in the cgs system?

See section 2-9, and examples 2-14 to 2-16, for discussion covered by the following.

2-40. What is the power in a circuit in which 90 V produces a current of 0.25 A?
2-41. If there is a current of 150 μA and a voltage of 450 V, what is the power dissipation?
2-42. A resistor is dissipating 25 W, with 250 mA flowing through it. What is the voltage across the resistor?
2-43. A 250-V, 500-W light bulb normally draws how much current?

See section 2-9, and examples 2-17 to 2-20, for discussion covered by the following problems.

2-44. A 5-kΩ resistor is dissipating 25 mW of power. What is the current?
2-45. A 12-Ω electric toaster wire has 10 A flowing through it. How much power is it consuming?
2-46. If 5 mA is flowing through a circuit, and 250 W is being dissipated, what is the resistance of the circuit?

See section 2-9, and examples 2-21 to 2-23, for discussion covered by the following problems.

2-47. A 20-kΩ resistor has 110 V across it. How much power is it dissipating?
2-48. If 12 V are being applied across a 30-Ω resistor, how much power is being consumed?
2-49. What is resistance of a 50-W, 110-V light bulb, operating normally?
2-50. A 300-Ω resistor is dissipating 12 W. What is the voltage across it?

chapter 3

Series Resistor Circuits

A *series* circuit is one where the current has only one possible path through which to flow. In the previous chapter's discussion of Ohm's law, only one resistor is used in each circuit. In this chapter, we discuss in detail series circuits containing more than just one resistor. Later, in the more advanced part of this chapter (section 3-10), we consider series circuits containing several *voltage* sources, as well as several resistors. Ohm's law is employed throughout in order to solve the problems presented by these series circuits.

3-1. Series Current

The circuit of schematic diagram shown in Fig. 3-1 is a series circuit, since the current, I (its direction is indicated by the arrows), flows from the negative end of the applied voltage, E_T, through resistors R_1 and R_2, to the positive end of E_T. This path through R_1 and R_2 is the only possible one for electrons to take to travel from the negative to the positive terminals of E_T.

As we pointed out in Chapter 2, the *same* electrons that leave the negative terminal of the applied voltage source E_T do not actually get around the circuit to the positive terminal. These electrons, coming out of the negative end of E_T, repel other electrons that are part of the atomic structure of the wires and the resistors. These repelled electrons move along only a tiny fraction of an inch, repelling still other electrons ahead of them. This action continues,

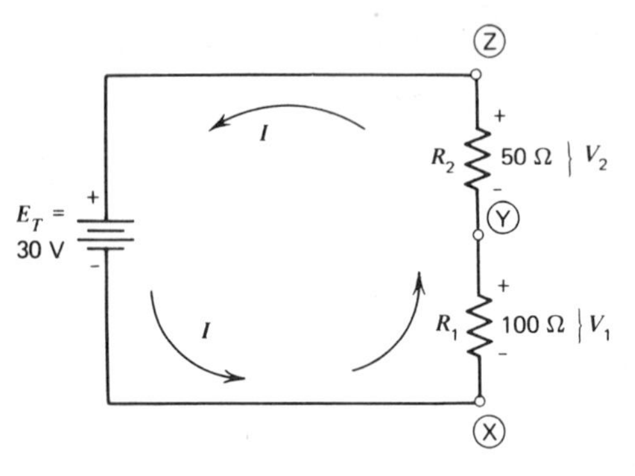

Figure 3-1. Schematic diagram of a series circuit with two resistors.

causing a motion of electrons throughout the circuit, resulting in an electron reaching the positive end of E_T for every electron that comes from the negative end of E_T.

This motion or drift of electrons, or current flow, is the same at *any* point in the circuit of Fig. 3-1. Current in a series circuit is uniform throughout the circuit. This current is usually referred to as I or I_T (I_{total}).

3-2. Series Resistance

The current in Fig. 3-1 flows through resistors R_1 *and* R_2, or a *total resistance*, R_T, equal to the *sum* of R_1 and R_2. This relationship is usually expressed as $R_T = R_1 + R_2$. In the circuit of Fig. 3-1, R_1 is 100 Ω and R_2 is 50 Ω. R_{total} is, therefore

$$R_T = R_1 + R_2$$
$$R_T = 100 + 50$$
$$R_T = 150\ \Omega$$

If there were *any* number of resistors connected in series, then the total resistance is simply the *sum* of all of them.

Example 3-1

Find the total resistance of the series circuit shown in Fig. 3-2.

Solution

$$R_T = R_1 + R_2 + R_3 + R_4$$
$$R_T = 10 + 20 + 30 + 40$$
$$R_T = 100\ \Omega$$

For Similar Problems Refer to End-of-Chapter Problems 3-1 to 3-2

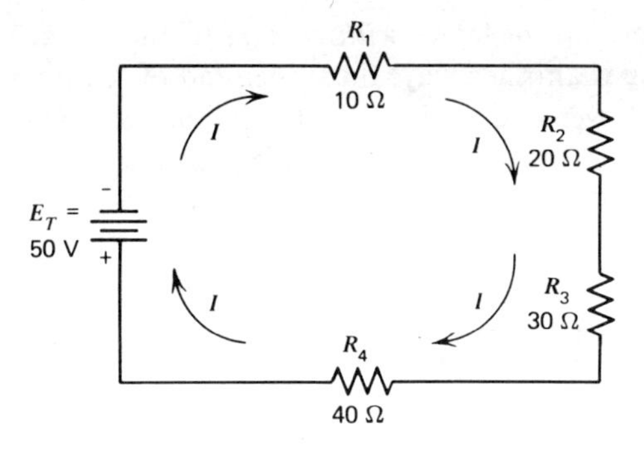

Figure 3-2. Schematic diagram of a series circuit with four resistors.

Figures 3-1 and 3-2 are *circuit diagrams* or *schematic diagrams*. The actual *physical* appearance of the four resistors and battery (E_T) of Fig. 3-2 is shown in Fig. 3-3. The resistors are often laid out one alongside the other on a board or chassis, and are strung between the pairs of mounting terminals or lugs. The connecting wires from the battery to the resistors, and between each resistor, are shown.

3-3. Series Voltage

In Fig. 3-1, electron flow, I, is shown by the arrows. Electrons only drift or flow from one point in a circuit to another if there is a difference of charge, or a voltage between the two points.

Since electrons always move from *negative* to *positive*, then a difference of charge, or a voltage, must be present *between* the ends of R_1, between points X and Y, and also *between* the ends of R_2, between points Y and Z. As shown in Fig. 3-1, electrons are moving up through R_1 and R_2. The lower end of R_1,

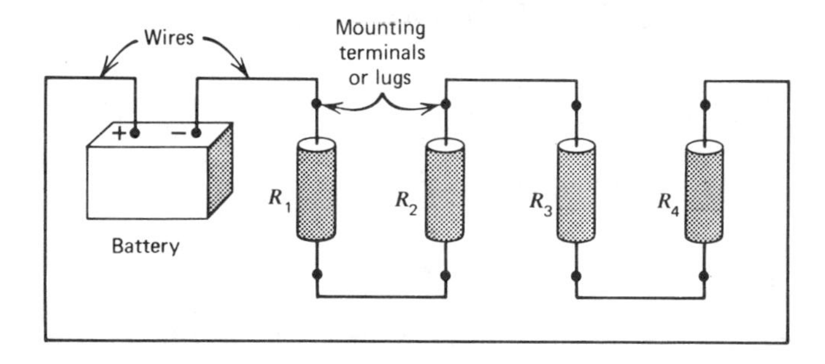

Figure 3-3. Physical appearance of the various parts of the schematic diagram of Fig. 3-2.

point X, is therefore negative *with respect* to the upper end of this resistor, point Y. Saying it another way, the upper end of R_1, point Y, is positive with respect to the lower end, point X. The $-$ and $+$ polarities, with *respect to each other*, are indicated at the ends of R_1, points X and Y.

Similarly, with electrons moving *up* through R_2, the lower end, point Y, is negative with respect to the upper end, point Z. Saying it the other way, point Z is positive with respect to point Y. These $-$ and $+$ polarities at the ends of R_2 are so indicated in Fig. 3-1.

The difference of charge between points X and Y is the voltage across R_1. This is usually called E_{R_1} or V_{R_1}, or more simply, V_1. The difference of charge between points Y and Z is, similarly, the voltage across R_2, or E_{R_2}, or V_2. To find E_{R_1} or E_{R_2}, the *current* must first be found. Then, using one of the forms of Ohm's law, $E = IR$, the voltage across R_1 or across R_2 may be calculated. This is illustrated in the following examples:

Example 3-2

Find the current in the circuit of Fig. 3-1.

Solution

$$I = \frac{E_T}{R_T}$$

$$I = \frac{30}{100 + 50}$$

$$I = \frac{30}{150}$$

$$I = 0.2 \text{ A}$$

This current flows through R_1 *and* R_2.

Example 3-3

Find the voltage across R_1 in Fig. 3-1.

Solution

$$E_{R_1} \quad \text{or} \quad V_1 = IR_1$$

$$V_1 = (0.2)(100)$$

$$V_1 = 20 \text{ V}$$

Example 3-4

Find the voltage across R_2 in Fig. 3-1.

Solution

$$E_{R_2} \quad \text{or} \quad V_2 = IR_2$$

$$V_2 = (0.2)(50)$$

$$V_2 = 10 \text{ V}$$

The voltage across R_1, (Example 3-3), and the voltage across R_2, (Example 3-4), of the circuit of Fig. 3-1 show a very impoprtant relationship when compared to the total applied voltage, E_T. This is explained in the following.

The applied voltage E_T of Fig. 3-1 is 30 V, While V_1 is 20 V (Example 3-3), and V_2 is 10 V (Example 3-4). Note that the sum of V_1 (20) and V_2 (10) is equal to E_T (30). This relationship is given by *Kirchhoff's voltage law* which states that the *sum* of the voltages in a *series* circuit is equal to the applied voltage, or

$$E_T = E_{R_1} + E_{R_2}$$

or

$$E_T = V_1 + V_2$$

Example 3-5

In the circuit of Fig. 3-2, find

(a) I
(b) E_{R_1} (V_1)
(c) E_{R_2} (V_2)
(d) E_{R_3} (V_3)
(e) E_{R_4} (V_4)

Solution

Note that in Fig. 3-2, R_{total} is now the sum of all four resistors.

(a) $$I = \frac{E_T}{R_T}$$

$$I = \frac{50}{R_1 + R_2 + R_3 + R_4}$$

$$I = \frac{50}{10 + 20 + 30 + 40}$$

$$I = \frac{50}{100}$$

$$I = 0.5 \text{ A}$$

$$
\begin{aligned}
&\text{(b) } E_{R_1} \quad \text{or} \quad & V_1 &= IR_1 \\
& & V_1 &= (0.5)(10) \\
& & &= 5.\ \text{V}
\end{aligned}
$$

$$
\begin{aligned}
&\text{(c) } E_{R_2} \quad \text{or} \quad & V_2 &= IR_2 \\
& & V_2 &= (0.5)(20) \\
& & &= 10\ \text{V}
\end{aligned}
$$

$$
\begin{aligned}
&\text{(d) } E_{R_3} \quad \text{or} \quad & V_3 &= IR_3 \\
& & V_3 &= (0.5)(30) \\
& & &= 15.\ \text{V}
\end{aligned}
$$

$$
\begin{aligned}
&\text{(e) } E_{R_4} \quad \text{or} \quad & V_4 &= IR_4 \\
& & V_4 &= (0.5)(40) \\
& & &= 20.\ \text{V}
\end{aligned}
$$

V_4 could also be found by an alternate method. After V_1, V_2, and V_3 have been found, then from Kirchhoff's voltage law:

$$
\begin{aligned}
E_T &= V_1 + V_2 + V_3 + V_4 \\
50 &= 5 + 10 + 15 + V_4
\end{aligned}
$$

and solving for E_{R_4} using Algebra by transposing the $+5$, $+10$, and $+15$ gives

$$
\begin{aligned}
50 - 5 - 10 - 15 &= V_4 \\
50 - 30 &= V_4 \\
20\ \text{V} &= V_4
\end{aligned}
$$

For Similar Problems Refer to End-of-Chapter Problems 3-1 to 3-6

3-4. Power Dissipation in Series Circuits

In the previous chapter, the *power* expended, measured in *watts*, due to a current flowing through a single resistor is given by the various forms of Watt's law (or power equations).

$$
\begin{aligned}
P \quad \text{or} \quad W &= EI \\
\text{or} \quad &= I^2R \\
\text{or} \quad &= \frac{E^2}{R}
\end{aligned}
$$

When there is more than one resistor in a series circuit, there are several powers or wattages. Each resistor dissipates some power, with the toal power

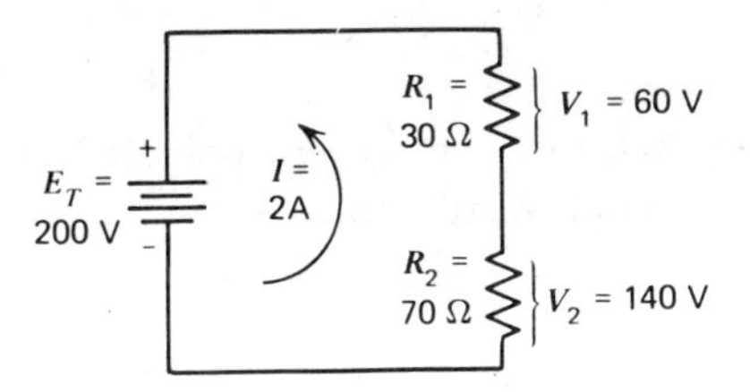

Figure 3-4. Series circuit used for Examples 3-6 to 3-8.

of the entire circuit being the sum of the individual wattages. If there are two resistors, R_1 and R_2, in a series circuit, as in Fig. 3-4, then the power dissipated in resistor R_1 is: $P_1 = V_1 I$, or $I^2 R_1$, or by the third form V_1^2/R_1, Note that the voltage, V_1, used, is just the voltage across resistor R_1. Similarly, the wattage in the second resistor R_2 is: $P_2 = V_2 I$, or $I^2 R_2$, or V_2^2/R_2. The total power consumed is then $P_T = E_T I$ or $I^2 R_T$, or E_T^2/R_T, or $P_T = P_1 + P_2$.

These examples illustrate the power that is dissipated in a series circuit.

Example 3-6

In the circuit of Fig. 3-4, find the power dissipated in resistor R_1 by using each of the three forms of Watt's law.

Solution

(a) $P_1 = V_1 I$

$= (60)(2)$

$= 120.$ W

(b) $P_1 = I^2 R_1$

$= (2^2)(30)$

$= 4(30)$

$= 120.$ W

(c) $P_1 = \dfrac{V_1^2}{R_1}$

$= \dfrac{(60)^2}{30}$

$= \dfrac{3600}{30}$

$= 120.$ W

Example 3-7

In the circuit of Fig. 3-4, find the power dissipated in resistor R_2 by using one of the three variations of Watt's law.

Solution

$$P_2 = V_2 I$$
$$= (140)(2)$$
$$= 280. \text{ W}$$

Example 3-8

In the circuit of Fig. 3-4, find the *total* power dissipated by employing one of the three variations of Watt's law.

Solution

$$P_T = E_T I$$
$$= (200)(2)$$
$$= 400. \text{ W}$$

NOTE

The power dissipated in R_1 (Example 3-6), is 120 W, while that in resistor R_2 (Example 3-7) is 280 W. The total power (Example 3-8) is 400 W. From this, the following relationship can be seen:

$$P_T = P_1 + P_2$$
$$P_T = 120 + 280$$
$$P_T = 400. \text{ W}$$

For Similar Problems Refer to End-of-Chapter Problems 3-7 to 3-10

MORE ADVANCED SERIES CIRCUITS

3-5. Voltages Across Series Resistors

In Fig. 3-5, which is the same as the circuit of Fig. 3-1, the 30 V is applied to R_1 (100 Ω) and R_2 (50 Ω) in series. From Example 3-2, the current was found to be 0.2 A, and from Examples 3-3 and 3-4, V_1 was found to be 20 V, while V_2 was 10 V. V_1 and V_2 were found using Ohm's law, $V = IR$.

Kirchhoff's voltage law showed that the *sum* of V_1 (20 V) and V_2 (10 V) equals E_T (30 V). In any series circuit, the total applied voltage, E_T, is "shared" or divided up by the resistors. The "share" that each resistor gets depends

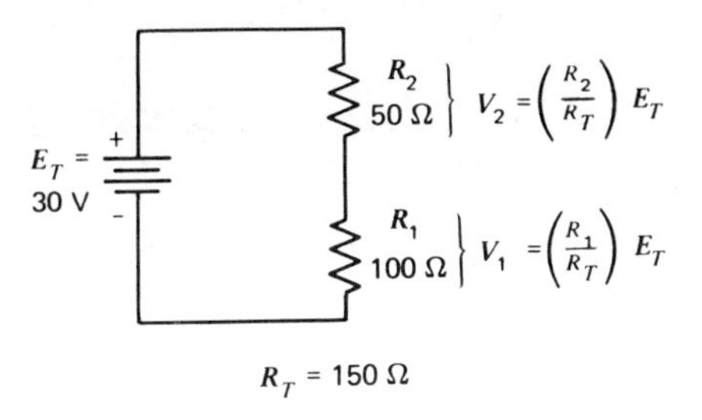

Figure 3-5.

on its size in relation to the total resistance. The voltage across *any* resistor in a *series* circuit is then simply a *fraction* of the total voltage.

$$V_{\text{across a resistor}} = \left(\frac{R_{\text{of that resistor}}}{R_{\text{total}}}\right) E_{\text{total}} \qquad (3\text{-}1)$$

From the above, it seems here that the voltage across a resistor is being calculated without using the current I. However, in referring again to the above equation, we note that E_{total} is actually being divided by R_{total}, and this E_T/R_T is really I. Then, this is being multiplied by the $R_{of\ that\ resistor}$. This is essentially multiplying I and R, although I is not readily apparent in the above equation.

This is illustrated by the following, which refers to the circuit of Fig. 3-5: Voltage across R_1, or

$$V_1 = \left(\frac{R_1}{R_T}\right) E_T \qquad (3\text{-}1)$$

$$V_1 = \left(\frac{100}{150}\right) 30$$

$$V_1 = 20\ \text{V}$$

Note that this agrees with the answer of Example 3-3. Similarly:
Voltage across R_2, or

$$V_2 = \left(\frac{R_2}{R_T}\right) E_T \qquad (3\text{-}1)$$

$$V_2 = \left(\frac{50}{150}\right) 30$$

$$V_2 = 10\ \text{V}$$

Note that this agrees with the answer of Example 3-4.

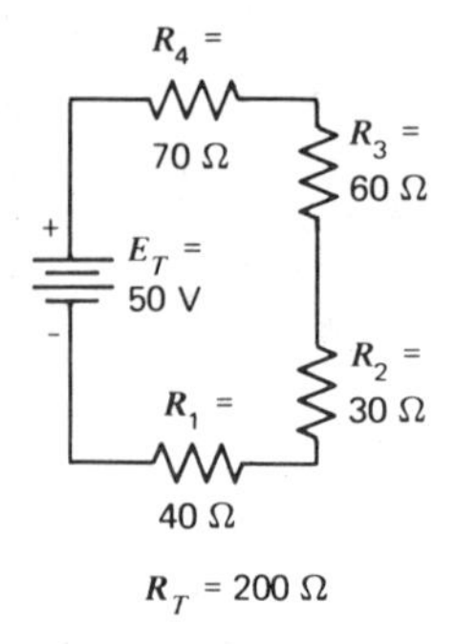

Figure 3-6. Series circuit used for Example 3-9.

Example 3-9

In the circuit of Fig. 3-6, find the voltages across resistors R_1 and R_2 using the fractional method, without first solving for the current.

Solution

(a) Voltage across R_1 or

$$V_1 = \left(\frac{R_1}{R_T}\right) E_T \tag{3-1}$$

$$V_1 = \left(\frac{40}{200}\right) 50$$

$$V_1 = 10 \text{ V}$$

(b) Voltage across R_2, or

$$V_2 = \left(\frac{R_2}{R_T}\right) E_T \tag{3-1}$$

$$V_2 = \left(\frac{30}{200}\right) 50$$

$$V_2 = 7.5 \text{ V}$$

For Similar Problems, See also Example 3-10 and Refer to End-of-Chapter Problems 3-11 to 3-13, and 3-15.

3-6. Ground, Negative, and Positive Voltages

A voltage, as has been shown, is simply a difference of charge between one point and another. It could be the difference between one terminal and the other terminal of a battery, or it could be the difference between one end of a resistor and the other end, with current flowing through the resistor.

In Fig. 3-7*a*, *b*, and *c*, the applied voltage of the battery is 100 V. This means, of course, that the difference between the ends of the battery is 100 V, with the upper end, in the drawing, being positive, and the lower end being negative. These + and − polarities are *with respect to each other*. R_1 and R_2 are assumed to be values such that the voltage across R_1 (or V_1) is 60 V, and the voltage across R_2 (or V_2) is 40 V, as shown.

Current flow, or electron drift, is up through the resistors, from the negative end of the battery to its positive end. Likewise, the voltage across R_1 (or V_1), shown as 60 V, is negative at the lower end, and positive at the upper end, since electrons move from negative to positive. The − and + polarities shown across R_1 in Fig. 3-7*a*, *b*, and *c* are only *with respect to each other*. The 40 V across R_2 is similarly marked − at the lower end, and + at the upper end. Again, these are only *with respect to each other*, and not with respect to anything else.

The connection to *ground* at the lower end of the circuit of Fig. 3-7*a*, is shown as a symbol with a few parallel, progressively smaller lines. This *ground* is simply a *reference* point from which all voltages can be measured. It is often only the metal chassis on which the circuit has been mounted. It could actually be a connection to the *earth*, but that is usually not the case. *Ground* in a circuit could even be a part of the wiring, or a *common* point called the *reference* point. By placing *ground* in the circuit, nothing is changed except that various points can be compared to this reference point.

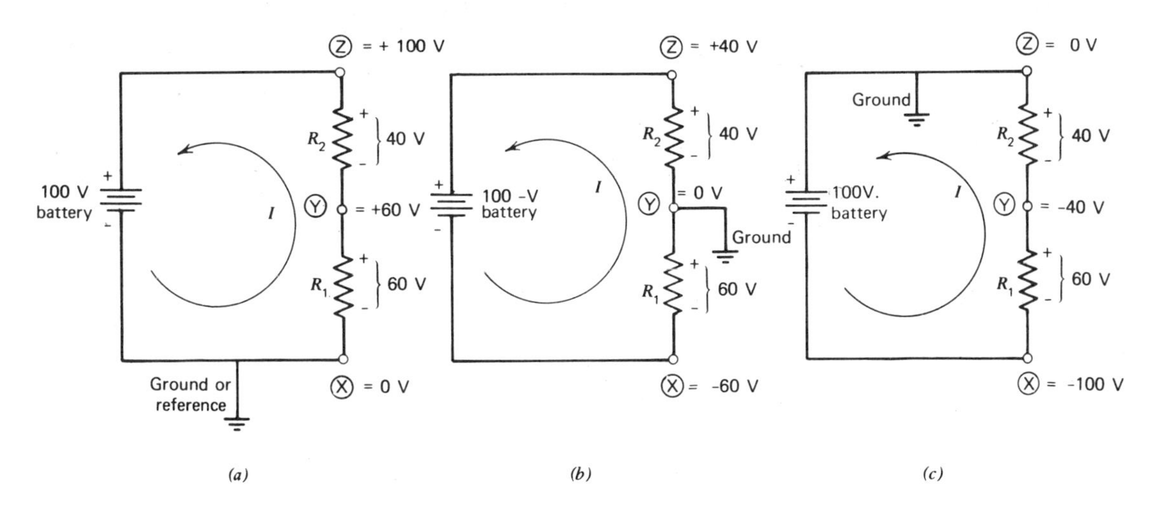

Figure 3-7. Ground, and negative and positive voltages. (a) Plus voltages with respect to ground. (*b*) Minus and plus voltages with respect to ground. (c) Minus voltages with respect to ground.

In Fig. 3-7*a*, this ground or reference point is shown connected to point X. Therefore, point X is said to be at *zero volts*, since it is the same as ground. The upper end of R_1 is 60 V positive with respect to its lower end. Therefore, point Y (the top of R_1) is said to be at +60 *V* with respect to ground (which is at the bottom of R_1).

Voltage across R_2 is 40 V, with point Z (the top of R_2) 40 V positive with respect to point Y (the bottom of R_2). Since point Y is +60 V, then point Z is +100 V, both with respect to ground (point X). The only time that a point in a circuit can be said to be +60 *V* or +100 *V* is when that point is measured with respect to a reference point or ground. With the *negative* side of the battery grounded, *positive* voltages with respect to ground are present at points Y (+60 V) and Z (+100 V).

In Fig. 3-7*b*, ground has been moved to point Y, and this point is now called *zero volts*. V_1 is still 60 V, and V_2 is still 40 V. The top of R_2, point Z, is still 40 V positive with respect to the bottom of R_2, point Y, shown by the polarities across R_2. Since point Y is zero volts, then point Z is now +40 V.

With V_1 still 60 V, then the bottom of R_1, point X, is 60 V negative with respect to the top of R_1, point Y, shown by the polarities across R_1. Since point Y is zero volts, then point X is now −60 V with respect to ground.

By grounding the circuit at point Y, *negative* and *positive* voltages, with respect to ground, are present: −60 V at point X, and +40 V at point Z. Note that the difference between point Z (+40 V) and point X (−60 V) is 100 V, or the battery voltage. Note also that point Z (+40 V) is 100 V *positive* with respect to point X (−60 V).

All *negative* voltages are present with *respect to ground* when the *positive* terminal of the battery is grounded. This is illustrated in Fig. 3-7*c*. Here,

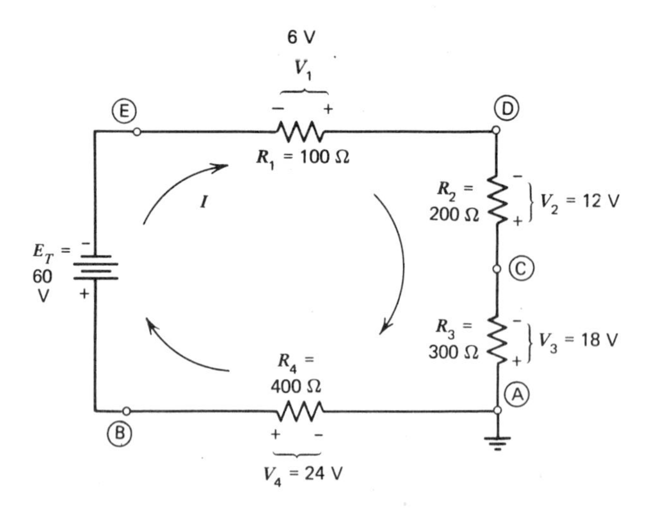

Figure 3-8 Examples 3-10 to 3-12.

point Z is grounded, and this point is said to be at zero volts. The lower end of R_2 (point Y) is still 40 V negative with respect to the top (point Z), shown by the polarities across R_2. Since point Z is zero volts, then point Y is -40 V.

Similarly, the lower end of R_1 (point X) is still 60 V negative with respect to its upper end (point Y), shown by the polarities across R_1. Since point Y is at -40 V with respect to ground, then point X is at -100 V with respect to ground.

Example 3-10

In the circuit of Fig. 3-8, find V_1 using the fractional method (equation 3-1), without solving for the current.

Solution

$$V_1 = \left(\frac{R_1}{R_T}\right) E_T \tag{3-1}$$

$$= \left(\frac{100}{1000}\right) 60$$

$$= 6 \text{ V}$$

By using the same method, now find V_2, V_3, and V_4. These values are shown in Fig. 3-8.

Example 3-11

From the values of V_1 V_2, V_3, and V_4 found in Example 3-10, in Fig. 3-8, determine the *voltage* and *polarity* between each of the following points and ground:

(a) Point A
(b) Point B
(c) Point C
(d) Point D
(e) Point E

Solution

The polarity of the applied 60 V, E_T, shown in Fig. 3-8, causes the current (electrons) to move clockwise as indicated by the arrows. The polarities of the voltage across each resistor is also shown.

(a) Point A is grounded, and is zero volts.

(b) Point B is at the positive end, the left end, of R_4. Since V_4, from Example 3-10, is 24 V, and point A is zero volts, then *point B is* $+24$ *V with respect to ground.*

(c) Point C is at the negative end, the top, of R_3. Since V_3, from Example 3-10, is 18 V, and point A is zero volts, then *point C is* -18 *V with respect to ground.*

(d) Point D is at the negative end, the top, of R_2. Since V_2, from Example 3-10, is 12 V, and point C is -18 V, then *point D* (which is 12 V more negative) *is* -30 *V with respect to ground.*

(e) Point E is at the negative end, the left end, of R_1. Since V_1, from Example 3-10, is 6 V, and point D is -30 V, then *point E* (which is 6 V more negative) *is* -36 *V with respect to ground.*

Example 3-12

From the results of Example 3-11 in Fig. 3-8, where point A = 0 V, B = +24 V, C = -18 V, D = -30 V, and E = -36 V, what is the difference and polarity between points:

(a) B and E ?
(b) B and D ?
(c) B and C ?
(d) E and C ?

Solution

(a) With point B = +24 V, and E = -36 V, the difference of charge, or voltage, between them is 60 V, with point B the + end, and E the − end. Note that this is the voltage across the applied E_T.

(b) With point B = +24 V, and D = -30 V, the difference of charge, or voltage, between them is 54 V, with point B the + end, and D the − end.

(c) With point B = +24 V, and C = -18 V, the difference of charge, or voltage, between them is 42 V, with point B the + end, and C the − end.

(d) With point E = -36 V, and C = -18 V, the difference of charge, or voltage, between them is 18 V, with C the + end, and E the − end.

For Similar Examples and Problems, See Examples 3-13 to 3-15 and 3-18, and Refer to End-of-Chapter Problems 3-14 to 3-18

3-7. Resistor Voltage Dividers, No Load

Often a voltage is required that is less than the available applied voltage. For example, as shown in Fig. 3-9, if a device requires only 30 V, then with the correct values of R_1 and R_2, the desired 30 V can be developed across R_1, even though the only available applied source of voltage is 100 V. If the device requiring the 30 V does not require current, then the only current flow in Fig. 3-9 is from the applied 100 V through R_1 and R_2. Certain devices in

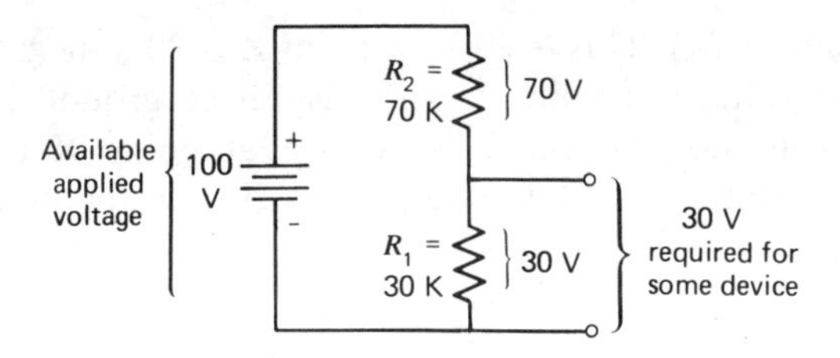

Figure 3-9. Simple resistor voltage divider.

electronics require a voltage to be present between two points, yet permit no current flow between these points. This is said to present *no load* to the applied voltage. One such example, which is of no real concern to the reader at this point, is the voltage called *bias* between the control grid and the cathode of a vacuum-tube amplifier. In this chapter, discussion of the resistor voltage divider will be limited to those circuits where no current flows to the device requiring the voltage. Later, in the chapter on more complex series-parallel circuits, we shall consider examples where current does flow to the external device, that is, where a *load* is presented to the applied voltage.

In Fig. 3-9, voltage across R_1 is the desired 30 V, since the values of R_1 and R_2 are 30,000 Ω (30 K) and 70,000 Ω (70 K), respectively. From the fractional voltage method presented in Section 3-5, voltage across R_1 in Fig. 3-9 is

$$V_1 = \left(\frac{R_1}{R_T}\right) E_T \tag{3-1}$$

$$= \left(\frac{30{,}000}{100{,}000}\right) 100$$

$$= 30 \text{ V}$$

Fig. 3-10 shows a resistor voltage divider where one end is grounded (point Y, the bottom of R_1), and where the other end is connected to -20 V (point

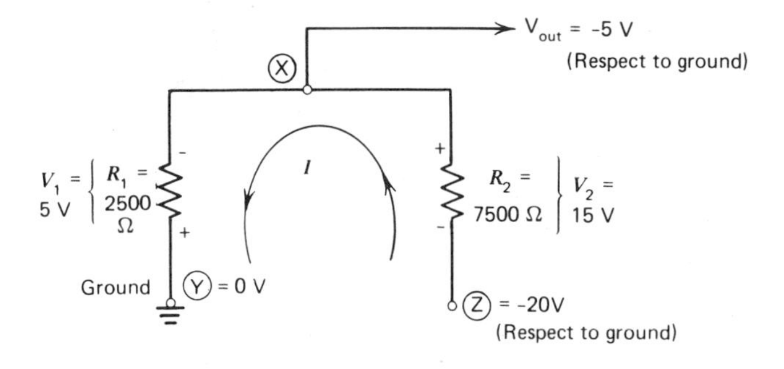

Figure 3-10. Voltage divider, Example 3-13.

Z, the bottom of R_2). This -20 V at point Z is 20 V negative with respect to ground. The voltage at point X with respect to ground is to be used by an external circuit, not shown. This voltage at point X is called the output voltage, V_{out}. Note that total voltage E_T is 20 V, the difference between point Z and ground.

Example 3-13

In the circuit of Fig. 3-10, find V_{out} with respect to ground.

Solution

Voltage across R_1, V_1 is

$$V_1 = \left(\frac{R_1}{R_T}\right) E_T \qquad (3\text{-}1)$$

$$= \left(\frac{2{,}500}{10{,}000}\right) 20$$

$$= 5 \text{ V}$$

Current flow, as shown, is from the -20 V at point Z, up through R_2, and down through R_1, producing the polarities across R_1 and R_2 shown. The upper end of R_1 is 5 V *negative* with respect to the lower end of R_1, point Y. Since point Y is ground or zero volts, then point X is -5 V.

An alternate method to find V_{out} is to find the voltage across R_2 (V_2). This is

$$V_2 = \left(\frac{R_2}{R_T}\right) E_T$$

$$= \left(\frac{7{,}500}{10{,}000}\right) 20$$

$$= 15 \text{ V}$$

The current flow up through R_2 means that the upper end of R_2 (point X) is positive 15 V with respect to the lower end (point Z). Since point Z is -20 V with respect to ground, then point X is -5 V (adding -20 and $+15$).

A resistor voltage divider may be connected between a *positive* voltage point and ground, or between a *negative* voltage point and ground (as shown in Fig. 3-10), or between *two positive* voltage points, or between *two negative* voltage points (as shown in Fig. 3-11), or between a *positive* voltage and a *negative* voltage point (Fig. 3-12).

In Fig. 3-11, resistors R_3 and R_4 make up a voltage divider connected between -100 V (point X) and -40 V (point Y). Ground is not shown, but the -100 V and -40 V are each with respect to ground. These negative voltages

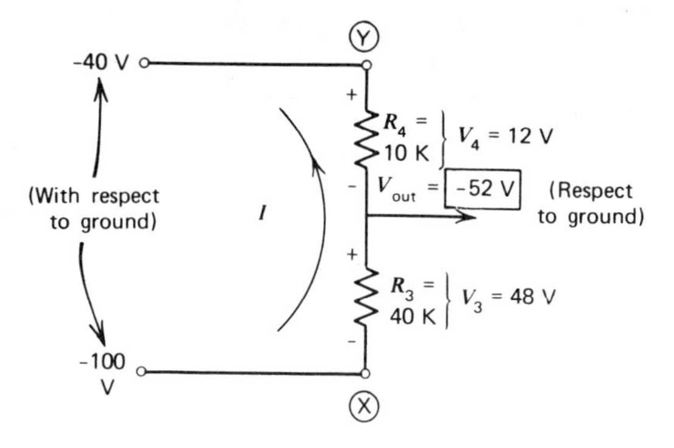

Figure 3-11. Voltage divider, Example 3-14.

are developed by a separate circuit which is not shown here, in which a ground connection is made, and with which we are not concerned at this point. A simple circuit producing the -100 V at point X and the -40 V at point Y can be seen in Fig. 3-7*c*. Later, the complete diagram is discussed in the chapter on series-parallel circuits.

Example 3-14

In the diagram of Fig. 3-11, find V_{out} with respect to ground.

Solution

The series resistors R_3 and R_4 are connected between the -40 V and the -100 V, for a total voltage difference between these points of 60 V. Point X (-100 V) is actually 60 V negative with respect to point Y (-40 V). Voltage across R_3 is, therefore,

$$V_3 = \left(\frac{R_3}{R_T}\right) E_T \tag{3-1}$$

$$= \left(\frac{40{,}000}{50{,}000}\right) 60$$

$$= 48 \text{ V}$$

Current flows as shown from the -100 V point X, up through R_3 and R_4, to the -40 V point Y. This produces the polarities across each resistor as shown. With 48 V across R_3, and the upper end (V_{out}) 48 V positive *with respect to the lower end* (point X = -100 V), then V_{out} is -52 V with respect to ground (adding -100 and $+48$).

Another method for finding V_{out} is first to solve for the voltage across R_4:

$$V_4 = \left(\frac{R_4}{R_T}\right) E_T$$

$$= \left(\frac{10{,}000}{50{,}000}\right) 60$$

$$= 12 \text{ V}$$

The polarities across R_4 are shown in Fig. 3-11. The lower end of R_4 (V_{out}) is 12 V negative *with respect to the upper end* (point Y = -40 V). Therefore, V_{out} is -52 V with respect to ground (adding -40 and -12).

In Fig. 3-12, series resistors R_1 and R_2 are connected between *a* $+100$ V (point B), and *a* -100 V (point A). The difference of voltage between points A and B is, therefore, 200 V. Depending on the values of R_1 and R_2, V_{out} is somewhere between $+100$ V and -100 V. V_{out} could be made somewhat *positive* with respect to ground, or somewhat *negative* with respect to ground, or even zero volts (same as ground), with the proper selection of R_1 and R_2.

Example 3-15

In Fig. 3-12, find V_{out} with respect to ground.

Solution

Current flows from the -100 V (point A), up through R_1 and R_2 to the $+100$ V (point B), producing the polarities across R_1 and R_2 shown. Voltage across R_1 is

$$V_1 = \left(\frac{R_1}{R_T}\right) E_T$$

$$= \left(\frac{20{,}000}{50{,}000}\right) 200$$

$$= 80 \text{ V}$$

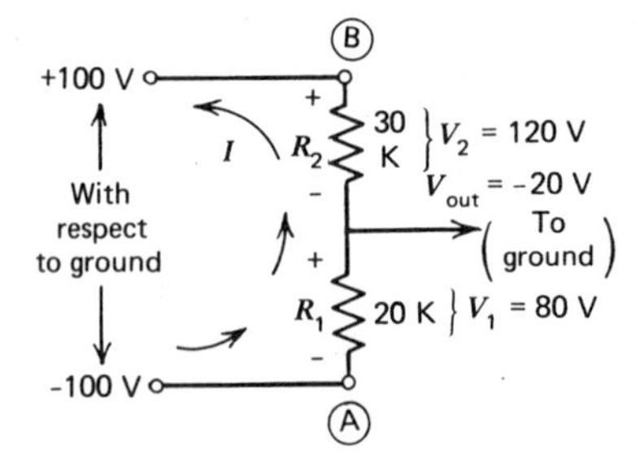

Figure 3-12. Voltage divider, Example 3-15.

The top of R_1 (V_{out}) is 80 V positive with *respect to the lower end* of R_1 (point A). Since point A is -100 V with respect to ground, then V_{out} is -20 V with respect to ground (adding -100 and $+80$).

For Similar Problems Refer to End-of-Chapter Problems 3-19 to 3-21

3-8. Series Filaments

Certain devices in electronics require that they be heated in order to operate. The vacuum tube is an example. Current is sent flowing through the *heater* or *filament* of the vacuum tube, causing it to become red hot. This heat produces an action within the tube called *electron emission* which must take place if the device is to do its job. In this chapter we do not discuss the operation of the vacuum tube, but only the circuit containing the filament portion of the tube.

To become red hot, the correct amount of current must flow through the filament wire. This wire acts as a resistor, and when its rated current flows through it, the rated voltage across it is produced. The tube manufacturer has designed the filament stating the correct rated values of voltage and current. If too much voltage appears across the filament, then the current through it becomes excessive, and the heater may quickly burn up and disintegrate, resulting in an "open" circuit. When the filaments each require the same value of current, they may be connected in series.

Figure 3-13*a* is the circuit of three series filaments, each requiring 0.2 A of current, with their resistances such that filament 1 has 20 V across it, filament 2 has 50 V, and filament 3 has 30 V. Note that, as discussed in Example

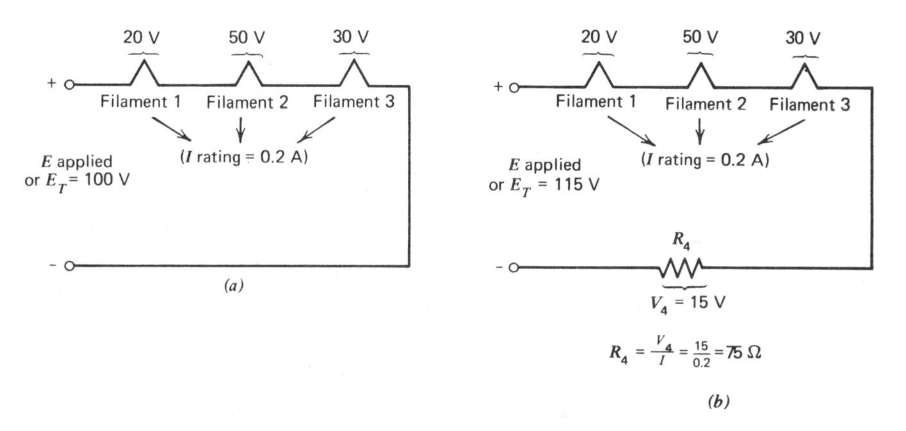

Figure 3-13. Series filaments. (*a*) Total filament voltage $= E_T$. (*b*) E_T is larger than total filament voltages, requiring resistor R_4.

3-4, Kirchhoff's voltage law holds true. That is, the sum of the voltages across each filament (20 + 50 + 30) is equal to the total applied voltage E_T (100 V).

When E_T is greater than the sum of the filament voltages, as shown in Fig. 3-13*b* then an additional resistor, R_4, must be added to the circuit. The only purpose of this resistor is to prevent too much voltage from being present across each filament, causing too much current through them. The sum of the filament voltages is 100 V. Since 115 V is applied in Fig. 3-13*b*, then without R_4, the "surplus" 15 V would divide up among the three filaments. This would mean that each filament would have more than its rated voltage.

The addition of R_4 of the correct value produces 15 V across R_4. Thus, the surplus of 15 V is "dropped" across R_4, leaving the remaining 100 V to be present across the series string of the three filaments as they require.

The correct value for R_4 is determined by the *voltage* that must appear or be "dropped" across it and the *current* that the filaments require.

$$R_4 = \frac{V_4}{I}$$

$$= \frac{15}{0.2}$$

$$= 75\ \Omega$$

Example 3-16

In Fig. 3-13*b* what is the power dissipated in resistor R_4?

Solution

Power dissipated in R_4 is

$$P_4 = V_4 I$$

$$= (15)\,(0.2)$$

$$= 3\ \text{W}$$

Example 3-17

Four filaments, each requiring 0.25 A are connected in series. If the filament voltages are 25, 50, 10, and 12, respectively, and the applied voltage is 120 V, draw the circuit showing any necessary additional parts and values.

Solution

Since the total filament voltage is 25 + 50 + 10 + 12, or 97 V, and the applied voltage is 120, then the "surplus" voltage of 120 − 97, or 23 V, must be "dropped" across a resistor that must be added to the four series filaments. The circuit would then be similar to that shown in Fig. 3-13*b*, but would contain four filaments and a resistor. The value of this added resistor is then

$$R = \frac{V_R}{I}$$

$$= \frac{23}{0.25}$$

$$= 92\ \Omega$$

and the power dissipated in the added resistor is

$$P = V_R I$$

$$= (23)\ (0.25)$$

$$= 5.75\ \text{W}$$

For Similar Problems Refer to End-of-Chapter Problems 3-22 to 3-23

3-9. Open Resistors in Series Circuits

A very common trouble in electronic circuits is the "open" reistor. Too much current through it causes it to overheat, resulting in its cracking or splitting apart. The resistor now acts as if it were simply two disconnected parts, or an *open* circuit.

Fig 3-14*a* shows four resistors in series, connected to a voltage source of 300 V. The values of the resistors are such that the 300 V is divided up among the resistors as shown, $V_1 = 25$ V, $V_2 = 50$ V, $V_3 = 100$ V, and $V_4 = 125$ V. (The student is urged to solve for these values, to check his understanding). Using ground as a reference point, then point A is -25 V with *respect to ground;* points B and C are the same as ground, or zero volt; point D is $+50$ V; point E is $+150$ V; and point F is $+275$ V. (The concepts of *ground* and *negative* and *positive* voltages are presented in Section 3-6.)

In Fig. 3-14*b*, the effects of an open resistor on the voltages at the various points A to E are shown. The discussion that follows explains the reasons for the voltages shown at each of these points.

In Fig. 3-14*b*, resistor R_1 is shown open. No current flows, and V_2, V_3, and V_4 are each zero volts. Since the lower end of R_2, point C, is grounded, it is zero volts with respect to ground. With $V_2 = 0$ V, then there is no difference of voltage between the upper end of R_2, point D, and the lower end. As a result, point D is also zero volts with respect to ground. Similarly, there is no difference of voltage between the ends of R_3 ($V_3 = 0$ V). The left end of R_3 point E, is therefore at the same potential as the right end, point D. Point E,

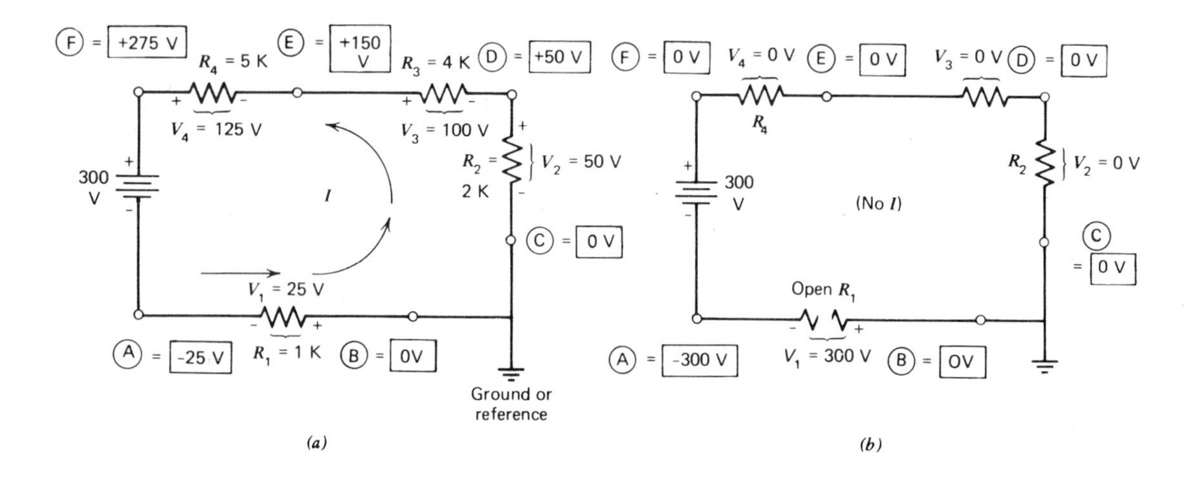

Figure 3-14. Effects of an open resistor in a series circuit. (a) Complete circuit. (b) Effect of an open R_1.

then, is also zero volts. The same is true of R_4, $V_4 = 0$ V, and point F is the same potential as point E, or zero volts with respect to ground. Note that with no current flowing throuth R_2, R_3, and R_4, the *positive* terminal of the applied 300 V, point F, is grounded through these resistors, and is the same as ground, or zero volts. The negative end of the applied voltage, point A, is therefore -300 V with respect to point F, or -300 V with respect to ground. This means that the voltage across R_1, V_1, is 300 V, as explained in the following.

With R_1 open, its resistance is practially infinite, or may actually be several billion ohms. In either case, it is so huge that current is so small that it may be considered to be zero. The applied 300 V divides up among the four resistors, with the huge resistance of the open R_1 getting practically all of the voltage, leaving nothing across the other resistors, V_1 is shown as 300 V. The right end of R_1, point B is grounded, and B is zero volts. With the very tiny current (practically zero) flowing through R_1 from left to right, the left end of R_1 is therefore -300 V with respect to ground. In a series circuit, the full applied voltage appears across the open resistor, with zero volts across the "good" resistors.

Example 3-18

In the circuit of Fig. 3-15, all voltages across the resistors are shown, as well as the voltages at points A, B, C, D, and E with respect to ground. As a review exercise, solve for V_1, V_2, V_3, and V_4, and verify the voltages at points A, B, C, D, and E (note that I is clockwise).

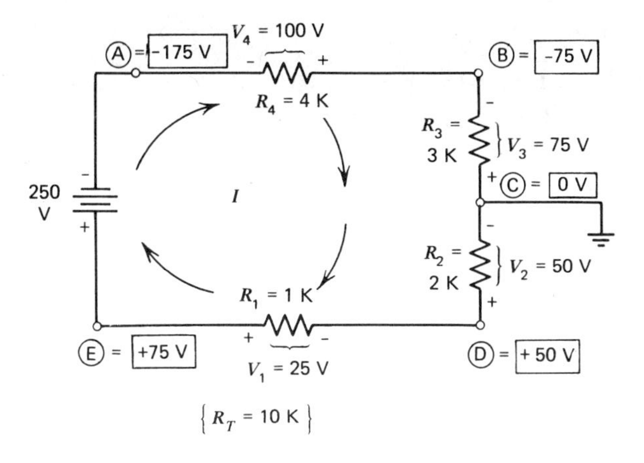

Figure 3-15. Examples 3-18 to 3-21.

Solution

All voltages are shown in Fig. 3-15. Verification is as follows, using R_1 as an example.

$$V_1 = \left(\frac{R_1}{R_T}\right) E_T \tag{3-1}$$

$$= \left(\frac{1\text{ K}}{10\text{ K}}\right) 250$$

$$= 25\text{ V}$$

Point C is grounded and is zero volts. Since $V_2 = 50$ V, point D is 50 V *positive* with respect to C. Point D is then +50 V with respect to ground.

Example 3-19

If R_1 in Fig. 3-15 opens, what is the voltage across each resistor, and at points A, B, C, D, and E with respect to ground?

Solution

With R_1 open, V_1 becomes 250 V. V_2, V_3, and V_4 each become zero volts. Points A, B, C, and D each become the same as ground, or zero volts. With 250 V across R_1, and point D = 0 V, then point E becomes +250 V with respect to ground.

Example 3-20

If R_2 in Fig. 3-15 opens, what is the voltage across each resistor, and at points A, B, C, D, and E with respect to ground?

Solution

With R_2 open, V_2 is now 250 V. V_1, V_3, and V_4 are now each zero volts. Point C is grounded and is zero volts. Since $V_3 = 0$ V and $V_4 = 0$ V, then point B is at the same potential as point C, or point B is also zero volts. Likewise, point A is the same as point C, or point A is also zero volts. Note that the *negative* end of the applied 250 V is at zero volts with respect to ground. The positive end of the applied voltage, point E, is therefore +250 V with respect to ground. With $V_1 = 0$ V, then there is no difference between points D and E, and point D is also +250 V with respect to ground. Observe that with point D at +250 V, and point C at zero volts, V_2 is 250 V.

Example 3-21

In Fig. 3-15, if R_4 opens up, what is the voltage across each resistor, and at points A, B, C, D, and E with respect to ground?

Solution

If R_4 opens, $V_4 = 250$ V, while V_1, V_2, and V_3 are each zero. Point C is grounded and is at zero volts. With $V_3 = 0$ V, point B is the same as point C, or point B is also zero volts with respect to ground. Since V_2 and V_1 are each zero, then point D is the same as point C (zero volts) and point E is the same as point D. Therefore, points D and E are also at zero volts with respect to ground. This means that the *positive* end of the applied 250 V is at zero volts, or ground. The *negative* end of the applied voltage, point A, is therefore −250 V with respect to ground. With point A = −250 V, and point B = 0 V, then V_4, across the open R_4 is 250 V, as expected.

For Similar Problems Refer to End-of-Chapter Problems 3-24 to 3-36

3-10. More Than One Applied Voltage

In more complex circuits there is often more than one applied voltage. These complex circuits usually contain several series circuits which are interconnected, forming a *mesh* or *network* circuit. In this section, we consider only *one* series portion of the mesh. In Chapter 6, networks, all the loops or series parts, are discussed.

Figure 3-16 shows a series circuit consisting of applied voltages E_1 and E_2 and resistors R_1 and R_2. E_1 should cause electrons to move counterclockwise,

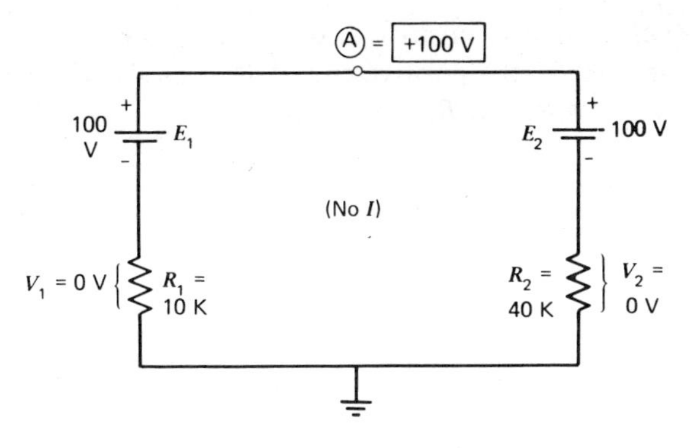

Figure 3-16. Applied voltages E_1 and E_2 are equal but opposing ($E_{\text{net applied}} = 0$ V).

while E_2 produces a clockwise movement of electrons. These voltages are said to be *opposing*. Since these applied voltages, E_1 and E_2, are equal, each being 100 V, the clockwise and counterclockwise currents completely cancel. No current flows, and voltage across resistor R_1, V_1, is zero, and across R_2, V_2 is also zero. Voltage at point A with respect to ground is the sum of E_1 and V_1, or the sum of E_2 and V_2. In either case, $100 + 0$ is 100 V. Therefore, point A is +100 V with respect to ground.

If the two applied voltages, E_1 and E_2 were *unequal*, as shown in Fig. 3-17, a current will flow as a result of the larger voltage. E_1 is 100 V while E_2 is only 70 V. The effect is as if E_1 were in the circuit alone and were 30 V. That is, E_1 and E_2 *oppose* one another. The difference between them is 30 V, and this difference acts as if it were the net total applied voltage, or E_T.

Current flows counterclockwise, in Fig. 3-17 because of the polarity of the larger applied voltage E_1. This current flows up through R_2, producing the

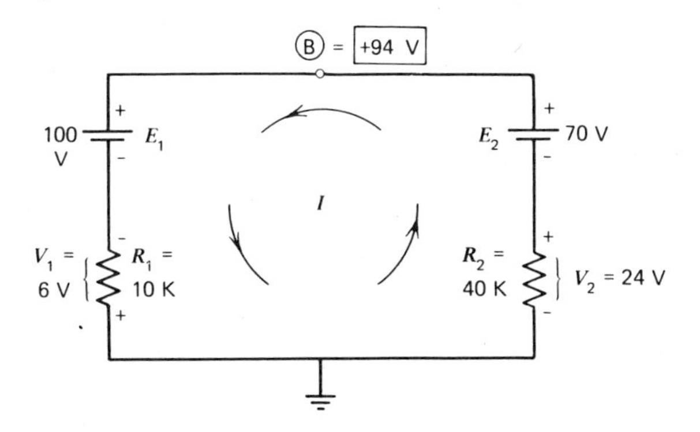

Figure 3-17. Applied voltages E_1 and E_2 are unequal and opposing ($E_{\text{net applied}} = 100 - 70 = 30$ V).

polarity of V_2 shown, positive at the upper end of R_2. Since this current flows down through R_1, the positive end of V_1 is as shown, at the lower end of R_1. Voltages across R_1 and R_2 are, respectively,

$$V_1 = \left(\frac{R_1}{R_T}\right) E_{\text{net applied}} \tag{3-1}$$

$$= \left(\frac{10\text{ K}}{50\text{ K}}\right) 30$$

$$= 6\text{ V}$$

and

$$V_2 = \left(\frac{R_2}{R_T}\right) E_{\text{net applied}}$$

$$= \left(\frac{40\text{ K}}{50\text{ K}}\right) 30$$

$$= 24\text{ V}$$

The voltage between point B and ground is the *algebraic* sum of E_2 and V_2, or the *algebraic* sum of E_1 and V_1. Observe that the E_2 and V_2 voltages are aiding each other. That is, the *negative* end of E_2 is connected to the *positive* end of V_2. As a result, their algebraic sum is simply the sum of 70 V and 24 V, or 94 V, positive at point B as shown.

This +94 V at point B with respect to ground is also the result of algebraically adding E_1 and V_1. Note that these voltages are opposing, since the *negative* end of E_1 is connected to the *negative* end of V_1. E_1 (100 V) is larger than V_1 (6 V) by 94 V, and their *sum in algebra* is actually their *difference*. Considering the *upper* ends of both voltages, E_1 could be called +100 V, while V_1 could be called −6 V. In algebra, *adding unlike signs* such as +100 and −6 really

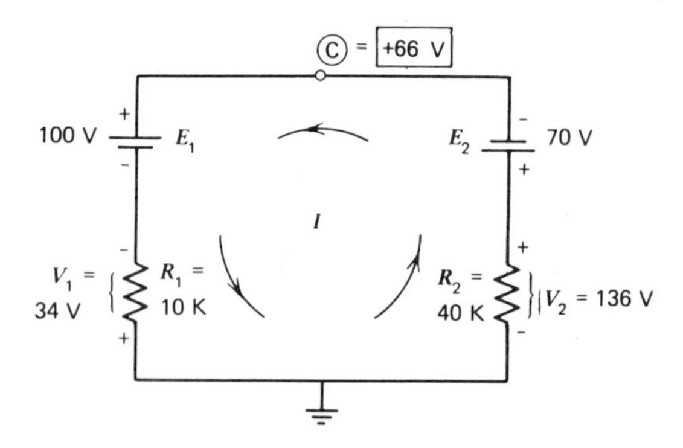

Figure 3-18. Applied voltages E_1 and E_2 are aiding ($E_{\text{net applied}} = 100 + 70 = 170$ V).

means *subtracting* the smaller number (6) from the larger one (100), giving the answer (94) the + sign of the larger number. Therefore, point B is +94 V with respect to ground.

In Fig. 3-18, the two applied voltages, $E_1 = 100$ V and $E_2 = 70$ V, are aiding each other (the *positive* of one is connected to the *negative* of the other). Both E_1 and E_2 will send current flowing counterclockwise. The total voltage causing current flow, or E_{net} applied, is the sum of E_1 and E_2, or $100 + 70 = 170$ V.

The counterclockwise current flows up through R_2 and down through R_1, producing the voltage polarities across the resistors as shown. These voltages are

$$V_1 = \left(\frac{R_1}{R_T}\right) E_{\text{net applied}} \tag{3-1}$$

$$= \left(\frac{10\text{ K}}{50\text{ K}}\right) 170$$

$$= 34\text{ V}$$

and

$$V_2 = \left(\frac{R_2}{R_T}\right) E_{\text{net applied}} \tag{3-1}$$

$$= \left(\frac{40\text{ K}}{50\text{ K}}\right) 170$$

$$= 136\text{ V}$$

Voltage at point C with respect to ground can be calculated by either algebraically adding E_2 and V_2, or E_1 and V_1. E_2 and V_2 are opposing (both positive ends are connected together), and their algebraic sum is (considering the upper ends of each): -70 and $+136$, or $+66$ V at point C.

Point C-to-ground-voltage is also the sum, in algebra, of the opposing E_1 and V_1, or $+100$ and -34, or $+66$ V.

In the following examples using Fig. 3-19, note there there are *four* applied voltages, E_1, E_2, E_3, and E_4. It must first be determined which are aiding and which are opposing. Then, the net applied voltage can be found, and the *direction* of current flow.

Example 3-22

In the circuit of Fig. 3-19, find (a) $E_{\text{net applied}}$, (b) voltage and polarity across each resistor, and (c) voltage at point X with respect to ground.

Solution

(a) Applied voltages, E_3, E_1, and E_2 each cause current to flow clockwise. Therefore, these voltages are aiding. Applied voltage E_4 causes a current flow

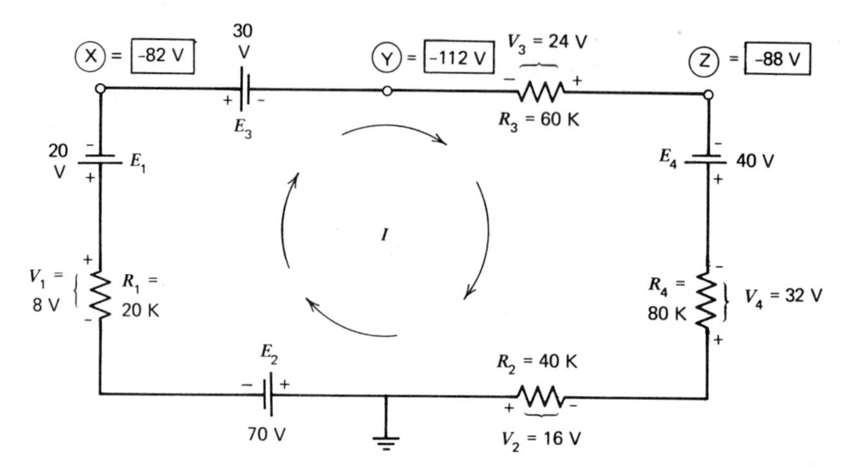

Figure 3-19. Examples 3-22 to 3-24.

$$E_{\text{net applied}} = \underbrace{E_1 + E_2 + E_3}_{\text{aiding}} - \underbrace{E_4}_{\text{opposing}}$$
$$= 20 + 70 + 30 - 40$$
$$= 80 \text{ V}$$

which is counterclockwise. E_4 then is bucking or opposing the others, and the net applied voltage is

$$E_{\text{net applied}} = E_3 + E_1 + E_2 - E_4$$
$$= 30 + 20 + 70 - 40$$
$$= 80 \text{ V}$$

(b) Since the sum of the E_3, E_1, and E_2 voltages (causing the clockwise current) is 120 V and is larger than the 40 V E_4 (which causes the counter-clockwise current), then the 80 V net produces a clockwise current shown in Fig. 3-19. Voltage polarity across each resistor is indicated in the diagram, with electrons, of course, going from − to +. Find the values of the resistor voltages by the method indicated for V_1.

$$V_1 = \left(\frac{R_1}{R_T}\right) E_{\text{net applied}} \tag{3-1}$$
$$= \left(\frac{20 \text{ K}}{200 \text{ K}}\right) 80$$
$$= 8 \text{ V}$$

(c) Voltage between point X and ground is the algebraic sum of E_1, V_1, and E_2. E_1 and E_2 are aiding voltages, but V_1 is opposing, since the *positive* end of V_1 is adjacent to the *positive* end of E_1, and the *negative* end of V_1 is adjacent to the *negative* end of E_2. By using the polarities of each voltage that is *toward* point X, the result is

$$\begin{aligned} V_{x\text{-to-ground}} &= -E_1 + V_1 - E_2 \\ &= -20 + 8 - 70 \\ &= -90 + 8 \\ &= -82 \text{ V} \end{aligned}$$

Point X is therefore -82V with respect to ground.

Example 3-23

Using the values found in the previous example, find the voltage at point Z with respect to ground in Fig. 3-19.

Solution

Voltage between point Z and ground is the algebraic sum of E_4, V_4, and V_2. These voltages are all aiding. Using their polarities nearer point Z, their sum is

$$\begin{aligned} V_{z\text{-to-ground}} &= -E_4 - V_4 - V_2 \\ &= -40 - 32 - 16 \\ &= -88 \text{ V} \end{aligned}$$

Point Z is therefore -88 V with respect to ground.

Example 3-24

Using the values found in the previous two examples, find the voltage at point Y with respect to ground in Fig. 3-19.

Solution

Point X is -82 V (from Example 3-22*c*). The 30 V of E_3 is between points X and Y, with the negative end of E_3 at point Y. Point Y is therefore 30 V *negative* with respect to point X (-82). Point Y then is -112 V with respect to ground.

Another solution for the voltage at point Y is that the voltage V_3 (24 V) is between points Z and Y with the negative end toward point Y. Point Y is then 24 V *negative* with respect to point Z. Since point Z has been found (Example 3-23) to be -88 V with respect to ground, then point Y is 24 V more negative, or Y is -112 V.

For Similar Problems Refer to End-of-Chapter Problems 3-37 to 3-38

PROBLEMS

See sections 3-1, 3-2, 3-3 for discussion covered by the following problems.

3-1. Four resistors, each 250-Ω, are connected in series with a 9-V battery. Find: (a) R_{total}, (b) I.

3-2. In the circuit for this problem, find: (a) R_{total}, (b) R_2.

3-3. Three resistors, a 4000-Ω, a 3000-Ω, and a 1000-Ω, are connected in series with a voltage source E_T. If the current flow is 0.5 mA, find: (a) voltage across the 4000-Ω resistor, (b) voltage across the 3000-Ω, (c) voltage across the 1000 Ω, and (d) E_T.

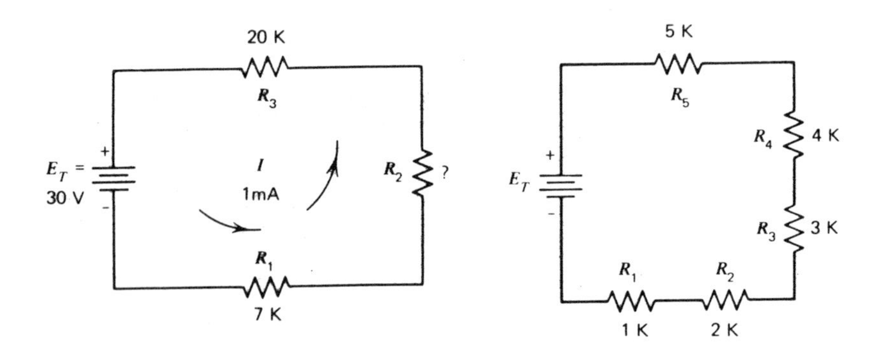

Problem 3-2.

Problem 3-4.

3-4. In the circuit shown, find R_{total}.

3-5. In the circuit of Problem 3-4, if V_3 is 6 V, find: (a) I, (b) V_1, (c) V_2, (d) V_4, (e) V_5, and (f) E_T.

3-6. In the circuit of Problem 3-4, if E_T is 120 V, find: (a) I, (b) V_1, (c) V_2, (d) V_3 (e) V_4, and (f) V_5.

See section 3-4 for discussion covered by the following problems.

3-7. In the circuit shown for this problem, find: (a) I, (b) V_1, (c) V_2, (d) P_1, (e) P_2, (f) P_T. (Solve for the power dissipated, using several methods.)

3-8. If 10 mA flows through a 20-, a 30-, and a 50-K resistor connected in series, find P_T.

3-9. A 2-K and an 8-K resistor are connected in series with a voltage source E_T. If the smaller resistor is dissipating 8 mW of power, find: (a) I, (b) V across the 2-K resistor, (c) power dissipated in the 8-K resistor, (d) E_T, (e) P_T.

3-10. Find the ohmic value of R_1 in the circuit shown for this problem.

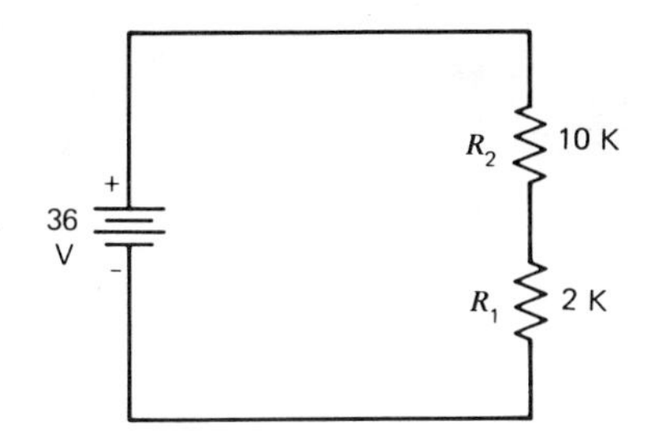

Problem 3-7.

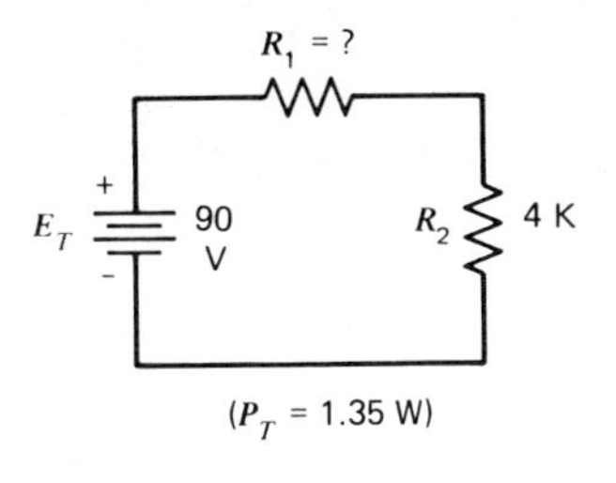

Problem 3-10.

More Challenging Problems: See section 3-5 for discussion covered by the following problems.

3-11. Three 100-Ω resistors are connected in series with a 30-V battery. Without finding the current, find the voltage across each resistor.

3-12. Three resistors, 1 K, 2 K, and 3 K, are connected in series with a 9-V source. Without solving for the current, find: (a) voltage across the 1-K resistor, (b) voltage across the 2-K, and (c) voltage across the 3-K.

3-13. In the circuit shown for this problem, find E_T without solving for the current.

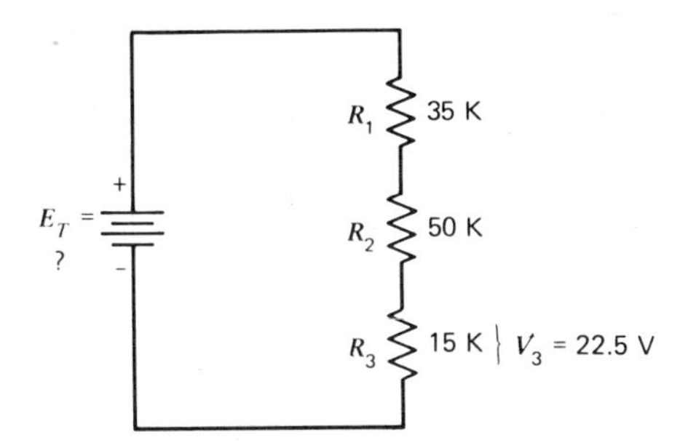

Problem 3-13.

See section 3-6 for discussion covered by the following problems.

3-14. In the circuit shown, (a) what is the voltage and polarity at point A with respect to ground, (b) what is the voltage at point B with respect to ground, and (c) what is E_T?

3-15. In the circuit shown for this problem, find the voltage and polarity of (a) V_1, (b) V_2, (c) V_3, and (d) V_4.

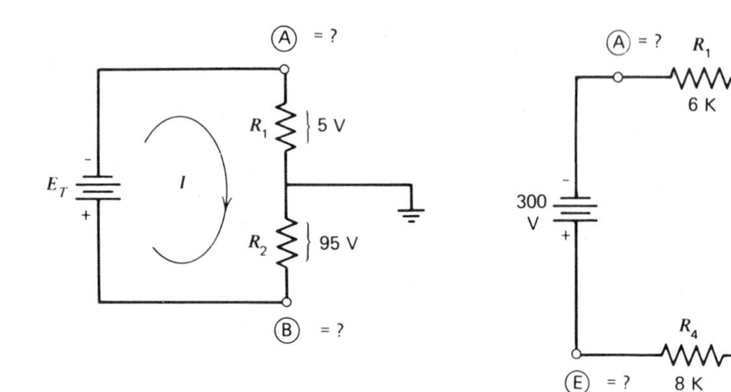

Problem 3-14.

Problem 3-15.

3-16. In the circuit of Problem 3-15, find the voltage and polarity with respect to ground at: (a) point A, (b) point B, (c) point C, (d) point D, and (e) point E.

3-17. If the applied 300 V were reversed in polarity in the circuit of Problem 3-15, find the voltage and polarity with respect to ground at: (a) point A, (b) point B, (c) point C, (d) point D, and (e) point E.

3-18. If, in some circuit, point X is -50 V with respect to ground; point Y is -10 V to ground; and point Z is $+200$ V to ground, what is the voltage between: (a) points X and Y, (b) points Y and Z, and (c) points X and Z.

See section 3-7 for discussion covered by the following problems.

3-19. In the circuit for this problem, find the voltage and polarity at point A with respect to ground.

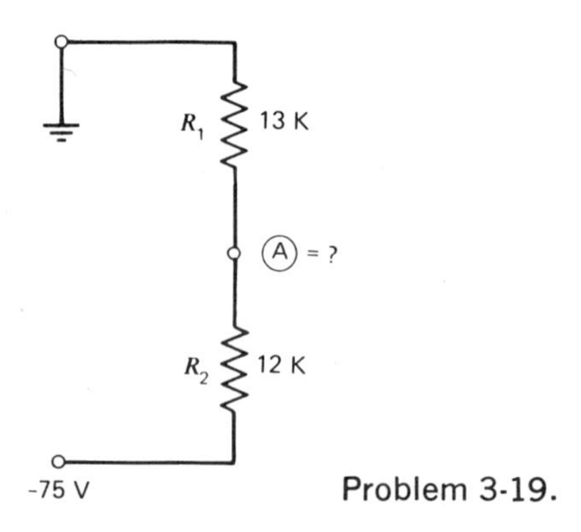

Problem 3-19.

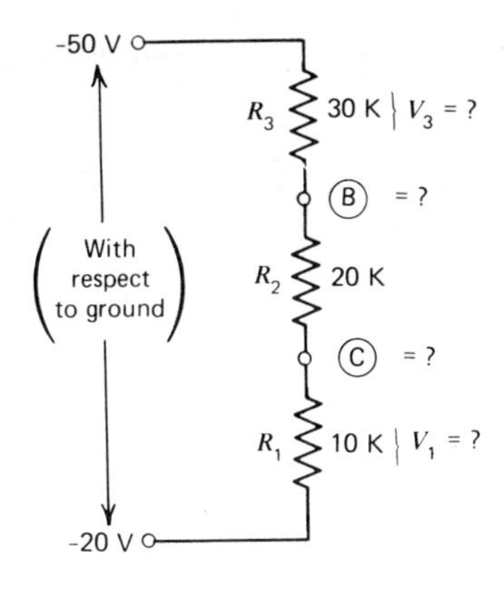

Problem 3-20.

-50 V
R_6
30 K
V_6 = ?
X = ?
With respect to ground
R_5
25 K
Y = ?
R_4
20 K
V_4 = ?
+100 V

Problem 3-21.

3-20. In the circuit shown, find: (a) V_3, (b) V_1 (c) voltage at point B with respect to ground, and (d) voltage at point C with respect to ground.

3-21. In the circuit shown, find: (a) V_6, (b) V_4, (c) voltage at point X with respect to ground, and (d) voltage at point Y to ground.

See section 3-8 for discussion covered by the following problems.

3-22. Redraw the circuit shown for this problem, adding any necessary resistors. Find: (a) power dissipated by each filament, (b) value of the required resistor that must be added, (c) power dissipated by this resistor, and (d) total power dissipated by the circuit.

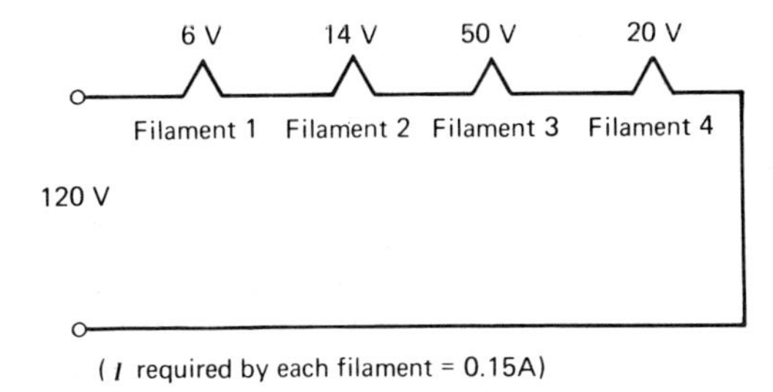

Problem 3-22.

3-23. What would probably occur in the circuit shown for Problem 3-22 if no additional resistor were added?

See section 3-9 for discussion covered by the following problems.

3-24. In the circuit of Problem 3-14, if R_1 opened, what would the voltage be with respect to ground at: (a) point A, (b) point B?

3-25. In the circuit of Problem 3-14, if R_2 opened, what would be voltage be with respect to ground at: (a) point A, (b) point B?

3-26. In the circuit of Problem, 3-15, what would the voltage be with respect to ground at points A, B, C, D, and E, if R_1 opened?

3-27. The same as Problem 3-26, but if R_2 opened?

3-28. The same as Problem 3-26, but if R_3 opened?

3-29. The same as Problem 3-26, but if R_4 opened?

3-30. In the circuit of Problem 3-20, what would the voltage be with respect to ground if R_1 opened, at points B and C?

3-31. The same as Problem 3-30, but if R_2 opened?

3-32. The same as Problem 3-30, but if R_3 opened?

3-33. If R_4 opened in the circuit of Problem 3-21, what would be the voltage at points X and Y with respect to ground?

3-34. The same as Problem 3-33, but if R_5 opened?

3-35. The same as Problem 3-33, but if R_6 opened?

3-36. In the circuit diagram of Problem 3-22, if filament 1 opened up, what would be the voltage across each individual filament?

See section 3-10 for discussion covered by the following problems.

3-37. In the circuit for this problem, find: (a) the net voltage that causes a current, (b) electron flow direction, clockwise or counterclockwise, (c) V_1

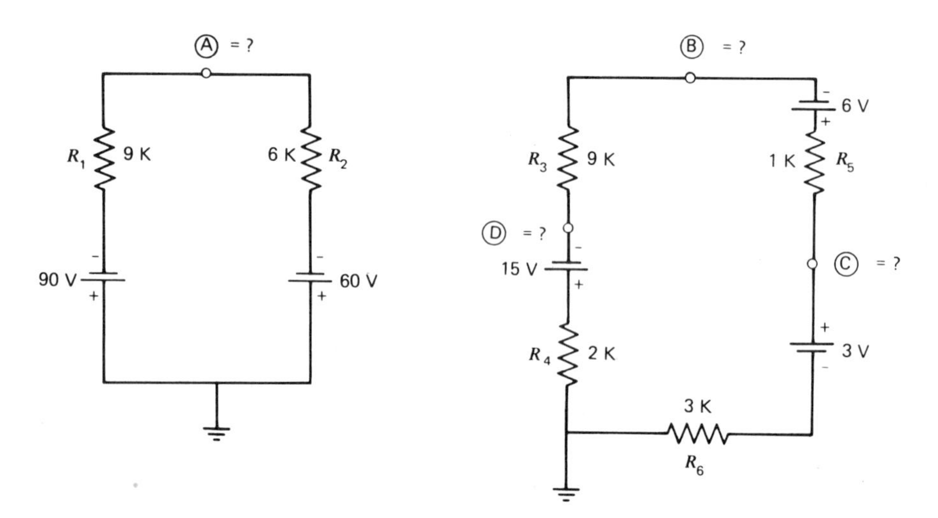

Problem 3-37.

Problem 3-38.

and its polarity, (d) V_2 and its polarity, and (e) voltage at point A with respect to ground.

3-38. In the circuit shown for this problem, find: (a) the net voltage that causes a current, (b) electron flow direction, clockwise or counterclockwise, (c) V_3 and its polarity, (d) V_4 and its polarity, (e) V_5 and its polarity, (f) V_6 and it polarity, (g) voltage at point B with respect to ground, (h) voltage at point C with respect to ground, and (i) voltage at point D with respect to ground.

chapter

4

Parallel Resistor Circuits

A *parallel* circuit is one where the current has more than one path to flow through. When resistors are connected in this parallel arrangement, less opposition is offered, since some current is bypassed through the additional path or paths. As a result, the *total* resistance is decreased when additional resistors are connected in parallel. This will become evident in the discussion of this chapter.

4-1. Two Resistors in Parallel

The diagrams of Fig. 4-1*a* to *d* illustrate the effect of connecting two resistors in a parallel circuit. Figure 4-1*a* shows an *open* circuit. That is, since both switches are in open positions, neither resistor is connected to the applied 120 V, and no current flows.

In Fig. 4-1*b*, switch 1 is closed, connecting R_1 across the applied 120 V. Since switch 2 is still open, R_2 is still disconnected. With R_1 alone in the circuit, current I_1 flows through it, making up a *series* circuit. This current I_1 is

$$I_1 = \frac{E_{\text{applied}}}{R_1}$$

$$I_1 = \frac{120}{60}$$

$$I_1 = 2 \text{ A}$$

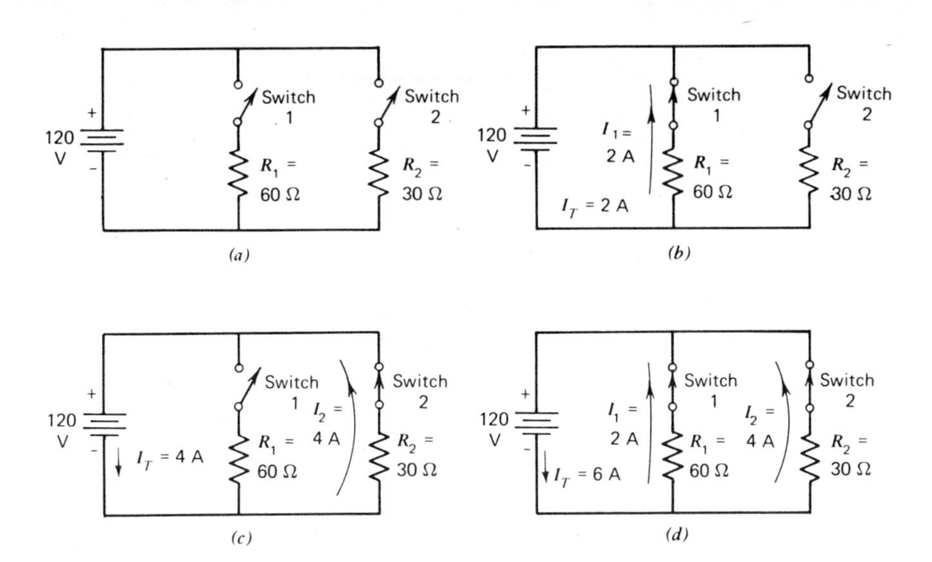

Figure 4-1. (*a*) Both switches open, no current. (*b*) Switch 1 closed, switch 2 open. Current flows through R_1. (*c*) Switch 2 closed, switch 1 open. Current flows through R_2. (*d*) Both switches closed. Current flows through R_1 and R_2.

In Fig. 4-1*c*, switch 2 is now closed, connecting R_2 across the applied 120 V. Switch 1 is now open, disconnecting R_1 from the circuit. With R_2 alone in the circuit, current I_2 flows through it, again making up a *series* circuit. Current I_2 is

$$I_2 = \frac{E_{\text{applied}}}{R_2}$$

$$I_2 = \frac{120}{30}$$

$$I_2 = 4 \text{ A}$$

When both switches are closed, as shown in Fig. 4-1*d*, resistors R_1 and R_2 are *both* connected across the applied 120 V. The 120 V E_{total} is also the voltage across R_1 (V_1) and also across R_2 (V_2). In a parallel circuit, therefore, $E_{\text{total}} = V_1 = V_2$. Now current flows through R_1 and also through R_2. Since current now has more than one path through which to flow, the diagram of Fig. 4-1*d* is a *parallel* circuit. One current flows in the left branch (through R_1), while a second current flows in the right branch (through R_2).

I_1 is still 2 A, and I_2 is still 4 A, with the *total* current I_T from the 120 V being the sum of the two branch currents. I_T is

$$I_T = I_1 + I_2$$

$$I_T = 2 + 4$$

$$I_T = 6 \text{ A}$$

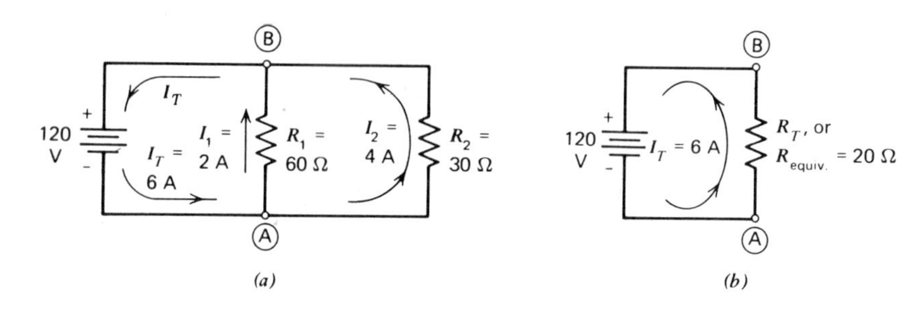

Figure 4-2. (*a*) Parallel circuit. (*b*) Equivalent circuit.

This relationship of the total current equaling the sum of the individual branch currents is called *Kirchoff's current law.*

The switches shown in Fig. 4-1*a* to *d* were included only to allow first R_1 alone, then R_2 alone, and finally both R_1 and R_2 to be placed in the circuit. The actual practical two-resistor parallel circuit is shown in Fig. 4-2*a*, without the switches. The total current, 6 A, from the negative end of the applied 120 V flows into point A. Leaving point A the current branches into two parts. One current, $I_1 = 2$ A, flows up through R_1 into point B, and a second branch current, $I_2 = 4$ A, flows up through R_2 into point B. At point B the two branch currents I_1 and I_2 join together to form the total current of 6 A.

Point A, where I_T first splits up into its two branches, and *Point B*, where the two branch currents join again to form I_T, are the ends of the parallel circuit. This means that R_1 and R_2 are in parallel or in *shunt* with each other. The *total* or equivalent resistance of the circuit can be found using Ohm's law:

$$R_T = \frac{E_T}{I_T}$$

$$R_T = \frac{120}{6}$$

$$R_T = 20\ \Omega$$

This means that between points A and B, R_1 (60 Ω) and R_2 (30 Ω) in parallel are the *equivalent* of a single 20-Ω resistor. This is shown in Fig. 4-2*b*. Note than I_T of 6 A *effectively* flows through the 20-Ω equivalent resistor. In a parallel circuit, R_{equiv} is the same as R_{total}. This *equivalent resistor is always less than the smaller of the two parallel resistors.* Another method of determining the *equivalent* resistance of two parallel resistors is as shown:

$$R_T \quad \text{or} \quad R_{\text{equiv}} = \frac{1}{\dfrac{1}{R_1} + \dfrac{1}{R_2}} \tag{4-1}$$

by taking the reciprocal of both sides of equation, it may be rewritten as:

$$\frac{1}{R_T \text{ or } R_{\text{equiv}}} = \frac{1}{R_1} + \frac{1}{R_2} \tag{4-2}$$

These equations are usually called the *reciprocal* method. Either equation 4-1 or 4-2 may be used in solving the following example.

Example 4-1

Find R_{equiv} of the circuit of Fig. 4-2*a*.

Solution

$$R_{\text{equiv}} = \frac{1}{\dfrac{1}{R_1} + \dfrac{1}{R_2}} \tag{4-1}$$

$$R_{\text{equiv}} = \frac{1}{\dfrac{1}{60} + \dfrac{1}{30}}$$

$$R_{\text{equiv}} = \frac{1}{\dfrac{1+2}{60}} \quad \text{(where 60 is a common denominator)}$$

$$R_{\text{equiv}} = \frac{1}{\dfrac{3}{60}}$$

$$R_{\text{equiv}} = 1 \times \frac{60}{3}$$

$$R_{\text{equiv}} = 20\ \Omega$$

as shown in Fig. 4-2*b*.

Example 4-2

(a) Find the power dissipated in R_1 in Fig. 4-2*a*.
(b) Find the power dissipated in R_2 in Fig. 4-2*a*.
(c) Find the *total* power dissipated in Fig. 4-2*a*.

Solution

(a)
$$P_1 = V_1 I_1 = (120)(2) = 240\ \text{W}$$

(b)
$$P_2 = V_2 I_2$$
$$= (120)(4)$$
$$= 480 \text{ W}$$

(c)
$$P_T = E_{\text{total}} I_{\text{total}}$$
$$= (120)(6)$$
$$= 720 \text{ W}$$

or
$$P_T = P_1 + P_2$$
$$= 240 + 480$$
$$= 720 \text{ W}$$

Note that in a parallel circuit, as in the series circuits of Chapter 3, the *total* power dissipation is equal to the *sum* of the wattages dissipated in each resistor.

For Similar Problems Refer to End-of-Chapter Problems 4-1 to 4-4

A simpler formula than equation 4-1 or 4-2 for solving the *equivalent* resistance of *two* parallel resistors can be derived. The equation is

$$R_{\text{equiv}} = \frac{R_1 R_2}{R_1 + R_2} \tag{4-3}$$

Equation 4-3 is also referred to as the *product divided by the sum* method of the two parallel resistors.

Example 4-3

Find the *equivalent* resistance of the 60-Ω (R_1) and the parallel 30-Ω resistor (R_2) of Fig. 4-2*a*, using equation 4-3.

Solution

$$R_{\text{equiv}} = \frac{\text{product}}{\text{sum}} \tag{4-3}$$
$$= \frac{(R_1)(R_2)}{R_1 + R_2}$$
$$= \frac{(60)(30)}{60 + 30}$$
$$= \frac{1800}{90}$$
$$= 20\ \Omega$$

Note that this answer of 20-Ω is the same found by using the method of Example 4-1.

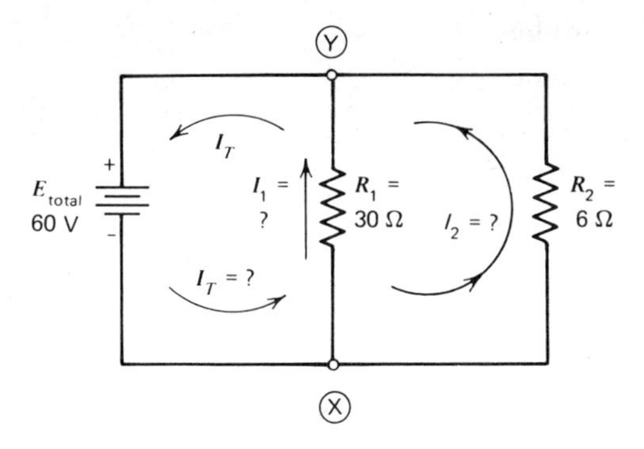

Figure 4-3.

Example 4-4

In the circuit of Fig. 4-3 find (a) V_1, (b) V_2, (c) I_1, (d) I_2, (e) I_T, and (f) R_{equiv} between points X and Y.

Solution

(a) Voltage across R_1, or $V_1 = E_{total}$

$$V_1 = 60 \text{ V}$$

(b) Voltage across R_2, or $V_2 = E_{total}$

$$V_2 = 60 \text{ V}$$

(c) $$I_1 = \frac{V_1}{R_1}$$

$$I_1 = \frac{60}{30}$$

$$I_1 = 2 \text{ A}$$

(d) $$I_2 = \frac{V_2}{R_2}$$

$$I_2 = \frac{60}{6}$$

$$I_2 = 10 \text{ A}$$

(e) $I_T = I_1 + I_2$ (Kirchhoff's current law).

$$I_T = 2 + 10$$

$$I_T = 12 \text{ A}$$

$$\text{(f)}\quad R_{\text{equiv}} = \frac{\text{product}}{\text{sum}} \tag{4-3}$$

$$R_{\text{equiv}} = \frac{R_1\, R_2}{R_1 + R_2}$$

$$R_{\text{equiv}} = \frac{(30)\,(6)}{30 + 6}$$

$$R_{\text{equiv}} = \frac{180}{36}$$

$$R_{\text{equiv}} = 5\ \Omega$$

The *equivalent* resistance between points X and Y is also the *total* resistance of the parallel circuit, and could also have been found using Ohm's law, $R_T = E_T/I_T$.

Example 4-5

If the 12-V filament of a vacuum tube should have 150 mA flowing through it, and it is connected in a circuit where 250 mA flows, what value of resistor (R_1) should be placed in shunt (parallel) with the filament? The circuit is shown in Fig. 4-4.

Solution

As shown in Fig. 4-4,

$$I_T = I_1 + I_{\text{fila}}$$

$$I_T - I_{\text{fila}} = I_1$$

$$250 - 150 = I_1$$

$$100\ \text{mA} = I_1$$

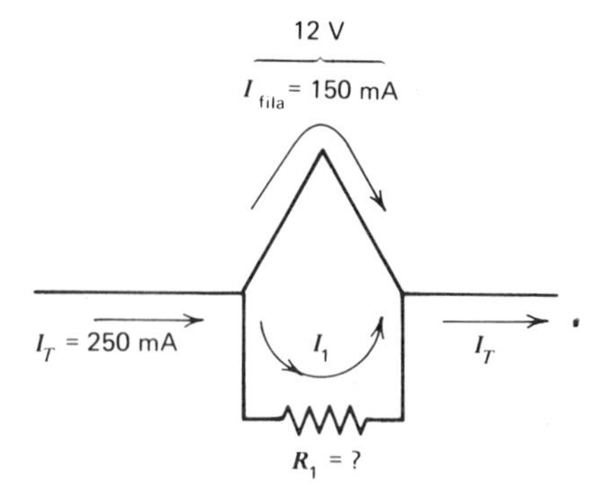

Figure 4-4.

With 12 V across the filament, then V_1 must also be 12 V, since R_1 is in parallel with the filament. Therefore

$$R_1 = \frac{V_1}{I_1}$$

$$R_1 = \frac{12}{100 \text{ mA}}$$

$$R_1 = \frac{12}{0.1 \text{ A}}$$

$$R_1 = 120\ \Omega$$

For Similar Problems Refer to End-of-Chapter Problems 4-5 to 4-11

4-2. Current Distribution in Two Parallel Resistor Circuits

In Fig. 4-2*a* it is shown that where R_1 (60 Ω) is in parallel with R_2 (30Ω), and each is connected across the applied 120 V, then I_2 (4 A) is twice the value of I_1 (2 A). That is, the *smaller* resistor (R_2 = 30 Ω) allows twice as much current to flow through it than flows through the larger resistor (R_1 = 60 Ω). The currents flowing through the parallel resistors are, therefore, said to be *inversely* proportional to the values of the resistors. That is illustrated in the derived equation 4-4 as shown:

$$\text{Voltage across } R_1 = \text{Voltage across } R_2$$

$$V_1 = V_2$$

$$I_1 R_1 = I_2 R_2$$

Dividing *both* sides of the equation by I_1 and also by R_2 yields:

$$\frac{R_1}{R_2} = \frac{I_2}{I_1} \tag{4-4}$$

Example 4-6

If R_1 is 60 Ω and is in parallel with R_2 of 30 Ω, and I_1 is 2 A (as in Fig. 4-2*a*), find I_2 by using equation 4-4.

Solution

$$\frac{R_1}{R_2} = \frac{I_2}{I_1}$$

$$\frac{60}{30} = \frac{I_2}{2}$$

$$\left(\frac{60}{30}\right)(2) = I_2$$

$$4 \text{ A} = I_2$$

This value of I_2 = 4 A agrees with that shown in Fig. 4-2*a*.

Example 4-7

In the circuit of Fig. 4-3, $R_1 = 30\ \Omega$ is in parallel with an $R_2 = 6\Omega$. If $I_2 = 10$ A (from Example 4-4), then find I_1 by using the current inverse proportional method of equation 4-4.

Solution

$$\frac{R_1}{R_2} = \frac{I_2}{I_1} \tag{4-4}$$

$$\frac{30}{6} = \frac{10}{I_1}$$

Cross multiplying gives

$$30\ I_1 = 60$$

$$I_1 = 2\text{ A}$$

This value of $I_1 = 2$ A agrees with that of Example 4-4c.

For Similar Problems Refer to End-of-Chapter Problems 4-12 to 4-13

Another method of solving for a *branch* current when R_1 and R_2 are known, as well as I_T, is given in the following equation 4-6:

$$I_1 = \left(\frac{R_2}{R_1 + R_2}\right) I_T \tag{4-6}$$

Equation 4-6 states simply that the current in one branch (I_1) is a *fraction* of I_{total}. This *fraction* is the resistance of the *other* branch divided by the sum of *both* branch resistors.

Similarly, the current in the other branch I_2 is

$$I_2 = \left(\frac{R_1}{R_1 + R_2}\right) I_T \tag{4-7}$$

Example 4-8

In Fig. 4-5, find: (a) I_1, using the fractional current method of equation 4-6, and (b) I_2, using equation 4-7.

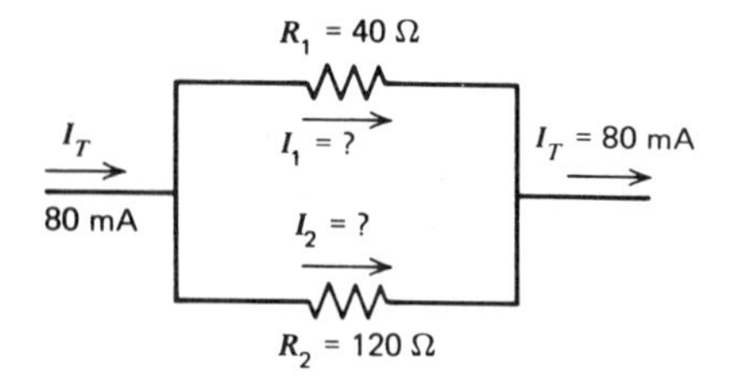

Figure 4-5.

Solution

(a) $$I_1 = \left(\frac{R_2}{R_1 + R_2}\right) I_T \qquad (4\text{-}6)$$

$$I_1 = \left(\frac{120}{40 + 120}\right) 80$$

$$I_1 = \left(\frac{120}{160}\right) 80$$

$$I_1 = 60 \text{ mA}$$

(b) $$I_2 = \left(\frac{R_1}{R_1 + R_2}\right) I_T \qquad (4\text{-}7)$$

$$I_2 = \left(\frac{40}{40 + 120}\right) 80$$

$$I_2 = \left(\frac{40}{160}\right) 80$$

$$I_2 = 20 \text{ mA}$$

Note that the sum of I_1 (60 mA) and I_2 (20 mA) is equal to I_T (80 mA).

For Similar Problems Refer to End-of-Chapter Problems 4-14 and 4-15

4-3. More Than Two Resistors in Parallel

When a third resistor is added in parallel with the other two, *additional* current now flows through this third possible path. As a result I_T increases. This means that R_T has decreased.

In the circuit of Fig. 4-6, three resistors are shown in parallel, each one being connected across the applied voltage source $E_T = 60$ V. The current through each branch is:

$$I_1 = \frac{E_T}{R_1}$$

$$= \frac{60}{30}$$

$$= 2 \text{ A}$$

$$I_2 = \frac{E_T}{R_2}$$

$$= \frac{60}{20}$$

$$= 3 \text{ A}$$

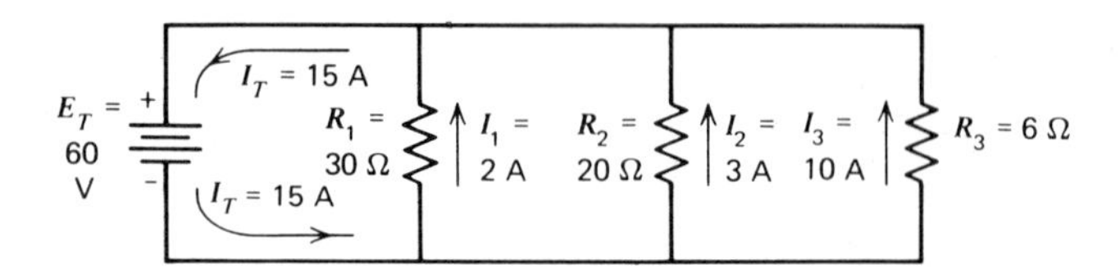

Figure 4-6.

$$I_3 = \frac{E_T}{R_3}$$

$$= \frac{60}{6}$$

$$= 10 \text{ A}$$

These currents are shown in Fig. 4-6. I_T is the sum of I_1, I_2, and I_3, or $I_T = 2 + 3 + 10 = 15$ A. *Total* Resistance, R_T, is now

$$R_T = \frac{E_T}{I_T}$$

$$= \frac{60}{15}$$

$$= 4 \ \Omega$$

An equation, similar to that used for two parallel resistors (equation 4-2), which is employed to find the equivalent resistance for three or more resistors in parallel is

$$\frac{1}{R_{\text{equiv}}} = \frac{1}{R_1} + \frac{1}{R_2} + \frac{1}{R_3} + \text{etc.} \tag{4-8}$$

or, another form, taking the reciprocal of both sides of equation 4-8 yields:

$$R_{\text{equiv}} = \frac{1}{\dfrac{1}{R_1} + \dfrac{1}{R_2} + \dfrac{1}{R_3} + \text{etc.}} \tag{4-9}$$

These equations are referred to as the *reciprocal* method:

Example 4-9

Find the equivalent resistance of the circuit of Fig. 4-6 using the equation 4-9.

Solution

$$R_{\text{equiv}} = \frac{1}{\dfrac{1}{R_1} + \dfrac{1}{R_2} + \dfrac{1}{R_3}} \tag{4-9}$$

$$= \frac{1}{\dfrac{1}{30} + \dfrac{1}{20} + \dfrac{1}{6}}$$

$$= \frac{1}{\dfrac{2 + 3 + 10}{60}} \quad \text{(where 60 is a common denominator)}$$

$$= \frac{1}{\dfrac{15}{60}}$$

$$= 1 \times \frac{60}{15}$$

$$= 4\ \Omega$$

This agrees with the result shown previously using Ohm's law, $R_T = E_T/I_T$. Note that R_{equiv} of the three parallel resistors, 30, 20, and 6 Ω, is 4 Ω and is always *less* than the *smallest* branch resistor.

For Similar Problems Refer to End-of-Chapter Problem 4-17

When only *two* resistors are connected in parallel, a simple method for finding the equivalent resistance of the combination is to use the product of the two divided by their sum (equation 4-3): $R_{\text{equiv}} = R_1R_2/(R_1 + R_2)$. When *three* resistors are connected in parallel, another method than that of Example 4-9 is to employ the *product-divided-by-the-sum* method (equation 4-3), but to do it *twice.* That is, use it first for *any two* of the resistors. Then combine this result with the *third* resistor using their product divided by their sum, as illustrated in the following example:

Example 4-10

Find R_{equiv} for the circuit of Fig. 4-6 using the *product/sum* method (equation 4-3).

Solution

First combine resistors R_1 and R_2 and find R_{equiv} of these two.

$$R_{1,\,2\,equiv} = \frac{R_1\,R_2}{R_1 + R_2} \tag{4-3}$$

$$= \frac{(30)\,(20)}{30 + 20}$$

$$= \frac{600}{50}$$

$$= 12\ \Omega$$

Now combine $R_{1,\,2\,equiv}$ with the third resistor R_3 to find R_{equiv} of the *entire* parallel circuit

$$R_T \quad \text{or} \quad R_{equiv} = \frac{(R_{1,\,2\,equiv})\,(R_3)}{R_{1,\,2\,equiv} + R_3}$$

$$= \frac{(12)\,(6)}{12 + 6}$$

$$= \frac{72}{18}$$

$$= 4\ \Omega$$

This agrees with the R_{equiv} for the same circuit found in Example 4-9.

Example 4-11

In the circuit of Fig. 4-7, find (a) V_1, (b) V_2, (c) V_3, (d) I_1, (e) I_2, (f) I_3, (g) I_T, (h) R_T, using Ohm's law, (i) R_T, using the reciprocal method (equation 4-9), and (j) R_T, using *product/sum* (equation 4-3) method *twice.*

Solution

(a) Voltage across R_1 is E_T. $V_1 = 120$ V.
(b) Voltage across R_2 is also E_T. $V_2 = 120$ V.
(c) Voltage across R_3 is also this same E_T.

$$V_3\ 120\ \text{V}$$

(d) $$I_1 = \frac{V_1}{R_1}$$

$$= \frac{120}{2000}$$

$$= 0.06\ \text{A}$$

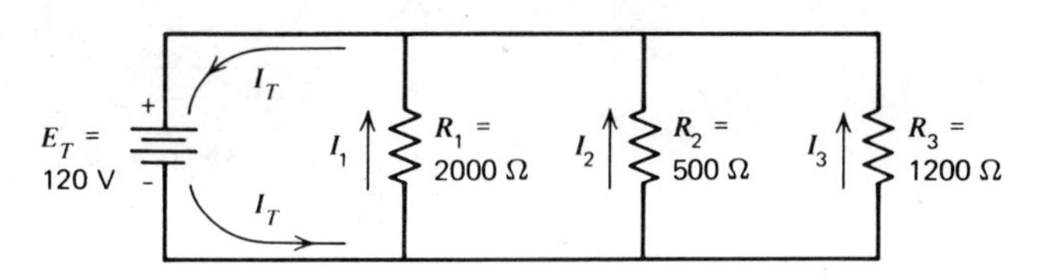

Figure 4-7.

(e) $I_2 = \dfrac{V_2}{R_2}$

$= \dfrac{120}{500}$

$= 0.24$ A

(f) $I_3 = \dfrac{V_3}{R_3}$

$= \dfrac{120}{1200}$

$= 0.1$ A

(g) $I_T = I_1 + I_2 + I_3$

$= 0.06 + 0.24 + 0.1$

$= 0.4$ A

(h) R_T, using Ohm's law

$$R_T = \frac{E_T}{I_T}$$

$$= \frac{120}{0.4}$$

$$= 300\ \Omega$$

(i) In a parallel circuit, R_T is the same as R_{equiv}. R_T, using equation 4-9

$$R_{\text{equiv}} \quad \text{or} \quad R_T = \frac{1}{\dfrac{1}{R_1} + \dfrac{1}{R_2} + \dfrac{1}{R_3}} \tag{4-9}$$

$$= \frac{1}{\dfrac{1}{2000} + \dfrac{1}{500} + \dfrac{1}{1200}}$$

$$= \frac{1}{\frac{3 + 12 + 5}{6000}}$$ (where 6000 is the common denominator)

$$= \frac{1}{\frac{20}{6000}}$$

$$= 1 \times \frac{6000}{20}$$

$$= 300\ \Omega$$

(j) R_T, using *product/sum* (equation 4-3) method *twice.*

First find R_{equiv} of *any two* resistors, say R_1 and R_2.

$$R_{1,\,2\ equiv} = \frac{R_1\ R_2}{R_1 + R_2}$$

$$= \frac{(2000)\ (500)}{2000 + 500} \quad \text{or} \quad \frac{(2 \times 10^3)\ (0.5 \times 10^3)}{(2 \times 10^3) + (0.5 \times 10^3)}$$

$$= \frac{1{,}000{,}000}{2500} \quad \text{or} \quad \frac{1 \times 10^6}{2.5 \times 10^3}$$

$$= 400\ \Omega \quad \text{or} \quad (0.4 \times 10^3\ \Omega)$$

Then find R_{equiv} (or R_T) of the complete circuit by combining the 400 Ω of $R_{1,\,2\ equiv}$ with the third parallel resistor R_3.

$$R_{equiv} \quad \text{or} \quad R_T = \frac{(R_{1,\,2\ equiv})\ (R_3)}{R_{1,\,2\ equiv} + R_3}$$

$$= \frac{(400)\ (1200)}{400 + 1200} \quad \text{or} \quad \frac{(4 \times 10^2)\ (12 \times 10^2)}{(4 \times 10^2) + (12 \times 10^2)}$$

$$= \frac{480{,}000}{1600} \quad \text{or} \quad \frac{48 \times 10^4}{16 \times 10^2}$$

$$= 300\ \Omega \quad \text{or} \quad (3 \times 10^2\ \Omega)$$

Observe that R_T or R_{equiv} is the same value of 300 Ω in each of the alternate methods of solution of parts (h), (i), and (j).

For Similar Problems Refer to End-of-Chapter Problems 4-16 to 4-19

Up to now, the values of all parallel resistors have been chosen for simplicity of calculations. In the reciprocal method, resistors such as 30 and 20 Ω can easily be combined, using 60 as their common denominator. More often, the resistor values are such that finding a common denominator for *four* parallel resistors such as 160, 125, 2700, and 1700 Ω is too time consuming. Using the reciprocal method (equation 4-9), each fraction (1/160, 1/125, 1/2700, and 1/1700) can be easily converted on the slide rule into its decimal equivalent. Then, adding up these decimals and taking the reciprocal of the sum gives the resulting R_{equiv} as shown in this example:

Example 4-12

Find the equivalent resistance of a circuit consisting of four parallel resistors, where $R_1 = 160\ \Omega$, $R_2 = 125\ \Omega$, $R_3 = 2700\ \Omega$, and $R_4 = 1700\ \Omega$.

Solution

$$R_{equiv} = \frac{1}{\frac{1}{R_1} + \frac{1}{R_2} + \frac{1}{R_3} + \frac{1}{R_4}} \tag{4-9}$$

$$R_{equiv} = \frac{1}{\frac{1}{160} + \frac{1}{125} + \frac{1}{2700} + \frac{1}{1700}}$$

$$= \frac{1}{\frac{1}{1.6 \times 10^2} + \frac{1}{1.25 \times 10^2} + \frac{1}{27. \times 10^2} + \frac{1}{17. \times 10^2}}$$

$$= \frac{1}{0.625 \times 10^{-2} + 0.8 \times 10^{-2} + 0.037 \times 10^{-2} + 0.0588 \times 10^{-2}}$$

$$= \frac{1}{1.5208 \times 10^{-2}}$$

$$= \frac{1 \times 10^2}{1.5208}$$

$$= 0.658 \times 10^2$$

$$R_{equiv} = 65.8\ \Omega$$

In the above, note that $1/R_1$ is 1/160 or 0.625×10^{-2}. From Chapter 2, recall that *conductance* (G) is the reciprocal of *resistance*, or $G = 1/R$. Therefore, Example 4-12 is really being performed by using conductances, and equation 4-9 may be rewritten as:

$$R_{equiv} = \frac{1}{\frac{1}{R_1} + \frac{1}{R_2} + \frac{1}{R_3} + \frac{1}{R_4}} \tag{4-9}$$

and, since $G_1 = 1/R_1$, and $G_2 = 1/R_2$, etc., then

$$R_{equiv} = \frac{1}{G_1 + G_2 + G_3 + G_4} \tag{4-10}$$

For Similar Problems Refer to End-of-Chapter Problem 4-19

4-4. Equal Resistors in Parallel

When resistors of the same ohmic value are connected in parallel, the equivalent resistance of the combination can be found by any of the previous methods. However, a simpler, faster method is shown in the following for *three* parallel resistors of equal value.

$$R_{equiv} = \frac{R}{3}$$

This means that where *three* resistors of equal value are connected in parallel, R_{equiv} is found by simply dividing the value of *one* resistor by the number *three* (the number of parallel branches). The equivalent resistance of any number (N) of equal parallel resistors is then

$$R_{equiv} = \frac{R}{N} \tag{4-11}$$

Example 4-13

(a) What is the equivalent resistance of *two* 12-kΩ resistors connected in parallel? (b) What is the equivalent resistance if a *third* 12-kΩ resistor is added in parallel with the other two? (c) If a *fourth* 12-kΩ is paralleled with the other three, what is the equivalent resistance?

Solution

(a) R_{equiv} for *two* 12-K resistors in parallel is

$$R_{equiv} = \frac{R}{N} \tag{4-11}$$

$$= \frac{12{,}000}{2}$$

$$= 6000\ \Omega \quad \text{or} \quad 6\ \text{k}\Omega$$

(b) R_{equiv} for *three* 12-K resistors in parallel is

$$R_{equiv} = \frac{R}{N} \tag{4-11}$$

$$= \frac{12{,}000}{3}$$

$$= 4000\ \Omega \quad \text{or} \quad 4\ \text{k}\Omega$$

(c) R_{equiv} for *four* 12-K resistors in parallel is

$$R_{equiv} = \frac{R}{N} \tag{4-11}$$

$$= \frac{12{,}000}{4}$$

$$= 3000\ \Omega \qquad \text{or} \qquad 3\ \text{k}\Omega$$

For Similar Problems Refer to End-of-Chapter Problems 4-20 to 4-22

4-5. Defective Resistors in Parallel Circuits

Very common troubles in electronic circuits are resistors that either increase or decrease in value. This is usually the result of excessive current through the resistor. An increase in value could be several times its normal value up to several megohms, or even an infinite value such as in an *open* resistor. A decreased value similarly could be a fraction of its normal value, even down to zero ohms in the case of a shorted resistor.

In Fig. 4-8, R_1 (300 Ω) is in parallel with R_2 (60 Ω). The equivalent resistance should be: $R_{equiv} = R_1R_2/(R_1 + R_2) = (300)(60)/(300 + 60) = 50\ \Omega$, and an ohmmeter placed between points A and B will indicate this amount. If the ohmmeter indicates a higher value than 50 Ω, it means that one (or possibly both) of the resistors is larger than normal. To measure R_1 or R_2 *alone*, the circuit must be opened up. That is, the single resistor to be measured must be disconnected from the parallel combination. If R_1 is to be checked with an ohmmeter, one end of R_1 must be disconnected, and then measured.

If R_2 is to be checked with an ohmmeter, placing the meter test leads between points C and D (in Fig. 4-8) actually measures both R_2 and R_1 connected in parallel. One end of R_2 must be disconnected, and the ohmmeter test leads must be placed between the ends of R_2.

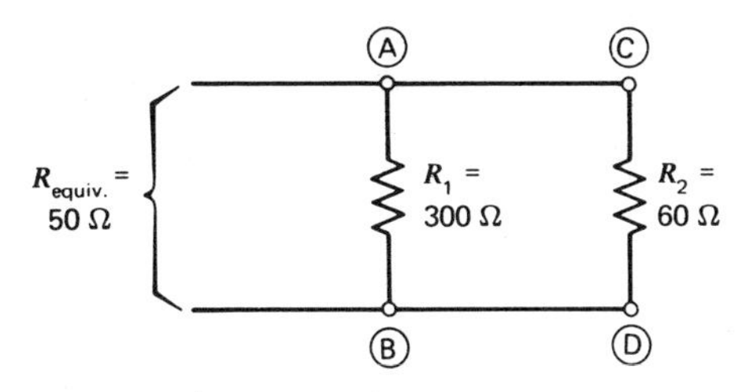

Figure 4-8.

If either R_1 or R_2 became much larger, R_{equiv} would no longer be 50 Ω, but would increase accordingly. If R_1 broke open (R_1 is now infinite), then R_{equiv} would become 60 Ω, or equal to R_2, which is now in the circuit alone. Similarly, if R_2 cracked open, R_{equiv} would now become 300 Ω (equal to R_1 alone).

If either resistor decreased value, such as becoming *zero* ohms (if it became shorted), then R_{equiv} would fall from its normal 50 Ω.

The same holds true for any number of parallel resistors, and to measure any one branch alone, that resistor must have one end disconnected.

If one resistor of a parallel circuit becomes defective, opening up or increasing its resistance, it causes the current through *that* branch to become zero or to decrease. This causes *total* current (of which this branch is a part) to decrease. Voltage, however, across each branch remains normal. The following example illustrates this.

Example 4-14

Three 6000-Ω resistors are connected in parallel and are connected to a 3-V applied voltage, as shown in Fig. 4-9*a*. (a) If everything is normal, find the equivalent resistance, current flow through each branch, and total current. (b) If R_1 becomes defective and opens up, find R_{equiv}, current through each branch, and I_{total}.

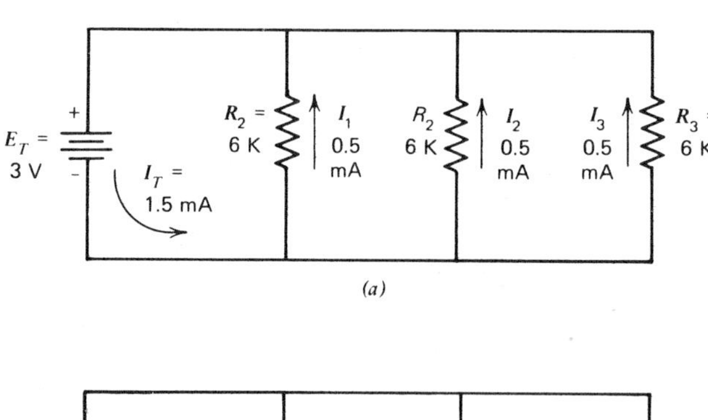

(*a*)

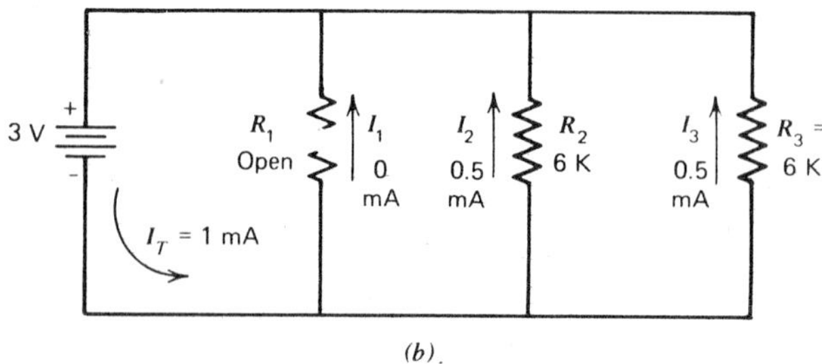

(*b*)

Figure 4-9. (*a*) Normal current. (*b*) Open R_1.

Solution

(a) Everything normal. R_{equiv} is

$$R_{equiv} = \frac{R}{N} \tag{4-11}$$

$$= \frac{6000}{3}$$

$$= 2000\ \Omega$$

$$I_1 = \frac{V_1}{R_1}$$

$$= \frac{3}{6000}$$

$$= .0005\ \text{A} \quad \text{or} \quad 0.5\ \text{mA}$$

Since R_2 and R_3 are each equal to R_1, then each branch current is 0.5 mA.

$$I_T = I_1 + I_2 + I_3$$

$$= 0.5 + 0.5 + 0.5$$

$$= 1.5\ \text{mA}$$

These currents are shown in Fig. 4-9*a*.

(b) If R_1 opens up, now only R_2 and R_3 are left in the parallel circuit, and R_{equiv} becomes

$$R_{equiv} = \frac{R}{N} \quad \text{where } N \text{ is now } \textit{two}$$

$$= \frac{6000}{2}$$

$$= 3000\ \Omega$$

Notice that since R_1 increased its value R_{equiv} also increased, going from 2000 to 3000 Ω.

I_1 is now zero, since R_1 is open or infinite.

I_2 and I_3 remain unchanged at 0.5 mA each, since R_2 and R_3 are still intact.

I_T now decreases from its normal value of 1.5 mA to:

$$I_T = I_1 + I_2 + I_3$$

$$= 0 + 0.5 + 0.5$$

$$= 1\ \text{mA}$$

These currents are shown in Fig. 4-9*b*.

For Similar Problems Refer to End-of-Chapter Problems 4-23 to 4-29

MORE ADVANCED PARALLEL CIRCUITS

Here we discuss current distribution in parallel circuits consisting of more than two branches. Also, a method of finding the value of a resistor that must be added in parallel with a known resistor in order to produce a desired value of equivalent resistance is presented.

4.6 Current Distribution In Two, Three, or More Parallel Resistors

An equation that can be used in a circuit containing *any* number of parallel resistors and is, therefore, of very great use in solving a branch current is

$$I_1 = \left(\frac{R_{equiv}}{R_1}\right) I_T \tag{4-12}$$

where I_1 is the current through one branch R_1, and R_{equiv} is the equivalent resistance of *any* number of parallel resistors. Similarly, current I_2 in a second branch R_2 is

$$I_2 = \left(\frac{R_{equiv}}{R_2}\right) I_T \tag{4-13}$$

and current I_3 in a third branch R_3 is

$$I_3 = \left(\frac{R_{equiv}}{R_3}\right) I_T \tag{4-14}$$

Equations 4-12, 4-13, and 4-14 are sometimes referred to as the *fractional* current method, since the current in any branch is a fraction of I_{total} as shown in the following:

$$I_{any\ branch} = \left(\frac{R_{equiv}}{R_{that\ branch}}\right) I_{total} \tag{4-15}$$

Observe that equation 4-15 is a general one, covering those of equations 4-12, 4-13, and 4-14.

The following example illustrates the usefulness of the general equation 4-15.

Example 4-15

As shown in Fig. 4-10, three resistors are connected in parallel. If the total current is 12 mA, find (a) I_1, (b) I_2, and (c) I_3.

Solution

(a) I_1 is a fraction of I_T, and may be found from

$$I_1 = \left(\frac{R_{equiv}}{R_1}\right) I_T \tag{4-12}$$

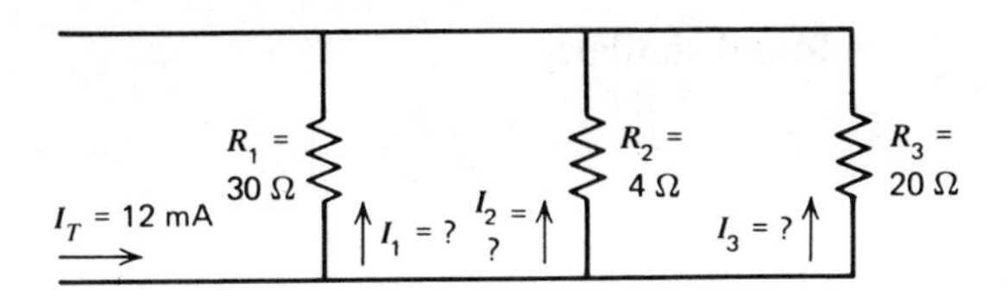

Figure 4-10.

First solve for R_{equiv} of the three resistors.

$$R_{equiv} = \frac{1}{\frac{1}{R_1} + \frac{1}{R_2} + \frac{1}{R_3}} \tag{4-9}$$

$$R_{equiv} = \frac{1}{\frac{1}{30} + \frac{1}{4} + \frac{1}{20}}$$

$$= \frac{1}{\frac{2 + 15 + 3}{60}}$$ (where 60 is a common denominator

$$= \frac{1}{\frac{20}{60}}$$

$$= 1 \times \frac{60}{20}$$

$$= 3\ \Omega$$

Now, I_1 may be found.

$$I_1 = \left(\frac{R_{equiv}}{R_1}\right) I_T \tag{4-12}$$

$$= \left(\frac{3}{30}\right) 12\ \text{mA}$$

$$= 1.2\ \text{mA}$$

(b) I_2 may be found as follows:

$$I_2 = \left(\frac{R_{equiv}}{R_2}\right) I_T \tag{4-13}$$

$$= \left(\frac{3}{4}\right) 12\ \text{mA}$$

$$= 9\ \text{mA}$$

(c) I_3 may be found similarly

$$I_3 = \left(\frac{R_{\text{equiv}}}{R_3}\right) I_T \qquad (4\text{-}14)$$

$$= \left(\frac{3}{20}\right) 12 \text{ mA}$$

$$= 1.8 \text{ mA}$$

Note that the sum of the three branch currents equals I_{total}, and therefore the answers check out.

$$I_1 + I_2 + I_3 = I_T$$

$$1.2 + 9 + 1.8 = 12$$

$$12 = 12$$

Example 4-16

Figure 4-11 shows three resistors, R_4, R_5, and R_6, in parallel. If $I_6 = 2$ mA, find (a) I_4, and (b) I_5.

Solution

(a) Use of the current and resistance ratio equation gives

$$\frac{I_4}{I_6} = \frac{R_6}{R_4} \qquad \text{(adaption of equation 4-4)}$$

$$I_4 = \left(\frac{R_6}{R_4}\right) I_6$$

$$I_4 = \left(\frac{60 \text{ K}}{120 \text{ K}}\right) 2 \text{ mA}$$

$$I_4 = 1 \text{ mA}$$

(b) Similarly, I_5 may be found as shown:

$$\frac{I_5}{I_6} = \frac{R_6}{R_5} \qquad \text{(adaption of equation 4-4)}$$

$$I_5 = \left(\frac{R_6}{R_5}\right) I_6$$

$$I_5 = \left(\frac{60 \text{ K}}{10 \text{ K}}\right) 2 \text{ mA}$$

$$I_5 = 12 \text{ mA}$$

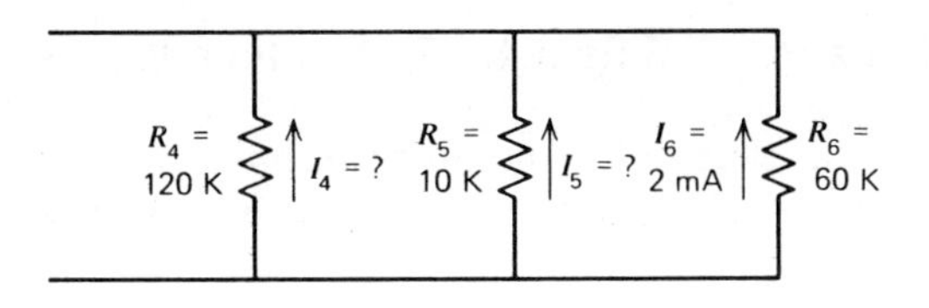

Figure 4-11.

I_4 and I_5 could also be found by using the basic Ohm's law, as shown in the following:

(a) Voltage across R_6, or

$$V_6 = I_6 R_6$$

$$V_6 = (2 \text{ mA})(60\text{K})$$

$$= (.002)(60{,}000) \quad \text{or} \quad (2 \times 10^{-3})(60 \times 10^3)$$

$$= 120 \text{ V}$$

Since R_4, R_5, and R_6 are in parallel, then $V_6 = V_4 = V_5$

$$V_6 = V_4 = V_5 = 120 \text{ V}$$

Then, solving for I_4 gives

$$I_4 = \frac{V_4}{R_4}$$

$$= \frac{120}{120{,}000} \quad \text{or} \quad \frac{120}{120 \times 10^3}$$

$$= .001 \text{ A} \quad \text{or} \quad 1 \text{ mA} \quad \text{or} \quad 1 \times 10^{-3} \text{ A}$$

Note that this is the same value found previously.

(b) And using Ohm's law to find I_5 gives

$$I_5 = \frac{V_5}{R_5}$$

$$= \frac{120}{10{,}000} \quad \text{or} \quad \frac{120}{1 \times 10^4}$$

$$I_5 = .012 \text{ A} \quad \text{or} \quad 12 \text{ mA} \quad \text{or} \quad 120 \times 10^{-4} \text{ A}$$

Note that this is the same value found previously.

For Similar Problems Refer to End-of-Chapter Problems 4-30 to 4-33

4-7. Adding a Parallel Resistor To Produce the Desired $R_{equivalent}$

Many times, the electronic laboratory technician has to add a resistor in parallel with an existing one to produce an equivalent resistance of some desired value. The equation that is used to determine the value of the resistor (R_1) that should be placed in parallel with a known resistor (R_2), resulting in the desired R_{equiv}, is derived in the following:

$$R_{equiv} = \frac{\text{product}}{\text{sum}} \tag{4-3}$$

$$R_{equiv} = \frac{R_1 R_2}{R_1 + R_2}$$

$$R_{equiv}(R_1 + R_2) = R_1 R_2$$

$$R_{equiv}(R_1) + R_{equiv}(R_2) = R_1 R_2$$

$$R_{equiv}(R_2) = R_1 R_2 - R_{equiv}(R_1)$$

Factoring out R_1 in the right side of the equation gives

$$(R_{equiv})(R_2) = R_1(R_2 - R_{equiv})$$

$$\frac{(R_{equiv})(R_2)}{R_2 - R_{equiv}} = R_1 \tag{4-16}$$

Where R_1 is the resistor to be placed in parallel with a known resistor R_2, in order to produce the desired amount of R_{equiv}.

Observe that equation 4–16 is simply the *product* of the two known quantities divided by their *difference*. This is similar to equation 4-3, which is the *product* of the two known items divided by their *sum*. The following examples illustrate the usefulness of equation 4-16.

Example 4-17

What value resistor, R_1, should be placed in parallel with a resistor, $R_2 = 40\ \Omega$, so that the combination R_{equiv} is $8\ \Omega$?

Solution

$$R_1 = \frac{\text{product}}{\text{difference}} = \frac{(R_2)(R_{equiv})}{R_2 - R_{equiv}} \tag{4-16}$$

$$R_1 = \frac{(40)(8)}{40 - 8}$$

$$R_1 = \frac{320}{32}$$

$$R_1 = 10\ \Omega$$

Example 4-18

A 24-kΩ resistor R_2 is to be paralleled with another resistor R_1 to produce an R_{equiv} of 6 kΩ. Find R_1.

Solution

$$R_1 = \frac{\text{product}}{\text{difference}} = \frac{(R_2)(R_{\text{equiv}})}{R_2 - R_{\text{equiv}}} \tag{4-16}$$

$$R_1 = \frac{(24{,}000)(6000)}{24{,}000 - 6000} \quad \text{or} \quad \frac{(24 \times 10^3)(6 \times 10^3)}{24 \times 10^3 - 6 \times 10^3}$$

$$R_1 = \frac{144{,}000{,}000}{18{,}000} \quad \text{or} \quad \frac{144 \times 10^6}{18 \times 10^3}$$

$$R_1 = 8000\ \Omega \quad \text{or} \quad 8 \times 10^3\ \Omega \quad \text{or} \quad 8\ \text{k}\Omega$$

The same procedure is followed for any number of parallel branches. For example, R_2 and R_3 are already in parallel. To produce a desired value of R_{equiv}, an unknown value resistor R_1 must be placed in parallel with the other two. To find the value of R_1, first find the value of the parallel combination R_2 and R_3. Then use this resultant of R_2 and R_3 called $R_{2,3}$ and the desired R_{equiv} in the equation 4-16 of their *product* divided by their *difference*

For Similar Problems Refer to End-of-Chapter Problems 4-34 to 4-38

PROBLEMS

See section 4-1 for discussion covered by the following problems.

4-1. R_1, a 240-Ω resistor is in parallel with R_2, an 80 Ω resistor. What is the equivalent resistance of this combination? (Use the reciprocal method.)

4-2. R_3, a 150-Ω resistor and R_4, a 300-Ω resistor are connected in parallel. What is their equivalent resistance? (Use the reciprocal method.)

4-3. If 12 V were connected to the parallel circuit of Problem 4-1, (a) what is V_1, and (b) what is V_2?

4-4. In Problem 4-3, find: (a) P_1, (b) P_2, (c) P_T.

4-5. In Problem 4-1, find R_{equiv} using the *product/sum* method.

4-6. In Problem 4-2, find R_{equiv} using the *product/sum* method.

4-7. If 9 V were connected to the parallel circuit of Problem 4-2, (a) what is V_3, and (b) what is V_4?

4-8. In Problem 4-7, find (a) P_3, (b) P_4, (c) P_T.

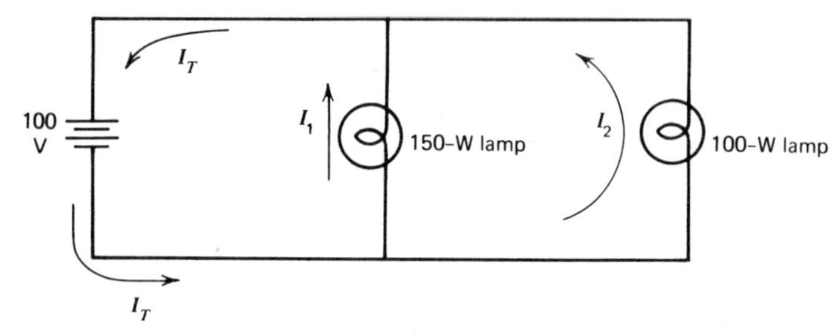

Problem 4-9.

4-9. The diagram for this problem shows a 150-W lamp and a 100-W lamp connected in parallel across 100 V. Using the appropriate form of Watt's law, find (a) I_1, and (b) I_2. Also find (c) I_T.
4-10. In Problem 4-9, find: (a) the resistance of the 150-W lamp, (b) the resistance of the 100-W lamp, and (c) the total or equivalent resistance.
4-11. The diagram for this problem shows the coil of a 1 mA meter, having a coil resistance of 180 Ω. R_1 and the meter coil are in parallel in a circuit where $I_{total} = 10$ mA. If $I_{coil} = 1$ mA, find: (a) V_{coil}, (b) I_1, (c) R_1, and (d) R_{total} or R_{equiv}.

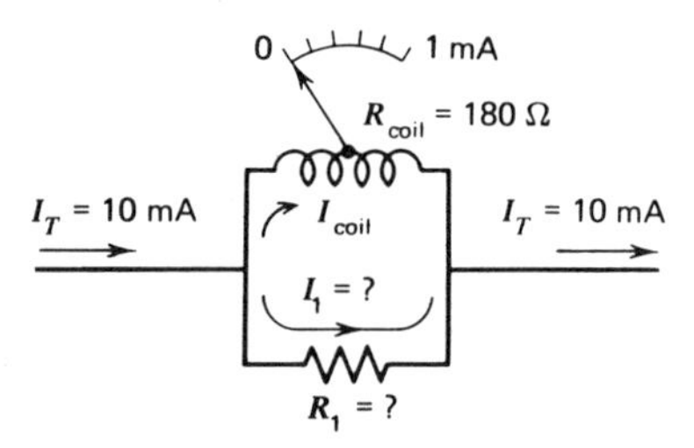

Problem 4-11.

See section 4-2 for discussion covered by the following problems.

4-12. In the diagram shown for this problem, find I_1 using the *current and resistor inverse ratio equation.*
4-13. In Problem 4-12, what is (a) V_2, and (b) V_1? Using Ohm's law, find (c) I_1, and check to see that it agrees with the I_1 value found in Problem 4-12.

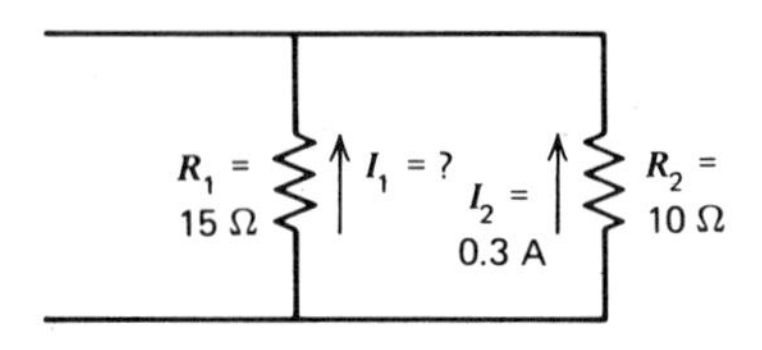

Problem 4-12.

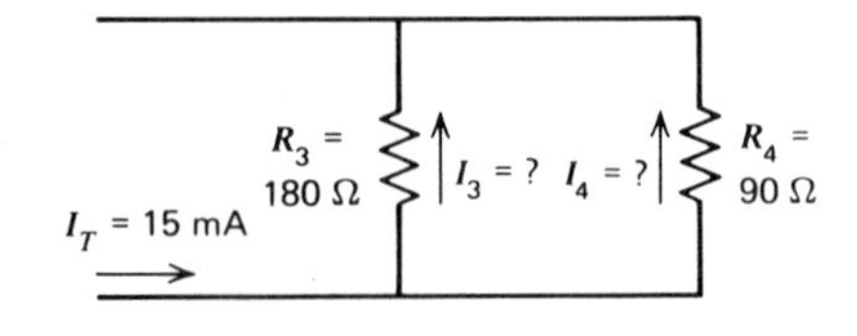

Problem 4-14.

4-14. In the diagram shown for this problem, find (a) I_3, and (b) I_4.

4-15. Two resistors, $R_1 = 600\ \Omega$ and $R_2 = 200\ \Omega$, are connected in parallel. If $I_{total} = 20$ mA, find (a) I_1, (b) I_2.

See section 4-3 for discussion covered by the following problems.

4-16. Three resistors, $R_1 = 60\Omega$, $R_2 = 120\ \Omega$, and $R_3 = 10\ \Omega$, are connected in parallel. Find R_{equiv} using the product/sum method *twice*, first solving for the combination of R_1 and R_2, and then combining this result with R_3 for the total or R_{equiv}.

4-17. In Problem 4-16 find R_{equiv} by using the reciprocal method.

4-18. In Problem 4-16, if 24 V were applied acorss the parallel circuit, then using Ohm's law find: (a) I_1, (b) I_2, (c) I_3 and, using Kirchhoff's current law, (d) I_T. Finally, again using Ohm's law, find (e) R_T or R_{equiv}.

4-19. Four resistors are connected in parallel. Find R_{equiv} using any desired method, where $R_1 = 125\ \Omega$, $R_2 = 200\ \Omega$, $R_3 = 300\ \Omega$, and $R_4 = 475\ \Omega$.

See section 4-4 for discussion covered by the following problems.

4-20. Two 150-Ω resistors are in parallel. What is R_{equiv}?

4-21. What is R_{equiv} if a *third* 150-Ω resistor were placed in parallel with the two of Problem 4-20?

4-22. Four 10 kΩ resistors are connected in parallel. Find R_{equiv}.

See section 4-5 for discussion covered by the following problems.

4-23. In Problem 4-20, what would R_{equiv} be if one of the two resistors opened up?

4-24. What happens to I_T in any parallel circuit if one branch resistor opens?

4-25. What is R_{equiv} in the circuit shown for Problem 4-12, (a) if R_1 opens, and (b) if R_2 (alone) opens?

4-26. If an ohmmeter were placed across R_4 in the circuit of Problem 4-14, what would the ohmmeter read?

4-27. In the circuit of Problem 4-14, what would an ohmmeter indicate if placed across R_3?

4-28. What should be done to the circuit of Problem 4-14 if it were desired to measure R_4 using an ohmmeter?

4-29. What should be done to the circuit of Problem 4-14 if it were desired to measure R_3 using an ohmmeter?

More Challenging Problems: See section 4-6 for discussion covered by the following problems.

4-30. In the circuit shown for this problem, find: (a) I_1, (b) I_2, (c) I_3, and (d) I_4, using the *fractional* method.

R_1 = 120 K, I_1 = ?, R_2 = 40 K, I_2 = ?, R_3 = 60 K, I_3 = ?, I_4 = ?, R_4 = 30 K, I_T = 36 mA

Problem 4-30.

4-31. In the circuit shown for Problem 4-30, find: (a) R_{equiv}, (b) $V_{R\,\text{equiv}}$, (c) V_1, (d) V_2, (e) V_3, (f) V_4, and using Ohm's law find: (g) I_1, (h) I_2, (i) I_3, (j) I_4. Check to see that the values for the branch currents are the same as those found in Problem 4-30.

4-32. In the circuit for this problem, $I_T = 30$ mA. Find: (a) I_1, (b) I_2, and (c) I_3.

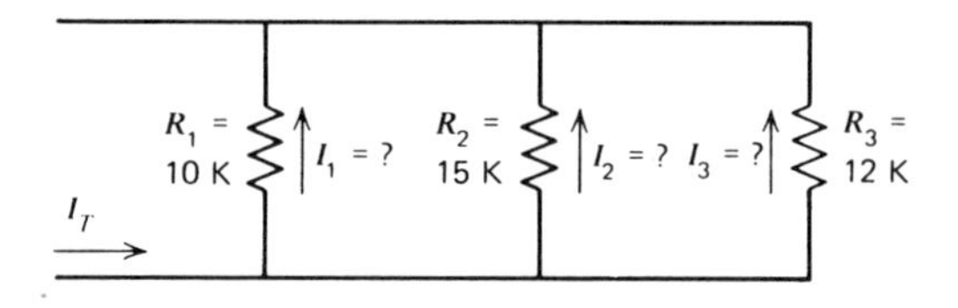

Problem 4-32.

4-33. In the circuit for Problem 4-32, if $I_2 = 2$ mA, find: (a) I_1 and (b) I_3, using the *current ratio = inverse of resistor ratio* method.

See section 4-7 for discussion covered by the following problems.

4-34. What value resistor R_1 should be paralleled with resistor R_2 (400 Ω) to produce an equivalent resistance of 80 Ω?

4-35. What value resistor R_3 should be paralleled with resistor R_4 (100 Ω) to produce an equivalent resistance of 90 Ω?

4-36. Resistor R_5 (560 Ω) is in parallel with R_6 (80 Ω). What value resistor R_7 should be added in parallel with the other two in order to produce an equivalent resistance of 60 Ω.

4-37. R_1 and R_2 are in parallel. If $R_2 = 9$ kΩ and $R_{equiv} = 6$kΩ, find the value of R_1.

4-38. Find the value of R_5 in the circuit shown for this problem.

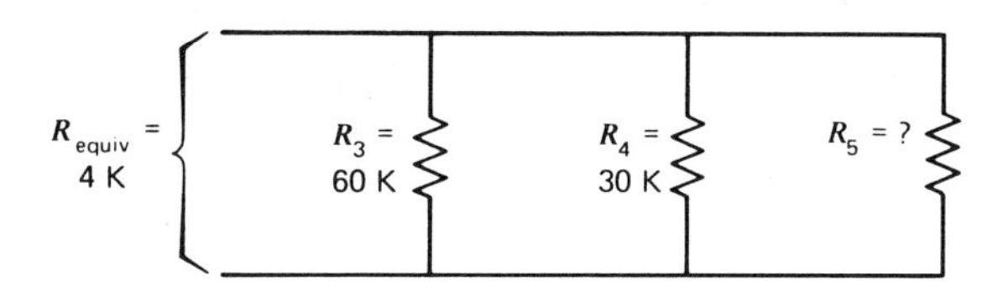

Problem 4-38.

chapter

5

Series-Parallel Combination Resistor Circuits

A circuit having one or more resistors through which *all* the current must flow is called a *series* circuit and is discussed in Chapter 3. A circuit where the current can split up and flow through two or more branches is called a *parallel* circuit and is discussed in Chapter 4. Combinations of these are called *parallel-series* or *series-parallel* circuits, and these various combinations are discussed in this chapter. Filaments of vacuum tubes are often connected in some combination circuit in order to operate properly. Power supplies of the larger electronic equipment such as radar, high power communication transmitters and receivers and the electronic apparatus of guided missiles and space vehicles usually are series-parallel circuits.

5-1. Parallel-Series Circuits

Figure 5-1*a* shows four resistors connected in a *parallel-series* circuit. Current from the negative end of the 60 V E_T flows to point A. Here the current divides into two branches, left and right. I_{left} flows up through R_2 and R_1 to point B. I_{right} flows up through R_4 and R_3 to point B. At point B the two branch currents join together, forming I_T, and flow to the positive end of E_T. Points A and B are the ends of a *parallel* circuit.

Since I_{left} flows through R_2 and R_1, then these two resistors are in *series* with each other. R_4 and R_3 are in *series* with each other also, since I_{right} flows through both of these. Since points A and B are the ends of a parallel circuit,

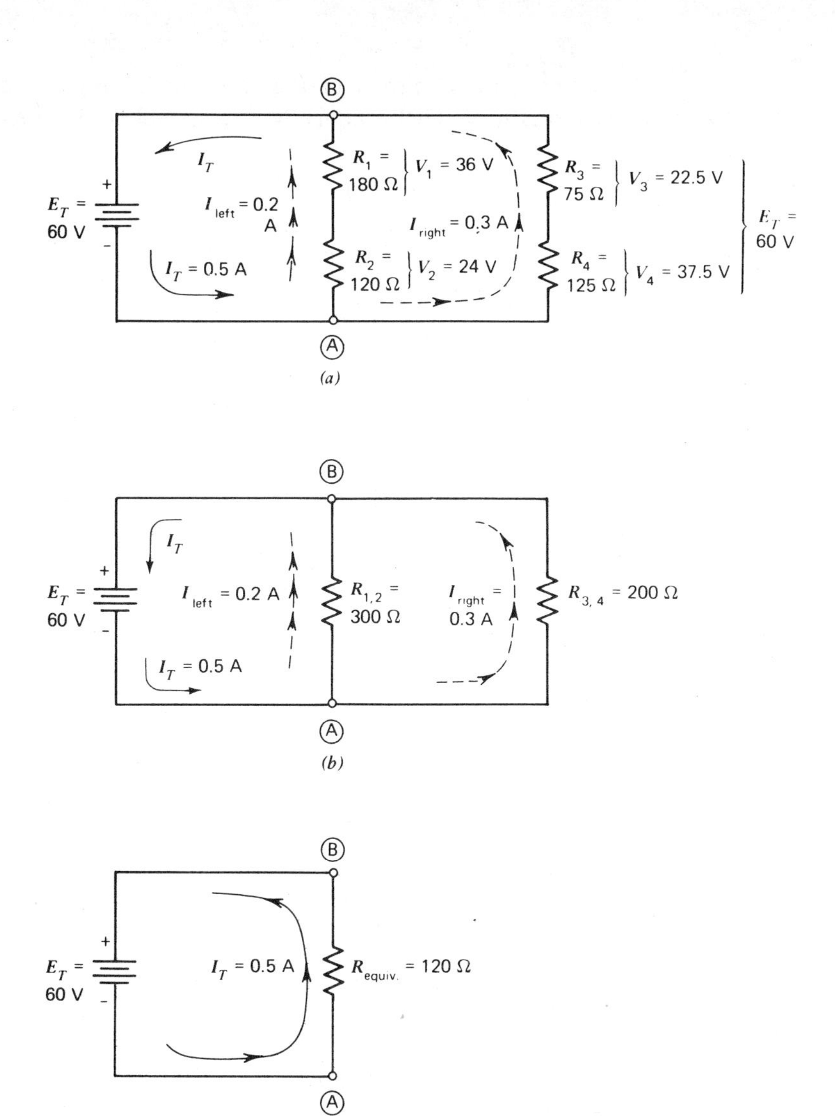

Figure 5-1. Parallel-series circuit analyzed. (*a*) Complete circuit. (*b*) More simplified circuit. (c) Equivalent circuit.

consisting of a left branch and a right branch, then the original circuit shown in Fig. 5-1*a* may be redrawn in a more simplified diagram as a simple *parallel* circuit shown in Fig. 5-1*b*. The left branch, R_{left}, shown as $R_{1,2}$, consists of the sum of R_2 (120 Ω) and R_1 (180 Ω), or $R_{1,2} = 300$ Ω. The right branch, R_{right}, shown as $R_{3,4}$, consists of the sum of R_4 (125 Ω) and R_3 (75 Ω), or $R_{3,4} = 200$ Ω.

The simple parallel circuit of Fig. 51-*b* made up essentially of 300-Ω and 200-Ω resistors in parallel is equivalent to a single resistor shown in Fig. 5-1*c*. This R_{equiv} is

$$R_{\text{equiv}} = \frac{\text{Product}}{\text{sum}} = \frac{(R_{1,2})\,(R_{3,4})}{R_{1,2} + R_{3,4}} \tag{4-3}$$

$$= \frac{(300)\,(200)}{300 + 200} \quad \text{or} \quad \frac{(3 \times 10^2)\,(2 \times 10^2)}{3 \times 10^2 + 2 \times 10^2}$$

$$= \frac{60{,}000}{500} \quad \text{or} \quad \frac{6 \times 10^4}{5 \times 10^2}$$

$$= 120\ \Omega \quad \text{or} \quad 1.2 \times 10^2\ \Omega$$

In Fig. 5-1*c* R_{equiv} is also the total resistance R_T.

Total current, $I_T = E_T/R_T = 60/120 = 0.5$ A, and is shown in Fig. 5-1*a*, *b*, and *c*. Current in each branch may be found in any of several ways. Using Ohm's law, $I_{\text{left}} = R_T/R_{1,2} = 60/300 = 0.2$ A. Similarly, $I_{\text{right}} = E_T/R_{3,4} = 60/200 = 0.3$ A. These branch currents are shown in Fig. 5-1*a* and *b*. Note that these results agree with Kirchhoff's current law, where $I_T = I_{\text{left}} + I_{\text{right}}$, $0.5\text{ A} = 0.2 + 0.3$ A.

Another method of solving for I_{left} as a fraction of I_T is

$$I_{\text{left}} = \left(\frac{R_{\text{right}}}{R_{\text{left}} + R_{\text{right}}}\right) I_T \qquad \text{(adaption of equation 4-6)}$$

$$= \left(\frac{200}{300 + 200}\right) 0.5$$

$$= \left(\frac{200}{500}\right) 0.5$$

$$= 0.2\ \text{A}$$

Another method for finding a branch current, say I_{right}, in terms of I_{left} and each branch resistor is

$$\frac{I_{\text{right}}}{I_{\text{left}}} = \frac{R_{\text{left}}}{R_{\text{right}}} \qquad \text{(adaption of equation 4-4)}$$

$$I_{\text{right}} = \left(\frac{R_{\text{left}}}{R_{\text{right}}}\right) I_{\text{left}}$$

$$I_{\text{right}} = \left(\frac{300}{200}\right) 0.2$$

$$I_{\text{right}} = 0.3\ \text{A}$$

This also agrees with the answer produced formerly using Ohm's law.

In Fig. 5-1*a*, I_{left} (0.2 A) is shown flowing through R_2 and R_1. These two resistors, as has been shown, are therefore in series with each other. Voltage across each resistor may be found by using Ohm's law. Voltage across R_1, or $V_1 = (I_{left})(R_1) = (0.2)(180) = 36$ V, shown in Fig. 5-1*a*. Voltage across R_2, or $V_2 = (I_{left})(R_2) = (0.2)(120) = 24$ V, shown in the diagram. The sum of these voltages is the voltage between points A and B, or E_{total}. This is Kirchhoff's voltage law, or $V_1 + V_2 = E_T$, $36 + 24 = 60$ V.

In the same manner, the sum of the voltages across R_3 and R_4 should be equal to E_{total}. $V_3 = (I_{right})(R_3) = (0.3)(75) = 22.5$ V, shown in Fig. 5-1*a*, and $V_4 = (I_{right})(R_4) = (0.3)(125) = 37.5$ V, also shown in the diagram. There sum is: $V_3 + V_4 = E_T$, $22.5 + 37.5 = 60$ V.

Example 5-1

The circuit diagram for this example is shown in Fig. 5-2*a*. Find the following: (a) which resistors are in *series* with which, and (b) which are in *parallel*, (c) R_{left}, (d) R_{right}, (e) I_{left}, (f) I_{right}, (g) I_T, (h) R_{equiv}, (i) V_1, (j) V_2, (k) V_3, (l) V_4, and (m) V_5.

Solution

By analyzing the circuit first, it can be seen that electrons coming from the negative end of E_T flow to point X. Here, the current divides into a left path and a right path. I_{left} flows through R_1 and R_2 to point Y. I_{right} flows through R_3, R_4, and R_5 to point Y. At point Y, I_{left} and I_{right} combine to form I_T which flows to the positive end of E_T.

(a) From the foregoing description of the circuit of Fig. 5-2*a*, R_1 and R_2 are in *series* with each other, forming R_{left}. Also, R_3, R_4, and R_5 are in *series* with each other, comprising R_{right}.

(b) The *parallel* circuit is made up of R_{left} in parallel with R_{right}. That is, the *series* circuit of R_1 and R_2 and the other *series* circuit of R_3, R_4, and R_5, are in *parallel* with each other.

(c) R_{left} consists of R_1 and R_2 in series, R_{left} or $R_{1,2} = R_1 + R_2 = 1\text{ K} + 2\text{ K} = 3\text{ K}$, shown in Fig. 5-2*b*.

(d) R_{right} consists of R_3, R_4, and R_5 in series. R_{right} or $R_{3,4,5} = R_3 + R_4 + R_5 = 3\text{ K} + 4\text{ K} + 5\text{ K} = 12\text{ K}$, shown in Fig. 5-2*b*.

(e) I_{left}, using Ohm's law, $= E_T/R_{1,2} = 24/3000 = .008$ A, or 8 mA.

(f) I_{right}, using Ohm's law, $= E_T/R_{3,4,5} = 24/12{,}000 = .002$ A, or 2 mA.

(g) I_T, from Kirchhoff's current law, $= I_{left} + I_{right} = 8\text{ ma} + 2\text{ mA} = 10$ mA.

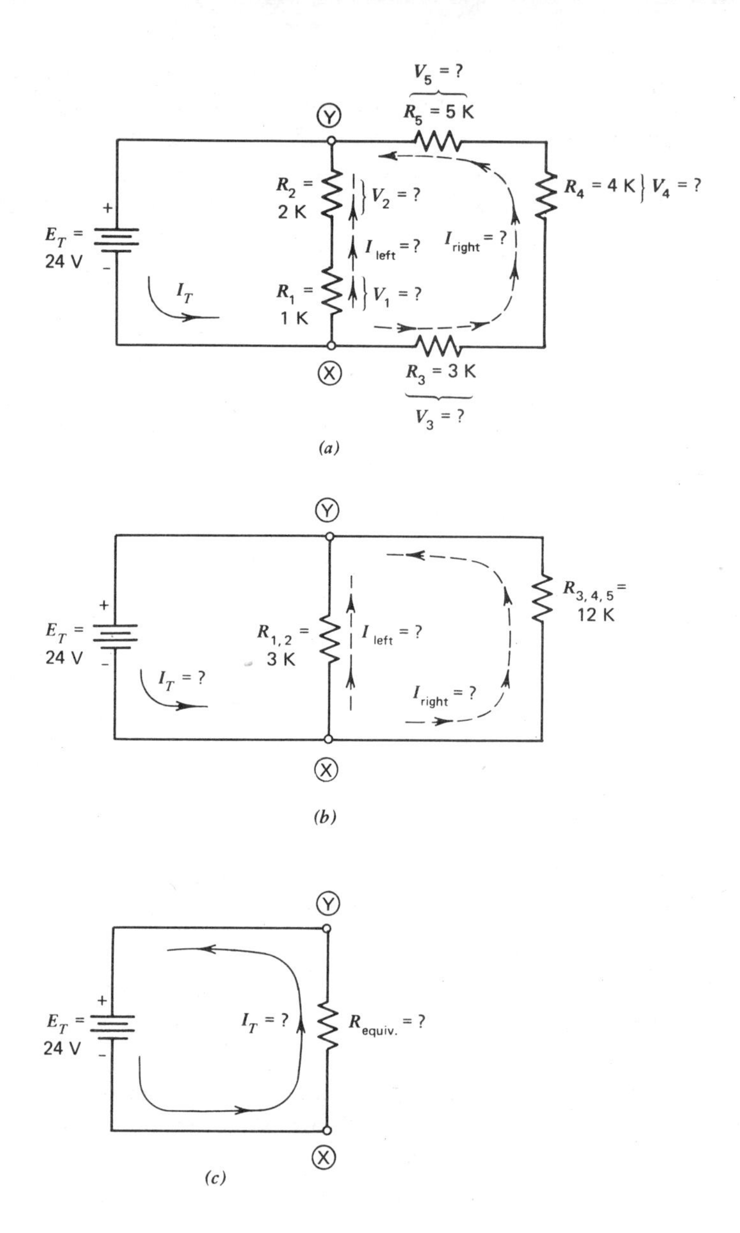

Figure 5-2. Parallel-series circuit of Example 5-1. (*a*) A complete circuit. (*b*) A more simplified circuit. (*c*) An equivalent circuit.

(h) R_{equiv}, which is also R_T in this circuit, or the resistance between points X and Y, $R_{xy} = E_T/I_T = 24/10$ mA $= 24/.01 = 2400\ \Omega$. Another method would be

$$R_{equiv} = \frac{\text{product}}{\text{sum}} = \frac{(R_{1,2})(R_{3,4,5})}{R_{1,2} + R_{3,4,5}} \qquad \text{(adaption of equation 4-3)}$$

$$= \frac{(3000)(12{,}000)}{3000 + 12{,}000} \quad \text{or} \quad \frac{(3 \times 10^3)(12 \times 10^3)}{3 \times 10^3 + 12 \times 10^3}$$

$$= \frac{36{,}000{,}000}{15{,}000} \quad \text{or} \quad \frac{36 \times 10^6}{15 \times 10^3}$$

$$= 2400\ \Omega \quad \text{or} \quad 2.4 \times 10^3$$

The simplified circuit showing R_{equiv} is shown in Fig. 5-2*c*.

(i) $V_1 = (I_{left})(R_1) = (8\text{ mA})(1\text{ K}) = (8 \times 10^{-3})(1 \times 10^3) = 8$ V.
(j) $V_2 = (I_{left})(R_2) = (8\text{ mA})(2\text{ K}) = (8 \times 10^{-3})(2 \times 10^3) = 16$ V.

Note that the sum of V_1 and $V_2 = E_T$, or $8 + 16 = 24$ V.

(k) $V_3 = (I_{right})(R_3) = (2\text{ mA})(3\text{ K}) = (2 \times 10^{-3})(3 \times 10^3) = 6$ V.
(l) $V_4 = (I_{right})(R_4) = (2\text{ mA})(4\text{ K}) = (2 \times 10^{-3})(4 \times 10^3) = 8$ V.
(m) $V_5 = (I_{right})(R_5) = (2\text{ mA})(5\text{ K}) = (2 \times 10^{-3})(5 \times 10^3) = 10$ V.

Note that the sum of V_3, V_4, and $V_5 = E_T$, or $6 + 8 + 10 = 24$ V.

For Similar Problems Refer to End-of-Chapter Problems 5-1 and 5-2

5-2. Series-Parallel Circuits

A *series-parallel* circuit is one having at least one resistor through which I_T flows (this is the *series* resistor), which is in *series* with at least two other *parallel* resistors. Figure 5-3*a* is an example of a simple series-parallel circuit.

Current flows from the negative end of E_T to point A. Here it splits into I_{left} through R_1 and I_{right} through R_2. The two branch currents join together at point B, and the combined current I_T flows through R_3 to the positive end of E_T. Points A and B, where I_T splits up and then joins together again, are the ends of a parallel circuit. This means that R_1 and R_2 are in parallel. Since I_T flows through R_3, then this resistor is the *series* resistor, being in series with the parallel combination of R_1 and R_2.

R_{equiv} of the parallel part of the circuit, or $R_{1,2}$ is shown in Fig. 5-3*b* and is: $R_{1,2}$ = product/sum $= (R_1)(R_2)/(R_1 + R_2) = (5)(20)/(5 + 20) = 100/25 = 4\ \Omega$.

As shown in Fig. 5-3*b*, $R_{1,2}$ is in series with R_3 so that $R_T = R_{1,2} + R_3 = 4 + 6 = 10\ \Omega$. Total current $I_T = E_T/R_T = 5/10 = 0.5$ A or 500 mA.

Voltage between points A and B, or across $R_{1,2} = (I_T)(R_{1,2}) = (0.5)(4) = 2$ V. $V_3 = I_T R_3 = (0.5)(6) = 3$ V. These are shown in Fig. 5-3*a* and *b*. The

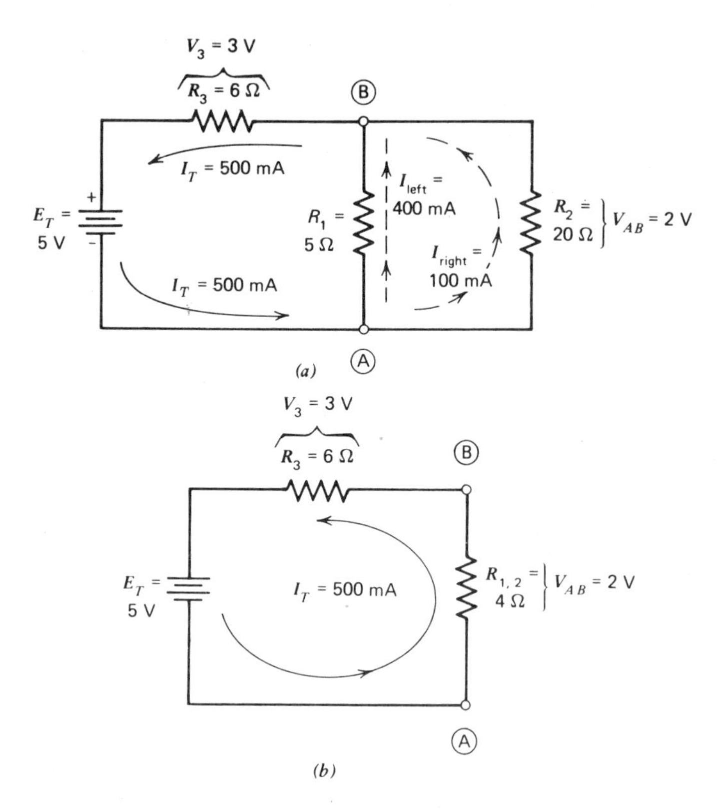

Figure 5-3. A series-parallel circuit. (*a*) A complete circuit. (*b*) An equivalent circuit.

voltage V_{AB} (2 V), is also the voltage across R_1 (2 V), and is also the same as across R_2 (2 V). Note that this voltage across the parallel portion (2 V) is not the entire applied voltage E_T, since there are 3 V *dropped* across R_3. With 5 V E_T, and 3 V across R_3, V_{AB} is 2 V, since $E_T = V_3 + V_{AB}$, or $5 = 3 + 2$.

I_{left} in Fig. 5-3*a* is a fraction of I_T, and may be found in several ways. Using Ohm's law: $I_{\text{left}} = V_{AB}/R_1 = 2/5 = 0.4$ A or 400 mA. I_{left} could also have been found from equation 4-6:

$$I_{\text{left}} = \left(\frac{R_2}{R_1 + R_2}\right) I_T \qquad (4\text{-}6)$$

$$= \left(\frac{20}{5 + 20}\right) 500 \text{ mA}$$

$$= \left(\frac{20}{25}\right) 500 \text{ mA}$$

$$= 400 \text{ mA}$$

I_{right} similarly could be found from Ohm's law: $I_{right} = V_{AB}/R_2 = 2/20 = 0.1$ A or 100 mA, or from equation 4-6.

The series-parallel circuits could become more complex than that shown in Fig. 5-3*a* by having two or more *series* resistors through which I_T flows, and also by having the parallel portions made up of *parallel-series* resistors. That is, the left branch and the right branch could each have two or more resistors in series with each other as shown in the circuits of Fig. 5-1*a* and 5-2*a*.

A complete and a more complex series-parallel circuit is shown in Fig. 5-4*a*. In analyzing the circuit, it can be seen that I_T flows from the negative terminal of E_T through R_1 to point C. At point C, the total current separates into I_{left} and I_{right}. I_{left} flows through R_2 and R_3. I_{right} flows through R_4, R_5, and R_6. I_{left} and I_{right} merge at point D, forming I_T, which flows through R_7 and to the positive terminal of E_T. Since I_T flows through R_1 and R_7, these resistors are the *series* resistors as shown in Fig. 5-4*b* and *c*. The *parallel* portion of the circuit is from point C to point D, and consists of a left branch (R_2 and R_3 in series) in parallel with a right branch (R_4, R_5, and R_6 in series).

R_{left} or $R_{2,3}$ is shown in Fig. 5-4*b*, and is the sum of $R_2 + R_3$ or $3\text{ K} + 1\text{ K} = 4\text{ k}\Omega$. R_{right} or $R_{4,5,6}$ is also shown in Fig. 5-4*b*, and is $R_4 + R_5 + R_6$ or $2\text{ K} + 4\text{ K} + 6\text{ K} = 12\text{ k}\Omega$.

At this point, find the following, comparing the answers with those shown in Fig. 5-4:

(a) $R_{equiv\ CD}$, (b) R_{total}, (c) I_T, (d) V_1, (e) V_{CD}, (f) V_7, (g) I_{left}, (h) I_{right}, (i) V_2, (j) V_3, (k) V_4, (l) V_5, and (m) V_6.

Another complex series-parallel circuit, shown in Fig. 5-5*a*, has two separate *parallel* portions, one between points A and B, and another between points C and D, and also two series resistors, R_3 and R_7.

Analyzing the circuit shows that electrons flow from the negative terminal of E_T to point A where the current divides into two paths, an upper and a lower. The *upper* current, I_{up}, flows through R_1 to point B, while the *lower* branch current, I_{lo}, flows through R_2 to point B. Points A and B are, therefore, the ends of a parallel circuit, the equivalent resistance of which, R_{AB} is shown in Fig. 5-5*b*.

Both currents, I_{up} and I_{lo}, combine at point B, forming I_T, which flows through R_3. R_3, then, acts as a series resistor. At point C, I_T again splits up into an upper and a lower current. I_{up} flows through R_6 and R_5 to point D, while I_{lo} flows through R_4 to point D. Points C and D are, therefore, also the ends of a *parallel* circuit. The upper branch resistance is made up of R_6 and R_5 in series with each other. The lower branch is R_4. The equivalent resistance, R_{CD}, is shown in the simplified circuit of Fig. 5-5*b*.

At point D, the two branch currents, I_{up} and I_{lo}, join together again, forming I_T, which flows through R_7 to the positive terminal of E_T. R_7 is, therefore, part of the *series* circuit shown in the simplified circuit of Fig. 5-5*b*.

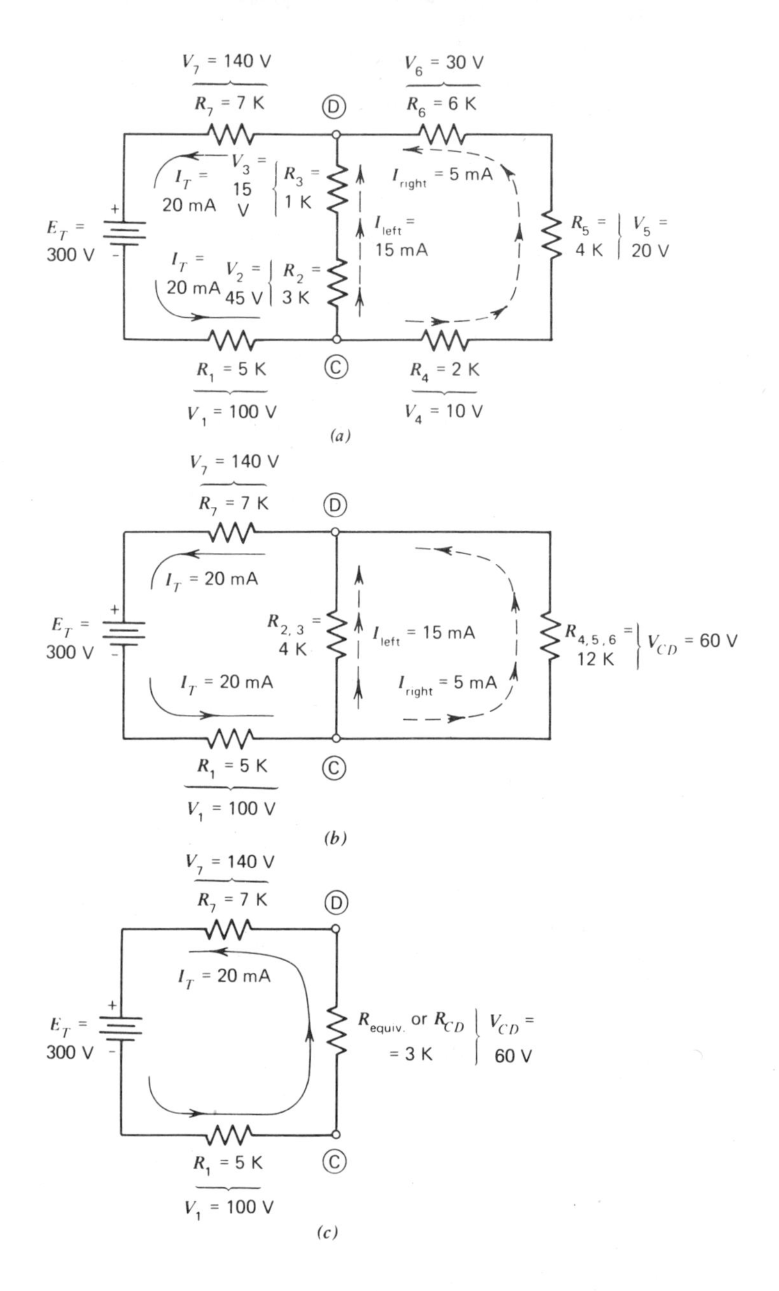

Figure 5-4. A more complex series-parallel circuit. (*a*) A complete circuit. (*b*) A more simplified circuit. (*c*) An equivalent circuit.

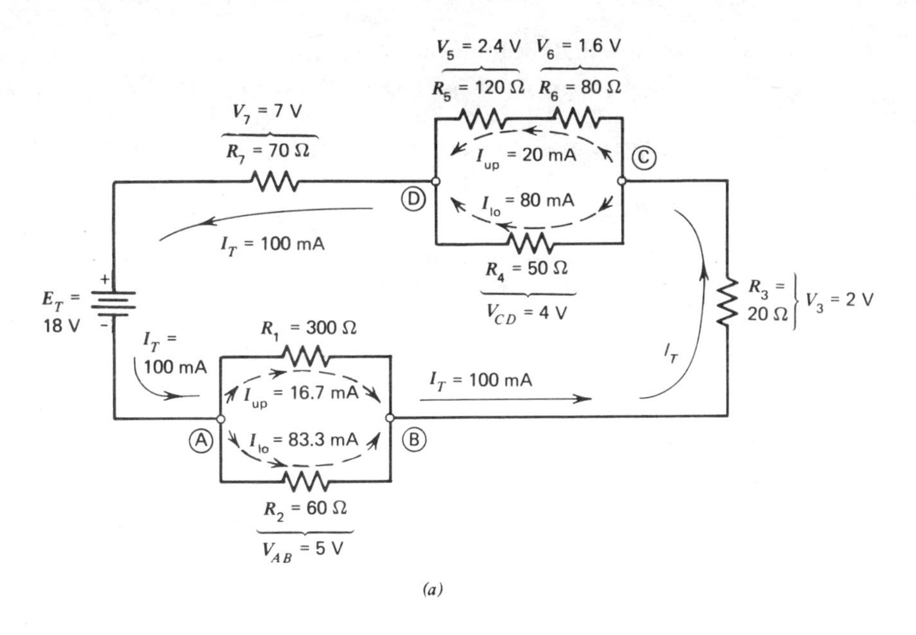

(a)

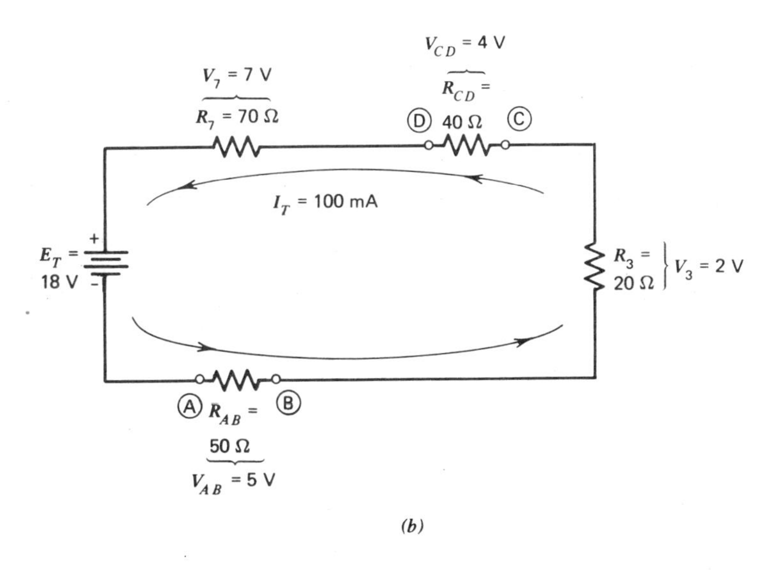

(b)

Figure 5-5. A complete series-parallel circuit. (*a*) A complete circuit. (*b*) A simplified equivalent circuit.

Now find the following, comparing the answers with those shown in Fig. 5-5: (a) $R_{\text{equiv } AB}$, (b) $R_{\text{equiv } CD}$, (c) R_T, (d) I_T, (e) V_{AB}, (f) V_3, (g) V_{CD}, (h) V_7, (i) $I_{\text{up } AB}$, (j) $I_{\text{lo } AB}$, (k) $I_{\text{up } CD}$, (l)$I_{\text{lo } CD}$, (m) V_5, and (n) V_6.

These examples are typical series-parallel circuits.

Example 5-2

Figure 5-6*a* shows an incomplete circuit consisting of four filaments, with the *required* current and voltage for each, and an applied $E_T = 110$ V. Figure 5-6*b* shows the completed diagram with three additional resistors. Explain the purpose of each of these resistors, and calculate the ohmic value of each.

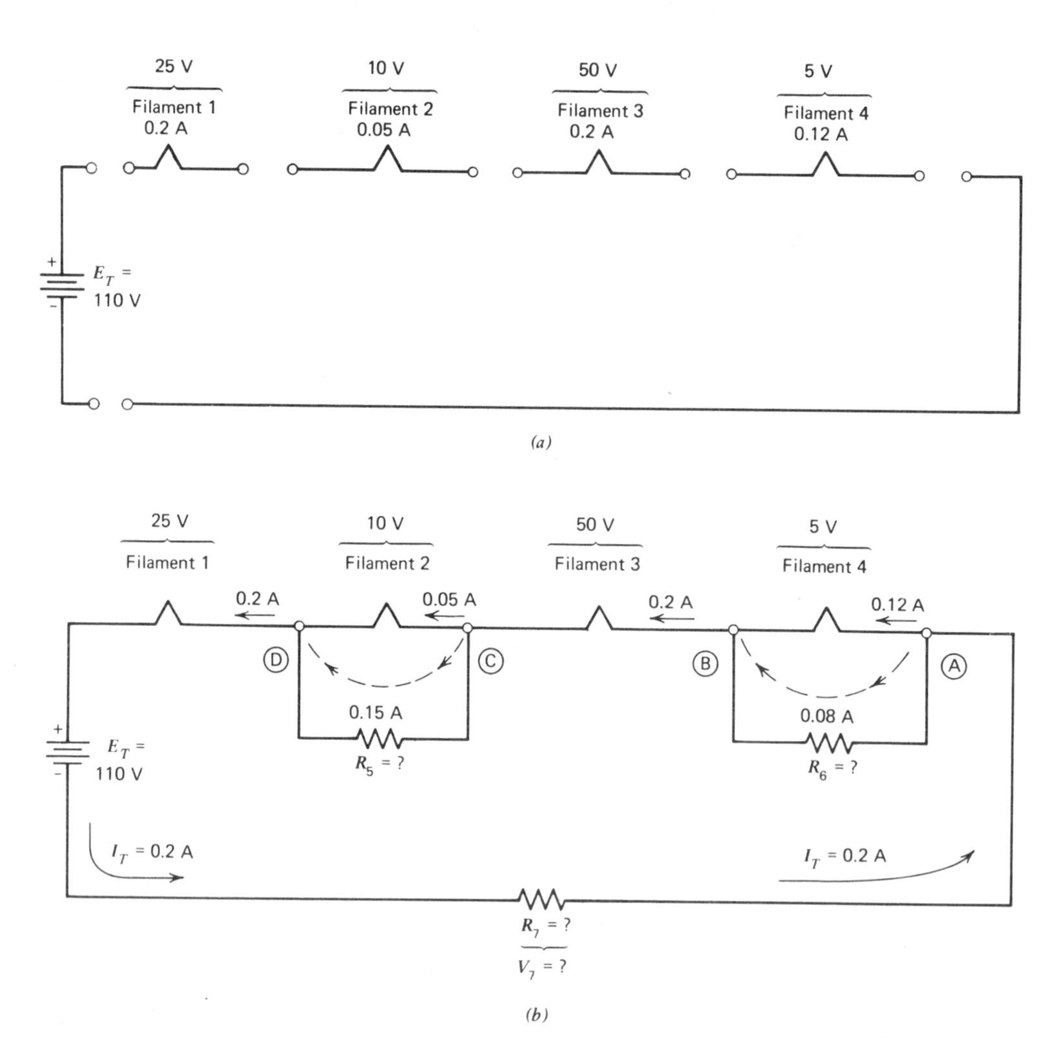

Figure 5-6. Example 5-2. (*a*) An incomplete circuit. (*b*) A complete circuit.

Solution

Filaments 1 and 3 require 0.2 A, while filament 2 requires a smaller current of 0.05 A, and filament 4 requires a current of 0.12 A. Since the largest current required is 0.2 A, then this current must flow in the circuit. With 0.2 A or 200 mA flowing, filaments 2 and 4 must be "shunted" by resistors so that not all the current will flow through filaments 2 and 4. Number 2 only requires 0.05 A or 50 mA. Therefore, the difference between 200 mA and 50 mA, or 150 mA must be given an alternate path that is through resistor R_5. This current 150 mA or 0.15 A is shown in Fig. 5-6*b* flowing through R_5.

Similarly, filament 4 should not have 200 mA flowing through it. Since it requires 0.12 A or 120 mA, then the difference between 200 mA and 120 mA, or 80 mA should flow through the alternate path, which is resistor R_6. This current 80 mA or 0.08 A is shown in Fig. 5-6*b* flowing through R_6.

The sum of the filament voltages of Fig. 5-6*a* is: $25 + 10 + 50 + 5 = 90$ V. With 110 V, E_T, applied and only 90 V required by the four filaments, then the difference between 110 and 90, or 20 V must be "dropped" across an additional resistor R_7, and V_7 is then 20 V. This resistor is added to the circuit shown in Fig. 5-6*b* simply as a voltage "dropping" device so that, from Kirchhoff's voltage law, the sum of each of the voltages will equal E_T. Note that the 10 V across filament 2 is also across its parallel resistor R_5 (between points C and D), and that the 5 V across filament 4 is also across its parallel resistor R_6 (between points A and B). Therefore, $V_{\text{fil 1}} + V_{CD} + V_{\text{fil 3}} + V_{AB} + V_7 = E_T$, or $25 + 10 + 50 + 5 + 20 = 110$ V.

To find the values of the resistors, the current *through* each, and the voltage *across* each must be determined in Fig. 5-6*b*. Voltage across R_5 is the same voltage as across filament 2 and is 10 V (or V_{CD}). Current through R_5 is 0.15 A as shown. Therefore, $R_5 = V_{CD}/I_{R5} = 10/0.15 = 66.7\ \Omega$.

Voltage across resistor R_6 is the same voltage as across filament 4 and is 5 V (or V_{AB}). Current through R_6 is 0.08 A as shown. Then, $R_6 = V_{AB}/I_{R6} = 5/.08 = 62.5\ \Omega$.

Voltage across R_7 is 20 V as has been described previously. Current through R_7 is $I_T = 0.2$ A as shown in Fig. 5-6*b*. Therefore, $R_7 = V_7/I_T = 20/0.2 = 100\ \Omega$.

Example 5-3

In the diagram of Fig. 5-7, if $V_4 = 6$ V, find: (a) I_{right}, (b) V_3, (c) V_{XY}, (d) I_{left}, (e) I_T, (f) V_1, (g) V_5, and (h) E_T.

Solution

As shown in the diagram, I_{total} flows from E_T through R_1 to point X. Here it becomes two parts, I_{left} and I_{right}. I_{left} flows through R_2 to point Y. I_{right} flows through R_3 and R_4 to point Y. Points X and Y, then, are the ends of a parallel

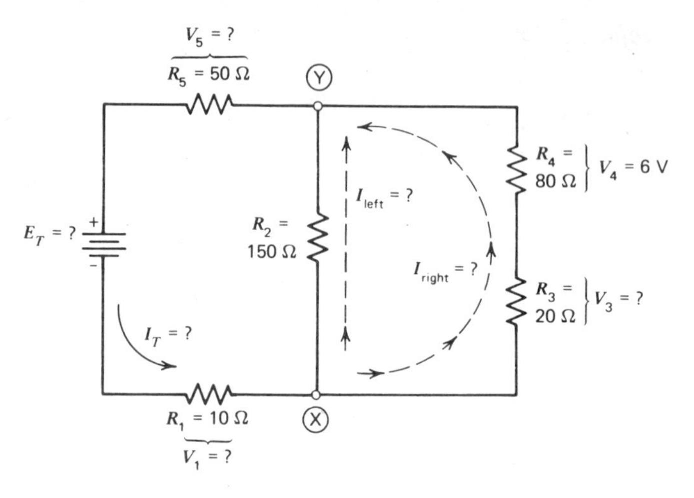

Figure 5-7. Example 5-4.

circuit consisting of R_2 in parallel with the series combination of R_3 and R_4. At point Y, I_{left} and I_{right} combine to again form I_T, which flows through R_5 to the positive end of E_T.

(a) Current through R_4 is I_{right} and is found from: I_4 or $I_{right} = V_4/R_4 = 6/80 = 0.075$ A of 75 mA.

(b) $V_3 = I_{right}\ R_3 = (0.075)\ (20) = 1.5$ V.

(c) $V_{XY} = V_4 + V_3 = 6 + 1.5 = 7.5$ V.

(d) V_{XY} is the voltage across the right-branch resistor, $R_3 + R_4$, and is also cross the left-branch resistor, R_2. Therefore, $I_{left} = V_{XY}/R_2 = 7.5/150 = 0.05$ A or 50 mA.

(e) From Kirchhoff's current law, $I_T = I_{left} + I_{right} = 50 + 75 = 125$ mA or 0.125 A.

(f) $V_1 = I_T\ R_1 = (0.125)\ (10) = 1.25$ V.

(g) $V_5 = I_T\ R_5 = (0.125)\ (50) = 6.25$ V.

(h) $E_T = V_1 + V_{XY} + V_5 = 1.25 + 7.5 + 6.25 = 15$ V.

For Similar Problems Refer to End-of-Chapter Problems 5-3 to 5-8

The following examples are practical ones using the actual circuit of a *shunt ohmmeter*, a device used to measure the value of an unknown resistor R_x. The actual circuit is shown in Fig. 5-8*a*, with the series-parallel equivalent in Fig. 5-8*b*. The resistance of the meter coil, R_{coil}, has been chosen as 1 kΩ for these examples for illustrative purposes, although R_{coil} is normally much smaller.

A brief circuit description is given here. The meter shown in Fig. 5-8*a* is a 1 mA type, requiring that much current to deflect the dial pointer full scale. If only 0.5 mA flows through the coil, the pointer only deflects half scale. The unknown resistor R_x to be measured is placed between the "test leads," or in shunt with the meter coil. Resistor R_1 is required to limit the current through the coil to a maximum of 1 mA, and is a variable resistor so that it may be decreased when the voltage of the battery, $E_T = 3$ V, falls due to aging.

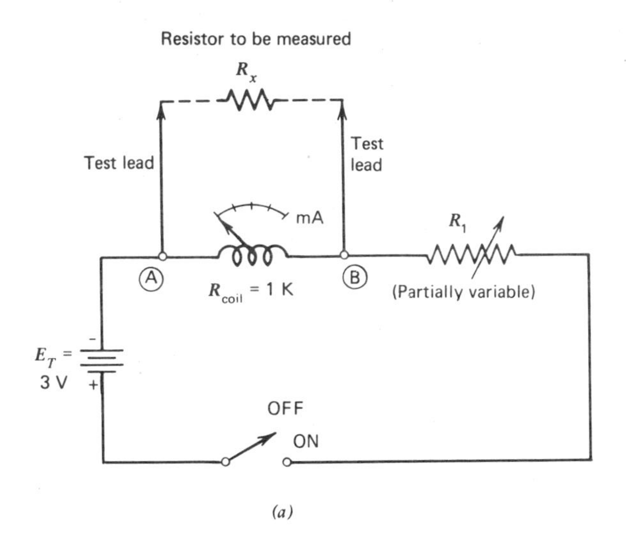

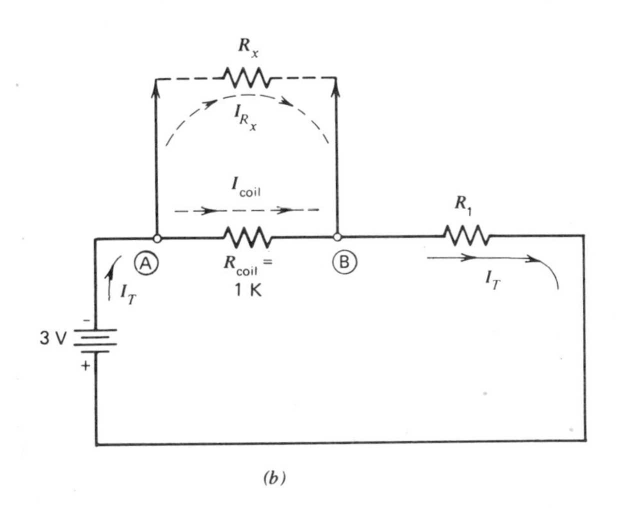

Figure 5-8. Examples 5-4 and 5-5. (*a*) Shunt-ohmmeter circuit. (*b*) Series-parallel circuit of shunt-ohmmeter.

Example 5-4

Find the value of R_1 if E_T is 3 V in Fig. 5-8*a*, if there is no R_x placed between the ends of the test leads, and it is desired to have 1 mA flowing through the meter coil when the switch is set to the ON position.

Solution

Without R_x, $R_T = R_{coil} + R_1$. Using Ohm's law to find R_T: $R_T = E_T/I_1 = 3/1 \text{ mA} = 3/.001 = 3000\ \Omega$.

And

$$
\begin{aligned}
R_T &= R_{coil} + R_1 \\
3000 &= 1000 + R_1 \\
3000 - 1000 &= R_1 \\
2000\ \Omega &= R_1
\end{aligned}
$$

Note that when R_x is omitted (test leads disconnected), 1 mA flows through R_{coil} and R_1, and the dial pointer deflects full scale to the dial reading of 1 mA. This point on the meter dial (1 mA) is, therefore, marked *infinity* ohms. If the test lead ends are shorted together ($R_x = 0\ \Omega$), current flows through R_1 only, since the meter coil is shorted across. The dial pointer does not deflect at all now and this point on the meter dial (0 mA) is marked *zero* ohms.

Example 5-5

What value R_x, placed between the ends of the test leads in Fig. 5-8 produces half-scale deflection of meter pointer, if $R_1 = 2\text{k}\Omega$ (from Example 5-4)?

NOTE

At first glance it would seem that if $R_x = R_{coil}$, or 1 K, meter current would become half-scale. This would be true where R_{coil} is very small compared to R_1. However, such is not the case here in this practice problem, and the value of R_x is found in the following solution.

Solution

With half-scale deflection of meter pointer, 0.5 mA flows through the coil. Voltage across the coil is: $V_{coil} = (I_{coil})(R_{coil}) = (0.5 \text{ mA})(1 \text{ K}) = (0.5 \times 10^{-3})(1 \times 10^{3}) = 0.5$ V.

$$
\begin{aligned}
E_T &= (V_{coil}) + (V_{R_1}) \qquad \text{(Kirchhoff's voltage law)} \\
3 &= 0.5 + V_{R_1} \\
3 - 0.5 &= V_{R_1} \\
2.5 \text{ V} &= V_{R_1}
\end{aligned}
$$

From Ohm's law, $I_{R_1} = V_{R_1}/R_1 = 2.5/2000 = 0.00125$ A or 1.25 mA. As shown in Fig. 5-8*b*, I_{R_1} is also I_T.

$$I_{\text{coil}} + I_{R_x} = I_T \qquad \text{(Kirchhoff's current law)}$$

$$0.5 + I_{R_x} = 1.25 \text{ mA}$$

$$I_{R_x} = 1.25 - 0.5$$

$$I_{R_x} = 0.75 \text{ mA}$$

Voltage across the coil (0.5 V) is also the voltage across resistor R_x. Therefore, $R_x = V_{R_x}/I_{R_x} = 0.5/0.75 \text{ mA} = 0.5/(0.75 \times 10^{-3}) = 0.667 \times 10^3 = 667\ \Omega$. This means that when a resistor, $R_x = 667\ \Omega$, is placed across the ends of the test leads, 0.5 mA flows through the meter coil. Therefore, at this point on the meter dial, the ohms scale would indicate 667 Ω.

For Similar Problems Refer to End-of-Chapter Problems 5-9 to 5-11

MORE ADVANCED SERIES-PARALLEL CIRCUITS

Here, more complex series-parallel circuits are analyzed, resistor voltage dividers with various loads connected to them are presented, the concepts of ground with negative and positive voltages are discussed, and the effects of an *open* resistor are shown.

5-3. More Complex Series-Parallel Circuits

Figure 5-9*a* shows a complex series-parallel circuit. By analyzing the circuit, it can be seen that I_T flows from the negative end of E_T through R_1 to point A. Since all the current flows through R_1, then this resistor is part of the *series* equivalent circuit and is shown in the more simplified diagrams of Fig. 5-9*b* and *c*.

At point A of Fig. 5-9*a*, I_T divides into a left branch flowing to point B through R_2 and into a right branch that also goes to point B, but first subdivides further. $I_{AB\ \text{right}}$ flows through R_3, and at point C it splits into $I_{CD\ \text{left}}$ through R_4 to point D, and $I_{CD\ \text{right}}$ through R_6, R_7, and R_8 to point D. Points C and D, therefore, are the ends of a parallel circuit made up of R_4 in parallel with the series resistors $R_6 + R_7 + R_8$. This parallel combination is $R_{CD\ \text{equiv}}$, shown in the simplified circuit of Fig. 5-9*b*.

At point D, the two branch currents, $I_{CD\ \text{left}}$ and $I_{CD\ \text{right}}$, join together forming $I_{AB\ \text{right}}$, and flow through R_5 to point B. Note that $I_{AB\ \text{right}}$ flows from point A to point B through R_3, $R_{CD\ \text{equiv}}$, and R_5. This is clearly shown in the more simplified circuit of Fig. 5-9*b*.

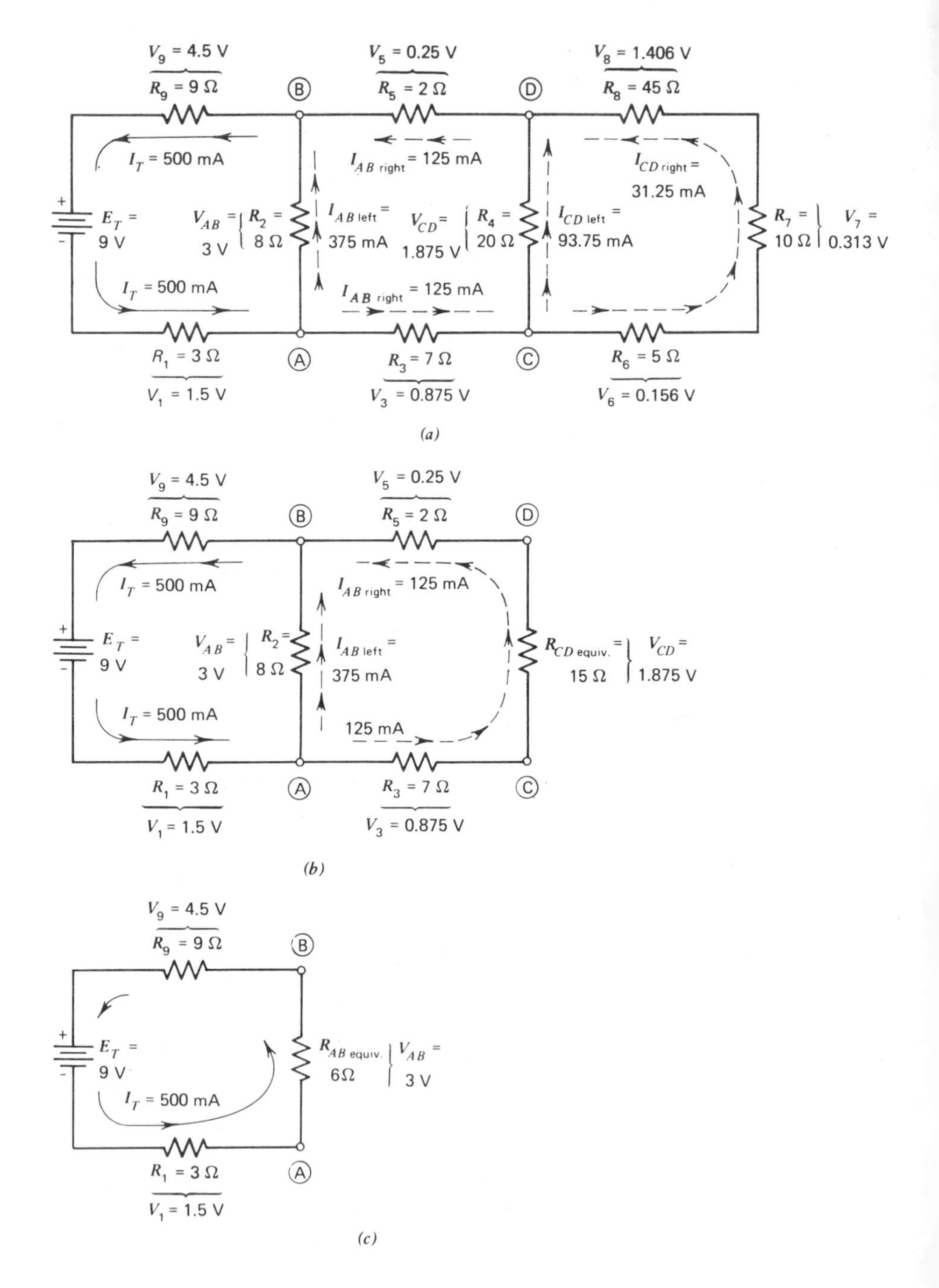

Figure 5-9. A complex series-parallel circuit analyzed. (*a*) A complete circuit. (*b*) A more simplified circuit. (*c*) An equivalent series circuit.

At point B, $I_{AB\,\text{right}}$ joins with $I_{AB\,\text{left}}$, forming I_T which flows through R_9 back to the positive terminal of E_T. Points A and B, therefore, are the ends of a large parallel circuit made up, as shown in diagrams (*b*) and (*c*), of R_2 in parallel with the series resistors $R_3 + R_{CD\,\text{equiv}} + R_5$. R_T finally consists of $R_1 + R_{AB\,\text{equiv}} + R_9$.

To solve for the voltage across each resistor and the current through each, R_T must be found. Since $R_{AB\,\text{equiv}}$ is part of R_T, $R_{AB\,\text{equiv}}$ must be found. However, $R_{CD\,\text{equiv}}$ is part of $R_{AB\,\text{equiv}}$, and R_{CD} *must* be found first. As shown in Fig. 5-9*a*, $R_{CD\,\text{equiv}}$ (or simply R_{CD}) consists of R_4 as a left-branch resistor in parallel with a right-branch resistor made up of R_6 in series with R_7 and R_8. The left-branch resistance of $R_4 = 20\ \Omega$. The right-branch resistance is the sum of $R_6 + R_7 + R_8 = 5 + 10 + 45 = 60\ \Omega$. Therefore, R_{CD} = product/sum = $(20)(60)/20 + 60 = 1200/80 = 15\ \Omega$. $R_{CD} = 15\ \Omega$ is shown in Fig. 5-9*b*.

Now $R_{AB\,\text{equiv}}$ (or simply R_{AB}) may be found. As illustrated in Fig. 5-9*b*, R_{AB} consists of R_2 as the left-branch resistor in parallel with the right-branch resistance. This right branch is made up of R_3, R_{CD}, and R_5 in series with each other, or $R_3 + R_{CD} + R_5 = 7 + 15 + 2 = 24\ \Omega$. Therefore, R_{AB} = product/sum = $(8)(24)/8 + 24 = 192/32 = 6\ \Omega$. This $R_{AB} = 6\ \Omega$ is shown in Fig. 5-9*c*.

R_T can now be found from Fig. 5-9*c*. $R_T = R_1 + R_{AB} + R_9 = 3 + 6 + 9 = 18\ \Omega$. $I_T = E_T/R_T = 9/18 = 0.5$ A or 500 mA. $V_1 = I_T R_1 = (0.5)(3) = 1.5$ V. V_1 may also be found by the fractional method of equation 3-1.

$$\text{Voltage across a series resistor} = \left(\frac{R_{\text{that resistor}}}{R_{\text{total}}}\right) E_T$$

$$V_1 = \left(\frac{R_1}{R_T}\right) E_T$$

$$V_1 = \left(\frac{3}{18}\right) 9$$

$$V_1 = 1.5\ \text{V}$$

$V_{AB} = I_T R_{AB} = (0.5)(6) = 3$ V; or, $V_{AB} = (R_{AB}/R_T)\, E_T = (6/18)\, 9 = 3$ V.

$V_9 = I_T R_9 = (0.5)(9) = 4.5$ V; or, $V_9 = (R_9/R_T)E_T = (9/18)\, 9 = 4.5$ V. These voltages, V_1, V_{AB}, and V_9, are shown in Fig. 5-9*c*. Note that their sum = $1.5 + 3 + 4.5 = 9\ \text{V} = E_T$, agreeing with Kirchhoff's voltage law.

Refer to Fig. 5-9*b* where $I_T = 500$ mA flows into point *A*, $I_{AB\,\text{left}}$ may be solved using Ohm's law, $I_{AB\,\text{left}} = V_{AB}/R_{AB\,\text{left}} = V_{AB}/R_2 = 3/8 = 0.375$ A or 375 mA. An alternate solution for $I_{AB\,\text{left}}$, using equation 4-15, is: $I_{AB\,\text{left}} = (R_{AB\,\text{equiv}}/R_2)\, I_T = (6/8)\, 500\ \text{mA} = 375$ mA.

Using Kirchhoff's current law, $I_{AB\,\text{right}}$ can now be found: $I_T = I_{AB\,\text{left}} + I_{AB\,\text{right}}$; $500 = 375 + I_{AB\,\text{right}}$, $500 - 375 = I_{AB\,\text{right}} = 125$ mA.

$V_3 = (I_{AB\ \text{right}})(R_3) = (0.125)(7) = 0.875$ V. V_3 could also be found by using the fractional method of equation 3-1. Referring to Fig. 5-9*b*, it can be seen that V_3 is a fraction of V_{AB}.

$$V_3 = \left(\frac{R_3}{R_{AB\ \text{right}}}\right) V_{AB} \quad \text{(adaption of equation 3-1)}$$

$$= \left(\frac{7}{7 + 15 + 2}\right) 3$$

$$= \left(\frac{7}{24}\right) 3$$

$$= 0.875 \text{ V}$$

$V_{CD} = (I_{AB\ \text{right}})(R_{CD}) = (0.125)(15) = 1.875$ V, or using the fractional method: $V_{CD} = (R_{CD}/R_{AB\ \text{right}})\ V_{AB} = (15/24)\ 3 = 1.875$ V.

$V_5 = (I_{AB\ \text{right}})(R_5) = (0.125)(2) = 0.25$ V. Note that in Fig. 5-9*b*, $V_3 + V_{CD} + V_5 = V_{AB}$, $0.875 + 1.875 + 0.25 = 3$ V.

With $V_{CD} = 1.875$ V, then the final two branch currents in Fig. 5-9*a*, $I_{CD\ \text{left}}$ and $I_{CD\ \text{right}}$ may be found. Using Ohm's law: $I_{CD\ \text{left}} = V_{CD}/R_{CD\ \text{left}} = 1.875/R_4 = 1.875/20 = 0.09375$ A or 93.75 mA.

$I_{CD\ \text{left}}$ could also be found using equation 4-15 where $I_{CD\ \text{left}}$ is a fraction of $I_{AB\ \text{right}}$: $I_{CD\ \text{left}} = (R_{CD\ \text{equiv}}/R_{CD\ \text{left}})\ I_{AB\ \text{right}} = (15/20)\ 125$ mA $= 93.75$ mA.

$I_{CD\ \text{right}}$ may be found from Ohm's law: $I_{CD\ \text{right}} = V_{CD}/R_{CD\ \text{right}} = 1.875/5 + 10 + 45 = 1.875/60 = 0.03125$ A or 31.25 mA. An alternate solution is, using equation 4-15: $I_{CD\ \text{right}} = (R_{CD\ \text{equiv}}/R_{CD\ \text{right}})\ I_{AB\ \text{right}} = (15/60)\ 125$ mA $= 31.25$ mA.

Both currents, $I_{CD\ \text{left}}$ and $I_{CD\ \text{right}}$, are shown in Fig. 5-9*a*. Note that the sum of these currents equal $I_{AB\ \text{right}}$, or 93.75 mA + 31.25 mA = 125 mA, agreeing with Kirchhoff's current law.

Voltages across resistors R_6, R_7, and R_8 can now be solved: $V_6 = (I_{CD\ \text{right}})(R_6) = (.03125)(5) = 0.156$ V (approximate). $V_7 = (I_{CD\ \text{right}})(R_7) = (.03125)(10) = 0.313$ V (approximate). $V_8 = (I_{CD\ \text{right}})(R_8) = (.03125)(45) = 1.406$ V (approximate). Note that the sum of these voltages equals V_{CD}, $V_6 + V_7 + V_8 = V_{CD}$, $0.156 + 0.313 + 1.406 = 1.875$ V, agreeing with Kirchhoff's voltage law.

Again, an alternate method of solving for V_6, V_7, and V_8, where each of these is a fraction of V_{CD}, is to use equation 3-1.

Example 5-6

In the circuit of Fig. 5-10*a*, all resistor values are given as shown, with $V_{11} = 0.24$ V. Find the following, comparing the answers with those shown in Fig. 5-10*a*, *b*, *c*, and *d*: (a) $I_{CD\ \text{right}}$, (b) V_{10}, (c) V_{12}, (d) V_{CD}, (e) $I_{CD\ \text{left}}$, (f) V_7, (g) V_8, (h) $R_{CD\ \text{equiv}}$, (i) $I_{AB\ \text{right}}$, (j) V_6, (k) V_9, (l) V_{AB}, (m) $I_{AB\ \text{left}}$, (n) V_2, (o) V_3, (p) $R_{AB\ \text{equiv}}$, (q) I_T, (r) V_1, (s) V_4, (t) V_5, and (u) E_T.

For Similar Problems Refer to End-of-Chapter Problem 5-12

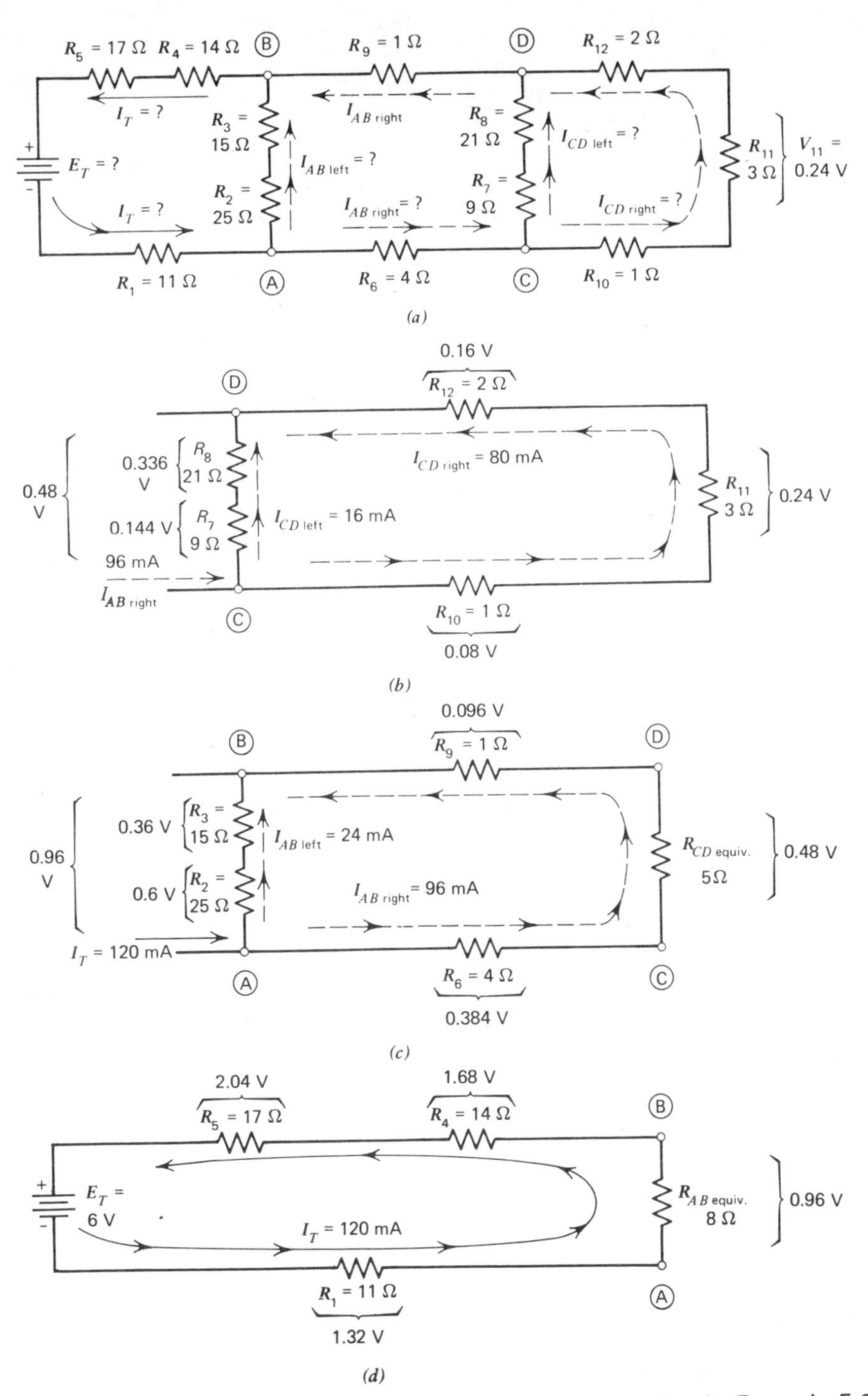

Figure 5-10. Example 5-6 and solution. (*a*) Complete circuit, Example 5-8. (*b*) Solution of *C-D* portion of circuit. (*c*) Solution of *A-B* portion of circuit. (*d*) Final solution.

5-4. Bridge Circuit

The simple parallel-series circuit of Fig. 5-11 is called a bridge circuit when a meter is connected between points X and Y, and is similar to a device called a Wheatstone bridge, used to measure the value of an unknown resistor. In Fig. 5-11, resistors R_1 and R_2 are in series with each other, and the series combination is connected across the 60-V E_T. Resistors R_3 and R_4 are also in series with each other, and this series combination is also connected across the 60-V E_T. Therefore, the two series circuits are in parallel with each other, with $R_1 + R_2$ being the left branch and $R_3 + R_4$ being the right brnach.

Voltages across each resistor may be found using equation 3-1: $V_1 = (R_1/R_1 + R_2)\ E_T = (1\text{ K}/1\text{ K} + 3\text{ K})\ 60\text{ V} = (1\text{ K}/4\text{ K})\ 60 = 15\text{ V}$. $V_2 = (R_2/R_1 + R_2)\ E_T = (3\text{ K}/1\text{ K} + 3\text{ K})\ 60\text{ V} = (3\text{ K}/4\text{ K})\ 60 = 45\text{V}$. Note that $V_1 + V_2 = E_T$; $15 + 45 = 60$. $V_3 = (R_3/R_3 + R_4)\ E_T = (25\text{ K}/25\text{ K} + 75\text{ K})\ 60\text{ V} = (25\text{ K}/100\text{ K})\ 60 = 15\text{ V}$. $V_4 = (R_4/R_3 + R_4)\ E_T = (75\text{K}/25\text{ K} + 75\text{ K})\ 60\text{ V} = (75\text{ K}/100\text{ K})\ 60 = 45\text{V}$. Note also that $V_3 + V_4 = E_T$, $15 + 45 = 60\text{ V}$.

These voltages also could have been found using Ohm's law. Current in the left branch is: $I_{\text{left}} = E_T/R_{\text{left}} = 60/1\text{ K} + 3\text{ K} = 60/4000 = .015\text{ A}$ or 15 mA. Then $V_1 = I_{\text{left}}\ R_1 = (.015)\ (1000) = 15\text{ V}$. $V_2 = I_{\text{left}}\ R_2 = (0.015)\ (3000) = 45\text{ V}$.

$I_{\text{right}} = E_T/R_{\text{right}} = 60/25\text{ K} + 75\text{ K} = 60/100{,}000 = 0.0006\text{ A}$ or 0.6 mA. $V_3 = I_{\text{right}}\ R_3 = (0.0006)\ (25{,}000) = 15\text{ V}$. $V_4 = I_{\text{right}}\ R_4 = (0.0006)\ (75{,}000) = 45\text{ V}$.

Ground is shown in Fig. 5-11 simply as a reference point. Point A which is grounded is therefore zero volts. Since the current I_{left} flows *up* through R_1 and R_2, the polarities across R_1 and R_2 are as shown. Point X, the top of R_1,

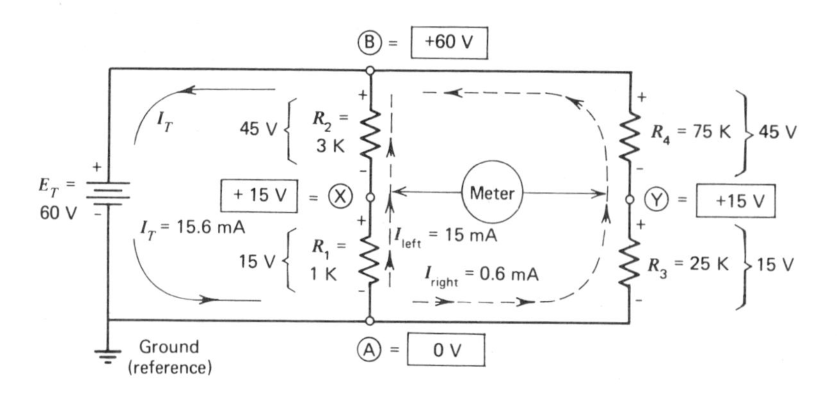

Figure 5-11. Bridge circuit, simple parallel-series circuit.

is + with respect to point A, the lower end of R_1. Since V_1 is 15 V, and point A is zero, then point X is +15 V *with respect to ground.* Point B is + with respect to point X as shown. Since V_2 is 45 V, and point X is +15 V, then point B is +60 V *with respect to ground.*

The current I_{right} flows *up* through R_3 and R_4. Point Y, the top of R_3, is + with respect to point A, the lower end of R_3. Since V_3 is 15 V, then point Y is +15 V *with respect to ground.* Observe that since point B is +60 V, and point Y is +15 V, the difference between these points is 45 V, which is V_4.

With the resistance values selected for Fig 5-11, points X and Y are at the same potential or voltage (+15 V) *with respect to ground.* The circuit is said to be *balanced* when there is no difference in voltage between points X and Y. Actually, ground could be eliminated from the circuit, and points X and Y would be at the same potential *with respect to point A*, and the circuit would still be balanced. For the circuit to be *balanced,* the ratio of the resistors of R_{left} must be the same as the ratio of the resistors of R_{right}, or

$$\frac{R_2}{R_1} = \frac{R_4}{R_3} \tag{5-1}$$

$$\frac{3\text{ K}}{1\text{ K}} = \frac{75\text{ K}}{25\text{ K}}$$

$$\frac{3}{1} = \frac{3}{1}$$

Another way of expressing this is that the ratio of the upper resistors must be the same as the ratio of the lower resistors, or

$$\frac{R_2}{R_4} = \frac{R_1}{R_3} \tag{5-2}$$

$$\frac{3\text{ K}}{75\text{ K}} = \frac{1\text{ K}}{25\text{ K}}$$

$$\frac{1}{25} = \frac{1}{25}$$

Actually, equation 5-2 may be algebraically derived from equation 5-1. The cross multiplication of numerators and denominators of both equations yields the same results: $R_2\ R_3 = R_1\ R_4$.

In an actual bridge circuit (discussed in Chapter 10, "Meters"), three of the four resistors are known and are adjusted so that the circuit is balanced. *Balance* is indicated when a sensitive current meter reads zero when connected between points X and Y. A simple bridge, similar to Fig. 5-11, is shown in Fig. 5-12. The fourth resistor is the unknown that is being measured by the

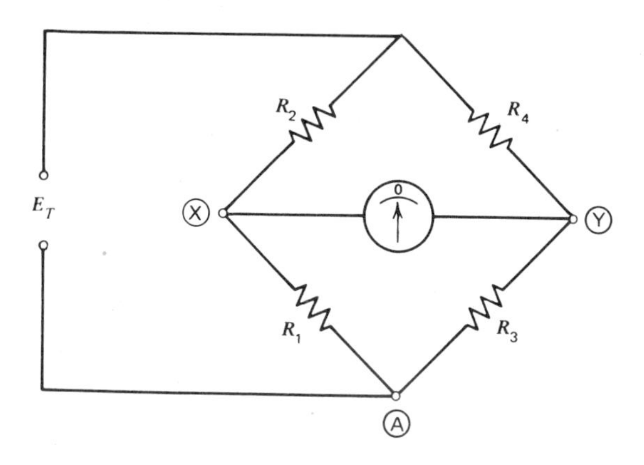

Figure 5-12. Balanced bridge circuit, Example 5-7 ($V_X = V_Y$ with respect to point A).

bridge. As an example, if R_1, R_2, and R_3 are each known, and the bridge is balanced, then the value of R_4 may be found from the following:

$$\frac{R_2}{R_1} = \frac{R_4}{R_3} \tag{5-1}$$

$$R_3\left(\frac{R_2}{R_1}\right) = R_4$$

The value of E_T has no effect on the *balance* of the bridge circuit.

Example 5-7

If the bridge of Fig. 5-12 is balanced, find R_4 if $R_1 = 500\ \Omega$, $R_2 = 2500\ \Omega$, and $R_3 = 120\ \Omega$.

Solution

Since the bridge is balanced (the meter between points X and Y reads zero), then the resistor ratio of equation 5-1 must be in effect: $R_2/R_1 = R_4/R_3$. Solving in terms of R_4 gives: $R_4 = (R_3)(R_2/R_1) = (120\ )(2500/500) = 600\ \Omega$.

For Similar Problems Refer to End-of-Chapter Problems 5-13 to 5-19

5-5. Resistor Voltage Dividers with Loads

In Section 3-7 of Chapter 3 a discussion of simple resistor voltage dividers without loads is given. As shown there, in Figures 3-9 to 3-12, the voltages produced by the series resistors are used by some external device. If this device does not require any current from the series-resistor voltage divider, then the device is said to present no load, and the voltage divider is unloaded. When the external device that is connected to the resistor voltage divider does have current flowing through itself, it does present a *load* to the voltage divider circuit.

In this section, our discussion is concerned with the more complicated voltage dividers with *loads*. To show the difference between the *unloaded* and the *loaded* voltage dividers, consider the following simple two-part example:

Example 5-8a

As shown in Fig. 5-13*a*, it is desired to produce an output voltage of +200 V with respect to ground from a 300-V power supply source by using a two-resistor voltage divider. The +200 volts is to be used by Device A, which requires no current. Find the values of the resistors, R_1 and R_2, if the current through the resistors, called *bleeder* current, is to be 1 mA.

Solution

Since voltage to Device A, between point X and ground, is +200 V, then voltage across $R_1 = 200$ V, and $V_{R_2} = 100$ V. From Ohm's law $R_1 = V_{R_1}/I_{R_1} = 200/1 \text{ mA} = 200/0.001 = 200{,}000\ \Omega$, and $R_2 = V_{R_2}/I_{R_2} = 100/1 \text{ mA} = 100/0.001 = 100{,}000\ \Omega$.

Example 5-8b

If the +200-V output from the 300-V power supply and the resistor voltage divider is to be connected to Device B, as shown in Fig. 13*b*, find the values

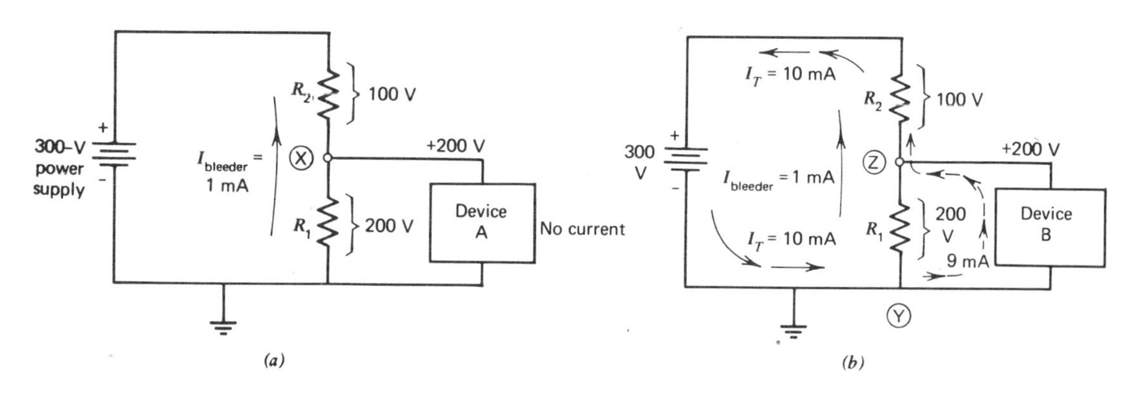

Figure 5-13. Voltage dividers. (*a*) No load, Example 5-8a. (*b*) Loaded, Example 5-8b.

of R_1 and R_2 where the *bleeder* current is to be 1 mA and where Device B requires 9 mA.

Solution

Since Device B requires +200 V, at point Z with respect to ground, then $V_{R_1} = 200$ V, and $V_{R_2} = 100$ V. Bleeder current, 1 mA, flows through resistors R_1 and R_2. The 9 mA of Device B flows through R_2 along with the 1 mA, making a total current $I_T = 10$ mA. I_T flows through R_2 and the power supply source, but not through R_1. At point Y, I_T separates into two parts. One part, $I_{\text{bleeder}} =$ 1 mA, flows through R_1, while the other part, $I_{\text{device B}} = 9$ mA, flows through Device B. At point Z, the two currents join together, forming $I_T = 10$ mA, and flow through R_2.

From Ohm's law then, $R_1 = V_{R_1}/I_{R_1} = 200/1 \text{ mA} = 200/0.001 = 200{,}000$ Ω, and $R_2 = V_{R_2}/I_{R_2} = 100/10 \text{ mA} = 100/0.01 = 10{,}000$ Ω. Device B acts as a simple resistor itself, with 200 V across it and 9 mA flowing through it. The resistance of this device then is: $R_{\text{device B}} = E/I = 200/9 \text{ mA} = 200/0.009$ $= 22{,}222$ Ω. Note that $R_{\text{device B}}$ is actually in parallel with R_1, and this combination is in series with R_2, making the entire circuit a series-parallel one.

In the circuit of Fig. 5-13*b*, Device B acts as a load on the resistor voltage divider. Device B in actual electronic equipment such as a television receiver or radar set is usually one or more amplifier stages. In the large electronic circuits there are many amplifier stages, some requiring different voltages and different currents than others. The following example illustrates this:

Example 5-9

Several amplifier stages, called *load A* in Fig. 5-14, require 150 V and draw 10 mA, while several other amplifiers, *load B*, require 250 V and draw 20 mA. The 250 V and 150 V are to be "tapped" off from a resistor voltage divider from a 300-V power supply as shown in Fig. 5-14. If bleeder current is to be 5 mA, find the values of resistors R_1, R_2, and R_3

Solution

To find the value of each resistor, voltage across each and current through each must be known. With +150 V with respect to ground required by the amplifiers of load A, then there is 150 V across R_1. As shown by the current arrow paths in Fig. 5-14, only the 5 mA I_{bleeder} flows through R_1. Therefore $R_1 = V_{R_1}/I_{R_1} = 150/0.005 = 30{,}000$ Ω.

Voltage required by the amplifiers of load B is 250 V. Therefore point B, the top of R_2, is +250 V with respect to ground. Point A, the lower end of R_2 is +150 V with respect to ground, since load A requires 150 V. Voltage across R_2 is the difference between the voltages at points B and A, or 100 V, the

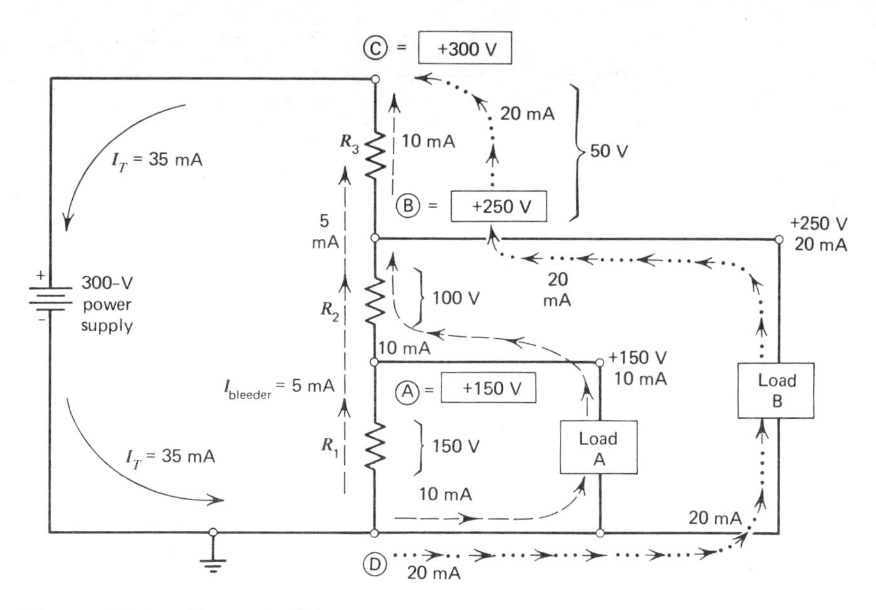

Figure 5-14. Example 5-9.

difference between +250 and +150. Current through R_2 consists of I_{bleeder} = 5 mA and load A current = 10 mA. I_{bleeder} flows through R_1, R_2, and R_3 as shown. $I_{\text{load A}}$ joins I_{bleeder} at point A, flowing through R_2, R_3, and the 300-V power supply, forming part of I_T. At point D, $I_{\text{load A}}$ flows to the right through load A. Current through R_2 then is I_{bleeder}, 5 mA + $I_{\text{load A}}$, 10 mA = 15 mA. Therefore, $R_2 = V_{R_2}/I_{R_2} = 100/15 \text{ mA} = 100/0.015 = 6667\ \Omega$.

Voltage across R_3 is 50 V, as shown, since point C, the top of R_3, is +300 V with respect to ground, while point B, the lower end of R_3, is +250 V. The difference between these points is 50 V, V_{R_3}.

Current through R_3 consists of I_{bleeder}, $I_{\text{load A}}$, and $I_{\text{load B}}$. As stated previously, I_{bleeder} and $I_{\text{load A}}$ each flow through R_3. At point B, $I_{\text{load B}}$ joins the other currents and also flows through R_3, as shown. $I_{\text{load B}}$ flows through R_3 and the 300-V power supply. At point D, this current flows to the right through load B. I_{R_3} is, therefore, = I_{bleeder} (5 mA) + $I_{\text{load A}}$, (10 mA) + $I_{\text{load B}}$ (20 mA) = 35 mA, or I_T.

$$R_3 = V_{R_3}/I_{R_3} = 50/35 \text{ mA} = 50/0.035 = 1428\ \Omega$$

Section 3-6 of Chapter 3 shows that both *negative* and *positive* voltages with *respect to ground* can be produced by grounding some point between the top and the bottom of the resistor voltage divider circuit. In the following example a *negative* voltage is to be produced for the part of the electronic equipment that does not draw any current and, therefore, does *not* present

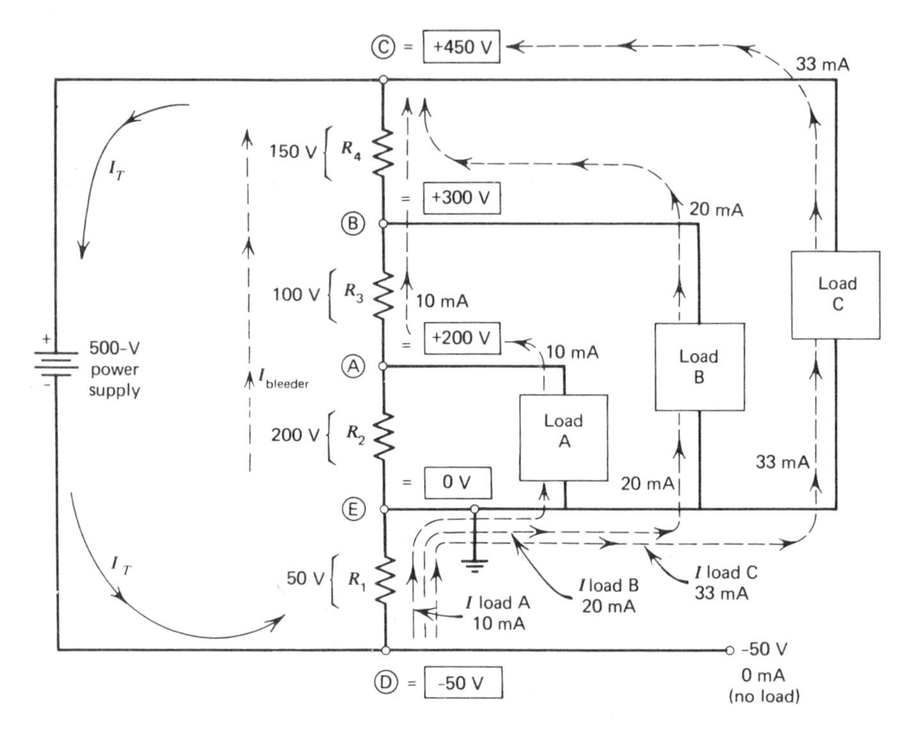

Figure 5-15. Voltage divider of Example 5-10, with negative voltage and positive voltages.

$$R_1 = \frac{V_1}{I_1} = \frac{50\ \text{V}}{70\ \text{mA}} = 714\ \Omega \qquad R_3 = \frac{V_3}{I_3} = \frac{100\ \text{V}}{17\ \text{mA}} = 5880\ \Omega$$

$$R_2 = \frac{V_2}{I_2} = \frac{200\ \text{V}}{7\ \text{mA}} = 28{,}600\ \Omega \qquad R_4 = \frac{V_4}{I_4} = \frac{150\ \text{V}}{37\ \text{mA}} = 4060\ \Omega$$

a load to the voltage divider. Also, three other loads (drawing current), each requiring different *positive* voltages, are connected to the various *taps* of the resistor voltage divider as shown in Fig. 5-15.

Example 5-10

The 500-V power supply and the resistor voltage divider is to apply the following: +450 V to several amplifier stages comprising load C, which draws 33 mA; +300 V to other amplifiers making up load B, which requires 20 mA; +200 V to the amplifiers of load A, which requires 10 mA; and −50 V with respect to ground to a part of the amplifiers requiring no current. The *bleeder current* flowing through each resistor should be 10% of the total current. Find (a) I_T, (b) I_{bleeder}, (c) current through each resistor, (d) voltage across each resistor, and (e) ohmic value of each resistor.

Solution

(a) Total current is the sum of *each* load current and bleeder current, or

$$I_T = I_{\text{load A}} + I_{\text{load B}} + I_{\text{load C}} + I_{\text{bleeder}}, \text{ and } I_{\text{bleeder}}$$

was stated to be 10% of I_T. Therefore,

$$I_T = 10 + 20 + 33 + 0.1\ I_T$$

$$I_T - 0.1\ I_T = 10 + 20 + 33$$

$$0.9\ I_T = 63$$

$$I_T = \frac{63}{0.9}$$

$$I_T = 70 \text{ mA}$$

(b) $I_{\text{bleeder}} = 10\%\ I_T$

$I_{\text{bleeder}} = (0.1)\,(70)$

$I_{\text{bleeder}} = 7 \text{ mA}$

Now find parts (c), (d), and (e) of this example and compare the answers with those shown listed in Fig. 5-15.

For Similar Problems Refer to End-of-Chapter Problems 5-20 to 5-23

5-6. Effects of Open Resistors in Parallel-Series Circuits

In Section 3-9 of Chapter 3 the effects of an *open* resistor in series circuits are discussed. If too much current flows through a resistor, causing a power dissipation that exceeds its maximum power capabilities, the resistor becomes excessively hot and may either increase in value substantially oe even crack open.

In a simple *parallel* circuit, an open resistor simply means that no current flows through this branch, while the other branch is unaffected. An example of this is the usual house wiring and connections. The 110 V from the power company is applied to *each* light bulb and electrical appliance (radio, electric refrigerator, toaster, television set, etc.) as a parallel circuit. If one light bulb burns out (opens up), each of the other bulbs and appliances are unaffected. Each of these other devices still has 110 V applied across it and still permits normal current through it. Only the defective lamp has no current flowing through it. The *total* current, of course, decreases.

In a simple *series* circuit, an open resistor prevents any current from flowing, since the circuit is not a complete path. With no current flow, voltage across each normal resistor is zero.

In a *parallel-series* circuit like that of Fig. 5-1*a*, an open resistor simply stops the current in its particular branch. Normal current and voltages continue to exist in the other branch. As an example, if R_1 (180-Ω resistor) in Fig. 5-1*a* opened up, I_{left} would become zero. I_{right} would continue to be its normal 0.3 A. I_{total} would now become equal to I_{right} alone, or 0.3 A. V_2, with no current flowing through R_2, now becomes zero volts. V_3 and V_4 remain normal at 22.5 V and 37.5 V, respectively. With R_1 open, the circuit becomes a simple *series* circuit consisting of R_3 and R_4 and E_T.

However with R_1 open, V_1 becomes 60 V instead of its normal 36 V. Reason for this is that the top of the open R_1 is still connected to point B which is the *positive* end of the 60 V E_T, while the lower end of R_1 connects through R_2 ($V_2 = 0$) to point A which is the *negative* end of the 60 V E_T. Therefore, voltage across the open R_1 = 60 V. Voltage across the *open* reistor always increases compared to its normal voltage.

For Similar Problems Refer to End-of-Chapter Problems 5-24 to 5-28

5-7. Effects of Open Resistors in Series-Parallel Circuits

The effect of an open resistor in a series-parallel circuit depends on whether the resistor is part of the parallel or series circuit. In Fig. 5-3*a*, an open R_3, the 6-Ω series resistor, results in a completely open circuit. Nó current flows, and V_1 and V_2 are each zero volts. V_3 is the full 5 V E_T.

A completely different effect would result if the 5-Ω R_1, part of the parallel section of Fig. 5-3*a* were to become open. Now the circuit would become a simple series type consisting of R_2 and R_3 in series with E_T. Current would flow through R_2 and R_3, but not through the open R_1. I_T would now become = $E_T/R_T = E_T/R_2 + R_3 = 5/20 + 6 = 5/26 = 0.192$ A or 192 mA instead of the original 500 mA. Note that with R_1 open, current through the remaining parallel branch, R_2, has now increased from its former value of 100 mA.

V_2 would now be $= I(R_2) = (0.192)(20) = 3.84$ V, instead of the normal value of 2 V shown. V_3 would be $= I(R_3) = (0.192)(6) = 1.16$ V, instead of the normal value 3 V shown. The voltage V_1 across the open resistor R_1 is the same as V_2, or 3.84 V instead of its normal value of 2 V.

A third effect would result if the 20-Ω resistor R_2 were to open in Fig. 5-3*a*. Again, the circuit would become a simple series type, this time consisting of R_1 and R_3 with E_T. I_T would now be $= E_T/R_T = E_T/R_1 + R_3 = 5/5 + 6 = 5/11 = 0.454$ A or 454 mA. Note that with R_2 open, current through the remaining parallel branch, R_1, has now increased from its former value of 400 mA.

V_1 is now $I(R_1) = (0.454)(5) = 2.27$ V instead of the normal 2 V shown. $V_3 = I(R_3) = (0.454)(6) = 27.3$ V instead of the normal value 3 V shown. Voltage V_2 across the open resistor R_2 is the same as V_1, or 2.27 V. Note that voltage across the open resistor has increased over its normal value of 2 volts.

Example 5-11

(a) If R_1 (5 K) opens up in Fig. 5-4*a*, what are the currents and voltages in the circuit? (b) If R_7 (7 K) opens in Fig. 5-4*a*, what are the currents and voltages in the circuit?

Solution

(a) and (b) R_1 and R_7 are *series* resistors in the series-parallel circuit of Fig. 5-4*a*. If either resistor opens, no current flows, and there is zero volts across every resistor in the circuit except the open one. The full 300 V E_T is measured across the open series resistor.

Example 5-12

What happens to the circuit of Fig. 5-4*a* if: (a) R_3 (the 1 K resistor) opens? (b) R_6 (the 6 K resistor) opens?

Solution

(a) If R_3 opens, the series-parallel circuit of Fig. 5-4*a* becomes a series circuit consisting of R_1, R_4, R_5, R_6, and R_7 and E_T. A current flows through these resistors, but none flows through the open branch made up of R_2 and R_3.

(b) An open R_6 (6 K resistor) in Fig. 5-4*a* again changes the circuit into a series type consisting of R_1, R_2, R_3, and R_7 and E_T.

For Similar Problems Refer to End of-Chapter Problems 5-29 to 5-40

PROBLEMS

See section 5-1 for discussion covered by the following problems.

5-1. Redraw the parallel-series circuit of Fig. 5-1*a*, changing all values to the following: $R_1 = 15$ K, $R_2 = 45$ K, $R_3 = 70$ K, and $R_4 = 50$ K, and $E_T =$

24 V. Find: (a) R_{left}, (b) R_{right}, (c) R_{equiv}, (d) I_{left}, (e) I_{right}, (f) I_T, (g) V_1, (h) V_2, (i) V_3, and (j) V_4.

5-2. Redraw the parallel-series circuit of Fig. 5-2*a*, but change all values to the following: R_1 = 15 K, R_2 = 25 K, R_3 = 20 K, R_4 = ?, R_5 = 30 K, and E_T = ?. If I_T = 4 mA, and V_1 = 45 V, find: (a) I_{left}, (b) V_2, (c) E_T, (d) I_{right}, (e) V_3, (f) V_5, (g) V_4, and R_4.

See section 5-2 for discussion covered by the following problems.

5-3. Redraw the series-parallel circuit of Fig. 5-3*a*, but all values are changed to the following: R_1 = 20 K, R_2 = 30 K, and R_3 = 8 K, and E_T = 100 V. Find: (a) R_{AB}, (b) R_T, (c) I_T, (d) V_3, (e) V_{AB}, (f)I_{R_1} (g) I_{R_2}, and (h) I_{R_3}.

5-4. Redraw the series-parallel circuit of Fig. 5-4*a* but change all values to the following: R_1 = 2 K, R_2 = 22 K, R_3 = 18 K, R_4 = 4 K, R_5 = 1 K, R_6 = 5 K, R_7 = 10 K, and E_T = 120 V. Find: (a) R_{CD}, (b) R_T, (c) I_T, (d) V_1, (e) V_7, (f) V_{CD}, (g) I_{left}, (h) I_{right}, (i) V_2, (j) V_3, (k) V_4, (l) V_5, and (m) V_6.

5-5. Redraw the series-parallel circuit of Fig. 5-5*a* but change all values to the following: R_1 = 10 K, R_2 = 15 K, R_3 = 3 K, R_4 = 7 K, R_5 = 20 K, R_6 = 22 K, R_7 = 25 K, and E_T = 400 V. Find: (a) R_{AB}, (b) R_{CD}, (c) R_T, (d) I_T, (e) V_{AB}, (f) I_{R_1}, (g) I_{R_2}, (h) V_3, (i) V_{CD}, (j) I_4, (k) I_{R_5, R_6}, (l) V_4, (m) V_5, (n) V_6, and (o) V_7.

5-6. Three filaments, A, B, and C, are to be connected to a supply voltage of 120 V. Filament A requires 35 V and 300 mA. Filament B requires 25 V and 300 mA. Filament C requires 30 V and only 100 mA. Draw the circuit showing any additional resistors required and their values. (see Fig. 5-6 for suggestions.)

5-7. Refer to Fig. 5-7 but change V_4 to 1.6 V, leaving all resistor values unchanged. Find: (a) I_{right}, (b) V_3, (c) V_{XY}, (d) I_{left}, (e) I_T, (f) V_1, (g) V_5, and (h) E_T.

5-8. Redraw Fig. 5-7 but change all values to the following: R_1 = 40 Ω, R_2 = 90 Ω, R_3 = ?, R_4 = 130 Ω, R_5 = 20 Ω, and E_T = ?. If V_1 = 4 V, and I_{R_2} = 66-2/3 mA, find: (a) V_{XY}, (b) I_{right}, (c) V_4, (d) V_3, (e) R_3, (f) V_5, and (g) E_T.

5-9. In the shunt-ohmmeter circuit of Fig. 5-8*a*, if the meter coil is a 100 microammeter with a coil resistance of 1 K (100 μA produces full-scale deflection), with an applied voltage of 3 V, find the value of R_1 that will permit full-scale deflection when the test leads are disconnected (R_x is infinite).

5-10. In the previous problem, what value of R_x, placed in shunt with the meter coil, reduces the meter coil current to half-scale deflection (50 μA)?

5-11. In Problem 5-9 how much current flows through the meter coil if R_x = 2 K?

More Challenging Problems: See section 5-3 for discussion covered by the following problems.

5-12. Redraw the circuit of Fig. 5-10a, but change all resistor values to the following: $R_1 = 1$ K, $R_2 = 2$ K, $R_3 = 4$ K, $R_4 = 4$ K, $R_5 = 5$ K, $R_6 = 6$ K, $R_7 = 7$ K, $R_8 = 8$ K, $R_9 = 14$ K, $R_{10} = 10$ K, $R_{11} = 11$ K, and $R_{12} = 9$ K. The only known voltage is $V_3 = 60$ V. Find: (a) R_{CD}, (b) R_{AB}, (c) R_T, (d) $I_{AB\,\text{left}}$, (e) V_2, (f) V_{AB}, (g) $I_{AB\,\text{right}}$, (h) V_6, (i) V_{CD}, (j) V_9, (k) $I_{CD\,\text{left}}$, (l) V_8, (m) V_7, (n) $I_{CD\,\text{right}}$, (o) V_{10}, (p) V_{11}, (q) V_{12}, (r) I_T, (s) V_1, (t) V_4, (u) V_5, and (v) E_T.

See section 5-4 for discussion covered by the following problems.

5-13. In the bridge circuit of Fig. 5-12, if $R_1 = 10$ K, $R_2 = 20$ K, $R_3 = 30$ K, $R_4 = 60$ K, and $E_T = 60$ V, what is: (a) Point X with respect to point A? (b) Point Y with respect to point A? (c) Point X with respect to point Y?

5-14. If the bridge circuit of Problem 5-13 (Fig. 5-12) is balanced: (a) What will the meter indicate between points X and Y? (b) if the applied voltage E_T is changed from 60 V to 120 V, in the balanced bridge, what will the meter indicate between points X and Y?

5-15. If the bridge of Fig. 5-12 is balanced, and $R_1 = 3$ K, $R_2 = 9$ K, and $R_3 = 800\ \Omega$, find R_4.

5-16. If the bridge of Fig. 5-12 is balanced, and $R_2 = 25$ K, $R_3 = 100$ K, and $R_4 = 500$ K, find R_1.

5-17. Find R_2 in the balanced bridge of Fig. 5-12 if $R_1 = 250\ \Omega$, $R_3 = 1$ K, $R_4 = 8$ K, and $E_T = 3$ V.

5-18. If the bridge of Fig. 5-12 is balanced, find R_3 if $R_1 = 1.5$ K, $R_2 = 9$ K, $R_4 = 120$ K and $E_T = 12$ V.

5-19. In the bridge circuit of Fig. 5-12, $R_1 = 4$ K, $R_2 = 2$ K, $R_3 = 250\ \Omega$, and $R_4 = 500\ \Omega$. If point X is +12 V with respect to point A: (a) What is E_T? (b) What is point Y with respect to point A? (c) What is point X with respect to point Y?

See section 5-5 for discussion covered by the following problems.

5-20. In the diagram for this problem, find: (a) I_T, (b) I_{R_1}, (c) I_{R_2}, (d) value of R_1 and its wattage dissipation, and (e) value of R_2 and its wattage dissipation.

5-21. In the diagram for this problem, find: (a) I_{bleeder}, (b) I_{R_3}, (c) I_{R_4}, (d) I_{R_5}, (e) R_3, (f) R_4, and (g) R_5.

5-22. In the circuit shown for this problem, note that I_{bleeder} is 10% of I_T. Find: (a) I_T, (b) I_{bleeder}, (c) I_{R_1}, (d) I_{R_2}, (e) I_{R_3}, (f) V_1, (g) V_2, (h) V_3, (i) E_T, (j) R_1, (k) R_2, and (l) R_3.

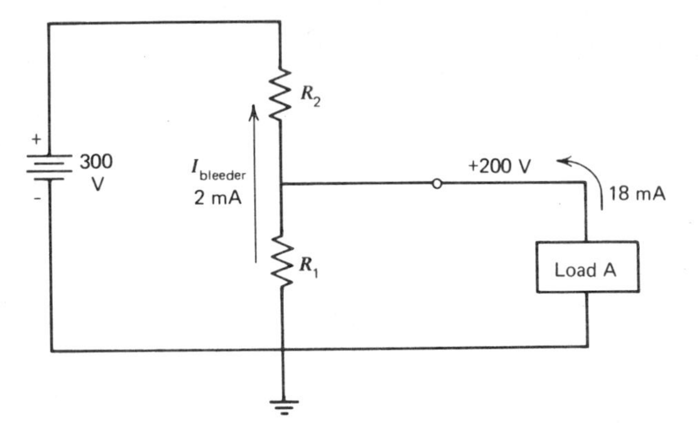

Problem 5-20.

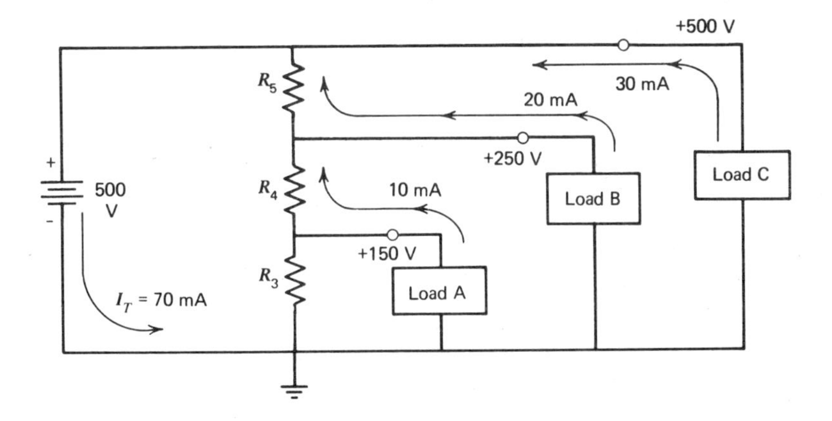

Problem 5-21.

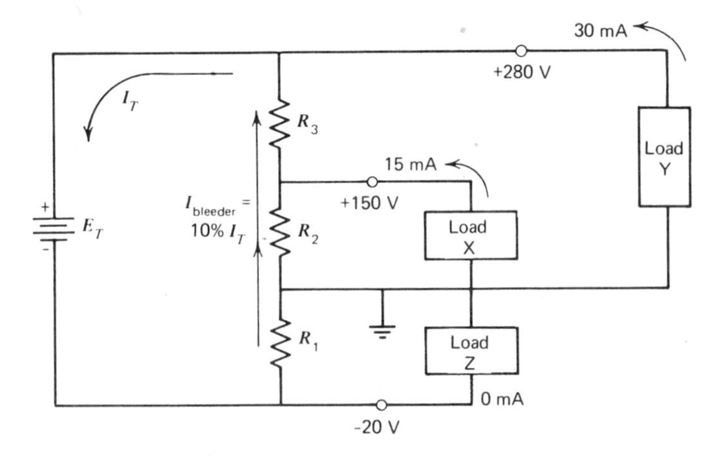

Problem 5-22.

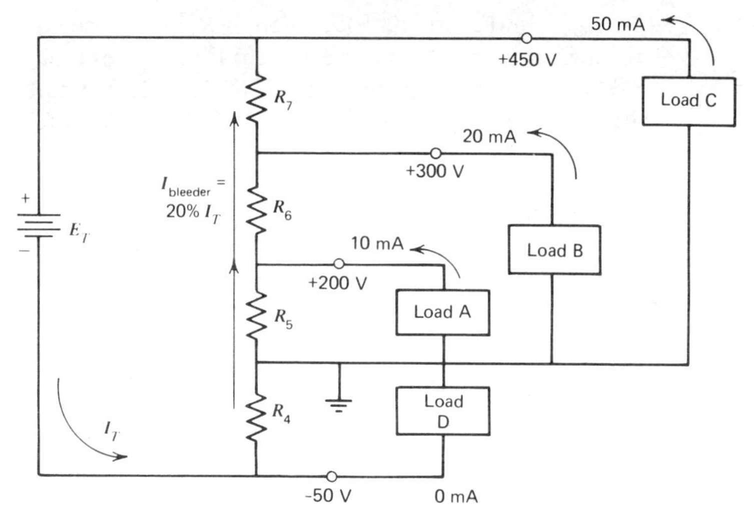

Problem 5-23.

5-23. I_{bleeder} is 20% of I_T in the circuit shown for this problem. Find: (a) I_T, (b) I_{bleeder}, (c) I_4, (d) I_5, (e) I_6, (f) I_7, (g) V_4, (h) V_5, (i) V_6, (j) V_7, (k) E_T, (l) R_4, (m) R_5, (n) R_6, (o) R_7.

See section 5-6 for discussion covered by the following problems.

5-24. In the parallel-series circuit of Fig. 5-1*a*, if resistor R_2 (120 Ω) opens, find: (a) I_{right}, (b) I_{left}, (c) I_T, (d) V_1, (e) V_2, (f) V_3, and (g) V_4.
5-25. In the parallel-series circuit of Fig. 5-1*a*, if resistor R_4 (125 Ω) opens, find: (a) I_{right}, (b) I_{left}, (c) I_T, (d) V_1, (e) V_2, (f) V_3, and (g) V_4.
5-26. In the circuit of Fig. 5-2*a*, if resistor R_1 (1 K) opens, find: (a) I_{right}, (b) I_{left}, (c) I_T, (d) V_1, (e) V_2, (f) V_3, (g) V_4, and (h) V_5.
5-27. In the bridge circuit of Fig. 5-11, if resistor R_2 (3 K) opens, find: (a) voltage at point X with respect to ground, and (b) voltage at point Y with respect to ground.
5-28. In the bridge circuit of Fig. 5-11, if resistor R_1 (1 K) opens, find: (a) voltage at point X with respect to ground, and (b) voltage at point Y with respect to ground.

See section 5-7 for discussion covering the following problems.

5-29. In the circuit of Fig. 5-5*a*, find all currents and voltages (a) if R_3 (20 Ω) opened, and (b) if R_7 (70 Ω) opened.

5-30. In the circuit of Fig. 5-5*a*, if R_1 (300 Ω) opens: (a) Should the current through R_2 increase or decrease? (b) Find I_{R_2}. (c) Find V_1.

5-31. In the complete series-parallel filament circuit of Fig. 5-6*b*, if *filament 3* (50 V and 0.2 A) opens, find: (a) $V_{fil\,1}$, (b) $V_{fil\,2}$, (c) $V_{fil\,3}$, (d) $V_{fil\,4}$, (e) V_7, and (f) I_T.

5-32. In the complete series-parallel filament circuit of Fig. 5-6*b*, if R_5 (the resistor shunting filament 2) opens, what should happen to the current flowing through filament 2 (do not actually solve)?

5-33. In the complex series-parallel circuit of Fig. 5-9*a*, if R_7 (10 Ω) opened: (a) Which resistors now form a parallel circuit? (b) Would R_T now increase or decrease (do not solve)?

5-34. In the complex series-parallel circuit of Fig. 5-9*a*, if R_2 (8 Ω) opened: (a) Which resistors now form a parallel circuit? (b) R_T now consists of which resistors (do not solve)?

5-35. In the complex series-parallel circuit of Fig. 5-9*a*, if R_4 (20 Ω) opened, which resistors now make up a parallel circuit?

5-36. In the complex series-parallel circuit of Fig. 5-9*a*, if R_3 (7 Ω) opens, which resistors now make up the circuit?

5-37. In the resistor voltage divider circuit of Fig. 5-14, if R_3 opened, what is: (a) V_3, (b) V_2, (c) V_1, (d) voltage at point B with respect to ground, (e) voltage at point A with respect to ground, and (f) I_T? (Note that with R_3 open, neither load A nor load B could operate.)

5-38. In the resistor voltage divider circuit of Fig. 5-14, if R_2 opened: (a) Would voltage at point B with respect to ground increase or decrease (do not solve)? (b) Would voltage at point A with respect to ground increase or decrease (do not solve)? (Note that with R_2 open, load A could not operate.)

5-39. In Fig. 5-14, if load B (+250 V, 20 mA) became inoperative: (a) Would V_3 increase or decrease? (b) Would voltage at point B with respect to ground increase or decrease?

5-40. In Fig. 5-15, if R_1 (between points D and E) opened, find voltages at each of the following points with respect to ground: (a) point A, (b) point B, (c) point C, (d) point D, and (e) point E.

chapter 6

Resistor Network Circuits

A *network* or *mesh* circuit is actually a complex series-parallel circuit that may contain more than one applied voltage source. To solve these more complicated circuits, other methods than those shown in Chapters 3, 4, and 5 have been developed. Solutions of network circuits are necessarily more complicated than those for series-parallel circuits. In the first part of this chapter, the various solution methods are demonstrated, using a simple series-parallel circuit having only one voltage source. This is done to aid the mastering of these methods. Later, these methods are applied to more complex circuits having *two or more voltage* sources.

The several methods of solution discussed in this chapter are: superposition, Maxwell's cyclic currents (also called *loop* currents), Thévenin's theorem, Norton's theorem, Millman's theorem, and nodal analysis. There are others too, but the half dozen examined here are adequate for practically all resistor networks. In every method, Ohm's law and either Kirchhoff's current or voltage law is employed. The discussion given in Chapter 5 on series-parallel circuits is also put to use.

6-1. Resistance Between Two Points

To more easily follow the discussions for some of the network solutions, an understanding of some simple circuits is first necessary. The three resistors of Fig. 6-1 represent a simple basic circuit without any applied voltage source.

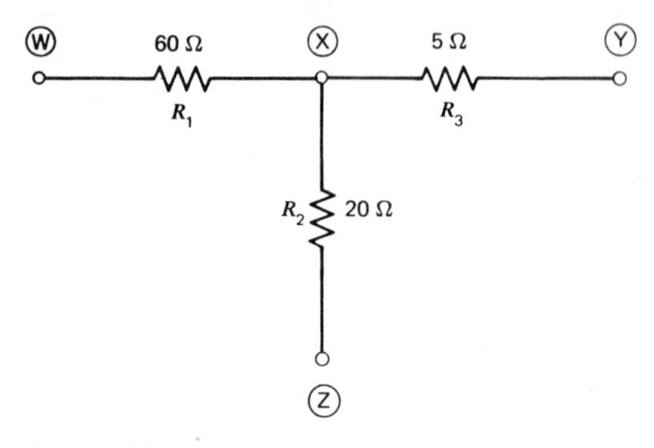

Figure 6-1.

Between points W and Z, the resistance consists of R_1 and R_2 in series, or $60 + 20 = 80\ \Omega$. If points Y and Z were shorted together, then the resistance between the *other* point X and point Z would consist of R_2 and R_3 in parallel, and $R_{XZ} = R_2 R_3/R_2 + R_3 = (20)(5)/20 + 5 = 100/25 = 4\ \Omega$. With points Y and Z still shorted together, the resistance between points W and Z is now R_1 in *series* with the *parallel* combination or R_2 and R_3. Therefore, $R_{WZ} = R_1 + R_{XZ} = 60 + 4 = 64\ \Omega$.

The following example is similar to the preceding discussion.

Example 6-1

In the resistor circuit of Fig. 6-2, without actually solving for the result, state the amount of resistance in *each* parallel branch: (a) between points A and B, and (b) between points B and D.

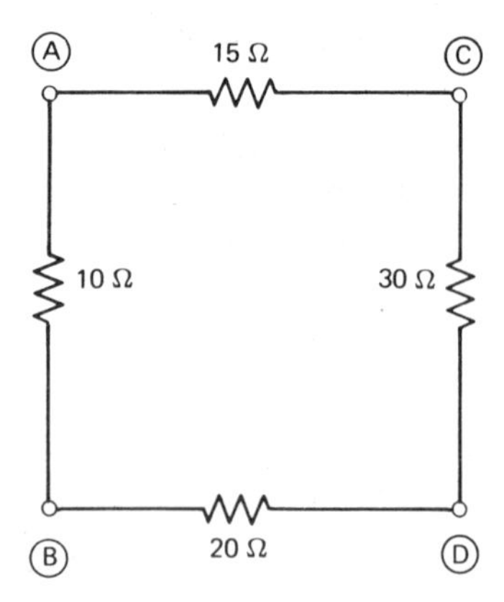

Figure 6-2. Example 6-1.

Solution

(a) Between points A and B, the 10-Ω resistor is in *parallel* with the *series* resistors of 15, 30, and 20 Ω, R_{AB} then consists of 10 Ω in parallel with 65 Ω.

(b) Between points B and D, the 20-Ω resistor is in *parallel* with the *series* resistors 10, 15, and 30 Ω. R_{BD} consists of 20 Ω in parallel with 55 Ω.

For Similar Problems Refer to End-of-Chapter Problems 6-1 to 6-4

6-2. Thévenin's Theorem

One method that is often employed to solve a network circuit is called Thévenin's theorem. This method, relating to a resistor network like that of Fig. 6-3*a*, states basically that there exists, between any *two* points in a circuit, an *equivalent* amount of resistance and voltage. For example in Fig. 6-3*a*, if the load resistor R_L were removed from the circuit, then between points A and B, some voltage and a single resistor could take the place of resistors R_1, R_2, R_3, R_4, and the applied voltage E_T. This *equivalent* R and E, called $R_{\text{Thévenin}}$ ($R_{\text{Thév}}$) and $E_{\text{Thévenin}}$ ($E_{\text{Thév}}$), could now be placed across R_{load}, between points A and B, with the same effect that the entire circuit would have on R_{load}.

Another way of explaining $E_{\text{Thév}}$ is what a voltmeter would read across terminals A and B with R_{load} open. Similarly, with R_{load} still open, $R_{\text{Thév}}$ is what an ohmmeter would read between terminals A and B, but with the voltage E_T removed and replaced by its internal resistance.

If it were desired to find the voltage across R_{load} and the current through it, R_{load} should first be opened and the following procedures followed:

To find $R_{\text{Thév}}$ between points A and B, remove R_{load} and replace the voltage source E_T with its internal resistance (assuming that the internal resistance of E_T is zero in this problem, E_T is replaced by a short) as shown in the first part of Fig. 6-3*b*. Between points C and D, R_4 is in parallel with R_1 and R_2 in series. R_{CD} is then $= [(30)(2 + 4)]/(30 + 2 + 4) = [(30)(6)]/36 = 5$ Ω. This 5 Ω now is the equivalent of R_1, R_2, and R_4, and is in series with R_3 as is shown in second part of Fig. 6-3*b*. An ohmmeter between points A and B in Fig. 6-3*b*, second part would read $R_{\text{Thév}} = 20$ Ω, which is $R_{CD} + R_3$.

To find $E_{\text{Thév}}$ between terminals A and B with R_{load} open, it is first necessary to find E_{CD}. This can be seen in the first part of Fig. 6-3*c*, where R_1, R_2, and R_4 are in series with the applied voltage E_T. Voltage across R_4 may be found from Ohm's law: $I = E_T/R_T = 36/R_1 + R_2 + R_4 = 36/2 + 4 + 30 = 36/36 = 1$ A; $V_{R_4} = I\,R_4 = (1)(30) = 30$ V as shown. Observe that since current (electron movement) is up through R_4, point D is + with respect to point C, as shown. This voltage V_{CD} is $E_{\text{Thév}}$, and together with $R_{CD} + (R_3)$ comprises the Thévenin *equivalent* circuit between points A and B shown in the second part of Fig. 6-3*c*.

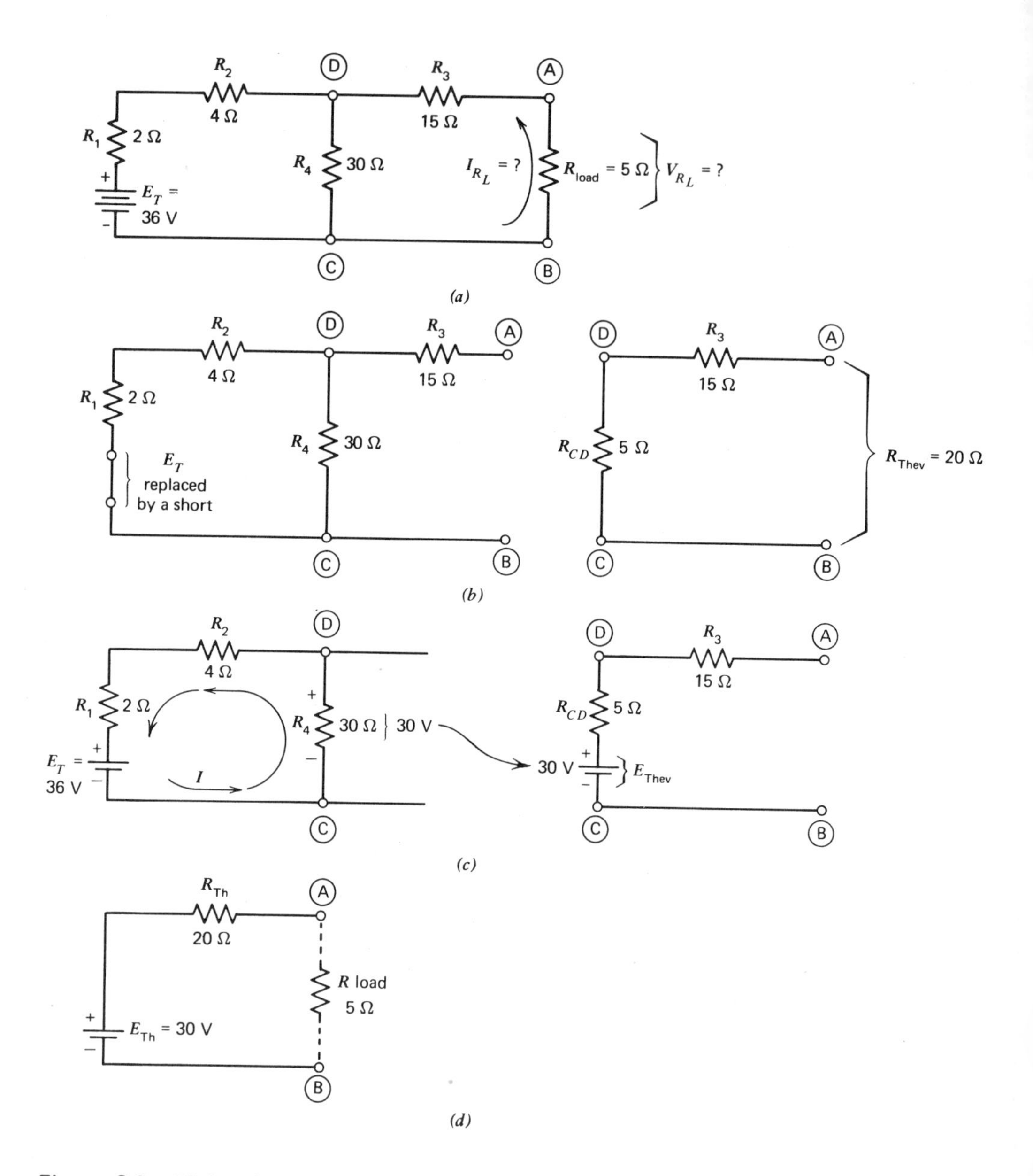

Figure 6-3. Thévenin solution. (*a*) Complete circuit. (*b*) $R_{\text{Thév}}$ (*R* load open). (*c*) $E_{\text{Thév}}$ (*R* load open). (*d*) Thévenin equivalent circuit (*R* load open).

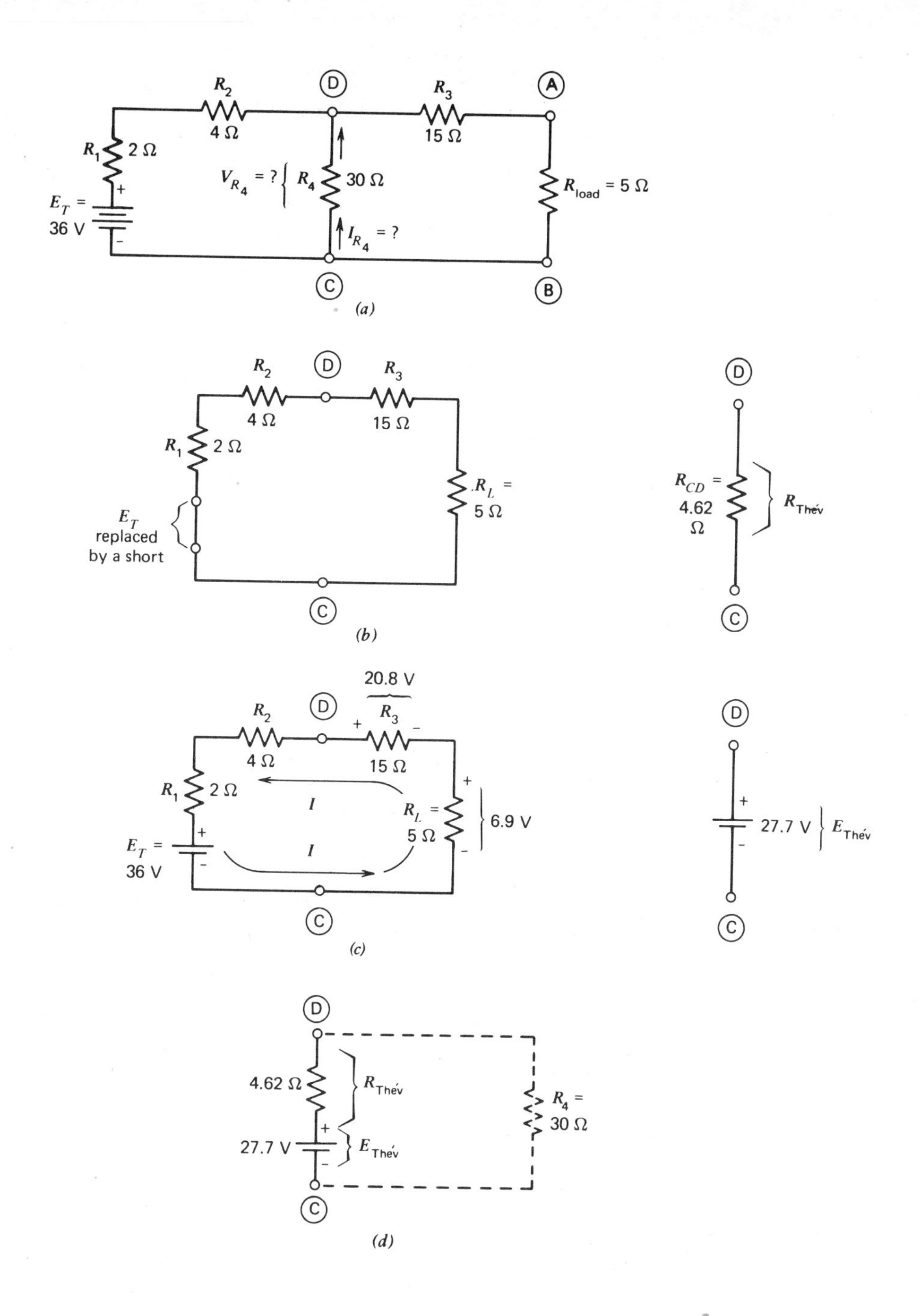

Figure 6-4. Problem 6-5 solutions. (*a*) Complete circuit. (*b*) $R_{\text{Thév}}$ (R_4 open). (*c*) $E_{\text{Thév}}$ (R_4 open). (*d*) Thévenin equivalent circuit (R_4 open).

Between points A and B resistor $R_3 = 15\ \Omega$, is in series with R_{CD} (5 Ω) and $E_{\text{Thév}}$ (30 V). As shown in Fig. 6-3*d*, the complete Thévenin equivalent circuit between points A and B, with R_{load} open, is 20 Ω ($R_{\text{Thév}}$) and 30 V ($E_{\text{Thév}}$) (point A is plus with respect to point B).

With the circuit simplified as shown in Fig. 6-3*d*, R_{load} (5 Ω) can now be placed in between points A and B. Recalling that the original problem was to find the current through, and voltage across R_{load}, the circuit of Fig. 6-3*d* is simply a series circuit. With R_{load} added to Fig. 6-3*d*, current through R_{load}, $I_{R_L} = E_{\text{Thév}}/R_T = 30/R_{\text{Thév}} + R_{\text{load}} = 30/20 + 5 = 30/25 = 1.2$ A. $V_{R_{\text{load}}} = (I_{R_L})(R_L) = (1.2)(5) = 6$ V. Solve for I_{R_L} and V_{R_L} in the series-parallel circuit of Fig. 6-3*a*, using the method of Chapter 5. If done correctly, I_{R_L} and V_{R_L} should agree with the Thévenin method just discussed.

For Similar Problems Refer to End-of-Chapter Problems 6-5 to 6-8

6-3. Norton's Theorem

Another method of solving network circuits is the use of *Norton's theorem.* This theorem, in brief, says that if the load resistor across the output terminals is shorted in a resistor-applied voltage network, the *current* flowing through the *short* may be depicted as a current-generating device. Then, with the short removed, the load resistor opened, and the applied voltage source replaced by its internal resistance, the *total resistance* between the *output terminals* acts as a resistor connected across the current generator. Finally, the load resistor is also placed across the current generator in order to find current through, and voltage across the load resistor.

Figure 6-5*a* is the same series-parallel circuit that was used in the discussion of Thévenin's theorem (Section 6-2). Again, it is desired to find I_{R_L} and V_{R_L}. In the Norton method, R_L, is first shorted, and the short-circuit current (I_N) is found. As shown in Fig. 6-5*b*, the short between terminals A and B, places R_3 and R_4 in parallel. This combination $R_{CD} = (15)(30)/15 + 30 = 450/45 = 10\ \Omega$. Total resistance to the applied voltage E_T is R_1 and R_2 in series with the parallel combination R_{CD}, or $R_T = R_1 + R_2 + R_{CD} = 2 + 4 + 10 = 16\ \Omega$. I_T (with R_L still shorted) $= E_T/R_T = 36/16 = 2.25$ A. Current through the short (between A and B) of Fig. 6-5*b* is the Norton current (I_N), and is a fraction of I_T. From Chapter 4, equation 4-6, $I_N = (R_4/R_3 + R_4)\, I_T = (30/15 + 30)\ 2.25\ \text{A} = (30/45)\ 2.25 = 1.5$ A. This $I_N = 1.5$ A is the Norton current generator shown in Fig. 6-5*b* and also in *d*.

Now, to find the Norton resistance (R_N), *remove* the short between A and B, *open* R_L, and *replace* E_T with a resistor equal to its internal resistance, as in Fig. 6-5*c*. Since, in Fig. 6-5*a* E_T has zero internal resistance, then E_T is simply replaced by a short. R_N is the resistance between terminals A and B. As shown in Fig. 6-5*c*, R_3 is in series with the parallel combination between points C and

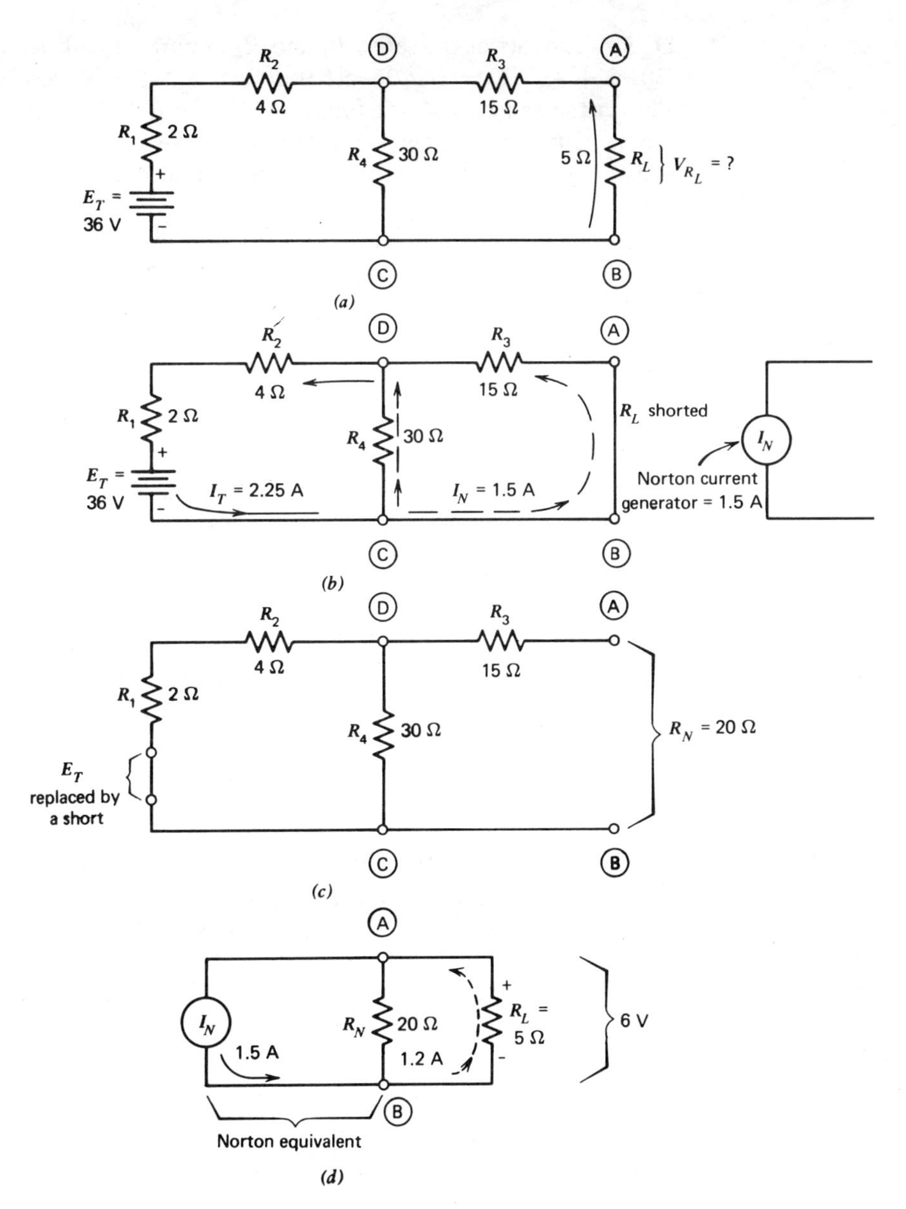

Figure 6-5. The Norton solution. (*a*) Complete circuit. (*b*) I_{Norton} (R_L shorted). (*c*) R_{Norton} (R_L open). (*d*) Norton equivalent circuit with R_L.

D. R_{CD} consists of the series R_1 and R_2 in parallel with R_4, or $R_{CD} = (2 + 4)(30)/2 + 4 + 30 = 180/36 = 5\ \Omega$. R_{AB} or $R_N = R_3 + R_{CD} = 15 + 5 = 20\ \Omega$. This is the same as $R_{\text{Thév}}$, found in Section 6-2. As shown in Fig. 6-5*d*, the Norton equivalent circuit consists of the 1.5-A current generator (I_N) with the resistor R_N connected across it. R_L is then also connected across the current generator.

To find I_{R_L} in Fig. 6-5*d*, I_{R_L} is a fraction of I_N (from equation 4-6). $I_{R_L} = (R_N/R_N + R_L)\, I_N = (20/20 + 5)\, 1.5 = (20/25)\, 1.5 = 1.2$ A. $V_{R_L} = I_{R_L}\, R_L = (1.2)\,(5) = 6$ V. These answers agree with those found using Thévenin's (Section 6-2).

The following example again refers to the same series-parallel circuit discussed previously, and refers to R_4 instead of R_L.

Example 6-2

In the circuit of Fig. 6-6*a*, find I_{R_4} and V_{R_4} employing Norton's theorem.

Solution

First short R_4 and solve for the short-circuit current. As shown in Fig. 6-6*b*, the shorted R_4 between points C and D also shorts out R_3 and R_L. As a result, only R_1 and R_2 are left in the circuit. $I_N = E_T/R_1 + R_2 = 36/2 + 4 = 6$ A. This is now I_N the Norton current generator.

To find the Norton equivalent resistance, R_N, between terminals C and D, remove the short, open R_4, and replace E_T by a resistor equal to its internal resistance. A *short* is used here, since the internal resistance of E_T has not been shown in Fig. 6-6*a*, and is therefore assumed to be zero.

As can be seen in Fig. 6-6*c*, R_N between terminals C and D consists of the series resistors R_1 and R_2 in *parallel* with the other series resistors R_3 and R_L. $R_N = (R_1 + R_2)(R_3 + R_L/R_1 + R_2 + R_3 + R_L = (2 + 4)(15 + 5)/2 + 4 + 15 + 5 = (6)\,(20)/26 = 120/26 = 4.62\ \Omega$.

The Norton equivalent circuit then consists of the current generator I_N connected to the Norton resistance R_N, as shown in Fig. 6-6*d*. R_4 is now also connected to the generator. I_{R_4} is a fraction of I_N. From equation 4-6 (Chapter 4), $I_{R_4} = (R_N/R_N + R_4)\, I_N = (4.62/4.62 + 30)\, 6\text{ A} = (4.62/34.62)\, 6 = 0.8$ A.

$$V_{R_4} = I_{R_4}\, R_4 = (0.8)\,(30) = 24\text{ V}$$

For Similar Problems Refer to End-of-Chapter Problems 6-9 to 6-11

6-4. Maxwell's Cyclic Currents, or Loop Currents

One method of solving a network is called *Maxwell's cyclic currents*, or *loop currents*. This involves assuming any *directions* of currents in each loop, where

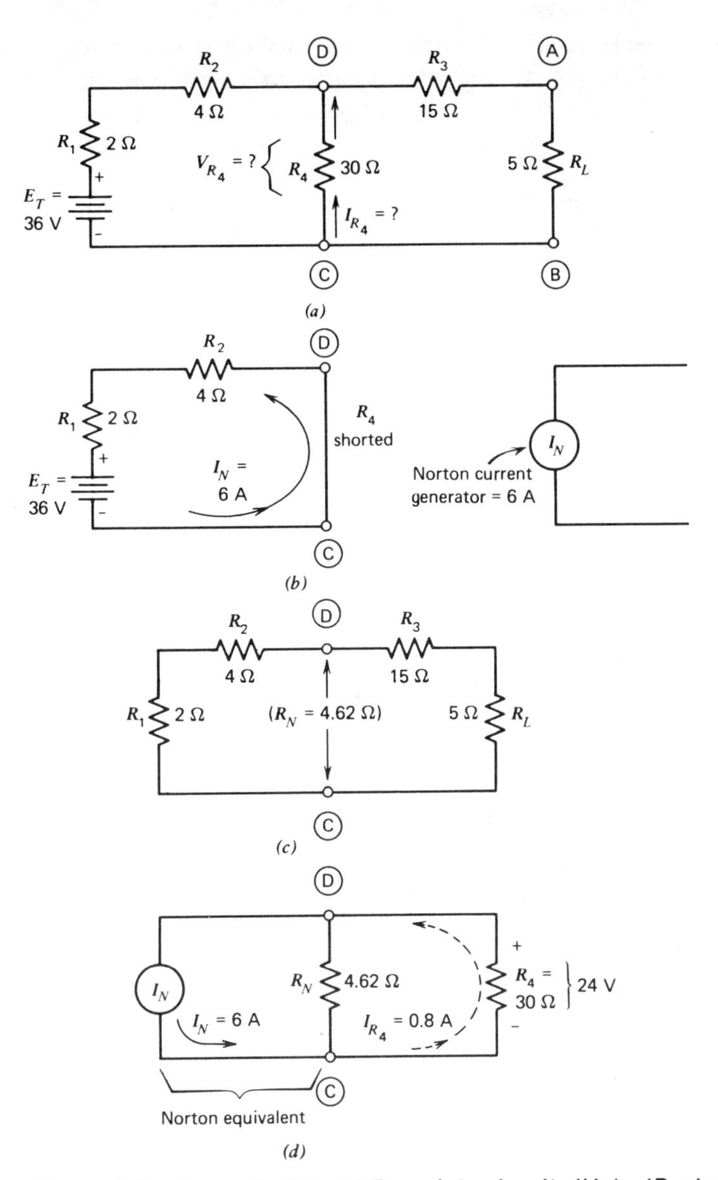

Figure 6-6. Example 6-2. (*a*) Complete circuit. (*b*) I_N (R_3 shorted). (*c*) R_{Norton} (R_4 open). (*d*) Norton equivalent circuit with R_4.

the paths of these currents must include *every* resistor. If the assumed *direction* of current is incorrect, a *negative* value will result, indicating this wrong direction. From Kirchhoff's voltage law, the voltages inside each loop are added and are equal to zero. An equation is written for each loop. Solving these simultaneous equations for the values of the currents then yields the true *directions* of these currents, and from their values, voltages across the resistors can be found.

Before solving a network, a few simple preliminary procedures will be discussed. The simple series circuit of Fig. 6-7*a* will be treated as a *loop*. Electron movement or current I is assumed to be flowing clockwise as shown. This is the correct direction, since the electron-flow theory states that electrons move from the negative end of the applied voltage source through the resistors to the positive end. Similarly, the voltage across each resistor is *negative* at the end where electrons *enter*, and is *positive* at the other end where electrons leave. These polarities across the resistor are with respect to each other and not with respect to something else. From Kirchhoff's voltage law, the sum of the voltages around the circuit is zero, that is, taking their polarities into account. Starting at point X in Fig. 6-7*a* "walk" around the circuit in the same direction as the assumed direction of I, or in the opposite direction. Give to *each* voltage

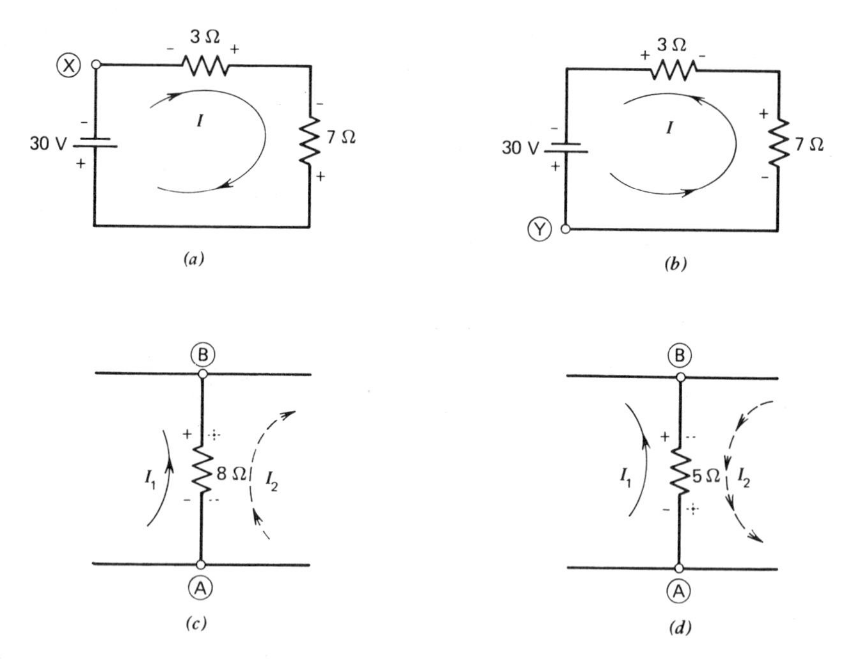

Figure 6-7. (*a*) Assumed direction of electron flow (I) is correct. (*b*) Assumed direction of electron flow (I) is incorrect. (c) Currents I_1 and I_2 both flow through the 8-Ω resistor in the same direction. (*d*) Currents I_1 and I_2 both flow through the 5-Ω resistor but in opposite directions.

the polarity seen as the voltage is approached. The *amount* of voltage across the 3-Ω resistor is unknown, but is actually IR, or (I) (3), or $3I$. Starting at point X and going clockwise with the current, the first voltage is therefore $-3I$. Going further, the next voltage is $-7I$. Still further, there is a known value of voltage, the $+30$. Going further around, completes the trip and brings us to our starting point X. Therefore, stating this as an equation gives

$$-3I - 7I + 30 = 0$$

$$-3I - 7I = -30$$

$$-10I = -30$$

$$I = 3 \text{ A}$$

The *positive* value of I means that the *assumed* direction is the *correct* direction.

In Fig. 6-7*b*, the current direction (electron movement) is purposely chosen to be incorrect. Starting at point Y and traveling completely around counterclockwise with the current yields

$$-7I - 3I - 30 = 0$$

$$-7I - 3I = +30$$

$$-10I = +30$$

$$I = -3 \text{ A}$$

The *negative* value of I simply indicates that the *assumed* current direction is actually just *backwards* from the true direction. Although it is evident in Fig. 6-7*b* that the assumed current direction has been reversed, it is often not apparent in the larger, more complex circuits.

In the partial circuit of Fig. 6-7*c*, two currents (electron movements) I_1 and I_2 are shown flowing through an 8-Ω resistor in the same direction. As a result, voltage across this resistor, "walking" from point A to point B, is $-8I_1 - 8I_2$.

If one of the currents flows *up*, while a second current flows *down* through a resistor, as shown in Fig. 6-7*d*, then the voltage across the 5-Ω resistor, "walking" from point A to B is $-5I_1 + 5I_2$. The *lower* end of the 5-Ω resistor as we travel from A to B is *negative due to* I_1, but is *positive due to* I_2.

The circuit of Fig. 6-8 is the same one used previously to illustrate Thévenin's and Norton's theorems to find I_{R_L} and V_{R_L}. Now, using loop currents, these unknowns will be found as shown in the following method.

In loop 1, made up of the components between points C, D, F, and E, a current I_1 is *assumed* to be flowing *counterclockwise* as shown. I_1 flows through the resistors R_4, R_2, and R_1, and the resulting voltage polarities across these resistors are as shown.

Loop 2 consists of the components between points B, A, D, and C. A current I_2 is *assumed* to be flowing in a *counterclockwise* direction in this loop, as shown by the dashed-line arrows. Resulting voltage polarities across

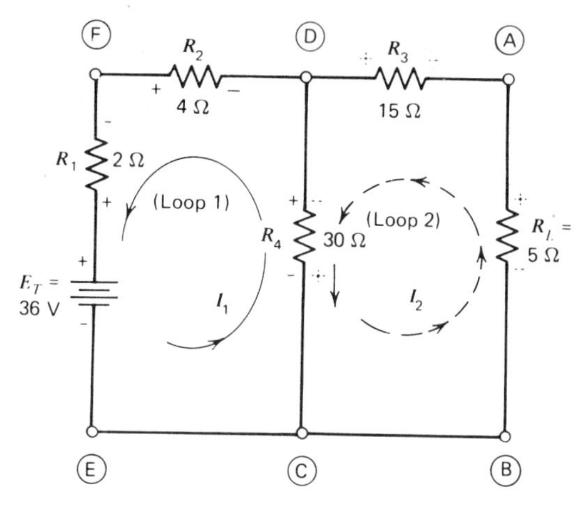

Figure 6-8. Loop currents.

resistors R_L, R_3, and R_4 are shown by the dotted-line plus and minus signs, and are determined by the assumed direction of current.

In Fig. 6-8, in loop 1, from Kirchhoff's voltage law, the following equation is set up, starting at point C and "walking" counterclockwise (ccw) completely around the loop:

$$-30I_1 + 30I_2 - 4I_1 - 2I_1 + 36 = 0 \tag{6-1}$$

Note that the 30-Ω resistor, at the lower end, is *negative* due to I_1, but is *positive* due to I_2.

Equation 6-1 above now becomes

$$-30I_1 - 4I_1 - 2I_1 + 30I_2 = -36$$

$$-36I_1 + 30I_2 = -36$$

Dividing the above by -6 to simplify, yields

$$6I_1 - 5I_2 = 6 \tag{6-2}$$

Similarly, in loop 2, starting at point B and "walking" ccw with I_2 completely around this loop gives

$$-5I_2 - 15I_2 - 30I_2 + 30I_1 = 0 \tag{6-3}$$

Note that the 30-Ω resistor, at the upper end, is *negative* due to I_2, but is *positive* due to I_1.

Equation 6-3 becomes

$$+30I_1 - 50I_2 = 0$$

Dividing the above by 10 to simplify, yields

$$3I_1 - 5I_2 = 0 \tag{6-4}$$

Rewriting both equations

$$6I_1 - 5I_2 = 6 \tag{6-2}$$

$$3I_1 - 5I_2 = 0 \tag{6-4}$$

Solving for I_1 by subtracting equation 6-4 from equation 6-2 in order to cancel one of the unknowns, I_2, yields

$$3I_1 = 6$$

$$I_1 = 2 \text{ A}$$

The positive value for I_1 means that the assumed ccw direction is the true direction. To find I_2, simply substitute for I_1 in equation 6-2:

$$6I_1 - 5I_2 = 6 \tag{6-2}$$

$$6(2) - 5I_2 = 6$$

$$12 - 5I_2 = 6$$

$$-5I_2 = 6 - 12$$

$$-5I_2 = -6$$

$$I_2 = 1.2 \text{ A}$$

The positive value of I_2 means that the ccw assumed direction is its true direction. Voltage across R_L, $V_{R_L} = I_2 R_L = (1.2)(5) = 6$ V. The answers of 1.2 A for I_2 (or I_{R_L}), and 6 V for V_{R_L} agree with those found using Thévenin's (Section 6-2) and Norton's (Section 6-3) theorems.

For Similar Problems Refer to End-of-Chapter Problems 6-12 to 6-14

6-5. Nodal Analysis

This method of solving circuits employs both Kirchhoff's voltage and current laws. Figure 6-9 is a simple series circuit that is being used here, at first, to introduce the nodal analysis method, also called *nodal voltage* method.

A *node* is the junction of two or more components (resistors or voltage sources). Point Y is called a *dependent node*, since the voltage at this point is fixed and depends on the amount of the applied voltage source E_T. Point X is called an *independent node*, since its voltage is a variable, controlled by the sizes of the resistors. Point Z is called the *reference node*, since the voltages at node X and at node Y are measured with *respect to node Z*. V_X means the voltage at node X with respect to node Z.

In this nodal method, currents are usually assumed to be flowing *into* the independent node X. As shown in the diagram, I_1 and I_2 are both flowing into point X. Obviously, one of these assumed current directions is wrong. The electron movement should be from the negative end of E_T, point Z, counterclockwise through R_2 and R_1. However, in this nodal method, the currents

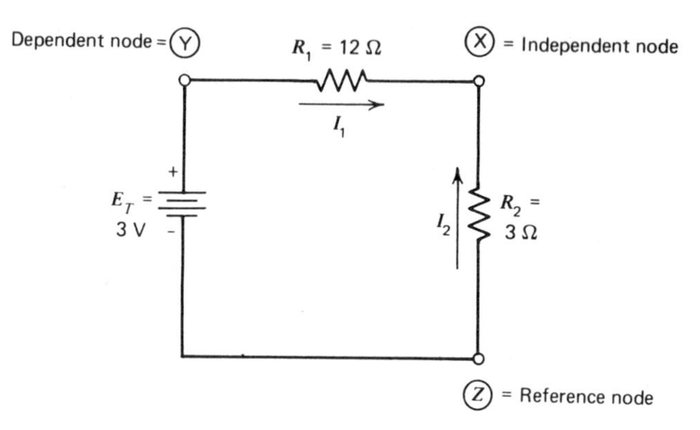

Figure 6-9. Simple nodal analysis.

are assumed to flow as shown. When solving for the currents, any *negative* value of current indicates that this current is actually flowing in an opposite direction from that which was assumed.

Kirchhoff's current law states that the sum of currents flowing *into* a point equals the sum of currents *leaving* that point. When *all* currents are assumed to be flowing *into* a point, then the sum of these currents is zero, or in Fig. 6-9,

$$I_1 + I_2 = 0 \tag{6-5}$$

The equation for the circuit, from equation 6-5 is

$$\frac{V_X - E_T}{R_1} + \frac{V_X}{R_2} = 0 \tag{6-6}$$

where the first fraction is actually I_1, and the second fraction is I_2. This equation 6-6 may or may not be apparent in referring to Fig. 6-9, and is derived in the following discussion.

Kirchhoff's voltage law, applied to Fig. 6-9, states that the sum of all the voltages equals zero, taking into consideration the polarities resulting from the assumed directions of currents. Starting at point Z and moving around the circuit in a clockwise (cw) direction, each voltage is given the polarity at the end that is first approached. The ends of *each* voltage that is approached first, traveling cw are: the lower end of E_T, which is its negative end; the left end of V_{R_1}, which is its negative end; and the upper end of V_{R_2}, which is its positive end. Expressed as an equation: $-E_T - V_{R_1} + V_{R_2} = 0$. Solving for V_{R_1} gives

$$V_{R_1} = V_{R_2} - E_T \tag{6-7}$$

Another term for V_{R_2} is the voltage at node X with respect to node Z, or simply V_X. Then

$$V_{R_1} = V_X - E_T \tag{6-8}$$

Repeating

$$I_1 + I_2 = 0 \tag{6-5}$$

$$V_{R_1} = V_X - E_T \tag{6-8}$$

and, from Fig. 6-9,

$$I_1 = \frac{V_{R_1}}{R_1}$$

And substituting for V_{R_1} in the above, from equation 6-8:

$$I_1 = \frac{V_X - E_T}{R_1} \tag{6-9}$$

And

$$I_2 = \frac{V_X}{R_2} \tag{6-10}$$

Then substituting for I_1 and I_2 in equation 6-5, gives the derived expression:

$$\frac{V_X - E_T}{R_1} + \frac{V_X}{R_2} = 0 \tag{6-6}$$

Solve for V_X (or V_{R_2}) using the values shown in Fig. 6-9:

$$\frac{V_X - 3}{12} + \frac{V_X}{3} = 0$$

Multiplying by 12, gives

$$V_X - 3 + 4\ V_X = 0$$

$$5\ V_X = 3$$

$$V_X = 0.6 \text{ V}$$

Solve for I_2 in equation 6-10:

$$I_2 = \frac{V_X}{R_2} \tag{6-10}$$

$$I_2 = \frac{0.6}{3}$$

$$I_2 = 0.2 \text{ A}$$

Solve for I_1 in equation 6-9:

$$I_1 = \frac{V_X - E_T}{R_1} \tag{6-9}$$

$$I_1 = \frac{0.6 - 3}{12}$$

$$I_1 = \frac{-2.4}{12}$$

$$I_1 = -0.2 \text{ A}$$

In the simple series circuit of Fig. 6-9, I_1 and I_2 are actually the same current and, as found above, are each 0.2 A. The *negative* sign for I_1, as stated previously, means that its *assumed* direction as shown in Fig. 6-9 is reversed from I_1 true direction. The *positive* value for I_2 denotes that the assumed direction of I_2 is its true direction. By reversing I_1 direction (which is a negative value), then, of course, both currents (which is really one and the same) flow in the same direction.

The advantage of the nodal method over that of the loop currents (or Maxwell's cyclic currents) method is that only one equation is required for a circuit containing one independent node.

The following example again uses a simple series circuit, but with E_T reversed from that of Fig. 6-9.

Example 6-3

In Fig. 6-10, using the nodal analysis method, find the current flowing in the circuit and the voltage and polarity across R_4.

Solution

Currents I_3 and I_4 are assumed to flow into node A, and from Kirchhoff's current law, $I_3 + I_4 = 0$. Voltage at node A (with respect to the reference node C) minus the *negative* E_T (voltage at node *B*), is V_{R_3}. Therefore, the nodal equation is

$$\frac{V_A - (-E_T)}{R_3} + \frac{V_A}{R_4} = 0 \tag{6-11}$$

(Note that the first fraction above is I_3, while the second fraction is I_4). The student should solve the above for V_A using the values of Fig. 6-10, and for I_3, I_4, and V_{R_4}, comparing his answers with those shown in the diagram.

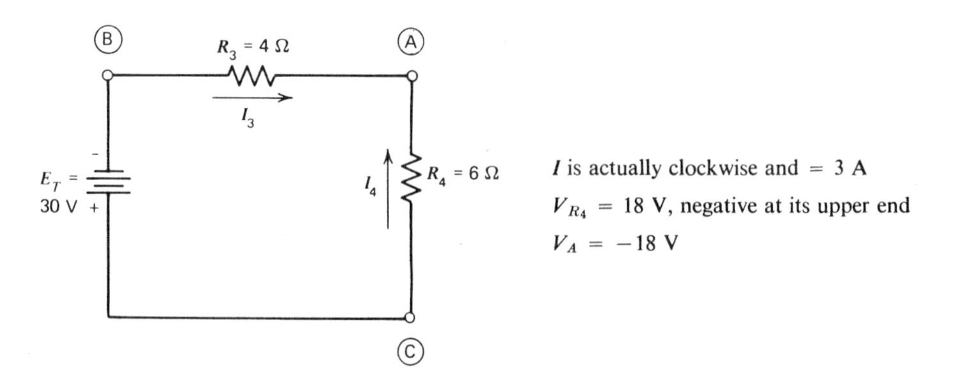

Figure 6-10. Simple nodal analysis, Example 6-3.

The next example again employs nodal analysis in the same series-parallel circuit used in the discussions of Thévenin (Section 6-2), Norton (Section 6-3), and loop currents (Section 6-4).

Example 6-4

In the circuit of Fig. 6-11, find the current through resistor R_L, the voltage across R_L, the current through R_4, and the voltage across R_4, using nodal analysis.

Solution

Assuming that the three individual currents flow into the independent node D, as shown, then: $I_{\text{left}} + I_{\text{center}} + I_{\text{right}} = 0$. Using point C as a reference node, then voltage across R_4 is the voltage at point D with respect to C, or can simply be termed V_D. Similarly, voltage at point E with respect to C, or actually E_T, may be termed V_E. Then: $(V_D - V_E)/(R_1 + R_2)$ is actually I_{left}, and V_D/R_4 is I_{center}. I_{right} is $V_D/(R_3 + R_L)$.

$$\frac{V_D - V_E}{R_1 + R_2} + \frac{V_D}{R_4} + \frac{V_D}{R_3 + R_L} = 0$$

$$\frac{V_D - 36}{2 + 4} + \frac{V_D}{30} + \frac{V_D}{15 + 5} = 0$$

$$\frac{V_D - 36}{6} + \frac{V_D}{30} + \frac{V_D}{20} = 0$$

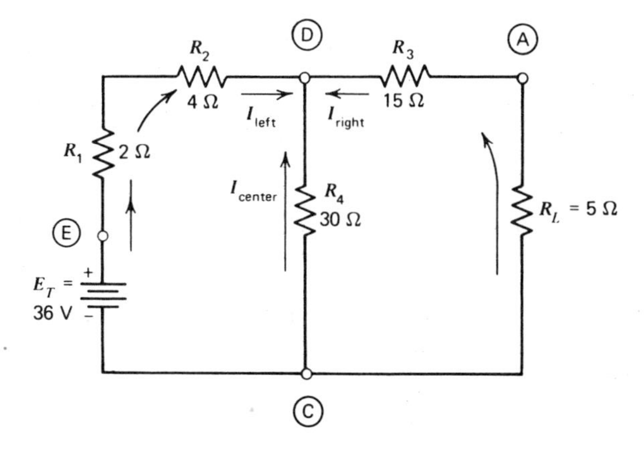

Figure 6-11. Nodal analysis, Example 6-4.

Multiplying by the common denominator 60 to clear fractions gives

$$10(V_D - 36) + 2V_D + 3V_D = 0$$
$$10V_D - 360 + 2V_D + 3V_D = 0$$
$$10V_D + 2V_D + 3V_D = 360$$
$$15V_D = 360$$
$$V_D = 24 \text{ V}$$

$$I_{\text{right}} = \frac{V_D}{R_3 + R_L}$$
$$= \frac{24}{15 + 5}$$
$$= \frac{24}{20}$$
$$= 1.2 \text{ A}$$

Note that I_{right} is the current that flows through R_L. This value of 1.2 A agrees with that found previously using the Thévenin, Norton and Loop current methods. Since the value of I_{right} is *positive* 1.2 A, then the assumed *direction* going *up* through R_L, as shown in Fig. 6-11, is its correct direction.

Voltage across R_L, $V_{R_L} = I_{R_L} R_L = (1.2)(5) = 6$ V, with point A the positive end. This, too, agrees with that found previously.

Current through R_4 is $I_{\text{center}} = V_D/R_4 = 24/30 = 0.8$ A. This also agrees with I_{R_4} found by using Thévenin in Example 6-3, and Norton in Example 6-4, and loop currents in Example 6-5. Since this current value is *positive* 0.8 A, then the assumed direction of I_{R_4} (or I_{center}), shown as flowing *up* through R_4, is its correct direction.

Voltage across R_4, $V_{R_4} = I_{R_4} R_4 = (0.8)(30) = 24$ V, with point D the positive end. This voltage and polarity agrees with that found previously.

For Similar Problems Refer to End-of-Chapter Problems 6-15 to 6-17

MORE COMPLEX RESISTOR NETWORKS

Here, resistor networks having more than one applied voltage source are solved by using the methods introduced previously. This includes Thévenin, Norton, loop currents, and nodal analysis as well as the superposition method and Millman's theorem, which were not included previously. In addition, unbalanced bridge networks are presented, introducing the delta to wye transformation.

6-6. Series Resistors, Two Voltage Sources

In Chapter 3, Section 3-10, series resistors with two voltage sources are discussed. A brief discussion of them will also be given here before the superposition theorem is introduced in the next section.

In Fig. 6-12 two resistors, R_1 and R_2, are in series with two applied voltages E_1 and E_2. It is desired to find V_{R_1}, V_{R_2}, and voltage between point X and ground. E_1 causes an electron movement that is counterclockwise, while E_2 produces a clockwise movement of electrons. The polarities of E_1 and E_2 are said to be *opposing* or *bucking*. If these voltage sources were equal, then no current would flow. As shown, E_2 is larger than E_1 by 6 V. As a result, current flows clockwise due to the net voltage of 6 V. This electron movement means that V_{R_1} and V_{R_2} have polarities, as shown, with the *positive* ends of these voltages at the *top* of R_1 and at the *bottom* of R_2.

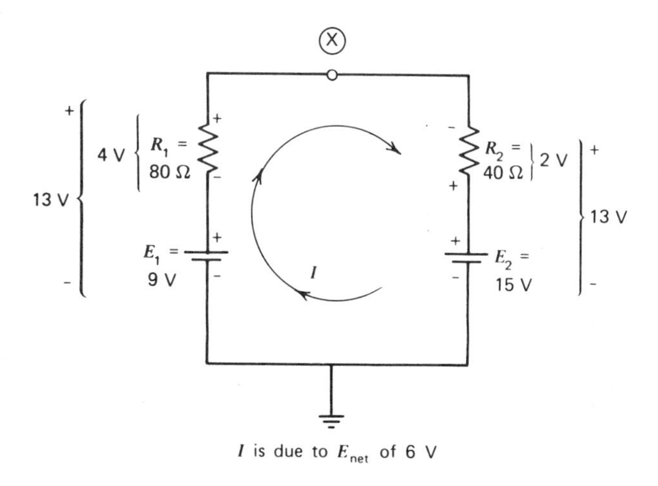

Figure 6-12. Series Circuit, Two-Voltage Sources.

Solving for V_{R_1} and V_{R_2}, using the fractional method of equation 3-1, gives: $V_{R_1} = [R_1/(R_1 + R_2)]\,E_{\text{net}} = [80/(80 + 40)]\,6 = (80/120)\,6 = 4$ V, as shown in Fig. 6-12. $V_{R_2} = [R_2/(R_1 + R_2)]\,E_{\text{net}} = [40/(80 + 40)]\,6 = (40/120)\,6 = 2$ V, as shown.

Voltage at point X with respect to ground may be determined by either adding algebraically E_1 and V_{R_1}, or E_2 and V_{R_2}. Note that E_1 and V_{R_1} are aiding, since their positive ends are at the upper ends of each. Therefore, $V_{x\text{-to-ground}} = +E_1 + V_{R_1} = +9 + 4 = +13$ V. If E_2 and V_{R_2} are to be added, it must be noted that their polarities are bucking; that is, the upper end of E_2 is positive, but the upper end of V_{R_2} is negative. Then, $V_{x\text{-to-ground}} = +E_2 - V_{R_2} = +15 - 2 = +13$ V.

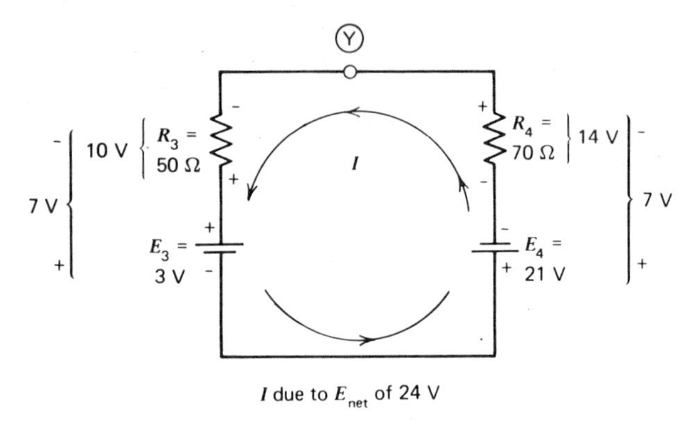

Figure 6-13. Series Circuit, Two-Voltage Sources.

Another simple series resistor circuit with two voltage sources is shown in Fig. 6-13. Again, it is desired to find V_{R_3}, V_{R_4}, and $V_{y\text{-to-ground}}$. The student should find these and compare his answers with those shown in the diagram.

For Similar Problems Refer to End-of-Chapter Problems 6-18 and 6-19

6-7. Superposition Theorem

This method of solution for networks having more than one voltage source uses *each* voltage source independently while the other sources are replaced by their internal resistances. For example, in the circuit of Fig. 6-14*a*, which is the same as that of Fig. 6-12, it is desired to find V_{R_1}, V_{R_2}, and $V_{x\text{-to-ground}}$. As shown in Fig. 6-14*b*, first the effect of E_1 alone is determined, with E_2 replaced by its internal resistance (zero ohms in this problem). With E_1 alone, current I_1 flows counterclockwise as shown. $V_{R_1} = [R_1/(R_1 + R_2)]\, E_1 = (80/120)\, 9 = 6$ V, with the polarity shown. $V_{R_2} = [R_2/(R_1 + R_2)]\, E_1 = (40/120)\, 9 = 3$ V, with the polarity shown.

In Fig. 6-14*c*, the effect of E_2 alone, this time with E_1 omitted. Now, current I_2 flows clockwise and $V_{R_1} = [R_1/(R_1 + R_2)]\, E_2 = (80/120)\, 15 = 10$ V, and $V_{R_2} = [R_2/(R_1 + R_2)]\, E_2 = (40/120)\, 15 = 5$ V, with polarities shown. Neither current I_1 nor I_2 actually flows. The only current that does, is the algebraic sum of the two.

In Fig. 6-14*d*, the effects of E_1 (from Fig. 6-14*b*), and the effects of E_2 (from Fig. 6-14*c*) are shown combined. Note that V_{R_1} consists of the 6 V (*negative* at the top of R_1 due to I_1), and the 10 V (*positive* at the top of R_1 due to I_2). The result of these two opposing voltages across R_1 is their algebraic sum, shown as 4 V positive at the top of R_1. $V_{x\text{-to-ground}}$ is then the algebraic sum of the two aiding voltages E_1 and $V_{R_1} = +9 + 4 = +13$ V.

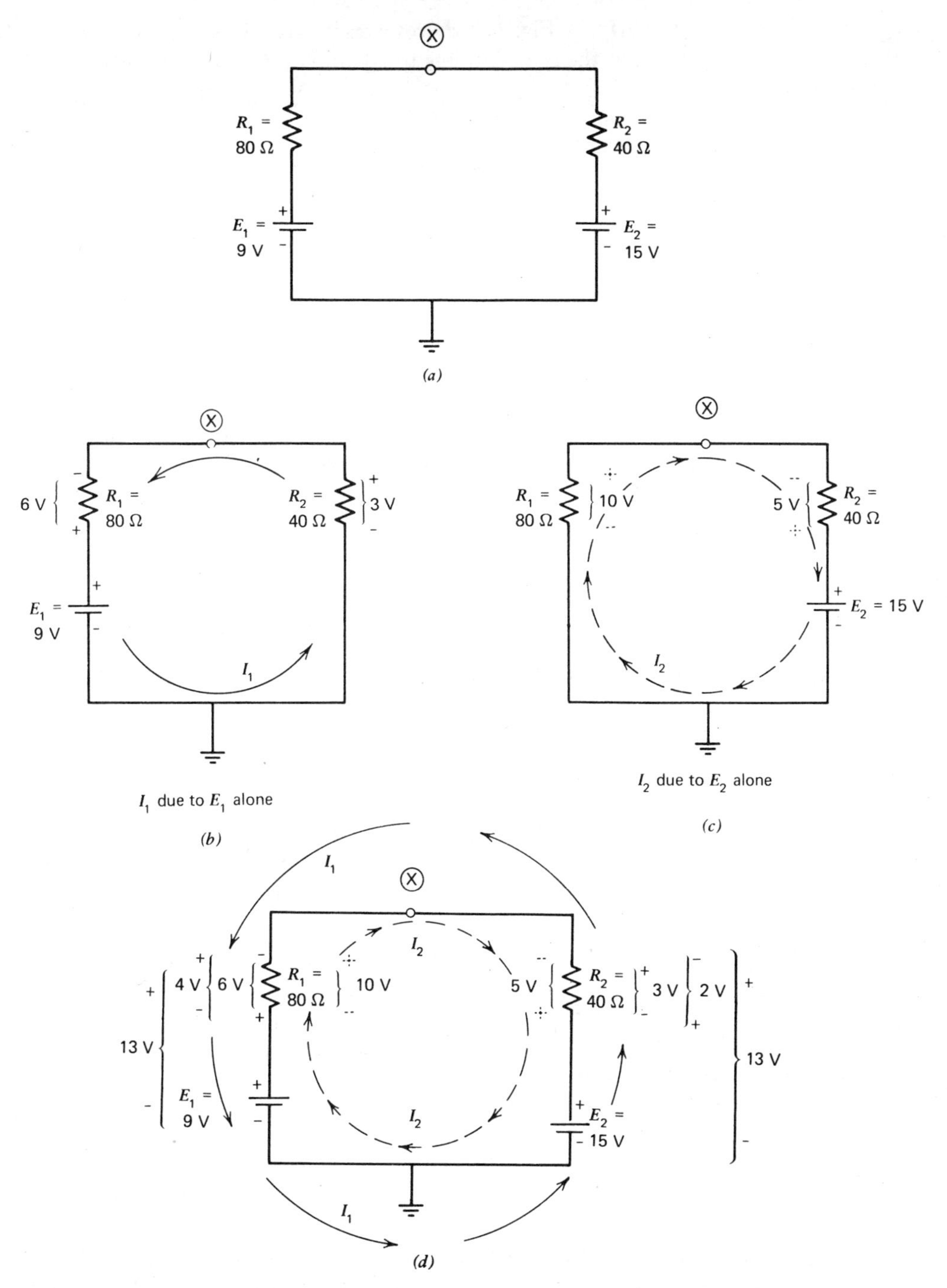

Figure 6-14. Superposition theorem. (*a*) Circuit. (*b*) Effect of E_1. (*c*) Effect of E_2. (*d*) Combined effects of E_1 and E_2.

Similarly in Fig. 6-14*d*, voltages across R_2 are also opposing, being 3 V (*positive* at the top of R_1 due to I_1) and 5 V (*negative* at the top of R_1 due to I_2). V_{R_2} is the algebraic sum, or $+3$ and $-5 = -2$ V, as shown, $V_{x\text{-to-ground}}$ is the algebraic sum of E_2 and $V_{R_2} = +15 - 2 = +13$ V.

Note that these answers shown in Fig. 6-14*d* agree with those shown in Fig. 6-12.

Example 6-5

In the circuit of Fig. 6-15*a*, find V_{R_3} and V_{R_4}, and $V_{y\text{-to-ground}}$ using the superposition theorem.

Solution

First solve for V_{R_3} and V_{R_4} using the applied voltage source E_3 alone, with E_4 replaced with its internal resistance (assumed to be zero in this example). This is shown in Fig. 6-15*b*, with a counterclockwise current, I_3. The reader should find these and compare with the results shown.

Now again find V_{R_3} and V_{R_4}, but this time with applied voltage source E_4 alone, while E_3 is replaced sith its internal resistance (zero in this example). This is shown in Fig. 6-15*c*. The student should again solve, and compare the answers with the results shown.

Figure 6-15*d* illustrates the combined effect of both E_3 and E_4. This is a combination of Fig. 6-5*b* and *c*. The student, once more, should find V_{R_3} and V_{R_4} due to both voltage sources, and finally, $V_{y\text{-to-ground}}$, comparing them with the answers shown.

For Similar Problems Refer to End-of-Chapter Problems 6-20 to 6-23

The circuit of Fig. 6-16*a* is another resistor network containing two voltage sources, but is more complex than those previously shown. Using the superposition theorem, it is desired to find I_{R_L} and V_{R_L}.

As shown in Fig. 6-16*b*, the effect of the applied voltage source E_1 alone, while E_2 is replaced by its internal resistance (zero in this problem), is found first. The circuit is actually a series-parallel one, with R_4 in parallel with the series resistors R_L and R_3, and this parallel circuit in series with R_2 and R_1. Resistance of the parallel part R_{CF} = product/sum = $(\mathrm{R}_4)\ (R_3 + R_L)/(R_4 + R_3 + R_L) = (30)\ (15 + 5)/(30 + 15 + 5) = (30)\ (20)/(30 + 20) = 600/50 = 12\ \Omega$. Total resistance of the E_1 circuit, $R_{T_1} = R_1 + R_2 + R_{CF} = 2 + 4 + 12 = 18\ \Omega$. Current due to E_1 alone is $I_1 = E_1/R_{T_1} = 84/18 = 4.667$ A, as shown in Fig. 6-16*b*.

From the fractional current equation 4-6, current through R_L due to E_1 alone is $I_{1\ \text{right}} = R_{\text{left}}/(R_{\text{left}} + R_{\text{right}})\ I_1 = R_4/[(R_4) + (R_3 + R_L)]\ I_1 = 30/(30 + 15 + 5)\ 4.667 = (30/50)\ 4.667 = 2.8$ A, as shown in Fig. 6-16*b*.

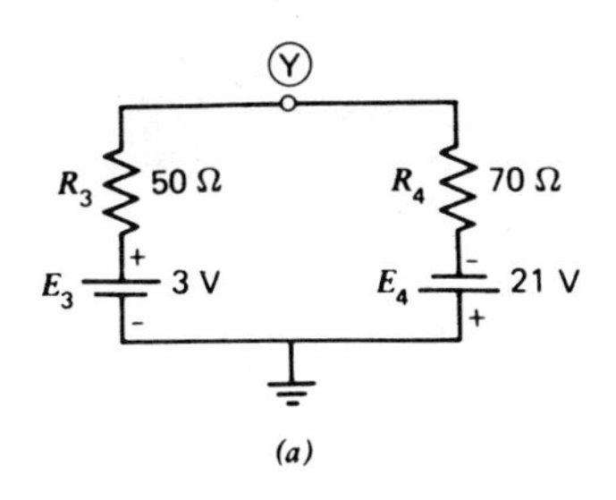

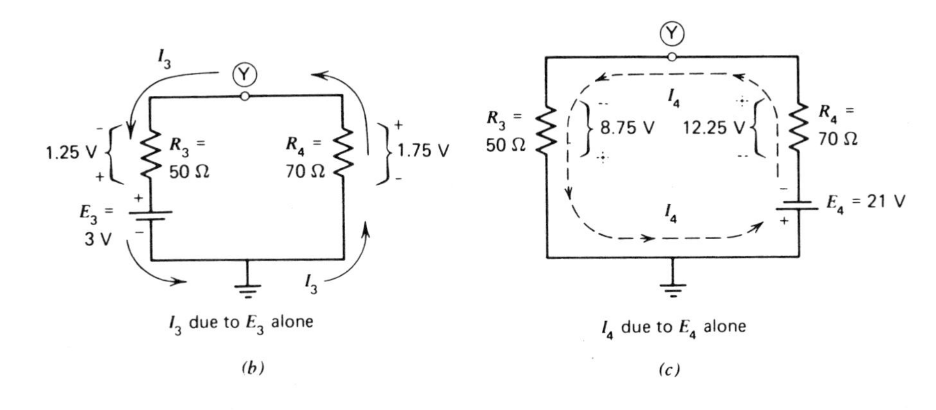

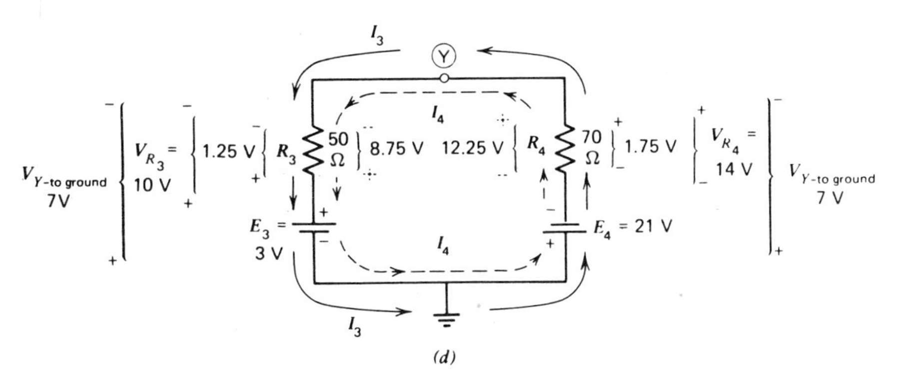

Figure 6-15. Example 6-5. (*a*) Circuit. (*b*) Effect of E_3. (*c*) Effect of E_4. (*d*) Combined effects of E_3 and E_4.

Now the effect of E_2 alone must be found, while the other voltage source E_1 is replaced by its internal resistance (zero, in this problem), as depicted in Fig. 6-16*c*. With E_2 the only voltage source, the circuit is again a series-parallel one, but different than that of Fig. 6-16*b*. Now, R_4 is in series with a parallel circuit R_{CF} made up of R_{left} and R_{right}. R_{left} consists of R_1 and R_2 in series, and R_{right} comprises R_L and R_3 in series. The parallel circuit is then R_{CF} =

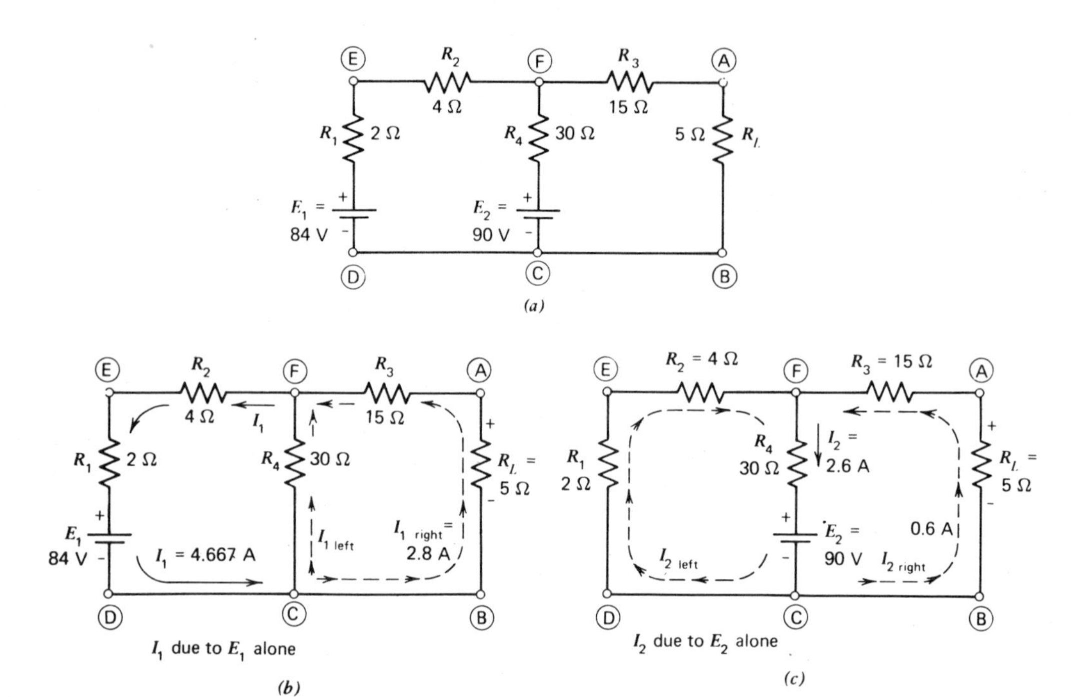

Figure 6-16. Superposition theorem. (*a*) Circuit. (*b*) Effect of E_1. (*b*) Effect of E_1. (*c*) Effect of E_2.

product/sum $= (R_{\text{left}})\,(R_{\text{right}})/(R_{\text{left}} + R_{\text{right}}) = (R_1 + R_2)\,(R_L + R_3)/[(R_1 + R_2) + (R_L + R_3)] = (2 + 4)\,(5 + 15)/[(2 + 4) + (5 + 15)] = (6)\,(20)/(6 + 20) = 120/26 = 4.62\ \Omega$.

Total resistance of the E_2 circuit is then $R_{T_2} = R_4 + R_{CF} = 30 + 4.62 = 34.62\ \Omega$. Current due to E_2 alone is $I_2 = E_2/R_{T_2} = 90/34.62 = 2.6$ A. From the fractional current equation 4–6, current through R_L due to E_2 alone is

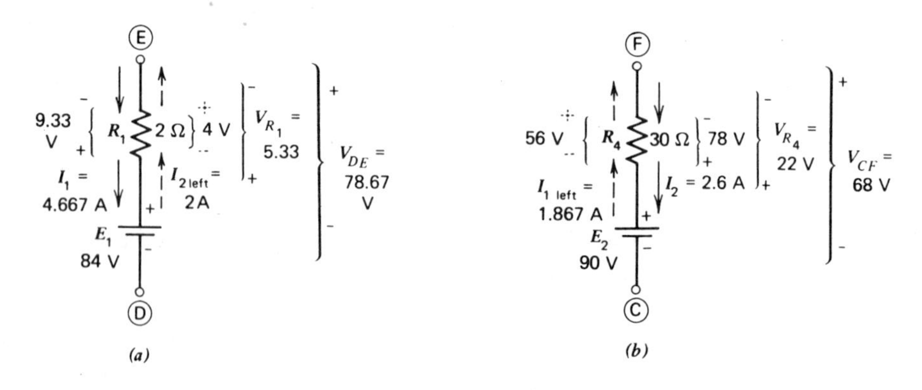

Figure 6-17. Example 6-6. (*a*) V_{R1} and V_{DE}. (*b*) V_{R4} and V_{CF}.

$I_{2\,right} = R_{left}/(R_{left} + R_{right})\, I_2 = (R_1 + R_2)/[(R_1 + R_2) + (R_L + R_3)]\, I_2 = (2 + 4)/[(2 + 4) + (5 + 15)]\, 2.6 = (6/6 + 20)\, 2.6 = (6/26)\, 2.6 = 0.6$ A, as shown in Fig. 6-16*c*.

As depicted in Fig. 6-16*b* and *c*, $I_{1\,right}$ (due to E_1) of 2.8 A, and $I_{2\,right}$ (due to E_2) of 0.6 A each flows *up* through R_L. Voltage across this resistor is, therefore, $V_{R_L} = (I_{1\,right} + I_{2\,right})\, R_L = (2.8 + 0.6)\, 5 = (3.4)\, 5 = 17$ V, with the polarity shown.

Example 6-6

In the circuit of Fig. 6-16*a* discussed above, find the voltage and polarity between (a) Points D and E, and (b) Points C and F, using the superposition theorem method, comparing them with answers shown in Fig. 6-17*a* and *b*.

For Similar Problems Refer to End-of-Chapter Problems 6-24 to 6-26

A resistor network containing three voltage sources is illustrated in Fig. 6-18*a*. Using the superposition theorem method, it is desired to find the voltage and polarity across R_L.

As shown in Fig. 6-18*b*, *c*, and *d*, the effect that *each* applied voltage source *alone* has on R_L is determined. Then, in Fig. 6-18*e*, these effects are noted, with the final result V_{R_L} as shown.

With E_5 the lone voltage source in Fig. 6-18*b*, the circuit is a series-parallel one. Resistors R_6, R_7, and R_L are in parallel with each other, and the combination is in series with R_5. The parallel resistors form a single equivalent one, $R_{XY} = 1/(1/R_6 + 1/R_7 + 1/R_L) = 1/(1/6 + 1/12 + 1/12) = 1/(2/12 + 1/12 + 1/12) = 1/(4/12) = (1)\,(12/4) = 3\ \Omega$. The total resistance that is presented to E_5, $R_{T_5} = R_5 + R_{XY} = 3 + 3 = 6\ \Omega$. Total current due to E_5 is $I_{T_5} = E_5/R_{T_5} = 12/6 = 2$ A. A fraction of this 2 A flows *up* through R_L. Using equation 4-15, $I_{R_L} = (R_{XY}/R_L)\, I_{T_5} = (3/12)\, 2 = 0.5$ A. Voltage across R_L due to E_5 alone is then $(I_{R_L})\, R_L = (0.5)\,(12) = 6$ V, with the polarity shown in Fig. 6-18*b*.

With E_6 now the lone voltage source in the circuit, as shown in Fig. 6-18*c*, the circuit is again a series-parallel combination. Observe that the polarity of E_6 is reversed compared to that of E_5 and E_7 (in the complete circuit of Fig. 6-18*a*). Total current from E_6 (in Fig. 6-18*c*), I_{T_6} flows up through R_6 and then branches out to flow *down* through R_5, R_7, and R_L. These branch currents then combine together to form I_{T_6} which enters the positive end of E_6. Therefore, R_5, R_7, and R_L form the parallel circuit, and their equivalent resistance, R_{XY}, is in series with R_6. R_{XY} is now $= 1/(1/R_5 + 1/R_7 + 1/R_L) = 1/(1/3 + 1/12 + 1/12) = 1/(4/12 + 1/12 + 1/12) = 1/(6/12) = (1)\,(12/6) = 2\ \Omega$. Total resistance presented to E_6 is $R_{T_6} = R_6 + R_{XY} = 6 + 2 = 8\ \Omega$. Total current due to E_6 alone is $I_{T_6} = E_6/R_{T_6} = 8/8 = 1$ A, as shown in Fig. 6-18*c*. A fraction of this current flows *down* through R_L. From equation 4-15,

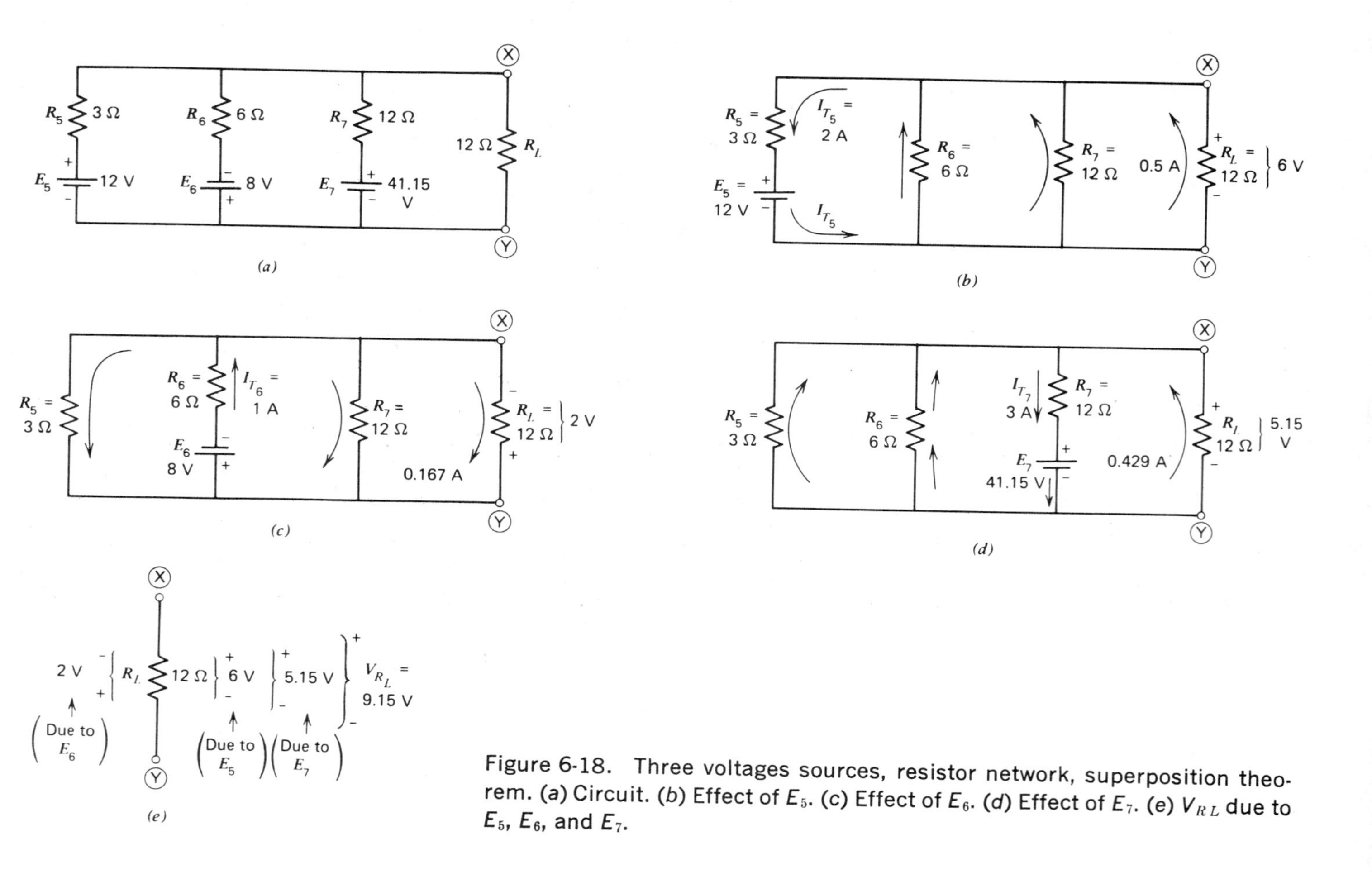

Figure 6-18. Three voltages sources, resistor network, superposition theorem. (a) Circuit. (b) Effect of E_5. (c) Effect of E_6. (d) Effect of E_7. (e) V_{RL} due to E_5, E_6, and E_7.

I_{R_L} is now $= (R_{XY}/R_L)\, I_{T_6} = (2/12)\, 1 = 0.167$ A, as shown. Voltage across R_L due to E_6 alone is then $= (I_{R_L})(R_L) = (0.167)(12) = 2$ V with the polarity shown.

Finally, as depicted in Fig. 6-18*d*, E_7 is the lone voltage source. Electrons coming from the negative end of E_7 branch out and flow *up* through R_5, R_6, and R_L. These branch currents then join to form I_{T_7}, which flows down through R_7 to the positive end of E_7. The parallel circuit, R_{XY}, is now R_5, R_6, and R_L in parallel. This combination is in series with R_7. R_{XY} is now $= 1/(1/R_5 + 1/R_6 + 1/R_L) = 1/(1/3 + 1/6 + 1/12) = 1/(4/12 + 2/12 + 1/12) = 1/(7/12) = (1)(12/7) = 1.715\ \Omega$. Total resistance "seen" by E_7 is $R_{T_7} = R_7 + R_{XY} = 12 + 1.715 = 13.715\ \Omega$. Total current due to E_7 is $I_{T_7} = E_7/R_{T_7} = 41.15/13.715 = 3$ A, as shown. A fraction of this current flows *up* through R_L. From equation 4-15, I_{R_L} is now $= (R_{XY}/R_L)\, I_{T_7} = (1.715/12)\, 3 = 0.429$ A. Voltage across R_L due to E_7 is $= (I_{R_L})(R_L) = (0.429)(12) = 5.15$ V, with the polarity shown in Fig. 6-18*d*.

Voltages across R_L and their polarities due to each of the voltage sources are shown in Fig. 6-18*e*. Note that the voltages due to E_5 and E_7 have the same polarities. These voltages across R_L (6 and 5.15 V) combine to form 11.15 V, positive at the top. The 2 V across R_L due to E_6 have the opposite polarity as shown. The final result, V_{R_L}, is the algebraic sum of the $+11.15$ V and the -2 V, or $V_{R_L} = 9.15$ V with polarity shown.

Example 6-7

In the complete circuit of Fig. 6-18*a*, find the voltage and polarity across R_5.

Solution

Referring to the previous discussion given for this diagram, find voltage across R_5 due to each voltage source individually, using Fig. 6-18*b*, *c*, and *d*. The reader should find V_{R_5}, referring to the results shown in Fig. 6-19.

For Similar Problems Refer to End-of-Chapter Problems 6-27 and 6-28

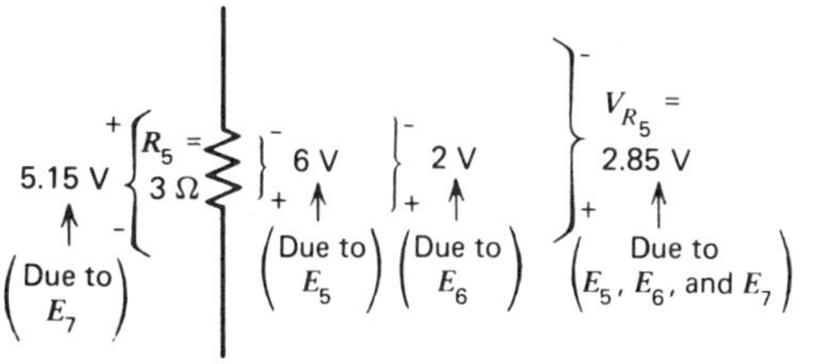

Figure 6-19. Example 6-7.

6-8. Thévenin Theorem, Two- and Three-Voltage Sources

In Section 6-2 a discussion for solving a resistor network containing only one applied voltage source is given, employing the Thévenin theorem. Here, in this section, two, and then three voltage sources in a resistor network are discussed.

The following discussion with all values illustrates the method of finding V_{R_L} in the complete circuit of Fig. 6-20*a*. Open R_L, and remove voltage sources E_1 and E_2, replacing them with their internal resistances, if any (assumed to be zero here), as shown in Fig. 6-20*b*. Find the resistance between points C and F, which consists of R_4 in parallel with the series resistors R_1 and R_2. R_{CF} = product/sum = $(R_1 + R_2)(R_4)/[(R_1 + R_2) + R_4] = (2 + 4)(30)/[(2 + 4)] + 30 = (6)(30)/[6 + 30] = 180/36 = 5\ \Omega$. Resistance between points A and B, with R_L still open, consists of R_3 and R_{CF} in series. $R_{AB} = R_3 + R_{CF} = 15 + 5 = 20\ \Omega$, as shown in Fig. 6-20*b*, second part. This is $R_{\text{Thév}}$.

With R_L still open, voltage sources E_1 and E_2 are placed back into the circuit as shown in Fig. 6-20*c*. E_1 and E_2 are bucking or opposing. That is, E_1 causes a counterclockwise current, while E_2 produces a clockwise movement of electrons. The difference between E_1 and E_2, or $E_{\text{net}} = 90 - 84 = 6$ V. E_{net} causes a clockwise current as shown. Voltage across R_4, from equation 3-1, $= (R_4/R_T)\,E_{\text{net}} = [30/(30 + 2 + 4)]\,6 = (30/36)\,6 = 5$ V, with polarity shown in Fig. 6-20*c*. Voltage between points C and F is the algebraic sum of E_2 and V_{R_4}. Note that these voltages have opposing polarities, with the upper end of E_2 being positive, while the upper end of V_{R_4} is negative. $V_{FC} = E_2 - V_{R_4} = +90 - 5 = +85$ V, with point F being the positive end, as shown. This is $E_{\text{Thév}}$.

The Thévenin equivalent circuit between points A and B, with R_L still open, consists of 20 Ω and 85 V. Placing R_L back between points A and B together with $R_{\text{Thév}}$ and $E_{\text{Thév}}$ is the final circuit shown in Fig. 6-20*d*. Voltage across $R_L = (R_L/R_T)\,E_{\text{Thév}} = [R_L/(R_L + R_{\text{Thév}})]\,E_{\text{Thév}} = [5/(5 + 20)]\,85 = (5/25)\,85 = 17$ V, with polarity shown in the diagram.

If we wished to find the *total* Thévenin equivalent resistance between points A and B, *with R_L in the circuit*, then, as shown in Fig. 6-20*b*, R_L is in parallel with $R_{\text{Thév}}$ (with the voltage source $E_{\text{Thév}}$ replaced by a short). $R_{\text{Thév total}} = (20)(5)/(20 + 5) = 100/25 = 4\ \Omega$.

The following example uses the same circuit of Fig. 6-20*a*, but this time we wish to find the Thévenin equivalent resistance and voltage "looking" in at the left side instead of that just discussed "looking" in at the right side.

Example 6-8

In the circuit of Fig. 6-20*a*, find the Thévenin equivalent resistance and voltage between points D and E.

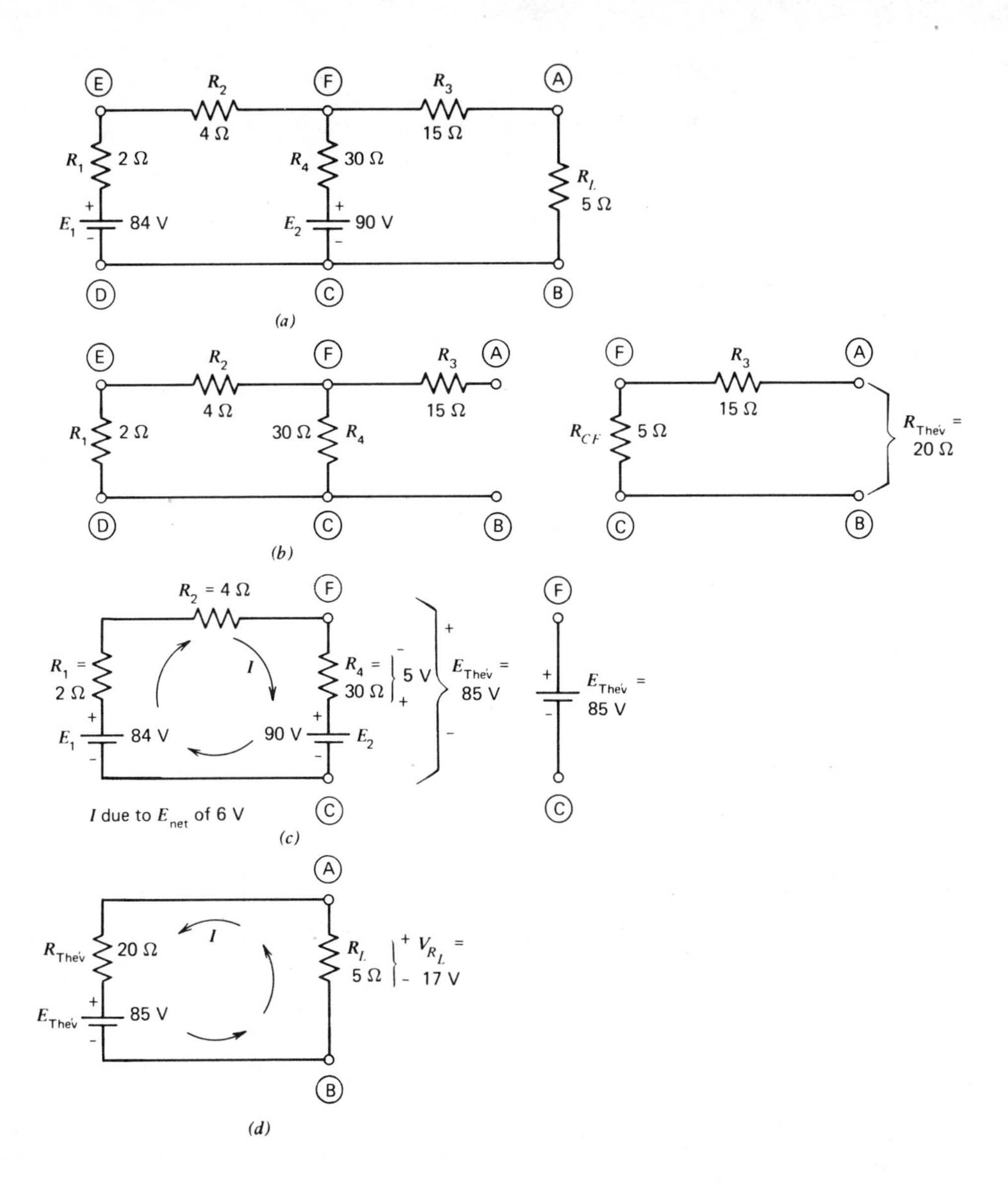

Figure 6-20. Thévenin theorem, two voltage source-resistor network. (*a*) Complete circuit. (*b*) $R_{\text{Thév}}$ (R_L open). (*c*) $E_{\text{Thév}}$ (R_L open). (*d*) R_L back in the circuit.

Solution

With R_1 open between points D and E, and with E_2 replaced by its internal resistance (assumed to be zero here), as shown in Fig. 6-21*a* first find R_{CF} and then $R_{\text{Thév}}$ between D and E (with R_1 still open). The student should continue solving this example using the results shown in Fig. 6-21*a* through *e*.

For Similar Problems Refer to End-of-Chapter Problems 6-29 to 6-32

Fig. 6-22*a* is a resistor network with three voltage sources. If it is desired to find the Thévenin equivalent voltage and resistance between points X and Y, it is necessary to first open R_9 between X and Y, and replace E_1 and E_2 by their internal resistances (assumed to be zero here). The circuit is now as shown in part (b). First solve for R_{WZ} which consists of R_5 and R_6 in series with each other, and in parallel with R_7. $R_{WZ} = (R_7)(R_5 + R_6)/[(R_7) + (R_5 + R_6)] = (60)(30 + 30)/(60 + 30 + 30) = (60)(60)/120 = 3600/120 = 30\Omega$.

The Thévenin resistance between points X and Y, with R_9 open, shown in part (c), consists of R_8, R_{WZ}, and R_{10} in series. $R_{\text{Thév partial } XY} = R_8 + R_{WZ} + R_{10} = 20 + 30 + 10 = 60\ \Omega$.

With R_9 still open, voltages E_3 and E_2 are again placed back into the circuit of Fig. 6-22*d*. Note that E_2 and E_3 are bucking or opposing. Their difference, E_{net}, is 20 V, which produces a counterclockwise current shown. Voltage across $R_7 = [R_7/(R_5 + R_6 + R_7)]\, E_{\text{net}} = [60/(30 + 30 + 60)]\, 20 = (60/120)\, 20 = 10$ V with the polarity shown in Fig. 6-22*d*. $E_{\text{Thév WZ}}$ is the algebraic sum of E_2 and V_{R_7}, which are each positive at their upper ends, and $= E_2 + V_{R_7} = 30 + 10 = 40$ V, with polarity shown.

Finally, in Fig. 6-22*e*, R_9 is connected back again, placing E_1 back into the circuit. $E_{\text{Thév WZ}}$ and E_1 are aiding each other, making $E_{\text{net}} = 40 + 20 = 60$ V, causing a counterclockwise current. $V_{R_9} = [R_9/(R_8 + R_9 + R_{10} + R_{WZ})]\, E_{\text{net}} = [90/(20 + 90 + 10 + 30)]\, 60 = (90/150)\, 60 = 36$ V, with the polarity shown. $E_{\text{Thév total XY}}$ is the algebraic sum of E_1 and V_{R_9}. Note that the upper end of E_1 is negative, while the upper end of V_{R_9} is positive. $E_{\text{Thév total XY}} = V_{R_9} - E_1 = 36 - 20 = 16$ V with the polarity shown in parts (e) and (g).

The Thévenin total equivalent resistance between points X and Y, with R_9 in the circuit, is shown in Fig. 6-22*f*. Again all voltage sources are replaced by their internal resistances (assumed to be zero here). The resistance between points X and Y consists of R_9 in parallel with the series resistors R_8, R_{WZ}, and R_{10}. Therefore, $R_{\text{Thév total XY}} = (R_9)(R_8 + R_{WZ} + R_{10})/(R_9 + R_8 + R_{WZ} + R_{10}) = (90)(20 + 30 + 10)/(90 + 20 + 30 + 10) = (90)(60)/(90 + 60) = 5400/150 = 36\ \Omega$, as shown in Fig. 6-22*f* and *g*.

The following example uses the same three-voltage-source resistor network as that just discussed, except that the Thévenin equivalent between two other points is to be found.

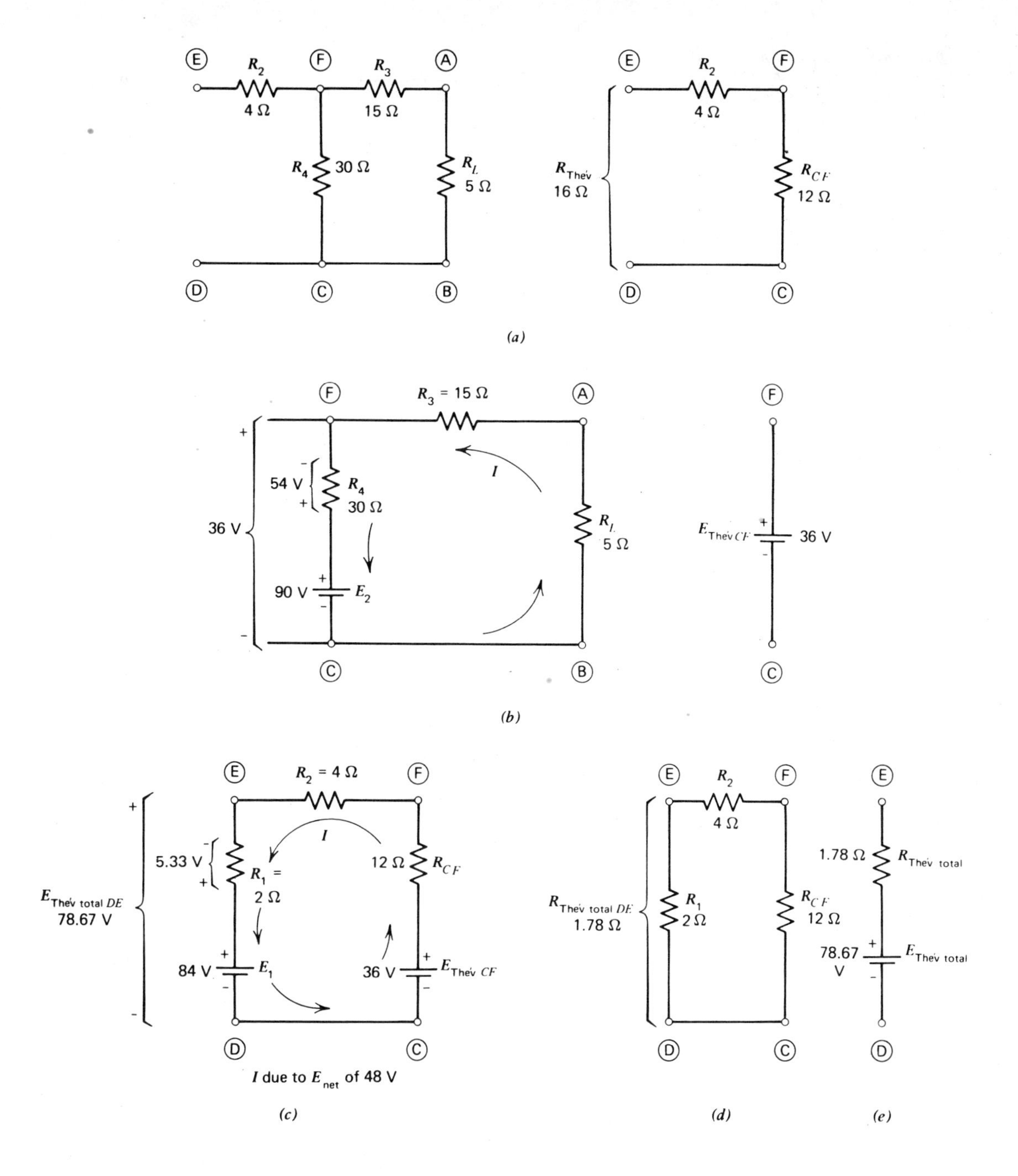

Figure 6-21. Example 6-8. (*a*) $R_{\text{Thév}\ DE}$ (R_1 open). (*b*) $E_{\text{Thév}\ CF}$ (R_1 open). (c) $E_{\text{Thév}}$ total *DE* (R_1 back in the circuit). (*d*) $R_{\text{Thév}}$ total *DE* (R_1 in circuit). (e) Thévenin total equivalent circuit *DE* (with R_1 in the circuit).

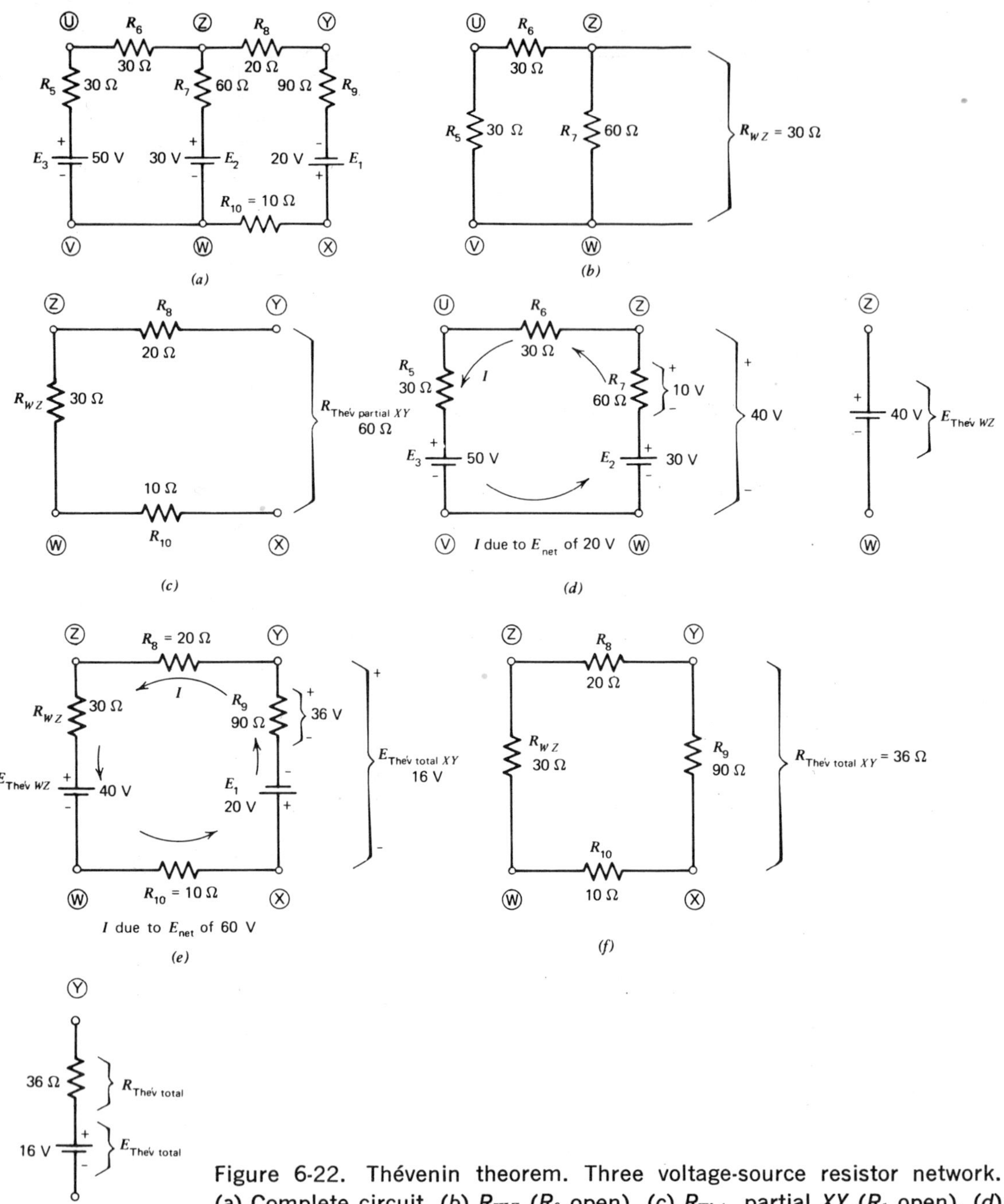

Figure 6-22. Thévenin theorem. Three voltage-source resistor network. (*a*) Complete circuit. (*b*) R_{WZ} (R_9 open). (*c*) $R_{Thév}$ partial *XY* (R_1 open). (*d*) $E_{Thév\ WZ}$ (R_9 open); (*e*) $E_{Thév}$ total *XY* (R_9 in the circuit. (*f*) $R_{Thév}$ total *XY* (R_9 in circuit). (*g*) Thévenin equivalent circuit.

Example 6-9

In the complete circuit of Fig. 6-22*a*, find the Thévenin equivalent resistance and voltage between points U and V.

Solution

Opening up R_5 and E_3 between points U and V in Fig. 6-22*a*, and replacing voltage sources E_2 and E_1 by their internal resistances (assumed to be zero here), yields the series-parallel resistor circuit of Fig. 6-23*a*. Solve this circuit using as a guide Fig. 6-23*a* to *f*.

For Similar Problems Refer to End-of-Chapter Problems 6-33 to 6-35

6-9. Norton Theorem, Two- and Three-Voltage Sources

In Section 6-3, Norton's theorem is introduced using a resistor network having one voltage source. Now, in this section, the solution of a resistor network having first two voltage sources, and later, three voltage sources is discussed using the Norton method.

In Fig. 6-24*a* a resistor network with two applied voltages is shown. It is desired to find the voltage across R_L, the 5-Ω resistor between points A and B. This same circuit was solved previously by using superposition in Section 6-7, Fig. 6-16, and also by using Thévenin in Section 6-8, Fig. 6-20.

A brief review of the Norton method of finding V_{R_L} in Fig. 6-24*a* is outlined first:

1. With R_L *shorted*, total current *through the short* due to both voltages E_1 and E_2 is found. This is the current from the generator, called $I_{N\ Total}$.
2. With R_L *open*, and all voltage sources replaced by their internal resistances, the resistance between points A and B is found next. This is resistor R_N, which is placed across the generator, I_{NT}.
3. Finally, R_L is also placed across the generator I_{NT}, and then current through R_L and voltage across R_L is found.

In the first step, R_L is shorted, as shown in Fig. 6-24*b*, and the effect of voltage source E_1 *alone* is found. The other voltage source E_2 is replaced by its internal resistance (assumed to be zero here), while the effect of E_1 is found. The circuit of part (b) is a series-parallel one consisting of resistor R_1 in series with R_2, and in series with the parallel portion between points C and F. R_{CF_1} is made up of R_4 in parallel with R_3. $R_{CF_1} = (R_4)(R_3)/(R_4 + R_3) = (30)(15)/(30 + 15) = 450/45 = 10\Omega$. Therefore, $R_{T_1} = R_1 + R_2 + R_{CF_1} = 2 + 4 + 10 = 16\ \Omega$. Total current $I_{T_1} = E_1/R_{T_1} = 84/16 = 5.25$ A, as shown in part (b).

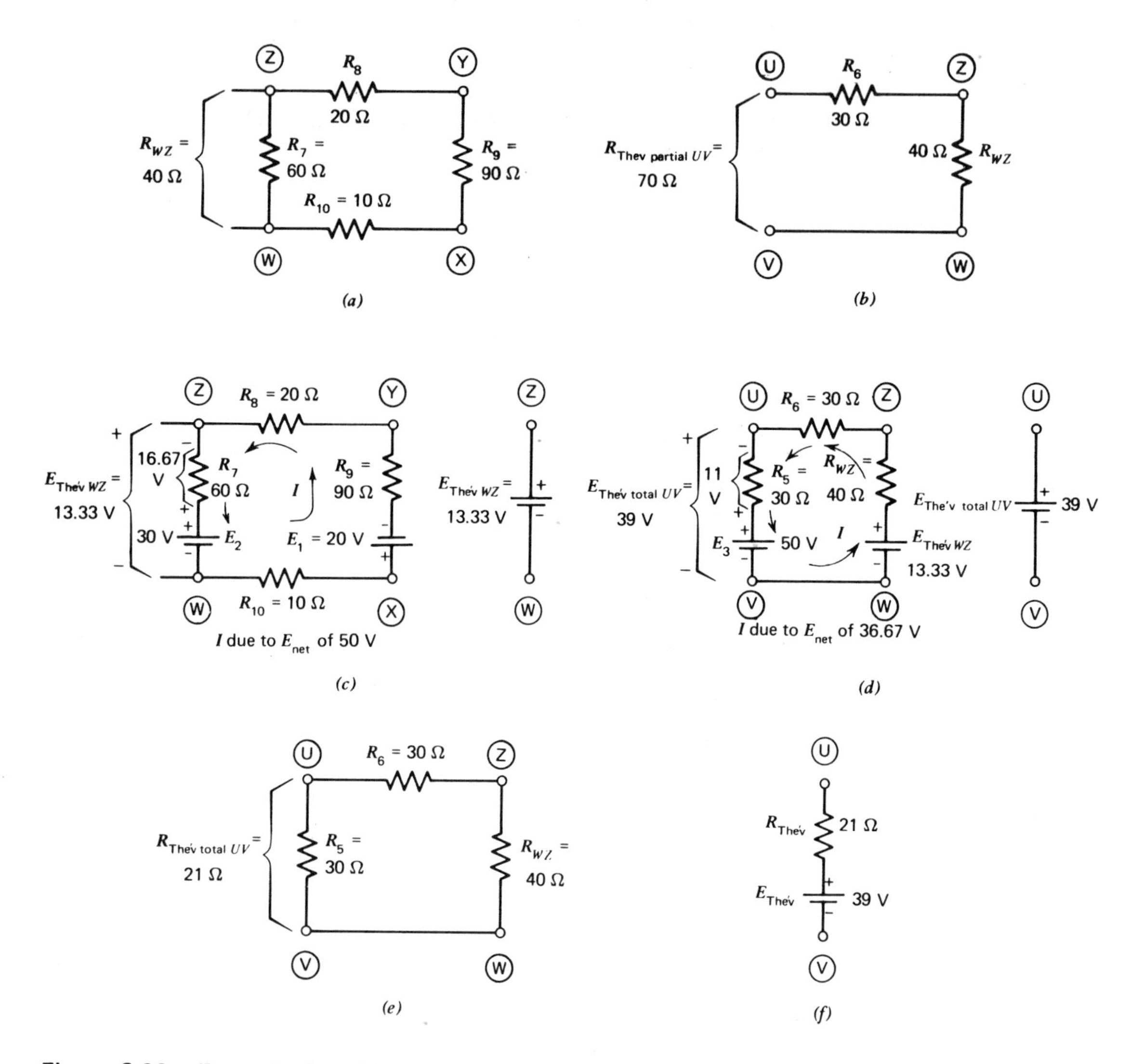

Figure 6-23. Example 6-9. (a) R_{WZ} (R_5 open of Fig. 6-22a. (b) $R_{\text{Thév}}$ partial UVS (R_5 open of Fig. 6-22a. (c) $E_{\text{Thév } WZ}$ (R_5 open). (d) $E_{\text{Thév}}$ total UV (R_5 in the circuit). (e) $R_{\text{Thév}}$ total UV (R_5 in the circuit). (f) Thévenin equivalent circuit.

Current I_{N_1} through the short between points B and A (the shorted R_L) may be found using equation 4-15, where $I_{N_1} = (R_{CF_1}/R_3)\,I_{T_1} = (10/15)\,5.25 = 3.5$ A as depicted flowing up from point B to point A.

With R_L still shorted, the effect of voltage source E_2 *alone* is found next. The other source E_1 is replaced by its internal resistance (assumed to be zero here), and the circuit is pictured in Fig. 6-24*c*. The circuit is a series-parallel

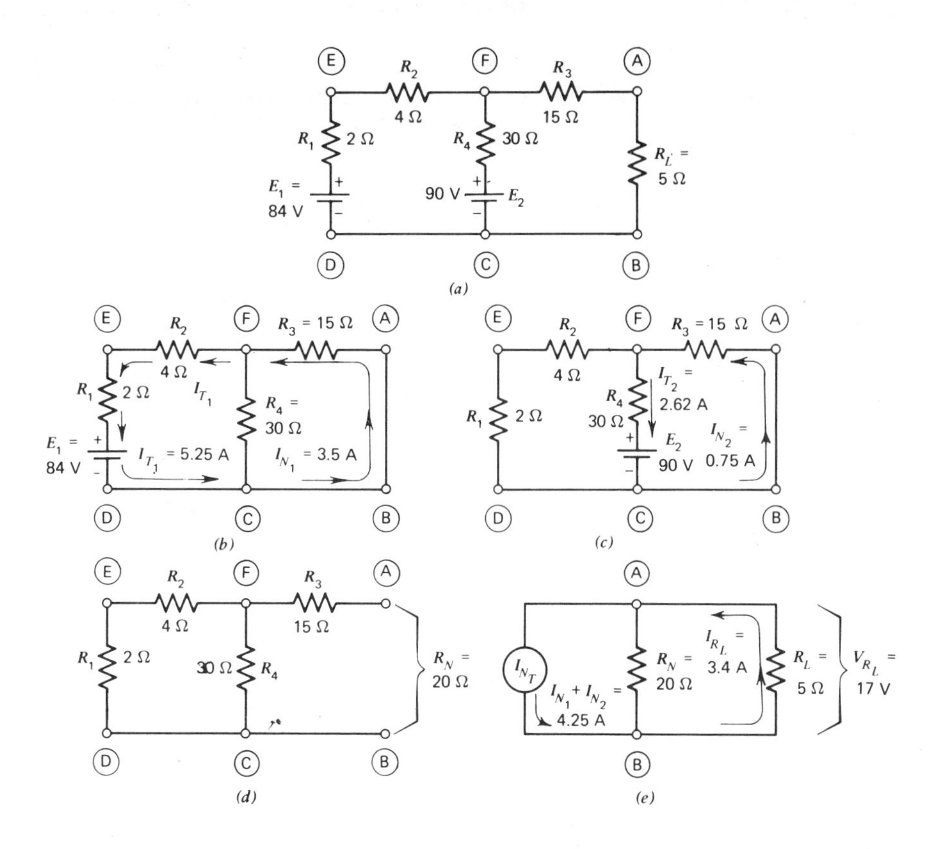

Figure 6-24. Norton theorem. Resistor network with two voltage sources. (a) Complete circuit. (b) Currents due to E_1 alone, R_L shorted. (c) Currents due to E_2 alone, R_L shorted. (d) R_N, between A and B, with R_L open. (e) Norton equivalent circuit with R_L.

combination, with R_4 in series with the parallel combination between points C and F. R_{CF_2} now consists of R_3 in parallel with the series resistors R_1 and R_2. $R_{CF_2} = (R_3)(R_1 + R_2)/[(R_3) + (R_1 + R_2)] = (15)(2 + 4)/[(15) + (2 + 4)] = (15)(6)/(15 + 6) = 90/21 = 4.29\ \Omega$. Total resistance in the circuit of (c), $R_{T_2} = R_4 + R_{CF_2} = 30 + 4.29 = 34.29\ \Omega$. Total current $I_{T_2} = E_2/R_{T_2} = 90/34.29 = 2.62$ A, depicted in diagram (c). Current I_{N_2} through the short between points B and A (the shorted R_L) may be found using equation 4-15, where $I_{N_2} = (R_{CF_2}/R_3)\, I_{T_2} = (4.29/15)\, 2.62 = 0.75$ A as shown flowing up from point B to point A. Note that I_{N_1} in part (b), and I_{N_2} in part (c) both flow *up* from point B to point A (the shorted R_L.) Therefore, the Norton generator I_{N_T} is a *current* device, where $I_{N_T} = I_{N_1} + I_{N_2} = 3.5 + 0.75 = 4.25$ A as shown in Fig. 6-24*e*. This is the combined effect of E_1 and E_2, and completes the first step in the previous review.

Now, as in step two of the review, R_L is *opened*, and all applied voltages, E_1 and E_2, are replaced by their internal resistances (zero in this discussion), as depicted in Fig. 6-24*d*. With R_L open, the resistance between points A and B is now found. This called the Norton resistance, R_N. R_N comprises R_3 in series with the parallel circuit of R_{CF}. This R_{CF_3} is made up of R_4 in parallel with R_1 and R_2 in series, or $R_{CF_3} = (R_4)(R_1 + R_2)/[(R_4) + (R_1 + R_2)] = (30)(2 + 4)/(30 + 2 + 4) = (30)(6)/(30 + 6) = 180/36 = 5\Omega$. Therefore, $R_N = R_3 + R_{CF_3} = 15 + 5 = 20\ \Omega$, and is shown in diagram (d), and also is connected across the Norton current generator I_{N_T} in part (e).

Finally, as stipulated in step three of the previous review, R_L is now inserted across the Norton current generator I_{N_T}, as shown in diagram (e). Current through R_L and voltage across R_L can now be found.

Note that R_N and R_L are in parallel. Resistance between points A and B, $R_{AB} = (R_N)(R_L)/(R_N + R_L) = (20)(5)/(20 + 5) = 100/25 = 4\ \Omega$. *Total* current from the generator, shown as $I_{N_T} = 4.25$ A, flows through this equivalent resistor R_{AB}. Therefore, $V_{R_{AB}} = (I_{N_T})(R_{AB}) = (4.25)(4) = 17$ V. This is also the voltage across R_L.

Current through R_L, $I_{R_L} = V_{R_L}/R_L = 17/5 = 3.4$ A as shown in diagram (e). These values agree with those found previously using superposition (Section 6-7) and using Thévenin (Section 6-8).

Example 6-10

In the same circuit of Fig. 6-24*a* just discussed, find V_{R_1} using Norton's theorem.

Solution

To solve for V_{R_1} using Norton's theorem, it is first necessary to short out R_1, and find the *total* current through this short due to all applied voltage sources. As shown in Fig. 6-25*a*, R_1 is shorted out, and the effect of source voltage E_1 alone is found, while E_2 has been replaced by its internal resistance (assumed to be zero here). The circuit of part (a) is a series-parallel one consisting of R_2 in series with the equivalent resistance between points C and F. Solve this example using the results shown in Fig. 6-25*a* to *d*.

For Similar Problems Refer to End-of-Chapter Problems 6-36 to 6-39

The circuit of Fig. 6-26*a* is a resistor network containing three voltage sources. The following discussion uses Norton's theorem to find the voltage across resistor R_9. This was previously solved in Section 6-11, Fig. 6-22, employing Thévenin's theorem.

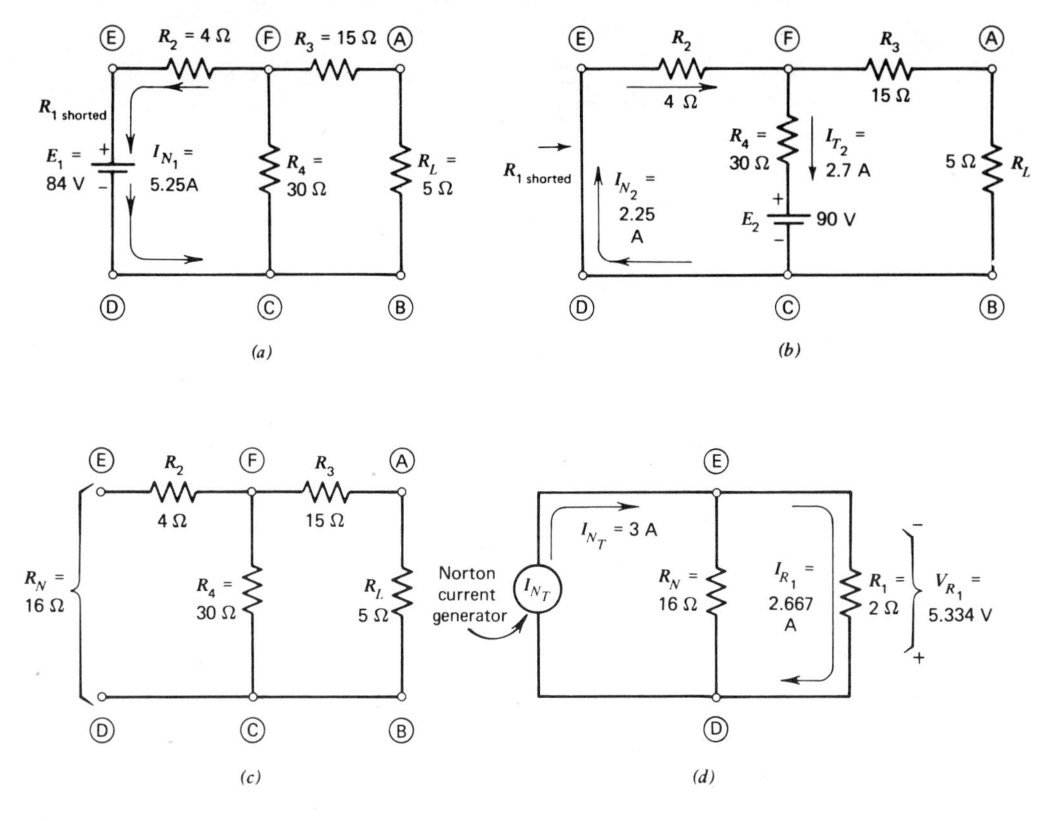

Figure 6-25. Example 6-10. (*a*) Current due to E_1 alone, R_1 shorted (from Fig. 6-24*a*). (*b*) Currents due to E_2 alone, R_1 shorted). (*c*) R_N (between *D* and *E*), R_1 open. (*d*) Norton equivalent circuit with R_1.

To find V_{R_9}, this resistor is first shorted. Current through the short is now found due to *each* of the applied voltage sources acting alone. Consider first the effect of E_1 *alone*, as shown in Fig. 6-26*b*. The other two voltage sources E_2 and E_3 are replaced by their internal resistances (assumed to be zero here). Diagram (b), with E_1 alone, is a series-parallel circuit. Resistors R_8 and R_{10} are in series with the parallel circuit between points W and Z. R_{WZ} consists of R_7 in parallel with resistors R_5 and R_6 in series. $R_{WZ} = (R_7)(R_5 + R_6)/[(R_7) + (R_5 + R_6)] = 60\ (30 + 30)/[(60) + (30 + 30)] = (60)\ (60)/(60 + 60) = 3600/120 = 30\ \Omega$.

$R_{T_1} = R_8 + R_{WZ} + R_{10} = 20 + 30 + 10 = 60\ \Omega$. Current flowing through the short (the shorted R_9) due to E_1 is $I_{N_1} = E_1/R_{T_1} = 20/60 = 0.333$ A, shown in diagram (b) flowing *up* through the short.

Consider next the effect of E_2 alone, with R_9 still shorted and with the other two voltage sources E_1 and E_3 replaced by their internal resistances (assumed

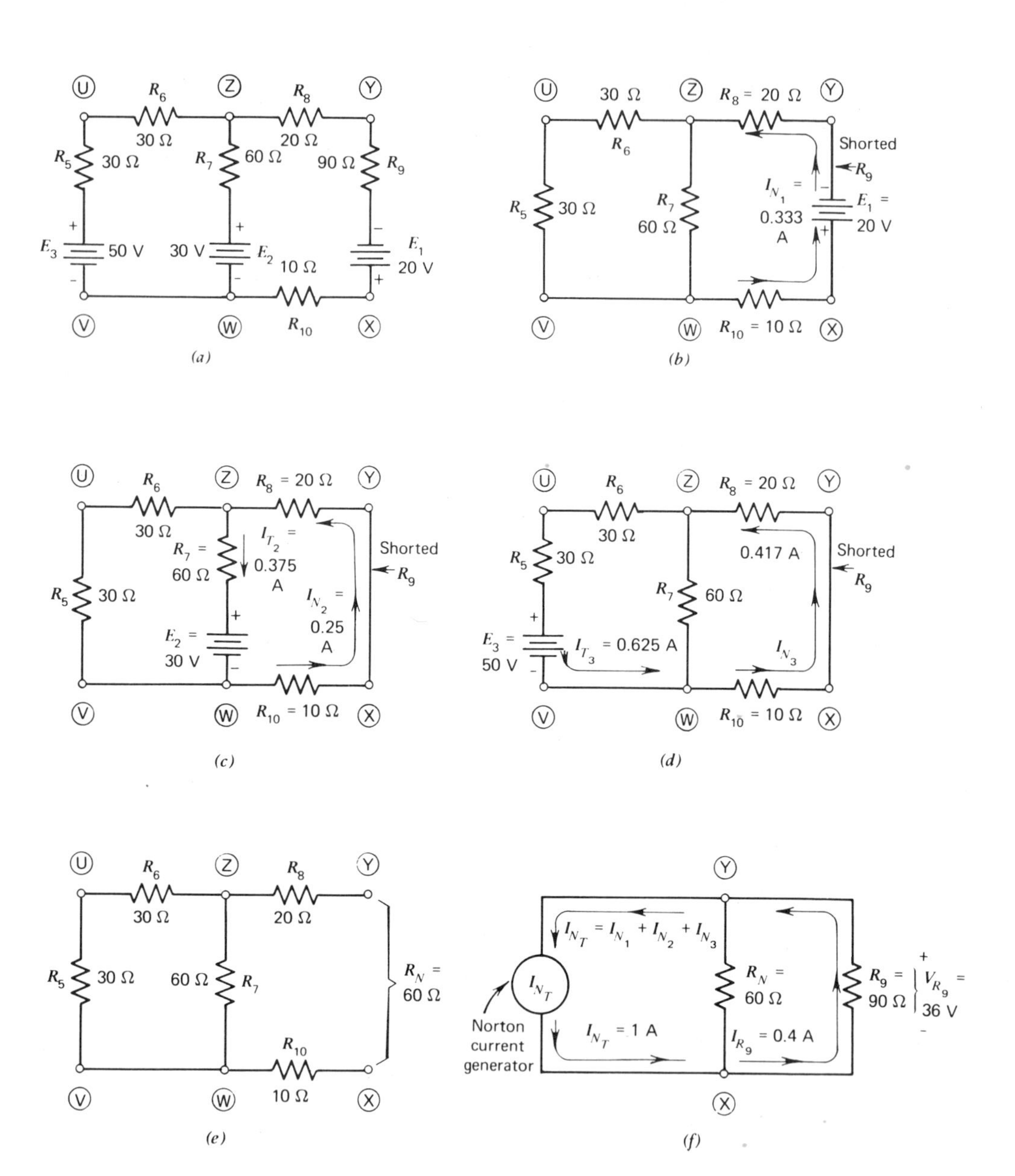

Figure 6-26. Norton theorem. Resistor network with three voltage sources. (*a*) Complete circuit. (*b*) Current due to E_1 alone, R_9 shorted. (*c*) Currents due to E_2 alone, R_9 shorted. (*d*) Currents due to E_3 alone, R_9 shorted. (*e*) R_N (between X and Y), R_9 open. (*f*) Norton equivalent circuit with R_9.

to be zero here), as depicted in diagram (c). Again, the circuit is a series-parallel one, but this time R_7 is in series with R_{WZ}. Now, R_{WZ} consists of series resistors R_5 and R_6 in *parallel* with series resistors R_8 and R_{10}. Therefore, $R_{WZ} = (R_5 + R_6)(R_8 + R_{10})/[(R_5 + R_6) + (R_8 + R_{10})] = (30 + 30)(20 + 10)/[(30 + 30) + (20 + 10)] = (60)(30)/(60 + 30) = 1800/90 = 20\ \Omega$.

Total resistance is now $R_{T_2} = R_7 + R_{WZ} = 60 + 20 = 80\ \Omega$. Total current due to E_2 alone is now $I_{T_2} = E_2/R_{T_2} = 30/80 = 0.375$ A as shown in part (c). Current through the short (the shorted R_9) and also through R_8 and R_{10} is a fraction of I_{T_2}, and from equation 4-6, $I_{N_2} = \{(R_5 + R_6)/[(R_5 + R_6) + (R_8 + R_{10})]\}\ I_{T_2} = \{(30 + 30)/[(30 + 30) + (20 + 10)]\}\ 0.375 = 60/(60 + 30)\ 0.375 = (60/90)\ 0.375 = 0.25$ A. This current is depicted in diagram (c) and is flowing *up* through the short.

The effect of voltage source E_3 alone is next to be considered in the circuit of part (d). R_9 is still shorted, and voltage sources E_1 and E_2 are now replaced by their internal resistances (assumed to be zero here). Now, the circuit consists of resistors R_5 and R_6 in series with the parallel circuit between points W and Z. R_{WZ} is made up of R_7 (60 Ω) in parallel with the series resistors R_8 and R_{10} (20 and 10 Ω). Therefore, $R_{WZ} = (R_7)(R_8 + R_{10})/[(R_7) + (R_8 + R_{10})] = (60)(20 + 10)/[(60) + (20 + 10)] = (60)(30)/(60 + 30) = 1800/90 = 20\ \Omega$.

Total resistance in part (d) is now $R_{T_3} = R_5 + R_6 + R_{WZ} = 30 + 30 + 20 = 80\ \Omega$. Total current I_{T_3} due to E_3 is $= E_3/R_{T_3} = 50/80 = 0.625$ A, as shown. The current I_{N_3} flowing through the short (the shorted R_9) is a fraction of I_{T_3}. Using equation 4-6, $I_{N_3} = \{R_7/[R_7 + (R_8 + R_{10})]\}\ I_{T_3} = [60/(60 + 20 + 10)]\ 0.625 = [60/90]\ 0.625 = 0.417$ A. Note that currents through the short, $I_{N_1} = 0.333$ A [diagram (b)], $I_{N_2} = 0.25$ A [diagram (c)], and $I_{N_3} = 0.417$ A [diagram (d)], all flow *up* from point X to point Y. Total current I_{N_T} through the short (the shorted R_9) is $I_{N_1} + I_{N_2} + I_{N_3} = 0.333 + 0.25 + 0.417 = 1$ A, and is depicted in part (f) as the Norton current generator.

Opening resistor R_9 in diagram (a), the resistance between points X and Y is depicted in part (e). R_8 and R_{10} are in series with R_{WZ}. Resistance between points W and Z consists of R_7 (60 Ω) in parallel with the series resistors R_5 and R_6 (30 and 30 Ω). $R_{WZ} = (R_7)(R_5 + R_6)/[(R_7) + (R_5 + R_6)] = (60)(30 + 30)/[60 + (30 + 30)] = (60)(60)/120 = 3600/120 = 30\ \Omega$. Resistance between points X and Y is then $= R_8 + R_{10} + R_{WZ} = 20 + 10 + 30 = 60\Omega$, as shown in diagram (e). This called the Norton resistance, R_N, and is placed across the Norton current generator I_{N_T}, as depicted in part (f).

Resistor R_9 is finally placed, along with R_N, across the Norton generator I_{N_T}, as shown. Current through R_9, from equation 4-6, is $I_{R_9} = [R_N/(R_N + R_9)]\ I_{N_T} = [60/(60 + 90)]\ 1 = (60/150)\ 1 = 0.4$ A, as shown in (f).

$V_{R_9} = (I_{R_9})(R_9) = (0.4)(90) = 36$ V, with the polarity shown in Fig. 6-26*f*. This agrees with that found using Thévenin, Fig. 6-22*e*.

The following example uses the same circuit, but employs Norton to find the voltage across resistor R_5.

Example 6-11

In the complete circuit of Fig. 6-26*a*, find V_{R_5} using Norton's theorem. This was found previously as part of Example 6-9, Fig. 6-23, using Thévenin.

Solution

Shorting resistor R_5, the current through the short due to each voltage

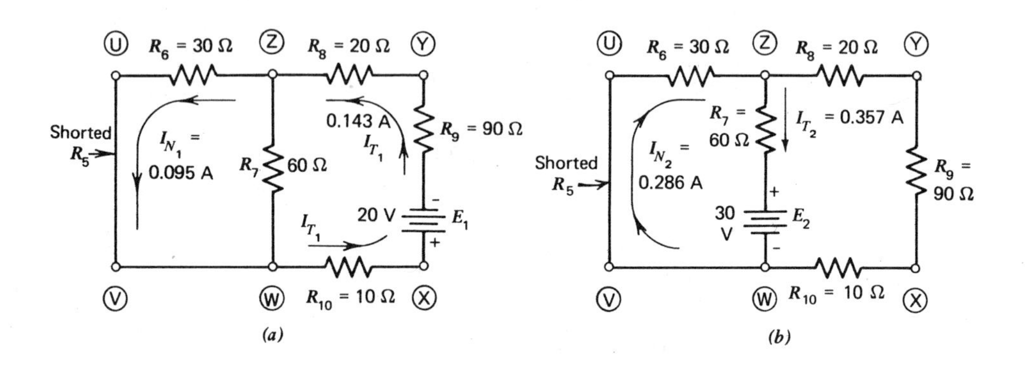

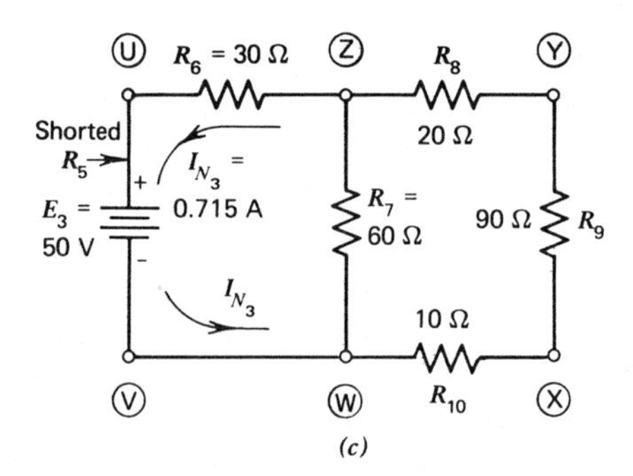

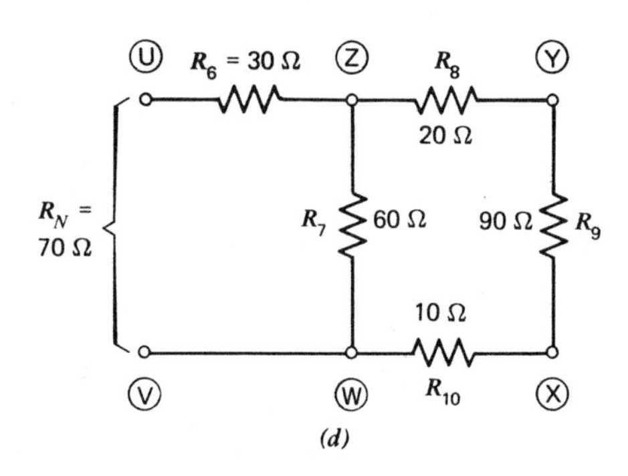

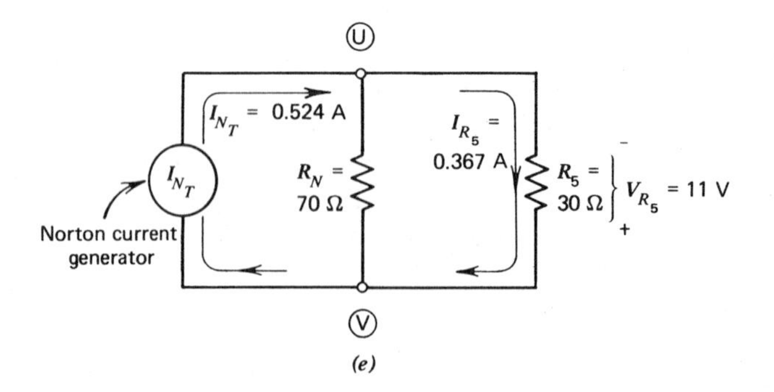

Figure 6-27. Example 6-11 (*a*) Currents due to E_1 alone, R_5 shorted (from circuit Fig. 6-26*a*). (*b*) Currents due to E_2 alone, R_5 shorted. (*c*) Current due to E_3 alone, R_5 shorted. (*d*) R_N (between points *U* and *V*), with R_5 open (*e*) Norton equivalent circuit with R_5.

source, E_1, E_2, and E_3, is found. As shown in Fig. 6-27*a*, E_1 is in the circuit alone, and R_5 is shorted. Solve this example using the results in Fig 6-27*a* to *e*.

For Similar Problems Refer to End-of-Chapter Problems 6-40 to 6-45

6-10. Maxwell's Cyclic Currents, Two- and Three-Voltage Sources

In Section 6-4, resistor networks containing only one voltage source are solved using Maxwell's cyclic currents or loop currents. In this section, two loop circuits result in two unknowns (I_1 and I_2) using two equations, and three loop circuits yield three unknown quantities, I_1, I_2, and I_3, using three equations.

Figure 6-28 depicts a resistor network with two voltage sources. We wish to find the values and polarities of V_{R_1}, V_{R_4}, and V_{R_L}. The currents I_1 and I_2 are assumed each to flow in a counterclockwise direction in their respective closed loops. I_1 is assumed to flow through loop 1 embracing points D, C, F, and E. Similarly, I_2 flows in loop 2 embracing points C, B, A, and F.

Loop 1 and I_1 yields the following equation, if point C is taken as the starting point, and a "walk" is taken around the loop in the direction of the assumed counterclockwise current. Also, the polarity of each voltage is that "seen" as the voltage is approached. Therefore in loop 1, starting at point C, the equation is: $-90 - 30I_1 + 30I_2 - 4I_1 - 2I_1 + 84 = 0$. Note that the voltage across the 30-Ω R_4 is *negative* at the *lower* end due to I_1, but is also *positive* at this same *lower* end due to I_2; hence, the voltage across R_4 is $\mathbf{-30I_1 + 30I_2}$. Combining and rewriting the previous equation yields: $-36I_1 + 30I_2 = 6$. Dividing both sides by 6, to simplify, gives: $\mathbf{-6I_1 + 5I_2 = 1}$ (equation 6-12).

Loop 2, starting at point C and traveling with the current I_2, yields the following: $-5I_2 - 15I_2 - 30I_2 + 30I_1 + 90 = 0$. Combining and rewriting

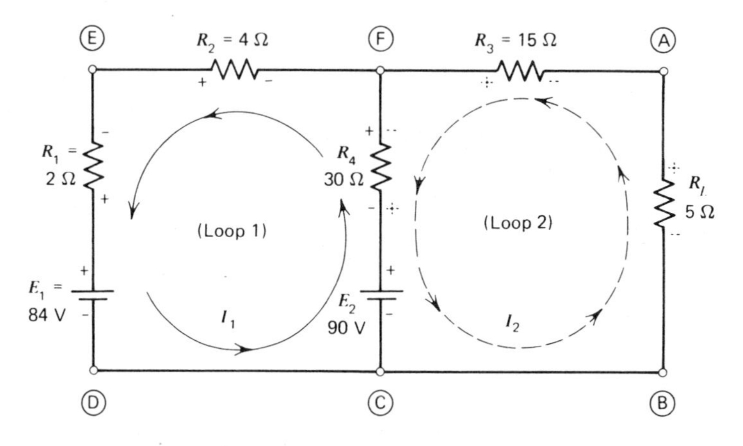

Figure 6-28. Loop currents, two voltage sources.

gives: $30I_1 - 50I_2 = -90$. Dividing both sides by 10, to simplify, yields: $\mathbf{3I_1 - 5I_2 = -9}$ (equation 6-13). Rewriting equations 6-12 and 6-13:

$$-6I_1 + 5I_2 = 1 \tag{6-12}$$

$$3I_1 - 5I_2 = -9 \tag{6-13}$$

Adding these equations eliminates one of the unknowns, I_2, as shown in the following:

$$-3I_1 = -8$$

$$I_1 = \frac{-8}{-3}$$

$$-3$$

$$I_1 = 2.667 \text{ A}$$

The positive answer for I_1 indicates that the assumed direction for I_1 is correct. Substituting the value 2.667 for I_1 in equation 6-12 yields

$$-6\,(2.677) + 5I_2 = 1 \tag{6-12}$$

$$-16 + 5I_2 = 1$$

$$5I_2 = 1 + 16$$

$$5I_2 = 17$$

$$I_2 = 3.4 \text{ A}$$

The positive value of I_2 indicates that its assumed direction is correct. $V_{R_1} = (I_1)\,(R_1) = (2.667)\,(2) = 5.334$ V, with the polarity shown in Fig. 6-28. This agrees with the results shown in Fig. 6-17*a*, using the superposition theorem; also, with the results shown in Fig. 6-21*c*, using Thévenin; and with the results shown in Fig. 6-25*d*, using Norton.

V_{R_4} is due to a net current made of up I_1 (2.667 A) flowing *up* through R_4, and I_2 (3.4 A) flowing *down* through R_4. I_{net} is the difference between these currents, or $3.4 - 2.667 = 0.733$ A, flowing *down*. $V_{R_4} = (I_{\text{net}})\,(R_4) = (0.733)\,(30) = 21.99$ V, or approximately 22 V, negative at the upper end. This agrees with the result shown in Fig. 6-17*b*, using superposition.

$V_{R_L} = (I_2)\,(R_L) = (3.4)\,(5) = 17$ V, with the polarity shown in Fig. 6-28. This agrees with the results shown in Fig. 6-20*d*, using Thévenin, and in Fig. 6-24*e*, using Norton.

Example 6-12

In the circuit of Fig. 6-29, using Maxwell's cyclic currents (loop currents), find (a) V_{R_1} and its polarity, (b) V_{R_3} and its polarity, and (c) V_{R_7} and its polarity.

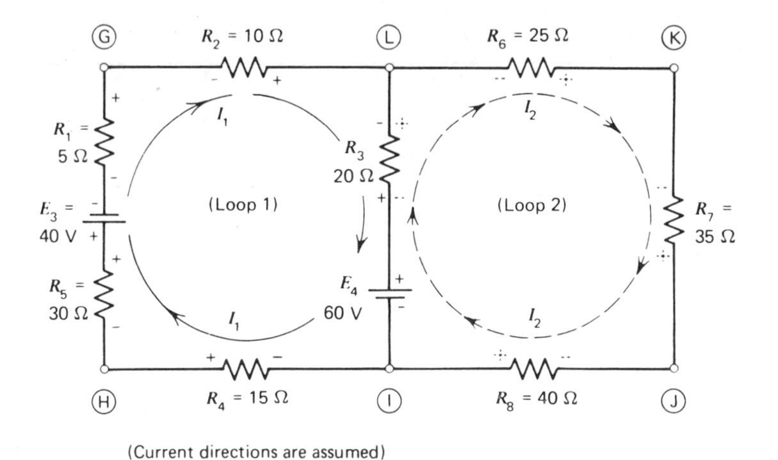

Figure 6-29. Example 6-22.

Loop 1; $-80I_1 + 20I_2 + 100 = 0$
Loop 2; $+ 20I_1 - 120I_2 - 60 = 0$

$I_1 = 1.175$ A
$I_2 = -0.304$ A

$V_{R1} = 5.88$ V (+ at top)
$V_{R3} = 29.6$ V (− at top)
$V_{R7} = 10.64$ V (+ at top)

Solution

Assume that current I_1 flows clockwise in loop 1 (points G, L, I, and H), and that current I_2 also flows clockwise in loop 2 (points L. K, J, and I). Now solve this example using the results shown in Fig. 6-29 as a guide.

For Similar Problems Refer to End-of-Chapter Problems 6-46 to 6-48

The circuit of Fig. 6-30 is a resistor network with three voltage sources and three loops. The following example employs Maxwell's cyclic currents to solve the circuit.

Example 6-13

In the diagram of Fig. 6-30, find the value and actual direction of currents I_1, I_2, and I_3.

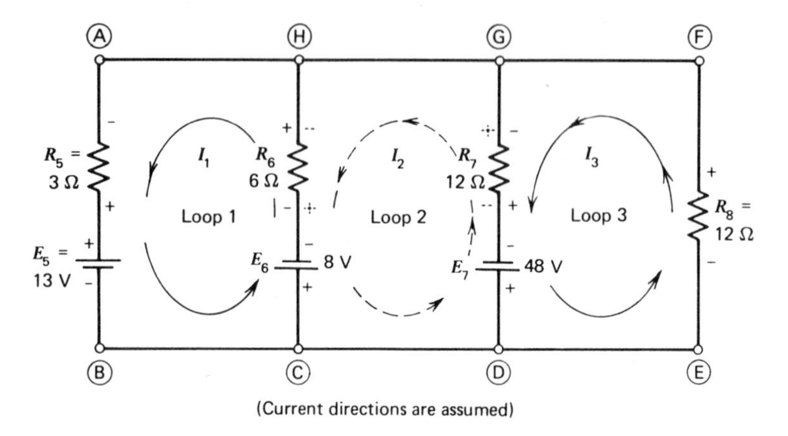

Figure 6-30. Three loop currents, Example 6-13. (Current directions are assumed.)

Loop 1; $-9I_1 + 6I_2 + 21 = 0$
Loop 2; $+6I_1 - 18I_2 + 12I_3 + 40 = 0$
Loop 3; $+12I_2 - 24I_3 - 48 = 0$

$I_1 = 4.84$ A
$I_2 = 3.75$ A
$I_3 = -0.125$ A

Solution

Assume that I_1 flows counterclockwise in the closed loop comprising points A, B, C, and H. Similarly, assume that I_2 also flows counterclockwise in its loop made up of points, H, C, D, and G. Finally, in the third loop of points G, D, E, and F, assume that I_3 also flows counterclockwise. Solve this example using the equations for the loops, and the final results, all listed in Fig. 6-30.

For Similar Problems Refer to End-of-Chapter Problems 6-49 and 6-50

6-11. Nodal Analysis, Two- and Three-Voltage Sources

In Section 6-5, the *nodal analysis* method of solving a simple resistor and single-voltage-source circuit is discussed. Refer to this previous section, as a review, before proceeding with the present discussion, which goes into the more complex resistor networks using two and three applied voltage sources, and two and three "loops."

In the circuit of Fig. 6-31, three currents, I_{left}, I_{center}, and I_{right}, are all *assumed* to be flowing *into* point F as shown. Actually, this is not possible since Kirchhoff's current law states that the sum of currents flowing *into* a point must equal that which flows *away* from that point. Therefore, one or two of

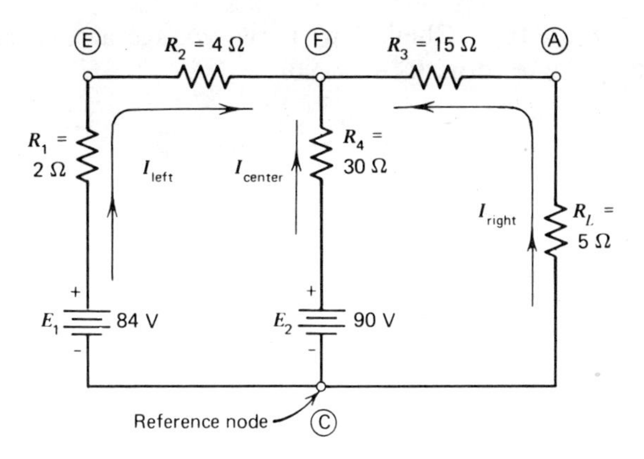

Figure 6-31. Nodal analysis. (Current directions are assumed.)

these currents are actually flowing in the opposite direction from that which is assumed. When solving for each of these currents, a negative value denotes that the assumed direction for that current is incorrect.

By assuming that all currents are entering point F, the sum of these currents must equal zero, or

$$I_{\text{left}} + I_{\text{center}} + I_{\text{right}} = 0 \tag{6-19}$$

The current shown in the diagram as I_{left} is simply, from Ohm's law, the voltage across resistors R_1 and R_2, divided by resistors R_1 and R_2, or

$$I_{\text{left}} = \frac{V_{R_1+R_2}}{R_1 + R_2} \tag{6-20}$$

Voltage across resistors R_1 and R_2 is simply the difference between the voltage at point F, V_F (with respect to the reference node, point C), and the voltage at the upper end of the applied voltage E_1 (also with respect to point C), or: $V_{R_1+R_2} = V_F - E_1$. Substituting for $V_{R_1+R_2}$ in equation 6-20 gives

$$I_{\text{left}} = \frac{V_F - E_1}{R_1 + R_2} \tag{6-21}$$

Similarly

$$I_{\text{center}} = \frac{V_{R_4}}{R_4} \tag{6-22}$$

Voltage across R_4 is the difference between the voltage at point F, V_F (with respect to point C), and the voltage at the upper end of the applied voltage E_2 (also with respect to point C). Therefore, $V_{R_4} = V_F - E_2$. Substituting for V_{R_4} in equation 6-22 yields

$$I_{\text{center}} = \frac{V_F - E_2}{R_4} \tag{6-23}$$

Current, I_{right}, from Ohm's law, is the voltage across resistors R_3 and R_L, divided by the resistors $R_3 + R_L$, or

$$I_{\text{right}} = \frac{V_{R_2+R_L}}{R_3 + R_L} \tag{6-24}$$

Voltage across these resistors, R_3 and R_L, is simply the voltage at point F (with respect to reference point C), or: $V_{R_3+R_L} = V_F$. Substituting for $V_{R_3+R_L}$ in equation 6-24 gives

$$I_{\text{right}} = \frac{V_F}{R_3 + R_L} \tag{6-25}$$

Finally, substituting in equation 6-19 for I_{left} (from equation 6-21), for I_{center} (from equation 6-23), and for I_{right} (from equation 6-25), gives the nodal analysis equation:

$$\frac{V_F - E_1}{R_1 + R_2} + \frac{V_F - E_2}{R_4} + \frac{V_F}{R_3 + R_L} = 0 \tag{6-26}$$

Replacing all terms with their values shown in Fig. 6-31, and solving for V_F yields

$$\frac{V_F - 84}{2 + 4} + \frac{V_F - 90}{30} + \frac{V_F}{15 + 5} = 0$$

$$\frac{V_F - 84}{6} + \frac{V_F - 90}{30} + \frac{V_F}{20} = 0$$

Multiplying all terms by 60 (common denominator) to clear all fractions, yields

$$10(V_F - 84) + 2(V_F - 90) + 3V_F = 0$$

$$10V_F - 840 + 2V_F - 180 + 3V_F = 0$$

$$15V_F = 840 + 180$$

$$15V_F = 1020$$

$$V_F = 68 \text{ V}$$

By knowing the value of voltage at point F ($V_F = 68$ V) with respect to point C, the values of the three currents in Fig. 6-31 may be found.

$$I_{\text{left}} = \frac{V_F - E_1}{R_1 + R_2} \quad \text{(from equation 6-21)}$$

$$I_{\text{left}} = \frac{68 - 84}{2 + 4}$$

$$I_{\text{left}} = \frac{-16}{6}$$

$$I_{\text{left}} = -2.667 \text{ A}$$

The *negative* sign of I_{left} indicates that this current is actually flowing in the reverse direction from that assumed and shown in Fig. 6-31. I_{left}, then, flows from point F to point E through R_2, and *down* through R_1.

$$I_{\text{center}} = \frac{V_F - E_2}{R_4} \qquad \text{(from equation 6-23)}$$

$$I_{\text{center}} = \frac{68 - 90}{30}$$

$$I_{\text{center}} = \frac{-22}{30}$$

$$I_{\text{center}} = -0.733 \text{ A}$$

The *negative* value for I_{center} indicates that this current, like I_{left}, actually flows in the opposite direction from that assumed and shown in Fig. 6-31. I_{center}, then, flows *down* through R_4.

$$I_{\text{right}} = \frac{V_F}{R_3 + R_L} \qquad \text{(from equation 6-25)}$$

$$I_{\text{right}} = \frac{68}{15 + 5}$$

$$I_{\text{right}} = \frac{68}{20}$$

$$I_{\text{right}} = 3.4 \text{ A}$$

This current, with its positive sign (no sign shown means it is positive), actually flows in the direction that was assumed and that is shown as *up* through R_L and from point A to point F through R_3.

Using the values and directions of the three currents in Fig. 6-31, voltages across the resistors and these polarities may now be found.

Voltage across R_L, between points A and C, $= (I_{\text{right}})(R_L) = (3.4)(5) = 17$ V, with point A positive with respect to point C. This agrees with the result using Thévenin for the same circuit used in Fig. 6-20, and also employing Norton in Fig. 6-24.

Voltage across R_1, with I_{left} flowing *down* through $R_1 = (I_{\text{left}})(R_1) = (2.667)(2) = 5.334$ V, negative at the upper end of R_1. Voltage between points E and C is made up of the 84 V E_1 and the opposing voltage V_{R_1} of 5.334 V. The algebraic sum is $+84 - 5.334$ or $+78.666$ V, positive at point E with respect to point C. This agrees with the result using Thévenin as depicted in Fig. 6-21*e*.

The following example employs the nodal analysis method in solving a resistor network containing three voltage sources. The circuit is identical to that used in Figs. 6-22 and 6-23 using Thévenin, and also employed in Norton in Figs. 6-26 and 6-27.

Example 6-14

In Fig. 6-32, using the nodal analysis method, find (a) the values and actual directions of the three currents shown, (b) voltage across R_5 and its polarity, and (c) voltage across R_9 and its polarity.

Solution

(a) Assuming that the three currents all flow into point Z as shown, then their sum is equal to zero, or

$$I_{\text{left}} + I_{\text{center}} + I_{\text{right}} = 0 \tag{6-27}$$

$I_{\text{left}} = V_{R_5+R_6}/(R_5 + R_6)$; but $V_{R_5+R_6} = V_Z - E_3$. Therefore

$$I_{\text{left}} = \frac{V_Z - E_3}{R_5 + R_6} \tag{6-28}$$

$I_{\text{center}} = V_{R_7}/R_7$; but $V_{R_7} = V_Z - E_2$. Therefore

$$I_{\text{center}} = \frac{V_Z - E_2}{R_7} \tag{6-29}$$

$I_{\text{right}} = V_{R_8+R_9+R_{10}}/(R_8 + R_9 + R_{10})$; but $V_{R_8+R_9+R_{10}} = V_Z - E_1$. Therefore

$$I_{\text{right}} = \frac{V_Z - E_1}{R_8 + R_9 + R_{10}} \tag{6-30}$$

The sum of the above currents (equations 6-28 to 6-30) is the nodal analysis equation:

$$\frac{V_Z - E_3}{R_5 + R_6} + \frac{V_Z - E_2}{R_7} + \frac{V_Z - E_1}{R_8 + R_9 + R_{10}} = 0 \tag{6-31}$$

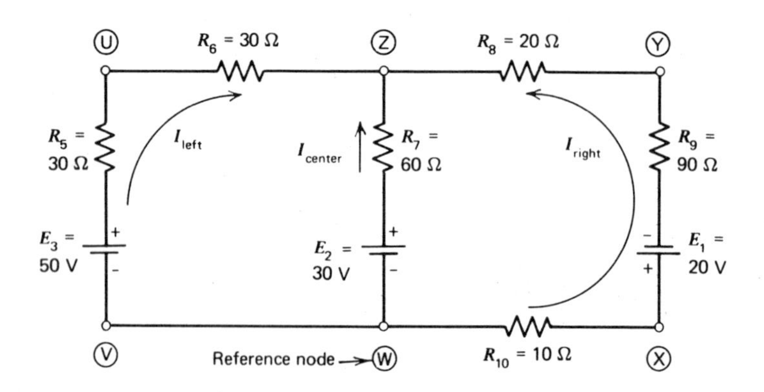

Figure 6-32. Example 6-14. (Current directions are assumed.)

Replacing all terms with their values shown in Fig. 6-32, solve for V_Z. (Note that E_2 and E_3 are *positive* at their upper ends, compared to point W, while E_1 is *negative* at its upper end.)

$$\frac{V_Z - 50}{30 + 30} + \frac{V_Z - 30}{60} + \frac{V_Z - (-20)}{20 + 90 + 10} = 0$$

$$\frac{V_Z - 50}{60} + \frac{V_Z - 30}{60} + \frac{V_Z + 20}{120} = 0$$

Multiplying all terms by 120 (common denominator) to clear fractions, gives

$$2(V_Z - 50) + 2(V_Z - 30) + V_Z + 20 = 0$$

$$2V_Z - 100 + 2V_Z - 60 + V_Z + 20 = 0$$

$$5V_Z = 100 + 60 - 20$$

$$5V_Z = 140$$

$$V_Z = 28$$

Solving for each current, by substituting for V_Z in equations 6-28, 6-29, and 6-30 gives

$$I_{\text{left}} = \frac{V_Z - E_3}{R_5 + R_6} \qquad \text{(from equation 6-28)}$$

$$I_{\text{left}} = \frac{28 - 50}{30 + 30}$$

$$I_{\text{left}} = \frac{-22}{60}$$

$$I_{\text{left}} = -0.367\ \text{A}$$

The *negative* sign for I_{left} indicates that this current actually flows in the reverse direction from that which is assumed and is shown in Fig. 6-32. I_{left}, therefore, flows from point Z through R_6 to point U, and *down* through R_5.

$$I_{\text{center}} = \frac{V_Z - E_2}{R_7} \qquad \text{(from equation 6-29)}$$

$$I_{\text{center}} = \frac{28 - 30}{60}$$

$$I_{\text{center}} = \frac{-2}{60}$$

$$I_{\text{center}} = -0.0333\ \text{A}$$

Again, the *negative* value for I_{center} indicates that this current also flows in the reverse direction from that which is assumed and is shown in Fig. 6-32.

Therefore, I_{center} actually flows *down* through R_7.

$$I_{right} = \frac{V_Z - E_1}{R_8 + R_9 + R_{10}} \qquad \text{(from equation 6-30)}$$

$$I_{right} = \frac{28 - (-20)}{20 + 90 + 10}$$

$$I_{right} = \frac{28 + 20}{120}$$

$$I_{right} = \frac{48}{120}$$

$$I_{right} = 0.4 \text{ A}$$

Since I_{right} has a positive value, its assumed direction is correct. This current, then, flows as depicted in Fig. 6-32.

(b) Voltage across R_5 is: $V_{R_5} = (I_{left})(R_5) = (0.367)(30) = 11.01$ V. Since I_{left} flows *down* through R_5, the upper end of R_5 is its negative end. This value and polarity of V_{R_5} agrees with the results found using Thévenin (Fig. 6-23) and Norton (Fig. 6-27).

(c) Voltage across R_9 is: $V_{R_9} = (I_{right})(R_9) = (0.4)(90) = 36$ V. Since I_{right} flows *up* through R_9, then the upper end of R_9 is its positive end. This value and polarity of V_{R_9} agrees with the results found using Thévenin (Fig. 6-22) and Norton (Fig. 6-23).

For Similar Problems Refer to End-of-Chapter Problems 6-51 and 6-52

Nodal Analysis of Resistor Network Containing "Three Loops." The following discussion of Fig. 6-33 employs the nodal analysis method for solving a resistor network made up of three "loops" and four applied voltage sources.

Example 6-15

In the circuit of Fig. 6-33, find the values and directions of all currents, and the voltages and polarities V_{AB}, V_{EB}, V_{DB}, and V_{CB} using nodal analysis.

Solution

Assume that the three currents $I_{5,9}$, I_6, and I_{DE} each flow into point E, an independent node. Also assume that the other independent node, point D, has three currents flowing into it. These currents, as shown, are I_{ED}, I_7 and $I_{8,11}$.

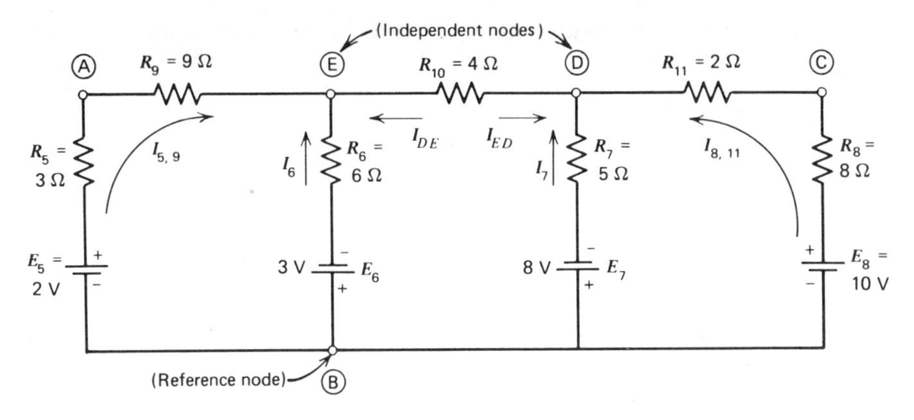

Figure 6-33. Nodal analysis, Example 6-15. (All current directions are assumed.)

Solving for the currents flowing into independent node point E gives

$$I_{5,9} + I_6 + I_{DE} = 0 \tag{6-32}$$

Current through R_5 and R_9, $I_{5,9}$ in equation 6-32 is the voltage across these resistors divided by the value of these resistors. Voltage across these resistors is the difference between the voltage at point E with respect to the reference node point B, and the voltage at the upper end of E_5, also with respect to point B. Therefore,

$$I_{5,9} = \frac{V_E - E_5}{R_5 + R_9} \tag{6-33}$$

Current through R_6, I_6 in equation 6-32, is the voltage across R_6 divided by this resistor. Voltage across R_6 is the differences between the voltage at point E (with respect to point B) and the voltage at the upper end of E_6 (also with respect to point B). Therefore

$$I_6 = \frac{V_E - E_6}{R_6} \tag{6-34}$$

Current through R_{10} from point D to point E, I_{DE} in equation 6-32, is the voltage across R_{10} divided by the R_{10} resistor. Voltage across R_{10} is the difference between the voltage at point E (with respect to reference point B) and the voltage at point D (also with respect to point B). Therefore

$$I_{DE} = \frac{V_E - V_D}{R_{10}} \tag{6-35}$$

Substituting in equation 6-32 for the currents (equations 6-33, 6-34, and 6-35) gives

$$\frac{V_E - E_5}{R_5 + R_9} + \frac{V_E - E_6}{R_6} + \frac{V_E - V_D}{R_{10}} = 0 \tag{6-36}$$

Replacing those terms whose values are given in Fig. 6-33 gives

$$\frac{V_E - 2}{3 + 9} + \frac{V_E - (-3)}{6} + \frac{V_E - V_D}{4} = 0$$

Solving yields

$$\frac{V_E - 2}{12} + \frac{V_E + 3}{6} + \frac{V_E - V_D}{4} = 0$$

Multiplying by 12 (common denominator) to clear all fractions yields

$$V_E - 2 + 2(V_E + 3) + 3(V_E - V_D) = 0$$

$$V_E - 2 + 2V_E + 6 + 3V_E - 3V_D = 0$$

$$6V_E - 3V_D + 4 = 0$$

$$6V_E - 3V_D = -4 \tag{6-37}$$

This equation will be solved by combining it with equation 6-43, derived in the following:

Solving for the currents flowing into the other independent node point D gives

$$I_{ED} + I_7 + I_{8,11} = 0 \tag{6-38}$$

Current from point E to point D, I_{ED} in equation 6-38, is the difference in voltage between these points divided by the resistance (R_{10}) between points E and D. Therefore

$$I_{ED} = \frac{V_D - V_E}{R_{10}} \tag{6-39}$$

Current through R_7, I_7 in equation 6-38 is the difference between the voltage at point D (with respect to reference node point B) and the voltage at the upper end of E_7 (also with respect to point B), divided by resistor R_7. Therefore

$$I_7 = \frac{V_D - E_7}{R_7} \tag{6-40}$$

Current through resistors R_8 and R_{11}, $I_{8,11}$ in equation 6-38, is the difference in voltage between point D (with respect to point B) and the voltage at the upper end of E_8 (also with respect to point B), divided by the resistors R_8 and R_{11}. Therefore

$$I_{8,11} = \frac{V_D - E_8}{R_8 + R_{11}} \tag{6-41}$$

Substituting in equation 6-38 for the currents (equations 6-39, 6-40, and 6-41) gives

$$\frac{V_D - V_E}{R_{10}} + \frac{V_D - E_7}{R_7} + \frac{V_D - E_8}{R_8 + R_{11}} = 0 \tag{6-42}$$

Replacing those terms whose values are given in Fig. 6-33 gives

$$\frac{V_D - V_E}{4} + \frac{V_D - (-8)}{5} + \frac{V_D - 10}{8 + 2} = 0$$

Solving yields

$$\frac{V_D - V_E}{4} + \frac{V_D + 8}{5} + \frac{V_D - 10}{10} = 0$$

Multiplying by 20 (common denominator) to clear all fractions yields

$$5(V_D - V_E) + 4(V_D + 8) + 2(V_D - 10) = 0$$

$$5V_D - 5V_E + 4V_D + 32 + 2V_D - 20 = 0$$

$$-5V_E + 11V_D + 12 = 0$$

$$-5V_E + 11V_D = -12 \tag{6-43}$$

Equations 6-37 and 6-43 are *simultaneous* equations and are solved using the determinant method as shown in the following:

$$6V_E - 3V_D = -4 \tag{6-37}$$

$$-5V_E + 11V_D = -12 \tag{6-43}$$

Solving for V_E using determinants gives

$$V_E = \frac{\text{numerator}}{\text{denominator}}$$

In the *denominator*, list in columns the coefficients for V_E and V_D as they appear in equations 6-37 and 6-43 (6 and -3, and below this, -5 and 11)

$$V_E = \frac{\text{numerator}}{\begin{vmatrix} 6 & -3 \\ -5 & 11 \end{vmatrix}}$$

Now, from the diagonal product of 6 and 11, *subtract* the other diagonal product of -5 and -3, giving

$$V_E = \frac{\text{numerator}}{(6)(11) - (-5)(-3)}$$

$$V_E = \frac{\text{numerator}}{66 - (15)}$$

$$V_E = \frac{\text{numerator}}{51}$$

In the *numerator* for V_E replace the V_E coefficients with the terms at the right side of the equations (-4 and -12) but again listing the V_D coefficients (-3 and 11), giving

$$V_E = \frac{\begin{vmatrix} -4 & -3 \\ -12 & 11 \end{vmatrix}}{51}$$

Again, from the first diagonal product of -4 and 11, subtract the second diagonal product of -12 and -3, yielding

$$V_E = \frac{(-4)(11) - (-12)(-3)}{51}$$

$$V_E = \frac{-44 - (36)}{51}$$

$$V_E = \frac{-80}{51}$$

$$V_E = -1.57 \text{ V}$$

Now, V_D may be similarly found. The *denominator* in the determinant method is the same for V_E and V_D.

$$V_D = \frac{\text{numerator}}{\text{same denominator as } V_E}$$

$$V_D = \frac{\text{numerator}}{51}$$

In the V_D *numerator* list the V_E coefficients (6 and -5) but replace the V_D coefficients with the terms from the right side of equations 6-37 and 6-43 (-4 and -12), giving

$$V_D = \frac{\begin{vmatrix} 6 & -4 \\ -5 & -12 \end{vmatrix}}{51}$$

Again, from the first diagonal product of 6 and -12, subtract the other diagonal product of -5 and -4, giving

$$V_D = \frac{(6)(-12) - (-5)(-4)}{51}$$

$$V_D = \frac{-72 - (20)}{51}$$

$$V_D = \frac{-92}{51}$$

$$V_D = -1.805 \text{ V}$$

From the values of V_E (-1.57 V), and V_D (-1.805 V), the various currents in Fig. 6-33 can now be found, using equations 6-33, 6-34, 6-35, 6-39, 6-40, and 6-41.

$I_{5,9}$ is:

$$I_{5,9} = \frac{V_E - E_5}{R_5 + R_9} \tag{6-33}$$

$$I_{5,9} = \frac{-1.57 - 2}{3 + 9}$$

$$I_{5,9} = \frac{-3.57}{12}$$

$$I_{5,9} = -0.297 \text{ A}$$

Note that the negative value of $I_{5,9}$ denotes that its actual direction is the opposite from the assumed direction shown in Fig. 6-33. Therefore, $I_{5,9}$ actually flows from point E toward point A, and *down* through R_5.

I_6 is:

$$I_6 = \frac{V_E - E_6}{R_6} \tag{6-34}$$

$$I_6 = \frac{-1.57 - (-3)}{6}$$

$$I_6 = \frac{-1.57 + 3}{6}$$

$$I_6 = \frac{1.43}{6}$$

$$I_6 = 0.238 \text{ A}$$

I_{DE} is:

$$I_{DE} = \frac{V_E - V_D}{R_{10}} \tag{6-35}$$

$$I_{DE} = \frac{-1.57 - (-1.805)}{4}$$

$$I_{DE} = \frac{-1.57 + 1.805}{4}$$

$$I_{DE} = \frac{0.235}{4}$$

$$I_{DE} = 0.05875 \text{ A}$$

I_{ED} is:

$$I_{ED} = \frac{V_D - V_E}{R_{10}} \tag{6-39}$$

$$I_{ED} = \frac{-1.805 - (-1.57)}{4}$$

$$I_{ED} = \frac{-1.805 + 1.57}{4}$$

$$I_{ED} = \frac{-0.235}{4}$$

$$I_{ED} = -0.05875 \text{ A}$$

The *negative* value for I_{ED} denotes that its actual direction is opposite to the assumed direction shown in Fig. 6-33. Note that I_{ED} actually flows from point D to point E, which agrees with that found in the previous equation 6-35 for I_{DE}.

I_7 is:

$$I_7 = \frac{V_D - E_7}{R_7} \tag{6-40}$$

$$I_7 = \frac{-1.805 - (-8)}{5}$$

$$I_7 = \frac{-1.805 + 8}{5}$$

$$I_7 = \frac{6.195}{5}$$

$$I_7 = 1.24 \text{ A}$$

$I_{8,11}$ is:

$$I_{8,11} = \frac{V_D - E_8}{R_8 + R_{11}} \tag{6-41}$$

$$I_{8,11} = \frac{-1.805 - 10}{8 + 2}$$

$$I_{8,11} = \frac{-11.805}{10}$$

$$I_{8,11} = -1.18 \text{ A}$$

The *negative* value of $I_{8,11}$ denotes that its actual direction is the reverse of the assumed direction shown. Therefore, $I_{8,11}$ actually flows from point D to point C, and *down* through R_8.

The voltages to be found, V_{AB}, V_{EB}, V_{DB}, and V_{CB} are: V_{AB}, the voltage at point A with respect to point B is the algebraic sum of V_{R_5} and the applied voltage E_5. $V_{R_5} = (I_{5,9})(R_5) = (0.297)(3) = 0.89$ V. As pointed out previously, the actual direction of $I_{5,9}$ is *down* through R_5, making the top of R_5 its negative end. Therefore, the polarity of V_{R_5} opposes that of the voltage source E_5. Voltage at point A (with respect to point B), $V_{AB} = V_{R_5} + E_5 = -0.89 + 2 = +1.11$ V.

Voltage at point E (with respect to point B), V_{EB}, or as it has been called previously V_E, is -1.57 V.

Voltage at point D (with respect to point B), V_{DB}, or as it has been called previously V_D, is -1.805 V.

Voltage at point C (with respect to point B), V_{CB} is the algebraic sum of V_{R_8} and the applied voltage E_8. $V_{R_8} = (I_{8,11})(R_8) = (1.18)(8) = 9.44$ V. As shown previously, the actual direction of $I_{8,11}$ is down through R_8, making the upper end of R_8 its negative end. The polarity of V_{R_8} opposes that of the applied voltage source E_8. Therefore, voltage at point C (with respect to point B), $V_{CB} = V_{R_8} + E_8 = -9.44 + 10 = +0.56$ V.

For Similar Problems Refer to End-of-Chapter Problem 6-53

6-12. Millman's Theorem, Two- and Three-Voltage Sources

Another method employed in solving a resistor network or a mesh circuit is called *Millman's theorem*. This is basically an adaptation of Norton's theorem (Section 6-9). In the simple circuit of Fig. 6-34, to find the voltage between point Y and X using Millman's theorem it is first necessary to find the current that would flow through a *short* placed between these points. This is the same as the Norton short-circuit current. This current is due, in part, to E_3 and R_3 and flows *up* through the short circuit from point X to point Y, and is also due to E_4 and R_4, and flows *down* through the short from point Y to point X. The current from E_3 flows *up* through the short toward point Y, making point Y *positive* with respect to point X. As a result, this current is given a *positive* sign. Current from E_4 flows *down* through the short toward point X, making point Y *negative* with respect to point X. Consequently, this current (from E_4)

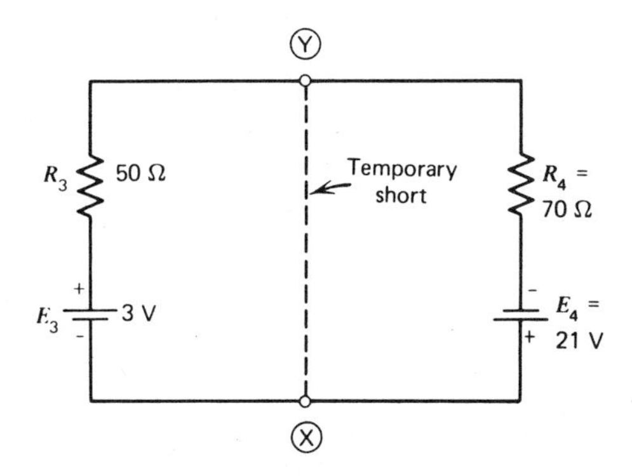

Figure 6-34. Millman's theorem, two voltage sources.

is given a *negative* sign. The currents through the short between X and Y are expressed in terms of Ohm's law as

$$\frac{E_3}{R_3} - \frac{E_4}{R_4}$$

the first term (E_3/R_3) is the current flowing *up* through the short, while the second term ($-E_4/R_4$) is the current flowing *down* through the short.

Removing the short between X and Y, and replacing the voltage sources E_3 and E_4 with their internal resistances (assumed to be zero here), the resistance between point X and Y consists of R_3 and R_4 in parallel. This is the same as R_{Norton} and also $R_{\text{Thévenin}}$.

Voltage between points Y and X is simply the product of the short-circuit current and the parallel resistors as shown in the following:

$$V_{YX} = (I_{\text{short-circuit}})(R_{\text{parallel}})$$

$$V_{YX} = \left(\frac{E_3}{R_3} - \frac{E_4}{R_4}\right)\left(\frac{1}{\frac{1}{R_3} + \frac{1}{R_4}}\right)$$

$$V_{YX} = \frac{\frac{E_3}{R_3} - \frac{E_4}{R_4}}{\frac{1}{R_3} + \frac{1}{R_4}} \tag{6-44}$$

Equation 6-44 is the Millman theorem formula. Using it with the values of Fig. 6-34 gives

$$V_{YX} = \frac{\frac{3}{50} - \frac{21}{70}}{\frac{1}{50} + \frac{1}{70}}$$

$$V_{YX} = \frac{\frac{21 - 105}{350}}{\frac{7 + 5}{350}}$$

$$V_{YX} = \frac{\frac{-84}{350}}{\frac{12}{350}}$$

$$V_{YX} = \left(\frac{-84}{350}\right)\left(\frac{350}{12}\right)$$

$$V_{YX} = -7 \text{ V}$$

The negative sign for V_{YX} means that point Y is negative 7 V with respect to point X. This agrees with the results of the previous identical circuit (Fig. 6-13), which was solved using loop currents.

The following example uses Millman's theorem in solving another two-voltage-source circuit.

Example 6-16

Find the voltage and its polarity between points F and C in Fig. 6-35.

Solution

Temporarily placing a short between points F and C produces currents flowing up through the short due to E_1 and E_2. These currents are: $E_1/(R_1 + R_2)$ and E_2/R_4. This is the same as the Norton current. Then, removing this short and replacing the voltage sources E_1 and E_2 by their internal resistances (assumed to be zero in this example), the resistance between points F and C consists of the three parallel branches. These are: $R_1 + R_2$, R_4, and $R_3 + R_5$. Voltage between points F and C, then, from equation 6-44 is

$$V_{FC} = \frac{\dfrac{E_1}{R_1 + R_2} + \dfrac{E_2}{R_4}}{\dfrac{1}{R_1 + R_2} + \dfrac{1}{R_4} + \dfrac{1}{R_3 + R_5}}$$

$$V_{FC} = \frac{\dfrac{84}{2 + 4} + \dfrac{90}{30}}{\dfrac{1}{2 + 4} + \dfrac{1}{30} + \dfrac{1}{15 + 5}}$$

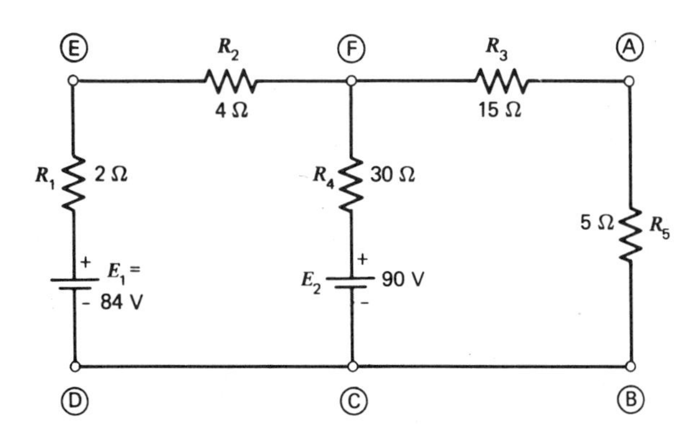

Figure 6-35. Millman's theorem, Example 6-16.

$$V_{FC} = \frac{14 + 3}{\frac{1}{6} + \frac{1}{30} + \frac{1}{20}}$$

$$V_{FC} = \frac{17}{\frac{10 + 2 + 3}{60}}$$

$$V_{FC} = \frac{17}{\frac{15}{60}}$$

$$V_{FC} = (17)\left(\frac{60}{15}\right)$$

$$V_{FC} = 68 \text{ V}$$

The 68 V is *positive* (since there is no negative sign), meaning that point F is positive with respect to point C. This agrees with the conclusion reached using loop currents in Section 6-10 for the identical circuit (Fig. 6-28).

For Similar Problems Refer to End-of-Chapter Problems 6-54 to 6-57

Three Voltage-Sources Using Millman. The diagram of Fig. 6-36 is a resistor network containing three sources of voltage. Note that the polarity of E_1 (20 V) is *negative* at its upper terminal, while the upper ends of E_2 (30 V) and E_3 (50 V) are *positive.* To find the voltage and polarity between points Z and W, it is first necessary to place a temporary short between these points, and to determine the currents and their directions through the short. Since the current due to E_1 flows *down* through the short from Z to W, then point Z is *negative* with respect to point W. Therefore, this current is to be

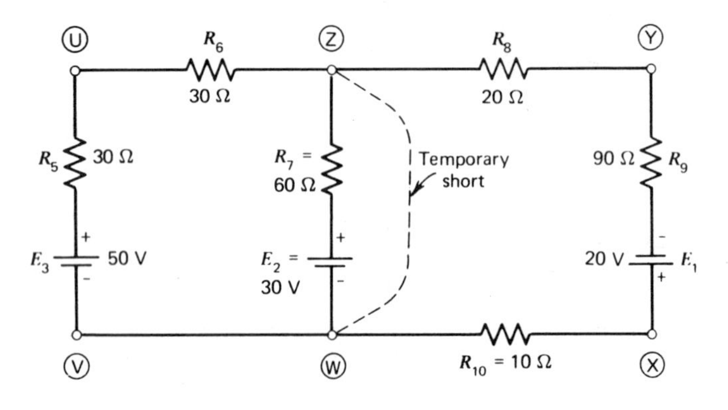

Figure 6-36. Millman's theorem, three voltage sources.

called *negative*. This current then is: $-E_1/(R_8 + R_9 + R_{10})$. The other two currents due to E_2 and E_3 each flow *up* through the short from W to Z. Therefore, point Z is *positive* with respect to point W, and these currents are termed *positive*. These currents are: E_2/R_7, and $E_3/(R_5 + R_6)$.

Removing the short, and replacing each voltage source with its internal resistance (none are shown, so these are assumed to be zero here), the resistance between points Z and W consists of three parallel branches. These are: $R_5 + R_6$ (the left branch), R_7 (the middle branch), and $R_8 + R_9 + R_{10}$ (the right branch).

Voltage between points Z and W, using a variation of equation 6-44, is then

$$V_{ZW} = \frac{\dfrac{-E_1}{R_8 + R_9 + R_{10}} + \dfrac{E_2}{R_7} + \dfrac{E_3}{R_5 + R_6}}{\dfrac{1}{R_8 + R_9 + R_{10}} + \dfrac{1}{R_7} + \dfrac{1}{R_5 + R_6}}$$

$$= \frac{\dfrac{-20}{20 + 90 + 10} + \dfrac{30}{60} + \dfrac{50}{30 + 30}}{\dfrac{1}{20 + 90 + 10} + \dfrac{1}{60} + \dfrac{1}{30 + 30}}$$

$$= \frac{\dfrac{-20}{120} + \dfrac{30}{60} + \dfrac{50}{60}}{\dfrac{1}{120} + \dfrac{1}{60} + \dfrac{1}{60}}$$

$$= \frac{\dfrac{-20 + 60 + 100}{120}}{\dfrac{1 + 2 + 2}{120}}$$

$$= \frac{\dfrac{140}{120}}{\dfrac{5}{120}}$$

$$= \left(\frac{140}{120}\right)\left(\frac{120}{5}\right)$$

$$= 28 \text{ V}$$

This 28 V, V_{ZW}, is positive (since there is no negative sign), meaning that point Z is positive with respect to point W. This agrees with one of the results of example 6-14, for the identical circuit (Fig. 6-32), using the nodal analysis method.

For Similar Problems Refer to End-of-Chapter Problems 6-58 to 6-61

6-13. Delta-to-Wye Transformation

Often, a complex resistor network may be reduced to a simpler series parallel circuit, and then solved more easily. This is possible when a portion of the network contains one or more *delta* (Δ) or *pye* (π) circuits. As shown in Fig. 6-37*a*, a pye (π) circuit simply has the resistors drawn in the shape of the Greek letter π. That is, the resistors in the diagram form the sides of this Greek letter. The identical circuit is shown in Fig. 6-37*b* in the shape of the Greek letter Δ (delta), by simply tilting R_1 and R_3 so that the three resistors now form a triangle or Δ. Fig 6-37*a* and *b* are therefore called π or Δ circuits, and are identical to each other. These circuits may be converted or transformed into an equivalent circuit called a tee (T) or wye (Y). The *T* and *Y*, which are equivalent to each other are shown in Fig. 6-38*a* and *b*. Note that the resistors R_x, R_y, and R_z form the shape of these letters.

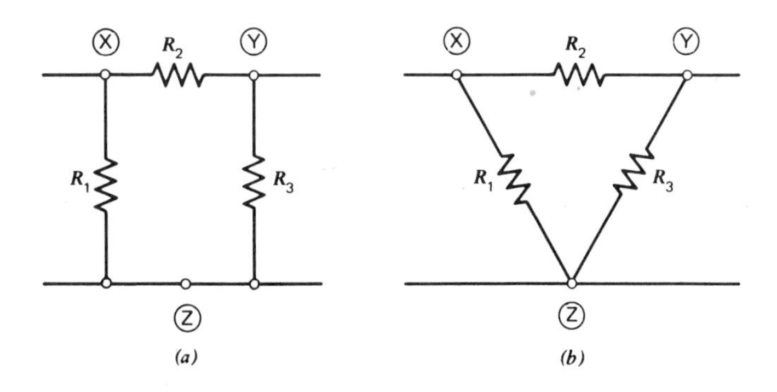

Figure 6-37. Pye or delta circuits are the same. (*a*) Pye (π) circuit. (*b*) Delta (Δ) circuit.

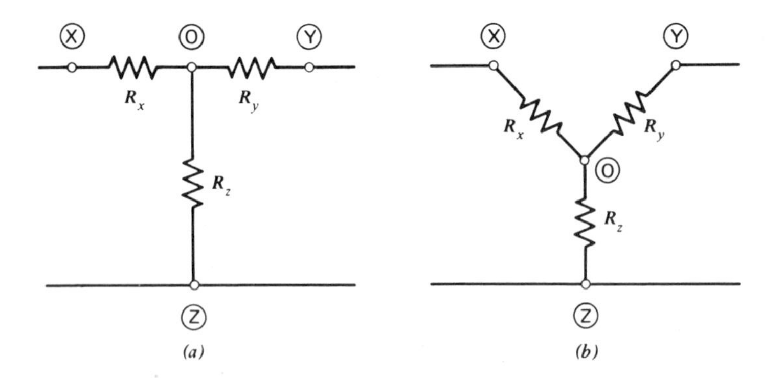

Figure 6-38. Tee or wye circuits are the same (*a*) T (tee) circuit. (*b*) Y (wye) circuit.

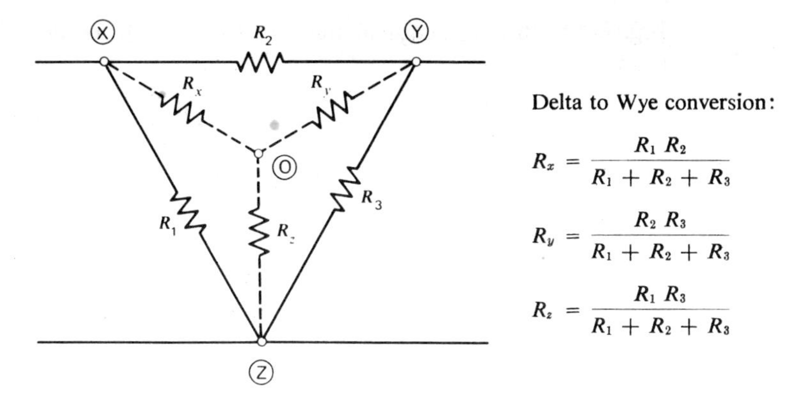

Figure 6-39. Delta-to-wye transformation.

As will be shown, it is often advantageous to convert a *delta* circuit into an equivalent '*wye* to more easily solve a resistor network. Figure 6-39 shows resistors R_1, R_2, and R_3 forming a Δ circuit. The equivalent Y circuit (which would replace the Δ) consists of the dotted-line portion, resistors R_x, R_y, and R_z. The delta-to-wye conversion equations for resistors R_x, R_y, and R_z in terms of the known values of R_1, R_2, and R_3 are shown in Fig. 6-39, and are repeated here:

$$R_x = \frac{R_1 R_2}{R_1 + R_2 + R_3} \tag{6-45}$$

$$R_y = \frac{R_2 R_3}{R_1 + R_2 + R_3} \tag{6-46}$$

$$R_z = \frac{R_1 R_3}{R_1 + R_2 + R_3} \tag{6-47}$$

Note that in the equations each Y resistor is equal to the *product* of the two Δ resistors to which the Y resistor is "connected," divided by the *sum* of all the Δ resistors. As an example, note that in Fig. 6-39, Y resistor R_x is shown connected to the junction of Δ resistors R_1 and R_2. As a result, R_x is equal to the *product* of R_1 and R_2 divided by the *sum* of all Δ resistors $R_1 + R_2 + R_3$. Similarly, R_y connects to the junction of R_2 and R_3. Therefore, R_y is equal to the *product* of R_2 and R_3 *divided* by the sum of all Δ resistors $R_1 + R_2 + R_3$.

Actually, in Fig. 6-39, Y resistor R_x is not connected to Δ resistors R_1 and R_2. R_x simply connects to point X, *replacing* resistors R_1 and R_2 in the conversion process. That is, Y resistors R_x, R_y, and R_z *replace* the original Δ resistors R_1, R_2, and R_3.

Now, equations 6-45 to 6-47 will be employed to convert the Δ circuit of

Fig. 6-40a into the equivalent Y circuit of Fig. 6-40b. The Y resistors are found as follows:

$$R_x = \frac{R_1 R_2}{R_1 + R_2 + R_3} \tag{6-45}$$

$$= \frac{(20)(30)}{20 + 30 + 50}$$

$$= \frac{600}{100}$$

$$= 6\ \Omega \qquad \text{as shown in Fig. 6-40}b$$

$$R_y = \frac{(R_2)(R_3)}{R_1 + R_2 + R_3} \tag{6-46}$$

$$= \frac{(30)(50)}{20 + 30 + 50}$$

$$= \frac{1500}{100}$$

$$= 15\ \Omega \qquad \text{as shown in Fig. 6-40}b$$

$$R_z = \frac{(R_1)(R_3)}{R_1 + R_2 + R_3} \tag{6-47}$$

$$= \frac{(20)(50)}{20 + 30 + 50}$$

$$= \frac{1000}{100}$$

$$= 10\ \Omega \qquad \text{as shown in Fig. 6-40}b$$

These values of the Y resistors R_x, R_y, and R_z.mean that the Y circuit replaces the original Δ resistors R_1, R_2, and R_3.

In the resistor network of Fig. 6-41a it can be seen that there are two delta circuits, an upper one (WXY), and a lower one (XYZ). If it is desired to find the total resistance that the network offers to the applied voltage source, it can be seen that the circuit cannot be solved by the ordinary series-parallel resistor method. The network of Fig. 6-41a is not a series-parallel type, since it cannot be determined which resistors are in series and which are in parallel.

However, if *either* delta circuit is replaced by its wye equivalent, then the network circuit can be reduced to a simpler type. In this discussion, the lower delta (XYZ), of Fig. 6-41a is converted to its wye equivalent (R_x, R_y, R_z) as shown in part (b) of this figure. Note that the XYZ delta of Fig. 6-41a is

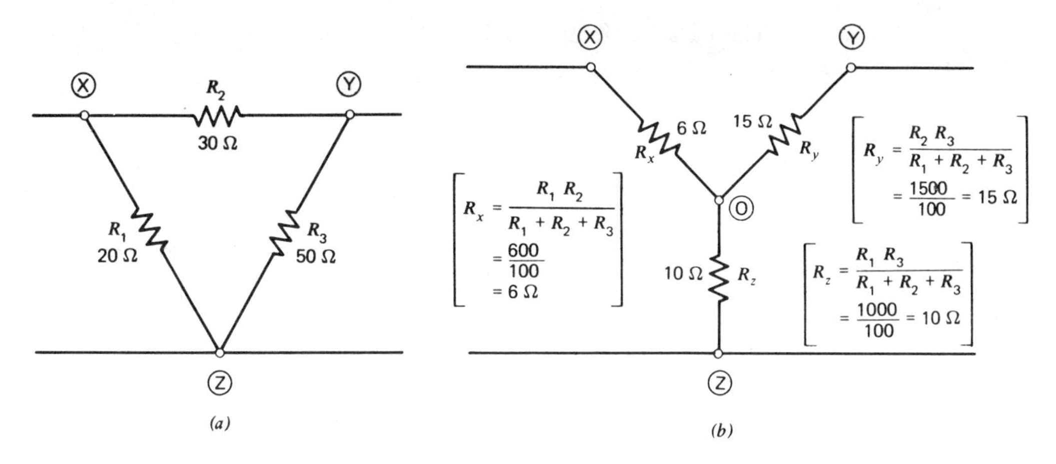

Figure 6-40. (*a*) Delta circuit. (*b*) Equivalent wye circuit.

identical to that of Fig. 6-40*a*. The wye equivalent was found previously and is shown in Fig. 6-40*b*. This is also the wye circuit (R_x, R_y, and R_z) of Fig. 6-41*b*. By inserting R_x, R_y, and R_z in place of R_1, R_2, and R_3, then the circuit of Fig. 6-41*b* should now be recognized as a series-parallel one. R_z is the series resistor, while the parallel circuit is between points O and W. This parallel circuit is comprised of a left branch consisting of $R_x + R_4$, and a right branch consisting of $R_y + R_5$.

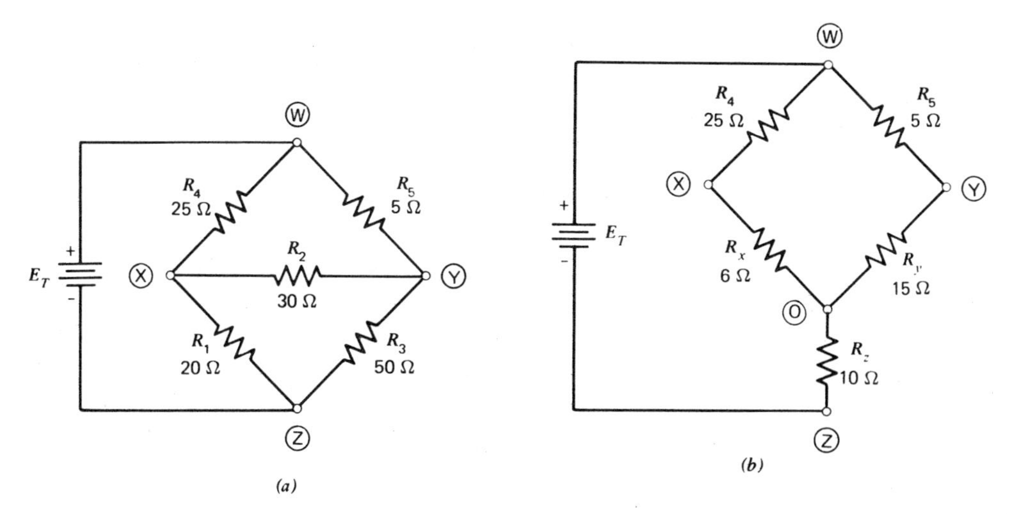

Figure 6-41. Converting a resistor network into a simpler series-parallel circuit by replacing the *XYZ* delta by the *XYZ* wye. (a) Resistor network comprising two Δ circuits (*XYZ* and *WXY*). (*b*) Equivalent series-parallel circuit.

The parallel circuit R_{ow} is then

$$R_{ow} = \frac{(R_x + R_4)(R_y + R_5)}{R_x + R_4 + R_y + R_5}$$

$$R_{ow} = \frac{(6 + 25)(15 + 5)}{6 + 25 + 15 + 5}$$

$$R_{ow} = \frac{(31)(20)}{51}$$

$$R_{ow} = \frac{620}{51}$$

$$R_{ow} = 12.16 \ \Omega$$

Total resistance R_T is

$$\begin{aligned} R_T &= R_z + R_{ow} \\ &= 10 + 12.16 \\ &= 22.16 \ \Omega \end{aligned}$$

The total resistance of the network of Fig. 6-41*a* could also have been found by replacing the *upper* delta WXY with its wye equivalent (Fig. 6-42) instead of converting the *lower* delta XYZ. Perform these steps using the results shown in Fig. 6-42 as a guide. Then find R_T of this series-parallel circuit. It should agree, within slide-rule accuracy, with the $R_T = 22.16 \ \Omega$ found previously.

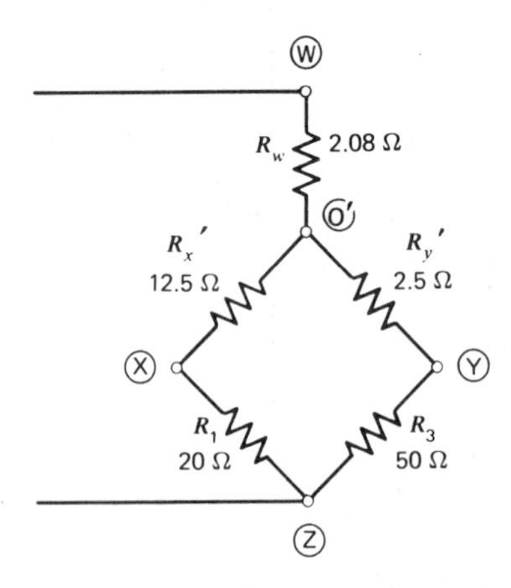

Figure 6-42. Converting the resistor network of Fig. 6-41a into a simpler series-parallel circuit by replacing the *WXY* delta by its equivalent *WXY* wye.

These examples follow the preceding discussion and employ the delta-to-wye transformation equations 6-45 to 6-47:

Example 6-17

Convert the *pye* or *delta* circuit of Fig. 6-43*a* to its equivalent *tee* or *wye* equivalent.

Solution

Note that the three "legs" of the Δ or π circuit of Fig. 6-43*a* are 40 Ω (between points A and C), 30 Ω (between points B and C), and 130 Ω (made up of the 80 + 50 Ω between points A and B). The three "legs" of the equivalent T or Y are shown in Fig. 6-43*b*, and their values are found in the following.

Resistor R_A connects to point A which, in Fig. 6-43*a*, is the junction of the 40- and 130-Ω "legs." Therefore, R_A is

$$R_A = \frac{(40)(80 + 50)}{40 + 80 + 50 + 30} \qquad \text{(from equation 6-45 to 6-47)}$$

$$= \frac{(40)(130)}{200}$$

$$= \frac{5200}{200}$$

$$= 26\ \Omega$$

shown in Fig. 6-43*b*.

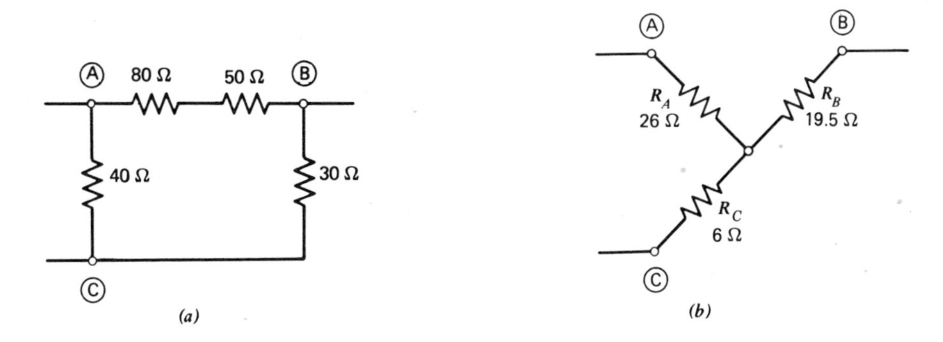

Figure 6-43. Example 6 17. (*a*) π or Δ circuit. (*b*) T or Y equivalent.

Resistor R_B connects to point B which, in Fig. 6-43*a*, is the junction of the 30- and 130-Ω "legs." Therefore, R_B is

$$R_B = \frac{(30)(80 + 50)}{40 + 80 + 50 + 30} \qquad \text{(from equation 6-45 to 6-47)}$$

$$= \frac{(30)(130)}{200}$$

$$= \frac{3900}{200}$$

$$= 19.5\ \Omega$$

shown in Fig. 6-43*b*.

Finally, resistor R_C connects to point C which, in Fig. 6-43*a*, is the junction of the 40- and 30-Ω "legs." Therefore R_C is

$$R_C = \frac{(40)(30)}{40 + 80 + 50 + 30} \qquad \text{(from equation 6-46 to 6-47)}$$

$$= \frac{1200}{200}$$

$$= 6\ \Omega$$

shown in Fig. 6-43*b*.

Example 6-18

Find the total resistance, R_T, between points A and D for the circuit of Fig. 6-44.

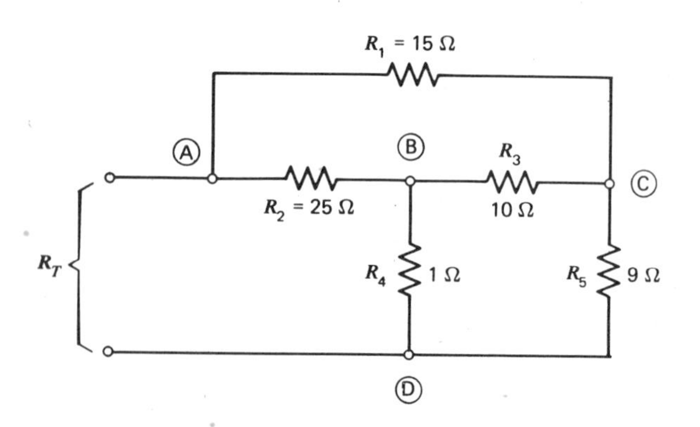

Figure 6-44. Circuit for Examples 6-18 and 6-19.

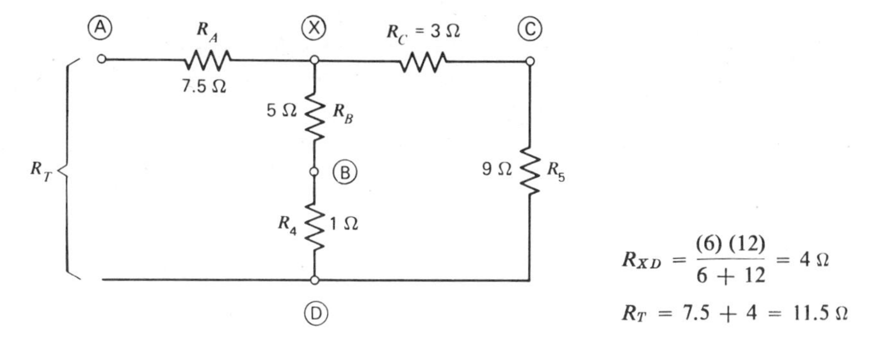

Figure 6-45. Solution for Example 6-18.

Solution

This resistor network contains two delta or pye circuits. One consists of resistors R_1, R_2, and R_3, between points A, B, and C. The other comprises R_3, R_4, and R_5, between points B, C, and D. The solution here will be to change the *ABC delta* into its *tee* equivalent, shown in Fig. 6-45. (In the next example, R_T will again be found, but by transforming, the *BCD* delta or pye.) Solve for the *tee* equivalent and R_T, comparing the answers with those shown in Fig. 6-45.

Example 6-19

As in the previous example, again find the total resistance R_T between points A and D of Fig. 6-44, but this time convert the *BCD delta* or *pye* into its *tee* or *wye* equivalent.

Solution

The *BCD* pye circuit made up of R_3, R_4, and R_5 can be converted into its equivalent tee (shown in Fig. 6-46). Solve for the tee equivalent and R_T, comparing the results with those shown in Fig. 6-46.

Note that this agrees with the result of the previous Example 6-18.

For Similar Problems Refer to End-of-Chapter Problems 6-62 to 6-67

6-14. Wye-to-Delta Transformation

In the previous Section 6-13, conversion from a delta (or pye) circuit to a wye (or Tee) is discussed. In many resistor networks it is sometimes useful to do just the reverse. That is, to change a wye configuration into its equivalent delta circuit. Figure 6-47 shows a wye circuit consisting of resistors R_x, R_y,

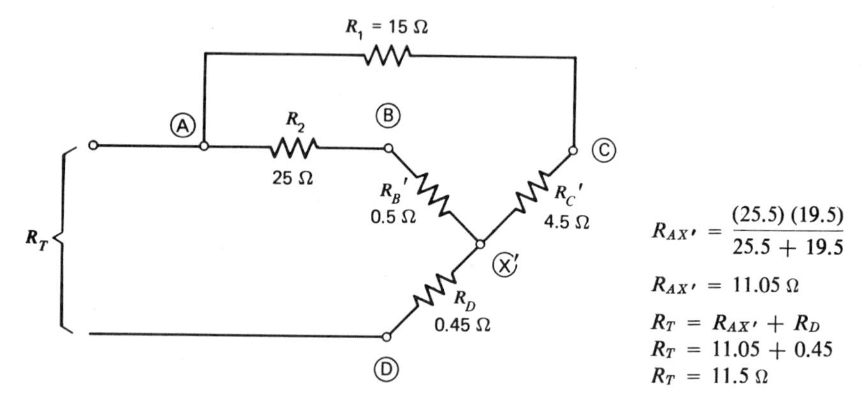

Figure 6-46. Solution for Example 6-19.

and R_z, and its *delta* replacement indicated by the dotted-line resistors R_1, R_2, and R_3. The Y-to-Δ transformation equations are also shown in Fig. 6-47 and are repeated here:

$$R_1 = \frac{R_xR_z + R_xR_y + R_yR_z}{R_y} \tag{6-57}$$

$$R_2 = \frac{R_xR_z + R_xR_y + R_yR_z}{R_z} \tag{6-58}$$

$$R_3 = \frac{R_xR_z + R_xR_y + R_yR_z}{R_x} \tag{6-59}$$

Note that in the equations each Δ resistor is equal to the sum of the products of each pair of Y resistors divided by the Y resistor, which is opposite the position of that Δ resistor.

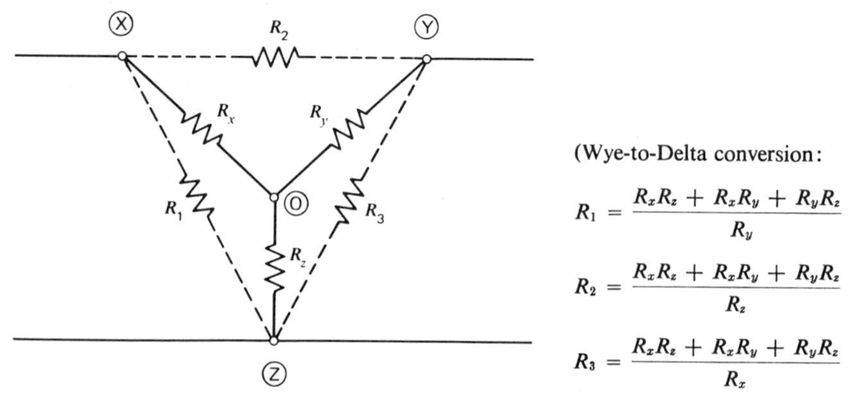

Figure 6-47. Wye-to-delta transformation.

As an example, note that in Fig. 6-47 delta resistor R_1 is equal to the *sum of the products* of each pair of wye resistors ($R_xR_z + R_xR_y + R_yR_z$), divided by the wye resistor (R_y), which is opposite the particular delta resistor (R_1).

Figure 6-48*a* shows a wye circuit with resistors R_x, R_y, and R_z. The equivalent delta circuit is shown in part (b) of this drawing. The values of the delta circuit resistors R_1, R_2, and R_3 are found by using the wye-to-delta transformation equations as shown:

$$R_1 = \frac{R_xR_z + R_xR_y + R_yR_z}{R_y} \tag{6-57}$$

$$= \frac{(6)(10) + (6)(15) + (15)(10)}{15}$$

$$= \frac{60 + 90 + 150}{16}$$

$$= \frac{300}{15}$$

$$= 20\ \Omega, \qquad \text{as shown in Fig. 6-48}b$$

$$R_2 = \frac{R_xR_z + R_xR_y + R_yR_z}{R_z} \tag{6-58}$$

$$= \frac{(6)(10) + (6)(15) + (15)(10)}{10}$$

$$= \frac{60 + 90 + 150}{10}$$

$$= \frac{300}{10}$$

$$= 30\ \Omega, \qquad \text{as shown in Fig. 6-48}b$$

$$R_3 = \frac{R_xR_z + R_xR_y + R_yR_z}{R_x} \tag{6-59}$$

$$= \frac{(6)(10) + (6)(15) + (15)(10)}{6}$$

$$= \frac{60 + 90 + 150}{6}$$

$$= \frac{300}{6}$$

$$= 50\ \Omega, \qquad \text{as shown in Fig. 6-48}b$$

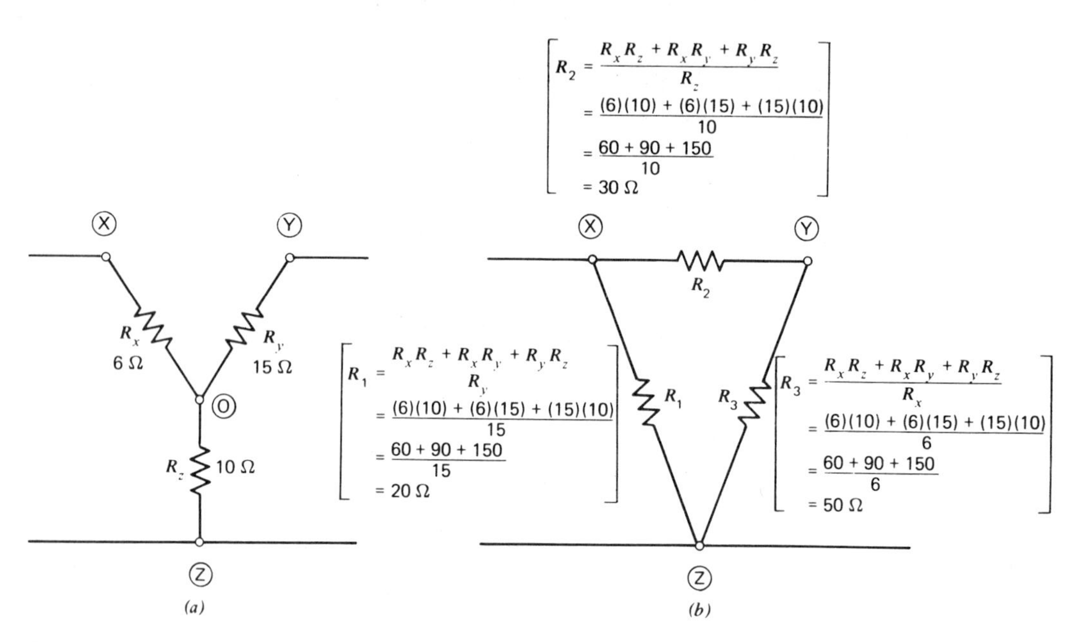

Figure 6-48. (*a*) Wye circuit. (*b*) Equivalaent delta circuit.

Note that the values of the delta resistors in this wye-to-delta conversion agree with those shown in the reverse operation (delta to wye) of Fig. 6-40.

Example 6-20

Figure 6-49*a* is a resistor network also referred to as a bridge circuit. It is desired to find the total resistance between points W and Z. employing the wye-to-delta transformation method.

Solution

Resistors R_1, R_2, and R_4 comprise a wye or tee configuration. Resistors R_3, R_2, and R_5 also form a wye. By converting either wye into its equivalent delta, the entire circuit now becomes a parallel-series circuit as shown in Fig. 6-49*c*.

The following solution changes the wye resistors R_1, R_2, and R_4 (Fig. 6-49*a*) into the equivalent delta resistors R_a, R_b, and R_c (Fig. 6-49*b* and *c*). The wye resistors R_1, R_2, and R_4 are shown removed from the complete circuit in part (c), with the delta resistors R_a, R_b, and R_c shown dotted in part (b).

$$R_a = \frac{R_1R_2 + R_2R_4 + R_1R_4}{R_2} \qquad \text{(adaptation of equations 6-57 to 6-59)}$$

$$= \frac{(20)(30) + (30)(25) + (20)(25)}{30}$$

$$= \frac{600 + 750 + 500}{30}$$

$$= \frac{1850}{30}$$

$$= 61.67\ \Omega,$$

as shown in Fig. 6-49*b* and *c*.

$$R_b = \frac{R_1R_2 + R_2R_4 + R_1R_4}{R_1} \qquad \text{(adaptation of equations 6-57 to 6-59)}$$

$$= \frac{1850}{20}$$

$$= 92.5\ \Omega,$$

as shown in Fig. 6-49*b* and *c*.

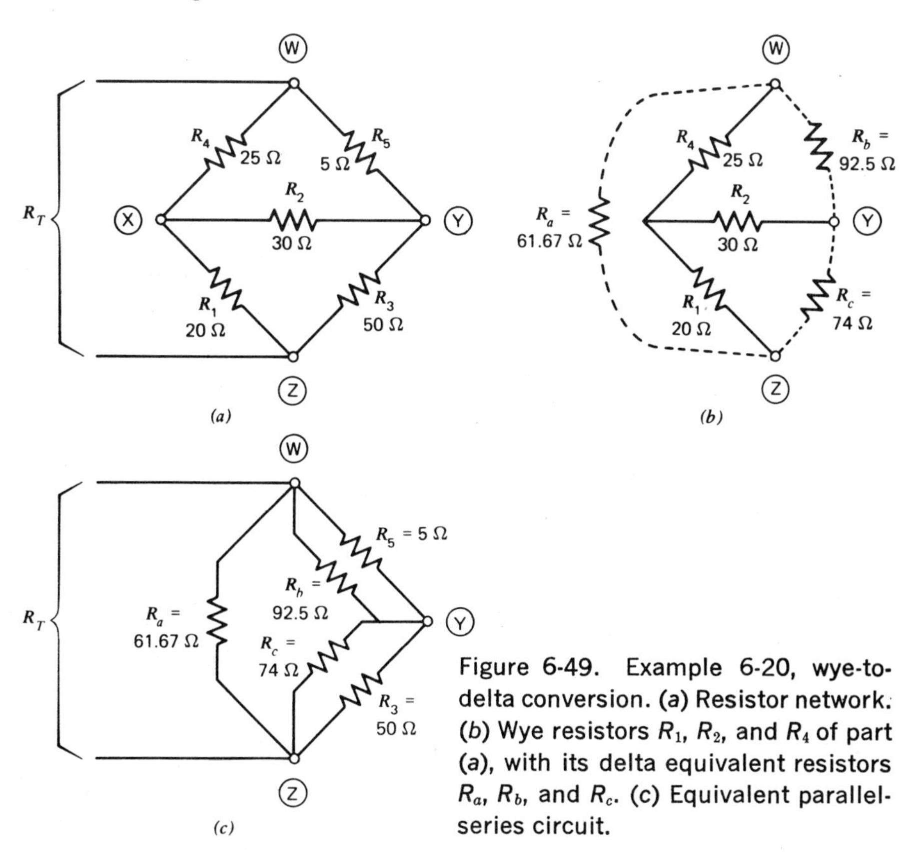

Figure 6-49. Example 6-20, wye-to-delta conversion. (*a*) Resistor network. (*b*) Wye resistors R_1, R_2, and R_4 of part (*a*), with its delta equivalent resistors R_a, R_b, and R_c. (*c*) Equivalent parallel-series circuit.

$$R_c = \frac{R_1R_2 + R_2R_4 + R_1R_4}{R_4} \qquad \text{(adaptation of equations 6-57 to 6-59)}$$

$$= \frac{1850}{25}$$

$$= 74\ \Omega,$$

as shown in Fig. 6-49*b* and *c*.

Delta resistors R_a, R_b, and R_c of Fig. 6-49*c* have replaced the wye resistors R_1, R_2, and R_4 of Fig. 6-49*a*. As shown in Fig. 6-49*c*, the circuit is a parallel-series type. R_b is in parallel with R_5, forming R_{wy}. R_c is in parallel with R_3, forming R_{yz}. Finally R_a is in parallel with $R_{wy} + R_{yz}$. Solving

$$R_{wy} = \frac{R_b\, R_5}{R_b + R_5}$$

$$= \frac{(92.5)\,(5)}{92.5 + 5}$$

$$= \frac{462.5}{97.5}$$

$$= 4.74\ \Omega$$

$$R_{yz} = \frac{R_c\, R_3}{R_c + R_3}$$

$$= \frac{(74)\,(50)}{74 + 50}$$

$$= \frac{3700}{124}$$

$$= 29.8\ \Omega$$

Finally

$$R_T = \frac{(R_a)\,(R_{wy} + R_{yz})}{R_a + (R_{wy} + R_{yz})}$$

$$= \frac{(61.67)\,(4.74 + 29.8)}{61.67 + (4.74 + 29.8)}$$

$$= \frac{(61.67)\,(34.54)}{96.21}$$

$$= \frac{2130}{96.21}$$

$$= 22.1\ \Omega$$

This agrees with the results shown previously for the identical circuit of Fig. 6-41*a* where, however, the delta-to-wye conversion was employed.

For Similar Problems Refer to End-of-Chapter Problems 6-68 to 6-72

PROBLEMS

See section 6-1 for discussion covered by the following problems.

6-1. In Fig. 6-2, what is the resistance between points A and D?
6-2. In Fig. 6-2, what is the resistance between points B and C?
6-3. In Fig. 6-2, if a short were connected between points A and C, what would the resistance become between points A and D?
6-4. In Fig. 6-2, if a short were connected between points B and C, what would the resistance become between points A and D?

See section 6-2 for discussion covered by the following problems.

6-5. In Fig. 6-4*a*, find the voltage (and its polarity) and resistance between points D and C, using Thévenin's theorem. (See parts *b*, *c*, and *d* of this figure for solutions.)
6-6. In the diagram for this problem find the voltage (and its polarity) and the resistance between points X and Y, using Thévenin.

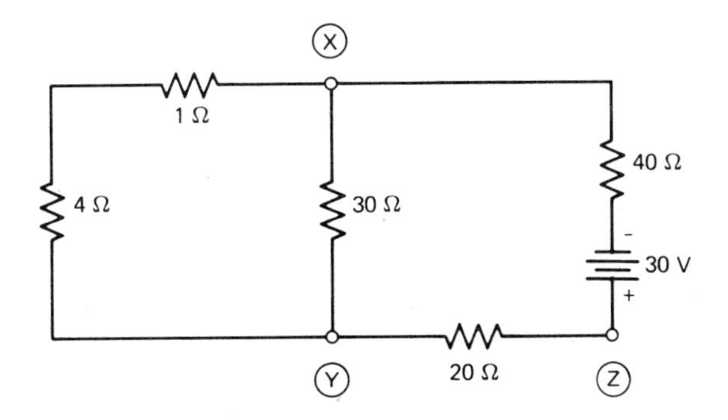

Problem 6-6.

6-7. In the diagram for Problem 6-6, find the voltage (and its polarity) and resistance between points X and Z, using Thévenin.
6-8. In the diagram for Problem 6-6, find the voltage (and its polarity) and the resistance between points Y and Z, using Thévenin.

See section 6-3 for discussion covered by the following problems.

6-9. In the diagram for Problem 6-6, find the voltage (and its polarity) and the resistance between points X and Y, using Norton's theorem.

6-10. In the diagram for Problem 6-6, find the voltage (and its polarity) and the resistance between points X and Z, using Norton's.

6-11. In the diagram for Problem 6-6, find the voltage (and its polarity) and the resistance between points Y and Z, using Norton's.

See section 6-4 for discussion covered by the following problems.

6-12. In the diagram for Problem 6-6, using Maxwell's cyclic current method, find the current and its direction through the 4-Ω resistor, and the voltage across it.

6-13. In the diagram of Problem 6-6, using Maxwell's cyclic current method, find the current through the 30-Ω resistor, and the voltage (and its polarity) across this resistor.

6-14. In this diagram for Problem 6-6, using Maxwell's cyclic current, find the current through the 20-Ω resistor, and the voltage (and its polarity) across the resistor.

See section 6-5 for discussion covered by the following problems.

6-15. Using the nodal analysis method in the diagram of Problem 6-6, find the voltage at point X with respect to point Y.

6-16. Using the nodal analysis method in the diagram of Problem 6-6, find the current (and its direction) through the 20-Ω resistor.

6-17. Using the nodal analysis method in the diagram of Problem 6-6, find the current (and its direction) through the 4-Ω resistor.

More Challenging Problems: See section 6-6 for discussion covered by the following problems.

6-18. In the diagram shown for this problem, find the voltage with respect to ground at (a) point A, and (b) point B, using only Ohm's law in a series circuit.

6-19. In the diagram shown for this problem, find the voltage with respect to ground at (a) point C, and (b) point D, using only Ohm's law.

See section 6-7 for discussion covered by the following problems.

6-20. In the diagram for Problem 6-18, using the superposition theorem, find: (a) voltage at point A with respect to ground due to applied voltage E_1

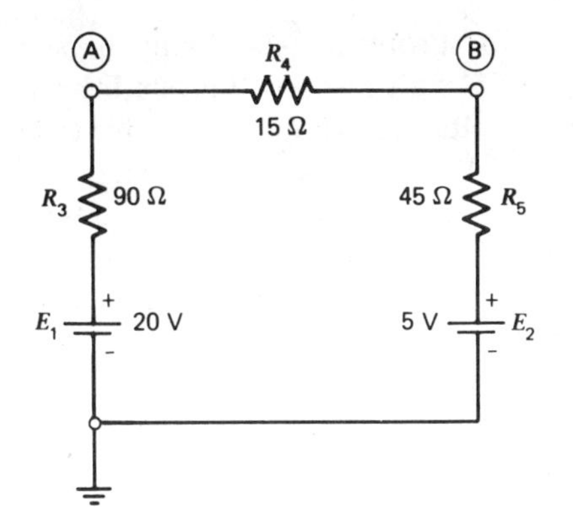

Problem 6-18.

Problem 6-19.

alone, (b) due to source voltage E_2 alone, and (c) due to both source voltages applied at the same time.

6-21. In the diagram for Problem 6-18, using the superposition theroem, find: (a) voltage at point B with respect to ground due to applied voltage E_1 alone, (b) due to source voltage E_2 alone, and (c) due to both source voltages applied at the same time.

6-22. In the diagram for Problem 6-19, using the superposition theorem, find: (a) voltage at point C with respect to ground due to applied source voltage E_3 alone, (b) due to source voltage E_4 alone, and (c) due to both source voltages.

6-23. In the diagram for Problem 6-19, using the superposition theorem, find: (a) voltage at point D with respect to ground due to applied voltage E_3 alone, (b) due to source voltage E_4 alone, and (c) due to both applied voltages.

6-24. In the circuit shown for this problem, using the superposition theorem, find: (a) voltage at point E with respect to ground due to applied voltage source E_5 alone, (b) due to voltage source E_6 alone, and (c) due to both applied voltages.

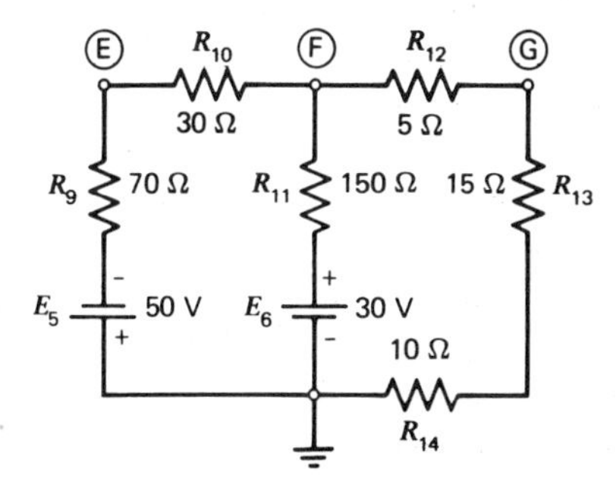

Problem 6-24.

6-25. In the circuit for Problem 6-24, using the superposition theorem, find: (a) voltage at point F with respect to ground due to applied voltage E_5 alone, (b) due to applied voltage E_6 alone, and (c) due to both applied voltages.
6-26. In the circuit for Problem 6-24, using superposition, find: (a) voltage at point G with respect to ground due to applied voltage E_5 alone, (b) due to voltage source E_6 alone, and (c) due to both voltage sources.
6-27. In Fig. 6-18*a* circuit, using superposition, find: (a) voltage and polarity across R_6 (6 Ω) due to source voltage E_5 alone, (b) due to source voltage E_6 alone, (c) due to source voltage E_7 alone, and (d) due to all three source voltages acting together.
6-28. In Fig. 6-18*a* circuit, using superposition, find: (a) voltage and polarity across R_7 (12 Ω) due to source voltage E_5 alone, (b) due to source voltage E_6 alone, (c) due to source voltage E_7 alone, and (d) due to all three source voltages acting together.

See section 6-8 for discussion covered by the following problems.

6-29. In Fig. 6-20*a*, using Thévenin's theorem, find the amount of resistance and voltage (and its polarity) between points C and F.
6-30. In the circuit for Problem 6-24, and using Thévenin's theorem, find the voltage (and its polarity) and the resistance between point E and ground.
6-31. In the circuit for Problem 6-24, and using Thévenin's theorem, find the voltage (and its polarity) and the resistance between point F and ground.
6-32. In the circuit for Problem 6-24, and using Thévenin's theorem, find the voltage (and its polarity) and the resistance between point G and ground.
6-33. In Fig. 6-18*a*, and using Thévenin's theorem, find the voltage (and its polarity) and the resistance between points X and Y.
6-34. In Fig. 6-22*a*, and using Thévenin's theorem, find the voltage (and its polarity) and the resistance between points W and Z.
6-35. In Fig. 6-30, omit all currents and resistor voltages shown, and using Thévenin's theorem, find the voltage (and its polarity) between points A and B.

See section 6-9 for discussion covered by the following problems.

6-36. In Fig. 6-24*a*, and using Norton's theorem, find the voltage (and its polarity) across R_4 (30 Ω).
6-37. In the circuit of Problem 6-24, and using Norton's theorem, find the voltage (and its polarity) across R_9 (70 Ω).
6-38. In the circuit of Problem 6-24, and using Norton's theorem, find the voltage (and its polarity) across R_{11} (150 Ω).
6-39. In the circuit of Problem 6-24, and using Norton's theorem find the voltage (and its polarity) across R_{13} (15 Ω).

6-40. In Fig. 6-26*a*, and using Norton's theorem, find the voltage (and its polarity) across R_7 (60 Ω).

6-41. In Fig. 6-18*a*, and using Norton's theorem, find the voltage (and its polarity) across R_6 (6 Ω).

6-42. In Fig. 6-18*a*, and using Norton's theorem, find the voltage (and its polarity) across R_7 (12 Ω).

6-43. In Fig. 6-30, omit all currents and resistor voltages shown, and using Norton's, find the voltage (and its polarity) across R_5 (3 Ω).

6-44. In Fig. 6-30, omit all currents and resistor voltages shown, and using Norton's, find the voltage (and its polarity) across R_6 (6 Ω).

6-45. In Fig. 6-30, omit all currents and resistor voltages shown, and using Norton's, find the voltage (and its polarity) across R_7 (12 Ω).

See section 6-10 for discussion covered by the following problems.

6-46. In the diagram for Problem 6-24, and using Maxwell's cyclic currents method, find the value of the current (and its direction) that flows through R_{10} (30 Ω).

6-47. In the diagram for Problem 6-24, and using Maxwell's cyclic currents method, find the value of current (and its direction) that flows through R_{12} (5 Ω).

6-48. In the diagram for Problem 6-24, and using Maxwell's cyclic currents method, find the value of current (and its direction) that flows through R_{11} (150 Ω).

6-49. In Fig. 6-18*a*, and using Maxwell's cyclic current method, find the value of current (and its direction) that flows through (a) R_5 (3 Ω), (b) R_6 (6 Ω), (c) R_7 (12 Ω), and (d) R_L (12 Ω).

6-50. In Fig. 6-33, omit all currents shown, and using Maxwell's cyclic currents method, find the value of current (and its direction) that flows through (a) R_5 (3 Ω), (b) R_6 (6 Ω), (c) R_{10} (4 Ω), (d) R_7 (5 Ω), and (e) R_8 (8 Ω).

See section 6-11 for discussion covered by the following problems.

6-51. In the circuit for Problem 6-24, and using the nodal analysis method, find the current (and its direction) through (a) resistor R_9 (70 Ω), (b) through R_{10} (30 Ω), (c) R_{11} (150 Ω), (d) R_{12} (5 Ω), (e) R_{13} (15 Ω), and (f) R_{14} 10 Ω).

6-52. In Fig. 6-16*a*, and using the nodal analysis method, find the current (and its direction) through: (a) R_1 (2 Ω), (b) through R_2 (4 Ω), (c) through R_4 (30 Ω), (d) through R_3 (15 Ω), (e) through R_L (5 Ω).

6-53. In Fig. 6-18*a*, and using the nodal analysis method, find the current (and its direction) through: (a) R_5 (3 Ω), (b) through R_6 (6 Ω), (c) R_7 (12 Ω), and (d) R_L (12 Ω).

See section 6-12 for discussion covering the following problems.

6-54. In the circuit for Problem 6-24, and using Millman's theorem find the voltage and its polarity with respect to ground at point E.
6-55. In the circuit for Problem 6-24, and using Millman's theorem, find the voltage and its polarity with respect to ground at point F.
6-56. In the circuit for Problem 6-24, and using Millman's theorem, find the voltage and its polarity with respect to ground at point G.
6-57. In Fig. 6-29, omit the currents shown, and also the voltage polarities across the resistors, and using Millman's theorem, find voltage and polarity across R_7 (35 Ω).
6-58. In Fig. 6-30, omit the currents shown and also the voltage polarities across the resistors, and using Millman's theorem, find the voltage and polarity across R_8 (12 Ω).
6-59. In Fig. 6-30, omit the currents shown and also the voltage polarities across the resistors, and using Millman's theorem, find the voltage and polarity across R_5 (3 Ω).
6-60. In Fig. 6-33, omit all currents shown, and using Millman's theorem, find the voltage and its polarity between points E and B.
6-61. In Fig. 6-33, omit all currents shown, and using Millman's theorem, find the voltage and its polarity between points D and B.

See section 6-13 for discussion covered by the following problems.

6-62. From diagram (a), the Δ circuit, shown for this problem, find the values of R_x, R_y, and R_z in (b), the equivalent Y circuit.
6-63. In Fig. 6-41*a*, change the values of resistors to the following: $R_1 = 1\,\Omega$, $R_2 = 6\,\Omega$, $R_3 = 3\,\Omega$, $R_4 = 4\,\Omega$, and $R_5 = 10\,\Omega$. Find the total resistance between points W and Z by first converting the *upper* delta (*WXY*) into its equivalent wye.

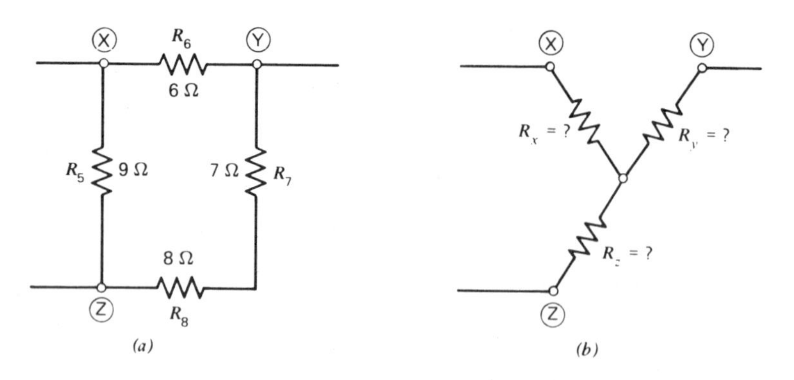

Problem 6-62. (a) Delta circuit. (b) Equivalent wye circuit.

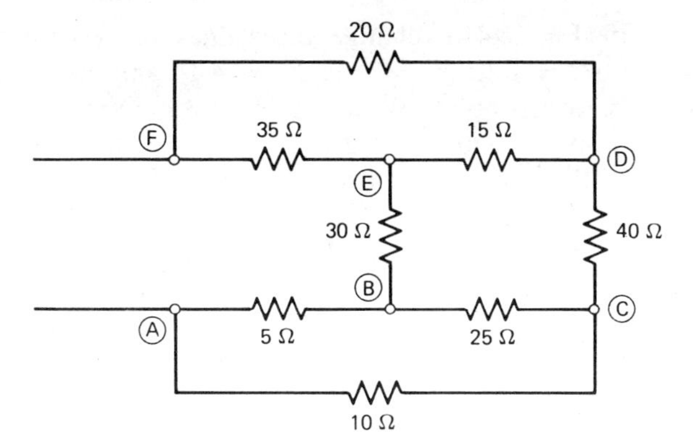

Problem 6-65.

6-64. In Fig. 6-41*a*, and using the resistor values given in Problem 6-63, find total resistance between points W and Z by first converting the *lower* delta (XYZ) into its equivalent wye.

6-65. In the circuit for this problem, find the values of the equivalent wye circuit resistors replacing the F, E, D delta, labeling these resistors R_f, R_e, and R_d.

6-66. In the circuit for Problem 6-65, find the values of the equivalent wye circuit resistors replacing the A,B,C delta, labeling these resistors R_a, R_b, and R_c.

6-67. In the circuit for Problem 6-65, find the total resistance between points A and F.

See section 6-14 for discussion covered by the following problems.

6-68. In Fig. 6-48*a*, change the values of wye resistors to the following: $R_x = 6\ \Omega$ (unchanged), $R_y = 18\ \Omega$, and $R_z = 3\ \Omega$, and find the new values of the equivalent delta circuit of Fig. 6-48*b*.

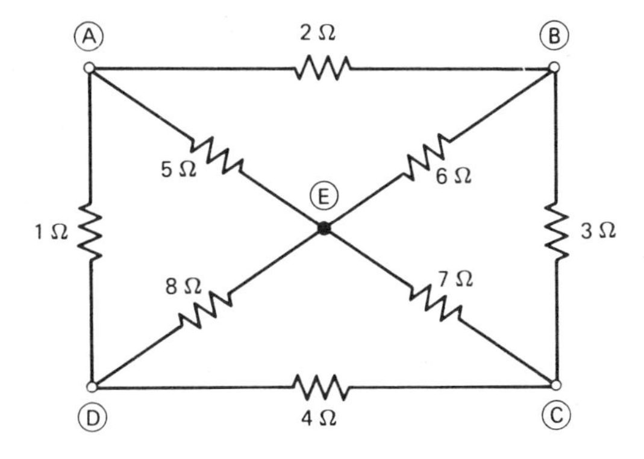

Problem 6-70.

6-69. In Fig. 6-41*a*, change the values of resistors to the following: $R_1 = 1\ \Omega$, $R_2 = 6\ \Omega$, $R_3 = 3\ \Omega$, $R_4 = 4\ \Omega$, and $R_5 = 10\ \Omega$. Find the total resistance between points W and Z by first converting the *wye* resistors R_1, R_2, and R_4 into its equivalent delta.

6-70. In the circuit shown for this problem, convert the wye resistors of *A*, *E*, *C*, and *D* (1, 8, and 4 Ω) into its equivalent delta circuit.

6-71. In the diagram for Problem 6-70, convert the wye resistors of *A*, *E*, *C*, and *B* (2, 6, and 3 Ω) into its equivalent delta circuit.

6-72. In the diagram for Problem 6-70, find the total resistance between points A and C, using the *wye-to-delta* transformations first (see the two preceding problems).

chapter 7

Batteries

A *battery* consists of two or more *cells*, each of which produces a pure dc voltage from a chemical action. The actual chemical changes are described in a chemistry text-book or in one specializing in batteries. The electronic technician should be familiar with what batteries are, how to connect them in circuits, and how best to determine whether they should or should not be replaced. This chapter discusses only what the electronic technician-engineer should know about batteries.

7-1. The Primary Cell

Various unlike metals, when submerged in an acid, slowly dissolve in the solution (called the *electrolyte*). As a result of a chemical action, *ions* are formed. These are charged atoms. Electrons are removed from the atoms of one metal, leaving positively charged ions there. Electrons are deposited on the other metal, producing negatively charged ions on that *electrode*. The difference in charge between the two electrodes is the dc voltage produced by the cell.

Long ago in 1799, Alessandra Volta, an Italian physicist, suspended a piece of zinc and a piece of copper in vinegar (acetic acid). The resulting chemical action placed a plus charge on the copper and a negative charge on the zinc, with a dc voltage between these electrodes of 1.1 V. This experimental device is known as the *Voltaic cell.*

This cell is called a *primary* cell since, eventually, either the zinc or the copper electrode dissolves into the acid (called the *electrolyte*), or the acid becomes too weak for further chemical action. The cell has now become useless and must be replaced. Some cells, made of other metals and electrolytes, need not be replaced, but can be recharged many, many times. These are called *secondary* or *rechargeable* cells. In this section, the primary cells are discussed. Although it is possible to "rejuvenate" some of the primary cells by using a "battery charger," their "new" life is only short-lived and the number of "recharges" possible is very small. With some *primary* cells such as the "mercury" cell and the "silver" cell, the danger of a violent rupture exists when attempting to recharge. Some manufacturers withdraw their battery warranty if an attempt has been made to recharge any primary cell.

Carbon-Zinc (or Leclanche) Cell. This is a primary cell producing *1.5 V*. It consists usually of a center *carbon* rod (the plus electrode) suspended inside a *zinc* can (the minus electrode), with a thick paste between the two. This material consists of, among other things, ammonium chloride (the electrolyte) and manganese dioxide, called the *depolarizer*. As the battery is used, hydrogen bubbles form around the plus electrode, increasing the *internal resistance* of the cell, and decreasing its output voltage. This is called *polarization*. To prevent it, the manganese dioxide (the depolarizer) combines with the hydrogen, forming water. The carbon-zinc cell is very popular in flashlights, toys, and the like because of its low cost. Although it is not designed to be rechargeable, this cell could be recharged a few times, depending on its condition.

Mercury Cell. This is another primary cell, producing about *1.35* to *1.4 V*, depending on the type of *depolarizer*. This cell consists of *mercuric-oxide* (the plus electrode and also the depolarizer), amalgamated *zinc* (the minus electrode), and *potassium hydroxide* (the electrolyte). Attempting to recharge this cell could cause it to burst dangerously.

The *Shelf life* of any battery is not limitless. Even without being used, some chemical action occurs as a result of impurities. This is called *local action*, and it results in a slow deterioration of the battery. The mercury cell has a very long shelf life.

Alkaline Cell. Depending on its construction, this cell could be either a primary type, or a rechargeable type. It produces *1.5 V* between the *zinc* (minus electrode) and the *manganese dioxide* (plus electrode), using *potassium hydroxide* as the electrolyte. This cell will provide up to ten times the service life of the carbon-zinc cell.

Silver-Oxide Cell. This is another type of primary cell, producing 1.5 *V* between its electrodes. *Silver oxide* is the positive electrode, also acting as the depolarizer. The negative electrode is *zinc*, with an *electrolyte* of either *potassium hydroxide* (in hearing aid batteries) or *sodium hydroxide* (in electric watch batteries). They have an excellent and long shelf life. Like the mercury cell, this cell could burst open if an attempt to recharge it were made.

Refer to End-of-Chapter Problems 7-1 to 7-5

Special Cells. The *solar cell* or *photovoltaic cell* produces about 0.25 V. It consists of a photosensitive element such as silicon, selenium, or cadmium which generates or produces a small voltage when exposed to light.

Another special type is the *fuel cell.* This cell does not consume or deteriorate the electrodes. Fuel such as alcohol, gasoline, hydrogen, and the like is fed in along with oxygen. These are consumed, not the electrodes. The fuel and oxygen are sent into separate areas of the cell, separated by the electrolyte. Ionization takes place at each electrode. Electrons from the fuel enter one electrode, making it negative. The oxygen removes electrons from the other electrode, making it positive. The difference of charge, or voltage, between the electrodes produces a current flow through the load which is connected between the electrodes. The ionized fuel and oxygen combine to form water and a gas, such as carbon dioxide. These are eliminated from the cell. The fuel cell only functions as long as fuel and oxygen are being pumped in and also only while the load is connected between the electrodes.

7-2. Secondary or Rechargeable Cells

When current from a cell flows through a load (light bulb, motor, etc.), the cell is said to be *discharging.* The electrodes, and possibly the electrolyte also, are being consumed or deteriorated. If the process can be reversed so that the electrodes and electrolyte are returned to their normal, former status, the cell is said to be *rechargeable*, or a *secondary cell.* This type can usually be recharged many, many times. Examples of them are given in this section.

Lead-Acid Cell. This *rechargeable* cell produces about 2.1 *V* between its electrodes. *Porous* or *spongy lead* is the negative electrode, while *lead dioxide* forms the positive electrode. The *electrolyte* is *sulfuric acid.* If recharged properly, the cell will operate staisfactorily each time, and also after numerous recharges. The *specific gravity* of the sulfuric acid electrolyte, measured with an *hydrometer*, indicates the status of the cell. Readings between 1.24 and 1.28 are usually indicative of a good cell.

Nickel-Cadmium Cell. This is another type of *secondary* or *rechargeable cell.* It produces about *1.25 V* between its electrodes. *Nickel hydroxide* is used as the positive electrode, while *cadmium* is the negative electrode. *Potassium hydroxide* is used as the electrolyte. This cell is exceptionally sturdy and can stand the abuse of heavy overload current drain, and being kept in a discharged condition for long periods.

Nickel-Iron Cell (Edison or Alkaline Cell). This *secondary* or *rechargeable cell* produces *1.4 V.* The positive electrode is *nickel*, while *iron oxide* is the negative electrode. *Potassium hydroxide* and *lithium hydroxide* make up the electrolyte. This electrolyte does not change during use or discharge of the cell, and measuring its specific gravity, therefore, does not indicate the cell condition.

Refer to End-of-Chapter Problem 7-6

7-3. Batteries: Cells Connected in Series and Parallel

Since a *battery* consists of two or more *cells*, this section discusses the effect of connecting cells in *series* and in *parallel.*

The *voltage* produced by a *cell* depends only on the materials used as the electrodes and electrolyte of the cell, and not on the size of these components. A carbon-zinc cell produces 1.5 V whether the carbon rod and zinc can are less than 1 in. in length and narrow in diameter, or are several inches long and have a wide diameter. The physically larger cell can supply current for a longer period of time than the smaller cell. This is called the *ampere-hour* (A-h) rating of the cell. A standard size carbon-zinc flashlight cell (size D) can supply 10 mA for 500 h, while the much smaller carbon-zinc pencil-type flashlight cell (size AAA) can supply 10 mA for only 45 h.

Cells in Series. Connecting cells in *series*, as shown in Fig. 7-1, consists of tieing the *negative* terminal of one cell to the *positive* of the next and taking the output between the remaining terminals. The *total* voltage is simply the *sum* of the *series* connected cells. A 9-V carbon-zinc battery is then made up of six 1.5 V cells connected in series. Similarly, a 12-V lead-acid "storage" battery, such as the type used in cars, is actually six of these 2.1-V cells connected in series.

Cells in Parallel. Connecting similar cells in *parallel*, as shown in Fig. 7-2, consists of tieing the *positive* terminal of one cell to the *positive* of the next,

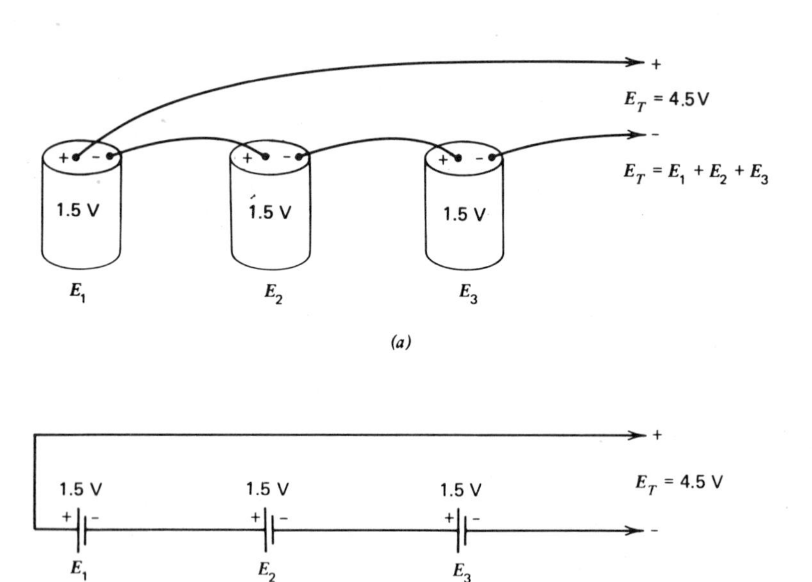

Figure 7-1. Cells in series. (*a*) Physical connections. (*b*) Circuit symbols.

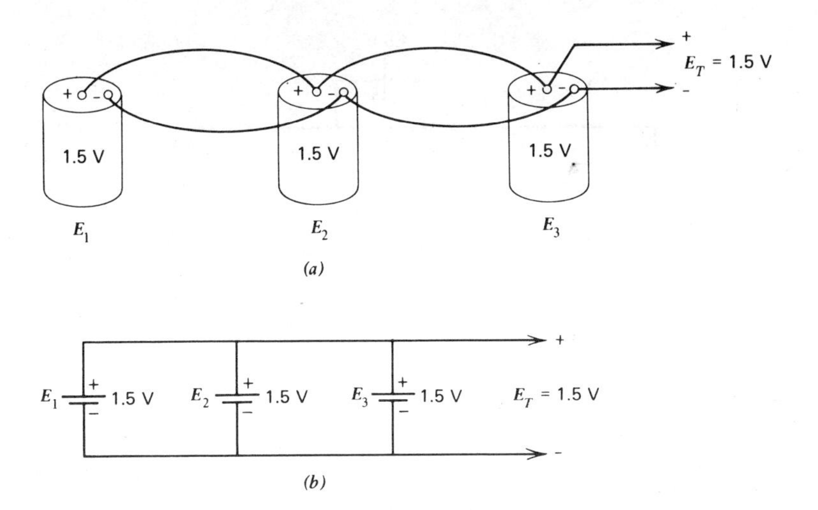

Figure 7-2. Cells in parallel. (*a*) Physical connections. (*b*) Circuit symbols.

and the *negative* terminal of one to the *negative* of the next. The output is taken between any pair of opposite polarity terminals. The result of parallel cells is that the *area* of each electrode adds to that of the others, making one large plus electrode and one large minus electrode. The total voltage is the same as any one of the individual cells alone. However, with the larger electrodes, the *current capacity* or *ampere-hours* has been increased.

If each cell of Fig. 7-2 is 1.5 V with a 5 A-h rating, then the parallel arrangement of the three cells will produce three times the current *capacity* or 15 A-h. If cells which are not similar, having different voltages, are connected in parallel, then the total voltage is not equal to any one of the cells, but can be found by solving the resulting mesh circuit (Chapter 6), similar to Figs. 6-12 and 6-18. The *internal resistance* of each cell must be known in this case. Internal resistance is discussed in the next section.

Example 7-1

In the circuit of Fig. 7-3, what should the milliammeter read when the switch is: (a) in position X, and (b) in position Y?

Solution

(a) When the switch is in position X, only the single 1.5-V cell is in the circuit. From Ohm's law, current is $I = E_T/R = 1.5/3000 = 0.0005$ A, or 0.5 mA.

(b) In position Y, two 1.5-V cells are in series. E_T is now 3 V. Current is: $I = E_T/R = 3/3000 = 0.001$ A, or 1 mA.

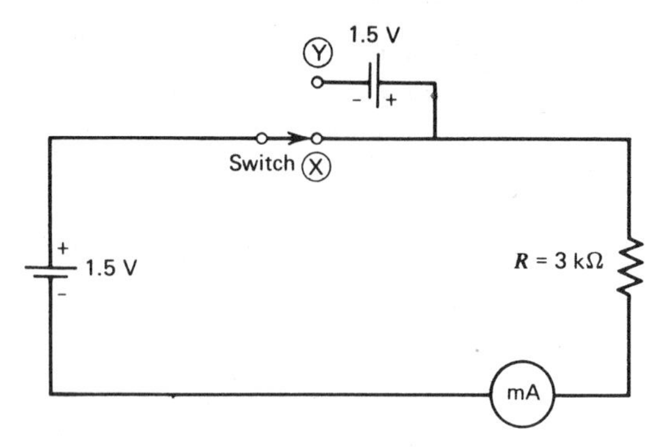

Figure 7-3. Example 7-1.

Example 7-2

In the circuit of Fig. 7-4, what should the milliammeter read (a) when the switch is open, and (b) when the switch is closed?

Solution

(a) When the switch is open, only the single 2-V cell is in the circuit. Current is: $I = E/R = 2/100 = 0.02$ A, or 20 mA.

(b) When the switch is closed, the second 2-V cell is placed in *parallel* with the first one. The voltage is still 2 V, and the current remains unchanged at 20 mA. The current *capacity* or *ampere-hours* has been doubled, however.

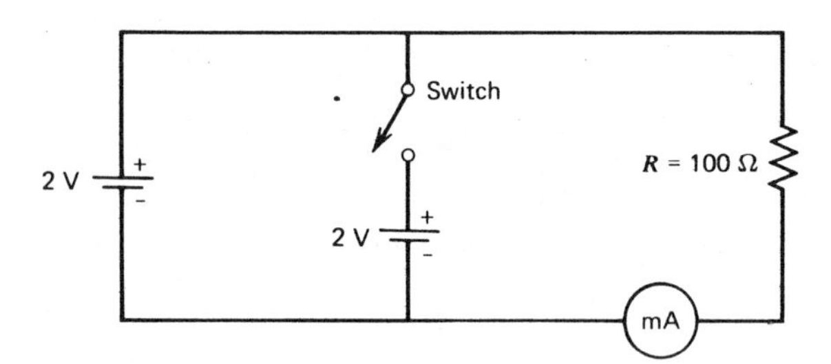

Figure 7-4. Example 7-2.

Cells in Series-Parallel. By connecting similar cells in a series-parallel combination, as shown in Fig. 7-5, the *voltages* of those in *series add,* while those in *parallel* keep their *voltage unchanged,* but their *current capacity,* (A-h) *add.* If each cell in Fig. 7-5 is 1.5 V having a 5-A-h rating, then the total voltage is 4.5 V, with a 10-A-h rating.

For Similar Problems Refer to End-of-Chapter Problems 7-7 to 7-11

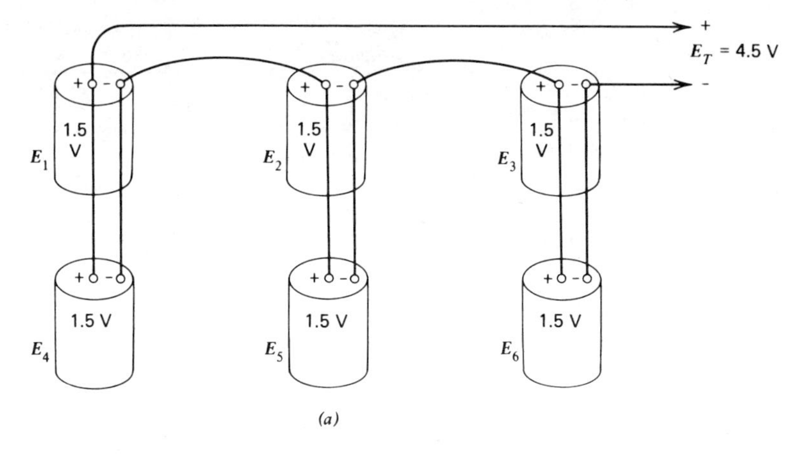

E_1 and E_4 are in *parallel* = 1.5 V
E_2 and E_5 are in *parallel* = 1.5 V
E_3 and E_6 are in parallel = 1.5 V

$\underbrace{E_1, E_4}_{1.5\text{ V}}$ is in *series* with $\underbrace{E_2, E_5}_{1.5\text{ V}}$ and is in *series* with $\underbrace{E_3, E_6}_{1.5\text{ V}}$ = 4.5 V

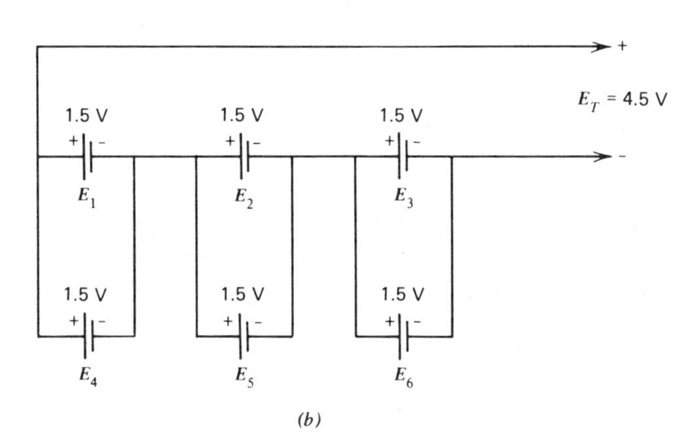

Figure 7-5. Cells in series-parallel. (*a*) Physical connections. (*b*) Circuit symbols.

7-4. Internal Resistance

The resistance inside a cell between the electrodes is called the *internal resistance*. When the cell or group of cells (battery) is connected to a load, as in Fig. 7-6*b*, $R_{int.}$, R_{load}, and the battery act as a simple series circuit. If the battery voltage *without* a load (switch open), as in Fig. 7-6*a*, is 30 V, then when a load is placed across the battery (switch closed) in Fig. 7-6*b*, current flows. There is

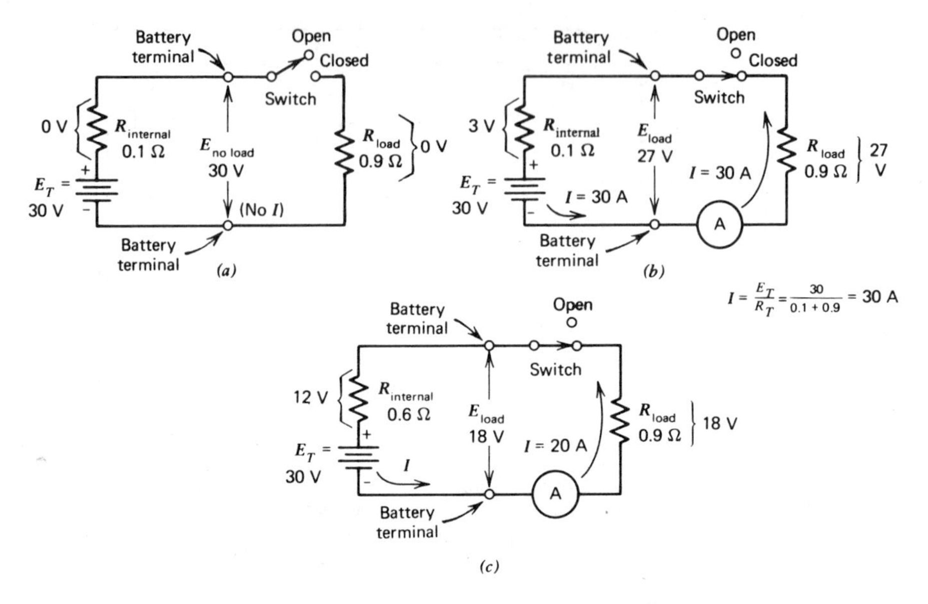

Figure 7-6. Effect of battery internal resistance. (*a*) Switch open, no load. (*b*) Switch closed, battery good (low R_{internal}).

a voltage drop ($I\ R_{\text{int.}}$) across $R_{\text{int.}}$, leaving less than the full 30 V across R_{load}. As the battery ages, the chemical changes occurring inside cause its $R_{\text{int.}}$ to increase. This increases the voltage drop across $R_{\text{int.}}$, *decreasing* the *voltage* available *across the load.* If the load operates satisfactorily at voltages between 30 and 25, then the *25* is called the *cutoff voltage*. This depends on the type of load (light bulb, toy motor, transistor, etc). If $E_{R\ \text{load}}$ becomes less than 25 V, the load fails to operate.

When no load is connected, as in Fig. 7-6*a*, no current flows, and there is no voltage across $R_{\text{int.}}$. Voltage between the battery terminals is the full 30 V. This is called the *no-load voltage*, $E_{no\ load}$.

As shown in Fig. 7-6*b*, a 0.9-Ω load resistor is connected between the battery terminals. Current is: $I = E_T/R_T = 30/(0.1 + 0.9) = 30/1 = 30$ A. Voltage across the load resistor is: $E_{R\ \text{load}} = I\ R_{\text{load}} = (30)\ (0.9) = 27$ V, as shown in Fig. 7-6*b*. $E_{R\ \text{load}}$ is also called battery terminal voltage with load, or E_{load}. Note that $E_{R\ \text{int.}}$ is 3 V, since $E_T = E_{R\ \text{int.}} + E_{R\ \text{load}} = 3 + 27 = 30$ V.

$R_{\text{int.}}$ cannot be measured with an ohmmeter, since the battery would cause a current to flow through the ohmmeter that could damage the ohmmeter. However, $R_{\text{int.}}$ can be calculated using Ohm's law, where $R_{\text{int.}} = E_{R\ \text{int.}}/I$. $E_{R\ \text{int.}}$ may be found, from Fig. 7-6*b*, as follows:

$$E_T = E_{R\ \text{int.}} + E_{R\ \text{load}}$$

$$E_T - E_{R\ \text{load}} = E_{R\ \text{int.}}$$

E_T may be called $E_{no\ load}$, and $E_{R\ load}$ may be called simply E_{load}. Therefore, $E_{R\ int.} = E_{no\ load} - E_{load}$, and $R_{int.}$ is

$$R_{int.} = \frac{E_{no\ load} - E_{load}}{I_{load}} \tag{7-1}$$

Then in Fig. 7-6*b*, $R_{int.}$ may be found using equation 7-1.

$$R_{int.} = \frac{30 - 27}{30}$$

$$R_{int.} = \frac{3}{30}$$

$$R_{int.} = 0.1\ \Omega$$

In Fig. 7-6*c*, $R_{int.}$ has increased, resulting in the current decreasing to 20 A, and E_{load} decreasing to 18 V. $R_{int.}$ is now: $R_{int.} = E_{no\ load} - E_{load}/I_{load} = 30 - 18/20 = 12/20 = 0.6\Omega$. Since the 30-V battery is now only delivering 18 V to the load, then with a 25-V *cutoff* value required by the load, the battery with its higher value of $R_{int.}$ (0.6-Ω instead of the previous 0.1 Ω) should be replaced.

The following example is a further illustration.

Example 7-3

A 45-V battery is used as a power source for a flashbulb. At the instant of flash, voltage across the bulb is 42 V, while the current is 6 A. Find the internal resistance of the battery.

Solution

$$R_{int.} = \frac{E_{no\ load} - E_{load}}{I_{load}} \tag{7-1}$$

$$R_{int.} = \frac{45 - 42}{6}$$

$$R_{int.} = \frac{3}{6}$$

$$R_{int.} = 0.5\ \Omega$$

For Similar Problems Refer to End-of-Chaper Problems 7-12 to 7-13

7-5. Testing a Battery

When the load connected to a battery does not function properly, the battery should be suspected, of course. Measuring the voltage across the battery terminals, with the battery removed from the equipment, could immediately indicate a defective battery if the voltmeter shows anything *less* than the full battery voltage. However, a full meter reading does *not* necessarily indicate a good battery.

The battery voltage should be read with the load connected to it. Then, depending on the condition of the battery, if its internal resistance is too large, the voltmeter across the load will indicate a lower voltage than the battery no-load voltage. Whether this lower voltage will adversely affect the load will depend upon the *cutoff voltage* of the load. Since this voltage is ordinarily not supplied with the load (light bulb, toy motor, electric knife, transistor radio, etc.), then any noticeable decrease in the battery voltage when connected to the load should cause the battery to be suspected. In most instances a, new battery that is known to be good should be substituted for the suspected battery.

A recommended test for any battery is to measure its voltage when a resistor is placed across its terminals that causes *one half* of the *maximum current drain* of the battery as stipulated by the manufacturer. If the battery is good, the voltmeter should read practically the full rated battery voltage.

A secondary battery such as the lead-acid type can also be checked by measuring the specific gravity of the electrolyte, using an hydrometer. Information as to the proper value at the particular temperature when measured must be available if this test is to be meaningful. When connecting a secondary battery to a charger, it is best to follow the recommendations of the battery manufacturer as to the amount and length of time of the charging current.

PROBLEMS

See section 7-1 for discussion covered by the following.

7-1. What is a battery?

7-2. Very briefly state how a cell produces a dc voltage.

7-3. What is a primary cell?

7-4. What is a secondary cell?

7-5. Name four different primary cells.

See section 7-2 for discussion covered by the following.

7-6. Name three different secondary or rechargeable cells.

See section 7-3 for discussion covered by the following.

7-7. Draw the circuit symbol diagram of three 1.25-V cells connected in series

7-8. What is the total voltage of the circuit of Problem 7-7?

7-9. Draw the circuit symbol diagram for three 1.5-V cells connected in parallel.

7-10. What is the total voltage of the parallel circuit of Problem 7-9?

7-11. A circuit consists of a single 1.4-V cell connected to a load resistor. Then, two more 1.4-V cells are placed in parallel with the first cell. (a) What remains unchanged? (b) What increases?

See section 7-4 for discussion covered by the following.

7-12. How can the internal resistance of a battery be measured with an ohmmeter?

7-13. A 9-V battery delivers 8.6 V to transistor amplifiers with 800 mA flowing. Find: (a) $E_{\text{no load}}$, (b) E_{load}, (c) $R_{\text{int.}}$.

chapter

8

Resistors and Conductors

Probably the two most generally used devices in electrical and electronic circuits are *resistors* and *conductors.* Resistors are components offering opposition to current flow and are placed in circuits either to limit the current to a desired value, or to make use of the voltage produced across the resistor. A conductor is simply a wire connected between an applied voltage source and resistors or tubes, transistors, and the like to complete a path through which current can flow. The conductor should have very little opposition or resistance to this current so as not to affect the circuit operation.

8-1. Resistors

A *resistor* is a device having the same opposition to direct and alternating currents. This opposition or *resistance* is measured in *ohms*, where the symbol Ω is used with a number to indicate the amount of resistance such as 10 Ω, 5000 Ω, and the like. In a circuit diagram, a resistor is drawn as a series of jagged sawteeth as shown in Fig. 8-1.

Refer to End-of-Chapter Problem 8-1

Types of Materials. Resistors are often made of a material composed of carbon granules and granules of an insulator (poor or nonconductor). These are molded together and encased in a sealed insulating tube with a tinned

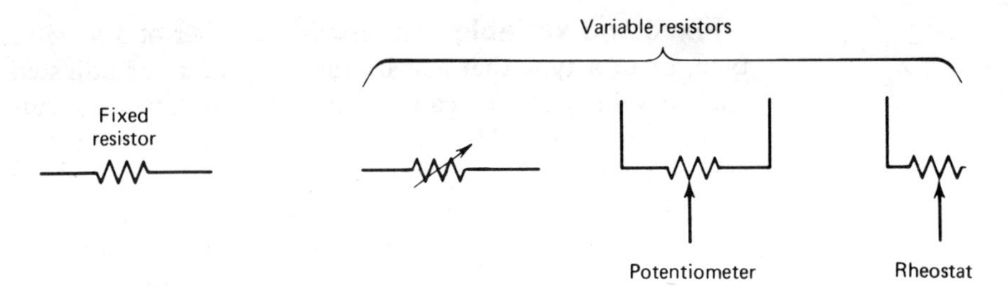

Figure 8-1. Resistor symbols.

copper wire extending from each end. This type is usually referred to as a *carbon resistor.*

Another common type, called a *wire-wound* resistor, consists of a wire such as manganin (alloy of copper, manganese, and nickel) or constantan (alloy of copper and nickel) wound around an insulating ceramic form, with the entire device enclosed in an insulating tube having a piece of tinned copper wire extending from each end. A wire-wound resistor is shown in Fig. 8-2. The higher wattage dissipation resistors are usually wire-wound types. A third type is called the *metal-film* resistor. A combination of metal and glass (or insulator) powders are fused onto a ceramic or glass form. The thickness of the film can be accurately controlled, resulting in a high degree of accuracy.

Refer to End-of-Chapter Problem 8-2

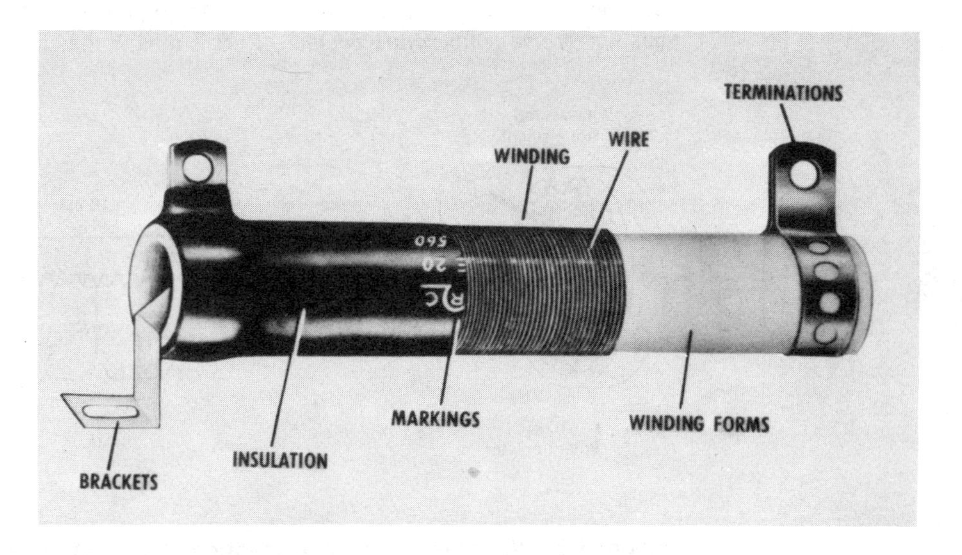

Figure 8-2. Wire-wound resistor (*courtesy IRC, Inc*).

Fixed and Variable. Resistors are either of a *fixed* value or nonvariable type, or of a type that can be varied in value or adjusted manually, or that change value automatically. Figure 8-1 shows the circuit symbols for *fixed* and manually *variable* resistors. Observe that the arrow denotes the resistor can be adjusted. The variable resistor may be either the carbon or the wire-wound variety. It consists of a movable or sliding arm that may be positioned along the resistive element, thereby using only a portion of it. In Fig. 8-3, a 10,000 Ω (10 kΩ) variable resistor or potentiometer is shown both in its physical appearance and as a circuit symbol. Note that the entire resistance of 10 kΩ is present between terminals A and B. The sliding contact or movable arm is connected to terminal C. At the position shown, if there is 4 kΩ between terminals A and C, then the remaining 6 kΩ is between B and C. The sliding center arm is either moved with a screwdriver inserted in a slot, or by manually turning a shaft or axle that is mechanically linked to the arm. If terminals A and B are each connected to the circuit, with a separate connection for terminical C, then the variable resistor is called a *potentiometer*, as shown also in Fig. 8-1. If only terminals A and C are used, then the resistor is called a *rheostat*. A common use of a potentiometer is as the *volume control* in a radio receiver or in an audio amplifier. If 2 V ac (the audio frequency *input* signal) is applied between terminals A and B (across the entire 10 kΩ), and an *output* signal voltage is taken between terminals A and C (across the 4 kΩ section), then the output voltage will be a portion of the 2 V, or actually 4/10 of 2 V, 4/5 V, or 0.8 V (from Chapter 3, Section 3-5).

The variable resistor just described is *manually* variable. Another type varies its resistance *automatically* whenever its temperature rises sufficiently, or whenever the voltage across it rises. One type is called a *thermistor*, and it has a *negative temperature coefficient*. This means that its *resistance decreases*

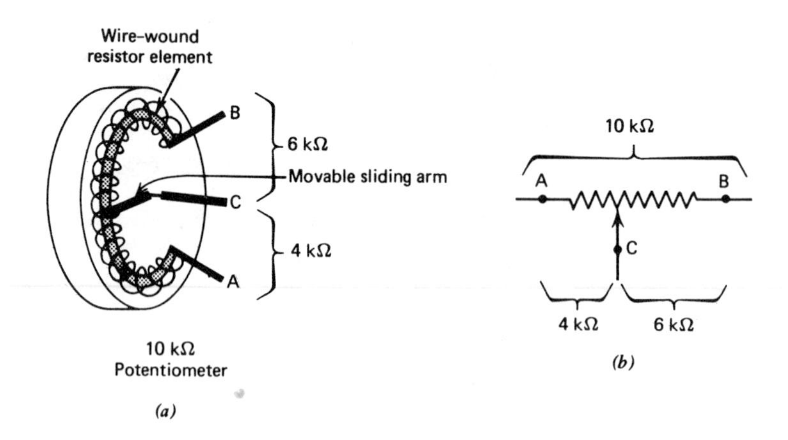

Figure 8-3. Variable resistor (potentiometer). (*a*) Physical appearance. (*b*) Circuit symbol.

as its *temperature rises.* Another type, called *thyrite,* also has a similar negative-acting characteristic. In this type, if the applied *voltage increases,* the *resistance* of the thyrite *decreases* even more rapidly, maintaining a more constant voltage across itself. Still another type is the *photoconductor* or *photoresistor.* This device decreases its resistance when exposed to light; the greater the light, the smaller its resistance becomes.

Refer to End-of-Chapter Problems 8-3 to 8-6

Linear and Nonlinear Resistors. When the resistance value of a device such as a thermistor, electron tube, transistor, or resistance wire *changes* in value, then its volt-ampere characteristic graph is not a straight line and the device is said to be a *nonlinear resistor.* This graph is shown in Fig. 8-4*a.* If the *resistor* value *remains constant* or unchanged over its operating range, then it is called a *linear resistor* since its volt-ampere characteristic graph is a straight line, as shown in Fig. 8-4*b.*

Refer to End-of Chapter Problems 8-7 and 8-8

Wattage Rating. When current flows through a resistor, power is dissipated as described in Chapter 2, Section 2-9. This power is dissipated in the form of heat in the resistor. The physical size (length and width) of the

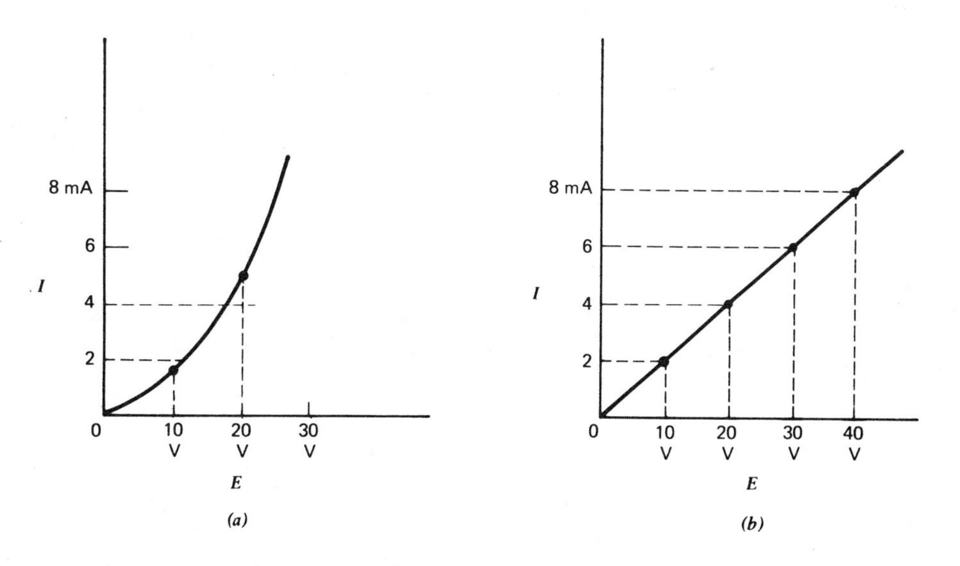

Figure 8-4. *I-E* characteristic graphs. (*a*) Nonlinear (curved) characteristic. (*b*) Linear (straight-line) characteristic.

resistor, and its construction materials, determine the ability of the resistor to dissipate heat without becoming too hot. Excessive heat could cause the resistor to either break apart, increase or decrease its ohmic value, or damage a nearby component such as a capacitor or transistor. Resistors, therefore, are rated as to their ability to dissipate power without causing damage. If a resistor has a power *rating* of 2 W, it means that this component will operate fairly warm if 2 W are actually being dissipated. If 4 or 5 W are being dissipated, this same resistor will operate very warm or even possibly very hot. The excessive heat could result in damage. If a resistor is actually dissipating 2 W, then one with *a rating* of about twice the dissipation, or 4 W rating, should be used. This resistor will operate fairly cool. If a resistor with a *rating* of 10 W is used here, where only 2 W are being dissipated, this component will run very cool. In a low-priced radio receiver, the manufacturer cuts costs by using an underrated resistor. That is, if a resistor dissipates 2 W, then instead of using a 4- or 5-W-rated resistor, a 1- or 2-W resistor is employed, with a resultant shortened life.

Refer to End-of Chapter Problems 8-9 to 8-11

Color Code. The amount of resistance in ohms, called the *ohmic value* of a resistor is usually indicated on the device as a series of color bands. This is shown in Fig 8-5*a* and in greater detail in Appendix VI. The most common resistor color code uses either three or four color bands. Each color shown listed in Fig. 8-5*b* and in Appendix VI has a numerical value from black (0), brown (1), red (2), orange (3), through to gray (8), and white (9).

The first color band is the band nearer one end. Its value is the *first* digit of the resistor size. The second color denotes the *second* digit. The third color-band value, if between the colors black through white (shown listed in Fig. 8-5*b* and Appendix VI) describe the *number of zeros* to be added after the first two digits. The following examples illustrate this.

Example 8-1

What ohmic values are indicated for each of the following resistor colors: (a) Green, brown, orange? (b) Yellow, red, red? (c) Blue, gray, brown? (d) Red, brown, black?

Solution

(a) Green, brown, orange = 51000 Ω; (b) yellow, red, red = 4200 Ω; (c) blue, gray, brown = 680 Ω; (d) red, brown, black = 21— Ω, or 21 Ω.

If there is also present a *fourth* band, usually silver or gold, it denotes the *percentage tolerance* of the resistor. This means how much *higher* or *lower* the resistor value may be off from the color-coded (or rated) value and still

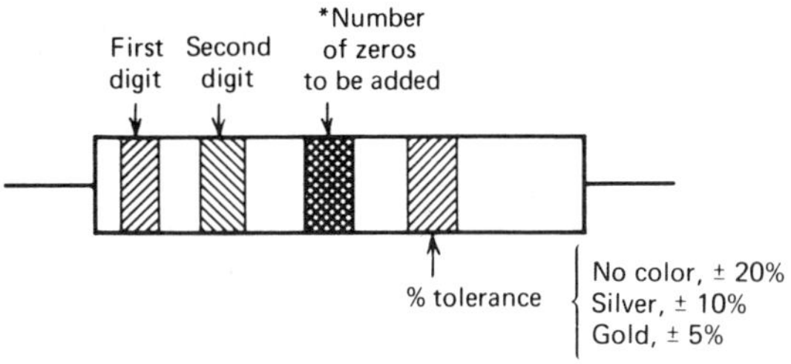

First Second digit digit

*Number of zeros to be added

% tolerance

No color, ± 20%
Silver, ± 10%
Gold, ± 5%

(a)

Resistor Color Code

Color	Value	Number of Zeros* To Be Added
Black	0	none
Brown	1	1
Red	2	2
Orange	3	3
Yellow	4	4
Green	5	5
Blue	6	6
Violet	7	7
Gray	8	8
White	9	9

(b)

* If third band is not one of the colors listed in the Color Code chart, but is either gold or silver, then do *not* add zeros. Instead, *multiply* the first two digits by: gold, 0.1; silver, .01.

Figure 8-5. Resistor color code. (*a*) Four-color-band resistor. (*b*) Color code chart.

be acceptable. The fourth color band of *silver* is a $\pm 10\%$ (read as + or -10%), tolerance, while *gold* is $\pm 5\%$ tolerance. The absence of a fourth band means $\pm 20\%$ tolerance. A 5000-Ω resistor with a $\pm 10\%$ tolerance would have the following color bands: green, black, red, silver. Its *maximum permissible* value would be: 5000 Ω + (10% of 5000), or 5000 + 500, or 5500 Ω. Its *minimum permissible* value would be: 5000 Ω − (10% of 5000), or 5000 − 500, or 4500 Ω. This 5000-Ω resistor with a tolerance of 10% would be considered good, that is, within its tolerance range, if it measured between 4500 and 5500 Ω. If it measured less than 4500 Ω or more than 5500 Ω, the resistor would be said to be out of its tolerance, and should be replaced.

Example 8-2

A resistor has the following four color bands: Red, black, orange, and gold. What is its: (a) Rated ohmic value? (b) Percentage of tolerance? (c) Maximum permissible value? (d) Minimum permissible value?

Solution

(a) 20,000 Ω, (b) ±5%, (c) 20,000 + (5% of 20,000), or 20,000 + 1000, or 21,000 Ω, (d) 20,000 − (5% of 20,000), or 20,000 − 1000, or 19,000 Ω.

As shown in Fig. 8-5 and also in the color code table of Appendix VI, if the *third* color band is *not* one of those listed between black (0) and white (9), but is either gold or silver, then do *not* add zeros after the first two digits. Instead, multiply these digits by either 0.01 for silver or 0.1 for gold. These are for resistors with small values of less than 10 Ω. The following example illustrates this.

Example 8-3

What are the rated values of the following color-coded resistors: (a) Yellow, blue, gold, silver? (b) Orange, black, silver, gold?

Solution

(a) Yellow, blue gold, silver = 46 × 0.1, ±10%, or 4.6 Ω ± 10%, (b) Orange, black, silver, gold = 30 × 0.01 ±5%, or 0.3 Ω ±5%.

Refer to End-of Chapter Problems 8-12 and 8-13

8-2. Conductors

To get the applied voltage across the load, which may be located several inches or feet away, wires (conductors) are used to connect the voltage source to its load. The resistance of the conductors should be as small as possible compared with the load so that practically all of the applied voltage appears across the load, with almost nothing "dropped" across the conductor. In a house, the voltage source from the electric power company must be connected to the various loads, for instance, the lights, the television receiver, the electric toaster, and the other electric appliances. If several appliances such as the toaster, the television set, and the lights are all operating at the same time, the large current flow could produce a voltage drop across improper or inadequate house wiring so that an applied 120 V could result in 30 V being dropped across the wires, leaving only 90 V for the loads. This could cause too little heat from the toaster, a small weak picture on the picture tube, and dim lights. Adequate house wiring uses heavier gauge or larger diameter wires. These conductors have less resistance, and, therefore, very little voltage is dropped across them.

Refer to End-of-Chapter Problems 8-14 and 8-15

Wire Sizes and Resistance. Table 8-1 lists the copper wire gage numbers, wire diameters, cross-sectional areas, and wire resistances per 1000 ft. This table is called the AWG (American Wire Gage) or B and S (Brown and Sharp). Observe that the *gage numbers* get *larger* for *smaller diameter* wires. Also note that the *resistance* of the wire *increases* as the *diameter decreases.* For example: a no. 10 wire has a diameter of 102 mils (102 milli inches) and a resistance of 1.02 Ω for 1000 ft, at a temperature of 25°C (Celsius). A no. 20 wire has a *smaller diameter* of only 32 mils, but a *higher resistance* of 10.4 Ω/1000 ft, at 25°C.

Also note that copper wire *resistance* approximately *doubles* for every increase of *three gauge* numbers. For example: no. 10 wire has 1.02 Ω, no. 13 has 2.04 Ω, no. 16 has 4.09 Ω, and no. 19 has 8.21 Ω. (these are per 1000 ft, at 25°C.)

Also note that the *wire resistance increases* approximately *tenfold* as *wire gage* numbers *increase by ten.* Examples: no. 2 wire has 0.159 Ω, no. 12 has 1.62 Ω, no. 22 has 16.5 Ω, and no. 32 has 167 Ω (these are per 1000 ft, at 25°C).

The diameter of a copper wire, as shown listed in Table 8-1, is in units called *mils.* These are actually *thousandths* of an inch or *milli* inch. The *cross-sectional area* of a round wire is in units called *circular mils.* This is simply the square of the diameter in mils. As shown in Fig. 8-6, a wire having a 0.001 in. (1 mil) diameter has a cross-sectional area of 1 cmil. In Table 8-1, a wire having a gage no. 36 has a diameter of *5 mils* (0.005 in.) and has a cross-sectional area of *25 cmil.*

Since the cross-sectional *area* in circular mils is the *square* of the *diameter* in mils, then a second wire having *twice the diameter* of a first wire has a cross-sectional *area four times* as large. For example, if one wire has a 3-mil diameter, its cross-sectional *area* is 9 cmil. If a larger wire has a diameter of 6 mils, or twice that of the smaller wire, the cross-sectional *area* of the larger is 36 cmil. Note that this is *four* times the area of the smaller wire.

The *resistance* of a wire is *inversely proportional* to the cross-sectional *area.* This means that the *resistance* becomes *smaller* as the *area* becomes *larger.* Therefore, if the *diameter* is *doubled,* the *area* becomes *four times* as large, but the *resistance* becomes *1/4* as large.

Refer to End-of-Chapter Problems 8-16 to 8-18

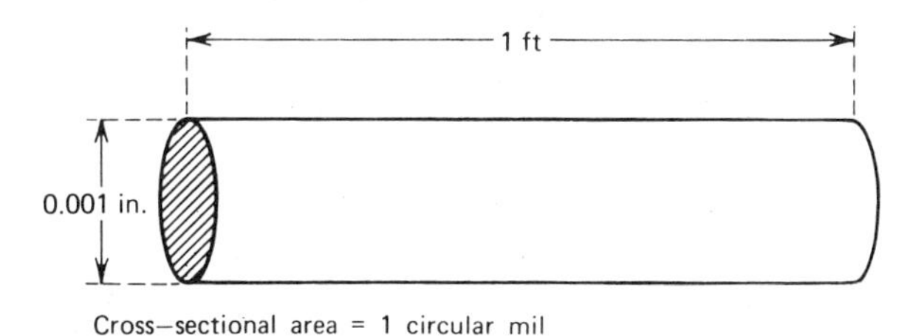

Figure 8-6. One circular-mil-foot.

TABLE 8-1 Copper Wire

(American Wire Gage or Brown and Sharp)

Gage Number	Diameter (mils)	Cross Section: Circular Mils	Cross Section: Square Inches	Ohms per 1000 ft: 25° C (= 77° F)	Ohms per 1000 ft: 65° C (= 149° F)
0000	460.0	212,000.0	0.166	0.0500	0.0577
000	410.0	168,000.0	.132	.0630	.0727
00	365.0	133,000.0	.105	.0795	.0917
0	325.0	106,000.0	.0829	.100	.116
1	289.0	83,700.0	.0657	.126	.146
2	258.0	66.400.0	.0521	.159	.184
3	229.0	52.600.0	.0413	.201	.232
4	204.0	41,700.0	.0328	.253	.292
5	182.0	33,100.0	.0260	.319	.269
6	162.0	26,300.0	.0206	.403	.465
7	144.0	20,800.0	.0164	.508	.586
8	128.0	16,500.0	.0130	.641	.739
9	114.0	13,100.0	.0103	.808	.932
10	102.0	10,400.0	.00815	1.02	1.18
11	91.0	8,230.0	.00647	1.28	1.48
12	81.0	6,530.0	.00513	1.62	1.87
13	72.0	5,180.0	.00407	2.04	2.36
14	64.0	4,110.0	.00323	2.58	2.97
15	57.0	3,260.0	.00256	3.25	3.75
16	51.0	2,580.0	.00203	4.09	4.73
17	45.0	2,050.0	.00161	5.16	5.96
18	40.0	1,620.0	.00128	6.51	7.51
19	36.0	1,290.0	.00101	8.21	9.48
20	32.0	1,020.0	.000802	10.4	11.9
21	28.5	810.0	.000636	13.1	15.1
22	25.3	642.0	.000505	16.5	19.0
23	22.6	509.0	.000400	20.8	24.0
24	20.1	404.0	.000317	26.2	30.2
25	17.9	320.0	.000252	33.0	38.1
26	15.9	254.0	.000200	41.6	48.0
27	14.2	202.0	.000158	52.5	60.6
28	12.6	160.0	.000126	66.2	76.4
29	11.3	127.0	.0000995	83.4	96.3
30	10.0	101.0	.0000789	105.0	121.0
31	8.9	79.7	.0000626	133.0	153.0

(Continued)

TABLE 8-1 *(continued)*

(American Wire Gage or Brown and Sharp)

		Cross Section		Ohms per 1000 ft	
Gage Number	Diameter (mils)	Circular Mils	Square Inches	25° C (= 77° F)	65° C (= 149° F)
32	8.0	63.2	.0000496	167.0	193.0
33	7.1	50.1	.0000394	211.0	243.0
34	6.3	39.8	.0000312	266.0	307.0
35	5.6	31.5	.0000248	335.0	387.0
36	5.0	25.0	.0000196	423.0	488.0
37	4.5	19.8	.0000156	533.0	616.0
38	4.0	15.7	.0000123	673.0	776.0
39	3.5	12.5	.0000098	848.0	979.0
40	3.1	9.9	.0000078	1,070.0	1,230.0

Specific Resistance. The amount of resistance in ohms offered by a material of a certain specified size is called its *resistivity* or *specific resistance* (at a specific temperature). The specified size is shown in Fig. 8-6 as being 1 ft in length, with a diameter of 1 mil (0.001 in.). Table 8-2 lists several materials starting with silver, the best conductor since it has the lowest resistance.

TABLE 8-2

Material	Specific Resistance (ρ) at 20°C (ohms per circular-mil-foot)	Temperature Coefficient (α) 20°C
Silver	9.8	0.0038
Copper	10.37	0.00393
Gold	14.7	0.0039
Aluminum	17.02	0.0039
Tungsten	33.2	0.0045
Nickel	50.	0.006
Iron	58.	0.0055
Steel (soft)	95.8	0.003
Manganin	270.	0.000006
Constantan	300.	0.000008
Nichrome	660.	0.00016
Carbon	20,000. (approximate)	−0.0005

The *specific resistance*, ρ (the symbol is the Greek letter, pronounced "rho"), is then actually *ohms per circular mil-foot.* The actual resistance of a round, tubular wire of some material may be found using the *specific resistance* ρ, the *length* l, and the *cross-sectional area* A. The *resistance* is *directly proportional* to the *length* (the greater the length, the larger is the resistance), but *inversely proportional* to the *cross-sectional area* (the *larger* the *area*, the *amaller* is its *resistance*). This is shown in the following formula for the resistance of a round, tubular wire of some material.

$$\text{resistance, in ohms} = (\text{specific resistance})\left(\frac{\text{length, in feet}}{\text{area, in circular mils}}\right) \quad (8\text{-}1)$$

$$R = (\rho)\left(\frac{l}{A}\right) \quad (8\text{-}1)$$

the following examples illustrate this.

Example 8-4

What is the resistance at 20°C of a round silver wire 2 ft long, having a diameter of 2 mils?

Solution

From Table 8-2, ρ is 9.8, and the cross-sectional area is the diameter squared or 4 cmil. Then

$$R = (\rho)\left(\frac{l}{A}\right) \quad (8\text{-}1)$$

$$R = (9.8)\left(\frac{2}{4}\right)$$

$$R = 4.9\ \Omega$$

Example 8-5

What is the resistance at 20°C of a 5-ft length of round copper wire 10 mils in diameter?

Solution

From Table 8-2, ρ is 10.37, and the cross-sectional area is the diameter squared, or 100 cmil. Then

$$R = (\rho)\left(\frac{l}{A}\right) \quad (8\text{-}1)$$

$$R = (10.37)\left(\frac{5}{100}\right)$$

$$R = 0.519\ \Omega$$

Note that this is approximately the same as the result if Table 8-1, the American Wire Gauge chart, had been used. Here, wire gage no. 30 has a cross-section area of approximately 100 cmil. At 25°C, the resistance is shown as 105 Ω/1000 ft, or 0.105 Ω/ft. Then for a 5-ft length, its resistance would be about 0.525 Ω. This is almost the same as the answer in Example 8-5.

For Similar Problems Refer to End-of-Chapter Problems 8-19 to 8-21

Temperature Coefficient. Most conductors undergo a change in their conductivity (the reciprocal of resistance) at different temperatures. The *temperature coefficient*, α, (Greek letter pronounced "alpha") is defined as the *amount of change in resistance* (ΔR) *per ohm per degree* temperature change Δt (centigrade or Celsius). The actual change in resistance, ΔR is

$$\Delta R = (R_{\text{original}})\,(\alpha)\,(\Delta t) \tag{8-2}$$

A *positive* α means that the resistance increases at higher temperatures, while a *negative* α signifies that the resistance varies inversely with the temperature. Actually, α itself varies slightly, but may be considered to be constant. The following examples illustrate this.

Example 8-6

A wire has 3 Ω of resistance at 20°C. (a) How much of a change in resistance results if the *temperature coefficient* α is + 0.005 and the temperature rises to 30°C? (b) What is the new resistance of the wire?

Solution

(a)
$$\Delta R = (R_{\text{original}})\,(\alpha)\,(\Delta t) \tag{8-2}$$
$$\Delta R = (3)\,(0.005)\,(10)$$
$$\Delta R = 0.15\ \Omega$$

(b) Since α is +, the change in resistance at the higher temperature, is *added* on to the original resistance.

$$R_{\text{new}} = R_{\text{original}} + \Delta R$$
$$R_{\text{new}} = 3 + 0.15$$
$$R_{\text{new}} = 3.15\ \Omega$$

Example 8-7

A 500-ft spool of copper wire has a resistance of 133 Ω at 20°C. (a) What change of resistance occurs if the temperature rises to 70°C? (b) What is the new resistance of the spool of wire?

Solution

(a) From Table 8-2, copper has an α of 0.00393, and Δt is 50°C.

$$R = (R_{\text{original}})(\alpha)(\Delta t)$$
$$R = (133)(0.00393)(50)$$
$$R = 26.1\ \Omega$$

(b) $R_{\text{new}} = R_{\text{original}} + \Delta R$

$R_{\text{new}} = 133 + 26.1$

$R_{\text{new}} = 159.1\ \Omega$

For Similar Problems Refer to End-of-Chapter Problems 8-22 to 8-24

PROBLEMS

See section 8-1, Resistors, for discussion covered by the following.

8-1. Is the ohmic value of a resistor the same for alternating current as it is for direct current?

See section, Types of Materials, for discussion covered by the following.

8-2. Or what materials are the two most common resistors composed?

See section, Fixed and Variable, for discussion covered by the following.

8-3. What is meant by (a) fixed resistor, and (b) manually variable resistor?
8-4. What is another name for a manually variable resistor where all three connections are used?
8-5. What is a thermistor?
8-6. What is a photoresistor?

See section, Linear and Nonlinear Resistors, for discussion covered by the following.

8-7. Why is a resistor that has a constant value called a linear resistor?
8-8. Why is a resistor that has a varying value called a nonlinear resistor?

See section, Wattage Rating, for discussion covered by the following.

8-9. If a resistor is dissipating 2 W, which is best to use: a 2-W-rated resistor or a 5-W-rated resistor?
8-10. If a 25-kΩ resistor dissipates 5 W, and it is replaced with a 25-kΩ 10 watter, how many watts is dissipated in the replacement resistor?
8-11. What could happen if the resistor in Problem 8-10 were replaced with a 2-W rated resistor?

See section on Color Code, for discussion covered by the following.

8-12. What are the rated ohmic values and tolerance percentage of each of the following color-band resistors: (a) Blue, green, brown, gold? (b) Brown, black, black, silver? (c) Green, black, green, silver? (d) Gray, black, gold, gold?
8-13. Since a resistor may be turned one way or the other, how do you determine which is the first color band?

See section 8-2, Conductors, for discussion covered by the following.

8-14. If 5-V is dropped across a wire leading to a load, and the applied voltage is 12-V, what is the load voltage?
8-15. With 120-V applied through connecting wires to a load, how many volts should be dropped across the wires in a properly operating circuit?

See section on Wire Sizes and Resistance, for discussion covered by the following.

8-16. From Table 8-1, the American Wire Gage chart, determine the wire diameter and its resistance per 1000 ft at 25°C for (a) no. 5 copper wire, and (b) no. 15 copper wire.
8-17. What is the cross-sectional area in circular mils for a wire having a diameter of 0.008 in.?
8-18. If the wire in the previous problem has a resistance of 167 Ω/1000 ft, what is the approximate resistance of another wire having half the diameter, or 0.004 in.?

See section on Specific Resistance, for discussion covered by the following.

8-19. Define specific resistance.
8-20. What is the resistance of 50 ft of round aluminium wire, no. 18 gage, at 20°C?

8-21. What is the resistance of 10 ft of round silver wire, no. 16 gage, at 20°C?

See section on Temperature Coefficient, for discussion covered by the following.

8-22. (a) What does a negative temperature coefficient mean? (b) What does a positive temperature coefficient mean?

8-23. A carbon-resistance wire used in an automobile is 300 Ω at a temperature of 90°C. What is its resistance when the temperature becomes 10°C? (Note from Table 8-2 that α is -0.0005.)

8-24. A copper wire spool has a resistance of 500 Ω at 65°C. What is its resistance at 20°C.

chapter
9

Magnetism and Electromagnetism

A simple explanation of magnetism is given at the beginning of this chapter, along with some practical uses of magnetism in electronic devices. Later in the chapter a more thorough discussion is presented.

9-1. Natural Magnets

For centuries man has observed the wonders of magnetism. The mysterious attraction that a particular type of rock has in making pieces of iron cling to it has long been known. The ancient Greeks believed that if their wooden ships sailed too close to certain islands that were rich in this magnetic stone, the iron nails would be pulled out of the planks and the ships would fall apart.

The Greeks named these stones *magnetite*, since the strongest of them were found in Asia Minor in an area called Magnesia. The ancient Chinese used a small magnetic stone placed on a piece of wood floating in a liquid-filled container. This was the forerunner of today's navigational compass. Since the floating-stone-device always aligned itself in a north-south position, it was called a *lodestone*, meaning "leading" stone. These stones are *natural* magnets. If a bar of steel is repeatedly rubbed against them, the steel bar exhibits the same attraction toward small pieces of iron that the magnetic rock has. The steel bar is called an *artifical* magnet.

Refer to End-of-Chapter Problems 9-1 to 9-3

9-2. Magnetic Field

For centuries the theory of a magnet's behavior was unknown. Even today, with all that is known, there is still some unexplained theory concerning magnets.

If tiny particles of iron, called iron filings or iron powder, are sprinkled over and around a plain steel bar, the iron powder covers the area in a random or disorderly fashion (Fig. 9-1*a*), much like salt and pepper being sprinkled over food on a plate. However, when the iron powder is sprinkled over and around a magnetized steel bar, the tiny iron particles always become arranged in the pattern shown in Fig. 9-1*b*. The orderly curved lines in which the iron particles arrange themselves show that there is present some invisible field around a magnetic steel bar that is not present around a plain steel bar. This invisible field (indicated by the lining up of the tiny iron particles) surrounding the bar magnet is called a *magnetic field* or *flux* and is assumed to be made up of *magnetic lines of force.* A strong magnet has more *magnetic lines per square area* than a weak magnet. This is called the *flux density.*

In the mks system (meter-kilogram-second), the term *weber* is used to denote a large number (10^8 or 100,000,000) of magnetic lines. The term *weber* honors Wilhelm E. Weber, a German physicist who, in 1833 to 1838, experimented with and successfully operated a telegraph device. The number of magnetic lines per square area, or the *flux density,* is measured in units called *tesla,* honoring Nicola Tesla (1856-1943), a Yugoslavian scientist who emigrated to the United States. He is known for his work on induction coils and motors and dynamos. A tesla is one weber per square meter, or 10^8 magnetic lines per square meter.

In the cgs system (centimeter-gram-second), the term *maxwell* denotes one magnetic line, and *flux density* is measured in units called *gauss.* A gauss is one *maxwell per square centimeter,* or one magnetic line per square centimeter.

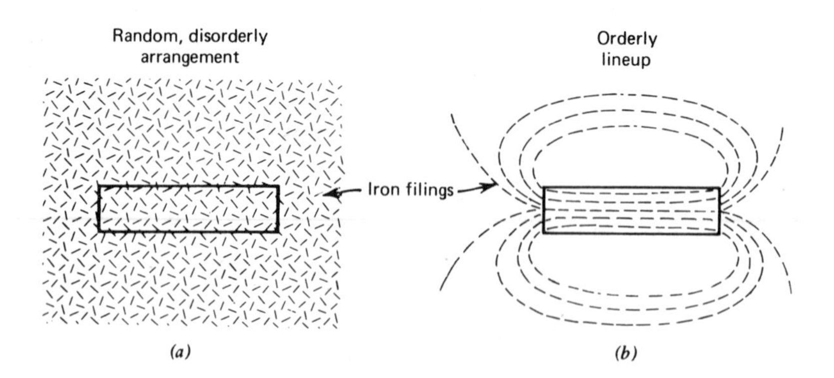

Figure 9-1. Effect of sprinkling iron filings over steel bars. (*a*) Plain steel bar. (*b*) Magnetic steel bar.

The gauss was named in honor of Karl F. Gauss a German scientist who, in 1833 to 1838, worked with Weber. The maxwell honors James C. Maxwell a Scotch physicist who, in 1864, first suggested the electromagnetic theory of light.

Note that in Fig. 9-1*b* the iron filing particles line up with a greater concentration (more per unit area) at the ends of the bar magnet. The ends are named *poles*; one end is called the north pole, and the other end is called the south pole. As far as is presently known, the magnetic lines may or may not have a direction. However, to explain certain happenings of magnetic devices, it has been assumed that these magnetic lines have a "direction," which is from the north pole of a magnet to the south pole of that magnet, in the area *outside* of the magnet. *Inside* the magnet itself, it is assumed that the lines go from south to north. This depicted in Fig. 9-2*a* for a bar magnet, and in Fig. 9-2*b* for a horseshoe magnet.

When two bar magnets are brought near each other, it is common knowledge that they will either attract one another and stick together, or that they will repel, depending on which ends of the magnets have been brought near each other. If the same polarity poles (north and north, or south and south) called *like* poles, are brought together, the bar magnets will not adhere, but will actually try to push each other apart, or repel one another. This repelling effect could easily be seen if one bar magnet were balanced at its center and suspended on a string so that magnet hung horizontally. When a second bar magnet is brought near the suspended magnet so that the adjacent poles of each magnet are the same, the suspended magnet will rotate away from the other. It is said, therefore, that *like poles repel.* Fig. 9-3*a* shows the appearance of the area between two like poles of a pair of bar magnets, A and B, if iron filings are sprinkled over them. Note that the magnetic lines of magnets A and B curve away from each other.

When opposite poles, or unlike poles (a north and a south) of magnets A and B are brought close to one another, there is an attraction between them. The magnetic lines (as indicated by the lineup of the iron filings) now exist

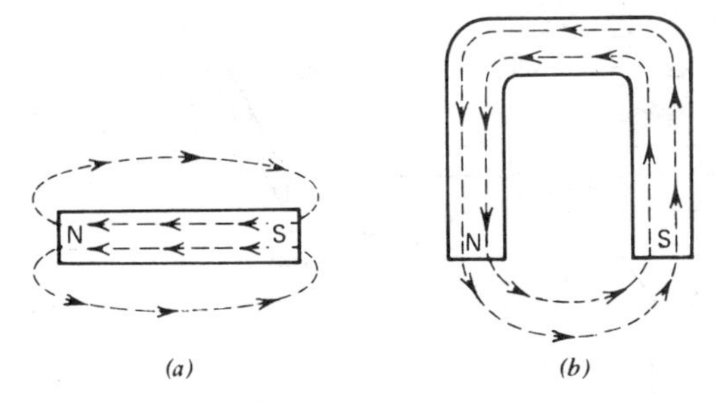

Figure 9-2. Assumed "direction" of magnetic lines of force. (*a*) Bar magnet. (*b*) Horseshoe magnet.

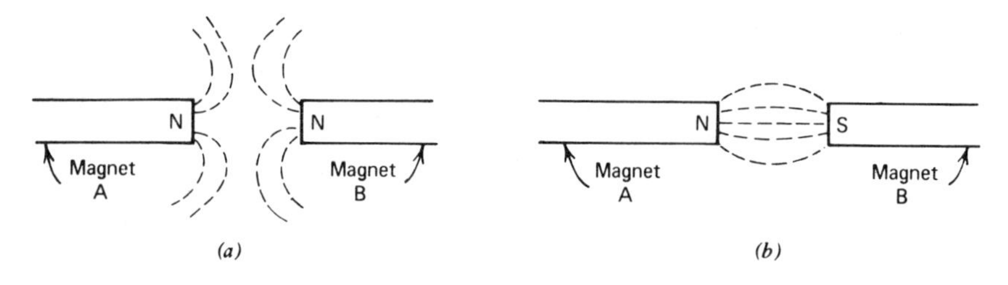

Figure 9-3. Effect of sprinkling iron filings between the ends of two bar magnets. (*a*) Like poles repel. (*b*) Unlike or opposite poles attract.

from the pole of magnet A to the opposite polarity pole of magnet B. This is shown in Fig. 9-3*b*. It is said, therefore, that *opposite poles attract.*

The effect that magnetic fields have on animals, including man, birds, insects, and fish is not known. However, experiments with birds and some insects indicate that these forms of life are aware of, and can sense magnetic lines.

In an interesting experiment involving insects, termite queens from South Africa were placed in a steel box. The steel acted as a magnetic shield (to be discussed later), so that no magnetic lines from nearby magnets could appear inside the box. The insects appeared throughout the box facing in all directions. When the insects were put in a wooden box, and a strong pair of magnets were placed outside on opposite sides of the box with unlike poles facing each other, the insects moved about and after a short time every termite faced in the same direction. The insects lined themselves up at right angles to the lines of force as depicted in Fig. 9-4.

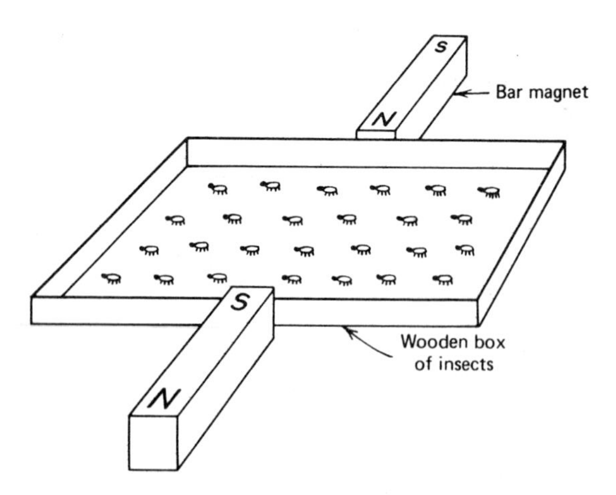

Figure 9-4. Termites all facing in the same direction when they are placed in a strong magnetic field.

It has also been observed that wild birds in migration change their flight paths to evade the strong magnetic fields of radio and radar transmitters. In laboratory experiments this has also been borne out. It is thought that the homing instincts which permit the salmon to locate its birthplace at spawning time, and the pigeon to find its way home are related to magnetic fields.

Refer to End-of-Chapter Problems 9-4 to 9-13

9-3. Magnetic Theory

The accepted explanation of the difference between an unmagnetized steel bar and one which is magnetized centers on the atoms and their electrons. The electrons rotating around the nucleus of each atom of an unmagnetized steel bar rotate in different directions. That is, some may move in clockwise orbits, while others travel counterclockwise. When more electrons travel in one direction than another, the material exhibits magnetic properties. Each atom seems to act like a tiny magnet with some having the reverse polarities of adjacent atoms, and some having the same polarities. This results in cancellations of magnetic fields in some areas, and in stronger fields in other areas. When practically all the electrons of all the atoms of the steel bar rotate in the same direction, the steel is magnetized at its maximum strength and is said to be magnetically *saturated.* This simple magnetic theory is depicted in Fig. 9-5. To support this theory of the "lining up" of the atoms, it has been found that the *length* of an iron bar actually changes when it is magnetized. This variation of size due to magnetization is called *magnetostriction.* To have a magnet, even that of a single atom, there must be moving electrons.

Refer to End-of-Chapter Problems 9-14 and 9-15

9-4. Magnetic Materials

Some materials such as iron may be strongly magnetized, while others like aluminum may only be weakly magnetized. Other substances such as glass,

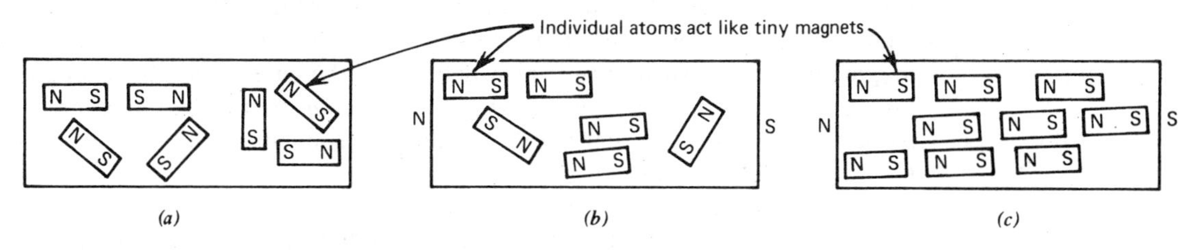

Figure 9-5. Simple magnetic theory. (*a*) Unmagnetized steel bar. (*b*) Weak bar magnet. (c) Strong bar magnet.

wood, and plastics cannot be magnetized at all. The ability to become magnetized depends on the *permeability* of the substance. *Permeability* is the ease with which magnetic lines of force are produced or travel along a material *compared to air*. The permeability of air, almost the same as a vacuum, is used as a reference and is given a value of 1. If a piece of iron has a permeability of 300, it means that a magnetic line would travel along the iron 300 times more easily than through the air. As a result, an iron or steel container acts as a magnetic *shield*, preventing most magnetic lines from getting inside the container, thus, keeping the contents of the container free from any external magnetic field. Figure 9-6*b* shows that the external magnetic field lines pass *along* the steel walls of the box rather than through the air. In Fig. 9-6*a* the magnetic field is practically unaffected by the presence of the wooden box. When it is desired to prevent a steady-strength, unvarying magnetic field from reaching some device, a magnetic shield of iron or steel is placed so that it completely encloses the device to be shielded. For example, a reel of magnetized tape, when stored away, must be shielded to prevent external magnetic fields from "washing away" the information stored on the tape.

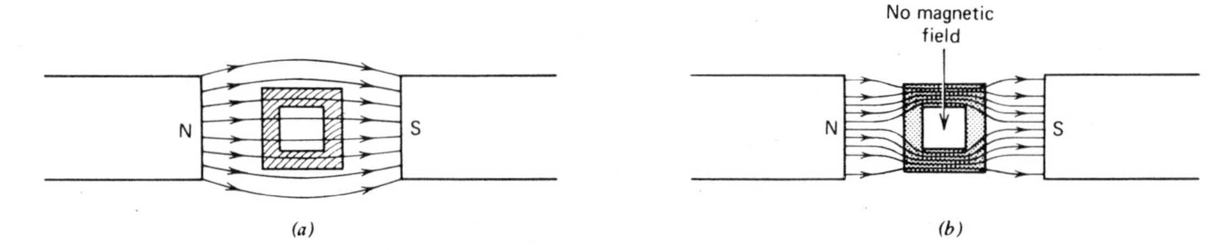

Figure 9-6. Magnetic shielding of steel box. (*a*) Wooden box, no effect. (*b*) Steel-walled box, shielding effect.

If a piece of iron were placed in a magnetic field, the magnetic lines (field) in the vicinity of the iron would bend toward the iron, trying to take the easier path through the iron rather than through the air. This is illustrated in Fig. 9-7, and is due to the high permeability of the iron. Note that the piece of wood which has about the same permeability as the air has no effect on the magnetic lines. *Permeability* is also defined as the ease with which magnetic lines are concentrated in a material. The *opposition* that a substance presents to the setting up of magnetic lines is called *reluctance*. A material that has a *high permeability* therefore has a *low reluctance*.

Materials such as iron, steel, and certain alloys consisting mainly of nickel and iron have permeability values from about 200 to well over 100,000. Some of these alloys are known by their commercial or trade names, for instance, *alnico*, *numetal*, and *permalloy*. When an unmagnetized iron bar is placed in a

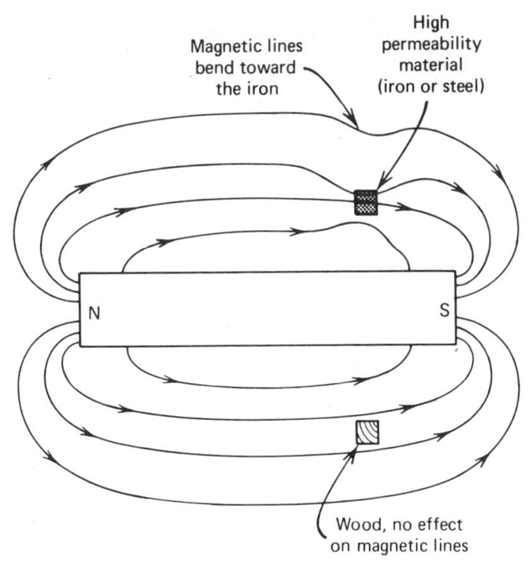

Figure 9-7.

strong magnetic field, the magnetic lines traveling through the iron cause some of the individual iron atoms to "line up" as shown in Fig. 9-5*b*. If the iron bar is tapped or struck several times while in the magnetic field, the vibrations of the metal shake up or loosen the atoms, allowing more of them to line up in the same directions. As shown in Fig. 9-8, the iron bar has become magnetized while under the influence of the strong magnetic field. Note that the direction of the magnetic lines from the permanent bar magnet "travel" from the north pole to the south pole *outside* the magnet, but go from the *south pole to the north pole inside* the magnet. These same lines will likewise "travel" *through* the iron bar, which has become magnetized, form its south to its north pole. The *left* end of the iron bar has become a south pole, which is the *opposite* polarity of the *left* end of the permanent bar magnet.

When a material such as iron becomes magnetized (Fig. 9-8) with opposite poles from the magnetizing force or bar magnet, the iron bar *field* is actually in the *same direction* as that of the bar magnet. The material is called *ferromagnetic*. These materials have high permeability values, and are attracted strongly by the magnet.

Certain substances such as aluminum, chromium, and platinum react similarly to iron, but to a much lesser degree. When these are placed in a strong magnetic field, they become very weakly magnetized, with the same magnetic poles as the iron bar of Fig. 9-8. These are called *paramagnetic*, and have a very low permeability, just a little more than that of air, and are only slightly attracted by the magnet.

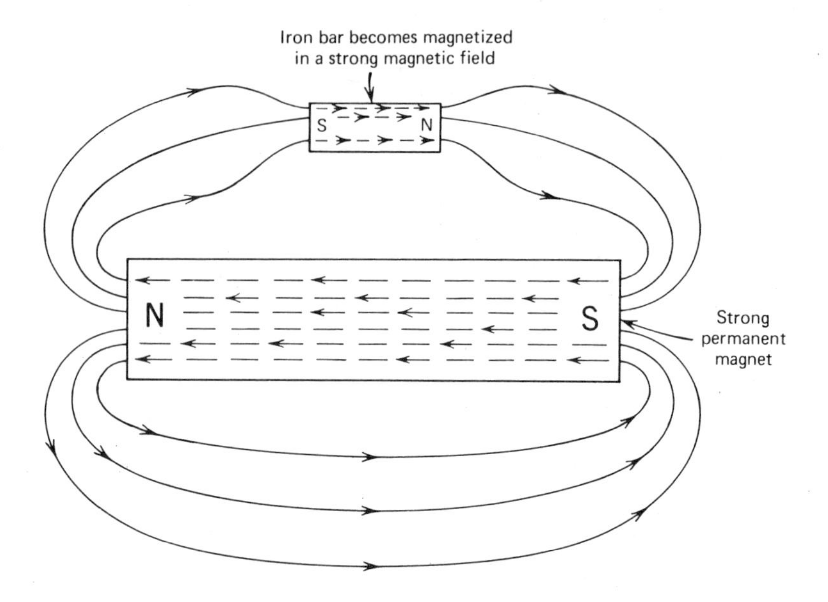

Figure 9-8.

A third class of magnetic substance is gold, silver, copper, zinc, and a few others. These have a permeability less than 1 (that of air). When these materials are placed in a strong magnetic field, they become very weakly magnetized, but react oppositely to the iron of Fig. 9-8. These materials become weakly magnetized with the same magnetic poles as that of the adjacent poles of the magnetizing force (the bar magnet). This means that their magnetic poles oppose those of the bar magnet. These materials are named *diagmagnetic*, and are weakly repelled by a magnet.

Substances such as wood, glass, water, and the like are not magnetic at all, and have the same permeability as air. They have no effect when placed in a magnetic field. A completely nonmagnetic material such as wood is shown in Figs. 9-6*a* and 9-7, having no noticeable effect.

When a piece of iron is itself magnetized by being placed in a strong magnetic field, as in Fig. 9-8, and then is removed from the field, the iron may or may not remain magnetized. If the iron is of a type called soft iron, it loses most of its magnetism almost immediately. On the other hand, hard steel, once magnetized, retains most of its magnetism for a long time and becomes a *permanent magnet*. However, with time, even the "permanent" magnets deteriorate. This weakening takes place at a decreasing rate. That is, when the magnet is new, it deteriorates more rapidly than when it is older. As a result, when the magnet is commercially manufactured, it is at first magnetized very strongly, approaching saturation. Then it is artificially and intentionally "aged" by being boiled in a liquid at a desired temperature for a desired time. When finished, the aged magnet will still deteriorate, but at a much slower rate.

If a magnet is dropped so that it is vibrated or jarred, some of its atoms are shaken up and become misaligned, resulting in a weakening of the magnet. The improper storage of a permanent magnet can also cause premature weakening. When magnets are to be stored away, they should be connected with opposite poles together so that one magnet does not demagnetize the other. Figure 9-9 shows how magnets should be stored. In each case, the magnetic lines travel *through* each magnet, from one to the other, in the normal direction (from south to north *inside* each magnet). As a result, there is almost no external field present in the surrounding air.

In Fig. 9-9*d* and *e* with only a single magnet, a soft iron bar, called a *keeper*, is connected across the magnet's poles, The iron bar becomes magnetized with its poles as shown, and again the magnetic lines travel through the large magnet and the small bar magnet, with almost no external field.

Refer to End-of-Chapter Problems 9-16 to 9-26

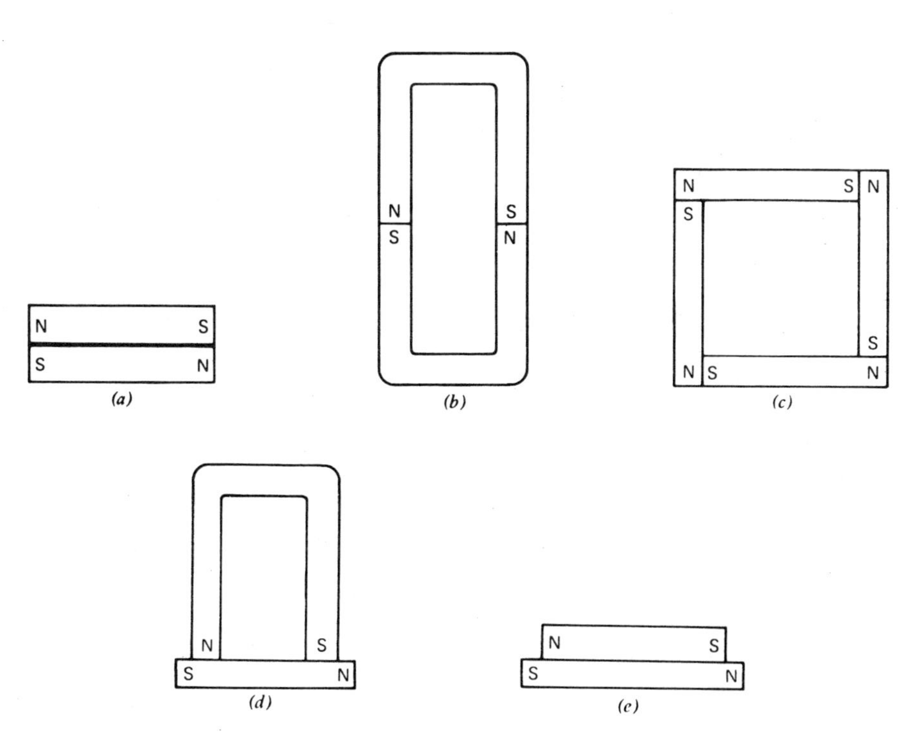

Figure 9-9. Proper storage of permanent magnets. (*a*) Pair of bar magnets. (*b*) Pair of horseshoe magnets. (*c*) Four bar magnets. (*d*) Iron bar acts as "keeper" for horseshoe magnet. (*e*) Iron bar acts as keeper for bar magnet.

9-5. Electromagnetism in a Wire

Section 9-3 describes a simple theory which states that each atom with electrons *moving* around the nucleus acts like a tiny magnet. When most atoms are so lined up that the electrons rotate in the same direction, the magnetic fields of these atoms add. As a result, the material exhibits magnetic properties. The *movement* of electrons seems to be required to produce a magnetic field.

The presence of a magnetic field due to the movement of electrons may be seen by sprinkling iron filings on a horizontally positioned cardboard through which passes a wire carrying current. This is shown in Fig. 9-10. In part (a) the switch is open and no current flows. The iron filings on the cardboard lie in a disorderly random arrangement. When the switch is closed, Fig. 9-10*b*, current flows through the wire. If the cardboard is tapped lightly, the tiny iron particles become lined up in an orderly arrangement as shown. The iron filings assume a circular pattern, with one circle within the next, all having a common center point, similar to the rings of a bullseye target. This pattern displayed by the iron filings indicates the presence of the invisible magnet lines of force that surround the wire while it is carrying a current.

The presence and direction of the magnetic lines seem to be indicated by the action of a magnetic compass needle. When this movable magnet is placed in a steady external magnetic field, it will rotate to a position where the magnetic lines can "travel" *through* the magnetic needle from its south to its north, as shown in Fig. 9-11. The direction of the magnetic lines is indicated by the arrows in Figs. 9-10*b* and 9-11. The *direction* of the magnetic lines surrounding

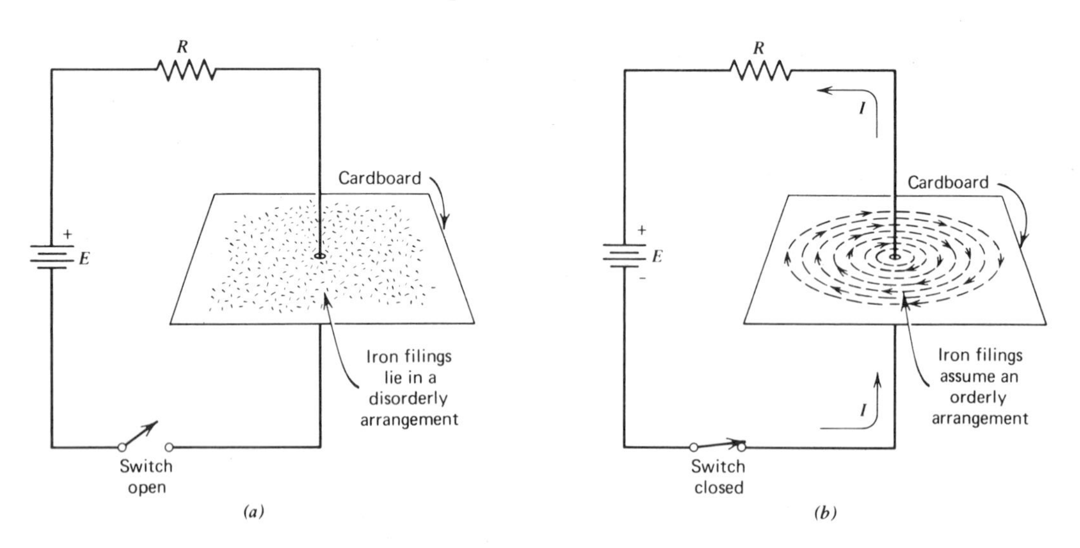

Figure 9-10. (*a*) No current, no magnetic field. (*b*) Current, magnetic field.

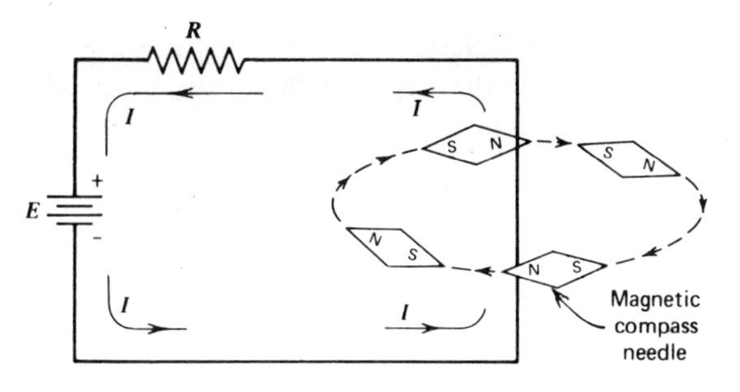

Figure 9-11. Magnetic compass needle lines up with magnetic field, which encircles wire.

a wire depends on the direction of the electron movement through it. In Fig. 9-12*a* current (electron movement) is shown going from left to right through a wire. The direction of the magnet lines may be predicted by a simple aid called the *left-hand wire rule*. If the *left* thumb is pointed along a wire in the direction of the electron movement, then the closed fingers of the left hand point in the direction of the magnetic lines. (The *right* hand is used if *conventional* current, from + to −, is followed, instead of the "electron current" from − to +.)

In Fig. 9-12*b* and *c*, *end* views of the current-carrying wire are shown. In part (b), the view is looking into the *right* end of the wire of part (a), with electrons coming *toward* the viewer. Notice that the *dot* shown in the center of the end view of the wire denotes the *point* of the arrow indicating the direction of electron movement. Using the left hand wire rule, hold the left thumb so that it points *out* from the paper of this page, in the direction of electron movement. The closed left fingers now point clockwise in the direction of the magnetic field that encircles the wire of Fig. 9-12*b*.

In Fig. 9-12*c*, the view is looking into the *left* end of the wire of part (a), with electrons going *away* from the viewer. Observe that the "X" or cross shown in the end view of the wire denotes the *tail* of the electron movement directional arrow. Again, the left-hand rule may be employed to determine

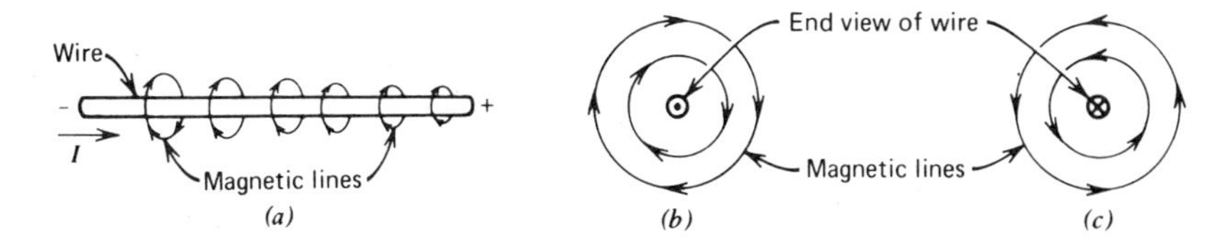

Figure 9-12. Magnetic field encircling current-carrying wire. (a) Magnetic lines. (*b*) Electrons coming toward viewer. (c) Electrons going away from viewer.

the direction of the magnetic lines encircling the wire. Hold the left thumb so that it now points *into* the paper of this page, in the direction of the electron movement. The closed left fingers now point counterclockwise in the direction of the magnetic field that encircles the wire of Fig. 9-12*c*.

The strength of the magnetic field (the number of lines per square unit area) depends on the current strength. A greater current in the wire produces more magnetic lines.

Refer to End-of-Chapter Problems 9-27 to 9-33

9-6. Electromagnetism in a Coil

If a piece of wire with current flowing through it is wrapped around a tubular-shaped cardboard form, as shown in Fig. 9-15*a*, then the magnetic field encircling one wire *adds* to the field of the adjacent wire, on the *same* side of the cardboard form. This is not too readily apparent, and can be explained by the following. Figure 9-13*a* shows the direction of the magnetic lines near the north pole of one magnet and the south pole of another, with the magnets a distance apart. Observe that in the *area between the poles*, one magnetic line points *upward*, while the other points *downward*. When the magnets are brought close together, as in Fig. 9-13*b*, the magnetic lines change their configuration to that shown, with the fields aiding and acting like one large bar magnet.

If two wires are placed parallel to each other, but a distance apart, with currents flowing in the *same* direction in each, then an end view of both wires

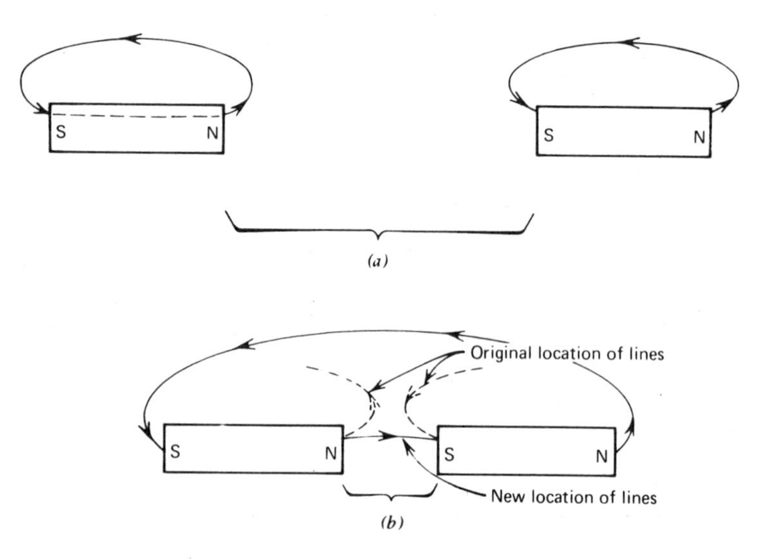

Figure 9-13. (*a*) Distance apart. (*b*) Close together.

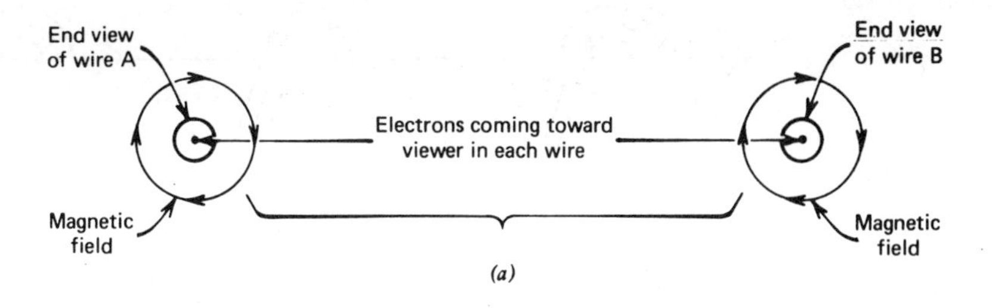

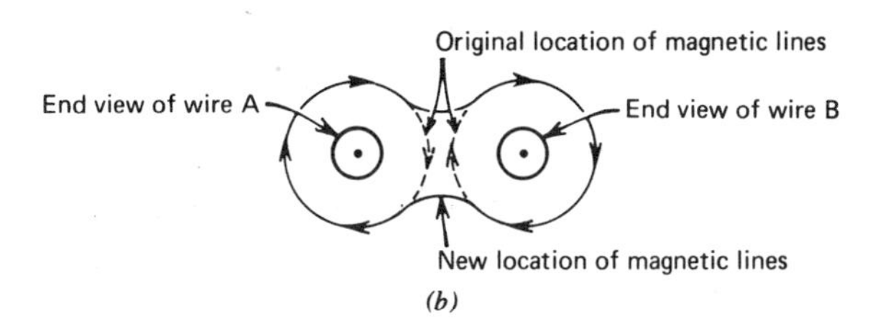

Figure 9-14. (*a*) Wires a distance apart. (*b*) Wires close together.

is as shown in Fig. 9-14*a*. If the electron flow is coming toward the viewer in both wires, then from the left hand rule, each wire is encircled by a clockwise magnetic field. Note that with the wires a distance apart, the *area between the wires* has one magnetic field pointing *upward*, while the other has a *downward* direction. When the wires are brought close together, as in Fig. 9-14*b*, the magnetic lines interact with each other and change their configuration to that shown, with the fields aiding. Note that the total magnetic field in Fig. 9-14*b* is still clockwise around both wires.

In Fig. 9-15*b*, a cross-sectional end view of the wires of the coil of part (a) is depicted. Current (electrons) flows *into* wires A, B, and C (away from viewer), and *out* of wires D, E, and F (toward the viewer). The magnetic fields encircling each wire are shown in part (b). If the coil wires are moved closer together, as shown in part (c), then the magnetic lines of each adjacent wire combine into the field shown. This is exactly the same as the field of a bar magnet, shown in Fig. 9-8. A coil of wire then, with a direct current flowing in it, acts as if it were a bar magnet, and is called an *electromagnet*.

Which ends are the north and south poles of a coil depend on the *direction* of current through the coil, and the *direction* that the coil is wound. The north and south poles of a coil or electromagnet may be predicted by a *left-hand coil rule*. This states that if the four fingers of the left hand are wrapped around the coil in the *direction* of the electron movement, then the extended left thumb will point to the north pole of the electromagnet.

In Fig. 9-16a the coil has *electrons* moving *downward* in the wires at the *front* of the coil (on the viewer's side of the coil). If the four fingers of the left

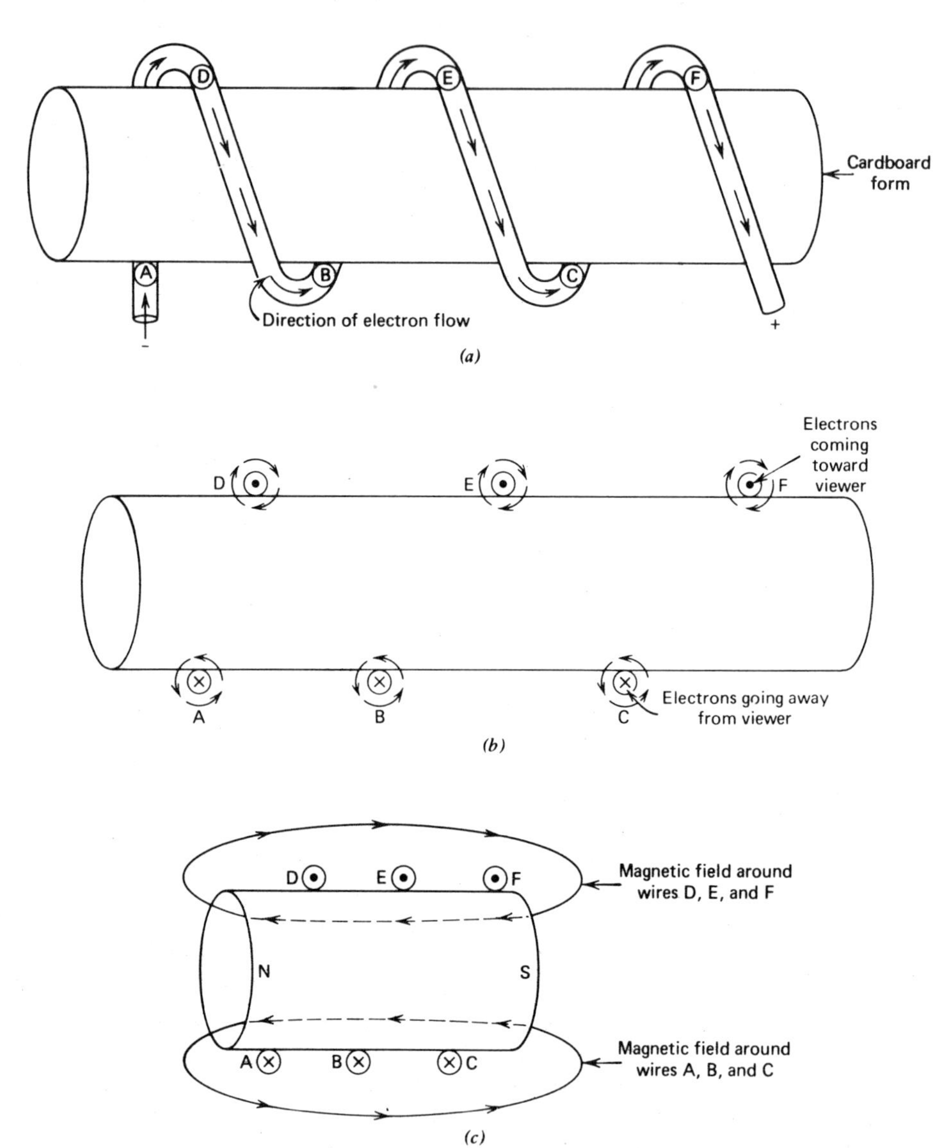

Figure 9-15. (*a*) Wire wrapped around a cardboard to form a coil. (*b*) Cross-sectional view of wires of coil, showing magnetic lines encircling each wire, wires widely spaced. (*c*) Cross-sectional view of wires of coil closely spaced; total magnetic field makes coil act like a bar magnet.

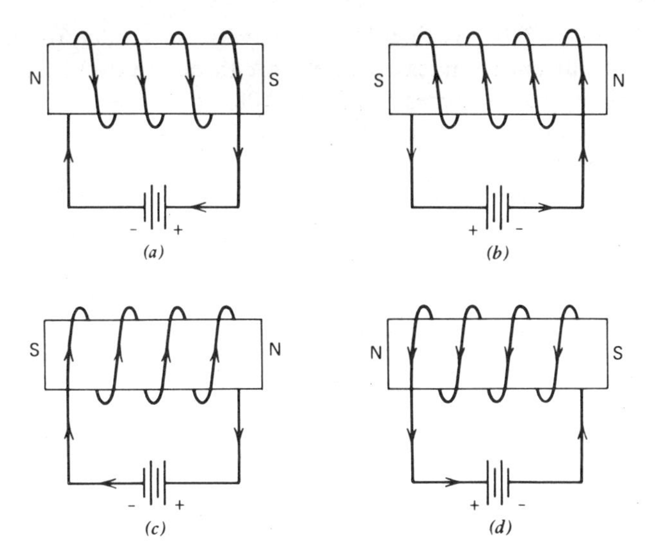

Figure 9-16. North and south polarities of electromagnets. Coils (*a*) and (*b*) are wound in the same direction, but have opposite direction currents. Coils (*c*) and (*d*) are wound in the same direction [but are reversed from coils in (*a*) and (*b*)], and also have opposite direction currents.

hand are wrapped around the coil so that they point downward in front of the coil, then the extended left thumb points to the *left* end of this coil. This is the north pole as shown.

In Fig. 9-16*b* the coil is *wound* in the same direction as the coil of part (a). However, the − and + voltage polarities have been reversed from part (a). Current (electron movement) is *upward* through the wires at the *front* of coil (b), as indicated. Following the left-hand coil rule, wrap the four fingers of the left hand so that they now point *up* in *front* of the coil. The extended left thumb now points to the *right* of the coil, and this is its north pole as shown in Fig. 9-16*b*.

Now use the left-hand rule to verify the magnetic poles as indicated in Fig. 9.16*c* and *d*.

Refer to End-of-Chapter Problems 9-34 and 9-35

9-7. Strength of an Electromagnet

In an electromagnet (a coil with current flowing through it), the strength of the magnetic field depends on the *number of turns* and the *amount of current*. This is called the *ampere-turns*. A coil having 100 turns of wire and 50 mA of current

(50 mA = 0.05 A) then has an *ampere-turns* value of 100 times 0.05, or 5. If the current is increased to 80 mA through this 100-turn coil, the ampere-turns product becomes 100 times 0.08, or 8.

The core around which the coil is wound also may affect the strength of the electromagnet. If the coil is wound on an iron core having a permeability of 300, the magnetic field becomes 300 times stronger than if the coil had been wound around a nonmagnetic material such as air or glass.

The ampere-turns is called the *magnetomotive force* (mmf), and is the total force that produces the magnetic lines. The *strength* of the magnetic field, or the *flux density* (see Section 9-2), is the number of magnetic lines per unit area. The strength of the field produced by the electromagnet is then actually dependent on *three* factors: *current*, *number of turns* (or ampere-turns), and the type of *core material.*

Refer to End-of-Chapter Problems 9-36 to 9-39

9-8. Practical Uses of Magnets and Electromagnets

Extensive use is made of magnets and electromagnets in electrical and electronic circuits. Among the more common are relays, meters, loudspeakers, transformers, electric motors and generators, and electron-beam deflection in a cathode-ray tube (the picture tube) in a television receiver or in a radar receiver.

In almost all these devices the basic idea is the same: current flowing through a coil produces a magnetic field. This electromagnetic field reacts with either a

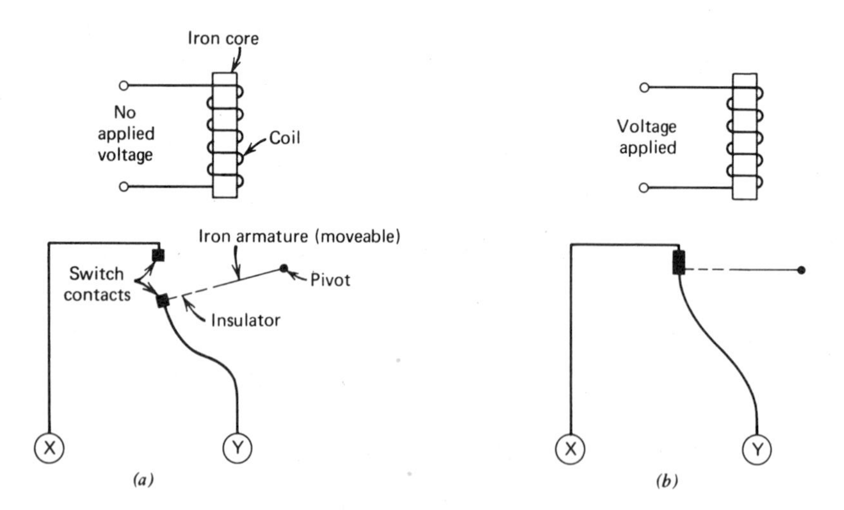

Figure 9-17. Simple relay. (*a*) Relay deenergized (switch normally open). (*b*) Relay energized (switch contacts held closed by the electromagnet).

ferromagnetic material, a permanent magnet, or another electromagnet, causing an action to occur.

The following preliminary discussions of some of these magnetic devices is given here to show the uses of magnetism. More thorough discussions are given in other chapters.

9-8A. The Relay

This is simply a switch that may be opened or closed by a current flowing through a coil. As shown in Fig. 9-17, when the coil is magnetized or *energized*, it attracts a moveable iron armature away from its previous spring-held position. The movement causes the switch contacts to close (make). When no current flows, the coil is unmagnetized (de-energized) and the contacts open (*break*). Figure 9-18 shows a relay circuit with two sets of switch contacts (*two poles*). (see Example 9-1.) Figure 9-19 shows a commercial relay. The reader is urged to see whether he can identify the coil, the armature, and the switch contacts in the photograph.

Example 9-1

Figure 9-18 shows a two-pole double-throw relay. Which alphabetically lettered terminals are making contact (connected together) when the relay is not energized, as drawn?

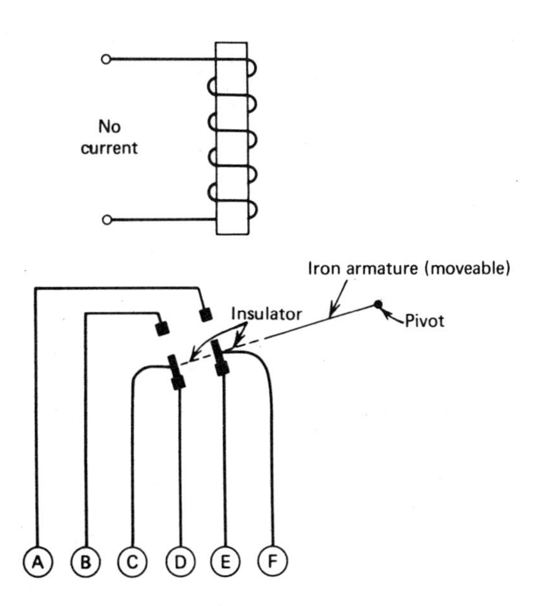

Figure 9-18. Two-pole, double-throw relay.

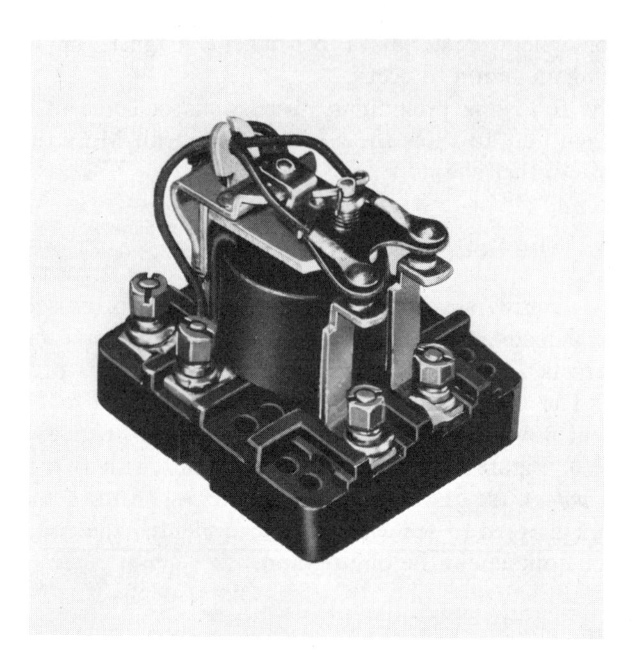

Figure 9-19. Relay (*Courtesy Ward-Leonard*).

Solution

With the iron armature in the downward position, the lower switch segments are making contact. Terminals C and D are connected together, and terminals E and F are also connected.

Refer to End-of-Chapter Problems 9-40 to 9-44

9-8B. The Meter

As shown in Fig. 9-20, when a current is allowed to flow through the meter coil, it becomes an electromagnet. If the current flows through the coil as shown, then the right side of the coil becomes its north pole, and it is attracted by the south pole of the permanent magnet. The coil now rotates clockwise around its pivot, away from its previous spring-held position. Movement of the coil causes the pointer or needle to swing along a graduated dial or scale. A more detailed description is given in Chapter 10.

Refer to End-of-Chapter Problems 9-45 to 9-47

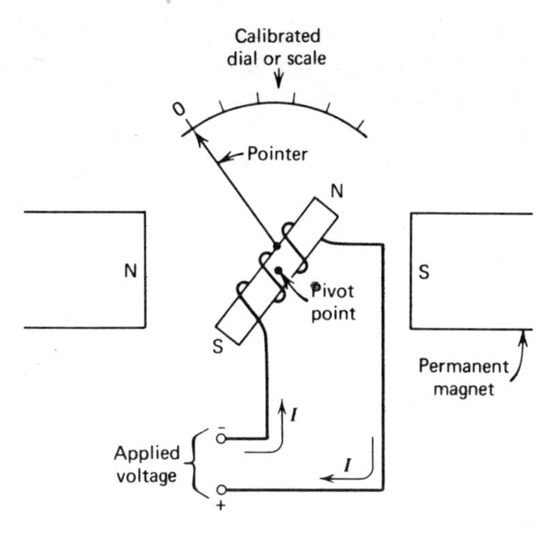

Figure 9-20. Basic moving-coil-type meter.

9-8C. The Loudspeaker

Another device employing magnetism is the loudspeaker shown in Fig. 9-21. A small, light, moveable coil called the *voice coil* is wound around a hollow cylindrically shaped cardboard core that is attached to a moveable large cardboard cone. An ac voltage called the audio signal is applied across the ends of the coil. The alternating current first has one polarity and then has the reverse polarity. The voice coil becomes magnetized at one instant so that it is

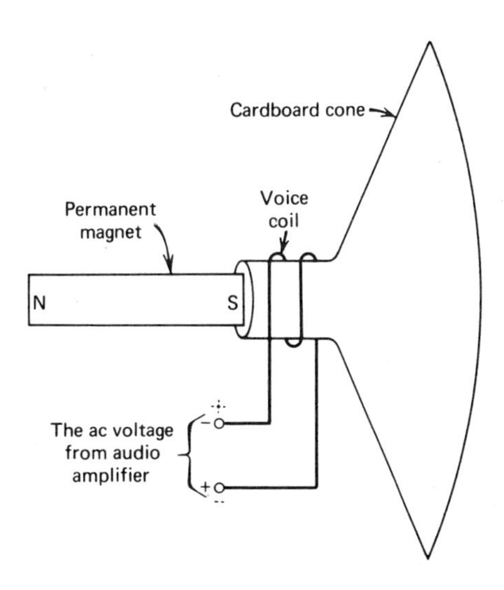

Figure 9-21. Loudspeaker.

repelled away from the permanent magnet, causing the large cone to move to the right, away from its normal spring-held position. At the next instant, with an opposite polarity voltage, the electromagnetic coil is attracted toward the permanent magnet, moving the large cone to the left. The movement of the large cone causes a large volume of air to vibrate, producing sound. The speaker shown is called a *permanent magnet (PM) dynamic* speaker. When a very large coil with a dc voltage across it is used instead of the permanent magnet, it is called an *electrodynamic* speaker.

Refer to End-of-Chapter Problems 9-48 to 9-52

9-8D. The Electric Motor Principle

A simple electric motor consists of a coil of wire mounted on an axle so that it is free to rotate, with the coil placed between the poles of a permanent magnet or strong electromagnet. This is similar to that of the meter shown in Fig. 9-20. When a voltage is applied to the ends of the coil, it becomes an electromagnet. Interaction between the permanent magnetic field and that of the

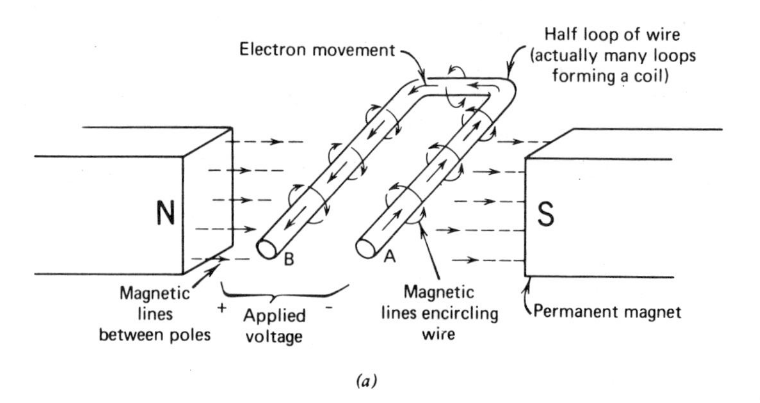

(a)

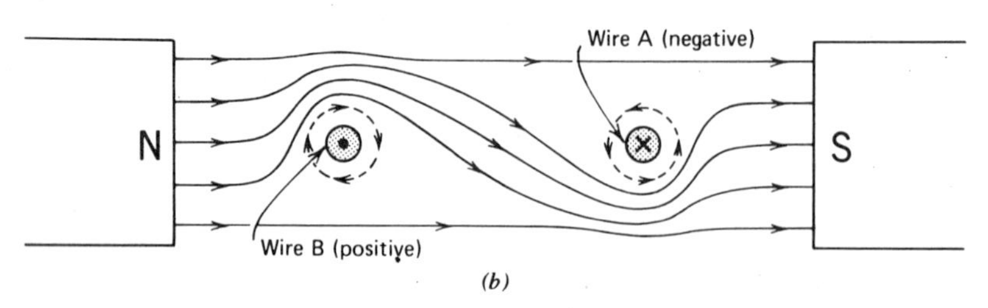

(b)

Figure 9-22. Electric motor principle. (*b*) End view of wires *A* and *B* of loop, with their encircling fields. Electrons flowing into *A* (going away from viewer), and out of *B* (coming toward viewer). *A* moves up; *B* moves down.

electromagnet causes a distortion of the permanent magnetic field, resulting in a movement of the coil.

A half-loop of wire, for simplicity, is shown in Fig. 9-22*a*. Current flow is into end A, through the wire, and out of end B. The magnetic lines encircling the coil wire are shown along with the lines between the poles of the permanent magnet. In part (b) of the drawing, it can be seen that the clockwise lines encircling wire B are in the same direction as the permanent magnetic lines above the wire, but are in the opposite direction below the wire. As a result, some of the permanent magnetic lines are distorted so that they pass above wire B where they have the same direction as the clockwise lines.

This produces a greater concentration of lines above wire B and a stronger field there. Since wire B is free to move, it will move downward toward the weaker magnetic field area. As B moves, it carries its encircling field downward, decreasing the concentration of lines in the area above the wire, as nature attempts to make the areas above and below wire B of equal magnetic strength. See Fig. 9-25 and the right-hand motor rule discussion in the next section.

The reader should determine how and why the other end of the coil, end A, moves up. A more thorough discussion of a dc motor is given in Chapter 10.

Refer to End-of-Chapter Problems 9-53 to 9-55

9-8E. Electromagnetic Deflection of the Electron Beam in a CRT

Another application of magnetism is the method of moving or deflecting the electron beam in a cathode-ray tube. The *picture tube* of a television receiver, and the *screen* or *indicator* of a radar set are a type of electron tube known as a *cathode-ray tube* (CRT). A simplified drawing of this tube is shown in Fig. 9-23. A portion of the tube, called the *electron gun*, emits a concentrated, narrow beam of electrons. These electrons strike the inside surface of the face or screen of the CRT, which is coated with a phosphor material. The

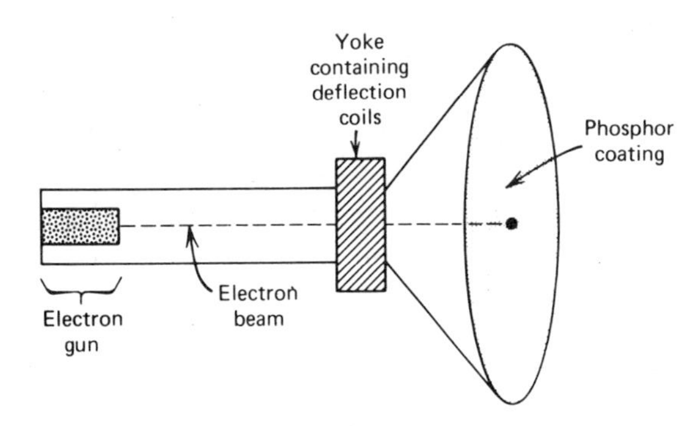

Figure 9-23. Simple cathode-ray tube.

phosphor glows at the point where it is struck by the electron beam, giving off a spot of light, which is seen by the person viewing the screen.

Around the narrow neck of the CRT is a *yoke*, inside of which are two pairs of coils. The electron beam passes between both pairs of coils. The *ac voltages* (voltages which keep reversing their polarity) are applied to each pair of coils. The reversing magnetic field between one pair of coils causes the electron beam to move vertically up and down. Similarly the field between the other pair of coils produces a horizontal deflection of the beam, left and right.

Figure 9-24*a* shows a front view of the CRT with the electron beam coming toward the viewer, as in wire B of the electric motor of Fig. 9-22*b*. A clockwise magnetic field encricles the beam as shown. When the ac voltage from the vertical amplifier has the polarity shown in Fig. 9-24*b*, the magnetic field between the vertical deflection coils is as shown. As discussed previously in Section 9-8D, Electric Motor Principle, the beam will move downward just as wire B did in Fig. 9-22*b*. An upward deflection of the beam is produced when the polarity of the applied ac deflection voltage reverses.

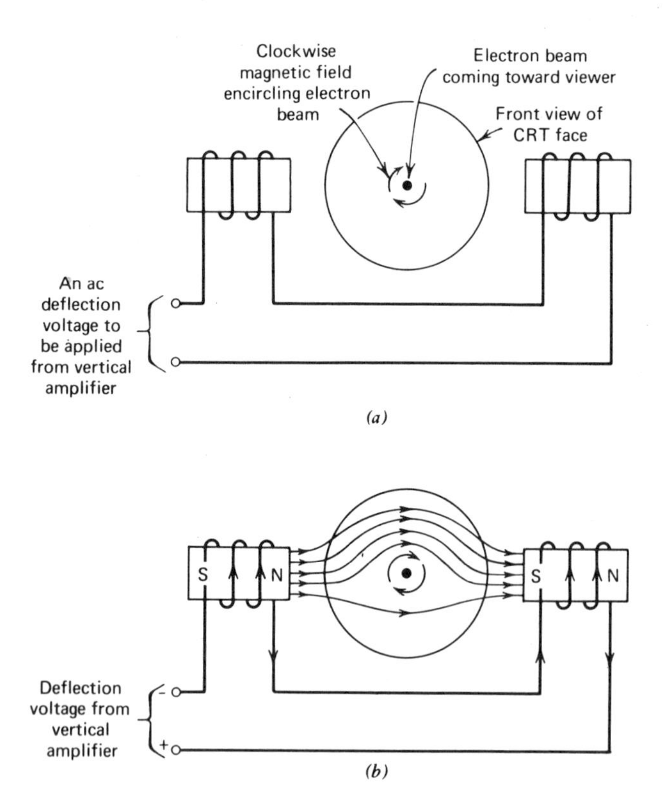

Figure 9-24. (*a*) Vertical deflection coils and cathode-ray tube. (*b*) Electron beam will be deflected downward.

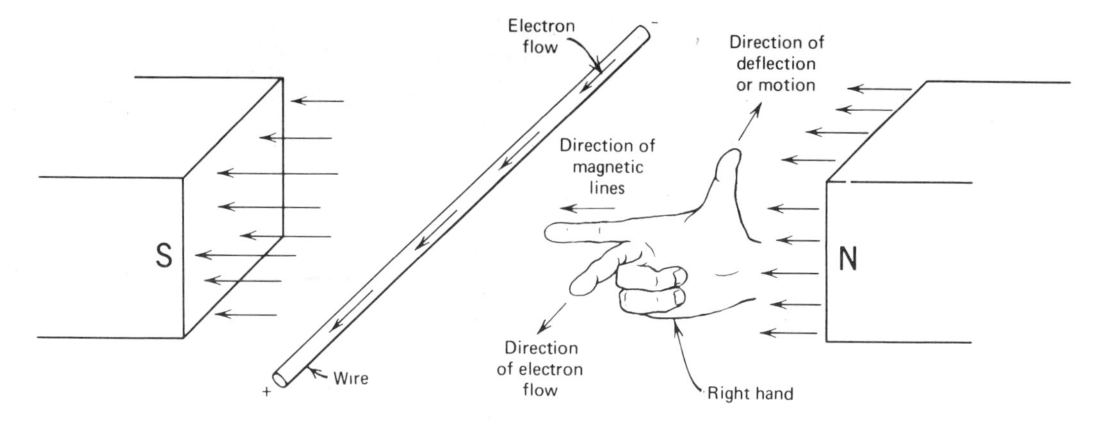

Figure 9-25. The right-hand motor rule.

The *direction* of the deflection or movement of the electron beam may be determined by the method described previously, or by another rule, called the *three-finger motor rule*, This involves the use of the thumb, index finger, and the middle finger of the right hand. As shown in Fig. 9-25, extend the right thumb, the right index finger, and the right middle finger so that they are at right angles to each other. Point the index finger in the direction of the magnetic lines, and the middle finger in the direction of the electron flow in the conductor. The thumb will now point in the *direction of deflection or movement* of the conductor that is carrying the electrons (in the case of a motor), or the electron beam (in the CRT).

Apply this three-finger (right-hand) motor rule for *each* wire of the simple motor of Fig. 9-22*b*, to determine whether wire A will deflect *upward*, while wire B deflects *downward*.

Refer to End-of-Chapter Problems 9-56 to 9-58

9-9. The Earth's Magnetism

The earth, spinning on its axis, acts as a huge magnet with its magnetic poles located at the upper and lower ends, near the geographic poles. The magnetism of the earth is due to the huge deposits of natural magnets in the earth and also, according to one explanation, results from the rotation of the earth.

Since the earth rotates from west to east, all parts of the earth are moving with it in this same direction. This movement of all parts of the earth (Fig. 9-26*b*) may be likened to the electron drift through the coil wires of Fig. 9-26*a*. In adapting the left-hand coil rule to the earth's rotation, the fingers should be

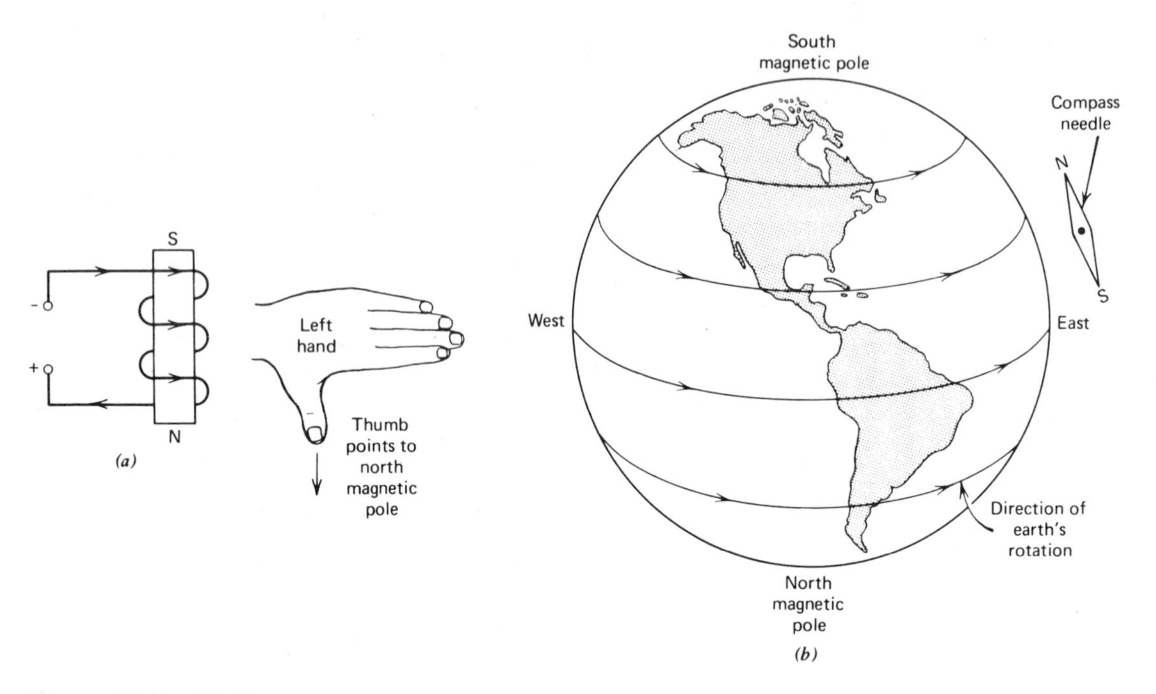

Figure 9-26. (*a*) Electromagnet. (*b*) Earth as a magnet.

pointed in the direction of the earth's rotation. Then, the extended left thumb points to the earth's *north* magnetic pole. Note that it is at the *bottom* end of the earth, where we might expect to find the earth's *south* magnetic pole!

This confusion is normal since, long ago, something was misnamed. The so-called *north* end of a bar magnet or a magnetic compass needle is actually a *south-seeking* pole. Observe that the magnetic compass needle shown in Fig. 9-26*b* has its *north* pole pointing to the *upper* end of the earth. Recalling that opposite poles attract, we see that this upper end of the earth must be the opposite of the north pole of the compass needle, or must be a *south* magnetic pole!

This confusing state of affairs is not helped by another, but opposite, explanation which holds that the upper end of the earth is, indeed, a north magnetic pole. The end of a magnetic compass needle (or bar magnet) which points to this upper end of the earth is then called a *north-seeking* pole, but is actually a *south* magnetic pole! However, since it points to the earth's (supposed) north pole, this north-seeking end of the compass needle is simply called a north pole—but, is, actually, the opposite of that of the earth.

Refer to End-of-Chapter Problems 9-59 and 9-60

MORE ADVANCED MAGNETISM

9-10. A Magnet

We previously pointed out in Section 9-3, Magnetic Theory, that the movement of electrons as they revolve around the nucleus of each atom produces a magnetic field. This is called the *electron orbital* motion, In an unmagnetized piece of iron, an equal number of electrons revolve around each nucleus in one direction as revolve in the opposite direction. As a result, the magnetic field produced by those electrons revolving, say, in a clockwise direction, cancel the field produced by those revolving in a counterclockwise direction. Another, and probably greater effect in producing a magnetic field is the *spinning* of the electrons on their axes. Remember that an electron resembles the action of the earth, which spins on its axis and also revolves around the sun. Similarly, the electron spins on its axis as it also revolves around its nucleus.

The magnetic field of numerous adjacent atoms add to one another, producing an area called a *domain*. In unmagnetized iron, the magnetic field of one domain is canceled by that of an adjacent domain. When more adjacent domains have magnetic fields which aid, the iron is magnetized. In a magnetically saturated iron, practically all domains are aligned with aiding magnetic fields.

If an iron ring in the shape of the letter "O" forms the core of an electromagnet (Fig. 9-27*a*), then when the current is increased, the magnetic field of the iron is increased. At saturation, however, increasing the current through the coil, produces no increase in the magnetic field. The *permeability* of this saturated iron, or its ability to permit the production of a larger magnetic field, decreases almost to that of a nonmagnetic substance. To prevent the iron from being magnetically saturated, the iron core may be changed from the shape of the letter "O" to the shape of the letter "C," or even to the shape of the numeral "1," as shown in Fig. 9-27*b* and *c*. The complete iron ring core is easily saturated. In Fig. 9-27*b* and *c*, an air space or *air gap* is introduced between the ends of the iron core. Since air cannot be magnetized, the air gap actually prevents the iron cores from easily becoming saturated.

Refer to End-of-Chapter Problems 9-61 to 9-63

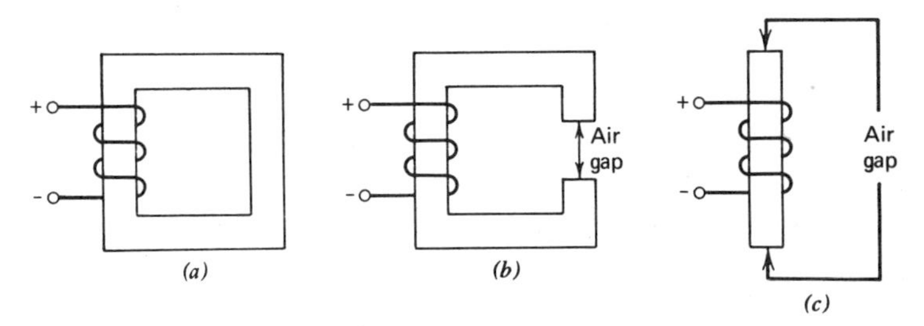

Figure 9-27.

9-11. Magnetic Units

Table 9-1 lists the terms and symbols used in magnetism. Each term is discussed more fully in the next five sections. These are *magnetic field* or *flux* (Section 9-12), the force or *magnetomotive force* (mmf) that produces the flux (Section 9-13), the *strength* of the magnetic field, called the *flux density* (B) (Section 9-14), the *magnetizing force* (H) (Section 9-15), *permeability* (μ) or the ease with which magnetic lines travel along a material compared with air (Section 9-16), and *reluctance* ($\mathcal{R}$) or the opposition of a material to the production of a magnetic field (Section 9-16).

Refer to End-of-Chapter Problems 9-64 to 9-70

9-12. Flux

The magnetic field or *flux* (ϕ) consists of numerous invisible flux lines, the presence of which is indicated by the pattern of iron filings that have been sprinkled over a magnet. In the mks system (meter-kilogram-second), flux is measured in *webers.* A weber is 10^8 lines, as listed in item one, Table 9-1. Since a weber is a large number of magnetic lines, micro webers are often used. In the cgs system (centimeter-gram-second) flux is measured in *maxwells.* One maxwell is one flux line. Note that one weber is, therefore, 10^8 maxwells, and a micro weber is 10^2 maxwells.

Some of the magnetic terms, *flux*, *magnetomotive force* (Section 9-13), and *reluctance* (Section 9-16, are interrelated, forming what is often called the magnetic Ohm's law. *Flux* (ϕ) is compared to the current in the electrical circuit; *magnetomotive force* (mmf) is compared to voltage; and *reluctance* ($\mathcal{R}$) is comparable to electrical resistance. Therefore, since

$$\text{current } (I) = \frac{\text{voltage } (E)}{\text{resistance } (R)} \qquad \text{(Ohm's law)}$$

then

$$\text{flux } (\phi) = \frac{\text{magnetomotive force (mmf)}}{\text{reluctance } (\mathcal{R})} \tag{9-1}$$

and

$$\text{mmf} = \phi\mathcal{R} \tag{9-5}$$

Refer to End-of-Chapter Problems 9-71 to 9-74

9-13. Magnetomotive Force

The force that produces the magnetic field is called the *magnetomotive force*, and the symbol is mmf. This is listed as item two in Table 9-1. The mmf is measured in units called *ampere-turns* in the mks system. An ampere-turn is the *product* of the *current* in a coil, in *amperes*, and the *number* of *tutns* of wire in the coil. *Ampere-turns* are usually abbreviated and written as *ampere-turn*, *AT*, or *NI* (where *N* is the *number* of turns, and *I*, the current). *Magnetomotive force* then is

$$\text{mmf} = NI \tag{9-2}$$

where mmf is in ampere-turns, N is the number of turns, and I is the current in amperes.

The following examples use equation 9-2.

Example 9-2

A coil with 200 turns has 300 mA flowing through it. Find the magnetomotive force or mmf in the mks system.

Solution

$$\text{mmf} = NI \quad \text{(from equation 9-2)}$$

$$\text{mmf} = (200 \text{ turns})\,(300 \times 10^{-3} \text{ A})$$

$$\text{mmf} = 60 \text{ At}$$

Example 9-3

If an mmf of 120 At is desired, how much voltage should be applied to a 3000-turn coil having a resistance of 500 Ω.

Solution

First the required current must be found.

$$\text{mmf} = NI \quad \text{(from equation 9-2)}$$

$$\frac{\text{mmf}}{N} = I$$

$$\frac{120}{3000} = I$$

$$0.04 \text{ A} = I$$

Now, using Ohm's law, find the required voltage.

$$E = IR$$

$$E = (0.04)\,(500)$$

$$E = 20 \text{ V}$$

For Similar Problems Refer to End-of-Chapter Problems 9-75 to 9-77

TABLE 9-1 Magnetic Terms, Units, and Equations

	Equation	mks System	cgs System
1. Magnetic field or *flux* lines (ϕ)	Flux = $\frac{\text{magnetomotive force}}{\text{reluctance}}$ $\phi = \frac{\text{mmf}}{\mathcal{R}} = BA$	*weber* 1 Wb = 10^8 lines	*maxwell* 1 Mx = 1 magnetic line (1 Wb = 10^8 Mx)
2. *Magnetomotive force* producing the field (mmf)	*Magnetomotive force* = (flux) (reluctance) mmf = $\phi\mathcal{R}$ mmf = NI	*ampere-turn*	*gilbert* (1 At = 1.256 Gbm Gi 0.7958 At = 1 Gbm Gi)
3. Magnetic field strength or *flux density*, is lines per unit area (B)	Flux density = $\frac{\text{flux}}{\text{area}}$ $B = \frac{\phi}{A}$	*tesla* 1 T = 1 Wb/m^2	*gauss* 1 G = 1 Mx/cm^2 (1 T = 10^4 G)
4. *Magnetizing force* or magnetomotive force per unit length (H)	Magnetizing force = $\frac{\text{magnetomotive force}}{\text{length}}$ $H = \frac{\text{mmf}}{l}$	*ampere-turns per meter*	*oersted* 1 Oe = 1 Gbm Gi/cm (1 At/m = 0.01256 Oe 79.58 At/m = 1 Oe)

5. *Permeability* or the ease of passing magnetic lines along a substance compared to air, indicates the ability to permit the production of a magnetic field. It is the permeance (see below) per unit length and per unit area (μ)	Permeability $= \dfrac{\text{flux density}}{\text{magnetizing force}}$ $\mu = \dfrac{B}{H}$ $\mu = \dfrac{l}{\mathcal{R}A}$ $\mu = \mathcal{P}\left(\dfrac{l}{A}\right)$	Permeability (μ) $= \dfrac{\text{tesla}}{\text{ampere-turn per meter}}$ For nonmagnetic materials, $\mu = (4\pi)(10^{-7})$ $\mu = (12.56)(10^{-7})$	Permeability (μ) $= \dfrac{\text{gauss}}{\text{oersted}}$ For nonmagnetic materials, $\mu = 1$
6. *Reluctance* is the opposition of a substance to the production of a magnetic field, similar to resistance in an electric circuit ($\mathcal{R}$)	Reluctance $= \dfrac{\text{magnetomotive force}}{\text{flux}}$ $\mathcal{R} = \dfrac{\text{mmf}}{\phi}; = \dfrac{l}{\mu A}$	Reluctance $= \dfrac{\text{ampere-turn}}{\text{weber}}$	Reluctance $= \dfrac{\text{gilbert}}{\text{maxwell}}$
7. *Permeance* is the reciprocal of reluctance; similar to conductance in an electric circuit. It is the ability of a substance to permit the production of a magnetic field ($\mathcal{P}$)	Permeance $= \dfrac{1}{\text{reluctance}}$ $\mathcal{P} = \dfrac{1}{\mathcal{R}}$	Permeance $= \dfrac{\text{weber}}{\text{ampere-turn}}$	Permeance $= \dfrac{\text{maxwell}}{\text{gilbert}}$

From item two, mmf, in Table 9-1, it may be seen that mmf is measured in units called *gilberts* in the cgs system (centimeter-grams-seconds). The following relationships convert from gilberts (Gbm Gi) (in the cgs system) to ampere-turns (in the mks system).

$$1 \text{ Gbm Gi} = 0.7958 \text{ At} \qquad (9\text{-}3)$$

$$1 \text{ At} = 1.256 \text{ Gbm Gi} \qquad (9\text{-}4)$$

The following examples employ equations 9-3 and 9-4.

Example 9-4

A 2500-turn coil having a resistance of 1000 Ω has 300 V applied to it. What is the mmf in gilberts (cgs system)?

Solution

Using Ohm's law, first find the current.

$$I = \frac{E}{R}$$

$$I = \frac{300}{1000}$$

$$I = 0.3 \text{ A}$$

Now find the mmf in ampere-turns (mks system).

$$\text{mmf} = NI \qquad (9\text{-}2)$$

$$\text{mmf} = (2500)\,(0.3)$$

$$\text{mmf} = 750 \text{ At}$$

Finally, convert ampere-turns to gilberts, using equation 9-4.

$$1 \text{ At} = 1.256 \text{ Gbm Gi} \qquad (9\text{-}4)$$

$$750 \text{ At} = (750)\,(1.256 \text{ Gbm Gi})$$

$$750 \text{ At} = 940 \text{ Gbm Gi}$$

For Similar Problems Refer to End-of-Chapter Problems 9-78 to 9-81

9-14. Flux Density

The *magnetic field* strength or *flux density* is the *number of magnetic lines or flux* (ϕ) *per unit area*, where the unit area is that cross-sectional area which is *perpendicular* to the lines. The letter B is used to denote *flux density*. As shown listed in item three of Table 9-1,

$$\text{flux density} = \frac{\text{flux}}{\text{area}} \qquad (9\text{-}6)$$

$$B = \frac{\phi}{A} \qquad (9\text{-}6)$$

In the mks system (meter-kilogram-second), flux density (B) is measured in a unit called *tesla* (T), where

$$1 \text{ tesla} = 1 \text{ weber (or } 10^8 \text{ lines) per square meter} \qquad (9\text{-}7)$$

In the cgs system, flux density (B) is measured in a unit called *Gauss* (G), where

$$1 \text{ gauss} = 1 \text{ maxwell (or 1 magnetic line) per square centimeter} \qquad (9\text{-}8)$$

Since a square meter (written as m^2) contains 100 centimeters times 100 centimeters, or 10,000 square centimeters (written as 10^4 cm^2) then, as listed in item three of Table 9-1,

$$1 \text{ T} = 10^4 \text{ G} \qquad (9\text{-}9)$$

The following examples for flux density (B) use equations 9-6, 9-7, 9-8, and 9-9.

Example 9-5

A magnetic field has 0.5 weber (Wb) in a perpendicular cross-sectional area of 2 m^2. What is the *flux density* (B) in *teslas* in the mks system?

Solution

$$B = \frac{\phi}{A} \qquad \text{(from equation 9-6)}$$

$$B = \frac{0.5 \text{ Wb}}{2 \text{ m}^2}$$

$$B = 0.25 \text{ T} \qquad \text{or} \qquad \text{Wb/m}^2$$

Example 9-6

How many *magnetic lines* are present if the flux is 0.01 Wb?

Solution

Since 1 Wb = 10^8 lines, then the flux (ϕ) of 0.01 Wb is

$$1 \text{ Wb} = 10^8 \text{ lines} \qquad \text{(from equation 9-7)}$$

and

$$0.01 \text{ Wb} = (0.01)(10^8 \text{ lines})$$
$$'' \text{ Wb} = (10^{-2})(10^8)$$
$$'' \text{ Wb} = 10^6 \text{ magnetic lines}$$
$$'' \text{ Wb} = 1{,}000{,}000 \text{ magnetic lines}$$

Example 9-7

Convert a flux density of 1250 G (cgs system) into tesla (mks system).

Solution

$$1 \text{ T} = 10^4 \text{ G} \qquad \text{(from equation 9-9)}$$
$$\frac{\text{T}}{10^4} = \text{G}$$
$$10^{-4} \text{ T} = \text{G}$$
$$(1250)\, 10^{-4} \text{ T} = 1250 \text{ G}$$
$$0.125 \text{ T} = 1250 \text{ G}$$

For Similar Problems Refer to End-of-Chapter Problems 9-82 to 9-91

9-15. Magnetizing Force

The *magnetomotive force* (mmf) (discussed in Section 9-13) per unit *length* is called the *magnetizing force* and the letter H is the symbol used. As shown listed in item four of Table 9-1,

$$\text{magnetizing force} = \frac{\text{magnetomotive force}}{\text{length}} \tag{9-10}$$

$$H = \frac{\text{mmf}}{l} \tag{9-10}$$

In the mks system the unit for magnetizing force is *ampere-turn per meter*. The ampere-turn is the mmf unit. In the cgs system the unit for H is *oersted*. An oersted (Oe) is a *gilbert per centimeter*. The gilbert is the mmf unit in the cgs system. To convert from the oersted to the ampere-turn per meter, or vice versa, the following equations are useful.

$$1 \text{ At/m} = 0.01256 \text{ Oe} \tag{9-11}$$

and

$$1 \text{ Oe} = 79.58 \text{ At/m} \tag{9-12}$$

In equation 9-10, l is the length of the *coil* in the case of an air core. When iron is used as the core of the coil, l is the length of the core. The following examples illustrate this.

Example 9-8

An air-core coil 30 cm long consists of 500 turns with 200 mA flowing through it. Find (a) the *magnetizing force* (H) in the mks system, and (b) the *magnetizing force* (H) in the cgs system.

Solution

(a) In the mks system,

$$\text{magnetizing force} = \frac{\text{magnetomotive force in } \textit{ampere-turns}}{\text{length in } \textit{meters}}$$

(from equation 9-10)

$$H = \frac{(200 \text{ mA})(500 \text{ turns})}{30 \text{ cm}}$$

$$H = \frac{(200 \times 10^{-3} \text{ A})(5 \times 10^{2})}{30 \times 10^{-2} \text{ m}}$$

$$H = 333.3 \text{ At/m}$$

(b) In the cgs system,

$$\text{magnetizing force} = \frac{\text{magnetomotive force in } \textit{gilberts}}{\text{length in } \textit{centimeters}}$$

(from equation 9-10)

To convert magnetomotive force (mmf) in ampere-turns into gilberts,

$$1 \text{ At} = 1.256 \text{ Gbm Gi} \qquad \text{(from equation 9-4)}$$

Therefore

$$\text{magnetizing force} = \frac{(1.256)(\text{ampere-turns})}{\text{length in centimeters}}$$

$$H = \frac{(1.256)(200 \text{ mA})(500 \text{ turns})}{30 \text{ cm}}$$

$$H = \frac{(1.256)(200 \times 10^{-3} \text{ A})(5 \times 10^{2})}{3 \times 10}$$

$$H = 4.19 \text{ Oe}$$

It should be pointed out that in an air-core coil many of the magnetic lines do not go through the entire coil length, and do not enter at one end and leave at the other. Since the air outside the coil has the same *permeability* (passes the lines with equal ease) as the air core inside, many lines enter and leave the *sides* of the coil, only going through the very center of it. As a result, the magnetizing force (H) found in Example 9-8*a* of 333.3 At/m, and in part (b) of 4.19 Oe is at the *center* of the coil only

When an iron core is used, most of the magnetic lines travel through the entire length of the iron, entering at one end and leaving at the other, since the iron core has much greater permeability than the air outside the coil. That is, the iron offers a much easier path for the magnetic lines than the air. As a result, the magnetizing force (H), is practically uniform throughout the iron core.

Example 9-9

An iron-core coil 8 cm long has a *core* length of 10 cm. If the coil has 800 turns with 50 mA flowing through it, find the magnetizing force (H) in (a) the mks system of ampere-turns per meter, and (b) the cgs system of oersteds.

Solution

Solve this example by the same method shown in Example 9-8, but use the *core* length of 10 cm. The answers are (a) 400 At, and (b) 5.024 Oe.

For Similar Problems Refer to End-of-Chapter Problems 9-92 to 9-99

9-16. Permeability, Permeance, and Reluctance

Refer to Sections 9-4 and 9-11 for introductory discussions of these topics. The ease with which a material allows a magnetic line to pass along that material compared to air is often referred to as its *relative permeability*. Permeability indicates the ability of a substance to permit the production of a magnetic field. The Greek letter μ (also used for the prefix *micro*) is the symbol for permeability.

A similar term that is very closely associated with permeability is called *permeance*. Permeance is the ability of a substance to permit the production of a magnetic field, whereas permeability indicates this ability. The symbol for permeance is the script letter $\mathcal{P}$. The relationship between the two terms is simply that *permeability* is the *permeance per unit length* and *per unit area*. This is shown listed in Table 9-1, item five, as

$$\text{permeability} = \text{permeance}\left(\frac{\text{length}}{\text{area}}\right) \tag{9-13}$$

$$\mu = \mathcal{P}\left(\frac{l}{A}\right) \tag{9-13}$$

permeability, μ, is also the ratio of the flux density (B) to the magnetizing force (H), as shown in Table 5-1 and in the following.

$$\text{permeability} = \frac{\text{flux density}}{\text{magnetizing force}} \tag{9-14}$$

$$\mu = \frac{B}{H} \tag{9-14}$$

Permeance ($\mathcal{P}$) is similar to *conductance* (G) in an electrical circuit, where conductance is the reciprocal of resistance, or $G = 1/R$. Similarly, as listed in item seven of Table 9-1.

$$\text{permeance} = \frac{1}{\text{reluctance}} \tag{9-15}$$

$$\mathcal{P} = \frac{1}{\mathcal{R}} \tag{9-15}$$

where *reluctance* ($\mathcal{R}$) is the opposition of a substance to the production of a magnetic field. *Reluctance* ($\mathcal{R}$) may be compared to the resistance of an electrical circuit. Similar to Ohm's law, where resistance = voltage/current, the following relationship occurs in magnetic circuits. This is from Section 9-12, equation 9-1, and Section 9-13, equation 9-5. These are also shown in Table 9-1 in items one and two.

$$\text{reluctance} = \frac{\text{magnetomotive force}}{\text{flux}} \tag{9-16}$$

$$\mathcal{R} = \frac{\text{mmf}}{\phi} \tag{9-16}$$

This is shown in Table 9-1 for item six.

Another useful equation involving *reluctance* ($\mathcal{R}$) and *permeability* (μ) is the following:

$$\text{permeability} = \frac{\text{length}}{(\text{reluctance})\ (\text{area})} \tag{9-17}$$

$$\mu = \frac{l}{\mathcal{R}A} \tag{9-17}$$

and solving in terms of reluctance ($\mathcal{R}$) gives

$$\mathcal{R} = \frac{l}{\mu A} \tag{9-18}$$

As shown in Table 9-1, item five, *permeability* (μ) in the mks system (meter-kilogram-seconds) is *tesla* (weber per square meter) per *ampere-turn per meter*. For nonmagnetic materials, the value of permeability is approximately that of a vacuum or air, and in the mks system it is 12.56×10^{-7}.

In the cgs system (centimeter-gram-seconds), *permeability* is *gauss* (maxwell per square centimeter) per *oersted* (gilbert per centimeter), and the value of permeability for a nonmagnetic material is 1 in the cgs system.

Reluctance ($\mathcal{R}$) as shown in Table 9-1, item six, in the mks system is *ampere-turn* per *weber*, while in the cgs system it is *gilbert* per *maxwell.*

Permeance ($\mathcal{P}$), the reciprocal of *reluctance*, is shown listed as item seven of Table 9-1. In the mks system, *permeance* is *weber* per *ampere-turn*. In the cgs system, it is *maxwell* per *gilbert*.

The following are examples of magnetic circuit problems that require the application of one or several of the equations shown listed in Tables 9-1 and 9-2.

Example 9-10

In the diagram of Fig. 9-28, find (a) *magnetomotive force* (mmf) in ampere-turns, (b) *reluctance* ($\mathcal{R}$), and (c) *flux density* (B).

Solution

(a) $$\text{magnetomotive force} = (\text{number of turns})\,(\text{current in amperes}) \quad (9\text{-}2)$$

$$\text{mmf} = NI$$

$$\text{mmf} = (500)\,(30 \times 10^{-3})$$

$$\text{mmf} = 15{,}000 \times 10^{-3}$$

$$\text{mmf} = 15 \text{ At}$$

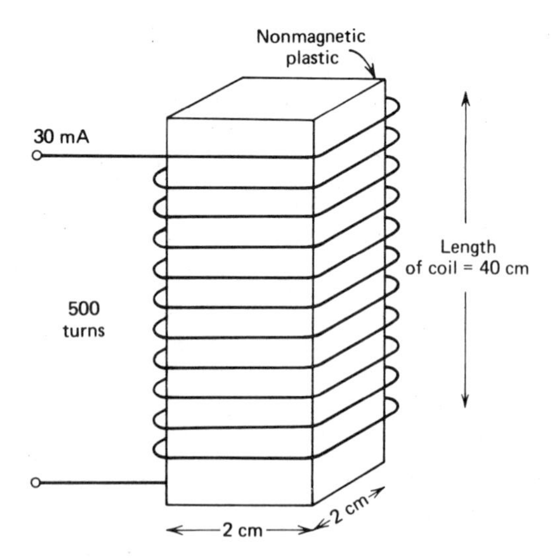

Figure 9-28. Example 9-10.

TABLE 9-2 Compilation of Magnetic Equations

Equation Number	Terms	Symbols
[a]9-1	flux $= \dfrac{\text{magnetomotive force}}{\text{reluctance}}$	$\phi = \dfrac{\text{mmf}}{\mathcal{R}}$
9-2	magnetomotive force = (number of turns) (current)	mmf $= NI$
9-3	1 Gbm Gi = 0.7958 At	
9-4	1 At = 1.256 Gbm Gi	
[a]9-5	magnetomotive force = (flux) (reluctance)	mmf $= \phi\mathcal{R}$
9-6	flux density $= \dfrac{\text{flux}}{\text{area}}$	$B = \dfrac{\phi}{A}$
9-7	1 T = 1 Wb (or 10^8 lines) per square meter	
9-8	1 G = 1 Mx/cm^2	
9-9	1 T = 10^4 G	
9-10	magnetizing force $= \dfrac{\text{magnetomotive force}}{\text{length}}$	$H = \dfrac{\text{mmf}}{l}$
9-11	1 At/m = 0.01256 Oe	
9-12	1 Oe = 79.58 At/m	
9-13	permeability = (permeance) $\left(\dfrac{\text{length}}{\text{area}}\right)$	$\mu = \mathcal{P}\left(\dfrac{l}{A}\right)$
9-14	permeability $= \dfrac{\text{flux density}}{\text{magnetizing force}}$	$\mu = \dfrac{B}{H}$
9-15	permeance $= \dfrac{1}{\text{reluctance}}$	$\mathcal{P} = \dfrac{1}{\mathcal{R}}$
[a]9-16	reluctance $= \dfrac{\text{magnetomotive force}}{\text{flux}}$	$\mathcal{R} = \dfrac{\text{mmf}}{\phi}$
9-17	permeability $= \dfrac{\text{length}}{\text{(reluctance) (area)}}$	$\mu = \dfrac{l}{\mathcal{R}A}$
9-18	reluctance $= \dfrac{\text{length}}{\text{(permeability) (area)}}$	$\mathcal{R} = \dfrac{l}{\mu A}$

[a] Magnetic "Ohm's law."

(b)
$$\text{reluctance} = \frac{\text{length in meters}}{\text{(permeability) (area in square meters)}} \tag{9-18}$$

$$\mathcal{R} = \frac{l}{\mu A}$$

where *permeability* (μ) of a nonmagnetic material is $(4\pi)(10^{-7})$, as listed in Table 9-1 item five under the mks system.

$$\mathcal{R} = \frac{40 \times 10^{-2}}{(4\pi)(10^{-7})(2 \times 10^{-2})^2}$$

$$\mathcal{R} = \frac{40 \times 10^{-2}}{(12.56)(10^{-7})(4 \times 10^{-4})}$$

$$\mathcal{R} = 0.796 \times 10^{9} \text{ At/Wb}$$

(c)
$$\text{flux density} = \frac{\text{flux}}{\text{area}} \tag{9-6}$$

$$B = \frac{\phi}{A}$$

but $\phi = \text{mmf}/\mathcal{R}$ (equation 9-1), and substituting for ϕ in equation 9-6 gives

$$B = \frac{\text{mmf}/\mathcal{R}}{A}$$

$$B = \frac{15/(0.796)(10^9)}{(2 \times 10^{-2})^2}$$

$$B = \frac{18.85 \times 10^{-9}}{4 \times 10^{-4}}$$

$$B = 4.71 \times 10^{-5} \text{ T} \qquad (\text{or Wb/m}^2)$$

Example 9-11

In Fig. 9-29, 5000 turns of wire are wrapped around an iron ring having a permeability of 150 with a circular cross section. If the inner diameter (D_1) is 6 cm, the outside diameter (D_2) is 8 cm, and 200 mA of current flows, find (a) reluctance ($\mathcal{R}$), (b) magnetic flux (ϕ), (c) flux density (B), and (d) magnetizing force (H).

Solution

The average length of the coil between the extreme upper and lower turns of wire (or the center circumference of the iron ring), and the cross-sectional *area* of the iron core must first be determined. The *average length* of the coil and its iron core is simply the average of the *inner* and *outer circumferences.*

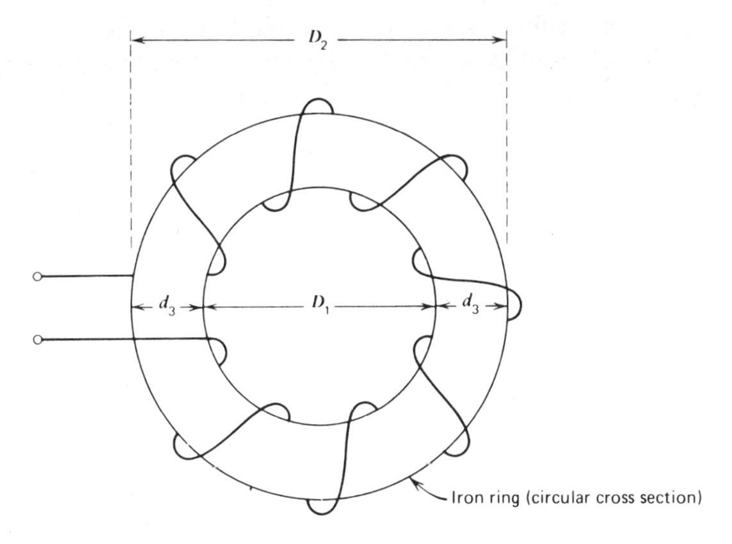

Figure 9-29. Toroid coil, Example 9-11.

The mathematical solution for the *circumference* of a circle is $2\pi r$, where r is the *radius*, and the *radius* is half of the *diameter*. The outer diameter D_2 is 8 cm, and the radius (r_2) of the outer circle is half of 8, or 4 cm. The inner diameter D_1 is 6 cm, and its radius (r_1) is half of 6, or 3 cm.

Therefore, the outer circumference in meters is

$$\begin{aligned}\text{Circum}_{\text{outer}} &= 2\pi r_2\\ &= (2)\,(3.14)\,(4 \times 10^{-2})\\ &= 25.12 \times 10^{-2}\text{ m}\end{aligned}$$

and

$$\begin{aligned}\text{Circum}_{\text{inner}} &= 2\pi r_1\\ &= (2)\,(3.14)\,(3 \times 10^{-2})\\ &= 18.84 \times 10^{-2}\text{ m}\end{aligned}$$

The average core length, or its average circumference is

$$\text{Circum}_{\text{avg}} \quad \text{or} \quad \text{core length} = \frac{\text{Circum}_{\text{outer}} + \text{Circum}_{\text{inner}}}{2}$$

$$\text{average length} = \frac{25.12 \times 10^{-2} + 18.84 \times 10^{-2}}{2}$$

$$\text{average length} = \frac{43.96 \times 10^{-2}}{2}$$

$$\text{average length} = 21.98 \times 10^{-2}\text{ m}$$

The width or thickness (d_3) of the iron ring itself may be determined from the values of $D_2 = 8$ cm and $D_1 = 6$ cm of Fig. 9-29. Note that the outer diameter D_2 is equal to the sum of $d_3 + D_1 + d_3$. Solving for d_3 gives

$$D_2 = d_3 + D_1 + d_3$$

$$8 \text{ cm} = d_3 + 6 \text{ cm} + d_3$$

$$8 - 6 = 2d_3$$

$$2 = 2d_3$$

$$1 \text{ cm} = d_3$$

The circular cross-sectional area of the iron ring is found by using the equation for the area of a circle $= \pi r^2$, where r is the radius (r_3) of the cross-sectional area. Since d_3 is 1 cm, then r_3 is half of that or 0.5 cm. Therefore, the circular *cross-sectional area* of the iron ring in square meters is

$$A = \pi r_3^2$$

$$A = (3.14)\,(0.5 \times 10^{-2})^2$$

$$A = (3.14)\,(0.25 \times 10^{-4})$$

$$A = 0.785 \times 10^{-4} \text{ m}^2$$

Now, knowing the core *length* (21.98×10^{-2} m) and the cross-sectional *area* (0.785×10^{-4} m^2), the main parts of this example are solved as follows.

(a)
$$\text{reluctance} = \frac{\text{length in meters}}{\text{(permeability) (area in square meters)}} \qquad (9\text{-}18)$$

$$\mathcal{R} = \frac{l}{(\mu)\,(A)}$$

$$\mathcal{R} = \frac{21.98 \times 10^{-2}}{(150)\,(0.785 \times 10^{-4})}$$

$$\mathcal{R} = \frac{21.98 \times 10^{-2}}{117.7 \times 10^{-4}}$$

$$\mathcal{R} = 0.187 \times 10^{2}$$

$$\mathcal{R} = 18.7 \text{ At/Wb}$$

(b)
$$\text{magnetic flux} = \frac{\text{magnetomotive force}}{\text{reluctance}} \qquad (9\text{-}1)$$

$$\phi = \frac{\text{mmf}}{\mathcal{R}}$$

but mmf $= NI$ (equation 9-2), and substituting for mmf in the preceding equation gives

$$\phi = \frac{NI}{\mathcal{R}}$$

$$\phi = \frac{(5000 \text{ turns})(200 \text{ mA})}{18.7}$$

$$\phi = \frac{(5 \times 10^{3})(200 \times 10^{-3})}{18.7}$$

$$\phi = 53.5 \text{ Wb}$$

(c)
$$\text{flux density} = \frac{\text{flux}}{\text{area in square meters}} \tag{9-6}$$

$$B = \frac{\phi}{A}$$

$$B = \frac{53.5}{0.785 \times 10^{-4}}$$

$$B = 68.2 \times 10^{4} \text{ T} \qquad (\text{or Wb/m}^2)$$

(d)
$$\text{magnetizing force} = \frac{\text{magnetomotive force}}{\text{length in meters}} \tag{9-10}$$

$$H = \frac{\text{mmf}}{l}$$

but mmf $= NI$ (equation 9-2), and substituting for mmf in above gives

$$H = \frac{NI}{l}$$

$$H = \frac{(5000 \text{ turns})(200 \text{ mA})}{21.98 \times 10^{-2}}$$

$$H = \frac{(5 \times 10^{3})(200 \times 10^{-3})}{21.98 \times 10^{-2}}$$

$$H = \frac{1000}{21.98 \times 10^{-2}}$$

$$H = 45.5 \times 10^{2}$$

$$H = 4550 \text{ At/m}$$

For Similar Problems Refer to End-of-Chapter Problems 9-100 to 9-110

9-17. B-H Magnetization Curve and Hysteresis

The *flux density* (B) of an air core, or nonmagnetic core, electromagnet is directly proportional to the *magnetizing force* (H). In other words, if the current through a coil is increased, the magnetic field strength also increases. In Fig. 9-30*a* an air-core coil is depicted having 3000 turns. The coil is 1 cm in diameter, and 10 cm in length between its upper and its lower turns. The current is varied in value from 0.5 to 2.5 mA in 0.5 mA steps. These are listed in the chart of Fig. 9-30*b*, shown as Case 1, Case 2, and so on.

The following calculations illustrate how the first line (Case 1) of the chart of Fig. 9-30*b* are found.

CASE 1

When the current is 0.5 mA,

$$\text{mmf} = NI \tag{9-2}$$

$$= (3000 \text{ turns}) (0.5 \text{ mA})$$

$$= (3000) (0.5 \times 10^{-3}\text{A})$$

$$= 1.5 \text{ At, as shown listed in Fig. 9-30}b$$

Magnetizing force,

$$H = \frac{\text{mmf}}{\text{length}} \tag{9-10}$$

$$H = \frac{1.5 \text{ At}}{10 \text{ cm}}$$

$$= \frac{1.5}{0.1 \text{ m}}$$

$$= 15 \text{ At/m, as shown listed in Fig. 9-30}b$$

Flux density

$$B = \frac{\phi}{A} \tag{9-6}$$

and

$$\phi = \frac{\text{mmf}}{\mathcal{R}} \tag{9-1}$$

and

$$\mathcal{R} = \frac{l}{\mu A} \tag{9-18}$$

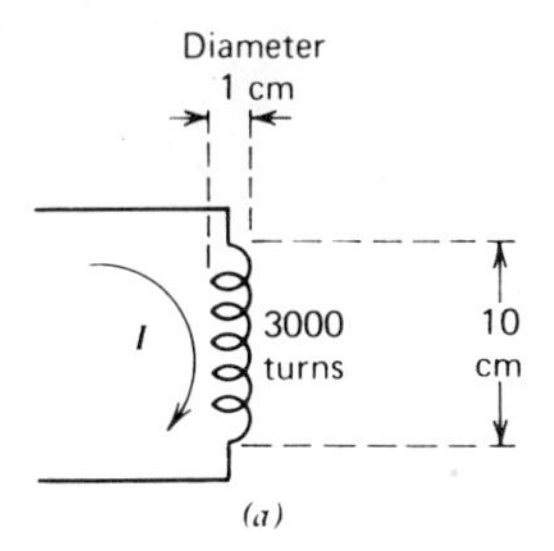

	I	mmf (ampere-turn)	Magnetizing Force (H) (ampere-turn per meter)	Flux Density (B) (microteslas)	Permeability μ
Case 1	0.5 mA	1.5	15	18.84	12.56×10^{-7}
Case 2	1. mA	3	30	37.68	12.56×10^{-7}
Case 3	1.5 mA	4.5	45	56.52	12.56×10^{-7}
Case 4	2. mA	6	60	75.36	12.45×10^{-7}
Case 5	2.5 mA	7.5	75	94.2	12.56×10^{-7}

(*b*)

Figure 9-30. (a) Air-core coil.

substituting for ϕ and for $\mathcal{R}$ in equation 9-6 yields

$$B = \frac{\dfrac{\text{mmf}}{l/\mu A}}{A}$$

Dividing by a fraction requires that the fraction be inverted and multiplied, as follows

$$B = (\text{mmf})\left(\frac{\mu A}{l}\right)\left(\frac{1}{A}\right)$$

Cancelling the A's gives

$$B = (\text{mmf})\left(\frac{\mu}{l}\right)$$

$$= (1.5)\left(\frac{12.56 \times 10^{-7}}{0.1 \text{ m}}\right)$$

$$= 188.4 \times 10^{-7} \text{ T (Wb/m}^2\text{)}$$

$$= 18.84 \times 10^{-6} \text{ T}$$

$= 18.84\ \mu\text{T}$ (micro webers/m^2), as shown listed in Fig. 9-30*b*

Permeability,

$$\mu = \frac{B}{H} \tag{9-14}$$

$$\mu = \frac{18.84 \times 10^{-6}}{15}$$

$$\mu = 1.256 \times 10^{-6}$$

$\mu = 12.56 \times 10^{-7}$ as shown listed in Fig. 9-30*b*.

CASE 2

When the current is 1 mA,

$$\text{mmf} = NI \tag{9-2}$$

$$= (3000 \text{ turns})\ (1 \text{ mA})$$

$$= (3000)\ (1 \times 10^{-3} \text{ A})$$

$= 3$ At, as shown listed in Fig. 9-30*b*.

Magnetizing force,

$$H = \frac{\text{mmf}}{\text{length}}$$

$$H = \frac{3}{10 \text{ cm}}$$

$$= \frac{3}{0.1 \text{ m}}$$

$= 30$ At/m, as shown listed in Fig. 9-30*b*

Flux density,

$B = (\text{mmf}) \left(\frac{\mu}{l}\right)$ (derived from equations 9-6, 9-1, and 9-18 as shown in the previous discussion for Case 1)

$$B = (3) \left(\frac{12.56 \times 10^{-7}}{0.1 \text{ m}}\right)$$

$$= 376.80 \times 10^{-7} \text{ T (Wb/m}^2\text{)}$$

$$= 37.68 \times 10^{-6} \text{ T}$$

$= 37.68\ \mu\text{T}\ (\mu\text{Wb/m}^2)$ as shown listed in Fig. 9-30*b*

Permeability,

$$\mu = \frac{B}{H} \tag{9-14}$$

$$= \frac{376.8 \times 10^{-7}}{30}$$

$= 12.56 \times 10^{-7}$, as shown listed in Fig. 9-30*b*

The reader is urged to verify the values of mmf, H, B, and μ for Cases 3, 4, and 5 in Fig. 9-30*b*, using the equations and method shown previously for Cases 1 and 2.

From the values of magnetizing force (H) and flux density (B) for Cases 1 to 5, shown listed in Fig. 9-30*b*, a *graph* is drawn, shown in Fig. 9-31. The resulting graph is simply a line drawn from the first dot to the second, and so on. Note that this is a *straight line*, and the graph is said to be *linear*. The graph of H and B is called the *B-H Magnetization curve*. Since the air core, or any nonmagnetic core, cannot be magnetized, then any increase in magnetizing force (H) results in a proportionate increase in flux density (B). Note that in the chart of Fig. 9-30*b*, permeability (μ) remains unchanged, since the *ratio* of B and H remains constant. The value for μ is 12.56×10^{-7}, in the mks system.

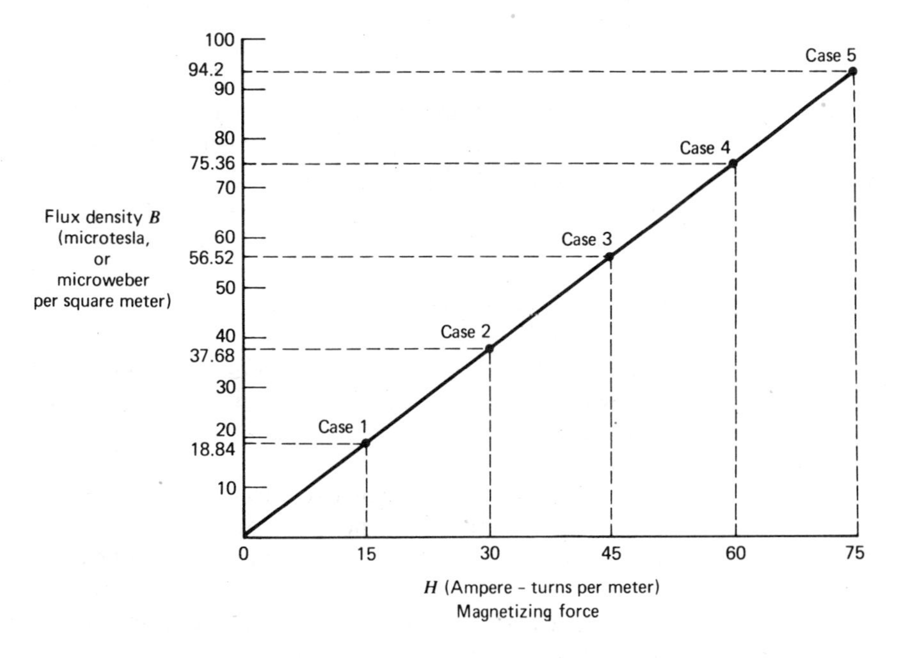

Figure 9-31. *B-H* magnetization curve for air-core coil.

When a magnetic substance such as iron or steel is used as the core of an electromagnet, the same is true up to a point. The graph of B-H rises *linearly* as H is increased, but only up to a certain amount of H. This is shown in Fig. 9-32 in the *ABC* region of the graph. As more and more of the electrons become aligned and spin in the same direction, the iron core approaches its magnetic saturation, in the *CDE* region. Increasing the magnetizing force (H) still further and by the same amounts as before, only results in the flux density (B) increasing by smaller amounts than before. This is in the *CDE* region of the graph of Fig. 9-32 where it is rising less steeply than before. The graph or curve is said to be "flattening out," and is no longer a straight line, but is now *nonlinear*.

From the values for B and H in the graph of Fig. 9-32, the permeability (μ) of the iron may be found, as shown in the following.

At point A, $H = 800$ At/m, and $B = 0.1$ T (Wb/m²). Therefore permeability

$$\mu = \frac{B}{H} \tag{9-14}$$

$$\mu = \frac{0.1}{800}$$

$$\mu = .000125 \qquad \text{or} \qquad 12.5 \times 10^{-5}$$

As saturation is approached, the permeability of the iron decreases. For example, at point E, permeability becomes, $B/H = 0.375/4000 = 9.375 \times 10^{-5}$.

In the linear graph of Fig. 9-31 for the air-core coil, flux density (B) will decrease along the same graph if the magnetizing force (H) is decreased. There is no magnetic field remaining when the force is reduced to zero. This is not true for an iron-core coil. If the magnetizing force is decreased back to zero, after having magnetized the iron, then some magnetic flux still remains in the iron. This is the *residual* magnetism which still resides in the iron even though there is no longer any magnetizing force. The iron retains some magnetism because of its *retentivity*.

This is illustrated in the graph of Fig. 9-33, called the *hysteresis* loop, where hystresis means a *lagging* behind or delay. This graph can best be explained as follows:

1. During time *a*, magnetizing force ($+H$) is increasing, resulting in flux density ($+B$) rising from point O to U.

2. During time *b*, magnetizing force ($+H$) is being decreased back to zero, but flux density ($+B$) only decreases from point U to point V, not reaching zero. Flux density is lagging behind the magnetizing force.

3. During time *c*, magnetizing force ($-H$) is being *increased* in the *opposite* direction. This coerces the iron to further decrease its flux density, and at

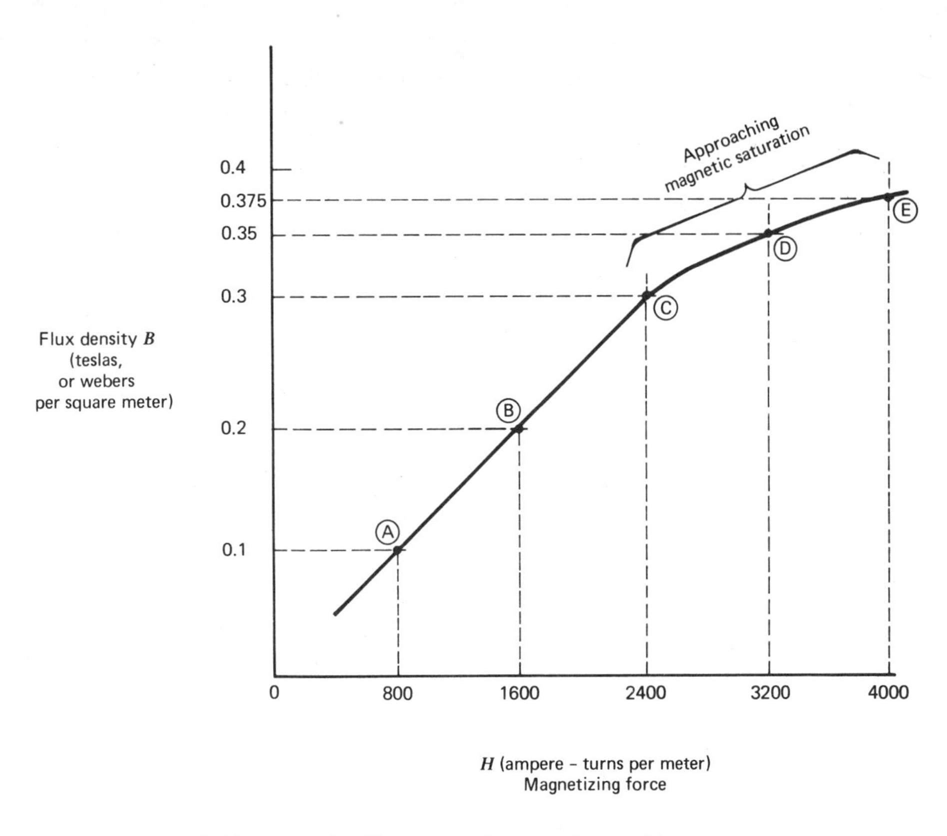

Figure 9-32. *B-H* magnetization curve for one type of iron.

point W, the iron has lost its magnetism. The magnetizing force ($-H$) required to demagnetize the iron is called the *coercive force*. As this opposite direction magnetizing force ($-H$) increases still further, the iron becomes *remagnetized* but with the *opposite* polarity. This is shown during the point W to point X interval, when flux density is $-B$.

The reader should follow the graph of Fig. 9-33 during times *d* and *e*.

The lagging of the flux density (B) behind the magnetizing force (H) is due to the *reluctance* of the iron to being magnetized, demagnetized, and remagnetized with opposite polarity. This produces some heat in the iron and is called hysteresis loss. Hard steel with high reluctance and high retentivity, therefore, has a greater hysteresis loss than soft iron, which has low reluctance and little retentivity.

Refer to End-of-Chapter Problems 9-111 to 9-128

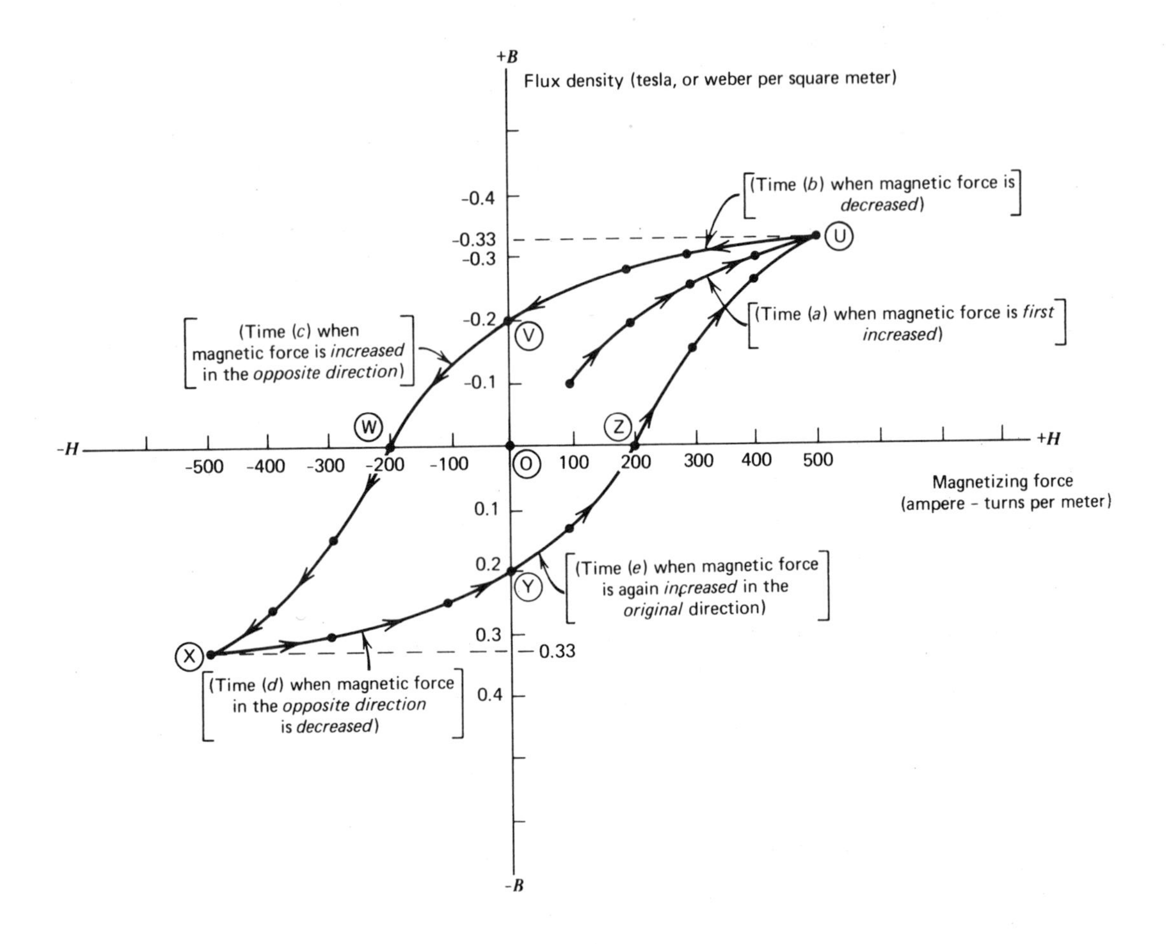

Figure 9-33. Hysteresis loop.

PROBLEMS

See section 9-1 for discussion covered by the following.

9-1. The attraction that certain stones and metals may have for iron or steel is called ______.

9-2. What are two names for a natural magnet?

9-3. What is an artificial magnet?

See section 9-2 for discussion covered by the following.

9-4. What seems to be indicated when iron filings are sprinkled around a bar magnet?

9-5. Define *flux.*

9-6. Define *flux density.*

9-7. Define *weber.*

9-8. Define *tesla.*

9-9. On a bar magnet, where is the greatest concentration of magnetic lines?

9-10. The names given to the extreme ends of a magnet are called the ______ and ______ poles.

9-11. It is usually assumed that the magnetic lines have a direction, outside the magnet, which is from the ______ to the ______ pole.

9-12. If two bar magnets are brought near to each other with their similar poles facing one another, what will the magnets do?

9-13. If the magnets of the preceding question had their opposite poles adjacent to one another, what would happen?

See section 9-3 for discussion covered by the following.

9-14. What is the basic difference getween the atoms of an unmagnetized steel bar and those of a bar magnet?

9-15. Describe the atoms of a saturated magnet.

See section 9-4 for discussion covered by the following.

9-16. What is permeability?

9-17. How can a steel box act as a magnetic shield for something inside the box?

9-18. What is reluctance?

9-19. How high in permeability values do certain alloys run?

9-20. What does a permeability of 5000 mean?

9-21. What happens to an iron bar when it is placed in a strong magnetic field?

9-22. If the north end of a bar magnet is brought near one end of an iron bar, what happens to this end of the iron?

9-23. Describe a ferromagnetic substance, and give an example of this type of metal.

9-24. Define paramagnetic material, and give an example of a metal in this category.

9-25. Define diamagnetic substance, and give an example of a metal in this category.
9-26. How should permanent magnets be correctly stored away?

See section 9-5 for discussion covered by the following.

9-27. What seems to be the cause of magnetism in a steel bar?
9-28. What seems to be the cause of magnetism in a wire?
9-29. Describe the magnetic lines that are present when a current occurs in a straight wire.
9-30. Describe two methods of proving that a magnetic field exists around a wire that has a direct current in it.
9-31. Describe a method of proving the *direction* of the magnetic lines around a wire that has a direct current in it.
9-32. Describe a simple rule that predicts the *direction* of the magnetic lines around a straight wire that has a direct current in it.
9-33. How could the magnetic field strength be increased around a straight wire?

See section 9-6 for discussion covered by the following.

9-34. Draw the end view of two parallel wires closely spaced, with electrons flowing into each wire, away from the viewer, showing the direction of the total magnetic field.
9-35. Determine the north and south poles of each of the electromagnets shown in parts (a), (b), and (c) of the illustration for this problem.

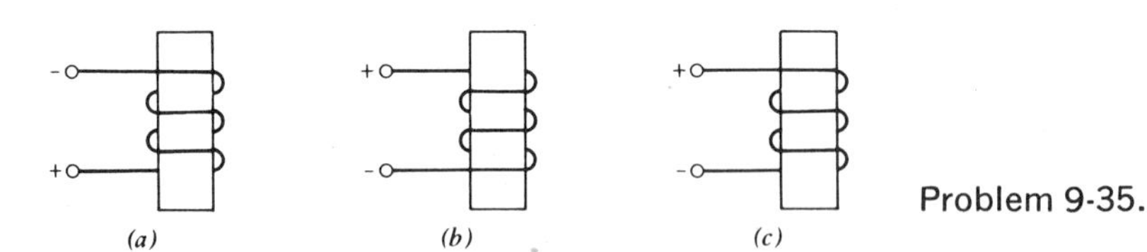

Problem 9-35.

See section 9-7 for discussion covered by the following.

9-36. If the current is decreased in a coil, what effect does it have on the strength of the magnetic field?
9-37. If a number of turns of wire on an electromagnet are stripped off, what effect will it have on the strength of the magnetic field, assuming no change in current?

9-38. What effect would it have on the strength of the magnetic field if the iron core were removed from an electromagnet?
9-39. What *three* things determines the strength of an electromagnet?

See section 9-8 for discussion covered by the following.

9-40. Name five electrical-electronic devices that employ magnetic principles.

See section 9-8A, The Relay, for discussion covered by the following.

9-41. When is a relay *energized*?
9-42. When is a relay *de-energized*?
9-43. What is a *double-throw* relay?
9-44. In Fig. 9-18, which alphabetically lettered terminals make contact when the relay is energized?

See section 9-8B, The Meter, for discussion covered by the following.

9-45. What are the main components of the moving-coil type of meter?
9-45. What determines the amount of coil movement?
9-47. What would occur if a voltage were connected to the coil terminals using the incorrect polarities?

See section 9-8C, The Loudspeaker, for discussion covered by the following.

9-48. What are the main parts of a loudspeaker?
9-49. What does the voice coil do?
9-50. What is an ac voltage?
9-51. What does *frequency* of the ac audio voltage mean?
9-52. How does an electrodynamic speaker differ from a PM speaker?

See section 9-8D, Electric Motor Principle, for discussion covered by the following.

9-53. What are the two main components of a simple electric motor?
9-54. In the illustration of this problem, no current flows through the wire, and the magnetic field is as shown. If current now flows *into* the wire (away from the viewer), draw *directional arrows* for the wire's magnetic field and the permanent magnet's new field.

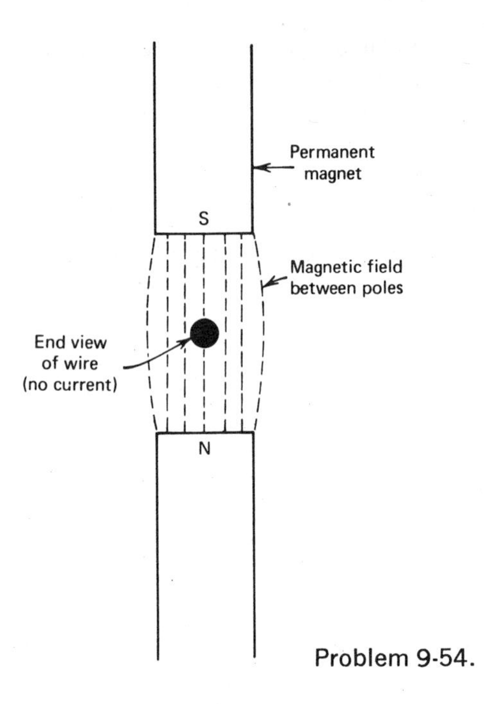

Problem 9-54.

9-55. In which direction will the wire of the previous problem attempt to move?

See section 9-8E, Electromagnetic Deflection in a CRT, for discussion covered by the following.

9-56. To determine the *direction of deflection* of a wire in which current is flowing, or the electron beam of a CRT, using the three-finger, right-hand rule, the index finger, middle finger, and right thumb should be pointed in which directions?

9-57. In the diagram for this problem, in which direction will the CRT electron beam be moved?

9-58. In the diagram for the previous problem, if the polarity of the deflection voltage were reversed from that shown, in which direction would the CRT electron beam be moved?

See section 9-9 for discussion covered by the following.

9-59. If the north magnetic pole of the earth is actually a *north* pole, then what is the magnetic pole of the end of a compass needle which points to this north pole of the earth?

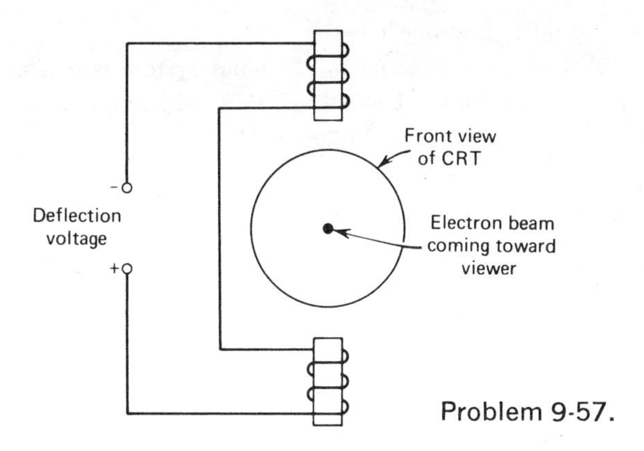

Problem 9-57.

9-60. Assuming that the south pole of a compass needle is truly a *south* magnetic pole, then what *magnetic polarity* is that end of the earth to which this compass end points?

See section 9-10 for discussion covered by the following.

9-61. What two things of an electron produce a magnetic field?
9-62. What are the areas comprising groups of magnetized adjacent atoms in iron called?
9-63. What effect does an air gap have on the iron core of an electromagnet?

See section 9-11 and Table 9-1 for discussion covered by the following.

9-64. What is the name and symbol for the *force* producing the magnetic field?
9-65. What is meant by *flux*, and what is the symbol used for it?
9-66. What is another term meaning magnetic field strength?
9-67. What is meant by *flux density*, and what symbol is used for it?
9-68. What is meant by *magnetizing force*, and what symbol is used?
9-69. How does *permeance* and *permeability* differ, and what symbols are used?
9-70. What is *reluctance* and what symbol is used for it?

See section 9-12 for discussion covering the following.

9-71. What is the unit of measurement of magnetic flux in the mks system?

9-72. What is a *weber*?

9-73. What is a *maxwell*, and in what system is it used?

9-74. What is the relationship of *flux*, *magnetomotive force*, and *reluctance* which is referred to as the magnetic Ohm's law?

See section 9-13 and examples 9-2 and 9-3 for discussion covered by the following.

9-75. Find mmf in the mks system if a coil with 5000 turns has 10 mA of current.

9-76. An mmf of 300 At is desired. What applied voltage should be used if the coil has 1500 turns and a resistance of 800 Ω?

9-77. How many turns (N) should a coil have if the current flowing through it is 50 mA and an mmf of 600 At is desired?

(*Note*: amperes must be used in equation 9-2, not milliamperes.)

See section 9-13 and examples 9-4 and 9-5 for discussion covered by the following.

9-78. The mmf of a coil in the cgs system is measured in what units?

9-79. If the mmf of a coil is 350 Gbm Gi (cgs system), what is the mmf in *ampere-turns* (mks system)?

9-80. A current of 150 mA flows through a coil having 300 turns. What is the mmf in *gilberts*?

9-81. What is the relationship between mmf, *flux* (ϕ), and *reluctance* ($\mathcal{R}$), often called the magnetic Ohm's law?

See section 9-14 for discussion covered by the following.

9-82. Define *flux density* (B).

9-83. What is the relationship between *flux density* (B), *flux* (ϕ), and *area* (A)?

9-84. What is the unit for *flux density* (B) in the mks system?

9-85. Define the term *tesla*.

9-86. What is the *flux density* (B) unit in the cgs system?

9-87. Define the term *gauss*.

9-88. A magnetic field of flux (ϕ) has 3000 μWb in an area 0.3 m by 0.05 m when the area is perpendicular to the flux. What is the *flux density* (B) in *teslas* in the mks system?

9-89. If a flux density (B) is 1.5 T, then in a perpendicular cross-sectional area 0.2 m by 0.05 m: (a) How many *webers* of flux (ϕ) are present, and (b) how many magnetic lines are present?

9-90. A magnetic field of flux (ϕ) consists of 13,500 Mx in a perpendicular area 1.5 cm by 1.5 cm. What is the *flux density* (B) in *gauss* in the cgs system?
9-91. Convert the following:
(a) 1.2 T (mks) into gauss (cgs).
(b) 6000 G (cgs) into tesla (mks).

See section 9-15 for discussion covered by the following.

9-92. What is meant by *magnetizing force*?
9-93. What is the symbol used for magnetizing force?
9-94. In the mks system, what is the unit of magnetizing force?
9-95. In the cgs system, what is the unit of magnetizing force?
9-96. Convert 100 At/m (mks system) into *oersteds* (cgs system), using equation 9-11.
9-97. Convert 10 Oe (cgs system) into *ampere-turns per meter* (mks system), using equation 9-12.
9-98. An air-core coil has 1000 turns and is 5 cm long. If a current of 300 mA flows through the coil, find the magnetizing force (H) in ampere-turns per meter (mks system) at the center of coil.
9-99. A coil is 20 cm long and consists of 500 turns wound on an iron core 25 cm long. If 10 mA flows through the coil, find the *magnetizing force* (H) in *oersteds* (cgs system) throughout the core.

See section 9-16 for discussion covered by the following.

9-100. What is meant by the term *permeability*?
9-101. What is the symbol for permeability?
9-102. Define *permeance*.
9-103. What symbol is used for permeance?
9-104. What is the relationship between permeability (μ), flux density (B), and magnetizing force (H)?
9-105. What is meant by *reluctance*?
9-106. What symbol is used for reluctance?
9-107. What is the relationship between permeance and reluctance?
9-108. 1500 turns of wire are wound around a round tubular plastic (nonmagnetic) form which has a diameter of 10 cm. If the distance between the upper and lower turns of wire of the coil is 200 cm, and 500 mA flows through the coil, find: (a) magnetomotive force (mmf) in ampere-turns, (b) reluctance ($\mathcal{R}$), and (c) flux density (B). (Refer to Examples 9-10 and 9-11 for similar solutions.)
9-109. If the *form* in the previous problem (9-108) on which the coil is wound were made of a magnetic substance having a permeability of 200, with

all other values the same as in Problem 9-108, find: (a) magnetomotive force (mmf), (b) reluctance ($\mathcal{R}$), and (c) flux density (B).

9-110. A coil is wound around an iron ring (see Fig. 9-29), having a permeability of 100. The coil has 300 turns of wire with 500 mA flowing through it, and the outer diameter (D_2) is 14 cm, and the inner diameter (D_1) is 10 cm. Find: (a) reluctance ($\mathcal{R}$), (b) magnetic flux (ϕ), (c) flux density (B), and (d) magnetizing force (H). (Refer to Example 9-11 for similar solution.)

See section 9-17 for discussion covered by the following.

9-111. In an air-core coil, if the magnetizing force (H) is doubled, what does the flux density (B) do?

9-112. What is the permeability (μ) of air in the mks system?

9-113. An air-core coil has a diameter of 2.5 cm and the distance between its upper and lower windings is 50 cm. If the coil has 5000 turns, and 30 mA flows through, find in the mks system: (a) mmf, (b) magnetizing force (H), and (c) flux density (B).

9-114. What is meant by the term "linear" graph?

9-115. The B-H magnetization curve for iron, shown in Fig. 9-32, is less steep in the CDE region than in the ABC region. What causes this?

9-116. What happens to permeability (μ) in the CDE region of the curve of Fig. 9-32 as compared to the steeper region region of ABC?

9-117. What is meant by the term "nonlinear" graph?

9-118. If the permeability of air in the mks system is 12.56×10^{-7}, and a steel rod has a *relative permeability* of 300, what is the permeability of the steel rod in the mks system?

9-119. If the μ of an iron bar decreases as saturation is approached, what happens to μ above saturation?

9-120. What is meant by the *retentivity* of iron?

9-121. What is *residual magnetism*?

9-122. When the magnetizing force (H) is decreased to zero in an iron-core coil, what does the flux density (B) become?

9-123. What must the magnetizing force (H) become, in an iron-core coil, to reduce the flux density (B) to zero?

9-124. What is the *lagging* of flux density (B) behind the magnetizing force (H) called?

9-125. What is *coercive force*?

9-126. What is *hysteresis*?

9-127. What is the loss of power called indicated by the heat in the iron when it is magnetized, demagnetized, and magnetized again with opposite polarity?

9-128. (a) If an ac voltage (one which keeps reversing its polarity) were to be applied to a metal-core coil, would soft iron or hard steel be more desirable in keeping any loss of power to a minimum? (b) Explain why.

chapter
10

DC Meters and Motor Principle

10-1. The Permanent-Magnet Moving-Coil Meter

When a current flows through a coil, the coil becomes an electromagnet with one end north, and the other end south. If this coil is placed between the poles of a permanent magnet so that *adjacent* poles of the coil and magnet are of *opposite* polarities, there is an attraction. If the coil is mounted on an axle or pivot, as shown previously in Fig. 9-20, so that it can turn or rotate, the magnetic attraction will cause the coil to do just that. A spiral hairspring keeps the coil in its normal position, but allows the coil to turn somewhat when the magnetic attraction *torque* (or twisting force) overcomes the restraining action of the spring. The amount of rotation of the coil, or its angle of deflection, is directly proportional to the amount of the current flowing in the coil. When the magnetic deflection torque is equal to the spring's restraining force, the coil stops its rotation and becomes stationary in its new deflected position.

A pointer is attached to the coil, and moves with it as the coil rotates. A calibrated or marked scale is positioned so that the pointer glides above the markings. The scale is graduated from zero to that value of current which causes the coil to rotate its maximum physical amount. This is called the *full-scale deflection current.*

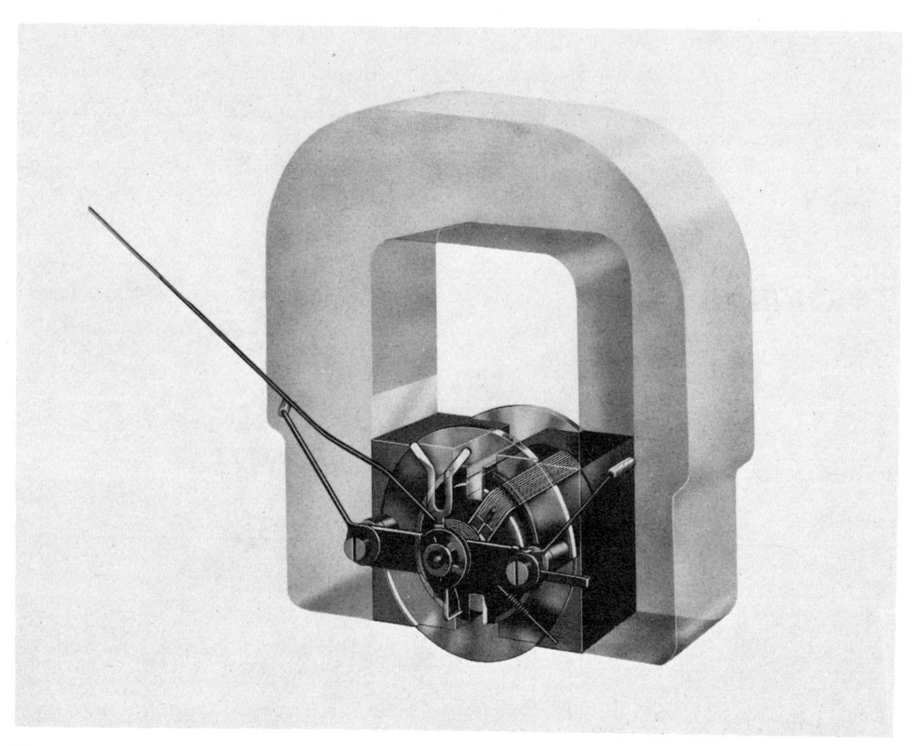

Figure 10-1. Permanent magnet moving coil meter movement (*courtesy Weston Instruments Inc*).

If the current becomes zero in the coil, the magnetic attraction force disappears, permitting the spiral hairspring to bring the coil back to its original or normal position, or zero. If a current were allowed to flow in the reverse direction through the coil, the north and south poles of the coil would reverse, causing the coil to turn in the opposite direction. If the coil's physical position is such that it should not rotate backward, then damage may result. However, in some meters, the coil is positioned so that its normal setting permits it to rotate either clockwise or counterclockwise. This is usually called a *galvanometer*, and has its *zero* marking on the graduated scale at the center.

Figure 10-1 shows the mechanism of the permanent-magnet moving-coil meter. This is referred to as the *D'Arsonval movement*, after its French inventor, Jacques D'Arsonval, who in 1881 first developed it as a fragile laboratory instrument. It is often also called a *Weston* movement, after Edward Weston, who in about 1887 changed the basic D'Arsonval movement into a more rugged commercial type of test instrument.

Refer to End-of-Chapter Problems 10-1 to 10-6

10-2. Ammeter, Multi-Range Conversions

When current flows through the coil of the moving-coil meter, it rotates as a result of the interaction between the permanent magnet and the electromagnet. The amount of current that causes the coil to rotate maximum, and the pointer to move to the extreme end of the graduated scale, is called the *full-scale deflection current* or the *current sensitivity* of the meter. A 1-mA meter, therefore, requires this amount of current to deflect the pointer to the extreme end of its graduated scale. Similarly, a 50-μA meter requires only this current for full-scale deflection.

To have greater utility, a meter that is to be used to indicate the amount of current in various circuits should have a wide variety of current ranges. For example, a 100-μA meter can only read up to a maximum value of 100 μA. If it is desired to measure the current in a circuit where 70 μA flows, or where 250 μA is flowing or, in another, where 7 mA (7000 μA) is flowing, it would be advantageous to use the same piece of test equipment to be able to measure the current in each of these circuits. Such a meter is said to have multi-ranges.

It is a simple task to enable a 1-mA meter to read larger currents, or a 50-μA meter to read larger values. All that is required is that a resistor of the correct value be placed in *parallel* or in *shunt* with the meter coil. If a 1-mA meter were placed in a circuit where 10 mA were flowing, the meter coil movement mechanism would become damaged. However, if a *shunt* resistor of the proper value were connected across the coil, then with 10-mA *total* current, 9 mA could flow through the resistor and 1 mA would flow through the meter coil. The addition of the shunt resistor is said to *convert* the meter to a higher range. Although the meter is still a 1-mA (for full-scale deflection) type, the shunt resistor permits the meter to be placed in a circuit where a greater current flows.

The value of the resistor depends on the voltage across it and the current through it (Ohm's law, $R = E/I$). Since the resistor is in parallel with the meter coil, the resistor voltage is equal to the coil voltage. *Voltage* across the coil is due to the current through it, and the resistance of the coil itself, or $E_{\text{coil}} = (I_{\text{coil}})(R_{\text{coil}})$. The shunt resistor is found using Ohm's law, $R_{\text{shunt}} = E_{R\ \text{shunt}}/I_{R\ \text{shunt}}$. The following example illustrates these points.

Example 10-1

Draw the circuit diagram of a 1-mA meter, having a coil resistance of 72 Ω, which is to be converted into a 10 mA range meter.

Solution

Figure 10-2 shows the meter coil and the required shunt resistor. Note that with 10 mA *total* current (the desired new range of the meter), and with the

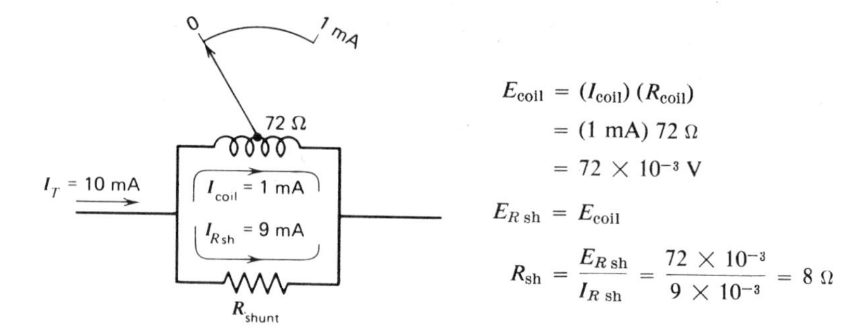

Figure 10-2. 1-mA meter converted into a 10-mA meter, Example 10-1.

coil only being capable of handling 1 mA (its full-scale deflection sensitivity), then the difference (10-1) of 9 mA flows through the shunt resistor. The value of this resistor is then

$$R_{sh} = \frac{E_{R\,sh}}{I_{R\,sh}} \qquad \text{(Ohm's law)}$$

but since R_{sh} is in parallel with the meter coil, then $E_{R\,sh} = E_{coil}$, and $E_{coil} = (I_{coil})\,(R_{coil})$ (Ohm's law)

then

$$R_{sh} = \frac{E_{coil}}{I_{R\,sh}} \qquad \text{(Ohm's law)}$$

also

$$I_{R\,sh} = I_{total} - I_{coil} \qquad \text{(Kirchhoff's current law)}$$

therefore,

$$R_{sh} = \frac{E_{R\,sh}}{I_{R\,sh}} \quad \text{or} \quad \frac{E_{coil}}{I_{R\,sh}} = \frac{(I_{coil})\,(R_{coil})}{I_T - I_{coil}} \tag{10-1}$$

$$R_{sh} = \frac{(1\text{ mA})\,(72\ \Omega)}{10\text{ mA} - 1\text{ mA}}$$

$$R_{sh} = \frac{(1 \times 10^{-3})\,(72)}{9 \times 10^{-3}}$$

$$R_{sh} = 8\ \Omega$$

In the circuit of Fig. 10-2, the coil and its shunt resistor act like a 10-mA meter. When 1-mA flows through the coil, 9 mA flows in R_{sh}, and I_{total} is, therefore, 10 mA. The full-scale reading of the meter is therefore read as 10 *mA*, instead of the actual value of 1 mA on the scale.

If the 1-mA meter were to be converted into a still *higher* range meter, as described in Example 10-2, then a *smaller* shunt resistor would be necessary.

The smaller the shunt resistor is, the greater is the fraction of I_T that flows through it. In Example 10-1, I_T is 10 mA, R_{sh} is 8 Ω, and 9 mA flows through it, while 1 mA flows through the meter coil. In the next example, the 1 mA meter is to be converted into a 100-mA type. By using a *smaller* value R_{sh} (see the solution to Example 10-2), then 99 mA will flow through it, while 1 mA goes through the meter coil.

Example 10-2

Draw the circuit diagram of a 1-mA meter having a 72-Ω coil, converted into a 100-mA range meter.

Solution

As shown in Fig. 10-3, if the meter is to be placed in a circuit where 100 mA is flowing, and the meter coil can only take a maximum of 1 mA, then the "surplus" 99 mA must go through the shunt resistor. Solve for the value of the required shunt resistor, using Fig. 10-3 as a guide.

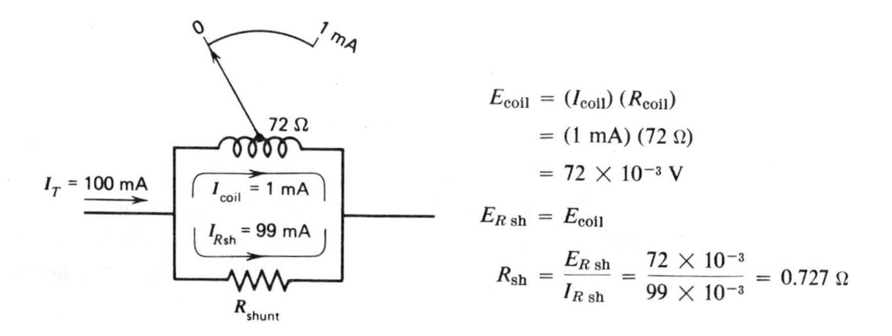

Figure 10-3. 1-mA meter converted into a 100-mA meter, Example 10-2.

Multi-Ranges. By switching in different values of shunt resistors across the coil of a meter, the basic meter, say a 1-mA type, could now be placed in a 10-mA circuit, or a 100 mA circuit, and so on, to indicate the amount of current flow. Figure 10-4 shows a 1-mA movement with a three-position *range switch.* With the switch as shown, there is no shunting resistor across the coil, and the meter now operates as a 1 mA range meter. In the other switch positions, either the 8 Ω or the 0.727-Ω resistor is in the circuit, making the meter a 10-mA or a 100-mA meter.

An important feature that this switch must have is a *make-before-break* or a *shorting* characteristic. This means that when the switch rotor is moved from one position to the next, there must not be a time, even for an instant, that

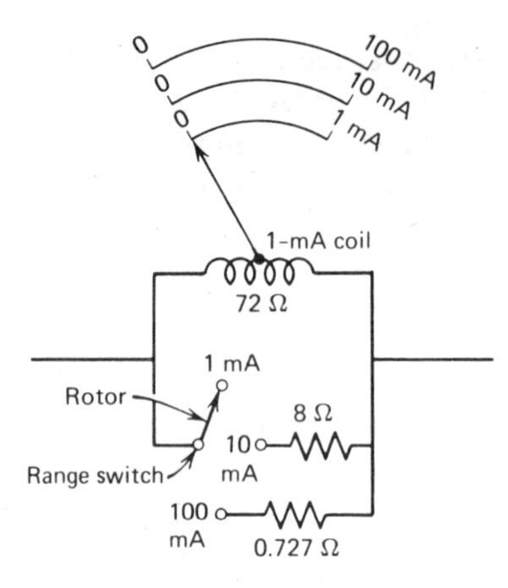

Figure 10-4. Multi-range current meter combining Examples 10-1 and 10-2.

the switch rotor is *open* or not making contact with, at least, one position. For example, if the switch rotor is on 100-*mA* range (the 0.727-Ω resistor is now in shunt with the coil), but if only about 5 or 6 mA is actually flowing, then the indication of this on the meter scale is quite small. To get a more accurate reading, the range switch should be turned to the 10 *mA* setting. At the instant that the switch rotor is between the 100 mA and the 10-mA positions, and if it is *not* contacting either, then there is no shunting resistor across the coil. At this instant, the meter is only a 1-mA range meter. If 5 or 6 mA were actually flowing, all this current would go through the meter coil, probably resulting in damage. A *shorting* switch, or a *make-before-break* type, simply momentarily shorts together two positions of the switch when the rotor is turned from one setting to the next. The rotor always *makes* the next contact before *breaking* the previous one. In other words, at no time is the meter coil left without a shunting resistor, while switching from one range to the next.

Refer to End-of-Chapter Problems 10-7 to 10-13

10-3. Using the Multi-Range DC Ammeter

The dc ammeter (or milliammeter, or microammeter) is, of course, used to read the current flow in some particular circuit. As such, it must be inserted, with the *correct polarity*, in *series* with that particular circuit. This means that the circuit must be opened up or disconnected so that the two leads of the current meter can be placed in the circuit.

Figure 10-5*a* depicts a simple series circuit, and Fig. 10-5*b* shows the physical connections with the resistors mounted on a board. Follow the wiring con-

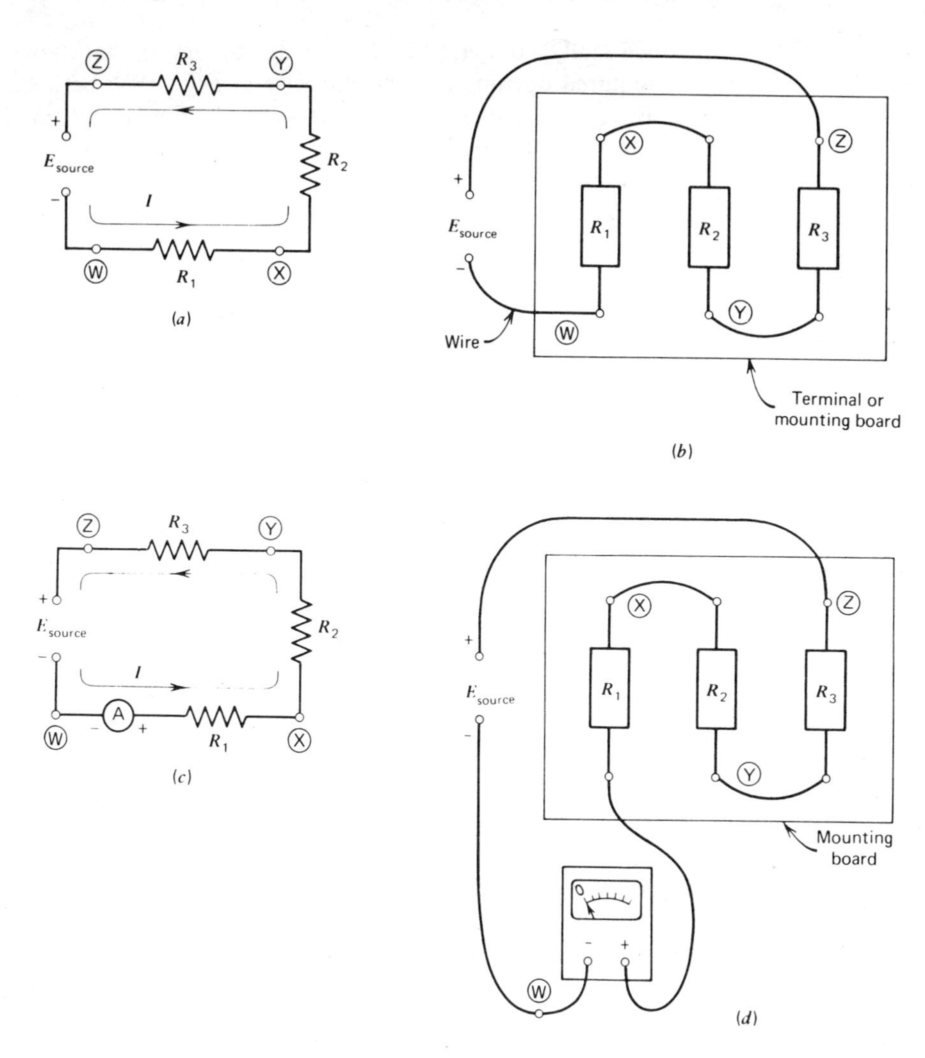

Figure 10-5. A dc current meter in a series circuit. (*a*) Schematic diagram of series circuit without meter. (*b*) Physical connections, or wiring of diagram of series circuit without meter. (c) Schematic diagram with meter. (*d*) Wiring diagram with meter.

nections of Fig. 10-5*b*, going from point W, to point X, to point Y and finally to point Z, and compare them to the schematic diagram of Fig. 10-5*a*.

Figure 10-5*c* is the series circuit of part (a) with the current meter added. Observe that the meter polarity is shown so that the current (electron movement) is from − to + through the meter. Now follow the wiring of part (d) with part (c), going from point to point.

Finally, compare Fig. 10-5*d* with (b), noting that to insert the current meter required opening the original wire of Fig. 10-5*b* between the *negative end of* E_{source} and the *end of* R_1. The meter in Fig. 10-5*d* simply replaces the wire between these two ends.

Figure 10-6*a* depicts a *series-parallel* circuit. Observe that R_4 is in *series* with the *parallel circuit* consisting of R_5, R_6, and R_7, and also in *series* with R_8.

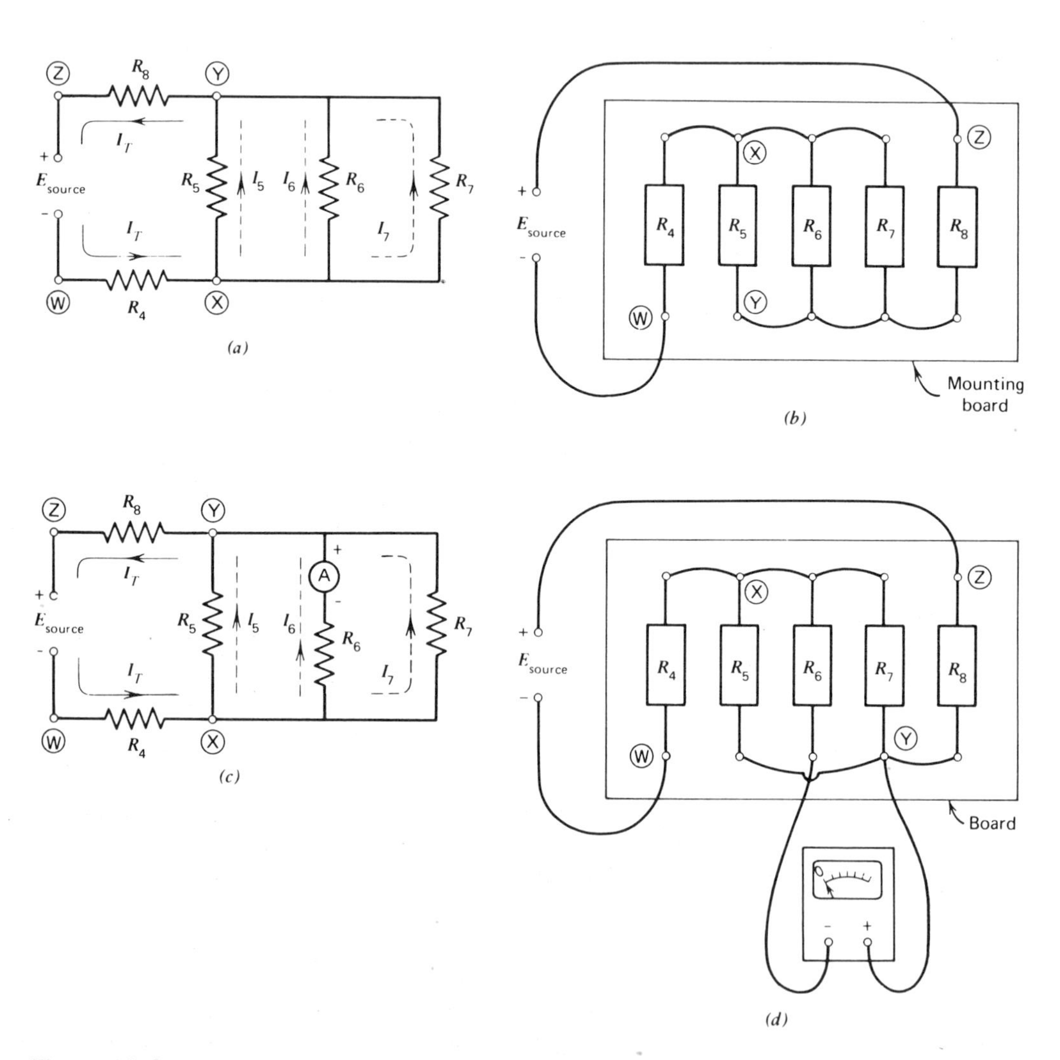

Figure 10-6. A dc current meter in a series-parallel circuit. (*a*) Schematic diagram of series-parallel circuit without meter. (*b*) Physical connections, or wiring diagram of series-parallel circuit without meter. (c) Schematic diagram with meter. (*d*) Wiring diagram with meter.

The wiring diagram is shown in Fig. 10-6*b*, and the reader should follow and compare each diagram (a) and (b), going from point W to point X to point Y, and finally to point Z.

If it is desired to measure the current in only *one* branch of the parallel part of the circuit, namely the current through resistor R_6, then the meter must be placed in *series* with this resistor only. This is shown in Fig. 10-6*c*. Note that the meter is shown so that the current I_6 (electron movement) flows through the meter from $-$ to $+$, with its negative lead connected to R_6, and its positive lead to *point Y* (the junction of R_5, R_7, and R_8).

The wiring diagram shown in Fig. 10-6*d* is identical to that of part (b) except that the lower end of R_6 in part (b) which connected to point Y (R_5, R_7, and R_8) is now, in part (d) connected *through* the *current meter* to this point Y. The reader is urged to follow the schematic diagram of (c), and the wiring diagram of (d), from point W, to point X, to point Y, and finally to point Z.

Ammeter Loading. When any piece of test equipment is used to indicate something about a circuit, the test equipment always has some effect on the circuit. This is called *loading*. It is desirable to have as little loading as possible. The milliammeter or microammeter with its parallel resistor (R_{sh}) has a combined resistance that is almost always very much smaller than the resistance of the circuit being measured. As a result, the circuit resistance is practically unchanged when the small resistance of the milliammeter circuit is added in series with the original circuit resistor. If, in that rare circuit where current is to be measured through *very small* resistors, the milliammeter resistance cannot be neglected because it does effect the circuit, then some other method has to be employed. One simple alternative method is to measure the *voltage* across one of the small resistors, using a *voltmeter* (described next, in Section 10-4). From Ohm's law, the current can be calculated, $I = E/R$.

Ammeter Precautions. As described previously, the dc current meter must be inserted in *series* with that resistor whose current is to be measured. The polarity of the meter must be observed, with the leads connected so that current (electron drift) goes through the meter from $-$ to $+$.

When the current through the circuit is completely unknown, the meter range switch should be on its highest setting at first. When the voltage is applied, the meter needle pointer should be watched closely. If it reads backward, the applied voltage should immediately be switched off. The meter leads have been connected backward, and should now be reversed.

If, when the voltage is applied, the meter pointer goes beyond the maximum deflection, it means that either the current is more than the meter's highest range, and cannot be measured with this particular meter, or that the range switch has not yet been turned to a sufficiently high range.

If the meter pointer is only deflected a small amount, turn the range switch to progressively lower ranges until a sizeable movement of pointer permits an accurate reading to be taken.

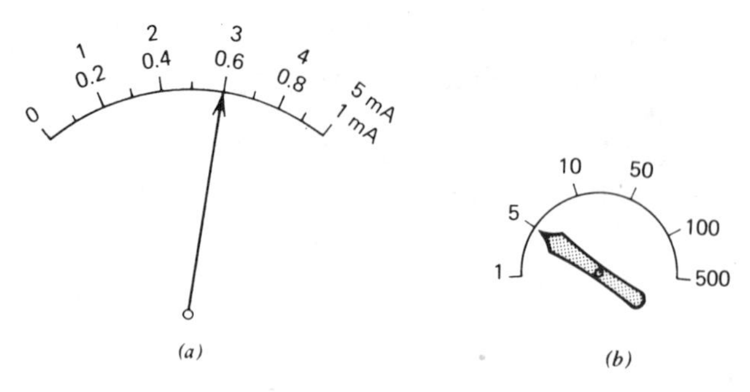

Figure 10-7. (*a*) Meter scale. (*b*) Range switch.

When reading the meter scale and pointer, care must be taken to note the setting of the range switch and to read the meter scale corresponding to the switch setting. The range *switch position* describes the *maximum* value of the meter scale.

Figure 10-7*a* is an example of a *meter scale*, and Fig. 10-7*b* shows some *range switch positions.* When the switch is on 5 mA, it means that the *maximum* value on the *scale* is also 5 mA. With the pointer as shown in (a), the meter should be read 3 mA.

If the switch were on 10 *mA*, then maximum scale reading would be 10 mA. The reading indicated now by the pointer position of (a) should be 6 mA.

Refer to End-of-Chapter Problems 10-14 to 10-23

10-4. Voltmeter, Multi-Range Conversions

The voltmeter is another piece of test equipment used frequently by the electronic technician. As its name implies, it measures or indicates the *voltage* or *difference in charge between two points* in a circuit. This could be the voltage *across* some part, such as a resistor, or it could be the voltage of a generator or a power supply.

The basic dc voltmeter is actually a current-operated meter as described in Section 10-1, *the permanent-magnet moving-coil meter* (also called the *D'Arsonval* or the *Weston* movement). The voltmeter is a current meter (milliammeter or microammeter) with a resistor added in *series* with the meter coil. The value of this resistor, which is added in series with the resistance of the meter coil, must be sufficient to limit the current through the meter coil to exactly full-scale deflection with the voltage theoretically applied. This voltage is the amount that is desired to produce full-scale current deflection, in a 100-*V range voltmeter*, or simply 100 *V*. In a 3-*V range volt meter*, it would be 3 *V*; in a 500-*V range voltmeter*, it would be 500 *V* and so on.

The resistor, then, determines the range of the voltmeter, and is called a *series multiplier*. The graduated scale of the meter actually, of course, indicates the current flowing through the meter coil, but the end of the scale is marked with the amount of *voltage* that produces a full-scale pointer deflection with the resistance in the circuit.

The following examples illustrate this.

Example 10-3

If a 1-*mA meter coil* having a *coil resistance of 150 Ω* were to be used as a 50-*V range voltmeter*, determine the value of the required *series multiplier resistor* that must be added in series with the coil. The circuit is shown in Fig. 10-8.

Solution

The desired 50-V range means that *if* 50 V were applied, a 1 mA (full-scale deflection) current should take place. If the 50 V were applied across the meter coil with only its 150 Ω, then an excessive current would flow: $I = E/R = 50\ \text{V}/150\ \Omega = 0.333\ \text{A}$, or 333 mA. To prevent this, the *series multiplier resistor* is added as shown in Fig. 10-8. The total resistance is, therefore,

$$R_T = \frac{E_{\text{range}}}{I_{\text{full-scale}}} \tag{10-2}$$

$$= \frac{50\ \text{V}}{1\ \text{mA}}$$

$$= \frac{50}{1 \times 10^{-3}\ \text{A}}$$

$$= 50\ \text{k}\Omega$$

$$R_T = \frac{E_{\text{range}}}{I_{\text{full-scale}}} = \frac{50\ \text{V}}{1\ \text{mA}} = 50\ \text{k}\Omega$$

$$R_T = R_{\text{ser.-mult}} + R_{\text{coil}}$$

$$50\ \text{K} = R_{\text{ser.-mult}} + 150$$

$$50\ \text{K} - 150 = R_{\text{ser.-mult}}$$

$$49{,}850\ \Omega = R_{\text{ser.-mult}}$$

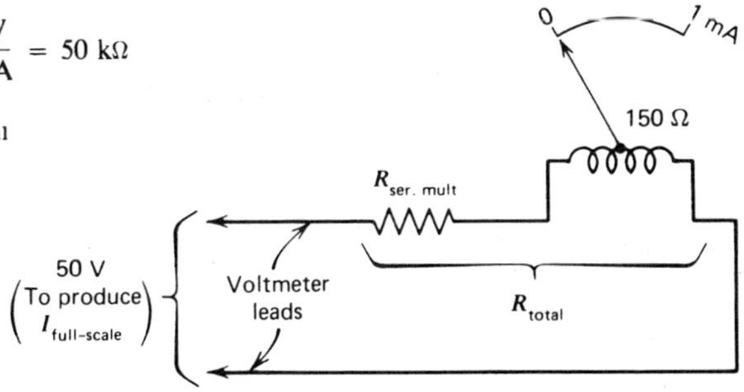

0
1 mA
150 Ω
$R_{\text{ser. mult}}$
50 V
(To produce $I_{\text{full-scale}}$)
Voltmeter leads
R_{total}

Figure 10-8. Converting a 1-mA meter into a 50-V range voltmeter, Example 10-3.

From Fig. 10-8, it can be seen that

$$R_{total} = R_{ser.\ mult} + R_{coil} \qquad \text{(resistors in series)}$$

and

$$50{,}000 = R_{ser.\ mult} + 150$$

$$50{,}000 - 150 = R_{ser.\ mult}$$

$$49{,}850\ \Omega = R_{ser.\ mult}$$

Therefore, by adding a resistor ($R_{ser.\ mult}$) of 49,850 Ω in series with the 150-Ω coil, full-scale deflection current (1 mA) flows if the meter leads are placed across two points with a 50-V difference between them. Since 50 V would produce a 1 mA reading, a voltage graduated scale with 50 V above the 1 mA marking is added to the meter scale as shown in Fig. 10-9.

Example 10-4

If the leads of the 50 V range voltmeter of Example 10-3 (Fig. 10-8) were connected between two points having only 10 V difference, what would the pointer indicate?

Solution

With 10 V applied and a total resistance of 50 kΩ (from Example 10-3), the current is

$$I = \frac{E}{R}$$

$$= \frac{10}{50\ \text{K}}$$

$$= \frac{10}{50 \times 10^{3}}$$

$$= 0.2 \times 10^{-3}\ \text{A} \qquad \text{or} \qquad 0.2\ \text{mA}$$

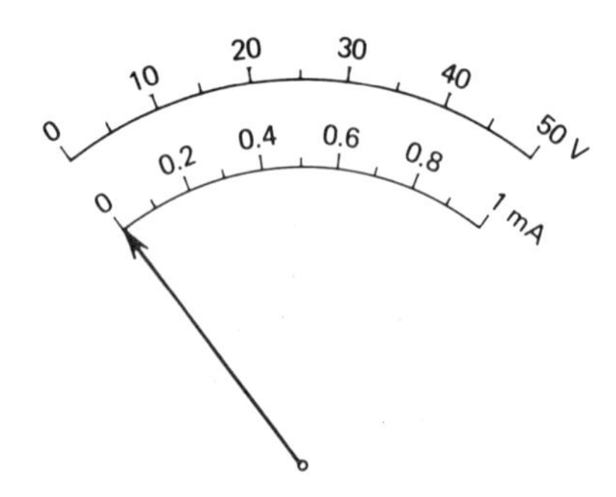

Figure 10-9. A 1-mA meter with a 50-V graduated scale.

The meter pointer therefore points to a 0.2 mA on the *current* scale (see Fig. 10-9), and since this is due to the 10 V applied, the corresponding figure on the voltage scale is 10 (see Fig. 10-9).

Any current meter having a resistor of the correct value in series with the coil becomes a voltmeter.

Example 10-5

A 200-μA meter with a coil resistance of 300 Ω is to be converted into a 500-V voltmeter. What is the value of the required *series multiplier resistor*? The circuit is shown in Fig. 10-10.

Solution

Since the range of this voltmeter is 500 V, it means that *if* 500 V were applied, it should produce a full-scale deflection of the pointer, which is 200 μA. Now solve for the value of $R_{\text{ser. mult}}$, using Fig. 10-10 as a guide.

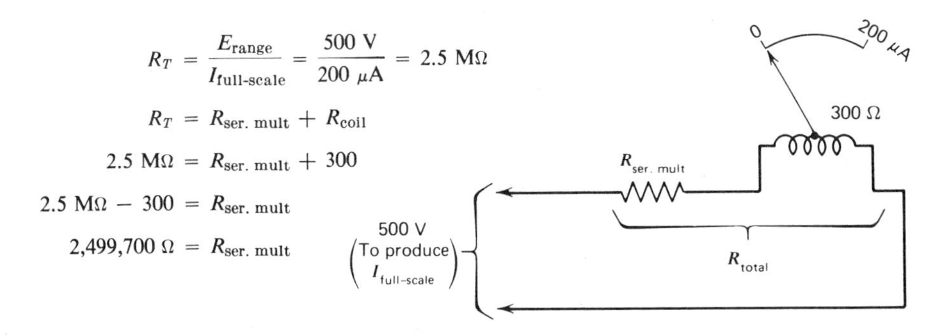

Figure 10-10. Converting a 200-μA meter into a 500-V range voltmeter, Example 10-5.

The Multi-Range Voltmeter. To be able to measure voltages over a wide range, the voltmeter has several ranges. Each one is selected by the range switch which simply connects a different value of $R_{\text{ser. mult}}$ in series with the meter coil. Figure 10-11 depicts the schematic diagram of a multi-range voltmeter. No values are given here, but this is one of the problems (see Problem 10-27) to be solved at the end of this chapter. Note that on the 5-V range, series-multiplier resistor R_1 is in series with the meter coil; on the 50-V range, R_2 is now in series with the coil; and so on.

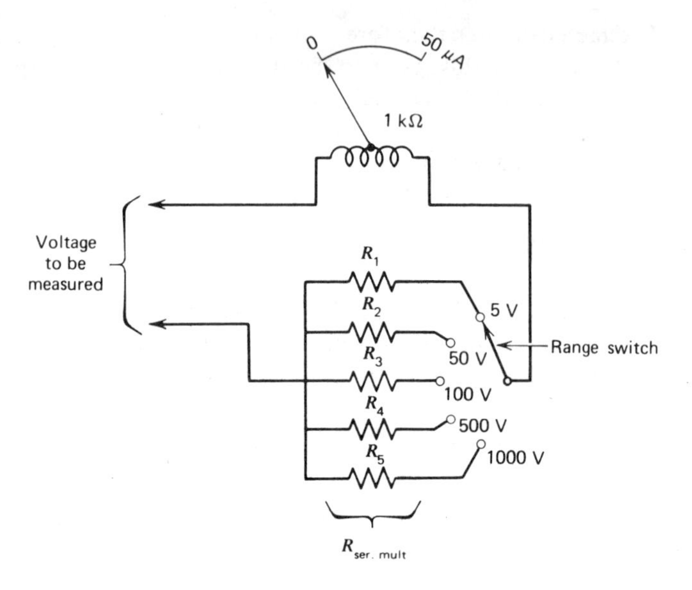

Figure 10-11. Multi-range voltmeter, Problem 10-27.

Ohms-per-Volt Sensitivity Rating. Any current meter, when converted into a voltmeter, has a characteristic called the *ohms per-volt sensitivity rating.* This is the *reciprocal* of $I_{\text{full-scale}}$, where *reciprocal* of any number is simply 1 over that number. (Reciprocal of 5 is 1/5; reciprocal of 100 is 1/100.) Therefore

$$\text{ohms-per-volt} = \frac{1}{I_{\text{full-scale}}} \tag{10-3}$$

or

$$\Omega/\text{V} = \frac{1}{I_{\text{full-scale}}}$$

Any 1 mA meter has an ohms-per-volt rating of

$$\Omega/\text{V} = \frac{1}{1 \text{ mA}} \tag{10-3}$$

$$\Omega/\text{V} = \frac{1}{1 \times 10^{-3} \text{ A}}$$

$$\Omega/\text{V} = 1 \times 10^{3}$$

$$\Omega/\text{V} = 1000$$

Similarly, any 50-μA meter has an ohms-per-volt rating of

$$\Omega/\text{V} = \frac{1}{50\ \mu\text{A}} \tag{10-3}$$

$$\Omega/\text{V} = \frac{1}{50 \times 10^{-6}}$$

$$\Omega/\text{V} = 0.02 \times 10^{6}$$

$$\Omega/\text{V} = 20{,}000$$

Very often a voltmeter has its Ω/V rating printed on the face of the meter. From this figure, two things may be determined: (a) the full-scale current required by the meter, and (b) the *total resistance* (R_T) of the voltmeter circuit.

To find the $I_{\text{full-scale}}$ when the Ω/V rating is known, use equation 10-3 (the reciprocal equation).

$$\Omega/\text{V} = \frac{1}{I_{\text{full-scale}}} \tag{10-3}$$

$$I_{\text{full-scale}} = \frac{1}{\Omega/\text{V}}$$

Example 10-6

A voltmeter is marked "100,000 Ω/V." What is the current required by this meter for full-scale deflection?

Solution

$$I_{\text{full-scale}} = \frac{1}{\Omega/\text{V}} \quad \text{(from equation 10-3)}$$

$$I_{\text{full-scale}} = \frac{1}{100{,}000}$$

$$I_{\text{full-scale}} = \frac{1}{1 \times 10^{5}}$$

$$I_{\text{full-scale}} = 1 \quad 10^{-5}\ \text{A}$$

$$I_{\text{full-scale}} = 0.00001\ \text{A} \quad \text{or} \quad 10\ \mu\text{A}$$

The Ω/V *rating* is actually the *ohms of resistance* for *each* volt of the *range* that the meter is operating on. From the Ω/V *rating* and the voltmeter *range*, the *total resistance* (R_T) of the voltmeter ($R_{\text{coil}} + R_{\text{ser. mult}}$) can be determined.

$$R_T = (\Omega/\text{V})\,(E_{\text{range}}) \tag{10-4}$$

From Example 10-3, it was found that a 1 mA meter operating on a 50-V range had a total resistance of 50 kΩ. This can also be found from the following (as shown previously, any 1-mA meter has 1000 Ω/V).

Then
$$R_T = (\Omega/\text{V})\,(E_{\text{range}}) \qquad (10\text{-}4)$$
$$R_T = (1 \times 10^3)\,(50\ \text{V})$$
$$R_T = 50 \times 10^3 \quad \text{or} \quad 50\ \text{k}\Omega$$

Example 10-7

Find the total resistance of a 200 μA meter that is operating as a 500-V range voltmeter (from example 10-5).

Solution

Solve for R_T in the method just described, comparing the result with that of Example 10-5.

For Similar Problems Refer to End-of-Chapter Problems 10-24 to 10-31

10-5. Using the Multi-Range DC Voltmeter

A dc voltmeter must always be placed *across* some component in a circuit to measure the voltage across that part. In a simple series resistor circuit, such as is shown in Fig. 10-5*a*, if it is desired to measure the total or source voltage, the leads of the voltmeter must be placed between *point W* and *point Z*, with the *negative* lead at *point W* and the *positive* lead at *point Z*. The test leads of a dc voltmeter must always be connected with the correct polarity, that is, the *negative* lead to the *negative point*, and the *positive* lead to the *positive point*. The terms *negative point* and *positive point* are relative. In Fig. 10-5*a*, voltage across R_1, is measured between points W and X, with point X positive with respect to W. Voltage across R_2 is measured between points X and Y, with point X negative with respect to Y. When measuring completely unknown voltages, always start with the voltmeter on its highest range.

Example 10-8

In Fig. 10-12 where would a voltmeter be connected as to its polarity to measure (a) V_{R_2} (b) V_{R_3} (c) V_{R_5} and (d) V_{R_6}?

Solution

Note the polarity of E_T, the applied voltage. Also note the direction of current flow (electron drift) as indicated by the arrows shown. Electrons

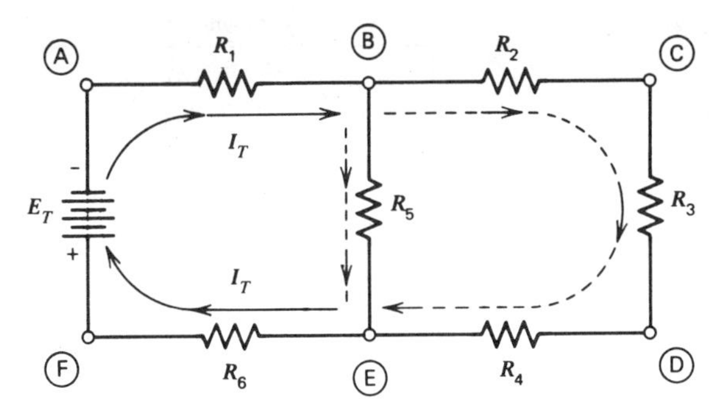

Figure 10-12. Example 10-8.

move from − to + through resistors, Therefore, (a) V_{R_2}, *negative* at point B, *positive* at point C, (b) V_{R_3}, *negative* at point C and *positive* at point D, (c) V_{R_5}, *negative* at point B, *positive* at point E, and (d) V_{R_6}, *negative* at point E and *positive* at point F.

Voltmeter as a Short-or-Resistance-Leakage-Indicator. Some components like capacitors (discussed in Chapters 13 and 14), and shielded cables used for microphones, audio, and transmission lines often develop a short or leakage between their leads only when a voltage is applied across them. The trouble, if it remains, even without the applied voltage, can easily be detected with an ohmmeter (see Section 10-6). However, if the defect is only present when sufficient voltage is applied, then a voltmeter is a reliable indicator.

The component to be checked may be connected in *series* with the voltmeter, as shown in Fig. 10-13*a* and *b* (this is not the normal method of connecting a voltmeter), and a voltage can be applied that is the required amount for the component in its normal circuit. If the capacitor in (a), or the cable in (b), is good, then these devices act as if they were an *open* circuit having infinitely large resistance. No current flows, and the meter indicates zero volts between points X and Y.

If the component is shorted (zero ohms) then the full applied voltage appears across points X and Y, and the voltmeter indicates this full voltage. Very often the trouble is due to some *leakage* in the insulating material of the capacitor or the cable, producing some resistance between the leads of the device, and the voltmeter reads some voltage; but it reads less than the full applied voltage.

Example 10-9

A small capacitor is suspected of being "leaky," yet checks perfectly normal (infinity) with the low voltage of an ohmmeter. When connected as shown in Fig. 10-13*a*, what would the voltmeter read if (a) capacitor were good, and (b) the capacitor were shorted?

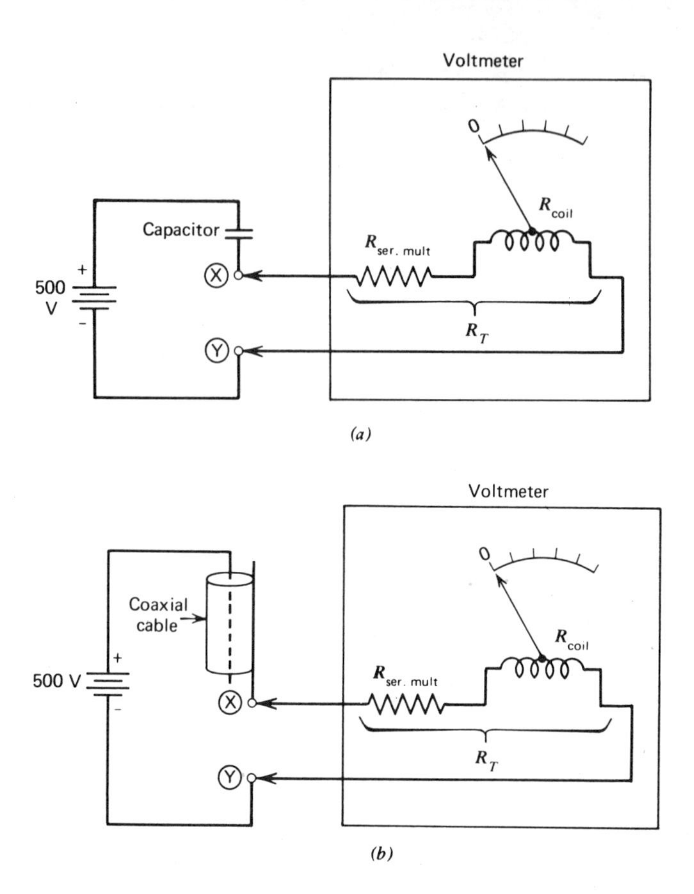

Figure 10-13. Voltmeter reads zero if component is good (not shorted or leaky). (*a*) Checking a nonelectrolytic capacitor with a voltmeter. (*b*) Checking a coaxial cable with a voltmeter.

Solution

(a) 0 V, and (b) 500 V.

Loading Effect of Voltmeter. Any test instrument that is inserted into an operating circuit has some effect on the circuit. This is called *loading* the circuit. It is desirable that the test equipment have as little effect as possible on the circuit. Since a voltmeter is put in *parallel* with the resistor (or other component) the voltage of which is to be measured, the voltmeter should have as *large* a resistance as possible compared to the resistor, so that the voltmeter does not *load* the circuit.

Figure 10-14 is a simple series-resistor circuit. Since R_1 and R_2 are equal, then voltage across R_1 is 3 V. This is the voltage across R_1 *without* a voltmeter. When a voltmeter is connected across R_1, as depicted in Fig. 10-15, the resist-

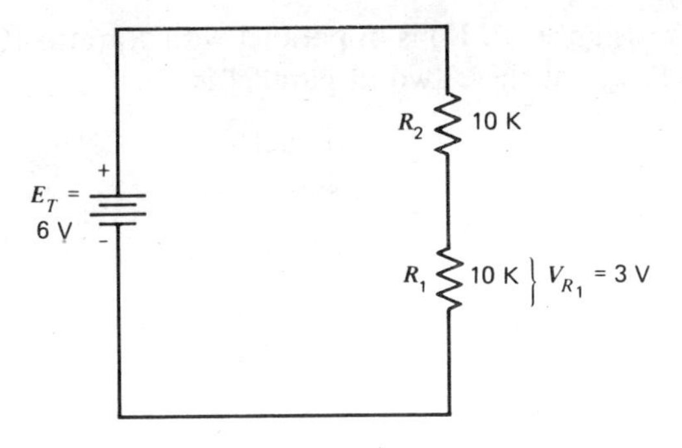

Figure 10-14.

ance of the voltmeter ($R_{\text{ser. mult}} + R_{\text{coil}}$) is now paralleling R_1. The following examples illustrate voltmeter *loading* of the circuit.

Example 10-10

If the voltmeter in Fig. 10-15 is a 1000 Ω/V type, and is on its 10-V range, find the voltage across R_1 with this meter connected across the resistor.

Solution

The meter, having a 1000 Ω/V rating and operating on its 10-V range, has a total voltmeter resistance of

$$R_{\text{meter}} = (\Omega/\text{V})\,(E_{\text{range}}) \tag{10-4}$$

$$R_{\text{meter}} = (1000)\,(10)$$

$$R_{\text{meter}} = 10\ \text{k}\Omega$$

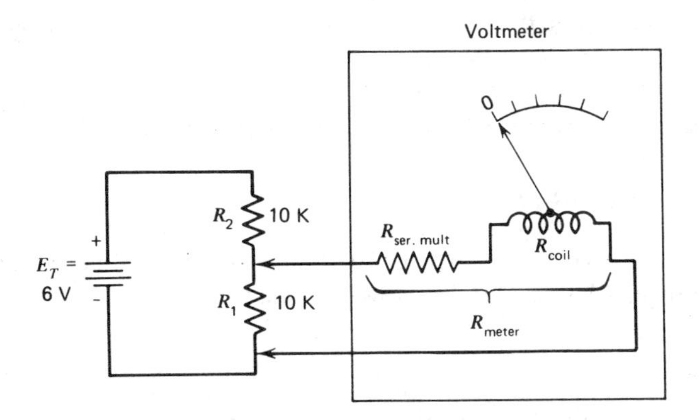

Figure 10-15. Voltmeter loading.

The meter resistance (10 K) is in parallel with R_1 (also 10 K). The equivalent resistance (R_{equiv}) of these two in parallel is

$$R_{equiv} = \frac{\text{product}}{\text{sum}} \tag{4-3}$$

$$R_{equiv} = \frac{(R_1)(R_{meter})}{R_1 + R_{meter}}$$

$$R_{equiv} = \frac{(10 \times 10^3)(10 \times 10^3)}{10 \times 10^3 + 10 \times 10^3}$$

$$R_{equiv} = \frac{100 \times 10^6}{20 \times 10^3}$$

$$R_{equiv} = 5 \times 10^3 \ \Omega \qquad \text{or} \qquad 5 \text{ k}\Omega$$

The total resistance of the circuit of Fig. 10-15 consists of R_2 in series with the R_{equiv} of R_1 and R_{meter}, or

$$R_T = R_2 + R_{equiv}$$

$$= 10 \text{ K} + 5 \text{ K}$$

$$= 15 \text{ k}\Omega$$

and

$$V_{R_1} \text{ now} = \left(\frac{R_{equiv}}{R_T}\right) E_T$$

$$V_{R_1} = \left(\frac{5 \text{ K}}{15 \text{ K}}\right)$$

$$V_{R_1} = 2 \text{ V}$$

Note that *without* the voltmeter, V_{R_1} of Fig. 10-14 is 3 V, and *with* the voltmeter across R_1 (Fig. 10-15), this voltage has become only 2 V. This change is due to *voltmeter loading.*

A more accurate result (closer to the actual 3 V) is achieved if the *voltmeter resistance is much larger* than R_1, as shown in the next example.

Example 10-11

If the voltmeter of Fig. 10-15 is a 20,000 Ω/V type, operating on its 10-V range, what would V_{R_1} now become?

Solution

Solve this example by using the method just shown. As a guide, use the following results: R of voltmeter = 200 K, R_{equiv} = 9.52 K, and V_{R_1} = 2.92 V.

Note that this meter (20,000 Ω/V) across R_1 results in 2.92 V across the parallel combination, compared to the previous meter (1000 Ω/V) of Example 10-10, which resulted in only 2 V, while the true voltage (without any meter to load the circuit) is actually 3 V. The percentage error may be found from the following equation:

$$\% \text{ error} = \frac{V_{R \text{ calculated without meter}} - V_{R \text{ read on meter}}}{V_{R \text{ calculated without meter}}} \text{ X}$$

Refer to End-of-Chapter Problems 10-32 to 10-41

10-6. The Series Ohmmeter

A third piece of test equipment used very often is the *ohmmeter*. As its name implies, it is used to measure the resistance, in ohms, between two points in a circuit. It is employed to measure certain components such as resistors, coils, transformer windings, and capacitors to determine which device in a circuit is defective. The ohmmeter indicates whether a wire or cable connecting two points in a circuit is good (continuity), or bad (open); whether a resistor is good (measures within its correct tolerances), or bad (open, or has changed value beyond its tolerance); whether a coil is good or not; whether a capacitor is good (charges up), or bad (shorted or leaky).

The simple ohmmeter is called a *series ohmmeter*, since the meter coil, the voltage source, and the device to be measured are all connected in series. A typical circuit is shown in Fig. 10-16. The voltage is usually a built-in battery, and is part of the ohmmeter circuit. Resistors R_1 and R_2 limit the current to that requiring full-scale deflection of the meter pointer, 1 mA in Fig. 10-16, when the test leads are shorted together. R_1 is a variable resistor, or potentiometer, which can be reduced in value as necessary to compensate for changes

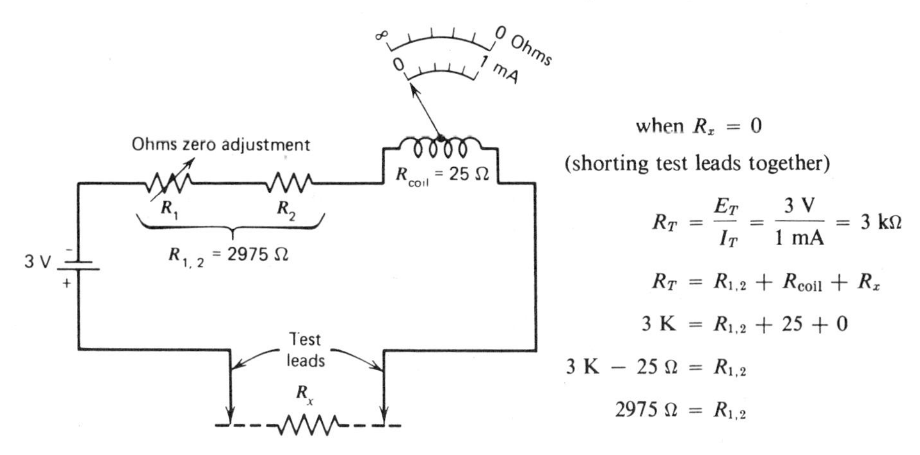

Figure 10-16. Series ohmmeter.

in the battery voltage and its internal resistance as the battery ages. Before the ohmmeter is used to measure some resistance, the operating procedure is to short the test leads together (meaning *zero* ohms between them). The pointer should deflect full scale to 1 mA, and on the ohmmeter scale, it should indicate *zero ohms*. R_1 should be adjusted so that the pointer does indicate this reading of zero ohms. The purpose of R_2 is to offer some protection to the meter if R_1 has been accidentally set to its minimum value, and the test leads shorted. R_2 helps to limit the current, with R_1 at minimum.

The combined value of R_1 plus R_2 is determined by the value of the battery voltage, R_{coil}, and the full-scale current of the meter. Then this combined value is divided up with approximately one half for R_1 and one half for R_2. If, as depicted in Fig. 10-16, the meter is a 1mA type with a coil resistance of 25 Ω, and the battery is 3 V, then with the test leads shorted, R_{total} is

$$R_T = \frac{E_T}{I_T} \qquad \text{(Ohm's law)}$$

$$R_T = \frac{3\text{ V}}{1\text{ mA}}$$

$$R_T = \frac{3}{1 \times 10^{-3}}$$

$$R_T = 3 \times 10^3\ \Omega \qquad \text{or} \qquad 3\text{ k}\Omega$$

and $R_1 + R_2$ will be called $\underline{R_{1,2}}$.

then

$$R_T = R_{1,2} + R_{coil}$$

$$3\text{ K} = R_{1,2} + 25$$

$$3\text{ K} - 25 = R_{1,2}$$

$$2975\ \Omega = R_{1,2}$$

This 2975 ohms is then divided up about equally between R_1 and R_2, with each about 1500 Ω, with R_1 being variable.

The 3-V battery and the 1 mA meter determine the total resistance (R_T = 3 kΩ). This also determines that value of the resistor (R_x), measured between the test leads, that produces a *half-scale current* deflection (or 0.5 mA). When $R_x = R_{1,2} + R_{coil}$, current will become half scale, as shown in the following. When R_x = 3 K, and $R_{1,2}$ = 2975 Ω, and R_{coil} = 25 Ω, then

$$R_T = R_{1,2} + R_{coil} + R_x$$

$$R_T = 2975 + 25 + 3000$$

$$R_T = 6000$$

and

$$I_T = \frac{E_T}{R_T}$$

$$I_T = \frac{3 \text{ V}}{6000 \ \Omega}$$

$$I_T = \frac{3}{6 \times 10^3}$$

$$I_T = 0.5 \times 10^{-3} \text{ A} \qquad \text{or} \qquad 0.5 \text{ mA}$$

Since this value of $R_x = 3\ K$ (plus $R_{1,2}$ and R_{coil}), resulted in 0.5-mA current flow, then the ohmmeter scale corresponding to this half-scale current is marked *3 K*, as depicted in Fig. 10-17.

The *ohms* scale is then calibrated for *each value* of R_x that produces a particular current. The following example illustrates this, with the results shown on the ohmmeter scale of Fig. 10-17.

Example 10-12

Calibrate the ohmmeter of Fig. 10-16 for the following values of R_x: (a) 100 Ω, (b) 500 Ω, (c) 1 kΩ, (d) 2 kΩ, and (e) 4 kΩ.

Solution

For each value of R_x, the resulting current must be found. This is calculated from $I_T = E_T/R_T$, where E_T is the 3-V battery, and $R_T = R_{1,2} + R_{\text{coil}} + R_x$. Therefore:

(a) When R_x *is 100* Ω:

$$R_T = R_{1,2} + R_{\text{coil}} + R_x$$

$$R_T = 2975 + 25 + 100$$

$$R_T = 3100$$

and

$$I_T = \frac{E_T}{R_T}$$

$$I_T = \frac{3}{3100}$$

$$I_T = \frac{3}{3.1 \times 10^3}$$

$$I_T = 0.968 \times 10^{-3} \text{ A}$$

$$I_T = 0.968 \text{ mA}$$

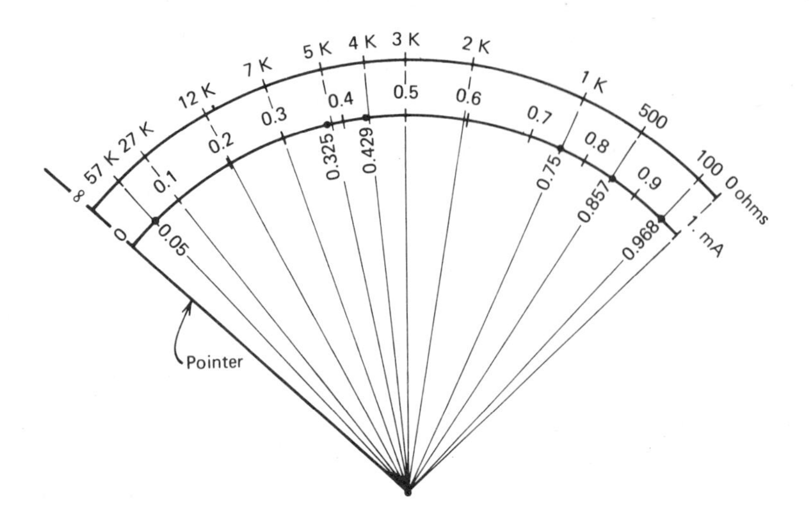

Figure 10-17. Calibrated ohmmeter scale of series ohmmeter of Fig. 10-16, Examples 10-12 and 10-13.

The value for $R_x = 100\ \Omega$ is now marked on the ohmmeter scale corresponding to 0.968 mA, as shown in Fig. 10-17.

(b) When R_x *is* $500\ \Omega$:

$$R_T = R_{1,2} + R_{\text{coil}} + R_x$$

$$R_T = 2975 + 25 + 500$$

$$R_T = 3500\ \Omega$$

and

$$I_T = \frac{E_T}{R_T}$$

$$I_T = \frac{3}{3500}$$

$$I_T = \frac{3}{3.5 \times 10^3}$$

$$I_T = 0.857 \times 10^{-3}\ \text{A}$$

$$I_T = 0.857\ \text{mA}$$

This value for $R_x = 500\ \Omega$ is now marked on the ohmmeter scale corresponding to 0.857 mA, as shown in Fig. 10-17.

Continue to solve this example, using the results shown in Fig. 10-17 as a guide.

Example 10-13

As a further exercise, calibrate the ohmmeter of Fig. 10-16 for the following values of R_x: (a) 5 kΩ, (b) 7 kΩ, (c) 12 kΩ, (d) 27 kΩ, and (e) 57 kΩ. Compare the answers with the results shown in Fig. 10-17.

Changing the Ohmmeter Range. The simple series ohmmeter shown in Figs. 10-16 and 10-17, and discussed previously in this section, may easily be made to read higher values of resistors. It is shown in Figs. 10-16 and 10-17 that by using a 3-V battery, an R_x (resistor to be measured) of 3 K produces a half-scale deflection, or 0.5 mA.

If a battery of 30 V (ten times the original) is substituted in place of the original 3 V, with the required larger value of $R_{1,2}$, then all ohmmeter scale readings are *multiplied by 10.*

Refer to End-of-Chapter Problems 10-42 to 10-48

10-7. Using the Ohmmeter

When one uses the ohmmeter, the adjustment for zeroing should be set first. With the test leads shorted together, the *zero adjust* control should be turned so that the meter pointer rests on *zero ohms.* Then, with the leads separated, the pointer should be on *infinity* (∞) *ohms.* Some ohmmeters have a second control called *ohms adjust.* This should be adjusted so that the pointer is on infinity when the test leads are separated.

Most ohmmeters have a range switch with positions such as: *R* × *1*, *R* × *10*, *R* × *100*, and so on. The ohmmeter scale is *nonlinear*, that is, the spaces or graduations have different values. By referring to Fig. 10-17, it can be seen that toward the left end of the ohmmeter scale, each one-half inch or one-quarter inch counts much more than do the same dimensions toward the right end. As a result, greater accuracy in reading may be achieved at the right end. Therefore, when attempting to read the ohmic value of some component, change the range switch setting so that the meter pointer is somewhere in that vicinity of the scale where greater accuracy results. However, it is important to remember to *multiply* the actual meter pointer reading by the setting of the range switch. For example, if the pointer indicates 2 K, but the range switch is on *R* × *10*, the value of the thing being measured is *20 kΩ.* Similarly a pointer reading of *500*, and the switch being on *R* × *100*, gives a value of *50,000* Ω (or 500 × 100), or *50* kΩ.

Since the ohmmeter uses its own voltage, usually a built-in battery (Fig. 10-16) then, when measuring the resistance of some component in a circuit, it is very important that there be *no other voltage in the circuit.* Any voltage other than the ohmmeter battery will produce an incorrect reading and will probably damage the meter. Before connecting the ohmmeter into a circuit, not only must the power supply or voltage be turned off, but any capacitor connected to the component being measured should first be discharged completely with a temporary *short* (piece of wire) placed across the capacitor.

In most circuits, one component is often in a parallel or a series-parallel arrangement with other parts. When placing the ohmmeter test leads across one component, it could actually read that part and others in parallel with it. Therefore, when this possibility exists, *one end* of the part to be measured should be disconnected from the circuit. The ohmmeter, now connected across that part, does not read anything else except the component being tested. This is especially true when measuring or testing a capacitor while that part is still connected in its circuit.

When one measures a capacitor, to repeat what was said previously, the capacitor must first be discharged to be sure that none of its voltage remains to damage the ohmmeter. The actual reading or indication that a capacitor produces on an ohmmeter depends on the value or size of the capacitor and, of course, on its condition as to good, shorted, or leaky. A capacitor should be checked by using the high-range of the ohmmeter.

A good capacitor of the smaller sizes (paper, or mica dielectric) gives only a slight and momentary movement of the ohmmeter pointer away from infinity (∞), but the pointer essentially *remains at infinity*. The momentary wiggle of the pointer is caused by the small capacitor's quickly charging up to the ohmmeter battery voltage.

A large good capacitor (the electrolytic type) causes the ohmmeter pointer to move quickly from infinity toward zero, and then slowly to creep back toward infinity, but usually never getting back to infinity. After possibly ten or twenty seconds with the ohmmeter still connected to the capacitor, the pointer may come to rest at some high resistance value. With the range switch on $R \times 100\ K$, the pointer may come to rest on *50*, giving a reading of $5\ M\Omega$ (or 50×100 K). A good but large capacitor of the electrolytic type has some normal leakage through the chemical dielectric or insulator, which is usually anywhere from a few hundred thousand ohms (400 K) to several megohms (10 or 20 MΩ), or larger—the larger, the better.

A poor or bad small capacitor (paper or mica dielectric) will produce some ohmmeter reading other than infinity. This could be zero up to several megohms, but the small capacitor should be replaced.

A poor or bad large capacitor (electrolytic type) will not show the charging effect of moving the meter pointer slowly back toward infinity, but the pointer will immediately swing to that amount of leakage. The lack of the slowly moving pointer, when the ohmmeter is on a high range, indicates a bad capacitor.

No movement of the pointer at all away from infinity, on the high ohmmeter range, indicates an *open* capacitor which should be replaced.

Refer to End-of-Chapter Problems 10-49 to 10-57

10-8. The DC Motor

In Section 9-8D, The Electric Motor Principle, we discuss a simple dc motor, showing its similarity to the permanent-magnet moving-coil meter. The student should now review Section 9-8D before continuing with this present section. Our discussion of dc motors here is sufficient to acquaint the *electronic student-technician* with the theory of electric motors. A more thorough presentation and explanation, such as is required by the *electrician-student*, may be obtained by referring to several of the strictly electrical textbooks.

Figure 10-18 shows the basic parts of the dc motor. Instead of a permanent magnet, a large electromagnet called the *field coil* is used. The *armature coil* is mounted on an axle so that it is free to rotate. The armature coil shown in Fig. 10-18 is only a single loop of wire for simplicity; actually, it consists of numerous turns of wire. Each end of the simplified *armature* coil is attached permanently to a separate metallic half-ring called the split-ring commutator.

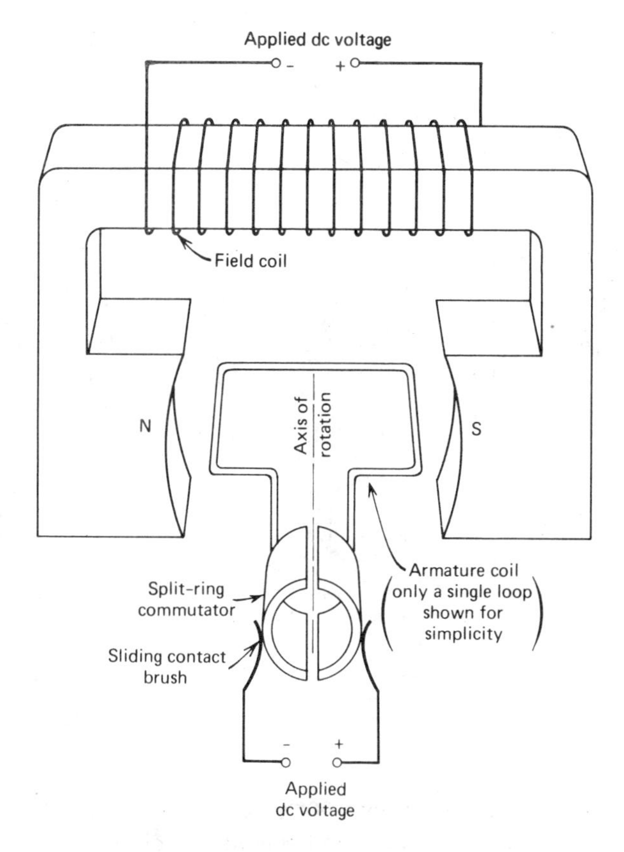

Figure 10-18. A simple dc motor.

This commutator ring is also mounted on the axle and is free to rotate with the armature coil. The applied dc voltage is connected to the moveable armature coil through a pair of stationary, but sliding contacts called *brushes*. These brushes are conductors and make contact with the metal split rings. If the armature and the split rings are rotated, then the brush that is permanently connected to the negative end of the applied dc voltage first makes contact with one split ring, and then as the armature and split rings rotate, this brush contacts the other split ring.

In Fig. 10-19*a* the dc voltage is applied through the brushes to the split-ring commutator. The brushes are permanently connected to the dc voltage; one brush is permanently negative, while the other is always positive. In the position of the coil and split rings shown in Fig. 10-19*a*, split ring *X* is connected to the negative brush, while *Y* is connected to the positive brush. Electron flow is from *X* to *Y* through the coil as indicated by the arrows on the coil itself. Observe that the coil is in a horizontal position, and that one end of the armature coil is permanently connected to the *X* split-ring, while the other end is always connected to *Y*. From the *three-finger right-hand rule for motors*, see Fig. 9-25, and also from Fig. 9-22, it may be found that the coil will start

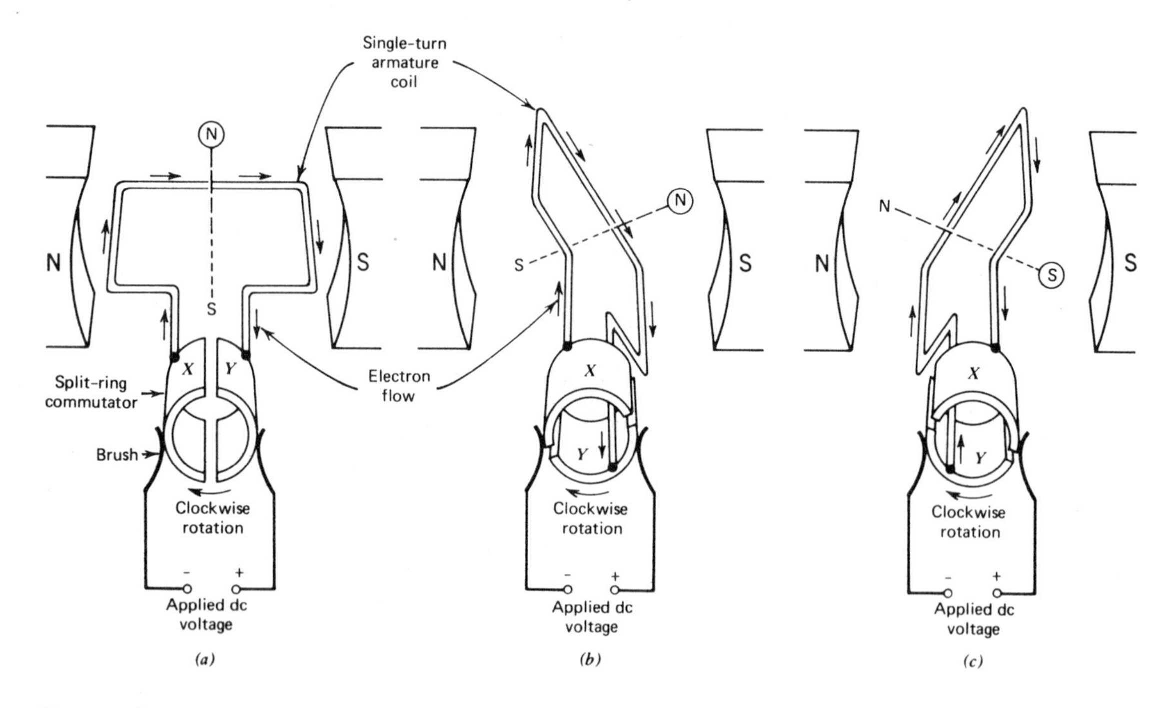

Figure 10-19. A dc motor operation. (*b*) Amost 90° rotation from (*a*). (c) About 95° rotation from (*a*), armature current and magnetic polarity has reversed.

rotating clockwise. This clockwise rotation may also be explained from the following. Observe also in Fig. 10-19*a* that from the *left-hand coil rule* (Section 9-6) the electromagnetic *armature* coil has it north pole above the coil, and its south pole below this coil. The *north* pole is shown *encircled.* The *encircled north pole of the armature* is now *repelled* by the *north pole of the field coil*, and *attracted* by the *south pole of the field coil*, producing a clockwise rotation of the armature coil and its split rings.

In Fig. 10-19*b*, the armature has rotated almost 90° clockwise from its position of (a). The wires of the armature coil are now almost in a vertical position, with the *X* wire (the one that is permanently connected to the *X* split ring) just about above the *Y* wire. The *X* ring is still connected to the negative terminal of the applied dc voltage through the negative brush, and electron flow is still from *X* to *Y* through the coil. From the *left-hand coil rule* (Section 9-6), it can be determined that the north pole of the armature coil again shown encircled, is now to the right of the coil, and is still attracted by the south pole of the field coil, continuing the clockwise rotation of the armature coil and its split rings.

If this condition continued, the coil would come to an abrupt stop when the *armature north pole* reached a point adjacent to the *field coil south pole.* However, as depicted in Fig. 10-19*c*, when the armature coil and the split rings have rotated a few degrees further, say about 5° beyond (b), or a total of about 95° from (a), ring *X* has moved to a position where it now contacts the *positive* brush, and *Y* now contacts the *negative* brush. This is the reverse of the previous connections shown in Fig. 10-19*a* and *b*. With split ring *Y* now connected to the negative brush, as shown in Fig. 10-19*c*, electron flow is from *Y* to *X* through the coil. Again, by using the *left-hand coil* rule, it can be determined that the armature north pole is now at the left, with its *south* pole now shown encircled, at the right. The *encircled pole is now a south pole*, whereas it had previously been a north pole. This south pole is now just slightly past the large field coil south pole, and is repelled, keeping the armature rotating clockwise.

The action of the split-ring commutator is to keep switching the negative and positive polarity from the brushes, first making *X negative* and *Y positive* (Fig. 10-19*a* and *b*), and then making *X positive* and *Y negative* (Fig. 10-19*c*). The commutator is, therefore, simply a mechanical device that causes the current to reverse through the armature coil, thus reversing its magnetic poles. This reversal occurs at the instant that the armature coil north pole is adjacent to the field coil south pole. The armature coil north then reverses to a south pole, causing a repelling action to occur, thus continuing the rotation of the armature coil.

A dc motor may be reversed in its direction of rotation by reversing the polarity of the applied dc voltage to *either* the armature coil or to the field coil, but not to both.

When the field coil and the armature coil are connected in series, it is called a *series* motor. When the two coils are connected in parallel, it is called a *shunt* motor. A third type is called a *compound* or *shunt-series* motor. Here, the field consists of two separate windings; one is connected in *shunt* with the armature, while the other is in *series* with the armature. Each type has different characteristics relating to its speed, its rotational or turning strength, called its *torque*, and its action with a varying load or weight. These characteristics and the reasons or explanations for them are not usually necessary for the electronic technician and have been omitted from this book. By referring to an electrical or motor textbook, this information can be obtained.

Refer to End-of-Chapter Problems 10-58 to 10-62

10-9. The Electrodynamometer

In the D'Arsonval or Weston moving-coil meter, described previously in Section 10-1, the moving coil rotated between the poles of a permanent magnet. Another version, called the *dynamometer* or *electrodynamometer*, employs a pair of stationary coils instead of the permanent magnet. In the diagram of Fig. 10-20, coils L_1 and L_2, the field coils, replace the large magnet of the Weston movement. Coil L_3 is the rotatable coil.

When the *dynamometer* is designed as an ammeter, stationary coils L_1 and L_2 are made of heavy, low-resistance wire. All the coils L_1, L_2, and L_3 are connected in series with the component whose current is to be measured. A shunt resistor (see Section 19-2), if necessary, is placed in parallel with the movable coil L_3.

When the *dynamometer* is designed to operate as a voltmeter, stationary coils L_1 and L_2 are constructed of a thin high-resistance wire. All three coils are connected in series with the multiplier resistor (Section 10-4), and the combination is connected across the component whose voltage is to be measured.

To measure the *power dissipation* or *wattage*, two regular meters of the D'Arsonval type are required. One is connected as a *voltmeter*, while the other is connected as an *ammeter*. Since power is the product of voltage and current ($P = EI$, Section 2-9), then by multiplying the readings of the voltmeter and ammeter, the power or wattage may be found. However, by employing the *dynamometer*, a single meter can read the wattage directly.

A *wattmeter* consists of the stationary coils, L_1 and L_2 of Fig. 10-20, of the *dynamometer* connected as an *ammeter* in *series* with the circuit being measured, while the movable coil L_3 with a suitable multiplier resistor (Section 10-4) is connected as a voltmeter across the component or circuit being measured. The amount of rotation or deflection of coil L_3, and its pointer, is directly proportional to the *current* flowing through field coils L_1 and L_2, and also to

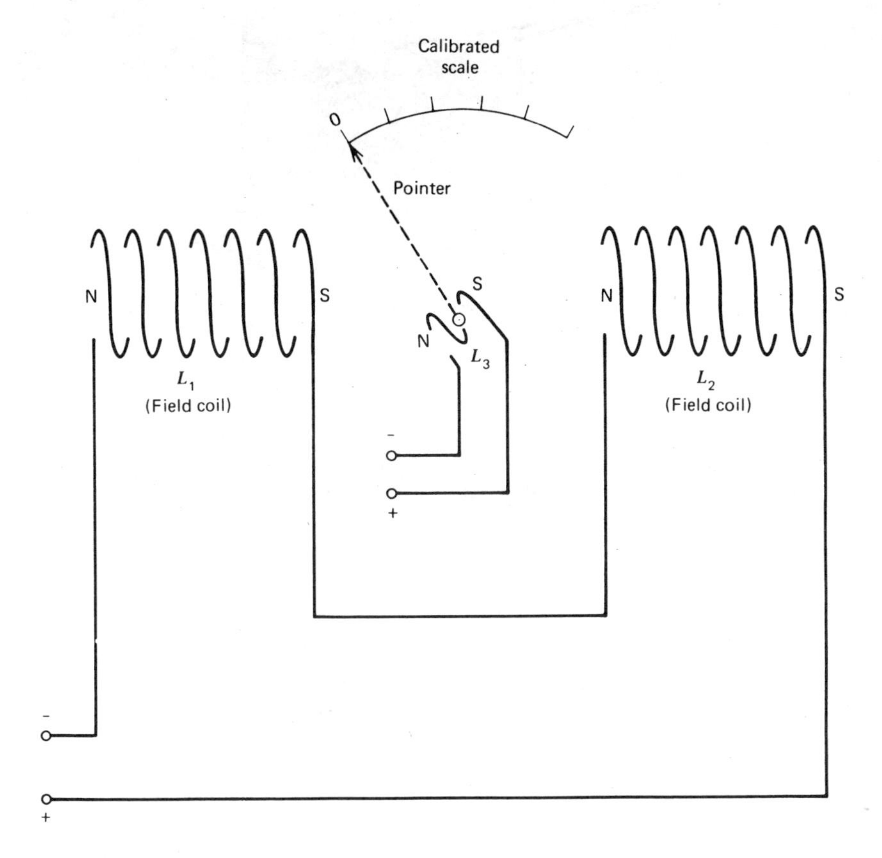

Figure 10-20. Electrodynamometer.

the *voltage* across (and its resultant current in) the movable coil L_3. The result is that the rotation of L_3 is actually proportional to the *product* of *voltage* and *current*, and thus the meter scale may be calibrated in *watts*.

Note that in Fig. 10-20, the negative and positive polarities applied to the field coils L_1 and L_2, and the movable coil L_3, produce the N and S magnetic polarities shown. Each coil has a north pole at its left end. As a result the adjacent ends of L_1 and L_3, and also L_3 and L_2, have opposite magnetic poles, producing an attraction. L_3, therefore, is rotated *clockwise* against the action of a spring (not shown).

If all voltage polarities were reversed from those shown in Fig. 10-20, then the magnetic poles of each coil would reverse. Now the right end of each coil would be its north pole. However, the adjacent ends of L_1 and L_3, and also L_3 and L_2, are still of opposite magnetic polarities, and movable coil L_3 is still attracted by the field coils L_1 and L_2. This again makes L_3 rotate clockwise. As a result, the *dynamometer* works whether the applied voltage has the polar-

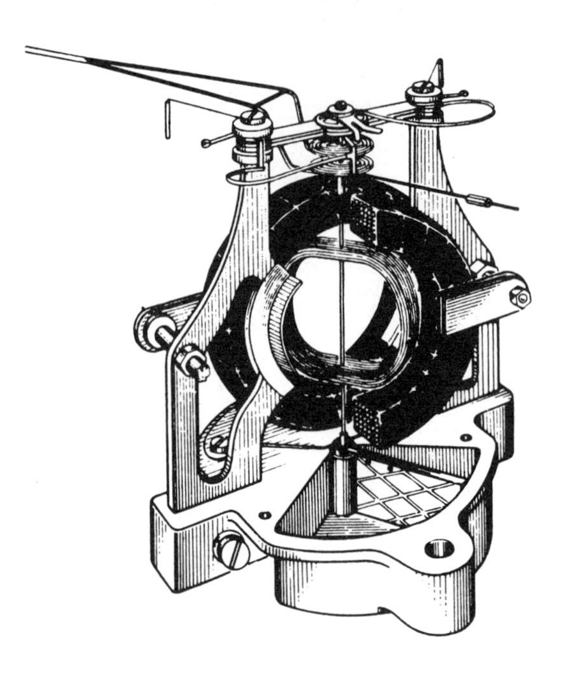

Figure 10-21. Electrodynamometer mechanism showing one field coil cut in half (*courtesy Weston Instruments, Inc*).

ities shown in Fig. 10-20 for a dc voltage, or whether the voltage polarities all keep reversing, such as for an ac voltage. This type meter, therefore, can be used for both dc and ac.

Figure 10-21 shows a detailed drawing of an electrodynamometer, with one of the field coils shown in a cutaway view. *Damping*, or the prevention of the moving coil rotating too quickly, is accomplished by a flat rectangular vane which moves with the coil. This damping vane may be seen at the bottom of Fig. 10-21. It moves in a close-fitting area, shown at the bottom of Fig. 10-21 and shaped like a large slice of pie. The vane cannot rotate too fast in its close fitting chamber because of the resistance offered by the air. Thus, the vane slows down the rotation of the moving coil.

Refer to End-of-Chapter Problems 10-63 to 10-67

10-10. Ayrton Shunt

In Section 10-2 it is shown that by placing a resistor of the correct value in *shunt* (parallel) with the movable coil of a meter, the meter has been converted into a higher range ammeter. By switching from one shunt resistor to another, the ammeter can now act as a multiple range current-reading device, as shown in Fig. 10-4. It is also pointed out that, when switching from one resistor to the next, the meter coil must never be without some shunt resistor. If it were, too much current could flow through the coil, damaging it. As a result, the switch must be of the *shorting* type or *make-before-break* type.

A more complex shunt-resistor circuit shown in Fig. 10-22 does not require this special type switch. The circuit is called the *Ayrton* shunt or *universal* shunt, and is also known as a *ring* shunt.

Note that in Fig. 10-22, the meter coil is *always* shunted by some resistance. On the 5 mA range, $R_1 + R_2$ acts as the shunt. As the range switch is moved to the 50 mA position, there is an instant when the switch is contacting neither position, but is open. The circuit is open, and no current flows at all, thus protecting the coil. When the range switch is turned to the 50 mA setting, R_1 is now in *series* with the meter coil, but R_2 is in *parallel* with the series circuit. The method of solving the required values of resistors R_1 and R_2 is a bit more complicated than the simpler circuit of Figs. 10-2, 10-3, and 10-4. The following example and detailed discussion illustrates one method of finding the values of R_1 and R_2 in Fig. 10-22.

Example 10-14

In the Ayrton shunt circuit of Fig. 10-22, find the values of R_1 and R_2 that will convert the 1 mA meter with a coil resistance of 150 Ω into a 5- and a 50-mA range meter.

Solution

On the 5-mA setting of the range switch, $R_1 + R_2$ form a total shunt resistor $R_{T\,sh}$. This is found in the same manner described previously in Section 10-2, and is shown in the following:

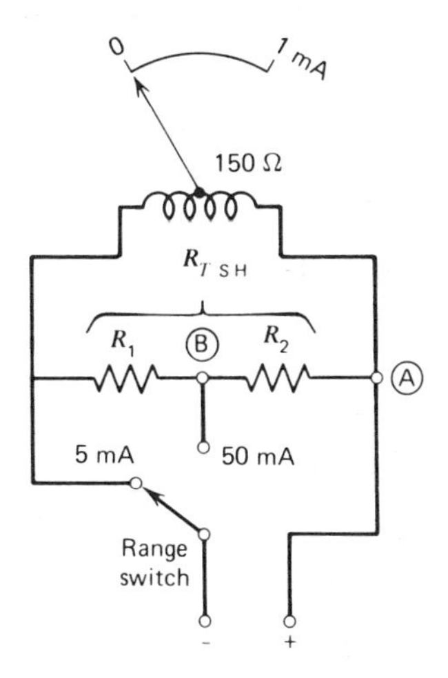

Figure 10-22. Ayrton shunt (universal or ring shunt), Example 10-14.

On the *5-mA range*, $I_{coil} = 1$ mA, and $I_{R_1R_2} = I_{Rt\ sh} = 4$ mA

$$R_{T\ sh} = \frac{V_{Rt\ sh}}{I_{Rt\ sh}} = \frac{\text{same as } V_{coil}}{I_{R_1R_2}} \qquad \text{(from equation 10-1)}$$

$$R_{T\ sh} = \frac{(1\ \text{mA})\ (150\ \Omega)}{4\ \text{mA}}$$

$$R_{T\ sh} = \frac{(1 \times 10^{-3})\ (150)}{4 \times 10^{-3}}$$

$$R_{T\ sh} = 37.5\ \Omega$$

On the *50-mA range* switch position of Fig. 10-22, current (electron drift) from the negative end of the circuit eneters point B, and splits up. Some current flows through R_2 to point A, while some current goes through R_1 and the coil in series to point A. Therefore, R_2 is in *parallel* with the series combination of R_1 and the coil. Since I_{total} is 50 mA, and I_{coil} (and I_{R_1}) is 1 mA, then I_{R_2} is 49 mA. Since the same voltage is across parallel branches, then

$$V_{R_2} = V_{R_1} + V_{coil}$$

and

$$(I_{R_2})\ (R_2) = I_{coil}\ (R_1 + R_{coil})$$

$$(49\ \text{mA})\ (R_2) = (1\ \text{mA})\ (R_1 + 150)$$

$$(49 \times 10^{-3})\ (R_2) = (1 \times 10^{-3})\ (R_1 + 150)$$

dividing both sides of the equation by 10^{-3} to simplify, yields

$$49\ R_2 = 1\ R_1 + 150 \qquad (10\text{-}6)$$

but $R_1 + R_2 = R_{T\ sh}$, and $R_{T\ sh}$ was found previously to be 37.5 Ω. Therefore

$$R_1 + R_2 = R_{T\ sh} \qquad (10\text{-}7)$$

$$R_2 = R_{T\ sh} - R_1$$

$$R_2 = 37.5 - R_1$$

now, substituting for R_2 in equation 10-6, gives

$$49\ (37.5 - R_1) = 1\ R_1 + 150$$

$$1835 - 49\ R_1 = 1\ R_1 + 150$$

Solving for R_1, $1835 - 150 = 1\ R_1 + 49\ R_1$

$$1685 = 50\ R_1$$

$$\frac{1685}{50} = R_1$$

$$33.7\ \Omega = R_1$$

and finally, solving for R_2

$$R_1 + R_2 = R_{T\,\text{sh}} \qquad (10\text{-}7)$$

$$R_2 = R_{T\,\text{sh}} - R_1$$

$$R_2 = 37.5 - 33.7$$

$$R_2 = 3.8\ \Omega$$

This method of solution which employs only Ohm's law, simple algebra, and a basic knowledge of series-parallel resistor circuits, will be used to find the resistor values of a more complicated Ayrton shunt circuit in the next example.

Example 10-15

In the Ayrton shunt diagram of Fig. 10-23, calculate the required values of resistors R_1, R_2, R_3, and R_4 to convert the 100-μA meter with a coil resistance of 2000 Ω into a meter with the following current ranges: 500 μA, 10 mA, 100 mA, and 500 mA.

Solution

On the *500 μA* range, the meter coil is shunted by $R_1 + R_2 + R_3 + R_4$. These resistors comprise the total shunt resistance, $R_{T\,\text{sh}}$. $R_{T\,\text{sh}} = R_1 + R_2 + R_3 + R_4$. The reader should verify that $R_{T\,\text{sh}}$ is found to be 500 Ω.

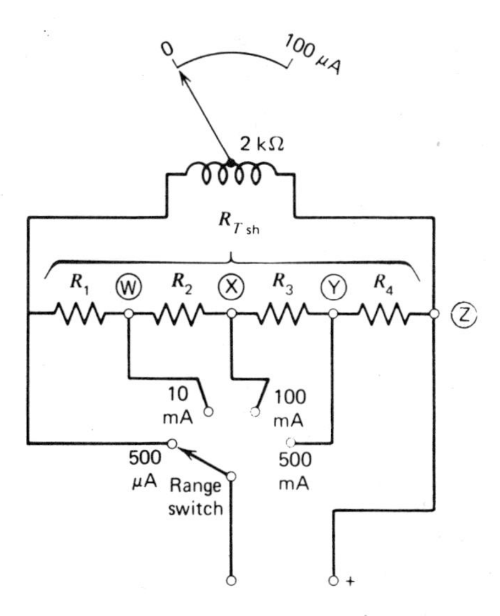

Figure 10-23. Ayrton shunt, Example 10-15.

Now find the value of each resistor, employing the method discussed in the previous example, and using the following as a guide.

On the *10 mA* range (or 10,000 μA) of Fig. 10-23, this current flows (electron movement) from the *negative* end of the circuit to point W and then splits up. 100μA flows through R_1 and the meter coil to point Z, the *positive* end of the circuit. The *difference* between 10 mA (or 10,000 μA) and 100 μA, or 9900 μA flows from point W through R_2, R_3, and R_4 to point Z. Therefore, $R_1 + R_{coil}$ is in parallel with $R_2 + R_3 + R_4$, and

$$V_{R_1,\,\text{coil}} = V_{R_2,R_3,R_4}$$

R_1 should be found to be 475 Ω.

When the range switch of Fig. 10-23 is set to the 100 *mA* position (or 100,000 μA), current from the negative terminal of the circuit flows to point X, and splits up. 100 μA flows through R_2, R_1 and the meter coil to point Z. The difference between the 100 mA and the 100 μA, or *99,900 μA* flows through R_3 and R_4 to point Z. Therefore, $R_2 + R_1 + R_{coil}$ is in parallel with $R_3 + R_4$, and

$$V_{R_1,R_2,R\text{ coil}} = V_{R_3,R_4}$$

R_2 should be found to be 22.5 Ω

When the range switch of Fig. 10-23 is turned to the *500-mA* position (or 500,000 μA), current from the negative end of the circuit flows to point Y and then splits up. 100 μA flows through R_3, R_2, R_1, and the meter coil to point Z. The difference between the 500 mA and 100 μA, or *499,900 μA* flows through R_4 to point Z. Therefore, $R_3 + R_2 + R_1 + R_{coil}$ is in parallel with R_4, and

$$V_{R_3,\,R_2,\,R_1,\,R\text{ coil}} = V_{R_4}$$

R_3 should be found to be 2 Ω, while R_4 should be 0.5 Ω.

For Similar Problems Refer to End-of-Chapter Problems 10-68 to 10-71

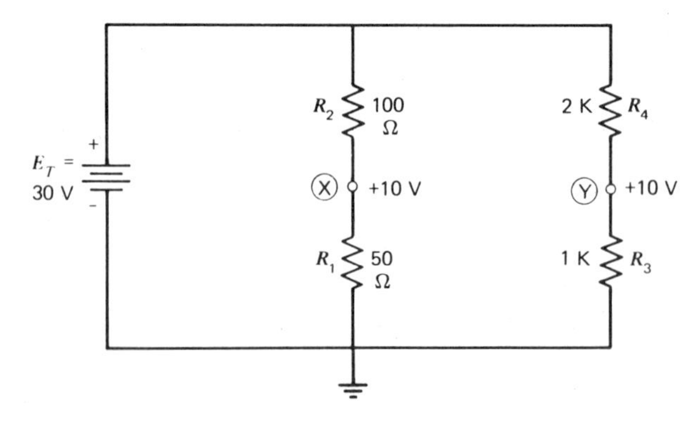

Figure 10-24. Wheatstone bridge principle.

10-11. Wheatstone Bridge

An instrument used to measure resistors accurately is the Wheatstone bridge. The basic principle of a bridge circuit may be explained by using the diagram of Fig. 10-24. Observe that it consists of two *parallel* branches. The left branch is comprised of R_1 and R_2 in *series*, while the right branch is made up of R_3 amd R_4 in series. Ground is shown, simply as a reference point, at the bottom of the diagram. Voltage at point X, and also at point Y, are found, with *respect to ground*, in the following.

Voltage at point X with *respect to ground* is actually the voltage across R_1. A simple method of determining it, from Section 3-5, is

$$V_{R_1} = \left(\frac{R_1}{R_1 + R_2}\right) E_T \qquad \text{(from equation 3-1)}$$

$$= \left(\frac{50}{50 + 100}\right) 30$$

$$= \left(\frac{50}{150}\right) 30$$

$$= 10 \text{ V}$$

Since current (electrons) flows *up* through R_1, then the upper end of R_1 (point X) is its positive end, and point X is, therefore, *+10V* with respect to ground, as shown on the diagram.

Voltage at point Y with *respect to ground*, or V_{R_3} is

$$V_{R_3} = \left(\frac{R_3}{R_3 + R_4}\right) E_T \qquad (3\text{-}1)$$

$$= \left(\frac{1000}{1000 + 2000}\right) 30$$

$$= \left(\frac{1000}{3000}\right) 30$$

$$= 10 \text{ V}$$

Since current also flows *up* through R_3, then the upper end of R_3 is its positive end, and point Y is, therefore, *+10 V* with respect to ground.

Note that points X and Y are at the same voltage, both being +10 V. The bridge circuit is now said to be *balanced*. If a current meter were connected between points X and Y, no current would flow through the meter, since there is no difference of potential, or no difference of voltage, between these points.

This *balanced* condition exists as long as the ratio of R_2 to R_1 is equal to the ratio of R_4 to R_3, or the ratio of the *left* branch resistors, *upper to lower*, is equal to the ratio of the *right* branch resistors, *upper to lower*.

Expressed as an equation, this is

$$\frac{R_2}{R_1} = \frac{R_4}{R_3} \tag{10-15}$$

Another similar relationship of the four resistors of Fig. 10-24 when the circuit is balanced is

$$\frac{R_2}{R_4} = \frac{R_1}{R_3} \tag{10-16}$$

or the ratio of the *upper* resistors, *left to right*, is equal to the ratio of the *lower* resistors, *left to right*.

Note that in either equation, *cross multiplication* (multiplying the left numerator by the right denominator, and the right numerator by the left denominator) yields the same resulting equation:

$$R_2 R_3 = R_1 R_4 \tag{10-17}$$

If the values of three of the resistors are known, then the value of the fourth may easily be found in the equation.

Figure 10-25 depicts a simple Wheatstone bridge circuit. Observe its similarity to Fig. 10-24. The four resistors of the bridge circuit are, as shown, usually drawn in this baseball-diamond shape. If the bridge is balanced, then points A and B are at the same potential, no difference exists between them, and the meter connected between them reads zero.

The meter is a *galvanometer*, that is, *zero* is at the *center* of the scale so that the pointer may deflect left or right, if necessary. Depending on the values of the four resistors when the bridge is *unbalanced*, point A may be lower than (or negative with respect to) point B. In this case, electrons would flow through the meter from point A to B. Again, depending on the resistor values, when the circuit is *unbalanced*, point B may be lower than (or negative with

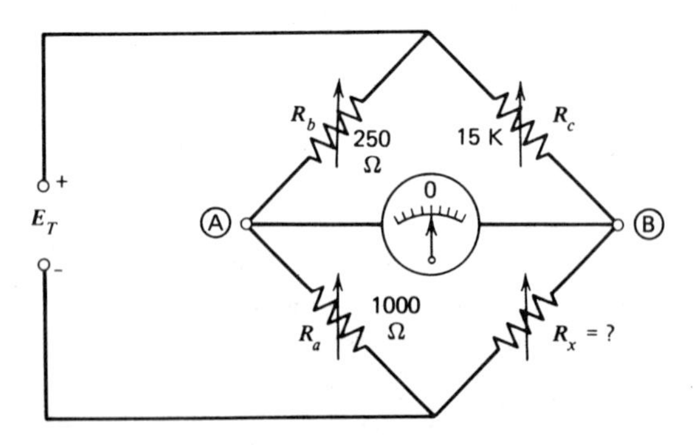

Figure 10-25. A simple wheatstone bridge.

respect to) point A. Electrons would now flow through the meter from point B to A. As a result, until the bridge is adjusted for a *balanced* condition, the meter pointer may deflect to the right or to the left, finally reading zero at the center of the scale when the bridge is balanced.

As it is shown in Fig. 10-25, the bridge is *balanced*, since the meter pointer reads zero. Assuming that R_x is an unknown value of resistor, and that the other three resistors, R_a, R_b, and R_c, have been varied until the meter indicates zero, R_x may now be determined in the following manner. (Note that the *value* of E_T, not shown in the diagram, has no effect on the bridge balance condition.)

$$\frac{R_b}{R_a} = \frac{R_c}{R_x} \quad \text{(from equation 10-15)}$$

Cross multiplying, yields

$$R_b\, R_x = R_a\, R_c \quad \text{(similar to equation 10-17)}$$

Solving for R_x by dividing both sides of equation by R_b, gives

$$\frac{R_b R_x}{R_b} = \frac{R_a R_c}{R_b}$$

Canceling R_b at the left side of equation, gives

$$R_x = \frac{R_a R_c}{R_b}$$

Solving for R_x by substituting the values of R_a, R_b, and R_c shown in Fig. 10-25, gives

$$R_x = \frac{(1000)\,(15000)}{250}$$

$$R_x = \frac{(1 \times 10^3)\,(15 \times 10^3)}{2.5 \times 10^2}$$

$$R_x = \frac{15 \times 10^6}{2.5 \times 10^2}$$

$$R_x = 6 \times 10^4$$

$$R_x = 60{,}000\ \Omega$$

Actually, any of the four resistor "legs" in the circuit of Fig. 10-25 could be the *unknown* resistor, with the other three resistor values known when the bridge is balanced (meter indicates zero). By using one of the previous equations, the unknown resistor could then be determined, as shown in the following example.

Example 10-16

In the Wheatstone bridge circuit of Fig. 10-25, assume that the bridge is balanced with the following resistor values. $R_a = 8.5\ k\Omega$, $R_b =$ unknown value, $R_c = 180\ \Omega$, $R_x = 3060\ \Omega$. Find the value of the unknown resistor, R_b.

Solution

For clarity redraw Fig. 10-25, substituting the values given in this example for those shown in the original. Then, using the cross-multiplication equation,

$$R_b R_x = R_a R_c \qquad \text{(similar to equation 10-17)}$$

Solving for R_b (the unknown resistor in this example), should verify that R_b is found to be 500 Ω.

For Similar Problems Refer to End-of-Chapter Problems 10-72 to 10-78

10-12. Using the Wheatstone Bridge

A commercial Wheatstone bridge is basically the same as that of the simple circuit of Fig. 10-25. To determine the value of the resistor being measured, R_x, it was necessary to know the values of the other three "legs," R_a, R_b, and R_c.

In the simplified commercial model of a Wheatstone bridge shown in Fig. 10-26, the unknown resistor, R_x, is connected as the lower right "leg" of the circuit. Resistor R_c, the upper right "leg" of the circuit, is actually a series of precision resistor with "taps" or connections to four separate switches. Each switch connects to a bank of 10 resistors, called *decades*. The top decade of R_c consists of ten 1000-Ω resistors in series. Its associated switch (called the 1000's) can connect in as part of R_c any value from 0 Ω up to 10,000 Ω in 1000-Ω steps.

This is connected to another decade of 100-Ω resistors, and its associated switch (called the 100's) can connect in as part of R_c any value from 0 to 1000 Ω in 100-Ω steps.

This, in turn, connects to still another decade consisting of 10-Ω resistors. The switch associated with this decade (called the 10's) can connect in as part of R_c any value from zero to 100 Ω, in 10-Ω steps.

Finally, a decade of 1-Ω resistors, works in conjunction with the *1's* switch, which can connect in as another part of R_c any value of zero to 10 Ω in 1-Ω steps.

Resistor R_c then depends on the settings of these four switches. As an example: if the *1000's* switch is set to *3*, and the *100's* is set to *8*, and the *10's* is set on *4*, while the *1's* is on *5*, the total value of resistor R_c is *3845* Ω.

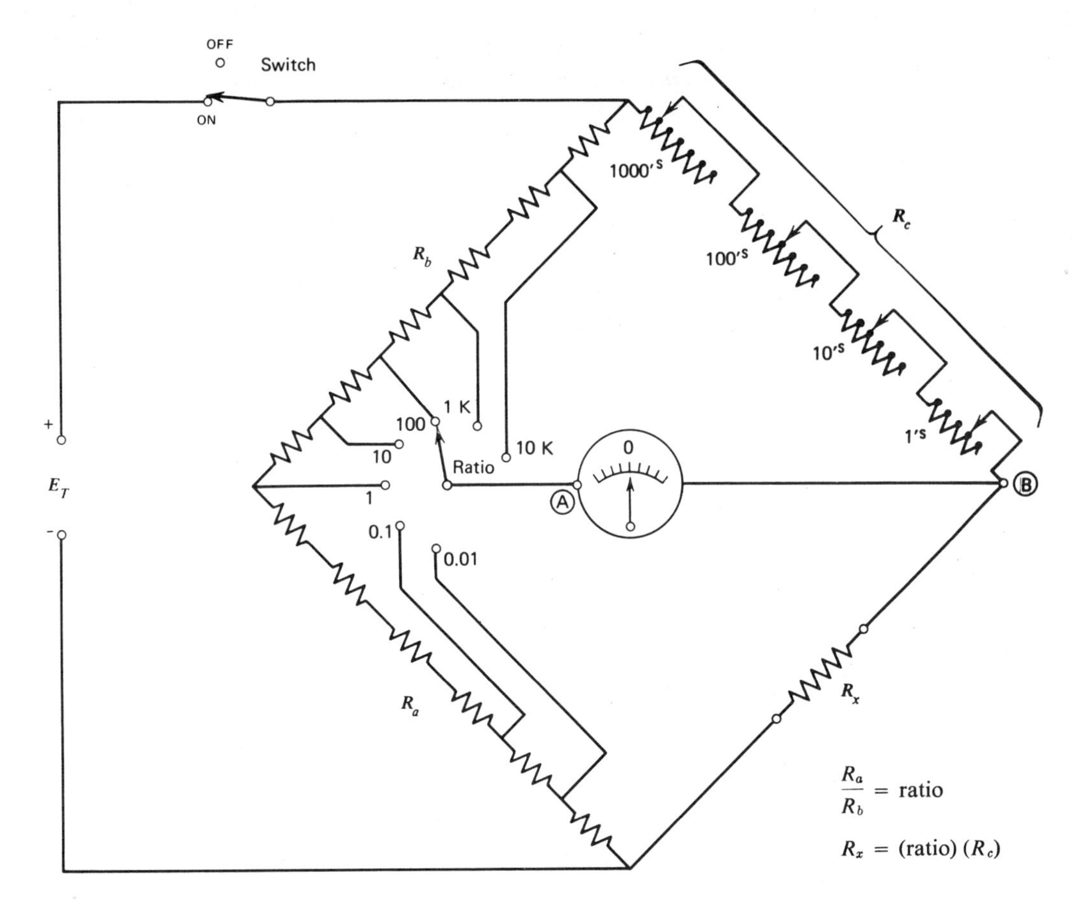

Figure 10-26. A simplified commercial wheatstone bridge.

When measuring an unknown resistor, R_x, it is connected to the terminals shown in Fig. 10-26. Then, with the ON-OFF switch momentarily closed, the four switches (*1000's*, *100's*, *10's*, and *1's*) and the *ratio* control knob are adjusted until the meter indicates zero, signifying that the bridge is balanced.

The *ratio* control knob, as shown in Fig. 10-26, is another switch that connects any one of the several "taps" between a series of resistors that comprises R_a and R_b, to one terminal of the meter. The settings of this switch are marked, *0.01, 0.1, 1, 10, 100*, and so on, and they indicate the *ratio* of the value of R_a to R_b. For example, when the *ratio* control is on *10*, it means that R_a is ten times as large as R_b; on *0.01* it means that R_a is $1/100$ of R_b, and so on.

The equation to find R_x is

$$R_x = \frac{R_a R_c}{R_b} \tag{10-18}$$

The actual values of R_a and R_b need not be known, only the *ratio* between them is necessary. Therefore, since

$$\frac{R_a}{R_b} \text{ is the } ratio$$

then replacing R_a/R_b in equation 10-18 with their ratio,

$$R_x = (\text{ratio})\,(R_c) \tag{10-19}$$

Determining the value of R_x, when the bridge is balanced, is then simply (equation 10-19) the *product* of the *ratio* knob setting and the value of R_c (see the photograph shown in Fig. 10-27).

The following example illustrates this:

Example 10-17

What are the values of R_x in each of the following, if the bridge is balanced with the following switch settings?

(a) *Ratio* control at 100; *1000's* at 8, *100's* at 6, *10'S* at 9, *1's* at 3.
(b) *Ratio* control at 0.01; *1000's* at 0, *100's* at 3, *10's* at 2, *1's* at 5.
(c) *Ratio* control at 10; *1000's* at 3, *100's* at 8, *10's* at 0, *1's* at 6.

Solution

$$R_x = (ratio)\,(R_c) \tag{10-19}$$

(a) $R_x = (100)\,(8693)$

$= 869{,}300\ \Omega$

(b) $R_x = (0.01)\,(0325)$

$= 3.25\ \Omega$

(c) $R_x = (10)\,(3806)$

$= 38{,}060\ \Omega$

Figure 10-27 shows a typical laboratory model Wheatstone bridge.

For Similar Problems Refer to End-of Chapter Problems 10-79 to 10-84

10-13. Shunt Ohmmeter

In Section 10-6, simple series ohmmeters are discussed. The circuit diagram of Fig. 10-16 depicts this simple series circuit consisting of the meter coil in

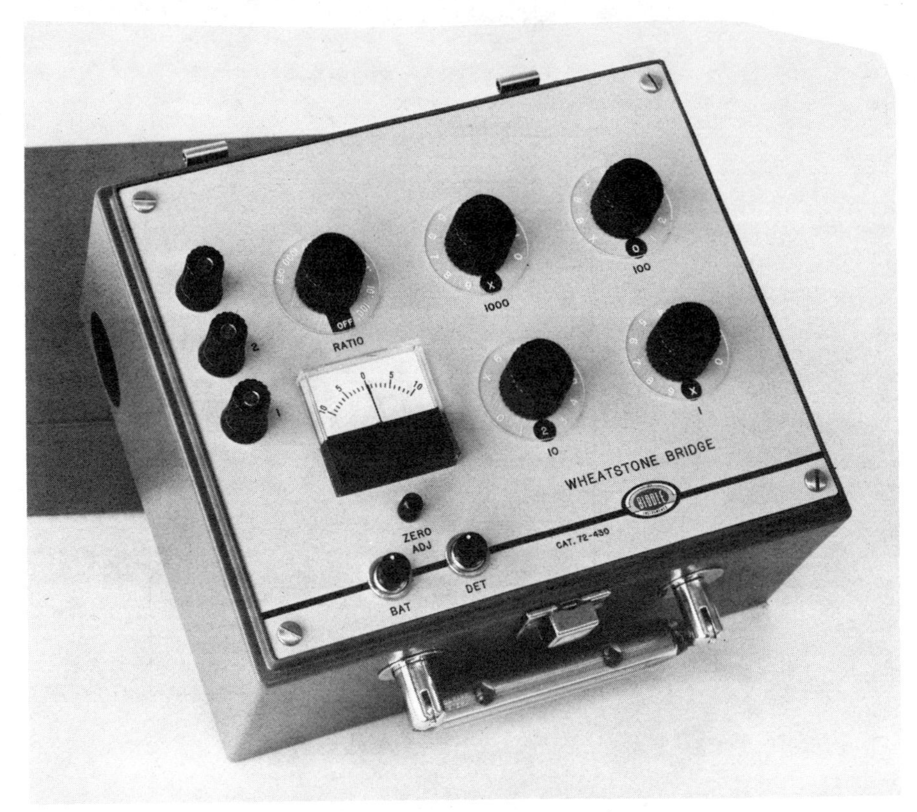

Figure 10-27. Wheatstone bridge (*courtesy James G. Biddle Co*).

series with the battery voltage source, the ohms zero adjustment, and the resistor to be measured. This type of ohmmeter circuit only has a limited resistance-measuring range. To increase it, say by ten times, it is necessary to use a battery ten times larger. It is therefore impractical to attempt to increase the range by 100, 1000, or 10,000 times, and so forth.

Other, more complex ohmmeter circuits have been developed that do not require progressively larger batteries but can be switched to higher resistance-measuring ranges, for example: R × 10, R × 100, R × 1000, R × 10 K, and R × 100 K.

In this section we discuss a typical ohmmeter circuit. Figure 10-28 shows a *shunt-type ohmmeter* having resistor-measuring capabilities that cover a wide range. Note that without the shunt resistors R_3 to R_8, the circuit is a simple *series* ohmmeter. The values of R_1 (the variable *zero adjust*) and R_2 depend on the battery voltage, the meter coil resistance, and the current required by the coil for full-scale deflection. The resistance of the *series* meter circuit (without

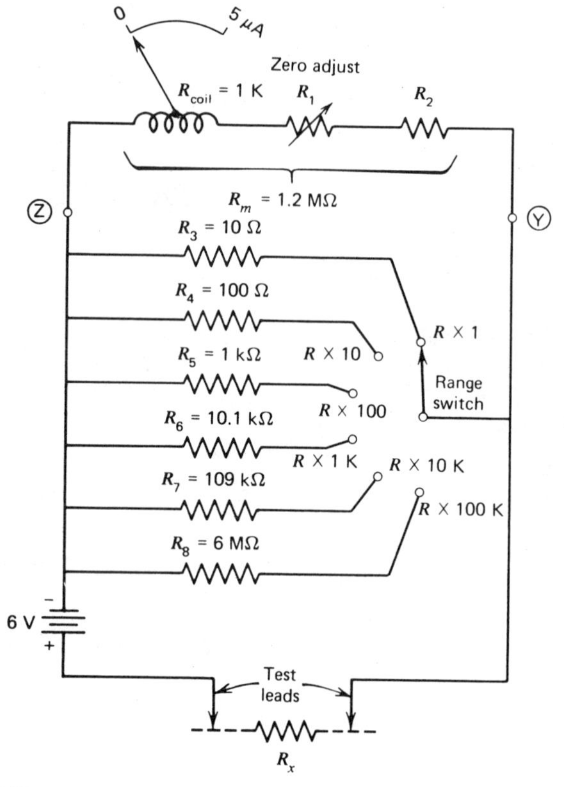

Figure 10-28. Shunt ohmmeter, multiple ranges.

R_3 to R_8), called here R_m, is $R_{coil} + R_1 + R_2$. When the test leads are shorted together ($R_x = 0\ \Omega$), then

$$R_m = \frac{\text{voltage of battery}}{I_{\text{full-scale}}}$$

$$R_m = \frac{6\ \text{V}}{5\ \mu\text{A}}$$

$$R_m = \frac{6}{5 \times 10^{-6}}$$

$$R_m = 1.2 \times 10^6\ \Omega \qquad \text{or} \qquad 1.2\ \text{M}\Omega$$

Since R_{coil} is 1000 Ω, then the *difference* (1.2 MΩ − 1 kΩ) of 1,199,000 Ω is usually split up with about one half for R_1 and one half for R_2.

The following detailed discussion derives the values of the *shunt* resistors R_3 through R_8 for the various ranges *R* × *1* through *R* × *100 K*.

On the lowest range switch setting, $R \times 1$, shunt resistor R_3 is connected as shown in Fig. 10-28. The value of R_3 depends primarily on how much resistance value of R_x (the resistor to be measured) is desired to produce *half-scale deflection* of the meter coil (or 2.5 μA). Assume that it is desired that *10* Ω of R_x should produce a half-scale deflection. Then, voltage across the coil $+ V_{R_1} + V_{R_2}$ (or V_{R_m}) is

$$\begin{aligned} V_{R_m} &= (I_{\text{half-scale}})\,(R_m) \\ &= (2.5\ \mu\text{A})\,(1.2\ \text{M}\Omega) \\ &= (2.5 \times 10^{-6})\,(1.2 \times 10^{6}) \\ &= 3\ \text{V} \end{aligned}$$

This is also the voltage between points Z and Y, or $V_{ZY} = 3$ V. Since E_{total} (the battery) is 6 V, and V_{ZY} is 3 V, then V_{R_x} is the difference, or also 3 V. Therefore, R_{ZY} must equal R_x, since they each have 3 V, or $R_{ZY} = 10\ \Omega$. R_{ZY} is R_3 in parallel with R_m (which is $R_{\text{coil}} + R_1 + R_2$). Then, to find the required value of R_3, use the two-parallel-resistor equation of their *product* divided by their *sum* (Chapter 4, equation 4-3).

$$\frac{(R_3)\,(R_m)}{R_3 + R_m} = 10$$

$$\frac{(R_3)\,(1.2 \times 10^6)}{R_3 + 1.2 \times 10^6} = 10$$

$$(R_3)\,(1.2 \times 10^6) = 10\,(R_3 + 1.2 \times 10^6)$$

$$1.2 \times 10^6\,R_3 = 10\,R_3 + 12. \times 10^6$$

$$1.2 \times 10^6\,R_3 - 10\,R_3 = 12. \times 10^6$$

the *10* R_3 is negligible, and may be omitted, giving

$$\begin{aligned} 1.2 \times 10^6\,R_3 &= 12. \times 10^6 \\ R_3 &= \frac{12 \times 10^6}{1.2 \times 10^6} \\ R_3 &= 10\ \Omega \end{aligned}$$

On $R \times 1$, with a desired value of 10 Ω for half-scale deflection, the necessary value of R_3 is therefore also 10 Ω, as shown in Fig. 10-28.

When the range switch is set to $R \times 10$, it means that a value of 100 Ω (or ten times the original 10 Ω) is now desired for half-scale current deflection. The necessary value for R_4 must now be found. Again, V_{R_m} remains the same as before, or 3 V, and V_{R_x} is still the other 3 V. Therefore, again $R_{ZY} = R_x$, which is now 100 Ω. R_{ZY} is R_4 in parallel with R_m (which is 1.2×10^6 Ω). Solving for R_4 gives

$$\frac{R_4 \, R_m}{R_4 + R_m} = 100$$

$$\frac{(R_4)\,(1.2 \times 10^6)}{R_4 + 1.2 \times 10^6} = 100$$

$$(R_4)\,(1.2 \times 10^6) = 100\,(R_4 + 1.2 \times 10^6)$$

$$1.2 \times 10^6 \, R_4 = 100 \, R_4 + 120. \times 10^6$$

$$1.2 \times 10^6 \, R_4 - 100 \, R_4 = 120 \times 10^6$$

Since *100* R_4 is negligible compared to $1.2 \times 10^6 \, R_4$, the equation becomes

$$1.2 \times 10^6 \, R_4 = 120 \times 10^6$$

$$R_4 = \frac{120 \times 10^6}{1.2 \times 10^6}$$

$$R_4 = 100 \, \Omega$$

This value for R_4 is shown in Fig. 10-28 on the $R \times 10$ range.

Now continue to find the values of shunt resistors R_5 to R_8 for ranges $R \times 100$ through $R \times 100$ K, using th : method just described, and using the results shown in Fig. 10-28 as a guide.

Note that on the highest range, $R \times$ *100 K*, the current flow between the *shorted* test leads is *minimum.* On this high range, current from the 6-V battery flows through the large shunt resistor R_8, *6 MΩ*, and also through the parallel R_m, *1.2 MΩ.*

On the lowest range, $R \times 1$, current flow between the *shorted* test leads is *maximum.* On the $R \times 1$ range, current from the 6-V battery flows through the small shunt resistor R_3, *10 Ω*, and also through the parallel R_m, *1.2 MΩ.*

Normally, when checking for the *continuity* of a wire, the lowest range $R \times 1$ is used. However, when checking for the continuity of a fuse that is rated at a tiny fraction of an ampere, it is quite possible to damage the fuse by excessive current from the ohmmeter. Therefore, fine fuses should be checked with the ohmmeter on its highest range, when only a tiny current flows.

Calibrating the Shunt Ohmmeter. The *shunt ohmmeter* of Fig. 10-28 with resistors R_3 to R_8 is designed with *10 Ω* of R_x between the test leads, producing half-scale deflection, or *2.5 μA.* This is shown on the calibrated scale of Fig. 10-29. Numerous other values of R_x would result in other values of meter current. For each assumed value of R_x, on the $R \times 1$ range, the resulting current is calculated, and the ohmmeter scale is then marked with that value of ohms opposite its associated current on the microampere scale.

First note, however, that on the $R \times 1$ range, shunt resistor R_3, *10 Ω*, is in parallel with R_m, *1.2 MΩ.* As a result, the parallel combination of these two

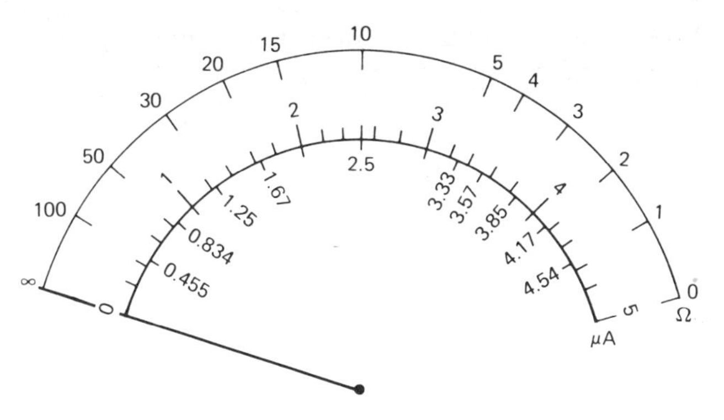

Figure 10-29. Calibrating the shunt ohmmeter of Fig. 10-28.

resistors, R_3 and R_m, is just about R_3 alone, or 10 Ω, or R_{ZY} will always be 10 Ω on $R \times 1$ range.

In the following meter scale calibrations for several values of R_x, first the voltage between points Z and Y is found. Then, by dividing V_{ZY} by R_m, current through the meter is determined. Each assumed value of R_x, and its resulting value of V_{ZY}, and finally the meter current, is shown listed in Table 10-1. Figure 10-29 shows the meter face with the value of ohms shown above its associated current.

When R_x *is zero*, V_{ZY} is the full 6-V battery voltage, and $I_{\text{meter}} = V_{ZY}/R_m = 6/1.2 \times 10^6 = 5 \times 10^{-6}$ A, or 5 μA, which is full-scale deflection.

If R_x *is 1* Ω, then from the circuit of Fig. 10-28, R_{ZY} (10 Ω) is in series with R_x (1 Ω) and with the 6-V battery. V_{ZY} is found

$$V_{ZY} = \left(\frac{R_{ZY}}{R_{ZY} + R_x}\right) E_T \qquad \text{(from equation 3-1)}$$

$$= \left(\frac{10}{10 \times 1}\right) 6$$

$$= \left(\frac{10}{11}\right) 6$$

$$= 5.45 \text{ V}$$

and

$$I_{\text{meter}} = \frac{V_{ZY}}{R_m}$$

$$= \frac{5.45}{1.2 \times 10^6}$$

$$= 4.54\ \mu\text{A}$$

These values are listed in Table 10-1, and the *1* Ω is shown above the *4.54 μA* in the scales of Fig. 10-29.

TABLE 10-1 Calibrating the Shunt Ohmmeter of Fig. 10-28 on $R \times 1$ Range

If R_x is (in Ohms)	then V_{ZY} is (in volts)	and I_{meter} is (in microamperes)
0	6	5
1	5.45	4.54
2	5	4.17
3	4.61	3.85
4	4.28	3.57
5	4	3.33
10	3	2.5
15	2.4	2
20	2	1.67
30	1.5	1.25
50	1	0.834
100	0.546	0.455
∞	0	0

When R_x *is 2* Ω, then V_{ZY} is found:

$$\begin{aligned} V_{ZY} &= \left(\frac{R_{ZY}}{R_{ZY} + R_x}\right) E_T \qquad \text{(from equation 3-1)} \\ &= \left(\frac{10}{10 + 2}\right) 6 \\ &= \left(\frac{10}{12}\right) 6 \\ &= 5 \text{ V} \end{aligned}$$

and

$$\begin{aligned} I_{\text{meter}} &= \frac{V_{ZY}}{R_m} \\ &= \frac{5}{1.2 \times 10^6} \\ &= 4.17\ \mu\text{A} \end{aligned}$$

These values are listed in Table 10-1, and the *2* Ω is shown above the *4.17* μA on the meter scales of Fig. 10-29.

Continue to find V_{ZY} and I_{meter} on the $R \times 1$ range for the following values of R_x: 3 Ω, 4 Ω, 5 Ω, 10 Ω, 15 Ω, 20 Ω, 30 Ω, 50 Ω, 100 Ω, and infinity. Use the method just described and, as a guide, Table 10-1, and the calibrated ohmmeter scale of Fig. 10-29.

For Similar Problems Refer to End-of-Chapter Problems 10-85 to 10-87

PROBLEMS

See section 10-1 for discussion covered by the following.

10-1. Briefly describe the construction of a permanent-magnet moving-coil meter.
10-2. Briefly describe the theory of operation of a permanent-magnet moving-coil meter.
10-3. In which direction will the coil and its pointer move if a reverse current is permitted to flow through the coil?
10-4. Describe the reason for your answer to the previous question.
10-5. What is the name for a meter that has the coil mounted so that *zero* is at the *center* of the graduated scale, and the coil is permitted to rotate in either direction?
10-6. By what other names is the permanent-magnet moving-coil meter known?

See section 10-2 for discussion covered by the following.

10-7. What must be done to a current-reading meter if it is desired to increase its range?
10-8. A 5-μA meter is to be converted into a 100 μA range meter. How much current should flow through the shunt resistor if 100 μA is actually flowing in the complete circuit?
10-9. A 100-μA meter (or 0.1 mA) has a resistor in parallel with the coil, converting it into a 10-mA range meter (or 10,000 μA). If 10 mA actually flows in the complete circuit, how much current flows through R_{sh}?
10-10. What size R_{sh} is necessary to convert a 50-μA meter with a coil resistance of 300 Ω, into a 1 mA range meter?
10-11. A 5-μA meter, having a coil resistance of 2 kΩ, is to be converted into a 10 mA meter. What value of R_{sh} is required?
10-12. When changing from one current range to another, why must the switch be of the *shorting* or *make-before-break* type?
10-13. What is a *shorting* or *make-before-break* switch?

Refer to section 10-3 for discussion covered by the following.

10-14. Should a current meter be placed in series or in parallel with the circuit being measured?
10-15. When inserting a current meter into a circuit, where should the negative and positive meter leads be connected?

10-16. If a current-reading meter were to be inserted into the series circuit of Fig. 10-5*c* between resistors R_2 and R_3, to which resistors should the negative and the positive meter leads be connected after opening the wire connecting these resistors?

10-17. A current-reading meter is to be inserted into the series-parallel circuit of Fig. 10-6*c* to read the current I_5 through R_5. If the meter were to be placed between point X and R_5, where should the negative lead and the positive lead of the meter be connected after opening the wire between point X and R_5?

10-18. What is meant by meter *loading* a circuit?

10-19. Explain why a current-reading meter will have little *loading* effect on most circuits.

10-20. What is indicated if the meter pointer moves backward off the graduated scale?

10-21. When measuring a completely unknown current, to what range should the meter be set?

10-22. What is indicated if the meter pointer moves all the way to the right, offscale?

10-23. Refer to the meter scale and range switch of Fig. 10-7*a* and *b*. How much current is indicated if the range switch is set to 500, and the meter pointer is at half scale?

Refer to section 10-4 for discussion covered by the following.

10-24. Where is the resistor connected with respect to the meter coil to convert a current-meter into a voltmeter?

10-25. How much voltage will produce full-scale deflection in (a) a 10-V range voltmeter, and (b) a 500-V range voltmeter?

10-26. What is the purpose of the series-multiplier resistor used in voltmeters?

10-27. Figure 10-11 shows a 50-μA meter with a coil resistance of 1 kΩ. The range switch connects in any one of five series-multiplier resistors for any of five different ranges. Calculate the value of the resistor required for the following ranges: (a) 5 V, (b) 50 V, (c) 100 V, (d) 500 V, and (e) 1000 V.

10-28. Refer to the meter scale shown in Fig. 10-9, and to the voltmeter diagram of Fig. 10-8. If this meter is basically a 1-mA type, operating as a 50 V voltmeter, how much current flows through the meter circuit if the voltage being measured is only (a) 10 V, and (b) 25 V?

10-29. What is the Ω/V sensitivity rating of a 200-μA meter?

10-30. What is the total resistance ($R_{coil} + R_{ser.\ mult}$) of the meter in the previous problem, operating as a 300-V voltmeter?

10-31. A voltmeter is marked "200,000 Ω/V." (a) What is the $I_{full\text{-}scale}$ of this meter, and (b) what is the total resistance ($R_{coil} + R_{ser.\ mult}$) of this meter on a 30-V range?

Refer to section 10-5 for discussion covered by the following.

10-32. In order to measure a voltage, how should a dc voltmeter be connected?

10-33. To measure V_{R_1} of Fig. 10-12, where should the negative and positive voltmeter leads be connected?

10-34. To what range should the voltmeter be switched when one is attempting to measure an unknown voltage?

10-35. To measure V_{R_4} of Fig. 10-12, where should the negative and positive voltmeter leads be connected?

10-36. If two resistors, each marked 1 kΩ, are connected in series with 120 V applied, and one resistor was *open*, what would a 20,000 Ω/V meter on its 150-V range read (a) across the good resistor, and (b) across the bad resistor? (*Hint*: the open resistor has an infinitely high value.)

10-37. Two resistors, each 200 kΩ, are connected in series with 10 V applied. The voltage across each is, therefore, 5 V. What would the voltage across one resistor become if a 5000 Ω/V meter on its 10-V range were connected across this resistor?

10-38. What is meant by *voltmeter loading*?

10-39. Should the resistance of a voltmeter be high or low in order not to affect a circuit?

10-40. A voltmeter loads a circuit and results in a voltage across a resistor of 100 V. Without the voltmeter, voltage across the resistor is calculated to be 125 V. What is the percentage error?

10-41. In Fig. 10-15, if a 100,000 Ω/V meter were used on its 10-V range and placed across R_1 (a) what would V_{R_1} become, and (b) what is the percentage of error if V_{R_1} without the meter is actually 3 V?

See section 10-6 for discussion covered by the following.

10-42. What is the purpose of an ohmmeter?

10-43. When measuring a piece of wire with an ohmmeter, what should the meter read if the wire is (a) open, and (b) good?

10-44. What is the purpose of the variable resistor often called *zero adjust*?

10-45. If the meter in Fig. 10-16 were a 50 μA meter, with a coil resistance of 100 Ω, and a 45-V battery were used, what would be the total resistance of the ohmmeter?

10-46. In the ohmmeter of the previous problem, what value of R_x (the resistor to be measured) would result in half-scale current deflection?

10-47. In calibrating the ohmmeter of Problem 10-45, what would each of the following values of R_x (resistor to be measured) correspond to on the microammeter scale; (a) 100 kΩ, (b) 1.1 MΩ, and (c) 2.1 Mg?

10-48. Describe a simple method of changing an ohmmeter range from its original scale readings to a higher range of $R \times 10$.

See section 10-7 for discussion covered by the following.

10-49. What adjustments are usually made when preparing to use the ohmmeter?
10-50. What does the ohmmeter indicate if the meter pointer is at 7.5 while the range switch is on $R \times 10$ K?
10-51. What important precaution must be taken to protect the ohmmeter before connecting it in a circuit?
10-52. To insure a true resistance reading when measuring some component in a circuit with an ohmmeter, what steps should be taken?
10-53. When measuring a capacitor with an ohmmeter, what important precaution must be taken to protect the ohmmeter?
10-54. When checking a small, good capacitor, what action should be observed on the high-range ohmmeter scale?
10-55. When checking a large, good capacitor, what action should be observed on the high-range ohmmeter scale?
10-56. What will a high-range ohmmeter read on a bad capacitor that is either shorted or leaky?
10-57. What high-range-ohmmeter action should be observed when measuring an "open" capacitor?

See section 10-8 for discussion covered by the following.

10-58. What are the main parts of a dc motor?
10-59. What is the purpose of the brushes?
10-60. What is the purpose of the split-ring commutator?
10-61. Why must the *direction* of current through the armature coil constantly be reversed?
10-62. What is the basic difference in the construction of a series motor and a shunt motor?

See section 10-9 for discussion covered by the following.

10-63. What is the main difference in the construction of the D'Arsonval or Weston movement and the electrodynamometer?
10-64. Redraw the dynamometer of Fig. 10-20 being used as an ammeter.
10-65. Redraw the dynamometer of Fig. 10-20 being used as a voltmeter.
10-66. When measuring the power dissipation of a circuit, using the *dynamometer* as a *wattmeter*, (a) where are the field coils L_1 and L_2 of Fig. 10-20 connected, and (b) where is the movable coil L_3 connected?
10-67. What would the dynamometer indicate when used as a wattmeter if: (a) polarities of all three coils were reversed, and (b) polarity of only movable coil L_3 were reversed?

See section 10-10 for discussion covered by the following.

10-68. In respect to the range switch mechanism, how is the Ayrton shunt circuit superior to the ordinary meter shunt circuit?

10-69. As shown in the diagram for this problem, find the values of resistors R_1, R_2, and R_3 for the Ayrton shunt that convert the 1-mA meter with a coil resistance of 100 Ω, into a multi-range meter of 10, 100, and 500 mA.

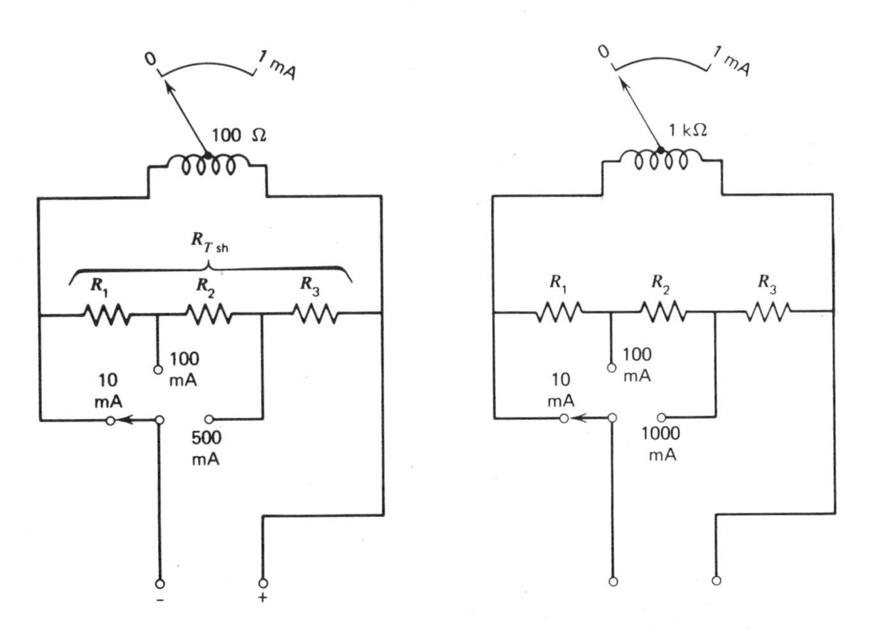

Problem 10-69. Problem 10-70.

10-70. In the Ayrton shunt circuit accompanying this problem, find the values of resistors, R_1, R_2, and R_3 that convert the 1-mA meter with a coil resistance of 1 kΩ, into a meter with 10-, 100-, and 1000-mA ranges.

10-71. In the Ayrton shunt circuit for this problem, find the values of resistors R_1, R_2, R_3, and R_4 that convert the 50-μA meter with a coil resistanec of 1000 Ω into the following ranges: 100 μA, 1 mA, 10 mA, and 50 mA.

See section 10-11 for discussion covered by the following.

10-72. In the simple Wheatstone bridge circuit of Fig. 10-25, set up a *fractional* equation using the letters R_a, R_b, and so on, where the ratio of the upper resistors is equal to that of the lower resistors.

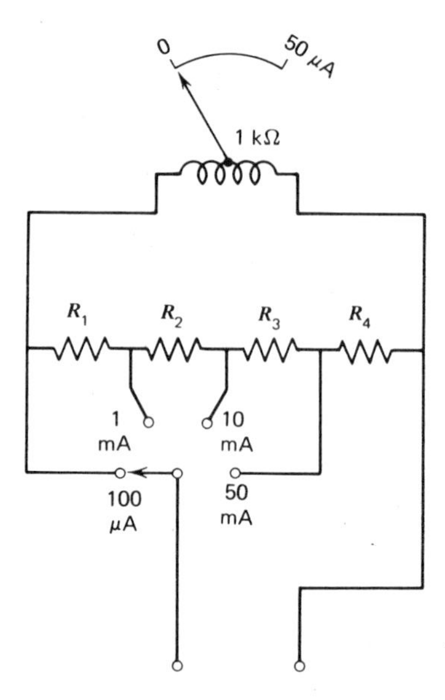

Problem 10-71.

10-73. In the simple Wheatstone bridge circuit of Fig. 10-25, set up an equation using the letters R_a, R_b, and so on, where *products* of diagonally opposite resistors are equal.

10-74. In the simple Wheatstone bridge circuit of Fig. 10-25, if $R_a = 50\ \Omega$, $R_b = 200\ \Omega$, $R_c = 4\ \text{k}\Omega$, and $R_x = 2\ \text{k}\Omega$, is the bridge balanced?

10-75. Is the simple Wheatstone bridge of Fig. 10-25 balanced if $R_a = 150\ \Omega$, $R_b = 200\ \Omega$, $R_c = 4\ \text{k}\Omega$, and $R_x = 2\ \text{k}\Omega$?

10-76. Find the value of R_a in Fig. 10-25, if $R_b = 350\ \Omega$, $R_c = 225\ \text{k}\Omega$, and $R_x = 45\ \text{k}\Omega$, and the bridge is balanced.

10-77. If the bridge of Fig. 10-25 is balanced, and $R_a = 552\ \Omega$, $R_b = 138\ \Omega$, $R_c =$ unknown, and $R_x = 84\ \text{k}\Omega$, find R_c.

10-78. What effect on bridge balance does the doubling of the applied voltage source have?

See section 10-12 for discussion covered by the following.

10-79. In respect to resistors, what is a decade?

10-80. In the commercial model of a Wheatstone bridge, as shown in Fig. 10-26, how is it possible to determine R_x (the unknown resistor) without knowing the actual value of R_a and R_b?

10-81. What is the value of R_c in Fig. 10-26 if the *1000's* switch is on *9*, the *100's* is on *2*, and the *10's* is set to *9* and the *1's* is set to *6*?

10-82. What is the minimum value of resistance that can be measured by the Wheatstone bridge circuit of Fig. 10-26? (Note the ratio range and R_c range.)

10-83. What is the maximum value of resistance that can be measured by the Wheatstone bridge circuit of Fig. 10-26? (Note the ratio range and R_c range.)
10-84. If the Wheatstone bridge circuit of Fig. 10-26 is balanced with the *ratio* control set at *10 K*, the *1000's* switch at *1*, the *100's* at *2*, the *10's* at *6*, and the *1's* at *2*, what is the value of R_x?

See section 10-13 for discussion covered by the following.

10-85. In the circuit of the shunt ohmmeter, Fig. 10-28 what value of R_3 is required to have a *1-Ω* value of resistor across the test leads result in a half-scale deflection of the meter, on the $R \times 1$ range?
10-86. In the *shunt ohmmeter* circuit of Fig. 10-28, what value of R_8 is required to have a 100 kΩ resistor across the test leads result in a half-scale deflection of the meter, on the $R \times 100$ K range?
10-87. In the *shunt ohmmeter* circuit of Fig. 10-28, if R_3 *is* 5 Ω, on the $R \times 1$ range, to have 5 Ω produce half-scale deflection, what *meter current* will flow when R_x, the resistor connected between the test leads, is: (a) 15 Ω, and (b) 100 Ω?

chapter

11

Induced Voltage, Inductance, and Alternating Current

The term ac stands for the words *alternating current*. This is a current that flows first in one direction and then in the opposite direction, periodically reversing or *alternating*. This type of current is due to a voltage that periodically reverses or alternates its positive and negative polarities. Such a voltage is called an *alternating voltage*. The term ac, although actually meaning alternating current, is often used as if it meant the single word *alternating*. Therefore, rightly or wrongly, through popular use, the term *ac current* has come to mean simply *alternating current* and not *alternating current current*! Similarly, an *alternating voltage* is commonly called an *ac voltage*, and this should not be taken to mean *alternating current voltage*! Following the popular usage, this text uses the "slang" expressions ac current and ac voltage.

An ac voltage is produced when a magnetic field and a conductor cut each other, first from one direction and then from the other.

In the first portion of this chapter, a simplified approach to induced voltage, inductance, and alternating current is presented. Then in the more advanced sections, a more detailed analysis is given.

11-1. Induced Voltage

If a bar magnet were moved into a coil of wire, as shown in Fig. 11-1*a*, the magnetic field of the bar magnet moves across the wires, cutting them. As a result, a voltage is produced or *induced* in the coil.

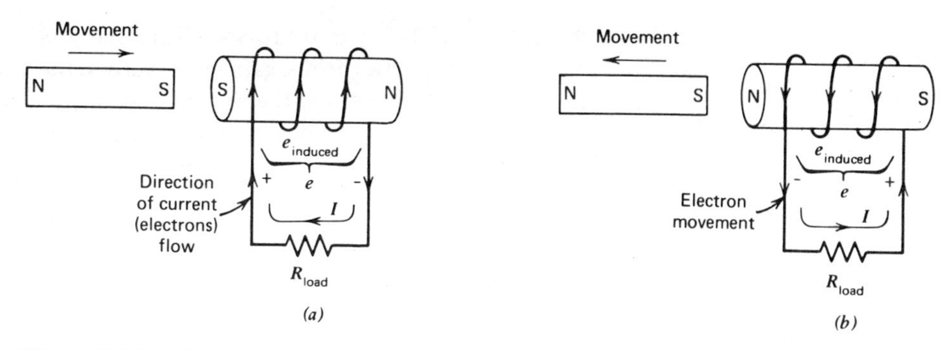

Figure 11-1. Induced voltage (*e*). (*a*) Magnet being moved toward coil. (*b*) Magnet being moved away from coil.

The *polarity* of this induced voltage in the coil depends on several things: the *direction of movement* of the bar magnet; the *magnetic polarity* of the *end* of the bar magnet near the coil; and the direction that the coil is wound (clockwise or counterclockwise). As a result of the voltage induced in the coil, a current will flow through the circuit consisting of the coil and the load resistor. This current is sometimes called the *induced* current. The polarity of the induced voltage, and the direction of the resultant current may be predicted from *Lenz's law* and the left-hand coil rule.

Lenz's Law. The *direction of current* flow due to an induced voltage will always produce a magnetic field that will *oppose* the original action that caused the induced voltage.

When we apply Lenz's law to Fig. 11-1*a*, the south end of the bar magnet is being moved toward the coil, inducing a voltage in the coil. The current flow (electron movement) in the coil produces an electromagnetic field that opposes the movement of the bar magnet. The left end of the coil in Fig. 11-1*a*, becomes a south pole in order to oppose the south end of the bar magnet's being moved toward the coil, while the right end of the coil becomes its north pole.

From the *left-hand coil rule* (see Section 9-6) with the left thumb extended towards the coil north pole, the enclosed fingers around the coil indicate the direction of electron flow. This is indicated by the arrows on the coil, and through the load resistor of Fig. 11-1*a*. Note that electrons are flowing through the load resistor from right to left. Since electrons move from negative to positive, then the right end of the resistor is its negative end, and the left end is its positive one, as shown in the drawing. Note that the polarity of the induced voltage at the ends of the coil acts the same as that of any applied voltage source with electrons moving from its negative to its positive end *through the load resistor* and, as in any voltage *source*, electron flow *through* the coil itself is from positive to negative.

When the bar magnet is held stationary, its magnetic field no longer cuts across the coil wires, and no voltage is induced. Current becomes zero.

When the magnet is being moved away from the coil, as shown in Fig. 11-1*b*, the moving magnetic field again cuts across the coil but from an opposite direction. Now determine the magnetic polarity of the coil, the current direction, and the voltage polarity. Be guided by the results shown in Fig. 11-1*b*. Note that the *direction* of current flow and the *polarity* of the induced voltage are the reverse in Fig. 11-1*b* from that of 11-1*a*. This is an ac voltage and current.

The following example illustrates the preceding discussion.

Example 11-1

Determine the direction of current flow and the polarity of the induced voltage in the coil when the bar magnet is moved down toward the coil of Fig. 11-2*a*, and then when it is moved up away from the coil of Fig. 11-2*b*.

Solution

Solve this example by using the results shown in the drawing as a guide.

Faraday's Law. The *amount* or *magnitude* of the induced voltage that is produced when a bar magnet is moved nearer or further away from a coil is determined by the *rate* at which the magnetic lines cut across the coil wires. This is called *Faraday's law* and depends on three factors: the *speed of movement* of the bar magnet, the *magnetic strength* of the bar magnet, and the *number of turns* of wire in the coil. The *rate* at which the magnetic lines cut the coil is directly proportional to these three factors.

For Similar Problems Refer to End-of-Chapter Problems 11-1 to 11-7

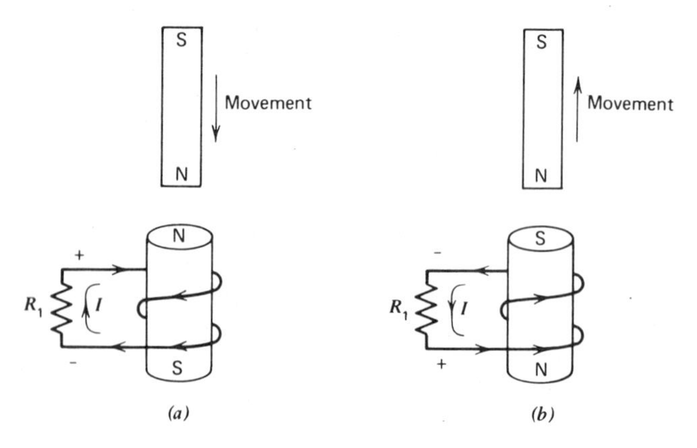

Figure 11-2. Example 11-1.

11-2. Simple AC Generator

The bar magnet being moved into, and out of, the coil in Fig. 11-1*a* and *b* producés an ac voltage. The coil and magnet make up a simple ac generator. An induced voltage also could be produced by moving the coil while the bar magnet is held stationary.

In Fig. 11-3 a piece of wire is shown between the poles of a stationary permanent magnet. If the wire is also held stationary so that it does not cut across any magnetic lines, then no voltage is induced. The same condition is true if the wire is moved horizontally to the left or to the right, but *parallel* to the magnetic lines so that the wire does not cut across any lines. Again, no voltage is induced.

When the wire is moved vertically upward, as shown in Fig. 11-3, so that it now cuts across magnetic lines, a voltage is induced in the wire. The *direction of the current* and the *polarity of the induced voltage* can be predicted by the *generator, left-hand, three-finger rule.* As illustrated in Fig. 11-3, extend the left thumb, index finger, and middle finger so that they are each perpendicular or at right angles to each other; point the *index finger* in the direction of *magnetic lines,* from north to south; point the *left thumb* in the direction of the *movement* of the wire; the *middle finger* now points in the *direction of the electron flow* in the wire.

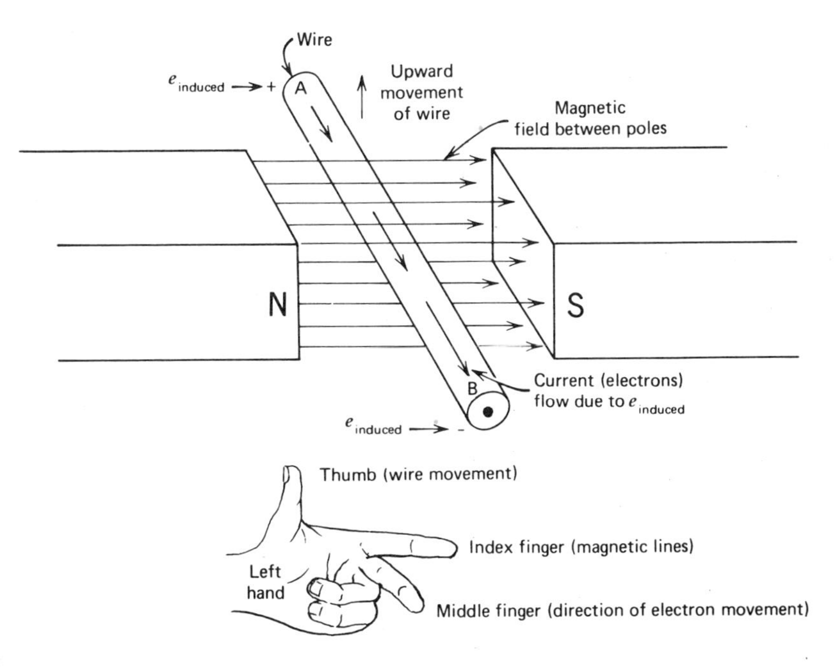

Figure 11-3. Wire being moved up in a magnetic field. Left-hand, three-finger rule for generator.

As shown in Fig. 11-3, current or electron flow in the wire is from point A to point B, toward the viewer. The dot at the near end of the wire is the symbol for the front or point of the directional arrow. With electrons leaving point A, in the wire, and moving to point B, the polarity of the induced voltage is *negative* at point B, and *positive* at point A, as shown. This electron flow is similar to the electrons moving *inside* any voltage source, from positive to negative.

An explanation as to why electrons will flow in the wire in the direction shown in Fig. 11-3, obeying the left-hand, three-finger rule, may be seen in the following.

Figure 11-4 depicts an end view of the wire in the magnetic field of the previous drawing of Fig. 11-3. When the wire is moved *upward*, it cuts across the magnetic lines, and a voltage is induced in the wire. This causes a current flow in the wire. If electrons flow *out* of the wire towards the viewer, as depicted in Fig. 11-4, then from the *left-hand wire rule*, the electromagnetic field encircles the wire in a *clockwise* direction. As shown in the drawing, the electromagnetic lines *aid* the permanent magnet lines *above* the wire, but *oppose* the permanent magnetic lines *below* the wire. As a result, the *weaker* magnetic area is *below* the wire. The conductor attempts to move toward this weaker magnetic area (from Section 9-8D and E). Thus the wire attempts to move *downward*, opposing the original action which consisted of moving the wire *upward*. This opposition obeys Lenz's law. Therefore, when the wire is moved up in the permanent magnetic field, as shown in Figs. 11-3 and 11-4, current (electrons) flows out of the near end of the wire, signifying that this is the negative end of the induced voltage.

In Fig. 11-5, the end view of the wire is shown in the field between the poles of a permanent magnet. If the wire is moved at a *constant speed* along a circular path starting at point A and proceeding to point B, then to point C, and

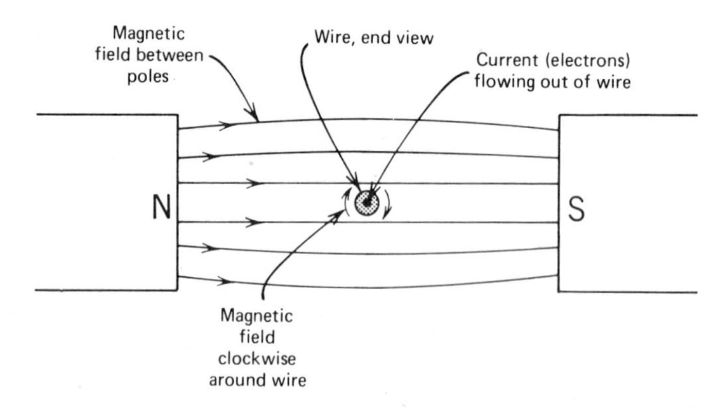

Figure 11-4. When wire is moved *up*, a current is produced flowing out of wire toward the viewer.

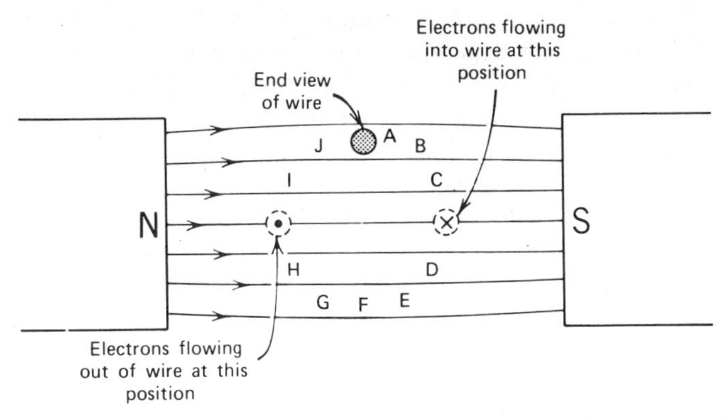

Figure 11-5. Moving a wire in a circular path in a magnetic field.

so on, a voltage will be induced in the wire whenever the conductor cuts across the magnetic field. For simplicity, only about seven magnetic lines are shown between the poles of the permanent magnet. The conductor is being moved completely around the circular path, or a total of 360°.

If the wire is started at position A and moved clockwise to position B, the wire is being moved parallel to the magnetic lines and is not cutting across the lines. As a result, no voltage is induced in the wire. This is illustrated in Fig. 11-6.

As the wire of Fig 11-5 is moved still further, and *down*, from position B to position C (1/8 rotation or 45°), it now cuts across one magnetic line. From

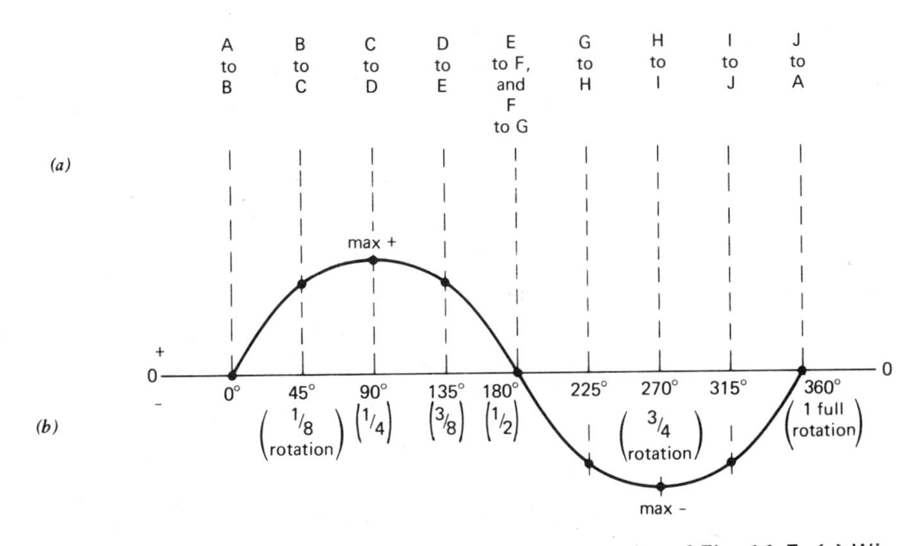

Figure 11-6. An induced ac voltage in the moving wire of Fig. 11-5. (*a*) Wire being moved from *A* to *B*, *B* to *C*, etc. (*b*) Voltage induced in wire.

Faraday's law, in the previous section, the amount of voltage that is induced in a conductor depends on the *rate* at which the conductor cuts across a magnetic field. When only one magnetic line is being cut, the rate is low and only a small voltage is induced in the coil. This is shown in Fig. 11-6*b*, where the induced voltage is small at the 45° position.

Rotating the conductor of Fig. 11-5 *down* further to the 90° position (1/4 rotation), or from point C to D, causes a maximum number of magnetic lines to be cut (three, in the simplified drawing). This produces a maximum voltage in the wire as shown in Fig. 11-6*b* at the 90° position. By employing the left-hand three-finger generator rule, the *direction* of the current flow in the wire may be determined. With the left thumb, index finger, and middle finger held at right angles to each other, point the thumb *downward* in the direction of the wire movement of Fig. 11-5, since the conductor moves *downward* from position C to D. Also point the left index finger in the direction of the magnetic lines of force, from the north to the south pole. The left middle finger will now point into the page of this book, indicating that electrons are moving into the wire of Fig. 11-5 at the position between positions C and D. This electron current flow is shown by the cross or "X" mark, indicating the "tail" of the current arrow, or that electrons are flowing into the wire, away from the viewer.

Now determine the amount of induced voltage as the wire is moved from position D to E, then from position E to F, and then from position F to G, using the results shown in Fig. 11-6*b* as a guide.

As the conductor of Fig. 11-5 is moved *upward* (for the first time) from position G to H, or at the 225° rotation, the wire again cuts across only one magnetic line, as it did going from position B to C and also from position D to E. As a result, the voltage induced is small. This shown in Fig. 11-6*b* at the 225° position. Note that the voltage at this point is below the zero axis, indicating that it is reversed in polarity from that at the 45° and 135° points. This, of course, is an ac voltage.

Again determine the amount of induced voltage as the wire continues to be moved upward from position H to I, then from position I to J, and finally from position J to A. As a guide, use the results shown in Fig. 11-6*b* at these positions.

The ac voltage induced in the wire of Fig. 11-5, as it was rotated in the magnetic field in its circular path, is called a *sine wave*. This is the voltage shown in Fig. 11-6*b*. It is called a *sine wave* because it is exactly the same form of graph that the values of the *sine* function of angles from 0 to 360° have in mathematics, but should not concern the reader here.

Example 11-2

In the diagram for this example, is the induced voltage *zero*, *small*, or *large* when the wire is being moved from (a) position W to X, (b) position X to Y, and (c) position Y to Z?

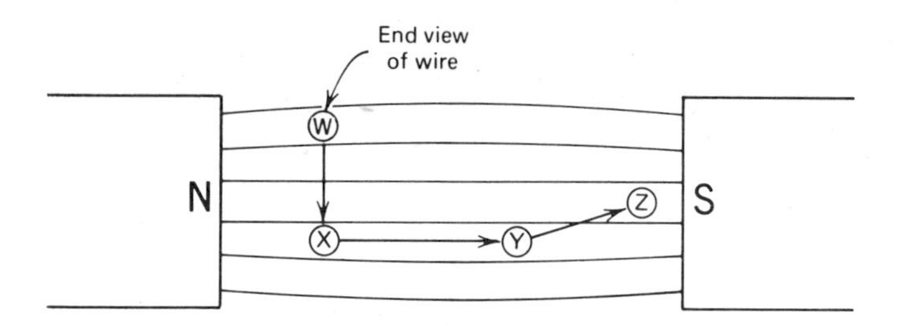

When wire is moved from:	$E_{induced}$ is:
W to *X*	Maximum
X to *Y*	Zero
Y to *Z*	Small, but of opposite polarity

Figure 11-7. Example 11-2.

Solution

Solve this example by using the results shown in Fig. 11-7 as a guide.

For Similar Problems Refer to End-of-Chapter Problems 11-8 to 11-10

11-3. Inductance

The property of any conductor's having a voltage *induced* in it when a magnetic field cuts across it, is called *inductance.* A piece of straight wire is an inductance, and so is the wiring of any circuit. A coil of wire is a much larger inductance than a single wire. The unit of inductance is called the *henry*, honoring the American scientist Joseph Henry who, in the mid 1800s, pioneered in knowledge of induction principles.

The value of an inductance or coil depends on several factors: its physical size (length and cross-sectional area), the number of turns of wire, and the permeability of its core material.

Refer to End-of-Chapter Problems 11-11 and 11-12

11-4. Self-Inductance

In Section 11-1 we show that when a bar magnet is moved near a coil, the magnetic lines cut across the coil wires, inducting a voltage in the coil. A voltage can also be induced in a coil without using a bar magnet.

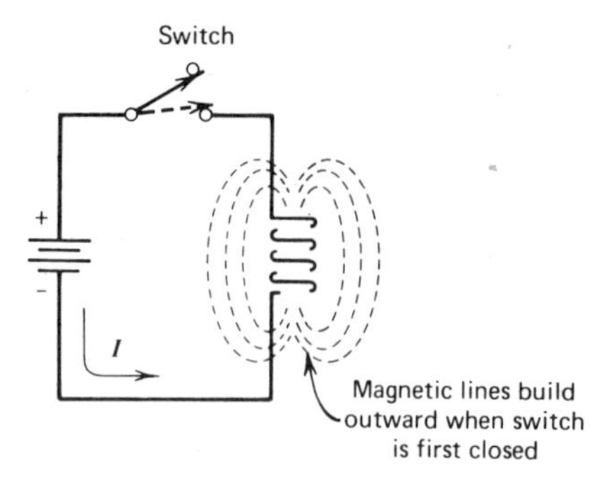

Figure 11-8. Self-inductance.

In Fig. 11-8, when the switch is *first* closed, the current starts flowing and a magnetic field starts building up. These magnetic lines emerge from the center of the coil and move outward. In so doing, the lines cut across the coil wires, inducing a voltage called a *counter electromotive force.* From Lenz's law (Section 11-1) the polarity of this induced voltage (e) in the coil is such as to oppose the original action (the current that started to flow when the switch was first closed). The coil is said to oppose a *change* of current, when the current went from zero (with the switch open), to some value (when the switch was closed).

The property of having a voltage induced in a conductor because of its *own* current's *changing* in it is called *self-inductance.* By definition, a coil has 1 henry (H) inductance if a current in it changes 1 A in 1 sec, inducing 1 V in the coil.

Refer to End-of-Chapter Problems 11-13 to 11-15

11-5. Mutual Inductance and Coefficient of Coupling

When current in a coil changes or varies, causing its magnetic field to move, the magnetic lines cut across the coil and also across any other nearby conductors. If a second coil is near the first, then some of the moving magnetic lines from the first coil cut across the second. A voltage is induced in the second coil as a result of a property called *mutual inductance.* That is, when a voltage is induced in a coil because of the moving magnetic field of *another* coil, the phenomena is called mutual inductance. This is shown in Fig. 11-9.

Coefficient of Coupling. When a voltage is induced in a second coil because of the moving magnetic field of a first coil, the coils are said to be magnetically *coupled.* The degree or amount of this coupling is called the

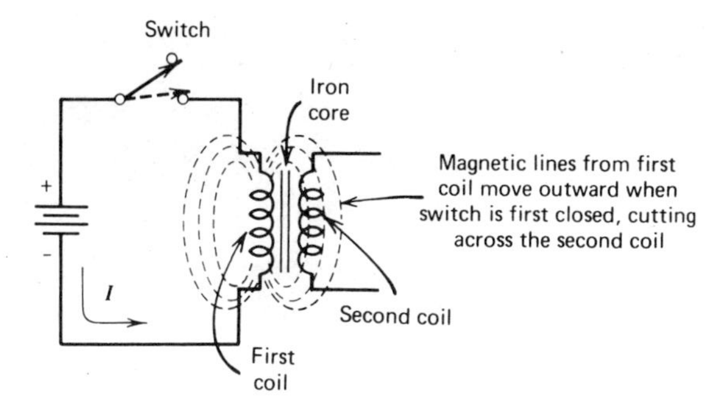

Figure 11-9. Mutual inductance.

coefficient of coupling, and it describes the proportion of the magnetic lines of the first coil that actually cut across the second coil. The maximum coefficient of coupling occurs when every magnetic line of the first coil cuts the second coil. In this case, the coefficient of coupling is at its maximum value of 1, which is also called *unity* coupling. When only one half the lines from the first coil cut across the second coil, the coefficient of coupling is 0.5.

The coupling between two coils may be increased by placing them parallel to each other (coils placed at a right angle, or perpendicular, to each other have almost zero coupling), or may be increased by bringing the coils closer together, even to winding one coil on top of the other. Maximum coupling (coefficient of coupling of 1) may be approached by winding both coils on one iron core. The schematic or circuit symbol for an iron core is shown in Fig. 11-9. Figure 11-10 is a photograph of iron-core inductances.

Refer to End-of-Chapter Problems 11-16 to 11-19

11-6. The Simple Transformer

A device that operates depending on mutual inductance and magnetic coupling is the *transformer*. In this device a varying current in the first coil, called the *primary*, causes its magnetic field to move. These magnetic lines cut across the wires of another coil called the *secondary*, inducing a voltage in it. The amount of voltage induced in the secondary depends on three main factors: (a) the voltage applied to the primary, (b) the number of turns of wire in the primary as compared to the secondary (this is called the *turns ratio*), and (c) the coefficient of coupling.

Figure 11-11 is a photograph of an iron-core transformer, while Fig. 11-12 shows a transformer circuit where the primary has 30 turns of wire and the secondary 150 turns. The *turns ratio* is simply the ratio of primary turns to

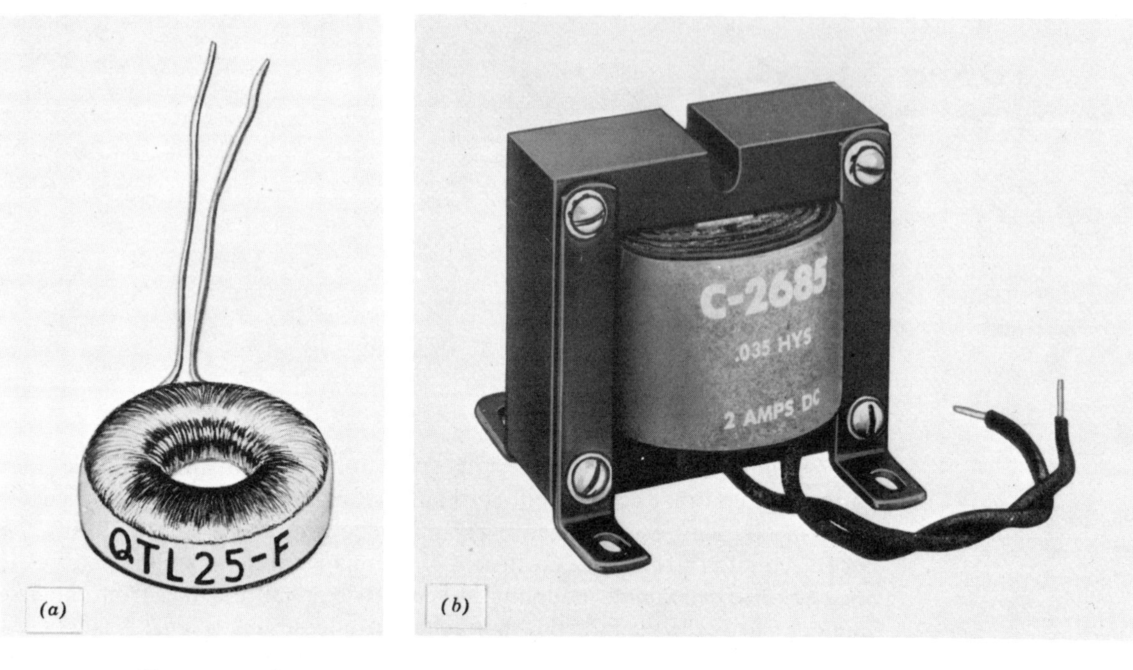

Figure 11-10. Iron-core inductances. (*a*) Toroid-wound coil (*courtesy Microtran Co*). (*b*) Iron-core coil (*courtesy Essex Stancor*).

secondary turns, or

$$\text{turns ratio} = \frac{\text{primary turns}}{\text{secondary turns}}$$

or

$$\text{turns ratio} = \frac{T_p}{T_s} \tag{11-1}$$

Example 11-3

What is the turns ratio of Fig. 11-12 where the primary consists of 30 turns and the secondary consists of 150 turns?

Solution

$$\text{turns ratio} = \frac{T_p}{T_s} \tag{11-1}$$

$$\text{turns ratio} = \frac{30}{150}$$

$$\text{turns ratio} = \frac{1}{5} \quad \text{or} \quad 1 \text{ to } 5$$

Figure 11-11. Iron-core transformer (*courtesy Essex Stancor*).

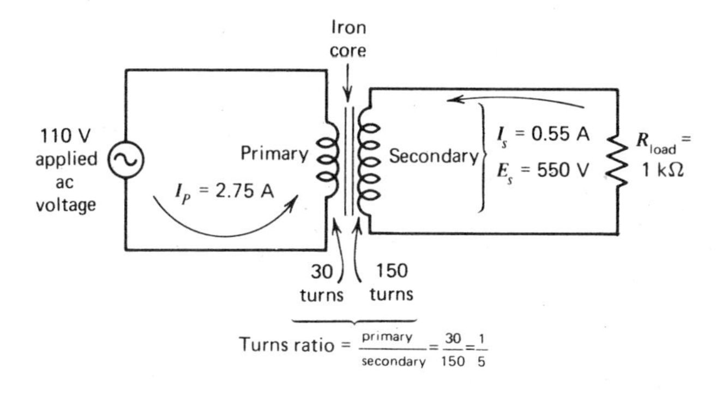

Figure 11-12. Voltage step-up transformer.

The voltage induced in the secondary depends on the *turns ratio* and the *voltage applied to the primary*. The ratio of primary voltage to secondary voltage is exactly the same as the turns ratio, assuming that the coefficient of coupling between the two coils is maximum, or 1. (With both coils wound on the same iron core, this is true.) Therefore

$$\text{turns ratio} = \text{voltage ratio}$$

$$\frac{T_p}{T_s} = \frac{E_p}{E_s} \tag{11-2}$$

Example 11-4

In the diagram of Fig. 11-12, find the voltage induced in the secondary if the voltage applied to the primary is 110 V, with 30 turns of wire in the primary, and 150 turns in the secondary, and assuming a coefficient of coupling of 1.

Solution

$$\text{turns ratio} = \text{voltage ratio}$$

$$\frac{T_p}{T_s} = \frac{E_p}{E_s} \tag{11-2}$$

$$\frac{30}{150} = \frac{110}{E_s}$$

To solve for E_s, cross multiply 30 by E_s and also 110 by 150, yielding

$$30\ E_s = (110)\ (150)$$

Now, dividing both sides of the equation by 30 gives

$$\frac{30\ E_s}{30} = \frac{(110)\ (150)}{30}$$

$$E_s = (110)\ (5)$$

$$E_s = 550\ \text{V}$$

Note that with a greater number of turns of wire in the secondary (150) than in the primary (30), the secondary voltage is 550 V, while the voltage applied to the primary is only 110 V. This is called a *voltage step-up transformer*. If the secondary has fewer turns than the primary, a smaller voltage is induced in the secondary. This type is called a *voltage step-down transformer*.

For Similar Problems Refer to End-of-Chapter Problems 11-20 to 11-22

Voltage Applied to a Transformer. To *induce* a voltage in the *secondary*, the magnetic field of the *primary* must be moving. This means that the current in the primary must vary. A steady dc type of current in the primary would produce only a stationary magnetic field that would not induce a secondary voltage. Therefore, an ac voltage is usually applied across the primary to permit a transformer to function.

However, a varying or fluctuating dc current in the primary would also produce a moving magnetic field that would cut across the secondary, inducing a voltage in it. An electron tube or a transistor is often connected to the primary of a transformer when it is desired to amplify an electronic signal (an ac voltage) and send it on (couple it) to the next amplifier stage. Signal current through an electron tube, and also through a transistor, is a varying direct current. As we show in a later section, varying direct current actually consists of two parts, a steady direct current and an alternating current. It is the ac component that permits the transformer to operate.

Voltage induced in the *secondary* coil is always ac voltage, as long as the voltage applied to the primary is either alternating current also, or is a varying direct current.

Refer to End-of-Chapter Problems 11-23 and 11-24

11-7. Transformer Currents

In the preceding section, we discuss voltages in the primary and in the secondary. The amount of current in each coil depends on the secondary voltage, the resistance in the secondary circuit, the turns ratio, and the power losses in the transformer.

In Fig. 11-12 a voltage step-up transformer circuit is shown. The primary has 30 turns and the secondary has 150 turns. Voltage applied to the primary is 110 V. From Example 11-4, secondary voltage is found to be 550 V. If a 1-kΩ load is placed across the secondary, as shown in Fig. 11-12, current in the secondary circuit can be found using Ohm's law, since voltage and resistance are known. This is shown in the following.

Example 11-5

Find the secondary current in Fig. 11-12.

Solution

$$I_{sec} = \frac{E_{sec}}{R_{\text{load in the sec}}} \qquad \text{(Ohm's law)}$$

$$I_{sec} = \frac{550}{1\ \text{k}\Omega}$$

$$I_{sec} = \frac{550}{1000}$$

$$I_{sec} = 0.55\ \text{A}$$

Power or wattage in the secondary circuit, using one of the power equations, is found as follows:

$$P_{sec} = E_{sec} I_{sec}$$

$$P_{sec} = (550 \text{ V}) (0.55 \text{ A})$$

$$P_{sec} = 302.5 \text{ W}$$

Transformer Power Losses. The power dissipated in the secondary circuit, as shown above, is actually supplied by the voltage source that is applied to the primary. If there were no power losses in the transformer, then the primary power would be the same as the secondary power. However, some power is wasted in the transformer as a heat loss. With proper construction, this power loss can be so small as to be negligible.

Power is lost or wasted in an iron-core transformer because of the following. As the magnetic field builds up and collapses, it induces a voltage in the iron core, itself. This produces undesirable currents inside the iron, causing it to heat up. These are called *eddy current* losses, which could become quite large if the core were a solid iron block. To reduce these currents, the iron core has high resistance built into it. One type of iron core is called *laminated* iron, and it consists of thin flat sheets of iron with insulation between each sheet. Another type called *powdered* iron, consists of tiny iron granules that are suspended in an insulating compound like liquid shellac. When the compound hardens, each tiny iron particles is insulated from the others. Voltage is still induced in the iron, but because of the high resistance between adjacent parts of the iron, very little current flows in the iron core, producing very little heat loss. Certain ceramics called *ferrites* have magnetic properties but are insulators. These cores keep the eddy currents down.

A second type of power loss in the iron core is called *hysteresis* loss. This is due to the reluctance or opposition of the iron to being magnetized when current flows in the coil, and then being magnetized with opposite polarity when the current reverses direction (ac) in the coil. To reduce this heat loss, the metal chosen as the core is a soft iron instead of a hard steel, since soft iron has low *reluctance* or little opposition to being magnetized and demagnetized, and then to being reverse magnetized.

A third loss is due to the resistance of the copper wire itself. This is simply called *copper* loss. By using wire of large diameter, the resistance is small and the copper loss is minimized.

The power losses in most transformers are usually so small as to be negligible. As a result, the power consumed by the *primary* from the voltage source is just about equal to the power dissipated in the *secondary* curcuit. The efficiency is said to be 100%. Since the primary power is equal to the secondary power, neglecting the small loss in the transformer, then the current in the primary may be found as shown in the following:

Primary Current. Assuming no power loss in the transformer, then

$$P_{\text{primary}} = P_{\text{secondary}}$$

and since power $= E\,I$, then

$$E_p\,I_p = E_s\,I_s \tag{11-3}$$

Dividing both sides of equation 11-3 by I_p and E_s yields

$$\frac{E_p\,I_p}{I_p\,E_s} = \frac{E_s\,I_s}{I_p\,E_s}$$

Canceling the I_p terms at the left side, and the E_s terms at the right side gives

$$\frac{E_p}{E_s} = \frac{I_s}{I_p} \tag{11-4}$$

Note that equation 11-4 shows that the current ratio, primary to secondary, is just the reverse of the voltage ratio, primary to secondary. In other words, if the *voltage* is increased, from primary to secondary, then the *current* is *decreased*, from primary to secondary, in the same proportion.

Since the voltage ratio is the same as the turns ratio shown previously in equation 11-2.

$$\frac{T_p}{T_s} = \frac{E_p}{E_s} \tag{11-2}$$

then the turns ratio (T_p/T_s) may be substituted for the voltage ratio (E_p/E_s) in equation 11-4, yielding

$$\frac{T_p}{T_s} = \frac{I_s}{I_p} \tag{11-5}$$

Note also that equation 11-5 states that when the *turns ratio* is *increased*, primary to secondary, then the *current is decreased*, going from primary to secondary. Equations 11-4 and 11-5 may only be employed when power losses in the transformer are negligible. In a later section of this chapter we discuss primary current, taking power losses into account.

Figure 11-12, and Examples 11-4 and 11-5 show a transformer having 30 turns of wire in the primary, 150 turns in the secondary, with 110 V applied to the primary. From Example 11-4, and by using equation 11-2 ($T_p/T_s = E_p/E_s$), secondary voltage is found to be 550 V. In example 11-5, where a load resistor in the secondary circuit is 1 kΩ, and by using Ohm's law, secondary current is found to be 0.55 A. The following examples illustrate how primary current may be found.

Example 11-6

In the circuit of Fig. 11-12 the primary has 30 turns, the secondary has 150 turns, and 0.55 A flows in the secondary. Assuming that power losses are negligible, find the primary current.

Solution

Solve for I_p using equation 11-5. (Find I_p, using the result shown in Fig. 11-12 as a guide.)

Example 11-7

In Fig. 11-13, 120 V ac is applied to a 500-turn primary. The load resistor in the 25-turn secondary circuit is 2 kΩ. Assuming a coefficient of coupling of 1, and negligible power loss, find: (a) E_s, (b) I_s, and (c) I_p.

Solution

Solve this example by using the results shown in Fig. 11-13.

For Similar Problems Refer to End-of-Chapter Problems 11-25 to 11-32

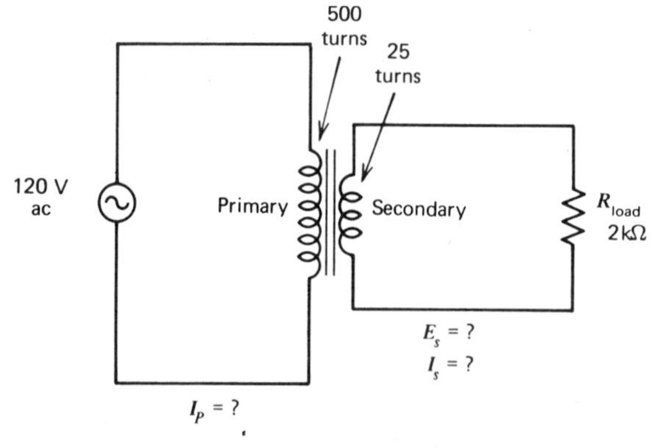

$$\frac{T_p}{T_s} = \frac{E_p}{E_s} \qquad I_s = \frac{E_s}{R} \qquad \frac{T_p}{T_s} = \frac{I_s}{I_p}$$

$$\frac{500}{25} = \frac{120}{E_s} \qquad I_s = \frac{6}{2\text{ K}} \qquad \frac{500}{25} = \frac{3\text{ mA}}{I_p}$$

$$E_s = 6\text{ V} \qquad I_s = 3\text{ mA} \qquad I_p = 0.15\text{ mA}$$

Figure 11-13. Example 11-7.

11-8. AC Voltage Sine Wave

In Section 11-2, a discussion of a simple ac generator is given. The result of moving a wire in a magnetic field (Figs. 11-3, 11-5, and 11-6) is described. As the wire is moved *down*, cutting magnetic lines, a voltage is induced in the wire. When the wire is moved *up*, cutting the magnetic field from the opposite direction, a reverse polarity voltage is induced in the wire. This reversing-polarity voltage is called an *alternating voltage*, or simply an ac *voltage*.

When the ac voltage is induced in the wire as a result of moving the wire in a circular path within the magnetic field (Fig. 11-5), then the ac voltage is that shown in Figs. 11-6*b* and 11-14. This type of ac voltage is called a *sine wave* because it has the same shape or form as the graph of the sine function of angles. (See Appendix 1, Mathematics for Electronics.) Observe in Fig. 11-6*b* that the induced ac voltage is actually a graph of voltage values plotted against the angle of rotation of the wire. The voltage graph is usually called a voltage *wave form* or *wave shape*. Note that it rises upward above zero, reaching a *maximum positive* value, and then dips down below zero, reaching a *maximum negative* value. The time that is required for the rotating wire of Fig. 11-5 to make one complete revolution, or 360°, is the time required to produce one complete waveshape or one ac-voltage *sine wave*.

The number of these sine waves produced in one second is called its *frequency*. For example, if the rotating wire of Fig. 11-5 revolves around its circular path at a rate of 60 times per second, then the induced ac-voltage sine wave of Fig. 11-6*b* has a *frequency* of 60 per second, usually expressed as 60 *cycles* per second (cps) or 60 *hertz*, where a *hertz* is a *cycle per second*).

Figure 11-14 shows two ac-voltage sine waves or 2 cycles. The voltage reaches a positive maximum value of +50 V, and a maximum negative value of −50 V. Each of these values is called the *maximum* or *peak voltage*. Going from the +50 V to the −50 V, for a *difference* of 100 V, this difference from the positive peak to the negative peak is called the *peak-to-peak* value, or $E_{p\text{-}p}$. In this case, it is 100 V peak-to-peak. Note that $E_{p\text{-}p}$ is twice E_{max}, and that E_{max} is one half of $E_{p\text{-}p}$.

It is often useful to be able to compare an ac voltage to a dc voltage. In Fig. 11-14 the ac sine-wave voltage reaches its 50-V peak value for only an

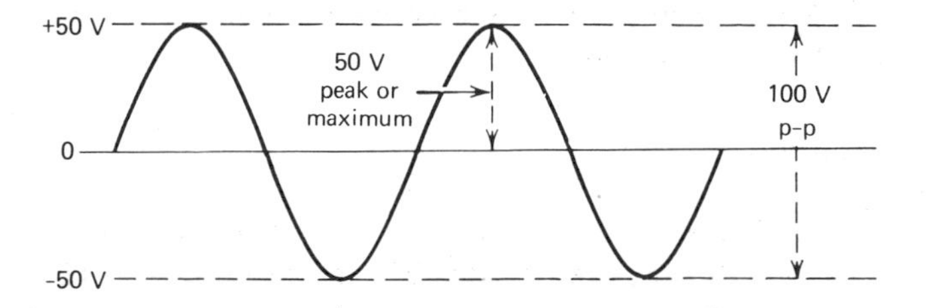

Figure 11-14. Two cycles of an ac voltage.

instant. Most of the time the ac voltage is much smaller than 50 V, even becoming zero for an instant. A dc voltage of 50 V has this value all the time. Therefore, 50 V dc has a much greater effect than 50 V of peak ac. To be able to compare an ac sine-wave voltage with a steady dc voltage, a certain value of alternating current is said to have the same heating *effect* as some amount of direct current. This value of alternating current is called its *effective* value. It is derived mathematically through a process called the *root mean square* or *rms*. This figure is *0.707 of the maximum value of the alternating current*. Therefore, the amount of an ac sine-wave voltage that has the same effect as a dc voltage is called the effective or rms voltage. Most ac voltmeters are calibrated to read this rms voltage. This is illustrated in the following equation and examples.

$$\text{Effective (or rms) voltage} = 0.707 \text{ maximum (or peak) voltage} \qquad (11\text{-}6)$$

Example 11-8

The 50-V *peak* ac sine wave of Fig. 11-14 would be read on an ac voltmeter as *effective* or *rms voltage*. This is the amount of steady dc voltage that has the same effect as the 50-V peak ac. What would the ac voltmeter read?

Solution

the voltmeter reads effective value.

$$E_{\text{effect}} \text{ (or rms)} = 0.707\, E_{\text{max}} \qquad (11\text{-}6)$$

$$E_{\text{effect}} = (0.707)\,(50)$$

$$E_{\text{effect}} = 35.35 \text{ V}$$

This means that 35.35 V of steady dc has the same effect as a 50-V peak ac sine wave voltage.

Example 11-9

What would an ac voltmeter read (rms) for an ac sine-wave voltage having 100 V peak-to-peak?

Solution

The 100-V peak-to-peak means that the peak value (or maximum) is actually one half of the peak-to-peak. This is shown in Fig. 11-14. With a 50-V peak, or maximum value, the effective value is the same as in the previous example, or 35.35 V, which is (0.707) (50).

Example 11-10

An ac sine-wave voltage is 338.4 V peak to peak. (a) What is its maximum voltage, and (b) what value of dc voltage has the same effect?

Solution

(a) The maximum or peak is one half of the peak-to-peak.

$$E_{max} = (\tfrac{1}{2})\, E_{peak\text{-}to\text{-}peak}$$

$$E_{max} = (\tfrac{1}{2})\, 338.4$$

$$E_{max} = 169.2 \text{ V}$$

(b) $$E_{effect \text{ or } rms} = (0.707)\, E_{max} \qquad (11\text{-}6)$$

$$E_{effect} = (0.717)\,(169.2)$$

$$E_{effect} = 120 \text{ V}$$

Note that this means that an ac sine-wave voltage having 338.4 V peak-to-peak amplitude, has a peak value of 169.2 V, and an effective or rms value (as read on an ac voltmeter) of 120 V.

For Similar Problems Refer to End-of-Chapter Problems 11-33 to 11-38

Maximum Value from Effective Voltage. Often it is useful to know the peak or maximum value, and also the peak-to-peak, from the ac voltmeter reading of the effective or rms value of a sine-wave voltage. Most ac voltmeters, therefore, have a peak-to-peak scale as well as an effective or rms scale.

When it is desired to know the maximum value of a sine wave from its effective value, equation 11-7 (derived below) is usually used.

$$E_{effect} = 0.707\, E_{max} \qquad (11\text{-}6)$$

Solve for E_{max} by dividing both sides by 0.707, giving

$$\frac{E_{effect}}{0.707} = E_{max}$$

The decimal 0.707 is equivalent to the fraction 707/1000, yielding

$$\frac{E_{effect}}{707/1000} = E_{max}$$

Dividing by the fraction 707/1000 requires *multiplying* by the *inverted* fraction, or

$$\left(\frac{1000}{707}\right)(E_{effect}) = E_{max}$$

1000/707 is approximately 1.41, producing

$$1.41\ E_{\text{effect}} = E_{\text{max}} \tag{11-7}$$

Note that the maximum voltage is 1.41 times the effective value.

Example 11-11

If an ac voltmeter reads 120 V rms for a sine wave, find: (a) the maximum voltage or peak, and (b) the peak-to-peak.

Solution

(a) $E_{\text{max}} = 1.41\ E_{\text{effect}}$ (11-7)

$E_{\text{max}} = (1.41)(120)$

$E_{\text{max}} = 169.2\ \text{V}$

(b) $E_{\text{p-p}} = (2)(E_{\text{max}})$

$E_{\text{p-p}} = 2\ (169.2)$

$E_{\text{p-p}} = 338.4\ \text{V}$

Note that these are the same figures as in Example 11-10.

Example 11-12

The secondary voltage of a transformer is 12 V ac. What is the (a) peak voltage, and (b) the peak-to-peak?

Solution

The term 12 V ac means the *effective* or *rms* value. Therefore

(a) $E_{\text{max}} = 1.41\ E_{\text{effect}}$ (11-7)

$E_{\text{max}} = 1.41\ (12)$

$E_{\text{max}} = 16.92\ \text{V}$

(b) $E_{\text{p-p}} = 2\ E_{\text{max}}$

$E_{\text{p-p}} = 2\ (16.92)$

$E_{\text{p-p}} = 33.84\ \text{V}$

For Similar Problems Refer to End-of-Chapter Problems 11-39 and 11-40

11-9. Inductance, More Advanced

In Section 11-3, the term *inductance* was described as the property of a conductor where a voltage is induced in it when a magnetic field cuts across it, or is cut by the conductor. A piece of wire, a coil of wire, and even a metal chassis all have this property, but the inductance of a piece of wire or the chassis is so small as usually to be negligible. The term inductance is generally reserved for a coil of wire.

Inductance, as previously mentioned, is measured in a unit called the *henry*. One henry is defined as the amount of inductance present where one volt is induced in a coil when a current change of one ampere in one second occurs. The value of an inductance is determined mainly by its physical size, such as the number of turns of wire, cross-sectional area, length, and also by the type of core material. The inductance in *henries* can be calculated using the following equations (*for single-layer coils only*):

$$\text{inductance} = \frac{(\text{number of turns})^2\,(\text{core permeability})\,(\text{cross-sectional area})}{\text{length}} \qquad (11.8)$$

or

$$L = \frac{N^2 \mu A}{l} \qquad (11\text{-}8)$$

where μ is the permeability of the core in *mks* units, and where μ of air is 12.56 times 10^{-7} (chapter 9, Table 9-1), and where μ of a magnetic material core becomes, in the equation, the core μ multiplied by the μ of air; A is the cross-sectional area of the core in *square meters*: and l is the average length of the core in meters.

The following examples illustrate the use of equation 11-8.

Example 11-13

Find the inductance of 5000 turns of wire wound on a round iron rod having a permeability of 300, if the rod is 15 cm long and 0.5 cm in diameter, as shown in Fig. 11-15.

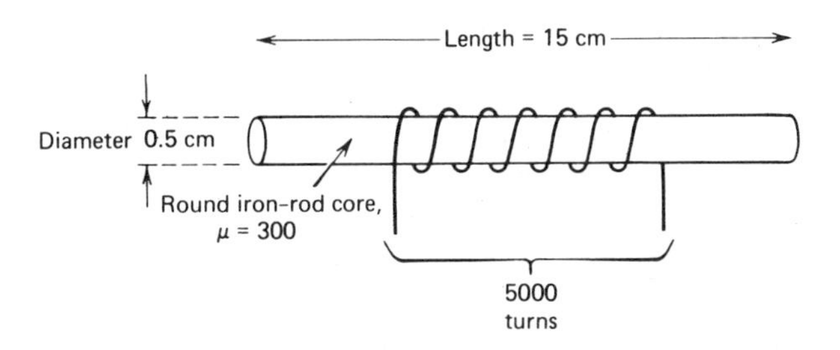

Figure 11-15. Example 11-13.

Solution

First find the cross-sectional area in square meters of the round iron rod with a diameter of 0.5 cm.

$$\text{radius} = \tfrac{1}{2}\ \text{diameter}$$

$$\text{radius} = \tfrac{1}{2}\ (0.5\ \text{cm})$$

$$\text{radius} = \tfrac{1}{2}\ (0.5 \times 10^{-2}\ \text{m})$$

$$\text{radius} = 0.25 \times 10^{-2}\ \text{m}$$

The area of a circle that is the cross-sectional area of the round rod is

$$\text{area} = \pi\ (\text{radius})^2$$

$$\text{area} = 3.14\ (0.25 \times 10^{-2})^2$$

$$\text{area} = 3.14\ (.0625 \times 10^{-4})$$

$$\text{area} = 0.196 \times 10^{-4}\ \text{m}^2$$

Note that in equation 11-8, the μ is that of the iron core (300) multiplied by the μ of air in the mks system (12.56×10^{-7}). The value of the inductance in henries is then

$$L = \frac{N^2 \mu A}{l} \tag{11-8}$$

$$L = \frac{(5000)^2\ (300)\ (12.56 \times 10^{-7})\ (0.196 \times 10^{-4})}{15\ \text{cm}}$$

$$L = \frac{(5 \times 10^3)^2\ (3 \times 10^2)\ (12.56 \times 10^{-7})\ (0.196 \times 10^{-4})}{15 \times 10^{-2}\ \text{m}}$$

$$L = \frac{(25 \times 10^6)\ (3 \times 10^2)\ (12.56 \times 10^{-7})\ (0.196 \times 10^{-4})}{15 \times 10^{-2}}$$

$$L = 12.3 \times 10^{-1}\ \text{H} \quad \text{or} \quad 1.23\ \text{H}$$

Example 11-14

Find the inductance of 200 turns of wire wound on a nonmagnetic tubular form, where the diameter is 10 cm and the coil length is 50 cm.

Solution

Solve for the inductance by using the method shown in the previous example. L should be found to be 791 μH.

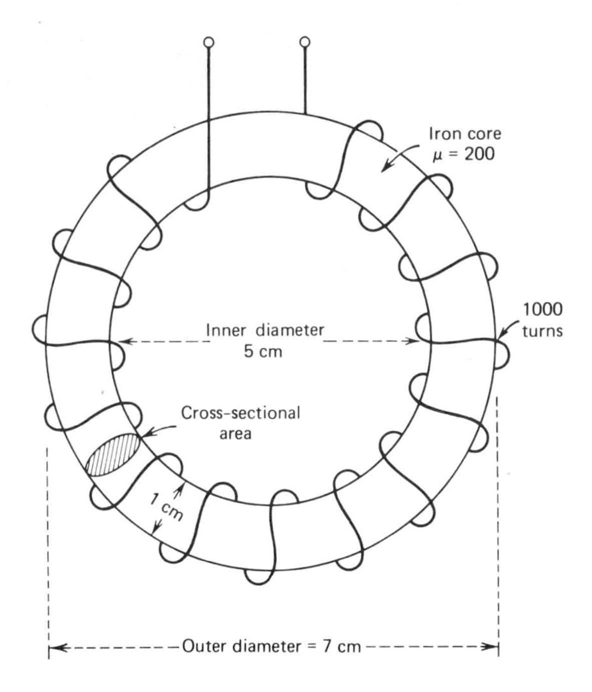

Figure 11-16. Toroid coil (doughnut-shaped core), Example 11-15.

Example 11-15

Find the inductance of 1000 turns of wire wound on a doughnut-shaped iron core, called a *toroid*, as shown in Fig. 11-16. The inner diameter is 5 cm, while the outside diameter is 7 cm. The iron has a permeability (μ) of 200.

Solution

The cross-sectional area of the ring core is found where the *diameter* or thickness of the ring core itself is 1 cm.

$$\text{cross-sectional area} = \pi\,(\text{radius})^2$$

$$\text{cross-sectional area} = (3.14)\left(\frac{1\text{ cm}}{2}\right)^2$$

$$\text{cross-sectional area} = (3.14)\,(0.5\text{ cm})^2$$

$$\text{cross-sectional area} = (3.14)\,(0.5 \times 10^{-2}\text{ m})^2$$

$$\text{cross-sectional area} = (3.14)\,(0.25 \times 10^{-4})$$

$$\text{cross-sectional area} = 78.5 \times 10^{-4}\text{ m}^2$$

The *average* length (or average circumference) of the doughnut-shaped core has a diameter that is the average between the 7 cm outer diameter and the 5 cm inner diameter, or a 6-cm diameter. This average length or circumference is then

$$\text{circumference} = 2\pi \text{ (radius)}$$

$$\text{circumference} = (2\pi)\left(\frac{\text{diameter}}{2}\right)$$

$$\text{circumference} = (6.28)\left(\frac{6 \text{ cm}}{2}\right)$$

$$\text{circumference} = (6.28)\ (3 \text{ cm})$$

$$\text{circumference} = (6.28)\ (3 \times 10^{-2} \text{ m})$$

$$\text{circumference} = 18.84 \times 10^{-2} \text{ m}$$

The inductance can then be found using equation 11-8. The μ in the equation is the μ of the iron core multiplied by that of air (12.56×10^{-7}) in the mks system (from Table 9-1).

$$L = \frac{N^2 \mu A}{l}$$

$$L = \frac{(1000)^2\ (200)\ (12.56 \times 10^{-7})\ (78.5 \times 10^{-4})}{18.84 \times 10^{-2}}$$

$$L = \frac{(1 \times 10^6)\ (2 \times 10^2)\ (985 \times 10^{-11})}{18.84 \times 10^{-2}}$$

$$L = \frac{1970 \times 10^{-3}}{18.84 \times 10^{-2}}$$

$$L = 104.5 \times 10^{-1} \text{ H} \qquad \text{or} \qquad 10.45 \text{ H}$$

The inductance of an iron-core coil, as shown in equation 11-8, depends on the permeability (μ) of the core material, as well as the physical dimensions of the coil. The amount of inductance is a fixed or constant value for a given coil as long as the iron core is not *magnetically saturated.* In normal operation, when the current is increased in the coil, the magnetic field becomes stronger. However, if sufficient current flows through the coil, there comes a point where the iron core is magnetized to its maximum. This is *magnetic saturation.* Increasing the current still further results in practically no increase in the strength of the magnetic field. The permeability of the iron core is reduced to that of a nonmagnetic material. With a smaller μ, the value of the inductance is decreased. In a special application, called the *swinging choke*, the inductance is permitted to decrease at higher currents.

For Similar Problems Refer to End-of-Chapter Problems 11-41 to 11-43

11-10. Self-Induced Voltage

A simple discussion of self-inductance is given in Section 11-4. It is stated there that when a current flowing in a coil changes value, the magnetic field moves, inducing a voltage in the coil. The amount of induced voltage e_L may be found from the following equation:

$$e_L = L\left(\frac{dI}{dT}\right) \tag{11-9}$$

or, stating it another way,

$$e_L = L\left(\frac{\Delta I}{\Delta T}\right) \quad \text{(another version of equation 11-9)}$$

In the above, L is in henries, dI or ΔI means a *change* of *current* (in amperes), and dT or ΔT denotes a change of time (in seconds).

The following example illustrates the use of equation 11-9.

Example 11-16

If a current of 12 mA increases steadily to 15 mA in 15 msec, how much voltage is induced in a 50 mH coil?

Solution

$$e_L = L\left(\frac{dI}{dT}\right) \tag{11-9}$$

$$e_L = (50 \text{ mH})\left(\frac{12 \text{ mA to } 15 \text{ mA}}{15 \text{ msec}}\right)$$

$$e_L = (50 \times 10^{-3})\left(\frac{3 \times 10^{-3}}{15 \times 10^{-3}}\right)$$

$$e_L = 10 \times 10^{-3} \text{ V} \quad \text{or} \quad 0.01 \text{ V}$$

Example 11-17

A current of 18 mA flows in a 25-H coil. During a 150-μsec period, the current rises steadily to 20 mA. Find the induced voltage during this short time interval.

Solution

The reader should solve for e_L, and should find it to be 333 V.

When the current *increases* in a coil, the magnetic field builds up, cutting the coil and inducing a voltage in the coil. When the current *decreases* in the coil, the magnetic field collapses, again cutting the coil but from the opposite direction. The induced voltage is now opposite in polarity from that which resulted when the current increased. This is shown in the following example.

Example 11-18

In Fig. 11-17 a current wave shape or graph is shown. In the first period, time T_1, of 20 μsec, the current goes from zero to 10 mA. In the second time period, T_2, also 20 μsec, the current remains at the 10-mA value. In the third time interval, T_3, again 20 μsec, the current rises from 10 mA to 15 mA. Finally, in the last time period, T_4, the current *decreases* from 15 mA to zero in 20 μsec.

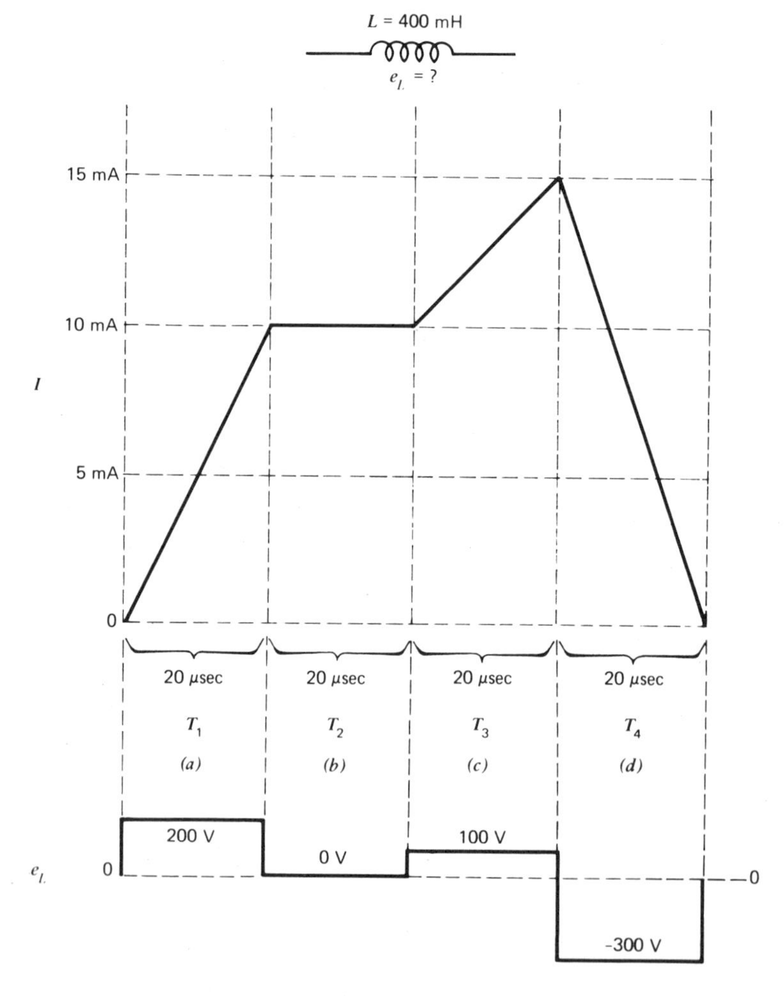

Figure 11-17. Example 11-18.

If this varying current flows through a 400-mA coil, find the amount of the induced voltage during: (a) time period T_1, (b) time period T_2, (c) time period T_3, and (d) time period T_4.

Solution

The reader should solve this example using the results shown in Fig. 11-17 as a guide. Note that in part (d), time period T_4, the current *decreased*. Therefore, e_L will be opposite in polarity from the voltages induced when the *current increased*.

For Similar Problems Refer to End-of-Chapter Problems 11-44 to 11-46

11-11. Mutual Inductance and Coefficient of Coupling (Advanced)

In Section 11-5, *mutual inductance* is described as the property of two coils where, when the current changes in one, the moving magnetic field of that coil cuts across the other coil, inducing a voltage in it. The coils are said to be *magnetically coupled.* The ratio of the number of magnetic lines that cut the second coil to the number of magnetic lines produced in the first coil is called the *coefficient of coupling*, or K. When every magnetic line from the first coil cuts the second coil, K is at its maximum value of 1, also called *unity coupling.*

The amount of mutual inductance (L_m) that is present depends on the value of each coil, and also on the coefficient of coupling (K). This is shown in the following:

$$L_m = K\sqrt{L_1 L_2} \qquad (11\text{-}10)$$

L_m, L_1, and L_2 are each in henries.

The following example illustrates this.

Example 11-19

Find the amount of mutual inductance (L_m) present in a transformer where the primary (L_1) is 2 H, the secondary (L_2) is 8 H, and the coefficient of coupling (K) is 0.8.

Solution

$$L_m = K\sqrt{L_1 L_2} \qquad (11\text{-}10)$$

$$L_m = 0.8\sqrt{(2)(8)}$$

$$L_m = 0.8\sqrt{16}$$

$$L_m = (0.8)(4)$$

$$L_m = 3.2 \text{ H}$$

Magnetically Coupled Coils In Series. When two coils that are magnetically coupled are connected in series, the total inductance (L_T) depends on the value of each coil, the amount of mutual inductance present, and the direction of winding of each coil.

Figure 11-18*a* shows two coils connected in series and wound in the same direction. Since a current flowing through both would produce magnetic fields which add, the coils are said to be connected *series-aiding*. the total inductance L_T is

$$L_{T\ \text{aiding}} = L_1 + L_2 + 2L_m \qquad (11\text{-}11)$$

The following example illustrates this.

Example 11-20

If the primary (L_1) of the transformer of Example 11-19 is 2 H, while the secondary (L_2) is 8 H, find the total inductance (L_T) if both coils are connected in series-aiding where the mutual inductance is 3.2 H.

Solution

$$L_{T\ \text{aid}} = L_1 + L_2 + 2L_m \qquad (11\text{-}11)$$

$$L_{T\ \text{aid}} = 2 + 8 + 2(3.2)$$

$$L_{T\ \text{aid}} = 2 + 8 + 6.4$$

$$L_{T\ \text{aid}} = 16.4 \text{ H}$$

Note, of course, that $L_{T\ \text{aid}}$ is greater than just the sum of L_1 and L_2.

Figure 11-18*b* depicts two coils connected in series, but wound in *opposite* directions. Since a current flowing through both would produce opposing magnetic fields, the coils are said to be connected *series-opposing*. Here, the total inductance (L_T) is

$$L_{T\ \text{oppos}} = L_1 + L_2 - 2L_m \qquad (11\text{-}12)$$

The following example shows the use of this.

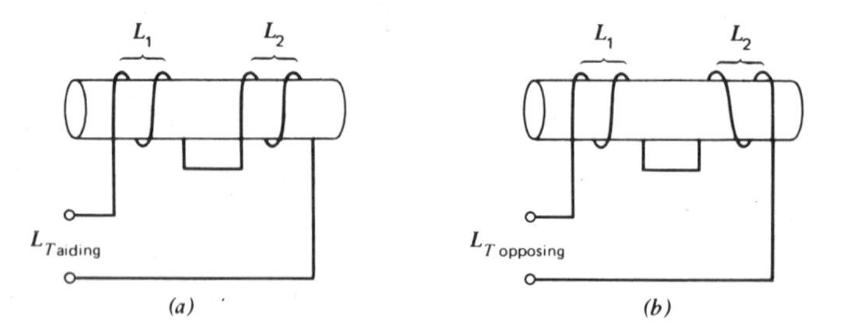

Figure 11-18. Coils in series, magnetically coupled. (*a*) Series aiding $L_{T\ \text{aiding}} = L_1 + L_2 + 2\ L_M$. (*b*) Series opposing $L_{T\ \text{opposing}} = L_1 + L_2 - 2\ L_M$.

Example 11-21

The transformer of Examples 11-19 and 11-20 consists of a primary (L_1) of 2 H, a secondary (L_2) of 8 H, and having mutual inductance of 3.2 H. Find the total inductance ($L_{T\text{ oppos}}$) if the two coils are connected series-opposing.

Solution

$$L_{T\text{ oppos}} = L_1 + L_2 - 2L_m \quad (11\text{-}12)$$

$$L_{T\text{ oppos}} = 2 + 8 - 2(3.2)$$

$$L_{T\text{ oppos}} = 2 + 8 - 6.4$$

$$L_{T\text{ oppos}} = 3.6 \text{ H}$$

Note that $L_{T\text{ oppos}}$ is *less* than the sum of L_1 and L_2.

By combining equations 11-11 and 11-12 another equation results which is useful for finding the mutual inductance (L_m) present in a transformer or between any two magnetically coupled coils. This is shown in the following:

From

$$L_{T\text{ aid}} = L_1 + L_2 + 2L_m \quad (11\text{-}11)$$

subtract

$$L_{T\text{ oppos}} = L_1 + L_2 - 2L_m \quad (11\text{-}12)$$

yielding

$$L_{T\text{ aid}} - L_{T\text{ oppos}} = L_1 - L_1 + L_2 - L_2 + 2L_m - (-2L_m)$$

which becomes

$$L_{T\text{ aid}} - L_{T\text{ oppos}} = +\ 2L_m + 2L_m$$

and finally

$$L_{T\text{ aid}} - L_{T\text{ oppos}} = 4L_m$$

and

$$\frac{L_{T\text{ aid}} - L_{T\text{ oppos}}}{4} = L_m \quad (11\text{-}13)$$

The use of equation 11-13 is illustrated in the following examples.

Example 11-22

The primary and secondary of the transformer of Examples 11-19, 11-20, and 11-21 are connected in series and measured on an inductance bridge which gives the total inductance of 3.6 H. Reversing the connections to *one* coil now produces a reading on the test instrument of 16.4 H. Find the mutual inductance which is present.

Solution

The total inductance values of 3.6 H and 16.4 H are obviously $L_{T\,\text{oppos}}$ and $L_{T\,\text{aid}}$, respectively. Therefore

$$L_m = \frac{L_{T\,\text{aid}} - L_{T\,\text{oppos}}}{4} \tag{11-13}$$

$$L_m = \frac{16.4 - 3.6}{4}$$

$$L_m = \frac{12.8}{4}$$

$$L_m = 3.2 \text{ H}$$

Note that this answer agrees with the L_m of Example 11-19, which it should, since the same transformer is being used in Examples 11-19 to 11-22.

From equation 11-10, $L_m = K\sqrt{L_1 L_2}$, the coefficient of coupling K may be found as shown in the following example.

Example 11-23

The transformer used in Examples 11-19 to 11-22 has a primary (L_1) of 2 H, and a secondary (L_2) of 8 H. If the mutual inductance (L_m) is 3.2 H, find the coefficient of coupling K.

Solution

$$L_m = K\sqrt{L_1 L_2} \tag{11-10}$$

Solving for K gives

$$\frac{L_m}{\sqrt{L_1 L_2}} = K$$

$$\frac{3.2}{\sqrt{(2)(8)}} = K$$

$$\frac{3.2}{\sqrt{16}} = K$$

$$\frac{3.2}{4} = K$$

$$0.8 = K$$

Note that this agrees with the K of Example 11-19.

When no magnetic coupling exists between two *coils in series*, then L_T is simply the *sum* of the coils, or $L_T = L_1 + L_2$, the same as resistors in series.

Coils in parallel, with no magnetic coupling, use a similar equation as parallel resistors, or $L_T = (L_1)(L_2)/(L_1 + L_2)$, or $L_T = 1/(1/L_1 + 1/L_2)$.

For Similar Problems Refer to End-of-Chapter Problems 11-47 to 11-51

11-12. Transformers (Advanced)

In Section 11-6, *The Simple Transformer*, and in Section 11-7, *Transformer Currents and Power Losses*, we point out in equation 11-2 that the *turns ratio* (primary to secondary) is equal to the *voltage ratio* (also primary to secondary). However, this is only true where the coefficient of coupling is at its maximum value of one. Similarly, in equation 11-5, it is shown that the *turns ratio* (primary to secondary) is equal to the *current ratio inverted* (that is, secondary to primary). This, too, is only true when the power losses are negligible so that the efficiency is 100%.

When the *coefficient of coupling* (K) of a transformer primary and secondary is less than one, then the voltage induced in the secondary is reduced by the lower value of K. Equation 11-2 of Section 11-6, $T_p/T_s = E_p/E_s$, must now be modified to

$$\frac{\text{turns, primary}}{\text{turns, secondary}} = \left(\frac{E_{\text{primary}}}{E_{\text{secondary}}}\right)(K) \qquad (11\text{-}14)$$

or

$$\frac{T_p}{T_s} = \left(\frac{E_p}{E_s}\right)(K) \qquad (11\text{-}14)$$

This means that if the secondary voltage would have been 360 V where K is 1, E_{sec} is reduced, where K is 0.5, to 360 times 0.5, or E_{sec} becomes 180 V. The following examples illustrate this.

Example 11-24

In Example 11-4, Section 11-6, the primary has 30 turns, the secondary 150 turns, applied voltage is 110 V ac, and the secondary voltage was found to be 550 V where K is 1. If K were only 0.8, what would E_{sec} be?

Solution

If the secondary voltage, where $K = 1$, is known to be 550 V, then with $K = 0.8$, E_{sec} would become (550) (0.8) or 440 V.

Example 11-25

A transformer has 800 primary turns, 38 secondary turns, 120 V ac applied to the primary and a coefficient of coupling (K) of 0.9. Find the secondary voltage.

Solution

By using the following equation, the reader should find that E_s is 5.13 V.

$$\frac{T_p}{T_s} = \left(\frac{E_p}{E_s}\right)(K) \tag{11-14}$$

For Similar Problems Refer to End-of-Chapter Problems 11-52 and 11-53

Tuned or Resonant Transformer Secondary. In some transformers the secondary is *tuned* or *resonant* to a particular frequency. Resonant circuits are discussed in Chapter 15. However, for the present it is sufficient to point out that a tuned secondary produces a step-up voltage at the output from the secondary circuit even though the coefficient of coupling K is much smaller than *one*. This voltage step-up is due to the nature of a resonant circuit, and will occur at a desired frequency only, producing *selectivity*. In this way, the tuner of a radio or of a television receiver can select the desired station and reject the others.

Transformer Efficiency. Power losses in transformers are described in Section 11-7. These losses show up as heat that is dissipated in the transformer. As a result of power losses, the power in the *secondary* circuit is always *less* than the power consumed from the applied voltage source in the primary. As an example, if 45 W are delivered to the secondary circuit, with 5 W being dissipated in the transformer iron core and wires, then there is 50 W (45 + 5) being consumed in the primary circuit from the applied voltage source.

If there were no power loss in the example just discussed, then with 45 W delivered to the secondary circuit, there would only be 45 W consumed in the primary circuit. With no power loss, that is, every watt consumed in the primary results in an equal number delivered to the secondary load, then the transformer is said to have an *efficiency* (η) of 100%. In actual practice, there is always some power lost or wasted in the transformer, resulting in less than 100% efficiency. The equation for efficiency in any device, including the transformer, is

$$\text{efficiency } (\eta) = \frac{P_{\text{output}}}{P_{\text{input}}} \tag{11-15}$$

where efficiency is a decimal number with a maximum value of 1; to convert this into percentage, *multiply by 100.*

For a transformer, the efficiency equation becomes:

$$\text{efficiency } (\eta) = \frac{P_{\text{sec}}}{P_{\text{pri}}} \tag{11-16}$$

The following examples illustrate the use of this equation.

Example 11-26

A transformer consumes 50 W from the applied voltage source to the primary, delivering 45 W to the secondary load circuit. What is the efficiency of the transformer?

Solution

$$\text{efficiency} = \frac{P_{\text{sec}}}{P_{\text{pri}}} \tag{11-16}$$

$$\text{efficiency} = \frac{45}{50}$$

$$\text{efficiency} = 0.9$$

The efficiency (or η) may be left as the decimal number 0.9 or, by multiplying by 100, the efficiency is expressed as 90% (0.9 × 100).

Example 11-27

If 17 W are delivered to the load in the secondary circuit, and the transformer efficiency is 85% find the power in the primary circuit.

Solution

By using the method just described, solve for the primary power, which should be found to be 20 W.

Example 11-28

A transformer is 92% efficient. If the primary consumes 60 W from the applied voltage source, how much power is delivered to the secondary load?

Solution

Solve for secondary power, which should be found to be 55.2 W.

For Similar Problems Refer to End-of-Chapter Problems 11-54 to 11-56

Transformer Currents. In Section 11-7, *Primary Current*, it is shown that with 100% efficiency the currents flowing in the primary and in the secondary depend on the turns ratio, the secondary voltage, and the load in the secondary circuit. Equation 11-3 ($E_pI_p = E_sI_s$), equation 11-4 ($E_p/E_s = I_s/I_p$), and equation 11-5 ($T_p/T_s = I_s/I_p$) are only correct for 100% efficiency, that is, when there are no losses in the transformer.

As shown in this present section, just previous to this, in *Transformer Efficiency*, the secondary power is less than the primary power, or P_{sec} = efficiency times P_{pri}. Therefore, to find the currents in a transformer, taking into account the power lost, the following procedures are recommended.

1. The secondary current is found in the usual manner by using Ohm's law $I_{sec} = E_{sec}/R_{load}$).
2. Then, the secondary power is found: $P_{sec} = E_{sec}\, I_{sec}$).
3. Next, the primary power is found using equation 11-16 (efficiency = P_{sec}/P_{pri}).
4. Finally, primary current is found: ($P_{pri} = E_{pri}\, I_{pri}$).

These procedures are followed in the next examples.

Example 11-29

A transformer has an efficiency of 90%, a coefficient of coupling of 0.85, and 120 V ac applied to a 300-turn primary. If the load resistor in the 1500-turn secondary is 50 kΩ, find: (a) E_{sec}, (b) I_{sec}, (c) P_{sec}, (d) P_{pri}, and (e) I_{pri}

Solution

$$\text{(a)} \quad \frac{T_p}{T_s} = \left(\frac{E_p}{E_s}\right)(K) \qquad \text{(11-14)}$$

$$\frac{300}{1500} = \left(\frac{120}{E_s}\right)(0.85)$$

Cross multiplying gives

$$300\ E_s = (1500)\,(120)\,(0.85)$$

$$E_s = \frac{(1500)\,(120)\,(0.85)}{300}$$

$$E_s = 510\ \text{V}$$

$$\text{(b)}\ I_{sec} = \frac{E_{sec}}{R_{sec}}$$

$$I_{sec} = \frac{510}{50\ \text{K}}$$

$$I_{sec} = \frac{5.1 \times 10^2}{5 \times 10^4}$$

$$I_{sec} = 1.02 \times 10^{-2}\ \text{A} \quad \text{or} \quad 0.0102\ \text{A} \quad \text{or} \quad 10.2\ \text{mA}$$

(c) $P_{sec} = (E_{sec})(I_{sec})$

$P_{sec} = (510)(0.0102)$

$P_{sec} = (5.1 \times 10^2)(1.02 \times 10^{-2})$

$P_{sec} = 5.2\ \text{W}$

(d) $\text{efficiency} = \dfrac{P_{sec}}{P_{pri}}$ (11-16)

$0.9 = \dfrac{5.2}{P_{pri}}$

$0.9\, P_{pri} = 5.2$

$P_{pri} = \dfrac{5.2}{0.9}$

$P_{pri} = 5.78\ \text{W}$

(e) $P_{pri} = (E_{pri})(I_{pri})$

$578 = 120\, I_{pri}$

$\dfrac{5.78}{120} = I_{pri}$

$0.0482\ \text{A} = I_{pri}$

Example 11-30

A transformer has a coefficient of coupling K of 0.833, an 80% efficiency, and 60 V ac applied to the 100-turn primary. If a 2.5 kΩ load is connected to the 10-turn secondary, find: (a) E_{sec}, (b) I_{sec}, (c) P_{sec}, (d) P_{pri}, and (e) I_{pri}.

Solution

Solve this example by using the method shown previously. As a guide, the following values should be found: $E_{sec} = 5$ V, $I_{sec} = 2$ mA, $P_{sec} = 10$ mW. $P_{pri} = 12.5$ mW, and $I_{pri} = 0.2083$ mA.

For Similar Problems Refer to End-of-Chapter Problems 11-57 to 11-59

Autotransformer. An *autotransformer* consists of one large coil with all, or some part of it, acting as the primary, and all, or some part of the coil, acting as the secondary. Figure 11-19*a* shows a coil with ends marked W and Z. The 110-V ac line voltage is applied to that part of the coil marked X and

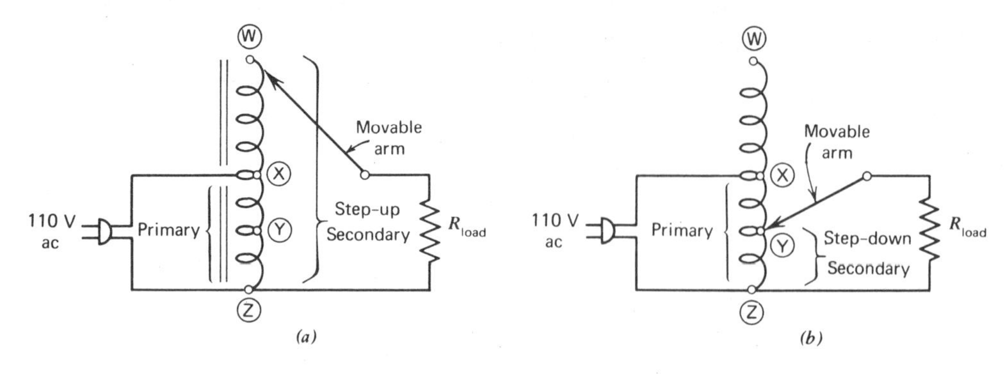

Figure 11-19. Variable autotransformer.

Z. This is the primary. The load resistor is connected in part (a) between *W* and *Z*, making the entire coil a voltage step-up secondary.

In part (b) of Fig. 11-19, the primary is still that portion between *X* and *Z*. Now, the movable arm is set at *Y*, connecting the load resistor between *Y* and *Z*, and making this a voltage step-down secondary. The movable arm could either be a switch, producing predetermined amounts of secondary voltages, or a slide, producing innumerable, continuously variable amounts of secondary voltages. Note that if the arm were set at point X, then the secondary would be the same as the primary, or a 1 : 1 turns ratio. Note also that point Z is always common to both the primary and the secondary. A conventional transformer with separate primary and secondary coils *isolates* the two circuits, while the autotransformer does not.

11-13. Transformer Impedance Matching

When a generator is connected to a load, often it is desired to produce maximum power in the load. Some examples of this are the loudspeakers of radio receivers and audio amplifiers, and the deflection coils of the picture and camera tubes of television receivers and transmitters. The source "generator" may be an electron tube or a transistor, while the load may be the voice coil of a speaker or the deflection coil of the television set. The generator has its own internal resistance, but the load usually has a different value of resistance. The term *resistance* used here is actually the *opposition* to alternating current, which is usually called *impedance*. The letter Z denotes *impedance*. Although the title of this section is "Transformer *Impedance* Matching," the terms *resistance* and *impedance* will be used here interchangeably.

When current fluctuates in the primary, the moving magnetic field induces voltage in the secondary. This secondary voltage and the load resistance determine the amount of secondary current. This varying *secondary* current causes a changing magnetic field which, in turn, induces voltage back into the primary. In other words, each coil induces voltage in the other. As a result, each coil *reflects* an impedance into the other.

Using equations 11-2 and 11-5, as shown in the following, results in a new equation that shows the *reflected impedances.*

$$\frac{T_p}{T_s} = \frac{E_p}{E_s} \qquad (11\text{-}2)$$

and

$$\frac{T_p}{T_s} = \frac{I_s}{I_p} \qquad (11\text{-}5)$$

Multiplying the above equations gives

$$\left(\frac{T_p}{T_s}\right)^2 = \left(\frac{E_p}{E_s}\right)\left(\frac{I_s}{I_p}\right) \qquad (11\text{-}17)$$

which may be rewritten as

$$\left(\frac{T_p}{T_s}\right)^2 = \left(\frac{E_p}{I_p}\right)\left(\frac{I_s}{E_s}\right) \qquad (11\text{-}17)$$

From Ohm's law, $R_p = E_p/I_p$, and $R_s = E_s/I_s$, and the reciprocal of R_s is $1/R_s = 1/(E_s/I_s)$ or $1/R_s = I_s/R_s$. Now, substitute R_p and $1/R_s$ in equation 11-17, yielding

$$\left(\frac{T_p}{T_s}\right)^2 = R_p\left(\frac{1}{R_s}\right) \qquad (11\text{-}18)$$

which may be rewritten as

$$\left(\frac{T_p}{T_s}\right)^2 = \frac{R_p}{R_s} \qquad (11\text{-}18)$$

where R_p is the *resistance* or *impedance* in the primary circuit, and R_s is the *resistance* or *impedance* in the secondary circuit.

Solving equation 11-18 in terms of the turns ratio by taking the square root of both sides of the equation gives

$$\frac{T_p}{T_s} = \sqrt{\frac{R_p}{R_s}} \qquad (11\text{-}19)$$

Maximum power is developed in a load when the load has a resistance equal to that of the generator. In the case of a generator (electron tube or transistor) with a much higher internal resistance than the load, a transformer with the correct turns ratio acts as the *impedance matching* device. That is, the high impedance generator "looks" into the *primary* and "sees" an R_p (in equation 11-18) of the correct desired high value. Similarly, the low impedance load looks into the *secondary* and sees an R_s (in equation 11-8) of the correct desired low value. In this way, the necessary maximum power is developed in the load. The following examples illustrate the use of equations 11-18 and 11-19.

Example 11-31

It is recommended that an electron tube be connected to a load of 15 kΩ for best operation. The actual load however is a 6-Ω voice coil of a speaker. What turns ratio should a transformer have to achieve best results? The electron tube is connected to the primary, and the speaker is in the secondary circuit.

Solution

R_p is the recommended value of resistance in the primary (15 K) that the tube should see. R_s is the value of the actual load (6 Ω) in the secondary.

$$\frac{T_p}{T_s} = \sqrt{\frac{R_p}{R_s}} \qquad (11\text{-}19)$$

$$\frac{T_p}{T_s} = \sqrt{\frac{15000}{6}}$$

$$\frac{T_p}{T_s} = \sqrt{2500}$$

$$\frac{T_p}{T_s} = 50$$

This means that the *turns ratio* (primary to secondary) should be 50:1, or 50 to 1.

Example 11-32

How much resistance does an electron tube "see," "looking" into the primary (R_p) of a 50:1 turns ratio transformer, if a 6-Ω load is connected to the secondary?

Solution

The reader should solve for R_p using equation 11-18, and should find that R_p is 15 kΩ.

Example 11-33

How much resistance does a loudspeaker see, looking back into the secondary (R_s) of a 50:1 turns ratio transformer, where the primary circuit (R_p) is 15,000 Ω?

Solution

Again, the reader should solve this example, this time for R_s in equation 11-18, and should find that R_s is 6 Ω.

For Similar Probelems Refer to End-of-Chapter Problems 11-60 to 11-62

11-14. Simple Rotating Generators

In Section 11-2 the simple ac generator is discussed. In Fig. 11-20 further details of the ac generator are shown. Here, the straight wire of Figs. 11-3 and 11-5 is formed into a single-loop coil. The coil is being rotated clockwise. When wire X is being moved down, the other end of the coil, wire Y, is moving upward. Each end, X and Y, is connected to a separate metal ring, called a slip-ring. The slip rings, also shown as X and Y, are mounted on the same mechanical axis as the coil and, therefore, rotate along with the coil. Two stationary carbon brushes make a sliding contact with the rotating slip rings. The brushes are shown as *A* and *B* in Fig. 11-20. A load, which, of course, is stationary, is connected between the brushes.

In Fig. 11-20*a*, the coil is shown being rotated with wire X moving down, and wire Y moving up in the magnetic field. As shown previously in Fig. 11-3, and using the left-hand three-finger generator rule, the wire moving *up* has electrons coming *out* of it. This is wire Y in Fig. 11-20*a*, shown as *negative*. At this instant, wire X is moving *down* and has electrons flowing *into* it, and is shown as *positive*. Brush B connects to the Y slip ring and is negative, while

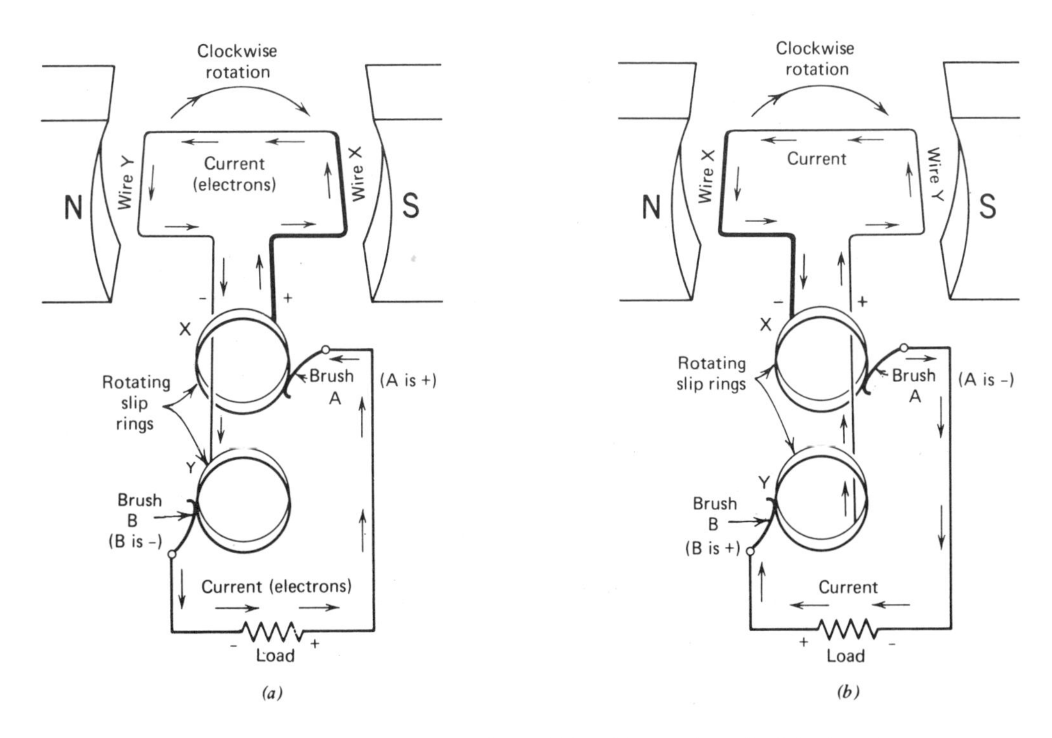

Figure 11-20. An ac rotating generator. (*a*) Wire *X* is +, brush *A* is +, wire *Y* is −, brush *B* is −. (*b*) Wire *Y* is now +, brush *B* is +, wire *X* is now −, brush *A* is −.

brush A connects to the X slip ring and is positive. Electron flow is from brush B to A, left to right through the load resistor.

In Fig. 11-20*b*, the coil has rotated 180°. Now wire X is moving *upward* and has electrons coming out of it. Wire X is now the *negative* end. Wire Y is moving *downward*, and it is now the *positive* end. Brush A, connecting to X, is negative, while brush B, connecting to Y, is now positive. As shown, electron flow is now from brush A to B, or right to left through the load. This is the reverse of the previous condition in Fig. 11-20*a* and is, therefore, an ac current and voltage.

DC Generator. By simply changing the slip rings of the ac generator of Fig. 11-20 to a single split ring called a *commutator*, as shown in Fig. 11-21, the output voltage between brushes A and B is changed into a dc voltage. The commutator is a single ring split into two halves. Brush B in Fig. 11-21*a* first connects to wire Y through its half slip ring, while wire Y is being moved upward in the magnetic field. Wire Y is *negative* at this time, as shown, with electrons coming out of its end, and therefore *brush B is negative.*

When the coil has rotated 180°, as seen in Fig. 11-21*b*, wire X is now moving upward, and it is now the negative end of the coil. Stationary brush B now con-

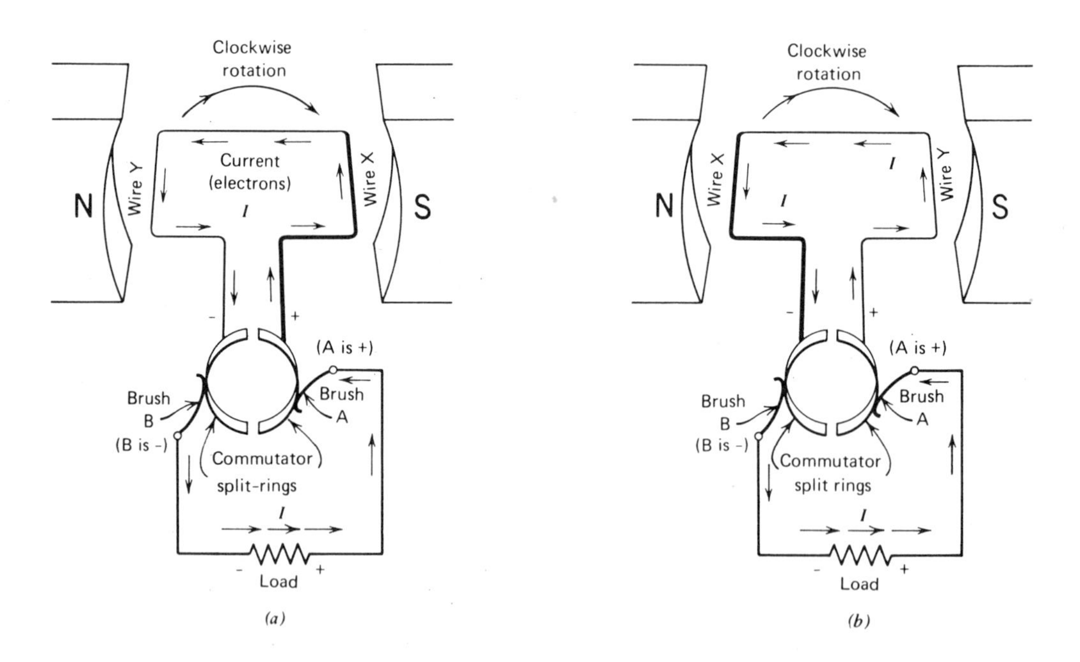

Figure 11-21. A dc rotating generator. (*a*) Wire *X* is +, brush *A* is +, wire *Y* is −, brush *B* is −. (*b*) Wire *X* is now a −, brush *B* is −, wire *Y* is now +, brush *A* is +.

tacts wire X through its half ring, which has now rotated so that it makes contact with this same brush. As a result, *brush B* is again the *negative* one. Note that *brush A* in Fig. 11-21*a* and *b* always contacts the wire that is the *positive* one, X in diagram (a), and then Y in part (b). Electron flow is *always* from brush B to A, or left to right through the load in both Fig. 11-21*a* and *b*. This is a dc current, and the voltage between the brushes is a fluctuating or varying dc voltage.

Refer to End-of-Chapter Problems 11-63 to 11-66

11-15. AC Sine Waves

In Section 11-2 it is shown that when a wire, or a coil, is rotated in a magnetic field, the voltage induced in the conductor has a sine-wave graph (Fig. 11-6, and repeated here as Fig. 11-22).

Frequency. One complete sine-wave induced voltage, shown in Fig. 11-22, is produced by one complete rotation (360°) of a coil in a magnetic field. If the coil is being rotated 100 times per second, then 100 sine waves are being produced in this time of one second. The 100-per-second is called the *frequency* of the sine waves. Frequency is measured in units called *hertz* (*Hz*), where a Hz is one cycle per second. The frequency of the previous sine wave is then 100 Hz. For higher frequencies the prefixes (kilo, mega, giga) are usually used.

The following example illustrates this.

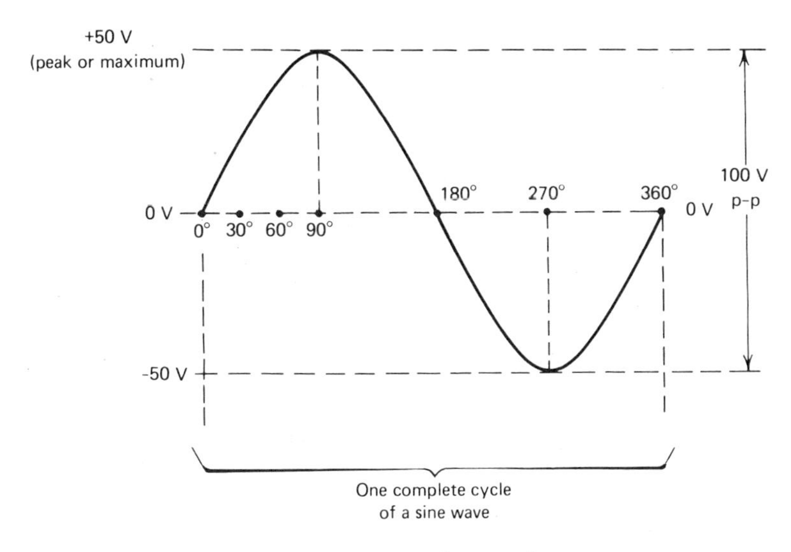

Figure 11-22. Sine wave, one complete cycle.

Example 11-34

Express the following frequencies in more convenient forms: (a) 20,000 Hz, (b) 1,600,000 Hz, (c) 66,000,000 Hz, and (d) 30,000,000,000 Hz.

Solution

(a) 20,000 Hz is a high *audio frequency* (af) and may be expressed as 20 kHz, or as 20×10^3 Hz.

(b) 1,600,000 Hz is a *radio frequency* (rf), being near the upper end of the *amplitude modulated* (AM) broadcast radio band. It may also be written as: 1600 kHz, or 1.6 MHz, or 1.6×10^6 Hz.

(c) 66,000,000 Hz is also called radio frequency (or rf) and is near the lower end of the television band. It may be expressed at 66 MHz, or as 66×10^6 Hz.

(d) 30,000,000,000 Hz is also referred to as radio frequency and is in the *superhigh frequency* (shf) region used in radar. It should be written for convenience as: 30,000 MHz, or 30 GHz, or as 30×10^9 Hz.

For a Similar Problem Refer to the End-of-Chapter Problem 11-67

Period. The *time duration* or *period* of a sine wave depends on the frequency, one being the *reciprocal* of the other.

$$\text{time or period} = \frac{1}{\text{frequency}} \tag{11-20}$$

or

$$T = \frac{1}{F} \tag{11-20}$$

and also

$$F = \frac{1}{T} \tag{11-21}$$

where F is in *hertz* or *cycles per second*, and T is in *seconds*.

In Fig. 11-23*a*, the frequency of the sine wave is 100 Hz, and the time of 1 cycle is 1/100 of a second, or 0.01 sec. In Fig. 11-23*b* the frequency is 200 Hz, and the time of a cycle is the reciprocal (equation 11-20) or 1/200 of a second, or 0.005 sec, or 5 msec.

Example 11-35

What is the *time* or *period* of a sine wave if its frequency is 25 MHz?

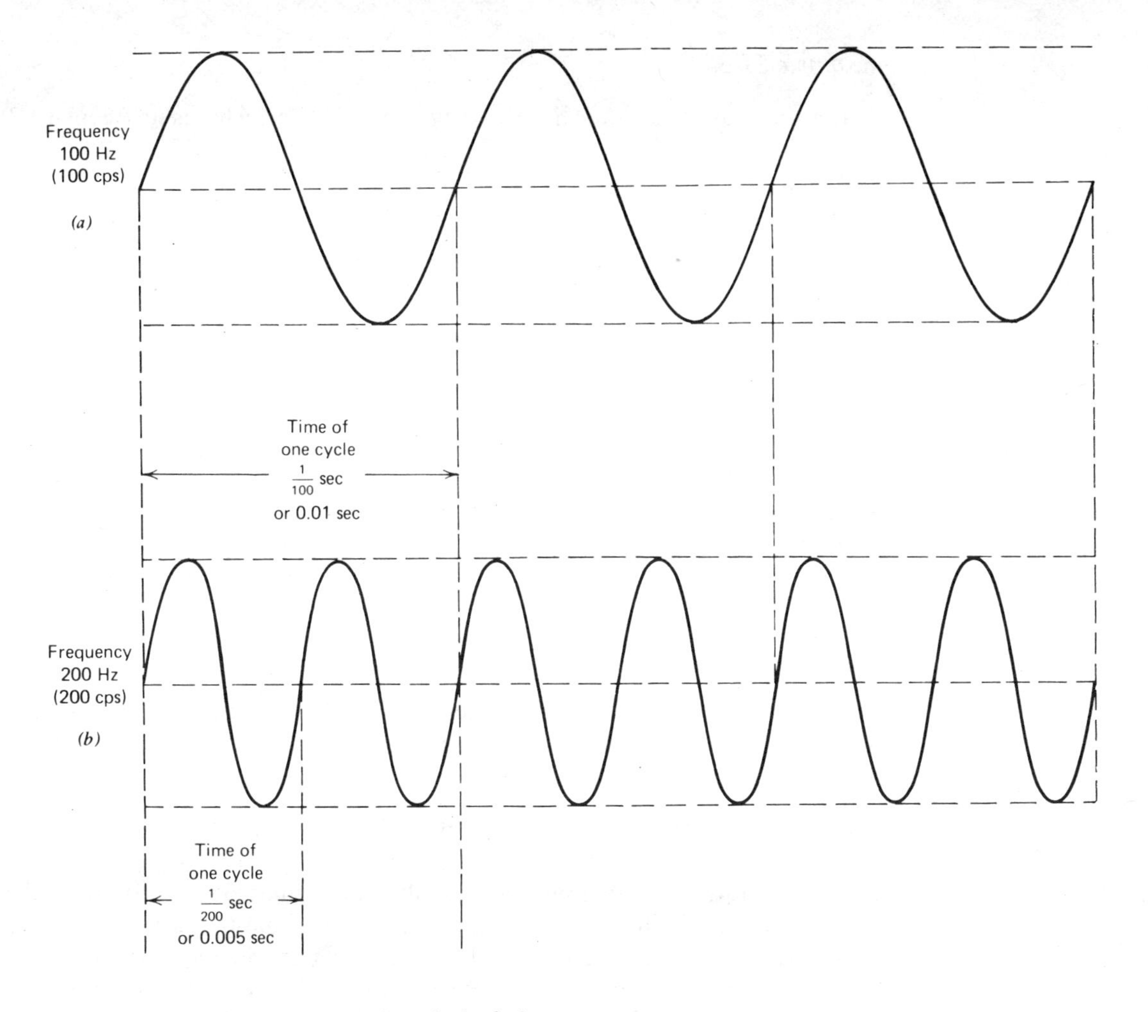

Figure 11-23. Frequency and period of sine waves.

Solution

$$T = \frac{1}{F} \tag{11-20}$$

$$T = \frac{1}{25 \text{ MHz}}$$

$$T = \frac{1}{25 \times 10^6}$$

$$T = 0.04 \times 10^{-6} \text{ sec} \quad \text{or} \quad 0.04\ \mu\text{sec}$$

Example 11-36

If a sine wave has a *time duration* or *period* of 5 μsec, what is its frequency?

Solution

$$F = \frac{1}{T} \tag{11-21}$$

$$F = \frac{1}{5\ \mu\text{sec}}$$

$$F = \frac{1}{5 \times 10^{-6}\ \text{sec}}$$

$$F = 0.2 \times 10^{6}\ \text{Hz} \quad \text{or} \quad 200{,}000\ \text{Hz} \quad \text{or} \quad 0.2\ \text{MHz}$$

Example 11-37

What is the frequency of a sine wave voltage if its time period is 20 *n*sec?

Solution

A *nanosecond* is 10^{-9} sec. The reader should solve for frequency by using the method shown previously, and he should find that F should be 50 MHz.

For Similar Problems Refer to End-of-Chapter Problems 11-68 to 11-71

Wavelength. Radio frequency signals are transmitted as variations of the electromagnetic field. A 1000-kHz (or 1 MHz) radio-frequency sine-wave signal is sent out from the transmitter as variations in the magnetic field, which changes in strength as the sine-wave amplitude becomes larger and smaller. The magnetic field travels through free space at the same speed or *velocity* as a light wave (also an electromagnetic field). This is approximately *186,000 miles per second* or *299,790 kilometers per second.* The approximate velocity is usually taken as 300,000 kilometers per second or 300,000,000 meters per second (300×10^{6} m/sec).

The *distance* that a radio-frequency sine-wave signal can travel in the *time period of one cycle* is called its *wavelength* (λ). It is also the horizontal *distance* from any point on 1 cycle to a corresponding point on the next cycle. This *wavelength* (λ) depends on the *velocity* and frequency, or

$$\text{wavelength} = \frac{\text{velocity}}{\text{frequency}} \tag{11-22}$$

$$\text{wavelength} = \frac{V}{F} \tag{11-22}$$

and also

$$F = \frac{V}{\lambda} \qquad (11\text{-}23)$$

where λ is in *meters*: *velocity* is 300×10^6 m/sec; and *frequency* is in Hz.
The following example illustrates the above discussion.

Example 11-38

What is the wavelength of a 1500-kHz AM radio broadcast transmitter?

Solution

$$\lambda = \frac{V}{F} \qquad (11\text{-}22)$$

$$\lambda = \frac{300 \times 10^6 \text{ m/sec}}{1500 \text{ kHz}}$$

$$\lambda = \frac{300 \times 10^6}{1500 \times 10^3}$$

$$\lambda = \frac{300 \times 10^6}{1.5 \times 10^6}$$

$$\lambda = 200 \text{ m}$$

For Similar Problems Refer to End-of-Chapter Problems 11-72 to 11-73

Instantaneous Values of a Sine Wave. The values or amplitudes of a sine wave vary from zero to its maximum value. The value at any instant depends on the *maximum* value and the *angle*, where the complete sine wave is 360°, one half sine wave is 180°, one quarter sine wave is 90°, and so on. This is shown in Figs. 11-22 and 11-24. The value of voltage at any *instant* (expressed as an angle) may be calculated from the following.

$$e_{\text{at angle } \theta} = (E_{\text{max}}) (\text{sin of angle } \theta) \qquad (11\text{-}24)$$

where the *sine* (abbreviated *sin*) of *any* angle (θ) is given in the trigonometric tables, or on the slide rule.
The following examples illustrate this, using Fig. 11-24.

Example 11-39

If the maximum value of voltage is 50 V (Fig. 11-24), find the instantaneous voltage of this sine wave at 30°.

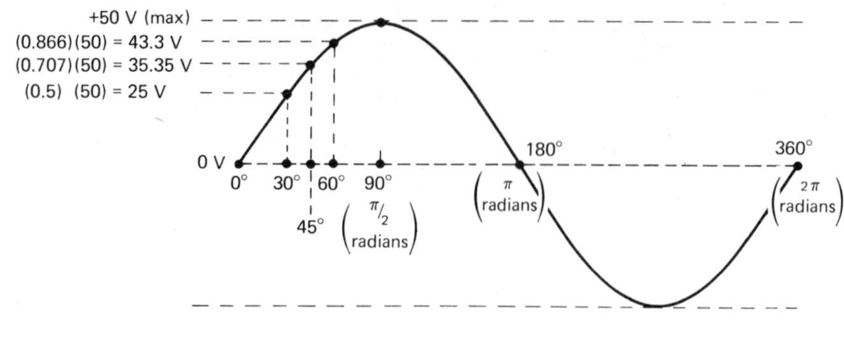

$e_{30°} = (\sin 30°)\,(E_{max})$ $\quad e_{45°} = (\sin 45°)\,(E_{max})$ $\quad e_{60°} = (\sin 60°)\,(E_{max})$

$e_{30°} = (0.5)\,(50)$ $\quad e_{45°} = (0.707)\,(50)$ $\quad e_{60°} = (0.866)\,(50)$

$e_{30°} = 25\text{ V}$ $\quad e_{45°} = 35.35\text{ V}$ $\quad e_{60°} = 43.3\text{ V}$

Figure 11-24. Instantaneous values of a sine wave.

Solution

From the trigonometric table, the sin of 30° is 0.5.

$$e_{30°} = (E_{max})\,(\sin 30°) \tag{11-24}$$

$$e_{30°} = (50)\,(0.5)$$

$$e_{30°} = 25\text{ V}$$

as shown in Fig. 11-24. Also shown are the values at 45° and 60°.

Example 11-40

Find the maximum value of a sine wave if the instantaneous value at 50° is 153.2 V.

Solution

Sin 50° is 0.766, from the trigonometric table

$$e_{50°} = (E_{max})\,(\sin 50°) \tag{11-24}$$

$$153.2 = (E_{max})\,(0.766)$$

$$\frac{153.2}{0.766} = E_{max}$$

$$200\text{ V} = E_{max}$$

Sometimes the *angular measurement* is not given, but instead a small period of time is stated after the sine wave has begun, as well as the frequency. From this, the actual angle can then be found. For example, if the frequency is

1 kHz, and it is desired to find the angle 100 μsec after the start of the cycle, then the time of 1 cycle must first be found.

$$T = \frac{1}{F} \tag{11-20}$$

$$T = \frac{1}{1 \text{ kHz}}$$

$$T = \frac{1}{1 \times 10^3 \text{ Hz}}$$

$$T = 1 \times 10^{-3} \text{ sec} \qquad \text{or} \qquad 0.001 \text{ sec} \qquad \text{or} \qquad 1000\ \mu\text{sec}$$

Since one complete sine wave contains 360°, then the short period of 100 μsec is equivalent to the following in degrees.

$$\text{degrees} = \left(\frac{100\ \mu\text{sec}}{1000\ \mu\text{sec}}\right)(360°)$$

$$\text{degrees} = \left(\frac{100}{1000}\right)(360)$$

$$\text{degrees} = \left(\frac{1}{10}\right)(360)$$

$$\text{degrees} = 36$$

This is a similar example:

Example 11-41

A 50-kHz sine wave has a peak or maximum value of 100 V. Find: (a) the angle, and (b) the voltage, 4 μsec after the start of the cycle.

Solution

The reader should solve this example using the method shown previously. He should find the following results: time = 20 μsec, angle = 72°, voltage = 95.1 V.

For Similar Problems Refer to End-of-Chapter Problems 11-74 to 11-77

Radian Measurement. Often, the rotation of a coil in a magnetic field or the angular measurement of points on the resultant sine wave may be described in another method other than by angles or in time. This other method is called *radian measurement.* Figure 11-25 shows a circle with a radius of length *OA*, and another radius *OB*. When the two radii are at an angle of 57.3° to each other, the arc *AB* is equal in length to the radius. The 57.3° angle is called a *radian.*

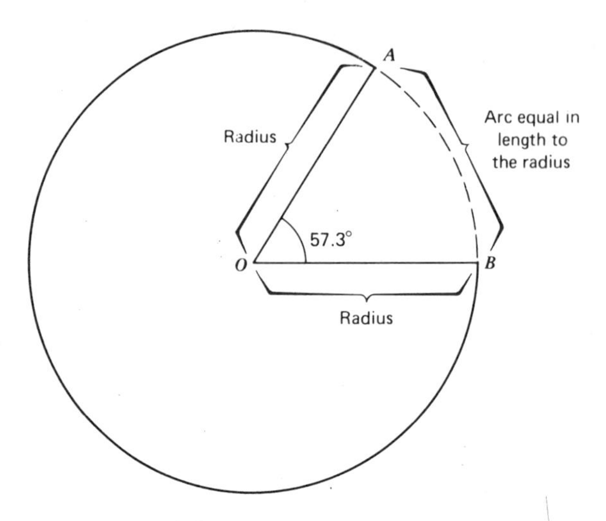

Figure 11-25. Circumference (360°) = 2 π radius; 57.3° = 1 radian.

The figure of 57.3° is derived from the fact that the complete circle contains 360°, and the *circumference* length is $2\pi R$, where $\pi = 3.14$ and R is the length of the radius, and also the length of the arc (AB of Fig. 11-25). Therefore, the 360° contains 2π radians, and a radian is found from:

$$2\pi \text{ radian} = 360°$$

$$(2)\,(3.14)\,(\text{radian}) = 360°$$

$$\text{radian} = \frac{360}{(2)\,(3.14)}$$

$$\text{radian} = 57.3°$$

As shown in the sine wave of Fig. 11-24, the complete cycle of 360° is marked *2π radians*. The half cycle, or 180°, is indicated as *π radians*, and the quarter cycle, or 90°, is shown as *$\pi/2$ radians*.

11-16. Average Value of Sine Wave

In Section 11-8, the *maximum* or *peak* value, and the *peak-to-peak* value, and the *effective* or *rms* value of a sine wave are discussed. The sine wave also has another value called its *average*. For a complete ac sine-wave voltage, going *above* the zero axis (meaning that the voltage has a certain polarity), and then going *below* the zero axis (meaning that the voltage has now reversed polarity), the average is on the zero axis itself. The *average* value is always at the point on the wave shape where a horizontal line drawn through this point has equal *areas* above and below this line. Note that in Fig. 11-26*a*, the average value for the ac sine wave is at zero, and that the shaded areas, A and B (above and below this average line), are equal.

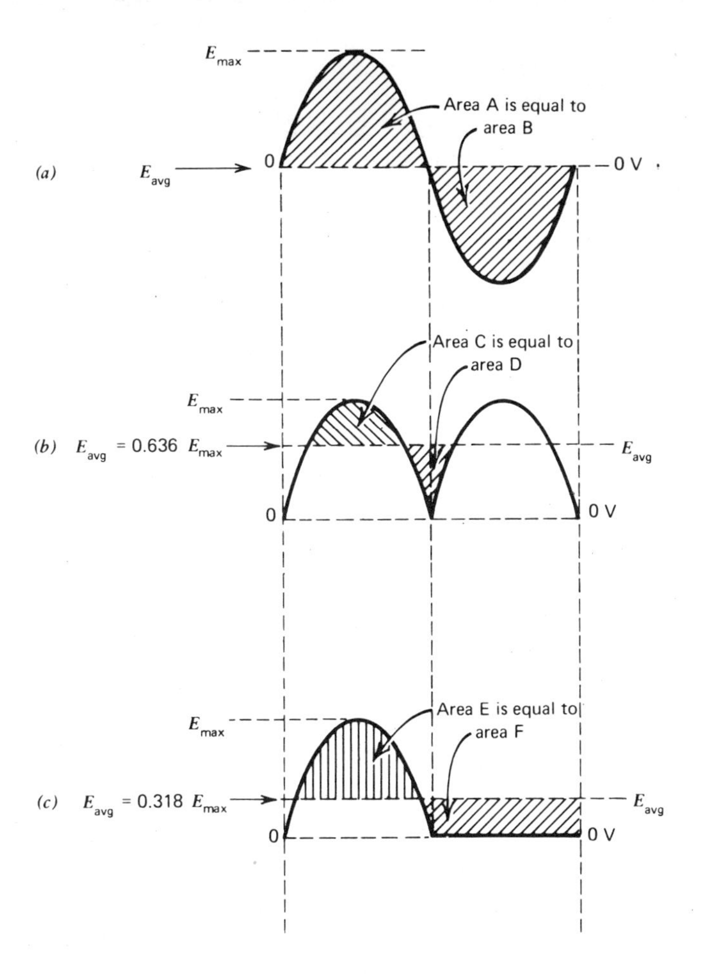

Figure 11-26. Average voltages.

Consecutive Half-Sine Waves. In Fig. 11-26*b*, two consecutive half-sine waves are shown. (This is produced at the output of a *diode full-wave rectifier*, discussed in a later chapter.) The *average* value of this is shown as *0.636 of the maximum* value. Note that the area (C) above this average line is equal to the area (D) below this line. The *average* is also called the *dc component*, or that which a dc meter would indicate.

The average value of these half-sine waves (or pulsating direct current) is derived simply by adding up a number of instantaneous values on the half cycle (three of these are shown in Fig. 11-24), and then dividing this sum by the number of values chosen, for the *average*. This, for a half-sine wave, is

$$E_{av} = 0.636\ E_{max} \qquad (11\text{-}24)$$

Example 11-42

What is the *average* voltage (the reading of a dc voltmeter) of a series of consecutive half-sine waves if the *peak* or *maximum* value is 70 V?

Solution

$$E_{av} = 0.636\ E_{max} \tag{11-24}$$

$$E_{av} = (0.636)(70)$$

$$E_{av} = 44.5\ \text{V}$$

Example 11-43

A dc voltmeter reads 57.2 V for a series of consecutive half-sine waves such as is shown in Fig. 11-26*b*. What is the *maximum* value of the sine wave?

Solution

The dc voltmeter indicates the *average* value of the half-sine wave.

$$E_{av} = 0.636\ E_{max} \tag{11-24}$$

$$57.2 = 0.636\ E_{max}$$

$$\frac{57.2}{0.636} = E_{max}$$

$$90\ \text{V} = E_{max}$$

For Similar Problems Refer to End-of-Chapter Problems 11-78 and 11-79

Alternate Half-Sine Waves. Figure 11-26*c* shows a half-sine wave followed by zero volts. (This is produced at the output of a *diode half-wave rectifier*, discussed in a later chapter.) As shown in the diagram, the *average* value is *0.318 of the maxium*. This is derived since the average of the *first* half cycle, which is *0.636* E_{max}, then becomes zero volts during the next half-cycle period. The complete average for the entire period is then one half of 0.636, or

$$E_{\text{average alternate half-sine wave}} = 0.318\ E_{max} \tag{11-25}$$

Example 11-44

If the maximum voltage is 80 V in Fig. 11-26*c* for alternate half-sine waves, what would a dc voltmeter read?

Solution

The dc voltmeter indicates the average voltage.

$$E_{\text{av alt half cycles}} = 0.318\ E_{\max} \tag{11-25}$$

$$E_{\text{av alt half cycles}} = (0.318)\ (80)$$

$$E_{\text{av alt half cycles}} = 25.4\ \text{V}$$

For Similar Problem See End-of-Chapter Problem 11-80

DC Sine Waves. Actually there is no such thing as a *dc sine wave* since a sine wave is always ac. However, it is possible to add a small ac sine wave onto a larger value of dc voltage. This is commonly found in electron tube and transistor amplifier circuits. The result of superimposing an ac voltage onto a dc voltage is a *varying* or *fluctuating* dc voltage. This is the *signal* voltage at the *plate* of an amplifier tube and at the *collector* of a transistor amplifier. This is discussed in a later chapter.

Figure 11-27*a* shows a 100-V dc voltage connected in series with a 10-V peak ac voltage. The result is a varying dc voltage shown in Fig. 11-27*b*. When the ac voltage is zero, the total voltage is 100 V. When the ac voltage is 10-V peak with the same polarity as the 100 V dc, the two voltages add, producing a total of 110 V. Finally, when the 10 V peak ac is opposing the polarity of the 100 V dc, the two voltages subtract, producing a total of 90 V.

The total voltage shown in Fig. 11-27*b* is a varying dc voltage fluctuating between 110 and 90 V. Note that the *average* value is at the center of the sine wave, at 100 V, where the shaded area above this line is equal to the area below the line. The 100 V is the dc component, and the 10-V peak (from 100

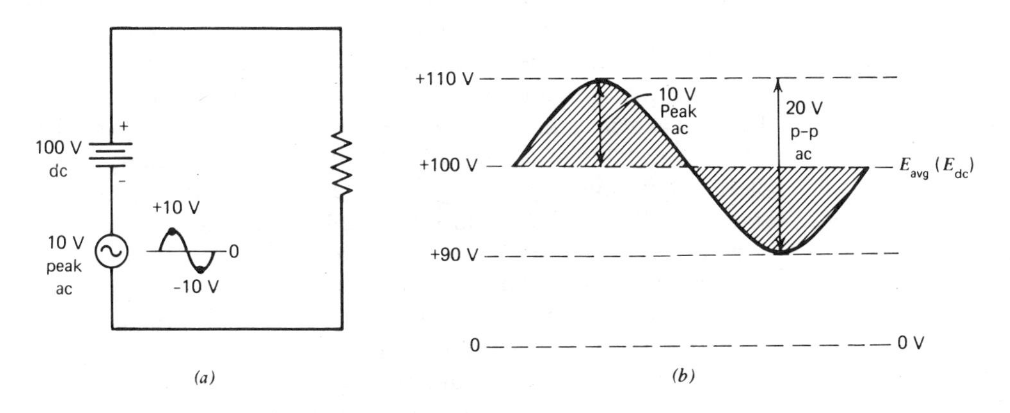

Figure 11-27. Sine wave varying dc voltage.

to 110 V, or from 100 to 90 V) is the maximum value of the ac component (or 20 V peak-to-peak).

The following example also illustrates this.

Example 11-45

If a sine-wave fluctuating dc voltage at the collector of a transistor varies between +9 and +5 V, what is its (a) dc or average voltage, and (b) peak ac voltage?

Solution

The average voltage or the dc component is the center of this sine-wave where there are equal areas above and below. This average is at +7 V, since this is 2 V below the upper limit of +9 V, and also 2 V above the lower limit of +5 V. This *+7-V average* can also be found from the fact that the dc voltage goes from +9 to +5, having a difference or *peak-to-peak ac value of 4 V*. The peak alternating current is then half of the peak-to-peak value, or the *peak* is $1/2 \times 4$, or *2 V*. The average dc is then the peak of 2 V either *added* to the +5 V *minimum*, or *subtracted* from the +9-V *maximum*, giving a +7-V average.

For Similar Problems Refer to End-of-Chapter Problem 11-81

11-17. AC Voltmeters

The *permanent-magnet moving-coil* meter (also called the *D'Arsonval* or the *Weston*), described in Section 10-1 and shown in Fig. 10-1, is a dc instrument. If alternating current were applied to it, the rapid polarity reversals would result in the moving coil and its pointer attempting to quickly rotate first in one direction, and then in the other. The weight and inertia of the coil and pointer would prevent this, with the actual result being no movement at all. The *dynamometer* described in Section 10-9, and shown in Figs. 10-20 and 10-21, is capable of ac operation as well as dc. The rapid polarity reversals of the alternating current simply reverse the north and south poles of the stationary field coils (L_1 and L_2) as well as the magnetic poles of the moving-coil (L_3). As a result, L_3 and its pointer deflect clockwise regardless of the polarity of the applied direct current, and do the same for alternating current. The dynamometer type is useful for alternating current which is either sinusoidal (sine wave) or nonsinusoidal.

The *permanent-magnet moving-coil* dc meter used with a *rectifier* is capable of ac operation. The rectifier, either a diode electron tube, or a diode semiconductor (selenium or germanium), or a copper-oxide type, changes the applied ac input into the dc output which the meter requires. The rectifier type is not accurate for nonsine waves.

An entirely different type of meter for ac operation is called the *moving iron-vane* meter. The alternating current is applied to a coil inside of which are two soft-iron plates. One is stationary, while the other, held lightly by a spring, is movable. The coil becomes an electromagnet due to the current flowing through it. Both iron plates inside the coil become magnetized with identical north and south poles, and as a result, repel each other. The rotation of the movable iron vane is mechanically coupled to the meter pointer, which deflects along the calibrated meter scale.

Two other ac meter types depend on *heat* to produce a meter pointer deflection. In one type, called the *hot-wire ammeter*, the alternating current flows through a silver-platinium wire, heating the wire. The movement caused by the expansion of the wire is mechanically transferred to the meter pointer, causing it to deflect along the calibrated meter scale.

A second type, also employing heat to produce a meter reading, is called the *thermocouple* meter. Two dissimilar metal strips such as *manganin* and *constantan* are joined together at one end. When this end is heated, a dc voltage is produced between the other open ends. The resultant dc voltage is connected to a conventional dc-type meter. The alternating current flows through a small *heater* resistor which heats up the thermocouple, producing an output dc voltage that is directly proportional to the amount of heat. This type is extremely useful at high frequency alternating current.

In both the *hot-wire* and *thermocouple* types, meter pointer deflection depends on the heat produced by the current flow. This is due to the power dissipation or I^2R. As a result, meter pointer deflection is proportional to the *square* of the current, following the *square law* principle. Meter scale is, therefore, not linear, with "crowding" of the numbers at the low end and "spreading" of the numbers at the high end of the calibrated scale.

All ac meter scales are calibrated as *effective* or *rms* values unless otherwise indicated. *Maximum*, *average*, or *peak-to-peak* values must be so marked.

Refer to End-of-Chapter Problems 11-82 to 11-84

PROBLEMS

See section 11-1 for discussion covered by the following.

11-1. When a magnetic field is stationary with respect to a coil, what is the voltage and the current?

11-2. What occurs when a magnetic field moves across the wires of a coil?

11-3. What takes place when the magnetic field cuts across the wires of a coil in the opposite direction?

11-4. If the bar magnet is moved away from the coil, as shown in the drawing for this problem: (a) Which end of the coil becomes its north pole, and (b) which end of the coil becomes the negative polarity?

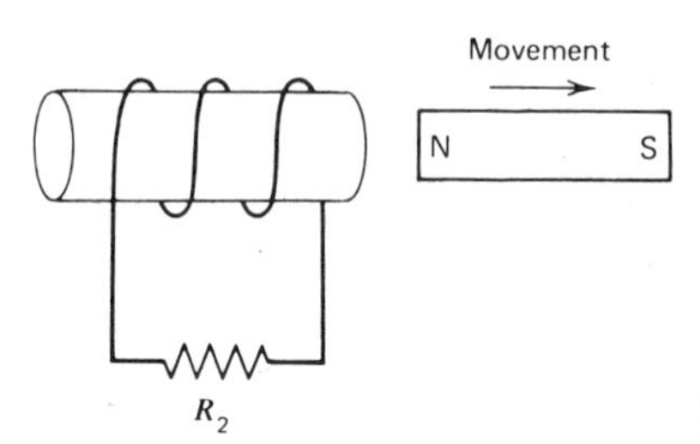

Problem 11-4.

11-5. What happens to the *amplitude* of the induced voltage if the bar magnet is moved more slowly toward a coil (see *Faraday's law*)?

11-6. What happens to the *amplitude* of the induced voltage if a stronger bar magnet is used in Fig. 11-1?

11-7. What happens to the *amplitude* of the induced voltage if the coil of Fig. 11-1 has more turns?

See section 11-2 for discussion covered by the following.

11-8. Refer to Example 11-2 and its diagram. How would the voltage induced in the coil compare with that of Example 11-2, if the wire were being moved (a) from position Z to Y, and (b) position Y to X, and (c) position X to W? (See Fig. 11-7.)

11-9. In the diagram of Fig. 11-3, if the north and south magnetic poles were reversed, and the wire were still moved upward, in which direction would electrons flow in the wire?

11-10. In Fig. 11-3, in which direction would electrons flow in the wire if the wire were being moved downward, and the north and south magnetic poles were reversed?

See section 11-3 for discussion covered by the following.

11-11. What is the property of an inductance?

11-12. What is the unit of inductance?

See section 11-4 for discussion covered by the following.

11-13. If current through a coil increases, describe what occurs.
11-14. What is the property of a conductor called when a voltage is induced in it due to a *change* of current?
11-15. How much inductance is present if 1 V is induced in a coil when the current goes from 0 to 1 A in 1 sec?

See section 11-5 for discussion covered by the following.

11-16. What is the name of the property of a circuit that describes a voltage being induced in one coil due to the moving magnetic field of another coil?
11-17. What is meant by saying that two coils are *magnetically coupled*?
11-18. What does a coefficient of coupling of 1, called unity coupling, mean?
11-19. Name four things that can be done to increase the coupling between two coils.

See section 11-6 for discussion covered by the following.

11-20. What is the turns ratio of a transformer which has 300 turns of wire in the primary, and 2100 turns in the secondary?
11-21. What voltage is induced in the secondary of an iron-core transformer where 120 V ac is applied to a 170-turn primary, while the secondary consists of 850 turns?
11-22. If 120 V ac were applied to an iron-core transformer primary consisting of 168 turns, what would the secondary voltage become if it had only 7 turns?

See section 11-6 for discussion covered by the following.

11-23. Why will a steady direct current applied to the primary of a transformer not cause any secondary induced voltage?
11-24. What kind of voltage would be induced in the secondary coil if: (a) Varying direct current were applied to the primary? (b) Alternating current were applied to the primary?

See section 11-7 for discussion covered by the following.

11-25. In the diagram for this problem, assume that the transformer has a coefficient of coupling of one. Find the secondary voltage.

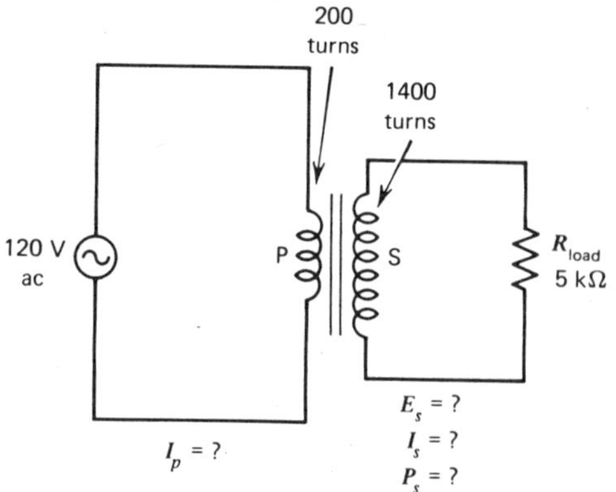

Problems 11-25 to 11-28.

11-26. In the diagram shown for this problem, and using the information found above, find the secondary current.

11-27. In the diagram shown for this problem and using the previously found answers, find the secondary power.

11-28. In the diagram shown for this problem, assume negligible power losses. Using the previously found answers, find the primary current.

11-29. Name three power losses that occur in an iron-core transformer.

11-30. (a) Describe eddy current loss, and (b) how it is kept to a minimum.

11-31. (a) What is hysteresis loss, and (b) how is it minimized?

11-32. (a) What is copper loss, and (b) how is it minimized?

See section 11-8 for discussion covered by the following problems.

11-33. A sine-wave ac voltage has a peak or maximum value of 200 V. What is its effective value?

11-34. (a) What is meant by the effective value of an ac voltage, and (b) the rms voltage?

11-35. What does an ac voltmeter usually read?

11-36. What is the peak-to-peak value of an ac sine-wave voltage if its peak is 75 V?

11-37. What is the maximum value of a sine-wave ac voltage if its peak-to-peak is 90 V?

11-38. An ac sine-wave voltage has a peak-to-peak of 20 V. What is its effective or rms value?

See latter part of section 11-8 for discussion covered by the following.

11-39. What is the peak or maximum value of an ac sine-wave voltage of 40 V rms?

11-40. An ac voltmeter reads 300 V across the secondary of a transformer. Find: (a) E_{max}, and (b) $E_{p\text{-}p}$.

More Advanced Problems: See section 11-9 for discussion covered by the following.

11-41. Find the value of inductance where 3000 turns of wire are wound on a square iron bar 10 cm long by 1 square cm. The coil itself is 7 cm long, and the iron has a μ of 150.
11-42. In the previous problem, if the iron bar were withdrawn from the 3000-turn coil, leaving it still in its original square shape, find the new value of inductance.
11-43. If 7000 turns of wire were wound on a doughnut-shaped iron core as a *toroid* (see Fig. 11-16), find the value of inductance where the outer diameter of the ring is 4 cm, the inner diameter is 3 cm, and the μ of the iron is 100.

See section 11-10 for discussion covered by the following.

11-44. Find the amount of voltage induced in a 500 μH coil if the current in it increases from 2 to 2.5 mA in 25 μsec.
11-45. If the current in the 500 μH coil of the previous problem now *decreases* from 2.5 mA to zero in 5 μsec, find the amount of voltage induced in this coil.
11-46. Compare the *polarity* of the induced voltage of Problem 11-44, when the current *increased*, to the voltage induced in Problem 11-45, when the current *decreased.*

See section 11-11 for discussion covered by the following.

11-47. Two coils are magnetically coupled and are connected in series. If one coil is 9 H, the second is 4 H, and the coefficient of coupling K is 0.9, find the amount of mutual inductance L_m present.
11-48. If L_1 is 5 H, L_2 is 6 H, and L_m is 3 H, find L_{total} if the coils are connected *series-aiding.*
11-49. If L_1 is 7 H, L_2 is 9 H, and L_m is 7.5, find L_{total} if the two coils are connected series-opposed.
11-50. Two magnetically-coupled coils are connected in series, and the total inductance is measured at 40 mH. By reversing the connections to one coil, the total inductance now measures 10 mH. Find the mutual inductance present.
11-51. Find the coefficient of coupling K in Problem 11-49.

See section 11-12 for discussion covered by the following.

11-52. A transformer has a coefficient of coupling K of 0.75, with 12 V ac applied to a 500-turn primary. Find the voltage induced in its 125-turn secondary.

11-53. With 60 V ac applied to a 25-turn primary, find the voltage induced in a 250-turn secondary if the coefficient of coupling K is 0.85.

See section 11-12 on transformer efficiency for discussion covered by the following.

11-54. Find the efficiency, in percentage, of a transfomer where the primary power is 80 W, and the secondary power is 76 W.

11-55. Find the power in the primary circuit where 90 W are delivered to the secondary circuit, and the transformer efficiency is 79%.

11-56. A transformer has an efficiency of 97%. If the power in the primary circuit is 12 W, find the power transferred to the secondary.

See section 11-12 on transformer currents for discussion covered by the following.

11-57. A transformer is 80% efficient. 120 V ac is applied to the primary and produces 1000 V in the secondary. If secondary current is 40 mA, find: (a) P_{sec}, (b) P_{pri}, and (c) I_{pri}.

11-58. If 12 V ac is induced in a secondary, and a 1-kΩ load is connected to the secondary, find: (a) I_{sec}, (b) I_{pri}, if 60 V ac is applied to the primary, and the transformer is 95% efficient.

11-59. The efficiency of a transformer is 90%, and its coefficient of coupling is 0.5. If 30 V ac is applied to the 70-turn primary, and a 500-Ω load is connected to its 280-turn secondary, find: (a) E_{sec}, (b) I_{sec}, (c) P_{sec}, (d) P_{pri}, and (e) I_{pri}.

See section 11-13 for discussion covered by the following problems.

11-60. A transistor device with an internal resistance of 10 kΩ is transformer coupled to a pair of 200-Ω (each) magnetic deflection coils of a television picture tube. What should the transformer turns ratio (primary to secondary) be to produce maximum power in the deflection coils?

11-61. A transformer has 500 turns in the primary and 100 turns in the secondary, and is used as coupling between a generator and a 400-Ω load. What impedance or resistance does the generator "see," "looking"" into the primary?

11-62. A transformer has a primary to secondary turns ratio of 5:1. If a generator in the primary circuit has 10-kΩ internal resistance, how much resistance does the load in the secondary circuit see?

See section 11-14 for discussion covered by the following.

11-63. What is the *commutator* of a dc generator?
11-64. What kind of voltage is induced in the rotating coil of a dc generator?
11-65. If there is ac voltage induced in the rotating coil of any generator, how is dc produced in the output of a dc generator?
11-66. Very briefly describe the operation of the commutator of a dc generator.

See section 11-15 for discussion covered by the following.

11-67. Express the following frequencies in one or more convenient forms: (a) 5000 Hz, (b) 660,000 Hz, (c) 105,000,000 Hz, (d) 150,000,000,000 Hz.

See section 11-15, Period, for discussion covered by the following.

11-68. What is the time duration of a sine wave if its frequency is 400 Hz.
11-69. A sine-wave voltage has a frequency of 2 GHz. What is the *period* of one cycle?
11-70. What is the frequency of a sine-wave voltage if its time period is 50 msec.
11-71. What is the frequency of a sine-wave voltage if its period is 250 nsec?

See section 11-15, Wavelength, for discussion covered by the following.

11-72. What is the wavelength of a 100-MHz sine wave?
11-73. What is the frequency if the wavelength is 150 cm?

See section 11-15, Instantaneous Values, for discussion covered by the following.

11-74. If E_{max} is 50 V, find the voltage of a sine wave at 60° (see Fig. 11-24).
11-75. If E_{max} is 150 V, find the instantaneous voltage of a sine wave at 20°.
11-76. Find the maximum value of a sine-wave voltage if its value at 49° is 25 V.

11-77. A 20-MHz sine wave has a maximum value of 40 V. Three nanoseconds after the start of the cycle find: (a) angle in degrees, and (b) value of the voltage. (*Hint*: 1 nsec = 1×10^{-9} sec).

See section 11-16 for discussion covered by the following.

11-78. What would a dc voltmeter read for consecutive half-sine waves if the maximum voltage is 30 V?

11-79. If the average voltage for consecutive half-sine waves is 12.7 V, what is its maximum value?

11-80. If E_{max} is 150 V, what is E_{av} for a series of alternate half-cycle sine waves such as is shown in Fig. 11-26*c*?

11-81. The voltage at the anode of an electron-tube amplifier is a sine-wave varying dc, going from a value of +205 V to a minimum of +195 V. What is (a) its dc average voltage, and (b) its peak ac voltage?

See section 11-17 for discussion covered by the following.

11-82. Name five types of meters capable of reading alternating current.

11-83. What type of voltage is produced by a thermocouple if: (a) Direct current is applied to the thermocouple heater? (b) Alternating current is applied to the thermocouple heater?

11-84. Most ac voltmeters are calibrated for what value of voltage (average, peak-to-peak, rms, or maximum)?

chapter 12

Inductors and Resistors in DC and AC Circuits

When a dc voltage is *first* applied to a coil (an inductance), a magnetic field starts to build up from the coil as it becomes an electromagnet. As this moving field cuts across the wires, a voltage is induced in the coil, resulting in a current in the coil which, from Lenz's law, opposes the original current flow. This opposition prevents the original current from becoming maximum immediately, as it would when a dc voltage is applied to a resistor.

To ac voltage, a coil reacts differently than to direct current. If the alternating current is a sine wave such as shown in Fig. 12-3*b*, then the current is almost always changing, producing a magnetic field that is also almost always moving, either building up or collapsing. This constantly moving field keeps inducing a voltage in the coil, not for only a short time as in direct current, but practically all the time. This induced voltage produces a current that, following Lenz's law, opposes the original current from the applied ac voltage source. This opposition is present all the time, since the magnetic field is constantly moving and constantly inducing a voltage in the coil. As a result, the inductance presents a much greater opposition (called *inductive reactance*) to alternating current than to direct current.

12-1. Inductor and Resistor in Series DC Circuit

When the switch in Fig. 12-1*a* is first closed, current just begins to flow, causing a magnetic field to start building up from the coil. This moving field

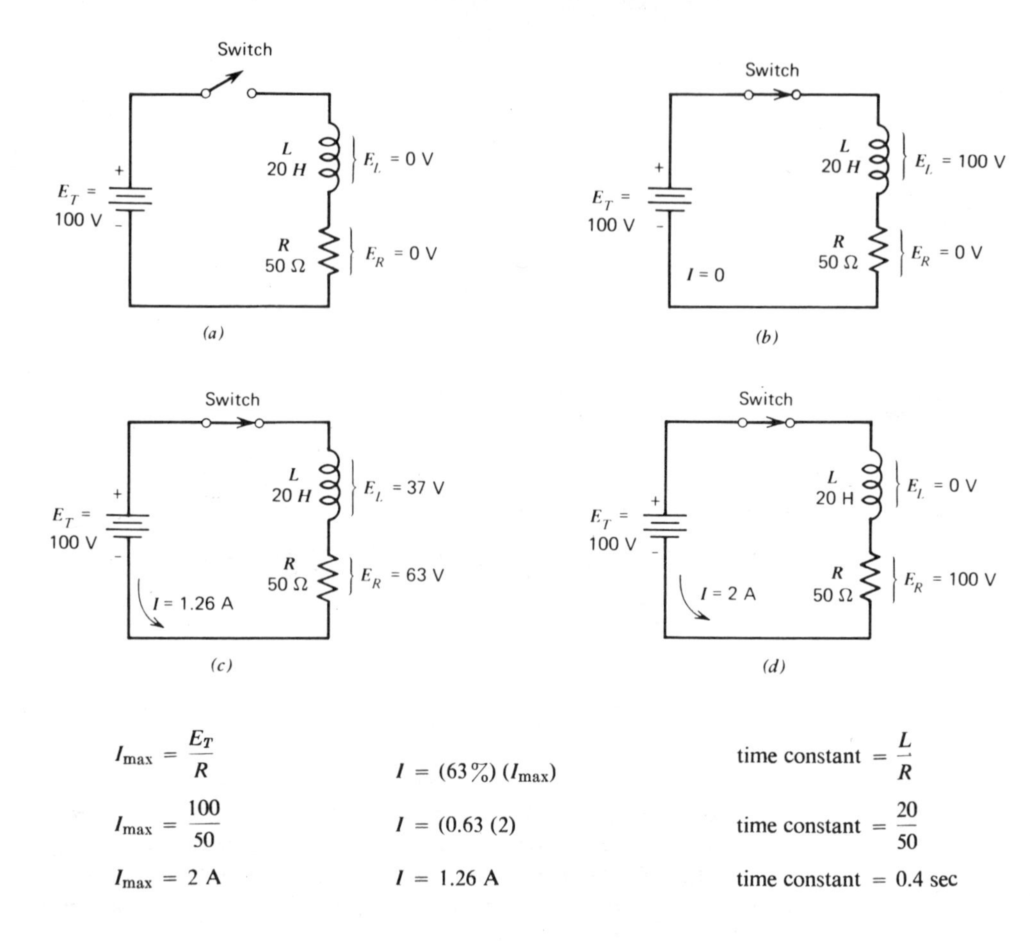

Figure 12-1. Inductor and resistor in dc circuit. (*a*) Open switch. (*b*) Instant that switch is closed. (c) One time constant after switch is closed. (*d*) Five time constants after switch is closed.

cuts across the coil wires, inducing a voltage in the coil, and producing a current. This current, sometimes called the "induced" current, must oppose the original current from the applied voltage source as stated in Lenz's law (Section 11-1). The induced current actually flows in the opposite direction from the original current. As a result of the opposition of the induced current, the original current cannot rise immediately from zero to its value of 2 A ($I = E/R = 100\ \text{V}/50\ \Omega = 2\ \text{A}$), as it would with only the 50-Ω resistor in the circuit.

Figure 12-1*b* shows the circuit at the instant the switch is closed. Although current from the voltage source has actually started, the opposition of the induced current results, at this first instant, in a delay of current so that it acts as if no current flows. Voltage across the resistor, E_R, is shown as zero volts, while voltage across the inductor, E_L, is 100 V. From Kirchhoff's voltage law (Section 3-3) the sum of E_R (0 V) and E_L (100 V) must equal the applied voltage (E_T = 100 V).

Time Constant. The graphs or curves shown in Fig. 12-2 show the values of E_R, E_L, and I at the instant that the switch is closed (time constant marked zero). The term *time constant* is a period of time equal to the value of the inductor (in henrys) divided by the value of the resistor (in ohms), or

$$\underset{\text{(in seconds)}}{\text{time constant}} = \frac{L \text{ (in henrys)}}{R \text{ (in ohms)}} \tag{12-1}$$

Example 12-1

Find the value of one time constant for the 20-H inductor and the 50Ω-resistor shown in Fig. 12-1.

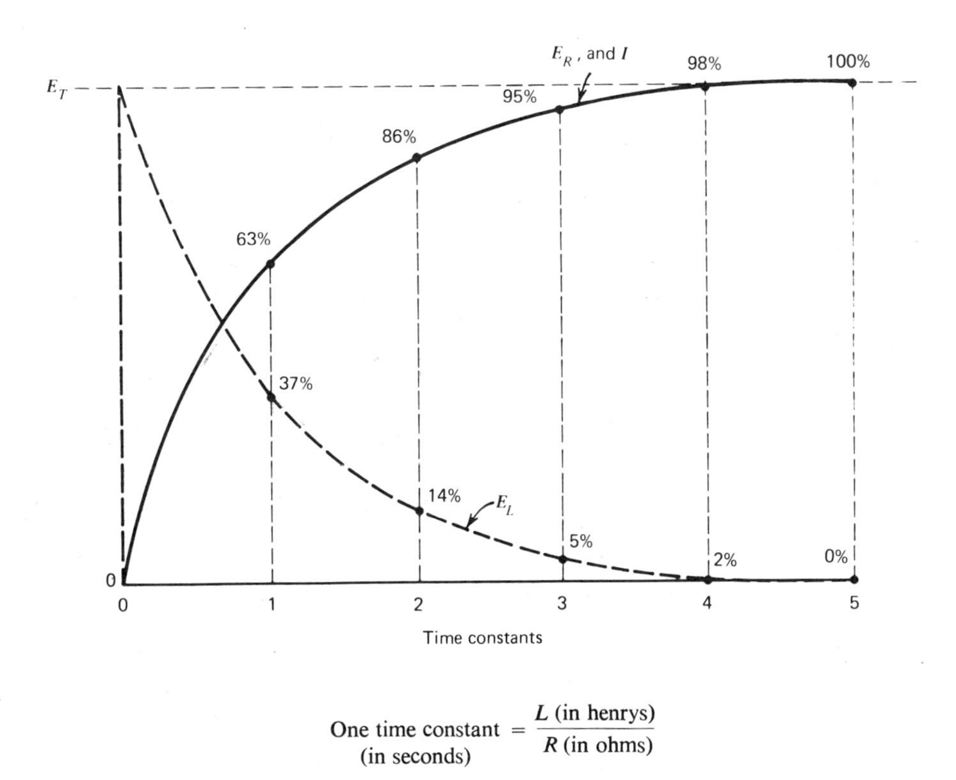

Figure 12-2. L/R time constant curves.

Solution

$$\text{time constant} = \frac{L}{R} \tag{12-1}$$

$$\text{time constant} = \frac{20}{50}$$

$$\text{time constant} = 0.4 \text{ sec}$$

As shown in the time constant curves of Fig. 12-2, the current I, and the voltage across the resistor E_R, rise from a zero value to 63% of their maximum value in the time of one time constant, or 0.4 sec in this example. The maximum value of current is $I = E/R = 100/50 = 2$ A. Therefore, at the end of the first time constant period, I is 63% of 2 A, or 1.26 A, as shown in the circuit of Fig. 12-1*c*.

Maximum value of voltage across the resistor is equal to the applied dc voltage source (100 V). As shown in the time constant graph of Fig. 12-2, E_R rises to 63% of its maximum value at the first time constant. E_R is then 63% of 100 V, or 63 V, shown in the circuit of Fig. 12-1*c*.

From Kirchhoff's voltage law, with 100 V E_T, and 63 V E_R, then E_L is the difference, or 37 V, as depicted in the circuit of Fig. 12-1*c*, and also shown on the E_L graph of Fig. 12-2 at the end of the first time constant.

The graphs of Fig. 12-2 show that at the end of the second time constant ($2 \times 0.4 = 0.8$ sec) the current has risen to 86% of its maximum value; E_R has also risen to 86% of its maximum; and E_L has decreased to 14%. This process continues, until about at the end of the fifth time constant ($5 \times 0.4 = 2$ sec), I has risen to about its maximum value ($I = E/R = 100/50 = 2$ A); E_R has reached its maximum of 100 V (equal to E_T); and E_L has decreased to zero, as shown also in Fig. 12-1*d*.

The following example shows the use of the L/R time constant curves of Fig. 12-2.

Example 12-2

A 150-mH inductor is in series with a 75-kΩ resistor, similar to the circuit of Fig. 12-1*a*. If 300 V dc were to be applied through a switch, find: (a) the time of one time constant, (b) I_{max}, (c) I, two time constants after switch has been closed, (d) E_R at this instant, (e) E_L at this instant, and (f) how much time, after the switch has been closed, it would take for I to reach its maximum value. Refer to the percentages shown in the graphs of Fig. 12-2.

Solution

Solve this example by using the following results as a guide: (a) time constant = 2 μsec, (b) I_{max} = 4 mA, (c) I at 2 μsec = 3.44 mA, (d) E_R at 2 μsec = 258 V, (e) E_L at 2 μsec = 42 V, and (f) 5 time constants = 10 μsec.

For Similar Problems Refer to End-of-Chapter Problems 12-1 to 12-5

12-2. Inductor in AC Circuit

Alternating current (ac) sine waves are discussed in Chapter 11, Section 8. The alternating current differs from a steady direct current in two main respects. The alternating current is almost always changing in value from one instant to the next, while the direct current remains at a constant value. The alternating current also reverses its direction, first flowing in one direction, and then in the opposite direction, while the direct current flows in only one direction.

The changing values of alternating current through a coil cause the magnetic field to vary in strength. This moving magnetic field keeps inducing a voltage in the coil, producing an "induced" current in the coil that *opposes* the original current from the applied voltage source. This opposition is present

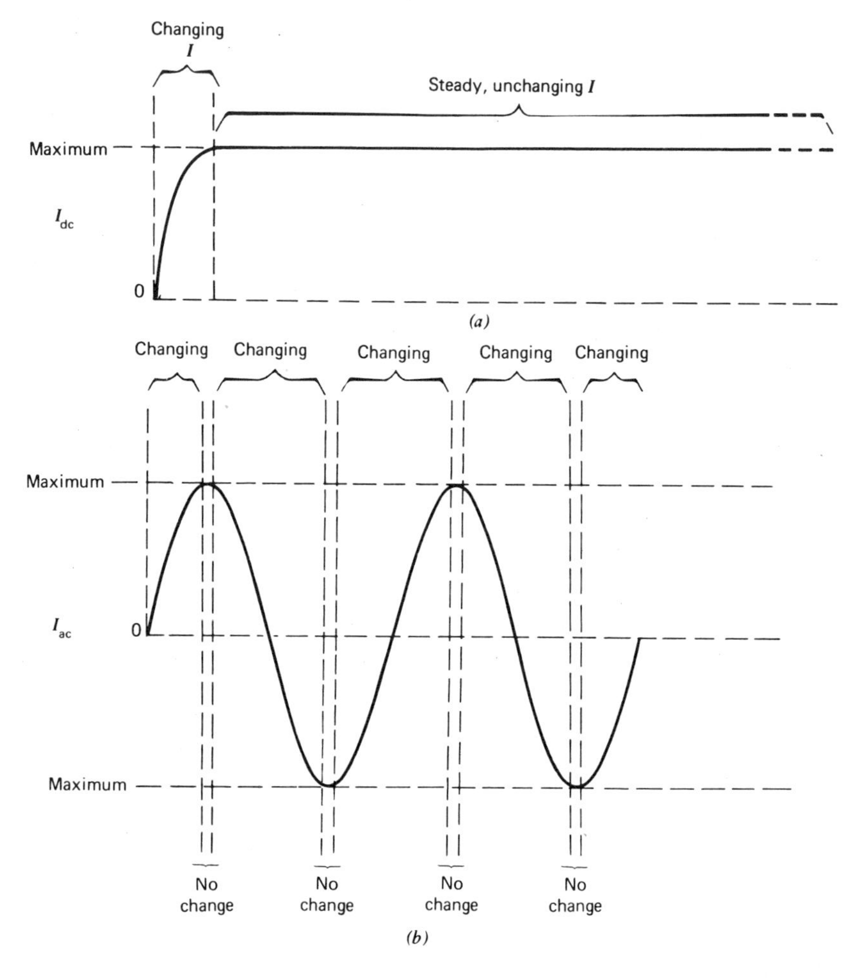

Figure 12-3. An ac 1.0 sine wave is changing value almost all the time.

practically all the time, as long as the ac current sine wave keeps changing in value, which is almost continuously, as may be seen in Fig. 12-3*b*. Note that the value of the ac sine wave is changing all the time except at those short instants where it is at maximum value. Therefore, unlike direct current where the current only changes until it has built up to its maximum and steady value, as shown in Fig. 12-3*a*, the opposition produced in a coil to alternating current is much greater. The coil *reacts* differently to alternating current than to direct current, and the *opposition* of the coil to alternating current is called *inductive reactance.*

Inductive Reactance. This is the opposition due to the induced voltage and its resultant current in the coil. *Inductive reactance* is usually written X_L and is read as "*X sub L.*" This opposition to ac current in a coil is measured in ohms, the same as a resistor. Inductive reactance (X_L) depends on the value of the inductor (in henrys) and the frequency of the alternating current (in hertz or cycles per second). The *frequency* describes how often a voltage is induced in the coil, while the value of the coil is a determining factor in the amount of this induced voltage. The amount of *X sub L* may be found from the following equation:

$$X_L = 2\pi f L \tag{12-2}$$

where X_L is in ohms, π is a mathematical symbol the value of which is 3.14 (2π is 6.28), f is the frequency of the ac sine wave in hertz, and L is the value of the coil in henrys.

The following examples illustrate the use of the X_L equation.

Example 12-3

Find the inductive reactance of a 5-H coil to an ac sine-wave voltage at a frequency of 100 Hz.

Solution

$$X_L = 2\pi f L \tag{12-2}$$
$$X_L = (2)(3.14)(100)(5)$$
$$X_L = (6.28)(500)$$
$$X_L = 3140\ \Omega$$

Example 12-4

What is the amount of opposition (X_L) offered by a 10-mH coil to a 300-kHz ac sine wave?

Solution

The reader should solve this example by using the previous solution as a guide, and should find that X_L is 18,840 Ω.

For Similar Problems Refer to End-of-Chapter Problems 12-6 and 12-7

12-3. Inductor and Resistor in AC Series Circuit

Any coil has some resistance in it, this resistance being primarily that of the coil wire itself. The resistance may be considered as a separate resistor acting as if it were in series with the inductive reactance of the coil. In Fig. 12-4 a coil having 60 Ω of resistance and 80 Ω of X_L is connected to an ac sine-wave voltage source. The *total* opposition of the coil is the sum of the resistance (60 Ω) and the inductive reactance (80 Ω), except that the two are not simply added together. They must be added by taking their *phasor* (or *vector*) sum. The reason for this is due to something called the *phase angle* or phase difference between the resistor and the inductive reactance. This is explained in the following discussion.

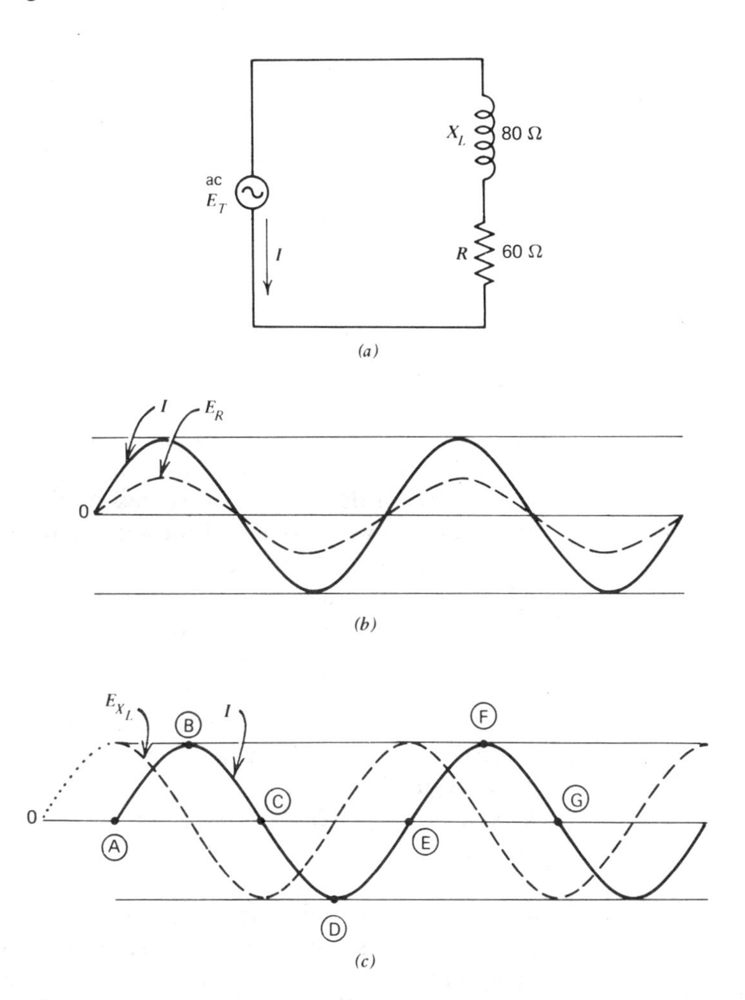

Figure 12-4. An ac circuit with sine waves showing phase relationships. (*b*) I and E_R in phase. (c) I lags E_{XL} by 90°.

Phase. When an ac current having a sine-wave graph, such as is shown in Fig. 12-3, flows through a resistor, the voltage across the resistor is also a sine wave and is *in phase* with the current. This means that when the current increases, E_R also increases. When I reaches its maximum value, E_R does the same. Finally, when I decreases back to zero, E_R follows exactly. This is depicted in Fig. 12-4*b*, where E_R is shown in phase with I. The *phase angle* between I and E_R is zero degrees.

When an ac sine-wave current, such as shown in Fig. 12-3, flows through a coil (assuming pure inductance, no resistance), voltage across the coil and current through it are not in phase. When the current has reached its maximum value and is not changing for a short instant, the magnetic field is stationary and does not cut the wires of the coil. As a result, no voltage is induced in the coil and E_{X_L} is zero. This is shown in Fig. 12-4*c* at points B, D, and F, where I is steady at its maximum values and E_{X_L} is at zero.

When the current rapidly *increases*, shown at points A and E of Fig. 12-4*c*, the rapidly increasing magnetic field induces a maximum voltage in the coil. Note that E_{X_L} is shown at its maximum value at points *A* and *E*.

When I rapidly *decreases* to zero, shown at points C and G of Fig. 12-4*c*, the rapidly collapsing magnetic field again induces a maximum voltage in the coil but one opposite in polarity from the voltage induced when the field was increasing. Note that in Fig. 12-4*c*, E_{X_L} is shown at maximum value at instant C and G, but that these maximum values of E_{X_L} are opposite in polarity to those at instants A and E.

The E_{X_L} sine wave of Fig. 12-4*c* is not in phase with the sine-wave I. Actually, I is said to be lagging behind E_{X_L} by 90°. If the E_{X_L} dashed-line graph is extended to the left of point A, as shown by the dotted-line section, it may be seen that when the E_{X_L} sine-wave has gone from zero to its maximum value (at instant A), which is 1/4 of a cycle or 90°, I is at zero (instant A), and is just starting to increase. Therefore, I is said to *lag* E_{X_L} by 90°.

Another way of depicting ac voltages in phase or 90° out of phase is by the use of *phasors*. These are discussed more fully in Appendix I, Mathematics for Electronics, but they are also discussed here. A student who is familiar with vectors and phasors may wish to omit the next subsection.

Phasors and Vectors. A force or a motion in some particular direction may be depicted by an arrow pointing in that direction. The amount of this force may be indicated by the size of the arrow. A drawing of this arrow is called a *vector*. A vector, then, is a straight line having *magnitude* and showing *direction*.

In Fig. 12-5*a*, the horizontal arrow or *vector* pointing to the right denotes a man swimming due east across a river at a speed of 4 miles per hour. Note that this vector is 4 units in length to indicate 4 mph. The vertical arrow pointing upward in Fig. 12-5*a* denotes the river current carrying the swimmer due north at a speed of 3 mph. Note that this vector is drawn 3 units in length, indicating 3 mph.

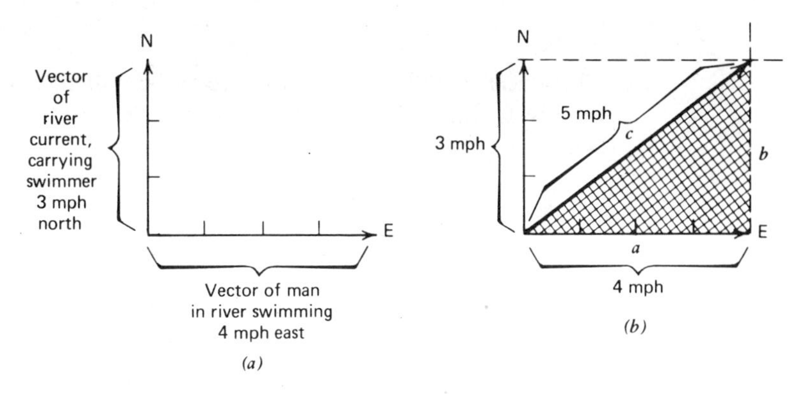

Figure 12-5. (*a*) Vectors. (*b*) Vector sum.

Obviously, the swimmer will move through the water with two forces acting on him; one force going east at 4 mph, and another force going due north at 3 mph. Therefore, the swimmer will be moving in a northeasterly direction which is the *vector sum* of the two vectors. This sum is shown in Fig. 12-5*b*. The vector sum can be found by drawing the 4 mph east vector 4 units (inches, centimeters, etc.) long, and at a right angle (90°) to it, and by drawing a second vector pointing north 3 mph, 3 units in length. Then a line can be drawn shown as the dashed lines in Fig. 12-5*b* at each arrow head vector, making it parallel to the other vector. The *vector sum* is the diagonal line joining the original connection point of the two vectors with the intersection point of the two dashed lines, as shown in Fig. 12-5*b*. If the length of this diagonal line is measured, it should indicate 5 units, or 5 mph. Note that the *vector sum* of 4 and 3 (at a right angle to each other) is 5, found by actual construction, or drawing. The sum may also be found by the following method.

The shaded area of Fig. 12-5*b* is a *right triangle* (a triangle containing a 90° angle), with the sides marked a, b, and c. A mathematical formula called the *Pythagoras theorem* states that the longest side of the triangle (c) is equal to the square root of the sum of the squares of the other two sides (b and c), or, simply

$$c = \sqrt{a^2 + b^2} \qquad (12\text{-}3)$$

In the shaded right triangle of Fig. 12-5*b*, side a is the 4 mph vector, and side b is equal in length to the 3 mph vector. Side c which is the vector sum is found from

$$c = \sqrt{a^2 + b^2} \qquad (12\text{-}3)$$

$$c = \sqrt{4^2 + 3^2}$$

$$c = \sqrt{16 + 9}$$

$$c = \sqrt{25}$$

$$c = 5 \text{ mph}$$

Example 12-5

Two vectors 90° out of phase are to be added. If one is 60 pointing horizontally to the right, while the other is 80 pointing vertically upward, find their sum.

Solution

Assume that the first vector (60) is side a of the right triangle and that the other vector (80) is side b, then the *vector sum* c is

$$c = \sqrt{a^2 + b^2} \tag{12-3}$$

$$c = \sqrt{60^2 + 80^2}$$

$$c = \sqrt{3600 + 6400}$$

$$c = \sqrt{10{,}000}$$

$$c = 100$$

For Similar Problem Refer to End-of-Chapter Problem 12-8

A vector can also represent a sine-wave voltage, or a sine-wave current at a particular frequency. In that case, the vector is assumed to be rotating like the sweep-second hand of a watch, except that the vector is rotated in the opposite direction or *counterclockwise*. A rotating vector is called a *phasor*. If the frequency of the ac voltage is 100 kHz, the *phasor* is assumed to be rotating counterclockwise at this speed of 100,000 rotations per second. The rapidly rotating vector or phasor would appear as a blur to the eye just as the spokes of a rapidly spinning wheel becomes invisible. To make use of the rotating vector, we shall always assume that it is being viewed under a *stroboscopic* light. This is a rapidly flashing (on and off) light source. If the light goes on and off at the same rate that the vector rotates, then the vector will appear lighted at the same position on each rotation, and will become dark or invisible during the remainder of its rotation. The stroboscopic light seems to "stop" the movement of the vector so that it *appears* stationary. A *phasor* then is actually a rapidly rotating vector that appears to be always in the same position.

E_R and E_{X_L} Addition. In Fig. 12-4*a*, a resistor (R) is shown in series with an inductive reactance (X_L) and is connected to an ac voltage (E_T). The same current (I) flows through R and X_L. In Fig. 12-4*b*, the sine wave I is shown in phase with E_R, while in Fig. 12-4*c* the sine wave I is shown lagging behind E_{X_L} by 90°.

The *phasors* of Fig. 12-6 show the same relationship as that of Fig. 12-4*b* and *c*. In the phasor diagram, note that I and E_R are in the same horizontal direction pointing to the right, denoting that I and E_R are in phase.

E_{X_L} is shown in Fig. 12-6 pointing vertically upward. Since all phasors are assumed to be rotating counterclockwise (but are shown at one instant as if

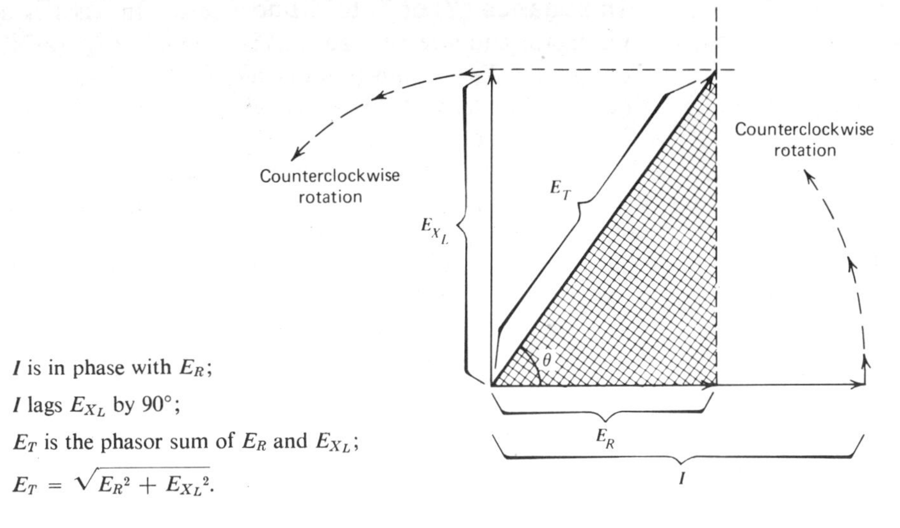

Figure 12-6. Phasors.

they were stationary), then the I phasor is 90° behind the E_{X_L} phasor. Phasors are added in the same way as stationary vectors. At each phasor arrowhead a line is drawn parallel to the other phasor. The sum (E_T) is the diagonal line shown in Fig. 12-6. Note that E_T is the longest side of the right triangle (the shaded area), with E_R and E_{X_L}, the other two sides. From the Pythagoras theorem, $c = \sqrt{a^2 + b^2}$ (equation 12-3), then

$$E_T = \sqrt{E_R^2 + E_{X_L}^2} \tag{12-4}$$

This sum of E_R and E_{X_L} is the total applied voltage E_T. The angle indicated as θ (theta) is the phase angle between E_T and I.

Voltage across a resistor (E_R), from Ohm's law, is simply the product of I and R. Voltage across an inductive reactance (E_{X_L}) is similarly the product of I and X_L.

Example 12-6

In Fig. 12-4*a*, if 200 mA flows through the X_L of 80 Ω in series with a resistor of 60 Ω, find: (a) E_{X_L}, (b) E_R, and (c) E_T.

Solution

Solve this example by using the following results as a guide: $E_{X_L} = 16$ V, $E_R = 12$ V, and $E_T = 20$ V.

For Similar Problems Refer to End-of-Chapter Problems 12-9 and 12-10

Impedance (Z) or Total Opposition. In a series ac circuit containing an inductor and a resistor such as is shown in Fig. 12-7*a*, the voltage phasors E_R, E_{X_L}, and E_T are shown in Fig. 12-7*b*. With $E_R = IR$, and $E_{X_L} = IX_L$, and since the same current flows through both R and X_L, then the voltage phasors (E_R and E_{X_L}) could be called simply R and X_L phasors as shown in Fig. 12-7*c*. Observe that the phasor R is drawn pointing horizontally to the right, while the phasor X_L is shown pointing vertically up, with the two phasors 90° out of phase with each other. R and X_L are each *oppositions* to current flow. The phasor sum of the two is the *total opposition* which is called *impedance* (Z).

As shown in Fig. 12-7*c*, Z is the longest side of the right triangle (the shaded area), with R and X_L the other two sides. The total opposition or impedance (Z) may be found by using the Pythagoras theorem,

$$Z = \sqrt{R^2 + X_L^2} \tag{12-5}$$

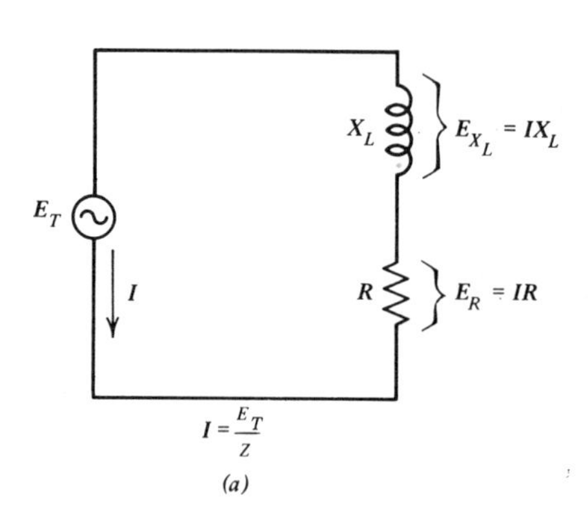

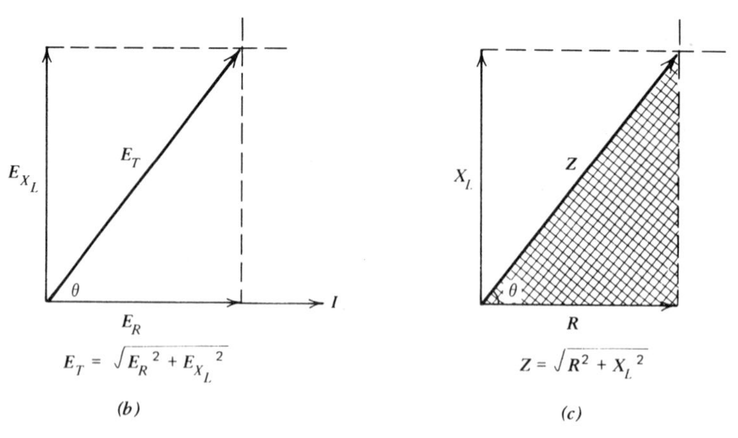

Figure 12-7. (a) An ac series circuit. (b) Voltage phasors. (c) R, X_L, and Z phasors.

If R and X_L are known, then their phasor sum is Z. This is shown in the following examples.

Example 12-7

As shown in Fig. 12-4*a*, X_L is 80 Ω, and the series R is 60 Ω. Find the total opposition or impedance (Z).

Solution

$$Z = \sqrt{R^2 + X_L{}^2} \tag{12-5}$$

$$Z = \sqrt{60^2 + 80^2}$$

$$Z = \sqrt{3600 + 6400}$$

$$Z = \sqrt{10{,}000}$$

$$Z = 100\ \Omega$$

Example 12-8

In a series ac circuit such as is shown in Fig. 12-7*a*, if E_T is 15 V, X_L is 200 Ω, and R is 150 Ω, find: (a) Z, (b) I, (c) E_{X_L}, and (d) E_R.

Solution

Solve this example by using the following as a guide: $Z = 250\ \Omega$, $I = E_T/Z = 0.06$ A, $E_{X_L} = 12$ V, and $E_R = 9$ V.

Example 12-9

In a series ac circuit such as is shown in Fig. 12-7*a*, E_T is 120 V at a frequency of 5 kHz. If the inductor L is 2 mH, having a resistance R of 50 Ω, find: (a) X_L, (b) Z, (c) I, (d) E_R, and (e) E_{X_L}.

Solution

Solve this example by using the following results as a guide: $X_L = 62.8\ \Omega$, $Z = 80.5\ \Omega$, $I = 1.49$ A, $E_R = 74.5$ V, $E_{X_L} = 93.5$ V.

For Similar Problems Refer to End-of-Chapter Problems 12-11 to 12-13

Inductive Reactances and Resistors in Series. When a coil L_1 and its resistance R_1 is connected in series with a second coil L_2 and its resistance R_2, as shown in Fig. 12-8*a*, both coils can be "paired," and both resistors can also be "paired." These are shown in Fig. 12-8*b*. Since each resistor voltage is

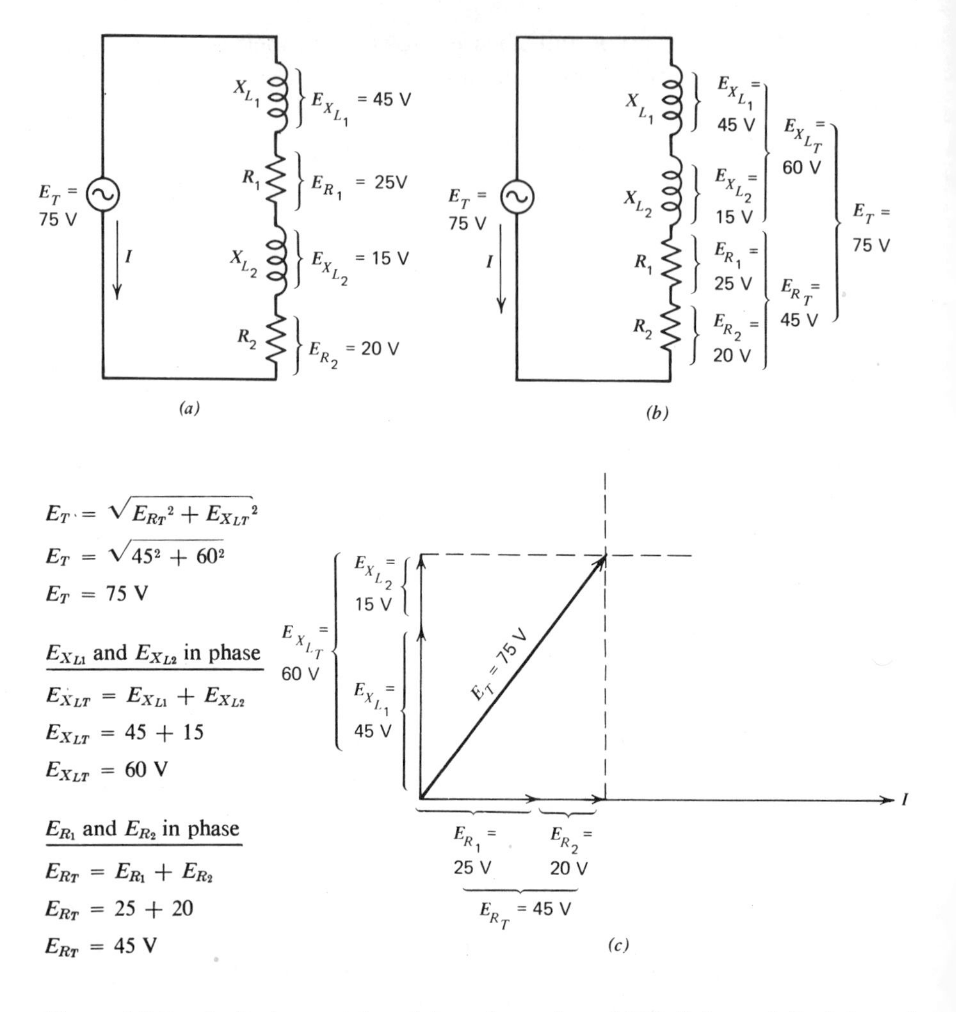

Figure 12-8. Inductors and resistors in series. (*a*) Coil L_1 and its internal resistance R_1 in series with coil L_2 and its internal resistance R_2. (*b*) E_{XL1} and E_{XL2} are "paired" together, and E_{R1} and E_{R2} are also paired. (c) Voltage phasors of series circuit of (*a*) and (*b*).

in phase with the current, as shown in the phasor diagram of Fig. 12-8*c*, then E_{R_1} is in phase with E_{R_2}. The *total* resistor voltage E_{R_T} is simply the arithmetic sum of the two, or, $E_{R_T} = E_{R_1} + E_{R_2} = 25 + 20 = 45$ V, as shown in Fig. 12-8*b* and *c*.

Also shown in Fig. 12-8*c*, each inductive reactance voltage *leads* the current by 90° (or I *lags* each inductive reactance voltage). Therefore, $E_{X_{L1}}$ and $E_{X_{L2}}$ are in phase with each other. The *total* inductive reactance voltage $E_{X_{LT}}$ is simply the arithmetic sum of the two, or $E_{X_{LT}} = E_{X_{L1}} + E_{X_{L2}} = 45 + 15 =$ 60 V, as shown in Fig. 12-8*b* and *c*.

The total voltage E_T is the *phasor sum* of the total resistor voltage E_{R_T} (45 V) and the total inductive reactance voltage $E_{X_{LT}}$ (60 V), or $E_T = \sqrt{E_{R_T}{}^2 + E_{X_{LT}}{}^2}$ $= \sqrt{45^2 + 60^2} = 75$ V, as shown in Fig. 12-8 *c*.

Similar to the voltage phasors of Fig. 12-8*c*, the phasors denoting opposition, R and X_L, are in the same positions as E_R and E_{X_L}. The following example illustrates this.

Example 12-10

Figure 12-9*a* shows the circuit of an ac voltage source applied to two coils connected in series. One coil has an inductive reactance ($X_{L_1} = 9$ kΩ) and a resistance ($R_1 = 5$ kΩ). The second coil has an inductive reactance ($X_{L_2} = 3$

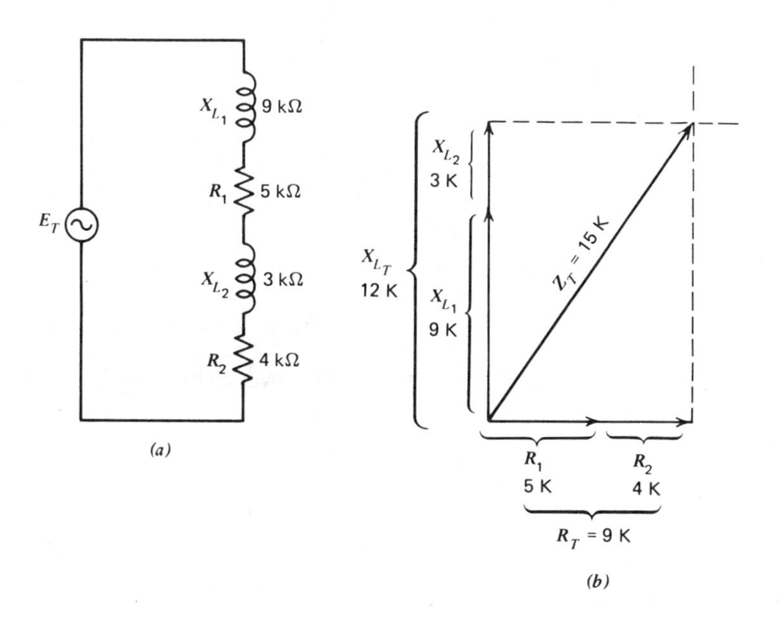

$R_T = R_1 + R_2$ $X_{LT} = X_{L_1} + X_{L_2}$ $Z_T = \sqrt{R_T{}^2 + X_{LT}{}^2}$

$R_T = 5\text{ K} + 4\text{ K}$ $X_{LT} = 9\text{ K} + 3\text{ K}$ $Z_T = \sqrt{(9\text{ K})^2 + (12\text{ K})^2}$

$R_T = 9\text{ k}\Omega$ $X_{LT} = 12\text{ k}\Omega$ $Z_T = 15\text{ k}\Omega$

EXAMPLE 12-10

Figure 12-9. Example 12-10. (*a*) Two coils, with their internal resistances connected in series (no magnetic coupling). (*b*) Phasor diagram.

kΩ) and a resistance (R_2 = 4 kΩ). Find: (a) R_{total}, (b) $X_{L\ total}$, and (c) total impedance Z_T.

Solution

Solve this example by using the results shown in Fig. 12-9 as a guide.

For Similar Problem Refer to End-of-Chapter Problem 12-14

Power in X_L and R Circuits. When an ac current flows through a resistor, energy is consumed and power is said to be dissipated in the form of heat in the resistor. This is not true with an inductor. In a coil, assuming that it is a pure inductor having no resistance, no power is consumed. The reason is quite apparent from the following. When the ac current *increases*, a magnetic field *builds up* and energy is *stored* in the field. When the ac current *decreases*, the magnetic field *collapses*, inducing a voltage in the coil, and the energy is returned to the circuit. As a result, the *pure* inductor does not dissipate power. Only the resistor consumes energy.

In a dc series circuit containing only resistors, the *total power* consumed is the product of total applied voltage and total current, or $P_T = E_T I_T$ (Chapter 2, equation 2-8, and also the Appendix III chart). In an ac series circuit containing X_L and R, this is not true. Here, $E_T I_T$ is called the *apparent* power whereas the actual or *true* power dissipated is

$$P_{true} = E_R I_T \tag{12-6}$$

since only the resistor dissipates power. This may be seen from the following example.

Example 12-11

In a series ac circuit, X_L = 200 Ω, R = 150 Ω, Z = 250 Ω, E_T = 15 V, I = 0.06 A, E_{X_L} = 12 V, and E_R = 9 V. Find: (a) the *apparent* power, and (b) the *true* power.

Solution

(a) the *apparent* power which is actually *not* being dissipated is

$$P_{apparent} = E_T I_T \tag{2-8}$$

$$P_{apparent} = (15)(0.06)$$

$$P_{apparent} = 0.9 \text{ W}$$

(b) The actual *true* power that is being dissipated is due to the resistor, therefore,

$$P_{\text{true}} = E_R I_T \tag{12-6}$$

$$P_{\text{true}} = (9)(0.06)$$

$$P_{\text{true}} = 0.54 \text{ W}$$

Note that the *true* power is *less* than the *apparent* power.

Power Factor. The *ratio* of the *true* power to the *apparent* power is called the power factor, or

$$\text{power factor (or PF)} = \frac{P_{\text{true}}}{P_{\text{apparent}}} \tag{12-7}$$

Since P_{true} is less than P_{apparent} in an X_L and R circuit, then power factor is always less than 1. Solving equation 12-7 in terms of true power gives

$$P_{\text{true}} = (P_{\text{apparent}}) \text{ (power factor)} \tag{12-8}$$

In any X_L and R circuit, the power factor is always *less* than 1. In a purely resistive circuit, the PF is exactly equal to 1, while in a purely inductive circuit (no resistor) the PF is zero. In an ac series circuit containing X_L and R then, the *power factor* may be defined as that factor that *reduces* the *apparent* power to the actual *true* power consumed. The following examples illustrate this.

Example 12-12

In the previous example, $P_{\text{true}} = 0.54$ W, and $P_{\text{apparent}} = 0.9$ W. Find the power factor of this circuit.

Solution

$$\text{PF} = \frac{P_{\text{true}}}{P_{\text{apparent}}} \tag{12-7}$$

$$\text{PF} = \frac{0.54}{0.9}$$

$$\text{PF} = 0.6$$

Example 12-13

If an X_L and R series ac circuit has a *power factor* of 0.75, find the true power that is dissipated if $E_{\text{total}} = 100$ V and $I_{\text{total}} = 2$ A.

Solution

$$P_{\text{true}} = (P_{\text{apparent}})\,(\text{PF}) \qquad (12\text{-}8)$$

$$P_{\text{true}} = (100\text{ V})\,(2\text{ A})\,(0.75)$$

$$P_{\text{true}} = (200\text{ W})\,(0.75)$$

$$P_{\text{true}} = 150\text{ W}$$

For Similar Problems Refer to End-of-Chapter Problems 12-15 to 12-18

12-4. Inductor and Resistor in AC Parallel Circuit

Figure 12-10*a* shows a coil in parallel with a resistor with 120 V ac applied across each. The applied voltage, E_T, is also E_{X_L} and also E_R. E_T is common

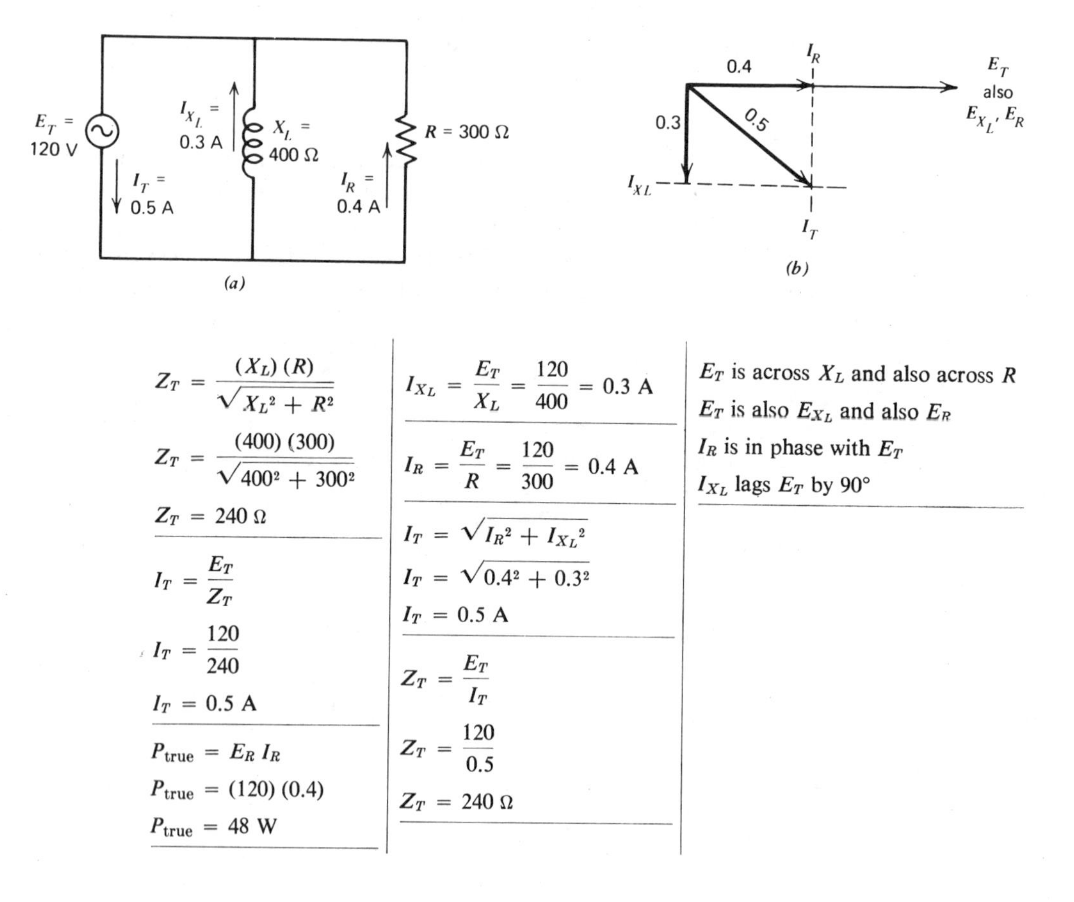

Figure 12-10. (*a*) Parallel ac circuit. (*b*) Phasors.

to X_L and R, and is used as the reference phasor. Current flow through the resistor branch and through the coil branch are illustrated by the phasors of Fig. 12-10*b*. I_R is in phase with the applied voltage across R, while I_{X_L} lags this same voltage. I_R and I_{X_L} are, therefore, 90° out of phase with each other.

With X_L in parallel with R, the total opposition or impedance Z_T may be found by using the "product over the sum" equation.

$$Z_T = \frac{\text{product}}{\text{sum}}$$

$$Z_T = \frac{(XL)\,(R)}{\sqrt{X_L^2 + R^2}} \tag{12-9}$$

Note that the sum of X_L and R (the denominator) is actually the phasor sum or $\sqrt{X_L^2 + R^2}$.

Total impedance Z_T could also be found using Ohm's law, $Z_T = E_T/I_T$. However, total current I_T in the circuit of Fig. 12-10*a* is the sum of the branch currents I_{X_L} and I_R. As shown in the phasor diagram of Fig. 12-10*b*, I_{X_L} and I_R are 90° out of phase with each other. To add them requires finding their *phasor sum*, or

$$I_T = \sqrt{I_{X_L}^2 + I_R^2} \tag{12-10}$$

The following examples solve the circuit of Fig. 10-12*a* using two different methods.

Example 12-14

In the circuit of Fig. 12-10*a*, find: (a) Z_{total}, (b) I_{total}, and (c) P_{true}.

Solution

Solve this example by using the methods and results shown in the text of Fig. 12-10 as a guide.

Example 12-15

By using a method other than shown in the previous example, solve the circuit of Fig. 12-10*a* for: (a) I_{X_L}, (b) I_R, (c) I_T, and (d) Z_T.

Solution

Solve this example by using the methods and results also shown in the text of Fig. 12-10 as a guide.

For Similar Problems Refer to End-of-Chapter Problems 12-19 to 12-20

Inductors and Resistors in AC Parallel Circuits. Figure 12-11*a* shows the circuit of two coils and two resistors connected in parallel with 24 V ac applied across each. Since the applied voltage E_T is across each branch, then E_T is used as a reference as shown in the phasor diagram of Fig. 12-11*b*. The currents in each branch are as shown with each resistor current (I_{R_1} and I_{R_2}) being in phase with E_T. The inductive branch currents ($I_{X_{L1}}$ and $I_{X_{L2}}$) each lag E_T by 90°.

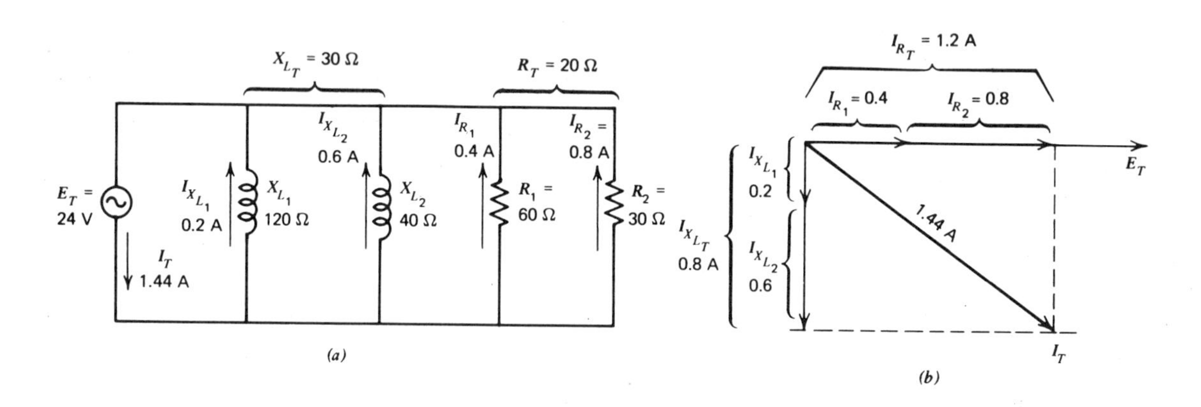

$$X_{LT} = \frac{(X_{L1})\,(X_{L2})}{X_{L1} + X_{L2}}$$

$$X_{LT} = \frac{(120)\,(40)}{120 + 40}$$

$$X_{LT} = 30\ \Omega$$

$$R_T = \frac{R_1\,R_2}{R_1 + R_2}$$

$$R_T = \frac{(60)\,(30)}{60 + 30}$$

$$R_T = 20\ \Omega$$

$$Z_T = \frac{(X_{LT})\,(R_T)}{\sqrt{X_{LT}^2 + R_T^2}}$$

$$Z_T = \frac{(30)\,(20)}{\sqrt{30^2 + 20^2}}$$

$$Z_T = 16.7\ \Omega$$

$$I_{RT} = I_{R1} + I_{R2}$$

$$I_{RT} = 0.4 + 0.8$$

$$I_{RT} = 1.2\ \text{A}$$

$$I_{XLT} = I_{XL1} + I_{XL2}$$

$$I_{XLT} = 0.2 + 0.6$$

$$I_{XLT} = 0.8\ \text{A}$$

$$I_T = \sqrt{I_{RT}^2 + I_{XLT}^2}$$

$$I_T = \sqrt{1.2^2 + 0.8^2}$$

$$I_T = 1.44\ \text{A}$$

E_T is also E_{XL1}, E_{XL2}, E_{R1} and E_{R2}

I_{R1} and I_{R2} are in phase with E_T

I_{XL1} and I_{XL2} each lag E_T by 90°

Figure 12-11. Coils and resistors in parallel. (*a*) Parallel circuit. (*b*) Phasors.

To find total current I_T, it is necessary to add I_{R_1} and I_{R_2}, resulting in I_{R_T}. $I_{X_{L1}}$ and $I_{X_{L2}}$ are also added, giving $I_{X_{LT}}$. Finally, the phasor sum of I_{R_T} and $I_{X_{LT}}$ results in I_T. Each individual branch current may be found by using Ohm's law where I_{branch} = voltage ÷ branch opposition.

Therefore, $I_{R_1} = E_{R_1}/R_1 = 24/60 = 0.4$ A, and $I_{R_2} = E_{R_2}/R_2 = 24/30 = 0.8$ A.

As shown in the phasor diagram of Fig. 12-11*b*, I_{R_1} and I_{R_2} are in the same direction, or in phase with each other. Their sum I_{R_T} is then: $I_{R_T} = I_{R_1} + I_{R_2} = 0.4 + 0.8 = 1.2$ A.

Similarly, $I_{X_{L1}} = E_{X_{L1}}/X_{L_1} = 24/120 = 0.2$ A, and $I_{X_{L2}} = E_{X_{L2}}/X_{L_2} = 24/40 = 0.6$ A.

The phasor diagram of Fig. 12-11*b* also shows that $I_{X_{L1}}$ is in phase with $I_{X_{L2}}$. Their sum $I_{X_{LT}}$ is then: $I_{X_{LT}} = I_{X_{L1}} + I_{X_{L2}} = 0.2 + 0.6 = 0.8$ A.

The total current I_T is the phasor sum of I_{R_T} and $I_{X_{LT}}$ which are 90° out of phase with each other as depicted in Fig. 12-11*b*, or

$$I_T = \sqrt{I_{R_T}^2 + I_{X_{LT}}^2} \tag{12-10}$$

$$I_T = \sqrt{1.2^2 + 0.8^2}$$

$$I_T = \sqrt{1.44 + 0.64}$$

$$I_T = \sqrt{2.08}$$

$$I_T = 1.44 \text{ A}$$

Using Ohm's law, the total impedance of the circuit of Fig. 12.11*a* may be found. $Z_T = E_T/I_T = 24/1.44 = 16.7\ \Omega$.

The following example shows a second method of finding the total impedance of the circuit of Fig. 12-11*a*.

Example 12-16

In the parallel circuit of Fig. 12-11*a*, find (a) Z_{total} and (b) I_{total}, using a different method than that just discussed.

Solution

(a) To find Z_T, X_{L_T}, and R_T must first be found, X_{L_T} consists of X_{L_1} and X_{L_2} in parallel, while R_T consists of R_1 in parallel with R_2. By using the *product divided by the sum* method, solve this problem by using the results shown in Fig. 12-11.

For Similar Problems Refer to End-of-Chapter Problems 12-21 and 12-22

INDUCTORS AND RESISTORS IN DC AND AC CIRCUITS, MORE ADVANCED

12-5. *L* and *R* in DC Series Circuit (More Advanced)

In Section 12-1 we discuss an inductor and resistor connected in series with an applied dc voltage source.

As shown in the time constant graphs of Fig. 12-2, current takes time to reach its maximum value, requiring about five time constants after the voltage is applied. Voltage across the resistor E_R similarly takes about five time constants to reach its maximum value. However, voltage across the inductor E_L immediately becomes maximum at the instant that the voltage is applied. Then, as current and E_R rise, E_L decreases. After approximately five time constants, E_L becomes zero when E_R rises to its maximum value, which is equal to the total applied dc voltage.

In Fig. 12-12*a*, when the switch is moved to the "closed" position, current starts flowing, and a magnetic field starts building up. This induces a voltage in the coil which produces the "induced" current that opposes the original current. As a result, the original current is delayed from immediately reaching its maximum or steady-state value ($I = E/R$). At this first instant, I is practically zero, E_R (or V_R) is zero, but E_L (or V_L) is at its maximum value, which is equal to the total applied voltage E_T. This is shown in the graphs in Fig. 12-12*b* at the instant when that switch is "closed."

As the original current rises more gradually or nonlinearly, the magnetic field increases more slowly, inducing a smaller voltage in the coil. As shown in Fig. 12-12*b*, the I and E_R graph rises, but at a progressively slower rate. This is called an *exponential* curve. During this time, E_L decreases exponentially as shown. When E_R becomes maximum (equal to E_T), E_L becomes zero.

The values of I, E_R, and E_L at any instant during this magnetic field rise time depend on the value of the applied dc voltage E_T, the values of L and R, and the period of time T that has transpired since the voltage was first applied (the switch being placed on "closed" position).

Time Constant. The values of L in henrys and R in ohms determine the *time constant* or τ (the Greek letter pronounced "tau"). This was expressed previously in equation 12-1 as

$$\text{time constant} = \frac{L}{R} \tag{12-1}$$

and will now be rewritten as

$$\tau = \frac{L}{R} \tag{12-1}$$

Voltages and Current Values During Magnetic Field Buildup or Rise. The values of I, E_R, E_L at some instant are written as the lower case or small letters i, e_R, *and* e_L. They may be found during the *buildup of the magnetic field*, as shown in Fig. 12-12*b* from the following.

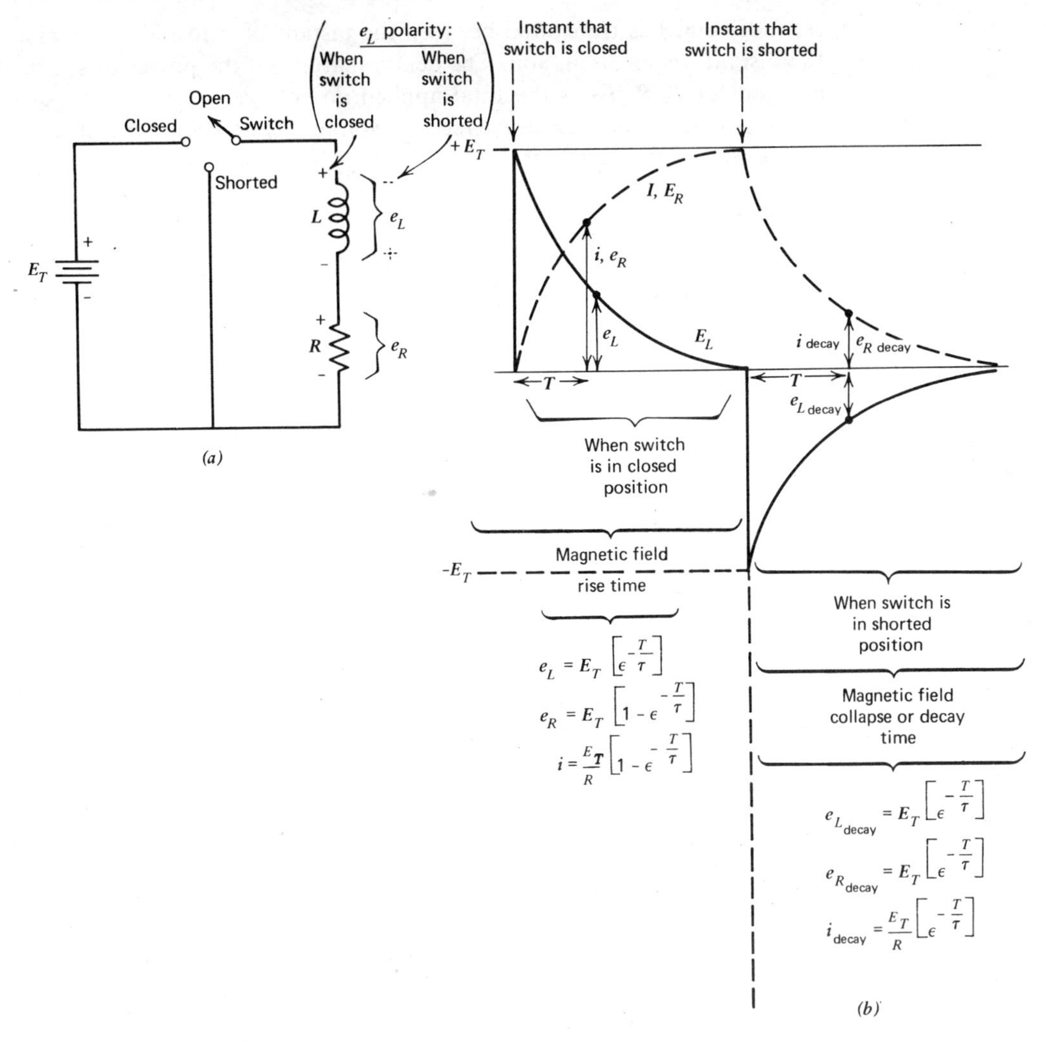

Figure 12-12. (*a*) An *LR* series circuit. (*b*) Time constant (τ) graphs and equations.

The value of the *decreasing* inductor voltage e_L at any instant during the magnetic field rise is

$$e_L = E_T\left(\epsilon^{-\frac{T}{\tau}}\right) \tag{12-11}$$

where ϵ (the Greek letter pronounced "epsilon") has a fixed mathematical value of 2.718, and the value of ϵ to its negative power is listed in the exponential function table in the Appendix. The T in the numerator is the time (in seconds) that has transpired between the instant of "closing" the switch and the instant under discussion. The denominator τ of the power of ϵ is the time constant or L/R. E_T is the total applied dc voltage.

The value of the *increasing* resistor voltage e_R at some instant during the rise of the magnetic field is

$$e_R = E_T\left(1 - \epsilon^{-\frac{T}{\tau}}\right) \tag{12-12}$$

The value of the *increasing* current i at some instant during the rise of the magnetic field is

$$i = \frac{E_T}{R}\left(1 - \epsilon^{-\frac{T}{\tau}}\right) \tag{12-13}$$

The following examples illustrate the use of equations 12-11 to 12-13.

Example 12-17

In the circuit shown for this example, where $E_T = 60$ V, $L = 20$ mH, and $R = 50\ \Omega$ (this could be the coil resistance), and the switch is placed in the "closed" position, find:

(a) time constant, τ.
(b) e_L 600 μsec after switch is closed.
(c) e_R, 600 μsec after switch is closed.
(d) i, 600 μsec after switch is closed.

Solution

(a) $$\tau = \frac{L}{R} \tag{12-1}$$

$$\tau = \frac{20\text{ mH}}{50}$$

$$\tau = \frac{20 \times 10^{-3}}{5 \times 10^{1}}$$

$$\tau = 4 \times 10^{-4}\text{ sec}$$

(b) $$e_L = E_T\left(\epsilon^{-\frac{T}{\tau}}\right) \tag{12-11}$$

$$e_L = 60\left(\epsilon^{-\frac{600\ \mu\text{sec}}{4 \times 10^{-4}\text{ sec}}}\right)$$

$$e_L = 60\left(\epsilon^{-\frac{600 \times 10^{-6}}{4 \times 10^{-4}}}\right)$$

$$e_L = 60\left(\epsilon^{-150 \times 10^{-2}}\right)$$

$$e_L = 60\left(\epsilon^{-1.5}\right)$$

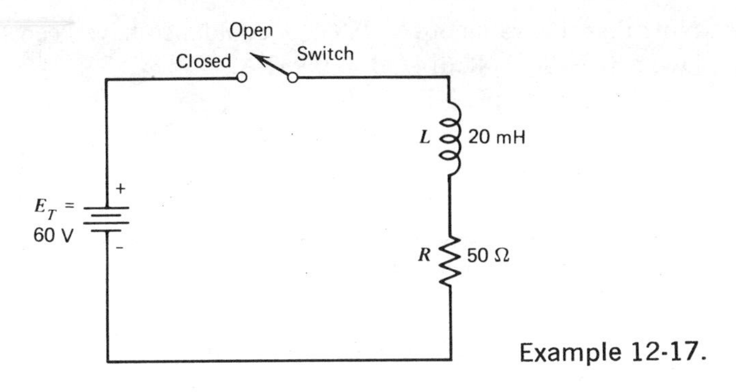

Example 12-17.

from the exponential function table in the Appendix:

$$\epsilon^{-1.5} \text{ has a value of } 0.2231$$

therefore

$$e_L = 60\ (0.2231)$$
$$e_L = 13.386 \text{ V}$$

(c) $e_R = E_T \left(1 - \epsilon^{-\frac{T}{\tau}}\right)$ (12-12)

$$e_R = 60 \left(1 - \epsilon^{-\frac{T}{\tau}}\right) \quad (12\text{-}12)$$
$$e_R = 60 \left(1 - \epsilon^{-\frac{600\ \mu\text{sec}}{4 \times 10^{-4}\ \text{sec}}}\right)$$
$$e_R = 60\ (1 - \epsilon^{-1.5})$$
$$e_R = 60\ (1 - 0.2231)$$
$$e_R = 60\ (0.7769)$$
$$e_R = 46.614 \text{ V}$$

Note that the sum of the voltage found, $e_L = 13.386$ V, and $E_R = 46.614$ V, must add up to the value of the applied E_T of 60 V.

(d) $i = \frac{E_T}{R}\left(1 - \epsilon^{-\frac{T}{\tau}}\right)$ (12-13)

$$i = \frac{60}{50}\left(1 - \epsilon^{-\frac{600\ \mu\text{sec}}{4 \times 10^{-4}\ \text{sec}}}\right)$$
$$i = 1.2\ (1 - \epsilon^{-1.5})$$
$$i = 1.2\ (1 - 0.2231)$$
$$i = 1.2\ (0.7769)$$
$$i = 0.932 \text{ A}$$

Note that this value of $i = 0.932$ A could also have been found using Ohm's law, $i = e_R/R = 46.614/50 = 0.932$ A.

Another example follows:

Example 12-18

If 50 V dc is applied to a 300-μH coil in series with a 1kΩ resistor (this could be the coil's resistance), find:

(a) time constant, τ.
(b) e_L, 0.75 μsec after the 50 V is applied.
(c) e_R, 0.75 μsec after 50 V is applied.
(d) i, 0.75 μsec after 50 V is applied.

Solution

Solve this example by using the method shown in the previous example, and compare the answers with these results: $\tau = 0.3$ μsec, $e_L = 4.1$ V, $e_R = 45.9$ V., and $i = 0.0459$ A.

For Similar Problems Refer to End-of-Chapter Problems 12-23 and 12-24

Time (T) during Magnetic Field Rise. Time (T) is shown in the graph of Fig. 12-12*b* as that period between the instant that the switch is first "closed" and that instant when e_L or e_R or i has reached its particular instantaneous value. If this instantaneous value, e_L, e_R or i is known, and total applied voltage E_T and the time constant τ are also known, then the elapsed time T may be found by using the appropriate formula (equations 12-11, 12-12, or 12-13). The following examples illustrate this.

Example 12-19

A dc voltage source of 15 V is applied through a switch to a 100-mH coil having 5 Ω resistance (the 5 Ω acts as a series resistor). Find: (a) time constant (τ), (b) time (T) required after switch is closed for e_L to decrease to 11 V from its starting value of 15 V.

Solution

$$\text{(a) time constant } \tau = \frac{L}{R} \qquad (12\text{-}1)$$

$$\text{time constant } \tau = \frac{100 \text{ mH}}{5\ \Omega}$$

$$\text{time constant } \tau = \frac{100 \times 10^{-3}}{5}$$

$$\text{time constant } \tau = 20 \times 10^{-3} \text{ sec} \quad \text{or} \quad 20 \text{ msec}$$

(b)
$$e_L = E_T \left(\epsilon^{-\frac{T}{\tau}} \right) \tag{12-11}$$

$$11 = 15 \left(\epsilon^{-\frac{T}{20 \times 10^{-3}}} \right)$$

$$\frac{11}{15} = \epsilon^{-\frac{T}{20 \times 10^{-3}}}$$

$$0.733 = \epsilon^{-\frac{T}{20 \times 10^{-3}}}$$

From the exponential function table in the Appendix, this value of 0.733 for ϵ^{-x} corresponds approximately to the exponent x of 0.31 (where the exponent $-x$ is actually the $-T/(20 \times 10^{-3})$. Therefore, if

$$0.733 = \epsilon^{-\frac{T}{20 \times 10^{-3}}}$$

then

$$0.31 = \frac{T}{20 \times 10^{-3}}$$

and

$$(0.31)(20 \times 10^{-3}) = T$$

$$6.2 \times 10^{-3} \text{ sec} = T$$

or

$$6.2 \text{ msec} = T$$

Example 12-20

In the previous example where $E_T = 15$ V, and $\tau = 20$ msec, find the time (T) required for e_R to rise to 4 V from its starting value of zero.

Solution

Time (T) should be the same answer as in the previous example (6.2 msec), since with a total applied dc voltage of 15 V, when e_L has become 11 V (previous example), e_R will be 4 V. Solve for T with the method shown in the previous example, but use equation 12-12.

For Similar Problems Refer to End-of-Chapter Problems 12-25 and 12-26

If the switch of Fig. 12-12*a* is now placed in the "shorted" position, E_T is removed and current I starts decreasing. This is depicted in the right half of the graphs of Fig. 12-12*b*. The magnetic field starts collapsing, inducing a voltage in the coil with the opposite polarity, shown as the dotted-line polarity e_L of Fig. 12-12*a*. Note that the reverse polarity E_L is also shown in the graph of Fig. 12-12*b*. This induced voltage is now the only voltage that keeps the current flowing, acting as if it were the applied voltage source. As the magnetic field collapses, E_L decreases. This causes the current to decrease, resulting in a decreasing E_R.

Note that in the graphs of Fig. 12-12*b*, during the time that the switch is in the shorted position, or during the magnetic field collapse time, E_R and E_L are equal to each other, but of opposite polarity, and are decreasing together. Note from the graphs that E_R has the same polarity (it is still *above* the zero axis) as it had during the time that the switch was in the "closed" position. Also note that the E_L graph is now *below* the zero axis, which is the reverse of what it was during the time that the switch was in the closed position.

At the instant that the switch is placed in the shorted position, after having been closed for about 5 time constants, the graphs of Fig. 12-12*b* show that E_R is equal to the $+E_T$, while E_L is equal to the reverse polarity or $-E_T$. Then, as time goes on, E_L and E_R both decrease. At any time T, as shown, the instantaneous values e_R and e_L are equal to each other, but are opposite in polarity; e_R is above the zero axis while e_L is below zero.

The instantaneous values of the *decreasing* e_L and e_R and i may be found by using similar equations to the decreasing e_L during the previous time when the switch was in the closed position [$e_L = E_T\,(\epsilon^{-T/\tau})$, eq. 12-11]. During the time that the switch is in the shorted position, the magnetic field is collapsing or decaying, and e_L, e_R, and i are d*ecreasing*; their values may be found from the following.

$$e_{L\text{ decay}} = E_T\left(\epsilon^{-\frac{T}{\tau}}\right) \tag{12-14}$$

and e_R which is equal to e_L is

$$e_{R\text{ decay}} = E_T\left(\epsilon^{-\frac{T}{\tau}}\right) \tag{12-15}$$

and

$$i_{\text{decay}} = \frac{E_T}{R}\left(\epsilon^{-\frac{T}{\tau}}\right) \tag{12-16}$$

The following examples illustrate the use of the above equations.

Example 12-21

In the circuit of Fig. 12-12*a*, E_T is 50 V, L is 25 mH, and R is 100 Ω. If the switch is in its closed position for a sufficient time about five time constants or longer), I and E_R reach their maximum values, and E_L decreases to zero after starting off at its maximum value (equal to E_T). Find (a) time constant (τ), (b) $E_{R\,max}$, (c) I_{max}, (d) e_L, 200 μsec after the switch is placed in the shorted position, (e) e_R at this same instant, and (f) i at this instant.

Solution

Solve this example by using the methods shown previously. As a guide, the following results are given: (a) time constant = 250 μsec, (b) $E_{R\,max}$ = 50 V, (c) I_{max} = 0.5 A, (d) e_L = 22.45 V, (e) e_R = 22.45 V, and (f) i = 0.2245 A.

Example 12-22

In the circuit of the previous example, the switch is first closed for sufficient time for E_R and I to rise to their maximum values. How much time (T) is required after the switch is moved to the shorted position for e_R to decrease from its maximum value of 50 V (equal to E_T) down to 4 V?

Solution

Find time (T) by using equation 12-15 and the method previously shown. As a guide, T should be found to be 633 μsec.

For Similar Problems Refer to End-of-Chapter Problems 12-27 and 12-28

12-6. Inductor and Resistor in AC Series Circuit (More Advanced)

In Section 12-2, we present the opposition of a coil to alternating current called *inductive reactance* (X_L), where $X_L = 2\,\pi fL$ (equation 12-2). Then, in Section 12-3, we discuss the series combination of X_L and R having a total opposition called *impedance* (Z), where Z is the *phasor sum* of X_L and R. Since these two oppositions to current (X_L and R) are 90° out of phase with each other, then $Z = \sqrt{R^2 + X_L^2}$ (equation 12-5).

Figure 12-13*a* shows a 3000-Ω resistor in series with a 4000-Ω inductive reactance. By using equation 12-5, $Z = \sqrt{R^2 + X_L^2}$, total impedance Z = 5000 Ω, as shown. Current, $I = E/Z$ = 120 V/5000 Ω = 24×10^{-3} A. Voltage across the resistor is: $E_R = IR = (24 \times 10^{-3})\,(3 \times 10^3) = 72$ V as shown. Voltage across the inductive reactance is: $E_{X_L} = (I)\,(X_L) = (24 \times 10^{-3})\,(4 \times 10^3) = 96$ V, as shown.

In Fig. 12-13*b*, the current and voltage phasors are shown. Note that since the same current flows through both R and X_L of Fig. 12-13*a*, the current phasor I is used as the reference phasor or zero-degree phasor. Voltage across the resistor, the E_R phasor, is in phase with I. Voltage across the coil or the inductive reactance, the E_{X_L} phasor, is *leading* I by 90° (or I *lags* E_{X_L} by 90°). Total voltage, E_T, is the phasor sum of E_R and E_{X_L}.

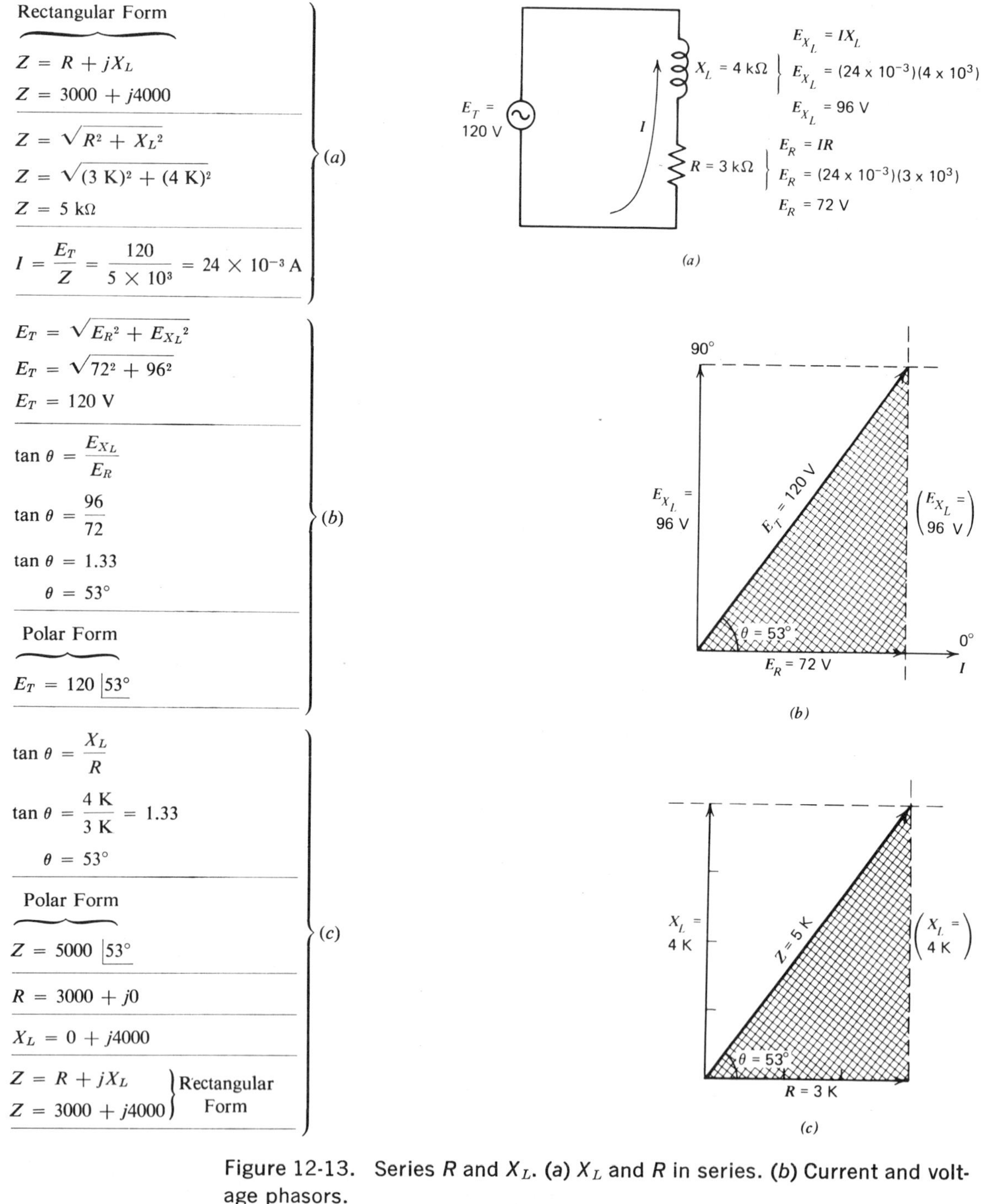

Figure 12-13. Series R and X_L. (*a*) X_L and R in series. (*b*) Current and voltage phasors.

Observe that the shaded area of Fig. 12-13*b* is a right triangle. One side (the base) is the E_R phasor (72 V). The right side is the same as the E_{X_L} phasor (96 V). The diagonal side or *hypotenuse* is the E_T phasor (120 V). The angle formed by the E_T and I phasors is called the *phase angle*, and the Greek letter θ (prounced "theta") is used to denote the phase angle. It is necessary to know this angle in more complex problems that involve power dissipation and power factor (discussed in simpler circuits in Section 12-3).

In the appendix or Mathematics for Electronics, the trigonometric functions of an angle in a right triangle are discussed. In this present chapter we now give a simple review that may be ommitted by the student who is familiar with this topic.

Trigonometric Functions. The functions of the angle θ in the right triangle of Fig. 12–14 that are of main concern to the electronics technician are the *sine*, *cosine*, and *tangent*. Each of these is simply the ratio of two particular sides of the right triangle. The *sine* is usually abbreviated as sin; *cosine* is simply written as *cos*; and *tangent* is written as *tan*. The *sine* of angle θ is the ratio of the *opposite* side to the *hypotenuse* or, as shown in Fig. 12-14,

$$\sin\theta = \frac{\text{opposite side}}{\text{hypotenuse}} \tag{12-17}$$

$$\sin\theta = \frac{X_L}{Z}$$

The *cosine* of angle θ is the ratio of the *adjacent* side to the *hypotenuse* or, as shown in Fig. 12-14,

$$\cos\theta = \frac{\text{adjacent side}}{\text{hypotenuse}} \tag{12-18}$$

$$\cos\theta = \frac{R}{Z}$$

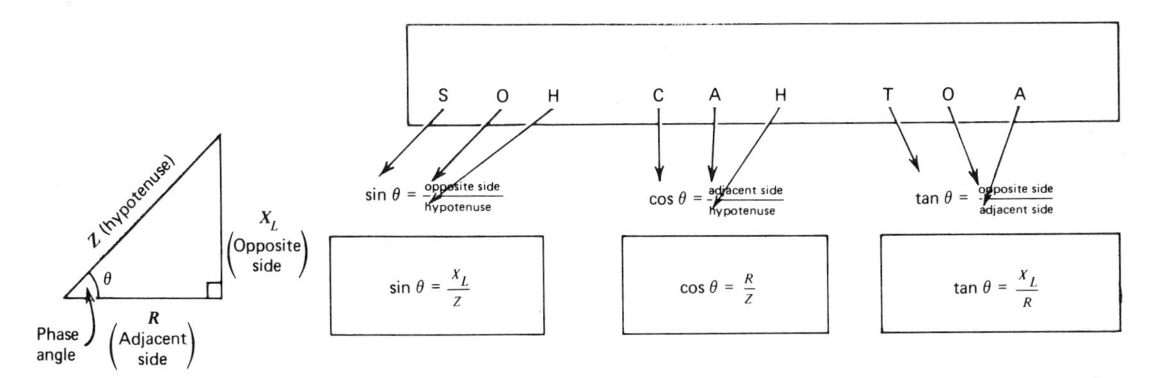

Figure 12-14. Trigonometric functions in a right triangle.

The *tangent* of angle θ is the ratio of the *opposite* side to the *adjacent* side or, as shown in Fig. 12-14,

$$\tan\theta = \frac{\text{opposite side}}{\text{adjacent side}} \tag{12-19}$$

$$\tan\theta = \frac{X_L}{R}$$

As indicated in Fig. 12-14, a helpful aid in learning the equations for sine, cosine, and tangent that students use is the coined word, or name, SOH-CAH-TOA. Some students pronounce it almost like "soak a toe"; others prefer that old popular expression, "sock it to her." Some, GI's or Navy veterans, even memorized a list of nine words, each starting with the letters of SOHCAH-TOA! At any rate, the student should know these three equations (12-17, 12-18, and 12-19).

In Fig. 12-13*b*, the shaded area is the right triangle having E_T (120 V) as the hypotenuse, E_R (72 V) as the *adjacent* side to the *angle* θ, and E_{X_L} (96 V) as the *opposite* side to θ. The angle θ may be found by using any one of the functions. Find the angle θ by using each function. The results should be: $\tan\theta = E_{X_L}/E_R = 96/72 = 1.33$, $\theta = 53°$; $\sin\theta = E_{X_L}/E_T = 96/120 = 0.8$, $\theta = 53°$; $\cos\theta = E_R/E_T = 72/120 = 0.6$, $\theta = 53°$.

In Fig. 12-13*c*, the phasors shown are R, X_L, and Z, instead of the voltage phasors (Fig. 12-13*b*) E_R, E_{X_L}, and E_T. The angle θ may again be found in Fig. 12-13*c* by employing any one of the functions, sine, cosine, or tangent. In this figure, the tangent function is shown. Note that θ is again found to be 53°.

The following discussion on complex numbers is covered more fully in the appendix on Mathematics for Electronics and may be omitted by the student who thoroughly understands it.

Complex Numbers. A method of representing an inductive reactance in series with a resistance uses a system called *complex numbers*. This employs a term called the *j factor*. The 4000-Ω X_L of Fig. 12-13*a* is written as $+j4000$, while the 3000-Ω R is simply written as *3000*. The two in series is, therefore, written as $3000 + j4000$. This form of a complex number is called the *rectangular form*. The first part (3000) represent the horizontal phasor ($R = 3000$) of Fig. 12-13*c*. The second portion ($+j4000$) is the vertical phasor ($X_L = 4000$) of Fig. 12-13*c*. The phasor sum (Z) is then $R + jX_L$, and $Z = \sqrt{R^2 + X_L^2}$.

Another form of a complex number is the *polar* form. This describes a phasor as having a magnitude at a certain phase angle. As shown listed in Fig. 12-13*c*, the total opposition or impedance Z is 5000 Ω at an angle of 53° and is written as $Z = 5000\,\underline{/53°}$.

To change from the rectangular form ($3000 + j4000$) to the polar form ($5000\,\underline{/53°}$), impedance is found from $Z = \sqrt{3000^2 + 4000^2}$. The phase angle θ is found as shown in Fig. 12-13*c* by using the tangent function, or $\tan\theta =$

4000/3000 = 1.33, and θ is then 53°. Therefore, the rectangular form (3000 + j4000) yields the polar form ($5000\angle 53°$).

The *polar* form of an impedance cannot be *added* to that of another impedance. To add complex numbers, it is necessary to use the *rectangular* forms. Therefore, it is convenient to change from the *polar* to the *rectangular form*. In Fig. 12-15, it may be seen that in the phasor diagram, Z is 5000 Ω at an angle of 53° (polar form, $Z = 5000\angle 53°$). The total impedance Z is made up of the horizontal phasor (R = 3000 Ω) and the vertical phasor (X_L = 4000 Ω). In Fig. 12-15, the trigonometric form and, then, the rectangular form are shown derived.

$$\text{trigonometric form} = Z(\cos\theta + j\sin\theta) \qquad (12\text{-}20)$$

By using the *trigonometric form* of a complex number, the *polar* form may then be converted into the *rectangular* form as shown in the following.

Example 12-23

In the phasor diagram of Fig. 12-15, change the polar form of impedance Z into its rectangular form.

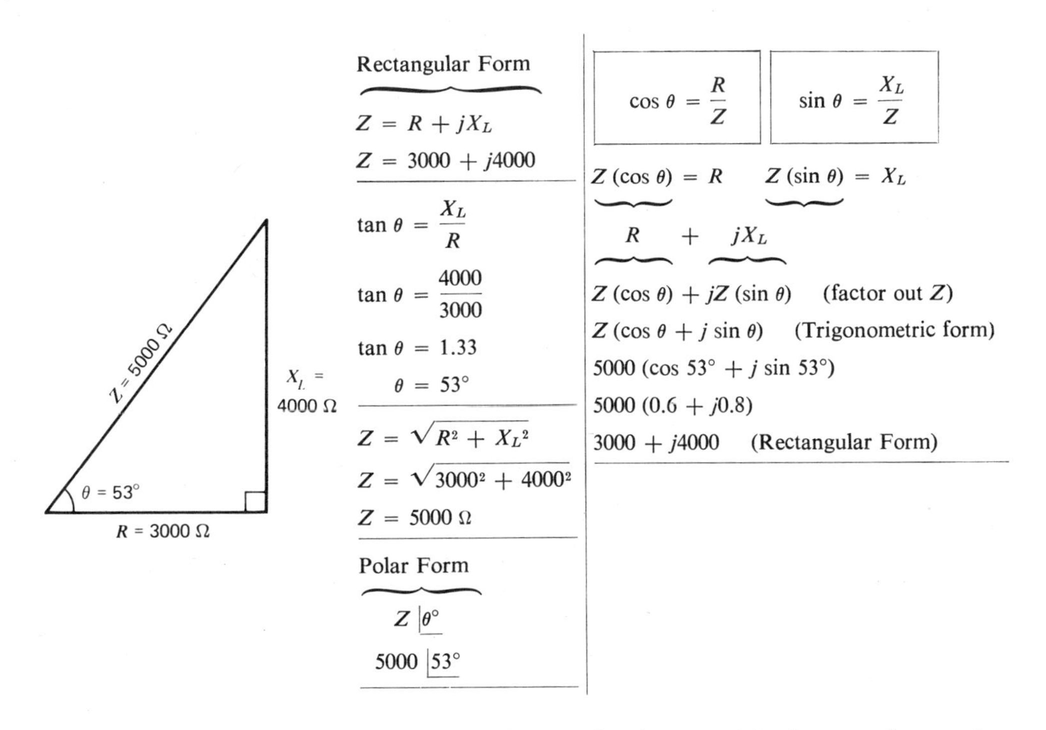

Figure 12-15. Rectangular, polar, and trigonometric forms of complex numbers.

Solution

The polar form of Z is

$$Z\underline{/\theta^\circ}$$

$$5000\underline{/53^\circ}$$

The *trigonometric* form becomes

$$Z(\cos\theta + j\sin\theta) \qquad (12\text{-}20)$$

$$5000(\cos 53^\circ + j\sin 53^\circ)$$

By using the trigonometric tables in the Appendix, where cos 53° is found to be 0.602, and sin 53° is found to be 0.799, then

$$5000(0.602 + j0.799)$$

Multiplying, gives the *rectangular* form:

$$3000 + j4000 \qquad \text{(approximately)}$$

For Similar Problems Refer to End-of-Chapter Problems 12-29 to 12-35

Complex Number Addition. When coils and resistors are in series, the inductive reactances (X_{L_1} and X_{L_2}) must be added. The resistors (R_1 and R_2) are also added. The rectangular forms are added by simply adding the R portions separately, and then adding the j portions separately. As an example, refer to Fig. 12-16. Here, a coil having an X_{L_1} of 300 Ω and a resistance R_1 of 70 Ω is connected in series with a second coil having an X_{L_2} of 500 Ω and a resistance R_2 of 130 Ω.

The first coil and its resistance have an impedance Z_1 expressed in rectangular form as *70 + j300*. The second coil and its resistance have an impedance Z_2 expressed in rectangular form as *130 + j500*. These are shown in Fig. 12-16*a*. Then, total impedance Z_T is the *sum* of the Z_1 and Z_2 rectangular forms. Adding R_1 (70 Ω) and R_2 (130 Ω) gives R_{total} of *200* Ω. Adding X_{L_1} (+j300) and X_{L_2} (+j500 Ω) gives a total inductive reactance X_{L_T} of *j800*.

Total impedance Z_T, expressed in rectangular form, is *200 + j800*, shown in Fig. 12-16*a*. The phasor representation of R_T, X_{L_T}, and Z_T is shown in Fig. 12-16*b*. The total impedance in its *rectangular* form, Z_T = *200 + j800*, is then converted to its *polar* form to find its magnitude and the phase angle, θ_T, as depicted in Fig. 12-16*b*, where $Z_T = 825\underline{/76^\circ}$. The value of Z_T is found by using the Pythagoras theorem for the right triangle of Fig. 12-16*b*:

$$Z_T = \sqrt{R_T^2 + X_{L_T}^2} \qquad (12\text{-}5)$$

$$Z_T = \sqrt{200^2 + 800^2}$$

$$Z_T = \sqrt{(2 \times 10^2)^2 + (8 \times 10^2)^2}$$

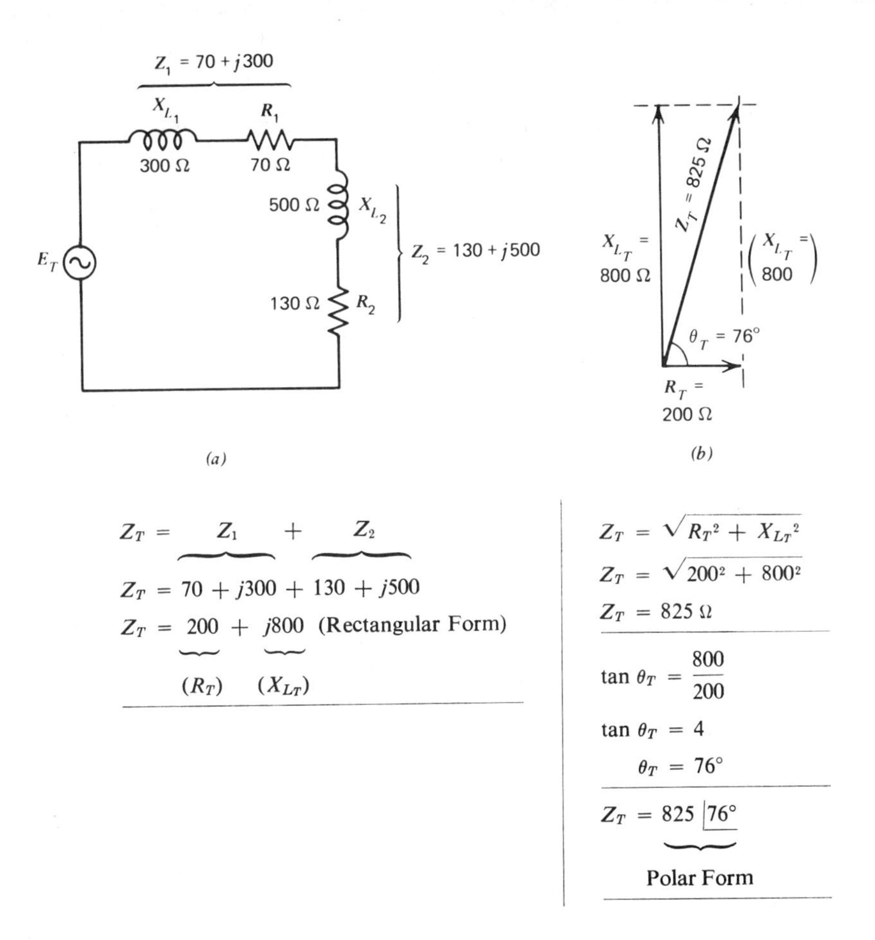

Figure 12-16. (*a*) Two coils and resistors in a series ac circuit. (*b*) Phasor diagram.

$$Z_T = \sqrt{4 \times 10^4 + 64 \times 10^4}$$

$$Z_T = \sqrt{68 \times 10^4}$$

$$Z_T = 8.25 \times 10^2$$

$$Z_T = 825\ \Omega$$

The phase angle θ_T of the right triangle of Fig. 12-16*b* may be found by using any of the trigonometric functions of the angle. The tangent gives: $\tan \theta_T = X_{LT}/R_T = 800/200 = 4$, $\theta_T = 76°$. Therefore, total impedance expressed in its polar form is $825\underline{/76°}$.

The following example illustrates the *addition* of complex numbers in an X_L and R series circuit.

Example 12-24

A coil has an inductive reactance X_{L_3} of 700 Ω and a resistance R_3 of 200 Ω. It is connected in series with a second coil that has an X_{L_4} of 800 Ω and a resistance R_4 of 300 Ω. Find: (a) Z_3 expressed in its rectangular form, (b) Z_4 in its rectangular form, (c) Z_T in its rectangular form, and (d) Z_T in its polar form.

Solution

Solve this example using the following results as a guide: (a) $Z_3 = 200 + j700$, (b) $Z_4 = 300 + j800$, (c) $Z_T = 500 + j1500$, and (d) $Z_T = 1580\angle 71.5°$.

For a Similar Problem Refer to End-of-Chapter Problem 12-36

Complex Number Multiplication. When one coil having X_{L_1} and R_1 is in *parallel* with a second coil having X_{L_2} and R_2, one method of finding the total impedance is to use the *product of the two divided by their sum*. Multiplying and dividing rectangular forms is more difficult than doing it by using polar forms. When *polar forms* of complex numbers are to be *multiplied*, the *values* are simply *multiplied* together, but the *angles* are *algebraically added.*

If $Z_1 = 15\angle 20°$ and is to be multiplied by $Z_2 = 30\angle 18°$, then $Z_{\text{product}} = 15$ *times* 30, or 450, while the angles of 20° and 18° are added, giving 38°. Z_{product} is then $450\angle 38°$. The following example illustrates this *multiplication* of *polar forms*.

Example 12-25

Multiply the following complex numbers and state the answers in polar forms.

(a) $(70\angle 30°)(50\angle 40°) =$

(b) $(150\angle 60°)(200\angle 50°) =$

(c) $(25\angle 70°)(30\angle 35°) =$

Solution

(a) $3500\angle 70°$

(b) $30{,}000\angle 110°$

(c) $750\angle 105°$

Complex Number Division. *Dividing polar forms* consists of *dividing* the values but *algebraically subtracting* the angles. If $Z_3 = 75\angle 30°$ is to be divided by $Z_4 = 25\angle 10°$, then the division is 75 ÷ 25 or 3, and the angle becomes 30° minus 10° or 20°. The result is then $3\angle 20°$. If the *divisor* angle (the term doing the division) is larger (and is positive) than the dividend (the term being divided), then the result is a *negative* angle. This is shown in the following. $120\angle 30°$ (*the dividend*) ÷ $40\angle 50°$ (the *divisor*) gives a result of 120 ÷ 40, or 3, while the resulting angle is 30° *minus* 50°, or −20°. The final result is then $3\angle -20°$.

The following example illustrates the *division* of *polar forms*.

Example 12-26

Perform the following polar-form divisions and state the answers in polar forms.

(a) $70\angle 30° \div 10\angle 70° =$

(b) $\dfrac{60\angle 50°}{12\angle 20°} =$

(c) $\dfrac{200\angle 80°}{25\angle 30°} =$

Solution

(a) $7\angle -40°$

(b) $5\angle 30°$

(c) $8\angle 50°$

For Similar Problems Refer to End-of-Chapter Problems 12-37 to 12-42

12-7. Inductors and Resistors in AC Parallel Circuits (More Advanced)

The circuit of Fig. 12-17*a* shows a coil having an X_{L_1} of 400 Ω and having a resistance R_1 of 300 Ω, comprising an impedance Z_1. This is in *parallel* with a second coil having an X_{L_2} of 600 Ω and having a resistance R_2 of 200 Ω, making an impedance Z_2. The two parallel branches Z_1 and Z_2 form a total impedance Z_T. To find Z_T, one method is to use the *product of* Z_1 *and* Z_2 *divided by their sum*, as shown in Fig. 12-17.

Z_1 expressed in rectangular form is $300 + j400$. The reader should change it into its polar form, as shown in the left column of Fig. 12-17. Z_1 in its polar form is $500\angle 53°$, as shown in Fig. 12-17.

Similarly, Z_2 should be converted from its rectangular form $(200 + j600)$ into its polar form, as shown in the center column of Fig. 12-17. Z_2, in its *polar form* is $633\angle 71.5°$.

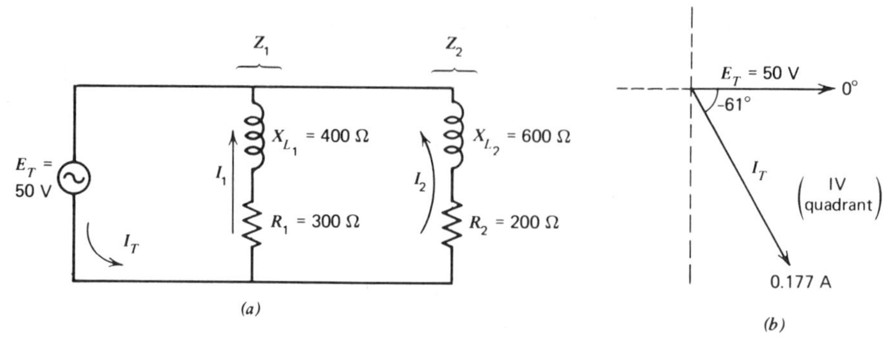

Left Branch	Right Branch	Total
$Z_1 = 300 + j400$ (Rectangular Form)	$Z_2 = 200 + j600$ (Rectangular Form)	$Z_T = \dfrac{(Z_1)(Z_2)}{Z_1 + Z_2}$
$Z_1 = \sqrt{300^2 + 400^2}$ $Z_1 = 500\ \Omega$	$Z_2 = \sqrt{200^2 + 600^2}$ $Z_2 = 633\ \Omega$	$Z_T = \dfrac{(500\angle 53^\circ)(633\angle 71.5^\circ)}{(300 + j400) + (200 + j600)}$
$\tan\theta_1 = \dfrac{400}{300} = 1.33$ $\theta_1 = 53^\circ$	$\tan\theta_2 = \dfrac{600}{200} = 3$ $\theta_2 = 71.5^\circ$	$Z_T = \dfrac{316{,}500\angle 124.5^\circ}{500 + j1000}$ $Z_T = \dfrac{316{,}500\angle 124.5^\circ}{1120\angle 63.5^\circ}$
$Z_1 = 500\angle 53^\circ$ (Polar Form)	$Z_2 = 633\angle 71.5^\circ$ (Polar Form)	$Z_T = 283\angle 61^\circ$ (Polar Form)
$I_1 = \dfrac{E_T}{Z_1} = \dfrac{50\angle 0^\circ}{500\angle 53^\circ}$ $I_1 = 0.1\angle -53^\circ$ A	$I_2 = \dfrac{E_T}{Z_2} = \dfrac{50\angle 0^\circ}{633\angle 71.5^\circ}$ $I_2 = 0.079\angle -71.5^\circ$ A	$I_T = \dfrac{E_T}{Z_T} = \dfrac{50\angle 0^\circ}{283\angle 61^\circ}$ $I_T = 0.177\angle -61^\circ$ (Polar Form)

$$I_T = I_1 + I_2$$

$$I_T = 0.1\angle -53^\circ + 0.079\angle -71.5^\circ$$

$$I_T = 0.1(\cos -53^\circ + j\sin -53^\circ) + 0.079(\cos -71.5^\circ + j\sin -71.5^\circ) \quad \text{(Trigonometric Form)}$$

$$I_T = 0.1(0.6 - j0.8) + 0.079(0.317 - j0.949)$$

$$I_T = 0.06 - j0.08 + 0.025 - j0.075$$

$$I_T = 0.085 - j0.155 \quad \text{(Rectangular Form)}$$

$$I_T = 0.177\angle -61^\circ \text{ A} \quad \text{(Polar Form)}$$

Figure 12-17. Inductors and resistors in a parallel ac circuit. (*a*) Parallel circuit. (*b*) Phasors.

Total impedance Z_T may now be found by using the product of Z_1 and Z_2, divided by their sum, as shown in the right column of Fig. 12-17. Solve for Z_T by using the method shown as a guide.

$$\text{total impedance } Z_T = 283\underline{|61^\circ}$$

This means that the circuit of Fig. 12-17*a* has a total impedance $Z_T = 283\Omega$. The phase angle between the total inductive reactance and the total resistance is 61°. This is also the phase angle between total voltage E_T and total current I_T.

I_T may be found from Ohm's law.

$$I_T = \frac{E_T}{Z_T}$$

$$I_T = \frac{50\underline{|0^\circ}}{283\underline{|61^\circ}}$$

In the parallel circuit of Fig. 12-17*a*, total voltage E_T is across Z_1 and also across Z_2. E_T is then used as a reference phasor, and its 50 V is then called zero degrees, as shown in Fig. 12-17*b*.

$$I_T = \frac{50\underline{|0^\circ}}{283\underline{|61^\circ}}$$

$$I_T = 0.177\underline{|-61^\circ} \text{ A}$$

The *negative* angle means that I_T is *lagging* 61° behind E_T which is the zero degree reference phasor, as shown in Fig. 12-17*b*.

Another method of solving for I_{total} and Z_{total} in the circuit of Fig. 12-17*a* is to first find each branch current I_1 and I_2, then add them for I_T. Z_T is then found by using Ohm's law. This second method is shown in the text of Fig. 12-17. Solve for I_1, I_2, I_T, and Z_T, using the procedures shown in the figure. The detailed solution of I_T is shown at the bottom of this figure, and the I_T components are shown in Fig. 12-18.

The following example has three parallel coils and resistors and will be solved in the same manner as is shown in the previous discussion.

Example 12-27

In Fig. 12-19, consisting of three inductors and three resistors in a parallel circuit, find the following: (a) the impedance in the left branch Z_3, (b) impedance of the center branch Z_4, (c) impedance of the right branch Z_5, (d) total impedance using the "product over the sum" method, and (e) total current I_T.

Solution

The reader should solve this example being guided by the procedures and results of Fig. 12-19.

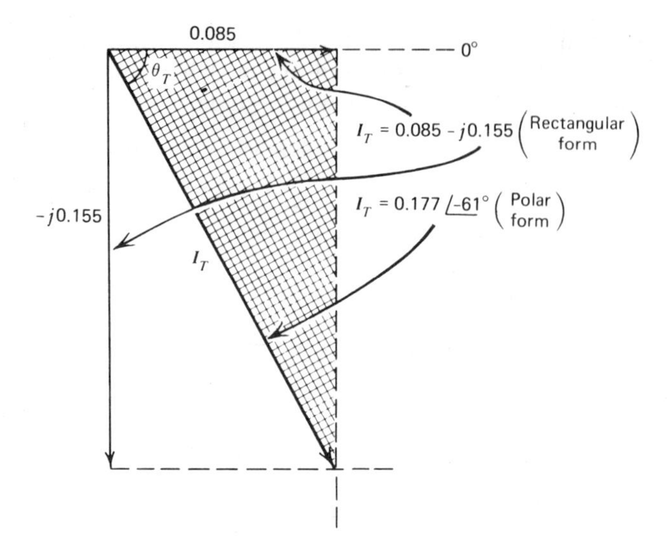

Figure 12-18. Phasor diagram of I_T and its components.

Example 12-28

In the three-branch parallel circuit of Fig. 12-19 (also used in the previous example) find: (a) left branch current I_3, (b) center branch current I_4 (c) right-branch current I_5, (d) total current I_T (compare with result in previous example), and (e) Z_T (compare with result in previous example).

Solution

Solve this example by using the method shown previously in the text of Fig. 12-17 for branch and total currents. As a guide, the following results should be used: (a) $I_3 = 0.485\angle{-76°}$, $= 0.117 - j0.47$; (b) $I_4 = 0.1635\angle{-78.5°}$, $= 0.0327 - j0.16$; (c) $I_5 = 0.116\angle{-54.5°}$, $= 0.0674 - j0.0944$; (d) $I_T = 0.2171 - j0.7244$, $= 0.757\angle{-73.5°}$; and (e) $Z_T = 13.2\angle{73.5°}$.

For Similar Problems Refer to End-of-Chapter Problems 12-43 to 12-46

Section 12-10, Inductance and Internal Resistance, discusses X_L and R (of coil) as a series circuit.

Power and Power Factor in L-R Series and Parallel Circuits, More Advanced. In Section 12-3, a discussion of the power dissipated in an inductor and resistor circuit is given. It is shown that power is only dissipated in the resistor, with none in the inductor. The reason for this is that when cur-

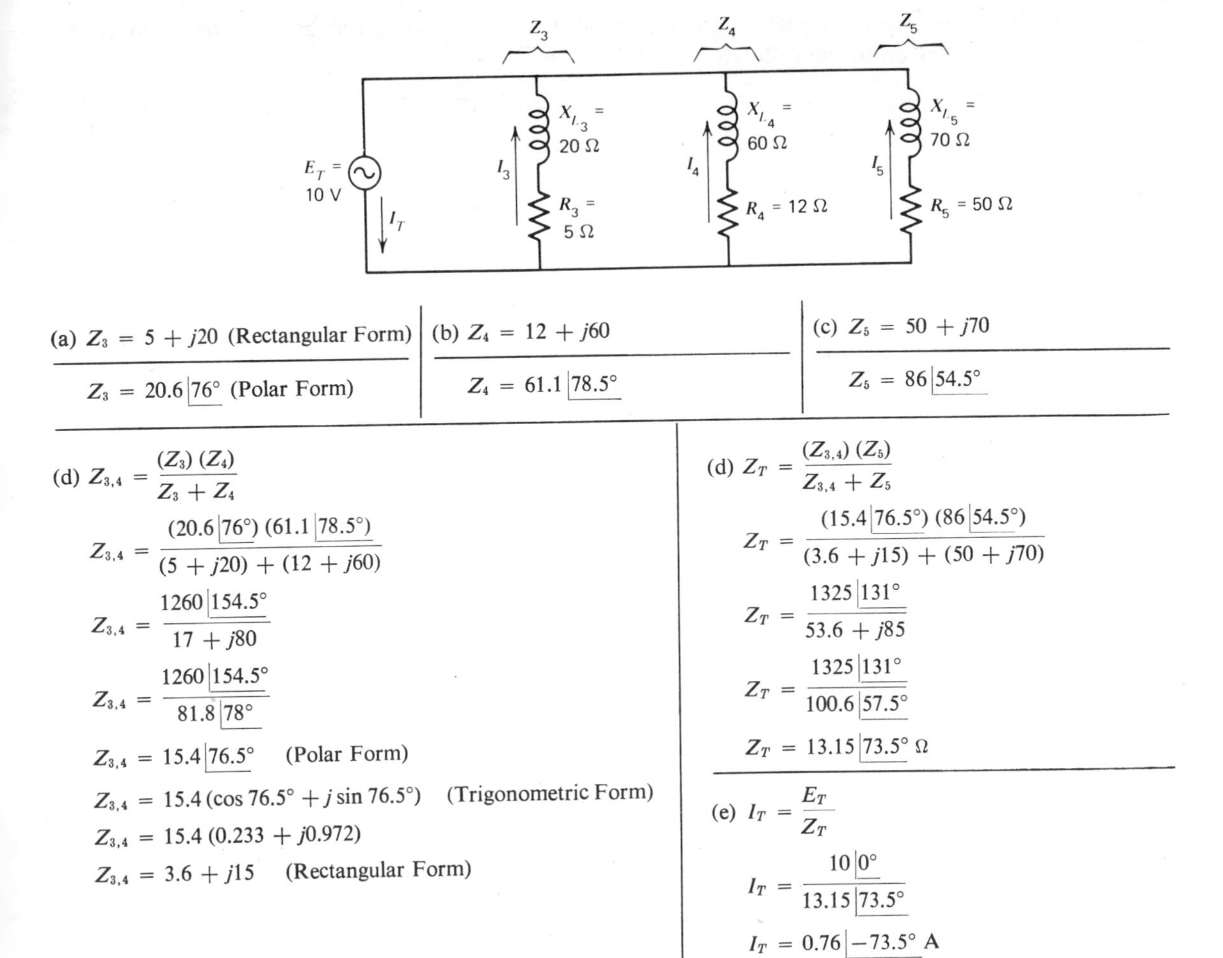

Figure 12-19. Example 12-27.

rent *increases* in the inductor, energy is stored in the magnetic field. When the current *decreases*, the collapsing magnetic field induces a voltage in the coil, thus returning energy back into the circuit.

In Section 12-3, it is shown that the *actual* or *true* power dissipated is the product of E_R and I (equation 12-6), while the product of E_T and I_T is called the apparent power. The apparent power in an LR ac circuit is not the power dissipated. The ratio of the *true* power to the *apparent* power is called the *power factor* (shown previously as equation 12-7).

Another method of finding the actual power dissipated or *true* power is given in the following.

$$P_{\text{true}} = (P_{\text{apparent}})\,(\text{power factor}) \qquad (12\text{-}21)$$

$$P_{\text{true}} = (E_T)\,(I_T)\,(\cos\theta_T) \qquad (12\text{-}21)$$

or

$$P_{\text{true}} = (E_T)\,(I_T)\left(\frac{R}{Z}\right) \qquad (12\text{-}21)$$

In the phasor diagram of Fig. 12-16*b*, the *cos* θ (adjacent side/hypotenuse) is R_T/Z_T, and is also called the *power factor*.

The following example shows a method of finding the true power dissipated in the series ac circuit of Fig. 12-16*a*.

Example 12-29

Assume that the applied voltage source $E_T = 33$ V in the circuit of Fig. 12-16*a*. (a) Find the total power dissipated or the *true* power by using equation 12-21. (b) Find the total power dissipated by using the power dissipated in each resistor.

Solution

(a) From the values shown in the text portion of Fig. 12-16, the total impedance is $Z_T = 825\angle 76^\circ\ \Omega$. With an $E_T = 33$ V, then the value of total current is

$$I_T = \frac{E_T}{Z_T}$$

$$I_T = \frac{33}{825}$$

$$I_T = 0.04\text{ A}$$

True total power is then

$$P_{\text{true}} = (E_T)\,(I_T)\,(\text{power factor}) \qquad (12\text{-}21)$$

$$P_{\text{true}} = (E_T)\,(I_T)\,(\cos\theta_T)$$

$$P_{\text{true}} = (33)\,(0.04)\,(\cos 76^\circ)$$

$$P_{\text{true}} = (33)\,(0.04)\,(0.242)$$

$$P_{\text{true}} = 0.32\text{ W}$$

Find the total true power by adding the true power dissipated in each resistor ($I_T^2\,R$). The following results should be used as a guide: $P_1 = 0.112$ W, $P_2 = 0.208$ W.

Parallel Circuits, Power. In the two-branch parallel circuit of Fig. 12-17*a*, total true power dissipated may be found in any one of several ways. The most direct approach is to employ the power factor method of equation 12-21 [$P_T = (E_T)(I_T)(\mathrm{PF})$]. A second method is to find the power dissipated in each resistor, and then take the sum of each branch power for the total. A third approach is to find the equivalent resistor (R_T) portion of Z_{total} and use the equation $P_T = (I_T^2)(R_T)$. The following example illustrates each of the three methods.

Example 12-30

In the two-branch circuit of Fig. 12-17*a*, and using the worked-out solutions shown in the text portion of Fig. 12-17, find: (a) total true power dissipated, using the power factor method of equation 12-21, (b) total true power dissipated by finding the power dissipated in each branch, and (c) total true power dissipated by finding the power dissipated in the equivalent or total resistance.

Solution

(a) From the text portion of Fig. 12-17, total current $I_T = 0.177\angle{-61^\circ}$ A. Solve for total true power, as shown previously, using the following result as a guide: $P_T = 4.28$ W.

(b) Again find P_T by adding the true power dissipated in each resistor. The results should be: $P_1 = 3$ W, $P_2 = 1.25$ W, and $P_T = 4.25$ W.

(c) From the text of Fig. 12-17, $Z_T = 283\angle 61^\circ\ \Omega$ (polar form). Changing Z_T into its *rectangular* form,

$$Z_T = 283\angle 61^\circ$$

$$Z_T = 283\ (\cos 61^\circ + j \sin 61^\circ)$$

$$Z_T = 283\ (0.485 + j0.875$$

$$Z_T = 137 + j248 \text{ (rectangular form)}$$

The $137 + j248$ means that the *total equivalent resistance* (R_T) of the circuit is *137* Ω, while the total inductive reactance (X_{L_T} is *248* Ω. Total true power is the power dissipated in the total equivalent resistance or

$$P_T = I_T^2 R_T$$

where I_T is shown in the text portion of Fig. 12-17 as 0.177 A, and

$$P_T = (0.177)^2\ (137)$$

$$P_T = (0.0314)\ (137)$$

$$P_T = 4.3 \text{ W}$$

Note that this answer of 4.3 W agrees, within slide-rule tolerance, with that part (a), 4.28 W, and that of part (b), 4.25 W.

For Similar Problems Refer to End-of-Chapter Problems 12-47 to 12-50

12-8. Energy in a Coil

When an alternating current (ac) flows through a coil, a magnetic field builds up and then collapses around the coil. Energy is stored in this field as it builds up, and energy is returned as this field collapses. When the coil is a pure inductance (theoretical) having no resistance, all the energy is stored in the magnetic field, since there is no resistance to dissipate any energy. Actually, every coil has some resistance, if only the ohmic resistance of the wire itself. The resistance produces some energy loss.

The energy stored in the magnetic field depends on the value of the inductance itself and the amount of current. This is shown in the following:

$$\text{energy} = \tfrac{1}{2}LI^2 \qquad (12\text{-}22)$$

where *energy* is in *joules* (a joule per second is a watt), L is in *henrys*, and I is in *amperes*.

The following examples illustrate this.

Example 12-31

When a current rises from zero to 5 mA through a 20-mH coil, how much energy is being stored in the magnetic field?

Solution

$$\text{energy} = \tfrac{1}{2}\, LI^2 \qquad (12\text{-}22)$$

$$\text{energy} = \tfrac{1}{2}\,(20\text{ mH})\,(5\text{ mA})^2$$

$$\text{energy} = \tfrac{1}{2}\,(20 \times 10^{-3})\,(5 \times 10^{-3})^2$$

$$\text{energy} = \tfrac{1}{2}\,(20 \times 10^{-3})\,(25 \times 10^{-6})$$

$$\text{energy} = 250 \times 10^{-9}\text{ J}$$

Example 12-32

When the switch is closed in the circuit shown, the 12 V is applied to the 250-μH coil in series with a 10-Ω resistor. Find (a) I, and (b) energy stored in the magnetic field.

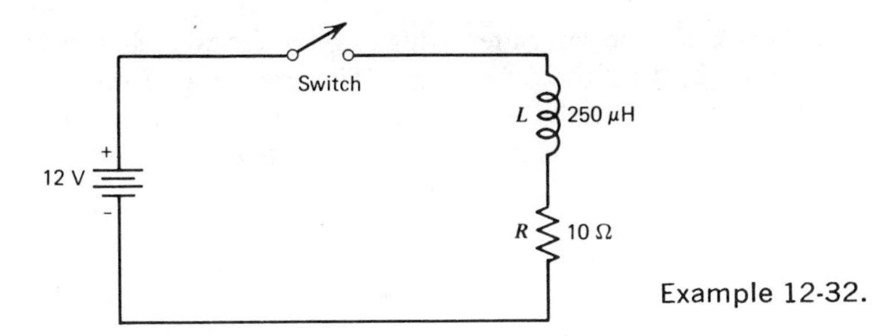

Example 12-32.

Solution

(a) $I = \frac{E}{R}$

$I = \frac{12}{10}$

$I = 1.2 \text{ A}$

(b) $\text{energy} = \frac{1}{2} LI^2$ (12-22)

$\text{energy} = \frac{1}{2} (250 \ \mu\text{H}) (1.2)^2$

$\text{energy} = \frac{1}{2} (250 \times 10^{-6}) (1.44)$

$\text{energy} = 180 \times 10^{-6} \text{ J}$

For Similar Problems Refer to End-of-Chapter Problems 12-51 and 12-52

Q of a Coil. The *ratio* of the amount of *energy stored* in the magnetic field around a coil, and the amount of *energy lost* because of the resistance of the coil is called the *quality* (Q) of the coil. It is also called the *figure of merit* of the coil. Depending on the coil construction (size of wire, core material, etc.), the Q of some coils may have a small ratio of about 5 (called a low Q), while other coils may run up to a large ratio of several hundred or a thousand or more (called a high Q).

The equation for Q is

$$Q = \frac{X_L}{R} \tag{12-23}$$

where Q is simply a ratio, having no unit; X_L is the inductive reactance in ohms; and where R in ohms is a combination of several things, each producing some energy loss, and is often called the *effective resistance*.

This resistance, R, consists of the following.

(a) The dc ohmic resistance of the wire itself.
(b) *Skin effect.* With direct current and also low frequency alternating current, electrons flow uniformly through the area of a wire. With high frequency alternating current, most of the current tends to flow along the outer surface

or skin of the wire, attempting to get as far away as possible from the magnetic lines in the center of the wire. This crowding of the electrons on the surface of the wire results in a greater opposition to the current. This is called *skin effect*. To reduce this skin effect opposition, a thicker wire having more surface area and, therefore, less opposition could be used. Since at these higher frequency ac currents, little or no current flows in the center portion of the wire, hollow tubing is often used instead of a solid, thick wire. Another method of reducing this skin effect is to use stranded wires insulated from each other and woven or braided into a cable-type conductor. Each wire strand is then, at some places, on the surface of the cable and, at other places, in the center. The strands are connected together at one end, acting like one heavy wire, and are also connected together at the other end. This is called *Litz* wire.

(c) Another portion of the effective resistance, R, of a coil is due to energy losses in the core itself. These are the *eddy currents* induced in the core by the varying magnetic field, and the *hysteresis* losses produced in the core each time the core is being magnetized, demagnetized, and then remagnetized with reverse poles.

This effective resistance cannot be measured directly, but may be found by knowing the X_L and Q of the coil. The value of Q of a coil at a particular frequency may be determined by using a laboratory instrument called a *Q-meter*. From equation 12-23,

$$Q = \frac{X_L}{R} \tag{12-23}$$

and solving in terms of R yields

$$QR = X_L$$

and

$$R = \frac{X_L}{Q} \tag{12-24}$$

The following examples illustrate the use of equations 12-23 and 12-24.

Example 12-33

A coil has an X_L of 5000 Ω at 150 kHz and an effective resistance R of 25 Ω. Find Q.

Solution

$$Q = \frac{X_L}{R} \tag{12-23}$$

$$Q = \frac{5000}{25}$$

$$Q = 200$$

Example 12-34

A 50-μH coil has an effective resistance of 20 Ω at a frequency of 20 MHz. Find (a) X_L, and (b) Q.

Solution

(a) $X_L = 2\,\pi FL$ (12-2)

$X_L = (6.28)\ (20\ \text{MHz})\ (50\ \mu\text{H})$

$X_L = (6.28)\ (20 \times 10^6)\ (50 \times 10^{-6})$

$X_L = 6280\ \Omega$

(b) $Q = \dfrac{X_L}{R}$ (12-23)

$Q = \dfrac{6280}{20}$

$Q = 314$

Example 12-35

A coil has an X_L of 3500 Ω at a frequency of 50 kHz. At this frequency, a Q-meter indicates a Q of 175. Find the effective resistance R.

Solution

$$R = \frac{X_L}{Q} \qquad (12\text{-}24)$$

$$R = \frac{3500}{175}$$

$$R = 20\ \Omega$$

For Similar Problems Refer to End-of-Chapter Problems 12-53 to 12-55

12-9. Voltage Square Waves and LR Circuits

A voltage *square wave* is simply the graph of voltage values (on the vertical axis) plotted against periods of time (on the horizontal axis). Figure 12-20*a* shows one cycle of a 50 kHz square wave dc voltage that goes from zero volts to 10 V. In many applications of advanced electronics such as computers, radar, space-vehicle guidance, and even television, square-wave voltages are employed.

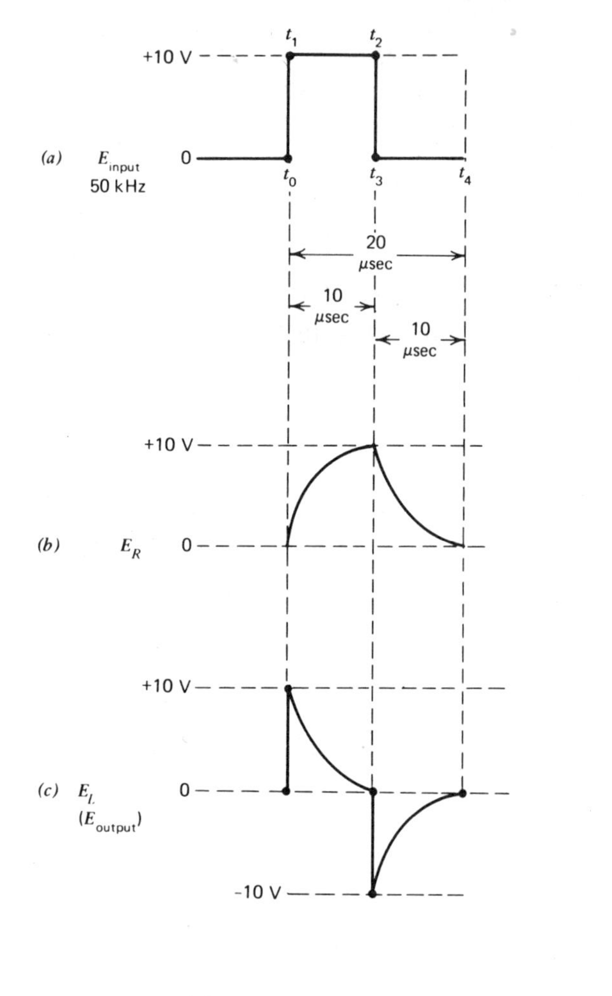

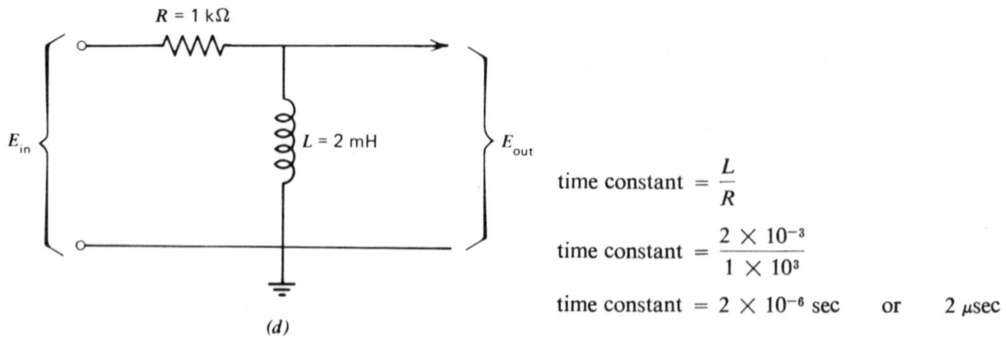

Figure 12-20. Voltage square wave applied to a short *LR* time-constant circuit (differentiating circuit).

In this section, an elementary discussion is given using a square-wave voltage that is applied to a coil and resistor as shown in the circuit of Fig. 12-20*d*.

Short Time-Constant, or Differentiating Circuit. The circuit of Fig. 12-20*d* consists of a 2 mH inductance in series with a 1 kΩ resistor. The time constant of a circuit, as discussed previously in Section 12-1, is: Time constant (in seconds) = L (in henrys)/R (in ohms) = 2 mH/ 1 K = $2 \times 10^{-3}/1 \times 10^{3} = 2 \times 10^{-6}$ sec, or 2 μsec.

The square-wave voltage shown in Fig. 12-20*a* has a frequency of 50 kHz, or 50,000 Hz. The time for one complete cycle is, therefore,

$$\underset{\text{(in seconds)}}{\text{time of one cycle}} = \frac{1}{\text{frequency (in hertz)}} \tag{12-25}$$

$$\text{time of one cycle} = \frac{1}{50 \text{ kHz}}$$

$$\text{time of one cycle} = \frac{1}{50 \times 10^{3}}$$

$$\text{time of one cycle} = 0.02 \times 10^{-3} \text{ sec} \quad \text{or} \quad 0.02 \text{ msec} \quad \text{or} \quad 20\ \mu\text{sec}$$

This is shown in Fig. 12-20*a* as the duration of one complete cycle, as the time from instant t_0 to instant t_4. A half-cycle is 1/2 of 20 usec, or 10 μsec, shown as the time between instants t_0 and t_3, and also as the time from instant t_3 to instant t_4.

The *LR* time constant of the circuit of Fig. 12-20*d* is 2 μsec, which is much less than the time duration of each half cycle (10 μsec). The L*R* time constant is therefore called a *short time constant.*

When a square-wave voltage is applied across a series *LR* circuit that is a short time constant, then the voltage produced across the coil (E_L) is a narrow spiked wave form, shown in Fig. 12-20*c*. This is called a *differentiated* wave, and the short time constant is called a *differentiating circuit.*

The following discussion explains the formation of these spiked voltage waves developed across the inductance. The *LR* time constant curves shown in the beginning of this chapter in Fig. 12-2, show that when a dc voltage is *first* applied to a series *LR* circuit, E_L immediately becomes equal to the applied voltage, while I and E_R are zero. Since the inductance opposes any *change* of current, the current attempts to flow, inducing voltage in L, which holds the current back. As a result, E_R is still zero, while E_L becomes equal to the applied voltage. This means that when the E input square wave of Fig. 12-20*a* goes from zero volts (at instant t_0) to +10 V (at instant t_1), E_R remains at zero, but E_L rises abruptly from zero to +10, shown in Fig. 12-20*b* and *c*.

Then E input (Fig. 12-20*a*) square wave remains at +10 V from instant t_1 to instant t_2 for 10 usec. This period of time is five times larger than the *LR* time constant of 2 μsec. In other words, there are *five* time constants during the 10 μsec period between instants t_1 and t_2. During this relatively long time (5 time constants), the current slowly rises to its maximum value, and a magnetic field builds up and then becomes stationary around the coil. As shown in the *LR* curves of Fig. 12-2, E_R rises gradually, becoming equal to the full applied voltage in about five time constants. During this time, E_L decreases gradually to zero.

As shown in Fig. 12-20*b* and *c*, E_R rises to +10 V at instant t_2, and E_L decreases from +10 V (at instant t_1) to zero volts (at instant t_2).

When *E* input square wave (Fig. 12-20*a*) changes abruptly from +10 V (at instant t_2) to zero volts (at instant t_3), the coil again opposes any *change* of current. With the applied voltage at zero (at instant t_3), current *attempts* to decrease. The magnetic field stored up around the coil starts collapsing, inducing a voltage in the coil that is opposite in polarity to the induced coil voltage when the magnetic field was building up. This opposite polarity coil voltage is shown at instant t_3 as −10 V (Fig. 12-20*c*). Since at this instant, current cannot yet change, then voltage across the resistor, E_R, remains at +10 V (Fig. 12-20*b*). This is also shown previously, in Fig. 12-12.

Finally, *E* input square wave (Fig. 12-20*a*) remains at zero from instant t_3 to instant t_4, a time of 10 μsec. This is five times as long as the *LR* time constant of 2 μsec. During this relatively long time (equal to five time constants), current has time to decrease to zero (shown previously in Fig. 12-12*b* during the "magnetic field collapse or decay time"). As the current decreases, E_R similarly decreases, and the weakening magnetic field induces smaller voltages in the coil.

After five *LR* time constants, *I* has become zero, and E_R and E_L have also become zero (at instant t_4), as shown in Fig. 12-20*b* and *c*.

The next cycle of the 50 kHz square wave starts at instant t_4 and the entire process is repeated.

Refer to End-of-Chapter Problems 12-56 to 12-58

12-10. Inductance and Internal Resistance

When the *inductive reactance* (X_L) of a coil is very much larger than the *resistance* of the coil, this *R* is usually ignored, and the coil may be thought of as being a *pure inductance*. If $X_L = 57\ \Omega$, while the coil resistance $R_L = 1\ \Omega$, then the coil impedance Z_L is just about the same as X_L, with the very small 1 Ω of resistance being negligible. The *polar form* of Z_L would be $57\underline{|89°}$ (approximate), while the *rectangular form* of Z_L is $1 + j57$.

If R_L is not insignificant compared to X_L, then the coil cannot be considered to be a pure inductance. In Fig. 12-13*a*, a coil is shown having an $X_L = 4000\ \Omega$. If the resistor shown, $R = 3000\ \Omega$, is the coil resistance, then the coil impedance, in its *polar form*, is shown in part (c) of this figure as $5000\angle 53^\circ$. By converting this into its rectangular form, Z of the coil $= 3000 + j4000$, where the 3000 is R of the coil, and the 4000 is X_L. The conversion from *polar* into *rectangular* is shown in Fig. 12-15 by first rewriting in the trigonometric form.

Figure 12-21*a* shows a coil having X_L, and with internal resistance R_L, connected in series with a resistor R_2. The total applied voltage is $E_T = 30$ V. Current is 300 mA, while voltage across R_2 is measured at 15 V, and voltage measured across the coil is 24.2 V (this includes E_{X_L} and E_{R_L}). The value of coil resistance R_L may be found from the following discussion.

The phasor diagram of Fig. 12-21*b* helps to visualize the equations in the discussion. The following values may be found by using Ohm's law and the circuit diagram of Fig. 12-21*a*.

$$R_2 = \frac{E_{R_2}}{I}$$

$$R_2 = \frac{15 \text{ V}}{300 \text{ mA}}$$

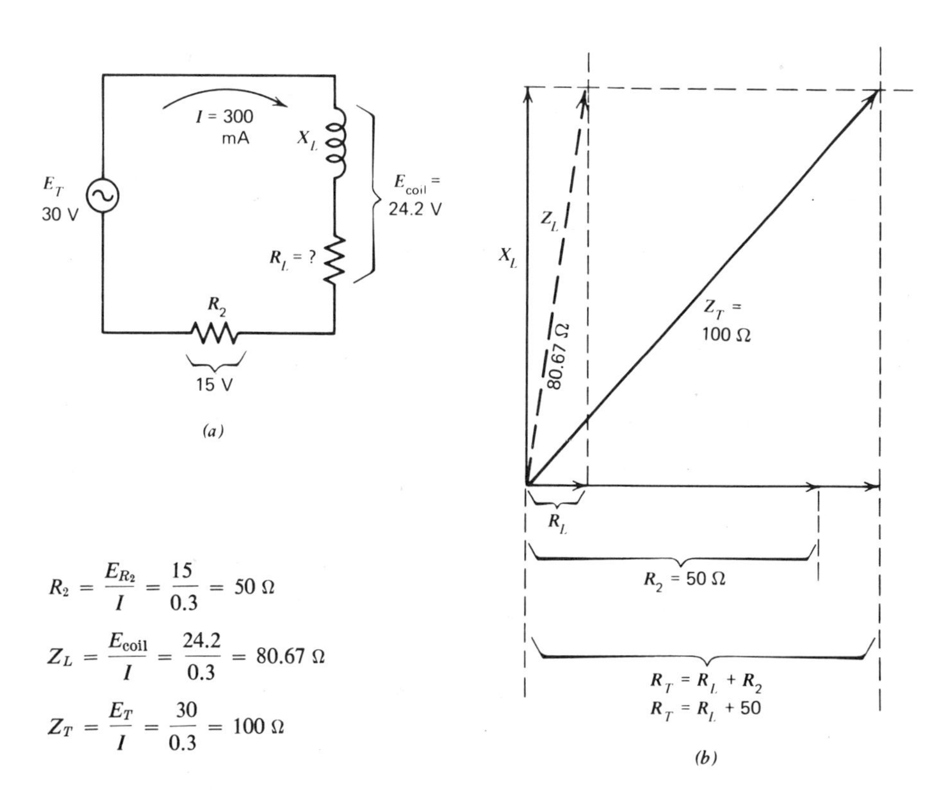

Figure 12-21. (a) Coil ($X_L + R_L$) in series with R_2. (b) Phasors.

$$R_2 = \frac{15}{0.3}$$

$$R_2 = 50\ \Omega$$

This is shown as the R_2 phasor in Fig. 12-21*b*.

Total impedance Z_T is

$$Z_T = \frac{E_T}{I}$$

$$Z_T = \frac{30\ \text{V}}{300\ \text{mA}}$$

$$Z_T = \frac{30}{0.3}$$

$$Z_T = 100\ \Omega, \quad \text{also shown in the phasor diagram}$$

Coil impedance Z_L is

$$Z_L = \frac{E_{\text{coil}}}{I}$$

$$Z_L = \frac{24.2\ \text{V}}{300\ \text{mA}}$$

$$Z_L = \frac{24.2}{0.3}$$

$$Z_L = 80.67\ \Omega, \quad \text{shown in the phasor diagram}$$

From Pythagoras' theorem, the following equations may be written. In the large right triangle of Fig. 12-21*b*, where Z_T is the hypotenuse,

$$Z_T^2 = X_L^2 + R_T^2 \tag{1}$$

and in the smaller right triangle where Z_L is the hypotenuse,

$$Z_L^2 = X_L^2 + R_L^2 \tag{2}$$

Subtracting equation 2 from equation 1 cancels the X_L term, yielding:

$$Z_T^2 - Z_L^2 = R_T^2 - R_L^2 \tag{3}$$

From the phasor diagram, R_T is the sum of R_L and R_2 (where R_2 has been found to be 50 Ω), or

$$R_T = R_L + 50$$

Substituting the term $R_L + 50$ in place of R_T in equation 3, gives

$$Z_T^2 - Z_L^2 = (R_L + 50)^2 - R_L^2 \tag{4}$$

Rewrite equation 4, putting in the values ($Z_T = 100$ and $Z_L = 80.67$).

$$100^2 - 80.67^2 = (R_L + 50)^2 - R_L^2 \tag{5}$$

$(R_L + 50)^2$ becomes $R_L^2 + 100\ R_L + 2500$, and equation 5 is now

$$100^2 - 80.67^2 = R_L^2 + 100\ R_L + 2500 - R_L^2$$

The R_L^2 terms cancel, and the equation becomes

$$10{,}000 - 6500 = 100\ R_L + 2500$$

Solving for R_L

$$10{,}000 - 6500 - 2500 = 100\ R_L$$

$$10{,}000 - 9000 = 100\ R_L$$

$$1000 = 100\ R_L$$

$$\frac{1000}{100} = R_L$$

$$10\ \Omega = R_L$$

For Similar Problems Refer to End-of-Chapter Problems 12-59 and 12-60

PROBLEMS

See section 12-1 for discussion covered by the following.

12-1. What is the value of the time constant where the inductor is 100 mH and the resistor is 200 Ω?

12-2. What is the time constant if $L = 200\ \mu H$ and $R = 50\ \Omega$?

12-3. If 6 V dc were applied to an inductor in series with a 300-Ω resistor, what would be the following at the instant the voltage was applied: (a) E_L, (b) I, and (c) E_R?

12-4. In the previous problem, after one time constant with the 6 V applied, what are the following values: (a) E_L, (b) I, and (c) E_R?

12-5. In Problem 12-3, after 5 time constants with the 6 V applied, what are the values of each of the following: (a) E_L, (b) I, and (c) E_R?

See section 12-2 for discussion covered by the following.

12-6. What is the amount of inductive reactance (X_L) of a 20-μH coil to an ac sine wave at a frequency of 75 kHz?

12-7. How much opposition (X_L) does a 25-mH coil offer to a 4-MHz ac sine wave?

See section 12-3 for discussion covered by the following.

12-8. What is the sum of two vectors 90° out of phase, if one is 9 pointing north, and the other is 12 pointing east?

See section 12-3, E_R and E_{X_L} Addition, for discussion covered by the following.

12-9. If 5 mA flows through a 4-kΩ resistor in series with an 8-kΩ X_L, find: (a) E_R, (b) E_{X_L}, and (c) E_T.
12-10. If 2 mA flows through a 5-kΩ resistor in series with a 15-kΩ X_L, find: (a) E_R, (b) E_{X_L}, and (c) E_T.

See section 12-3, Impedance or Total Opposition, for discussion covered by the following.

12-11. Find the total opposition or impedance (Z) if an X_L of 700 Ω is in series with an R of 300 Ω.
12-12. In a series ac circuit R is 40 Ω, X_L is 200 Ω, and the total applied voltage source E_T is 150 V. Find: (a) Z, (b) I, (c) E_R, and (d) E_{X_L}.
12-13. An ac voltage of 100 V at a frequency of 250 kHz is applied to a 400-μH coil having a resistance of 300 Ω. Find: (a) X_L, (b) Z, (c) I, (d) E_R, and (e) E_{X_L}.

See section 12-3, X_L's and R's in Series, for discussion covered by the following.

12-14. Two coils L_3 and L_4 are connected in series to an ac voltage, similar to that shown in Fig. 12-9*a*. If X_{L_3} is 300 Ω and R_3 is 100 Ω, while X_{L_4} is 500 Ω and R_4 is 150 Ω, find: (a) R_{total}, (b) $X_{L\ \text{total}}$, and (c) Z_{total}.

See section 12-3, Power in X_L and R circuits, for discussion covered by the following.

12-15. A 5000-Ω resistor is connected in series with a 7000-Ω resistor. If 180 V alternating current is applied, find the power factor.
12-16. A coil having an X_L of 10,000 Ω to an ac applied voltage of 200 V has practically no resistance. Find: (a) power factor, (b) P_{true} dissipated.
12-17. An X_L and a series R are connected to an ac source of 250 V. If $E_{X_L} = 200$ V, $E_R = 150$ V, and $I = 50$ mA, find: (a) P_{apparent}, (b) P_{true}, and (c) power factor.
12-18. In a series X_L and R circuit, 50 V ac is applied and the current is 60 mA. If the power factor is 0.8, find (a) P_{apparent}, and (b) P_{true}.

See section 12-4 for discussion covered by the following.

12-19. If 30 V ac were applied to an X_L of 60 Ω in parallel with an R of 20 Ω, find: (a) Z_{total}, (b) I_{total}, and (c) P_{true}, using $Z_T = (X_L)(R)/\sqrt{X_L^2 + R^2}$, (equation 12-9).

12-20. For the same circuit as Problem 12-19, find: I_T by using $I_T = \sqrt{I_{X_L}^2 + I_R^2}$ (equation 12-10). (The value found for I_T should agree with that of the previous problem.)

See section 12-4, Inductors and Resistors in Parallel AC Circuits, for discussion covered by the following.

12-21. In the parallel circuit shown for the diagram accompanying the problem, find: (a) $I_{X_{L1}}$, (b) $I_{X_{L2}}$, (c) I_{R_1}, (d) I_{R_2}, (e) $I_{X_{LT}}$, (f) I_{R_T}, (g) I_{total} (use: $I_T = \sqrt{I_{X_{LT}}^2 + I_{R_T}^2}$) and (h) Z_{total} (use Ohm's law).

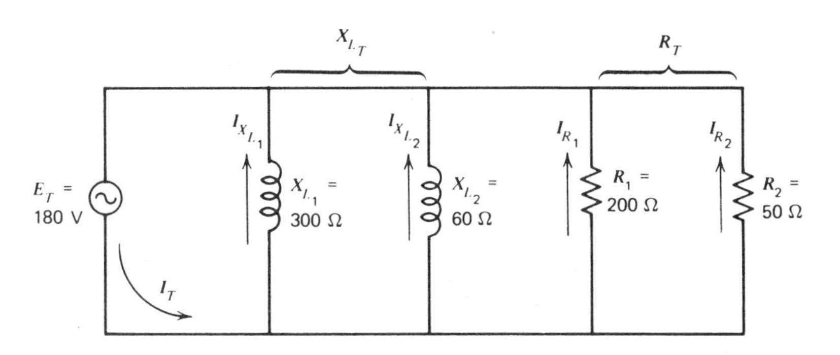

Problem 12-21.

12-22. In the circuit shown for Problem 12-21, find: (a) $X_{L\text{ total}}$, (b) R_{total}, (c) Z_{total} (use: $Z_T = (X_{L_T})(R_T)/\sqrt{X_{L_T}^2 + R_T^2}$), and also find (d) I_{total} (use Ohm's law).

See section 12-5, Voltages and Current Values During Magnetic Field Rise, for discussion covered by the following.

12-23. 100 V dc is applied to a 0.2 H coil in series with 10 Ω of resistance (this 10 Ω could be the resistance of the coil). Find (a) the time constant (τ), and the following values 10 msec after the voltage is applied: (e) e_L, (c) e_{R_1}, and (d) i.

12-24. A 50-mH coil is connected in series with a 250-Ω resistor with 10 V dc applied. Find (a) the time constant (τ), and the following values 700 μsec after the voltage is applied: (b) e_L, (c) e_{R_1}, and (d) i.

See section 12-5, Time During Magnetic Field Rise, for discussion covered by the following.

12-25. If 30 V dc is applied to a 500-mH coil in series with a 200-Ω resistor, find (a) time constant (τ) and, (b) the time (T) required after the voltage is first applied for e_L to fall to 6 V from its starting value of 30 V.

12-26. In the circuit described in the previous problem, find the time (T) required after the 30 V is applied for e_R to rise to 27 V.

See section 12-5, Voltages, Current, and Time During Magnetic Field Collapse or Decay, for discussion covered by the following.

12-27. In the circuit of Fig. 12-12*a*, assume that the switch has been in the "closed" position for sufficient time for I to reach its maximum value. If $E_T = 100$ V, $L = 300$ μH, and $R = 200$ Ω, find (a) time constant, and (b) current (i_{decay}), 2 μsec after the switch has been moved to the "shorted" position.

12-28. In the previous problem using the circuit of Fig. 12-12*a*, find the time (T) required after the switch is moved to the shorted position for e_R to fall from its maximum value (equal to E_T of 100 V) down to 50 V.

See section 12-6 for discussion covered by the following.

12-29. In the phasor diagram for this problem, find the angle θ where X_L is 100 Ω and Z is 200 Ω. (*Hint*: since the *O*pposite side and *H*ypotenuse are involved, the *S*ine function should be used: *SOH* CAH TOA).

12-30. In the phasor diagram shown, find the angle θ where R is 360 Ω and Z is 400 Ω (SOH *CAH* TOA).

12-31. In the phasor diagram shown, find the angle θ where R is 600 Ω and X_L is 2400 Ω (SOH CAH *TOA*).

12-32. In the phasor diagram shown where $\theta = 30°$, $R = 866$ Ω, $X_L = 500$ Ω, and $Z = 1000$ Ω, express Z in its *rectangular* form.

12-33. In the previous problem, express Z in its *polar* form.

12-34. Convert the rectangular form of the following complex number into its polar form: $100 + j200$.

12-35. Convert the *polar* form, $50\angle 30°$, into its *trigonometric* form, and then into its *rectangular* form.

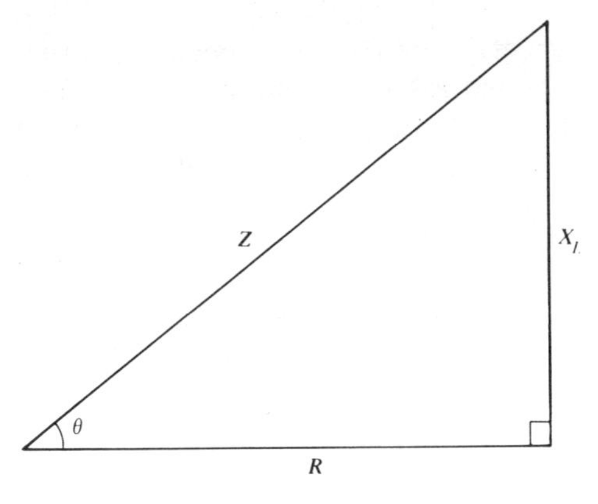

Problems 12-29 to 12-33.

See section 12-6, Complex Number Addition, for discussion covered by the following.

12-36. A coil having an $X_{L_1} = 2\ \text{k}\Omega$ and having an $R_1 = 600\ \Omega$ is connected in series with a second coil having an $X_{L_2} = 3\ \text{k}\Omega$ and an $R_2 = 400\ \Omega$. Find: (a) Z_1 in rectangular form, (b) Z_2 in rectangular form, (c) Z_{total} in rectangular form, and (d) Z_{total} in polar form.

See section 12-6, Complex Number Multiplication and Division, for discussion covered by the following.

12-37. $(12\angle 30^\circ)(60\angle 40^\circ) =$

12-38. $(40\angle 60^\circ)(20\angle 40^\circ) =$

12-39. $(300\angle 15^\circ)(250\angle 50^\circ) =$

12-40. $300\angle 15^\circ \div 250\angle 50^\circ =$

12-41. $40\angle 60^\circ \div 20\angle 40^\circ =$

12-42. $\dfrac{(60\angle 70^\circ)(30\angle 15^\circ)}{40\angle 50^\circ} =$

See section 12-7 for discussion covered by the following.

12-43. A coil having an $X_{L_1} = 30\ \Omega$ and an $R_1 = 10\Omega$ is in parallel with a second coil having an $X_{L_2} = 40\ \Omega$ and an $R_2 = 20\ \Omega$. The applied voltage source E_T is 50 V. Find: (a) impedance of first branch Z_1, (b) impedance of second branch Z_2, (c) total impedance Z_T, using "product over the sum," and (d) total current I_T, using Ohm's law.

12-44. In the circuit of the previous problem, find: (a) current in first branch I_1, (b) current in second branch I_2, (c) total current I_T, using the sum of the branch currents, and (d) total impedance Z_T, using Ohm's law. Compare the answers for I_T and Z_T with those of the previous problem.

12-45. A third coil having an $X_{L_3} = 50\ \Omega$ and an $R_3 = 25\ \Omega$ is connected in parallel with the other two coils of Problem 12-43. Find (a) the impedance of this third branch Z_3, (b) total impedance Z_T, by using the "product over the sum" method, and (c) total current I_T, using Ohm's law.

12-46. In the previous problem containing all three parallel branches, find (a) current in third branch I_3, (b) total current I_T, using the sum of the branch currents, and (c) Z_T, using Ohm's law. Compare the answers for I_T and Z_T with those of the previous problem.

See section 12-7, Power and Power Factor in LR Circuits, for discussion covered by the following.

12-47. In the three-branch parallel circuit of Fig. 12-19, find the true power (P_3) dissipated in the first branch Z_3, (a) by using equation 12-21, $P_3 = E_T I_3 \cos \theta_3$ and also (b) by using $P_3 = (I_3)^2 R_3$.

12-48. In the same circuit (Fig. 12-19), find the true power (P_4) dissipated in the middle branch Z_4, (a) by using equation 12-21, $P_4 = E_T I_4 \cos \theta_4$, and also (b) by using $P_4 = (I_4)^2 R_4$.

12-49. In the same circuit (Fig. 12-19), find the true power (P_5) dissipated in the right-hand branch Z_5, (a) by using equation 12-21, $P_5 = E_T I_5 \cos \theta_5$, and also (b) by using $P_5 = (I_5)^2 R_5$.

12-50. In the same circuit (Fig. 12-19), find the total *true* power (P_T) dissipated (a) by using equation 12-21, $P_T = E_T I_T \cos \theta_T$, and also (b) by using $P_T = (I_T)^2 R_T$, where R_T is the total equivalent resistance in the circuit. (R_T may be found by converting the *polar* form of Z_T into its rectangular form.

See section 12-8 for discussion covered by the following.

12-51. Find the energy in joules stored in the magnetic field of a 10-H coil if the current rises from zero to 200 mA.

12-52. When 30 V dc is applied to a 2-H coil having 10 Ω of resistance, find the energy stored in the magnetic field.

See section 12-8, Q of a Coil, for discussion covered by the following.

12-53. A coil has an X_L of 500 kΩ at a frequency of 5 kHz. Find the Q of this coil if its effective resistance is 10 kΩ.

12-54. A 500-μH coil has a measured Q of 157 at a frequency of 20 kHz. Find (a) X_L, and (b) effective resistance R.

12-55. A coil has an X_L of 150 Ω at a frequency of 0.15 MHz. If its measured Q is 100, what is its effective resistance R?

See section 12-9 for discussion covered by the following.

12-56. A voltage square wave has a frequency of 1 MHz. What is the time duration of (a) 1 cycle, in μsec, and (b) 1 half cycle, in μsec?

12-57. A 5-μH coil is connected in series with a 100-Ω resistor. What is the LR time constant?

12-58. If the square-wave voltage of Problem 12-56 went from zero volts to +30 V, and were applied to the circuit of Problem 12-57 (similar to that shown in Fig. 12-20), what are the following values: (a) E_R at instant t_1 (b) E_L at t_1, (c) E_R at t_2, (d) E_L at t_2, (e) E_R at t_3, (f) E_L at t_3, (g) E_R at t_4, and (h) E_L at t_4?

See section 12-10 for discussion covered by the following.

12-59. A coil consists of its X_L and its resistance R_L. If the coil impedance is $Z_L = 200\underline{|30°}$, find (a) R_L, and (b) X_L.

12-60. 100 V ac is applied to a coil in series with a resistor R_3. The coil has X_L and R_L, as shown in Fig. 12-21*a*. If E_{R_3} is 60 V, E_{coil} is 63.2 V (across both X_L and R_L), and I is 20 mA, find (a) R_3, (b) Z_L, (c) Z_T, and (d) R_L.

chapter 13

Capacitors and Resistors in DC and AC Circuits

A commonly used device in most electronic circuits is the *capacitor*, so named because it has the ability or capacity to store up a difference of charge, or a voltage. The capacitor was formerly called a *condenser*, and is still known by this name in the automotive field, where certain components of the distributor device are called the condenser and points.

The capacitor has unique properties which are used to advantage in electrical and electronic circuits. One unusual characteristic of a capacitor is that it prevents a current from flowing when the capacitor is in series with a resistor in a dc circuit, yet the same series circuit permits a current flow when an ac voltage is applied.

Another unusual feature that is peculiar to the capacitor is its ability to acquire a voltage across itself that becomes, with sufficient time, equal to the applied dc voltage. This is called the *charging* of the capacitor, which takes time. Then, when the applied dc voltage is reduced to zero, the capacitor voltage decreases, eventually becoming zero. This is called the *discharge* of the capacitor, which also takes time. As a result of this time requirement, a capacitor is often used as an electrical *filter*, where rapid variations of dc voltage do not give the capacitor sufficient time to either charge or discharge, resulting in a constant or steady dc voltage across the capacitor. Thus the varying dc voltage becomes a steady value. This process is called *filtering*.

A third use of capacitors is as part of a *tuned* or *resonant* circuit. In a radio receiver, either AM (amplitude modulation), or FM (frequency modulation), or television (which uses AM for the picture and FM for the sound), the transmitted signal of one station is desired, while all others must be suppressed. This desired transmitter produces a high frequency ac voltage, while the undesired stations produce ac voltages at other frequencies. The capacitor C in conjunction with an inductor L act as a tuning circuit. This combination of C and L produces a maximum voltage at the frequency of the desired station, but very much smaller voltages at other frequencies. Thus the desired station signal is "tuned in," while others are rejected. By changing the value of either the capacitor or the inductor, a different frequency signal can be tuned in. Such a capacitor is called a variable or tuning capacitor.

13-1. The Capacitor

A capacitor consists basically of two conductors separated by an insulator. The diagram symbol for a capacitor is shown in Fig. 13-1*b*. The two conductors are usually called the capacitor *plates*, and the insulator is called the *dielectric*. In a small simple capacitor, there are two flat silver-coated copper plates

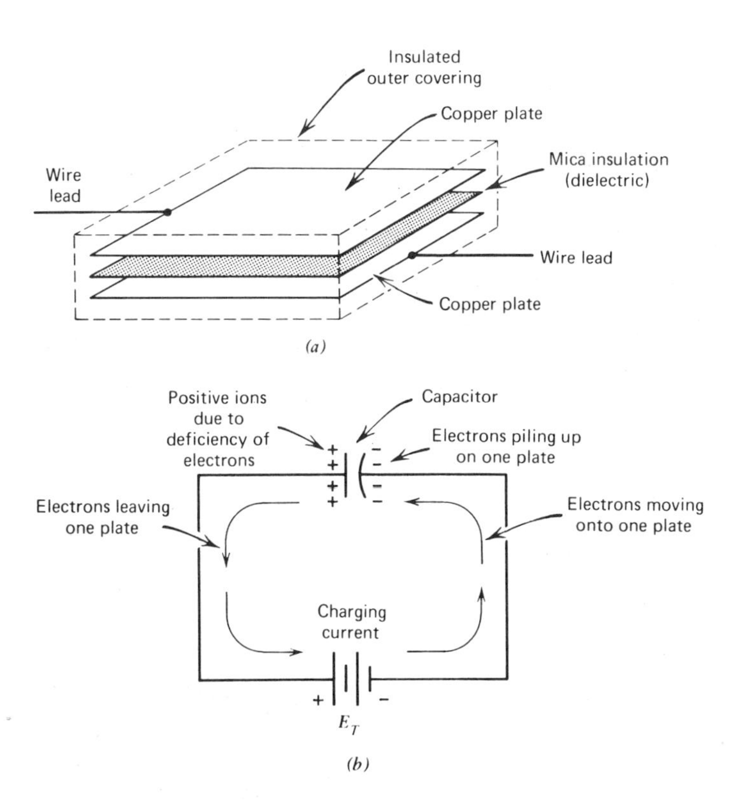

Figure 13-1. Capacitor. (*a*) Simple capacitor. (*b*) Capacitor circuit with applied dc voltage.

facing each other with a thin sheet of mica (an insulator) sandwiched in between, as shown in Fig. 13-1*a*. A larger capacitor consists of two long strips of thin metal such as tinfoil rolled up with insulating paper strips between the metallic strips.

When a dc voltage is applied to a capacitor, as shown in Fig. 13-1*b*, electrons move from the negative terminal of the voltage source and pile up on one capacitor plate. The accumulation of electrons on this plate makes this plate negative. This repels electrons off the other plate, leaving this second plate with a deficiency of electrons. This second plate now becomes positively charged. The difference in charge between the two plates is the voltage across the capacitor. After a short period of time, the capacitor voltage becomes equal to the applied voltage E_T. The capacitor is now said to be fully charged. When this occurs, the current stops.

Refer to End-of-Chapter Problems 13-1 to 13-3

Factors Affecting Capacitor Value. Capacitors are rated in units called *farads*, named after the English scientist Michael Faraday (1791-1867). The unit farad is very large, and capacitor values are usually in much smaller but more practical values such as *millionths of a farad*, called *microfarads*, or *millionths of a millionth of a farad*, called *micro microfarads*, also called *picofarads*. Microfarad is usually abbreviated as μF or sometimes mF.

The value or size of a capacitor is dependent on three factors: the *area of plates*, the *distance between the plates*, and the *type of insulator or dielectric* between the plates.

Refer to End-of-Chapter Problems 13-4 and 13-5

Area of Plates. A larger area plate can hold more electrons, resulting in a higher value capacitor. The simple capacitor drawn in Fig. 13-1*a* could have plates about 1/2 in. wide and 1/2 in. long, for a total area of each plate of 1/4 sq in. A typical value of such a small capacitor may be 0.00025 μF. A tubular shaped, much larger value capacitor about 0.25 μF has two tinfoil strips rolled up, each about 1 in. wide by 24 in. long, for a total area of each plate of 24 sq in.

The area of each plate may be large without resulting in an excessively long or physically large capacitor, as discussed previously, by rolling up the strips of tin foil. In a variable capacitor, pictured in Fig. 13-2, each plate consists of several half-moon-shaped flat metal discs that are connected together. Each plate may actually have half a dozen or more small sections connected together, resulting in the equivalent of two large plates. To be effective, the plates must be facing each other; that is, their flat surfaces must be adjacent to one another. In the type of variable capacitor pictured in Fig. 13-2, one set of

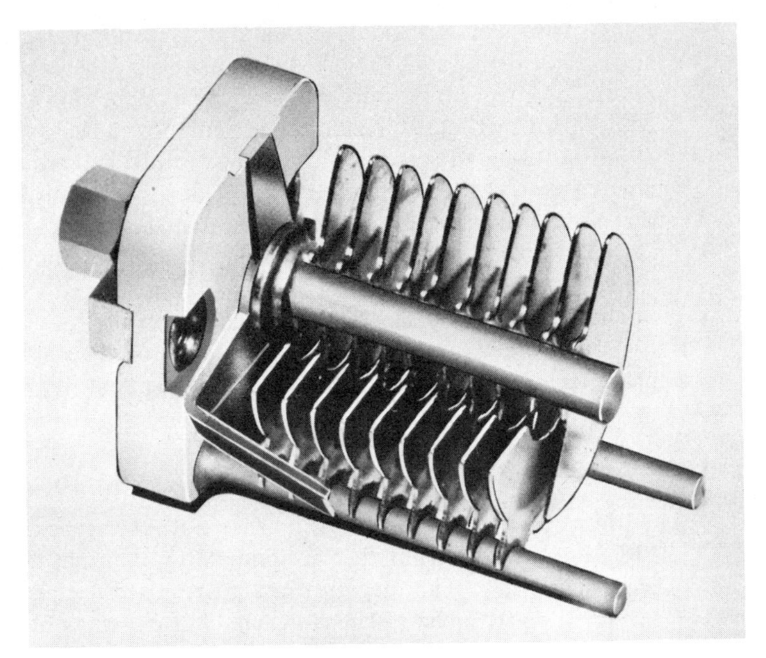

Figure 13-2. Variable capacitor (*courtesy of E. F. Johnson Co., Waseca, Minn*).

plates, shown as the 10 half-moon sections, can be rotated. This plate is called the *rotor*. The other plate, shown as the 9 section half-moons, is stationary and is called the *stator*. Note that by turning the rotor plates, they can be enmeshed in between the stators. In the picture the plates are about half way in mesh.

In Fig. 13-3*a*, a drawing of a single *stator* plate (plate 1), and a single *rotor* plate (plate 2), is shown with the plates completely unmeshed or out of mesh.

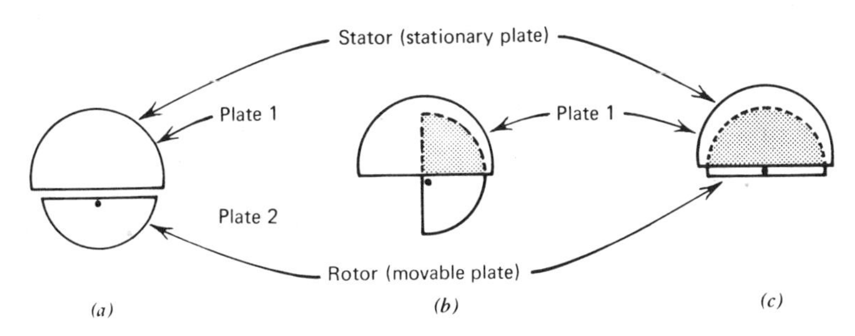

Figure 13-3. Variable capacitor, plates in and out of mesh. (*a*) Minimum capacitance plates completely out of mesh. (*b*) Approximately half of maximum capacitance, plates half in mesh. (c) Maximum capacitance, plates fully in mesh.

In this case, the area of the plates are practically meaningless, since the flat surfaces of the plates are not adjacent to each other; that is, not facing each other. In this position, only the small area of the thin edges of the plates are adjacent to each other, resulting in minimum capacitor value.

In Fig. 13-3*b*, the rotor plate is shown turned half way in mesh. The shaded portion shows the active areas of each plate, that is, the areas of plates 1 and 2 that are adjacent to each other. Note that this is about one half the area of each plate, resulting in a capacitor value that is about one half of the maximum value.

The rotor plate is shown turned all the way in mesh behind the stator in Fig. 13-3*c*. Now the entire area of each plate is adjacent to or facing the other plate, resulting in a maximum value of capacitance.

Refer to End-of-Chapter Problem 13-6

Distance Between Plates. When a dc voltage is applied to a capacitor, Fig. 13-1*b*, the electrons that accumulate on the right-hand plate charge that plate negatively. This negative charge repels electrons off the other plate, leaving a positive charge on this plate. The closeness of the two plates determines how easily electrons moving onto one plate can repel electrons from the other plate. If the plates were several inches apart (not practical), then the negative plate being so far away could barely have any repelling effect on electrons on the other plate. Such a capacitor would be extremely tiny in value.

The spacing between the plates of the smaller capacitors such as the mica type (about 0.00025 μF), and the paper tubular type (about 0.25 μF), is the thickness of the mica sheet, or of the paper strip. The very large capacitors, called *electrolytics* (about 50 μF), have an extremely small spacing between plates. This type of capacitor depends on a chemical action during its manufacture. A microscopically thin layer of insulating material such as aluminum oxide acts as the dielectric, allowing the plates to be extremely close together, giving the capacitor its large value. A type of variable capacitor, called a ***trimmer***, permits the spacing between the plates to be varied.

Refer to End-of-Chapter Problems 13-7 and 13-8

Dielectric or Insulator Between Plates. A capacitor exists only where the capability of storing a difference of charge between the plates is possible. This storage of a charge is only possible if electrons can move onto one plate, making the plate negative, and can repel electrons off the other plate. Without being able to repel electrons off this other plate, electrons can never move onto the first plate, and a capacitance will not be present. With air as the insulator between the two plates, as in a variable capacitor, electrons on one plate repel electrons off the other plate because of the closeness of the plates.

This repelling effect is increased when an insulator called the *dielectric* is placed between the two plates. An electron that has moved onto one plate can now more easily repel an electron from the other plate via the effect on the dielectric. This is illustrated in Fig. 13-4. In part (a) of this drawing, an uncharged capacitor is shown, with an atom of the dielectric having the normal circular orbit of its electrons. In Fig. 13-4*b*, an electron enters plate 1, making it negative. This now repels an electron on the atom of the insulating dielectric. Since this material is an insulator, electrons cannot easily be forced to leave the atom of the dielectric. However, the electron on this atom will be repelled away from the electron on plate 1, resulting in an eliptically distorted electron orbit on the dielectric atom.

The electron elongated path around the nucleus of the dielectric atom (Fig. 13-4*b*, brings this electron nearer to plate 2 than it would be if it had continued on a circular path. As a result, the electron of the dielectric atom, moving closer to plate 2, helps to repel an electron away from this plate. The net result is that the electrons moving onto one plate of a capacitor, using an insulatior between the plates, have a much greater repelling effect on electrons at the other plate than if air were used as the dielectric.

The increase in this repelling effect increases the value of the capacitor by a factor called the *dielectric constant*. For example, the insulator *mica* has a *dielectric constant* of about five. This means that if mica were substituted for air between the plates of a capacitor, with no other change in the capacitor construction, then the capacitor value would become five times as large.

Another example is worth noting. A ceramic insulator, *barium-strontium titanate*, has a dielectric constant of about 7500. With this material substituted for the air between the capacitor plates and with no other change in the capacitor construction, the capacitor value becomes 7500 times larger. Table 13-1 lists several insulating materials with the dielectric constant of each. When the electron orbit path is distorted in the dielectric as shown in Fig. 13-4, the molecules are said to be *polarized molecules*.

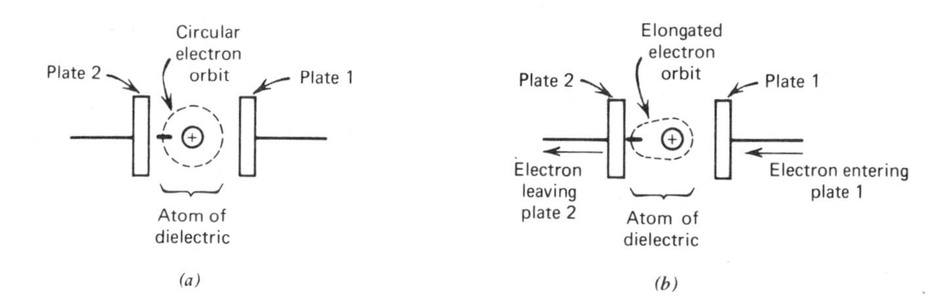

Figure 13-4. Effect of dielectric. (*a*) Uncharged capacitor. (*b*) Capacitor charging.

TABLE 13-1 Dielectric (Insulator) Materials Characteristics

Dielectric	Dielectric Constant (Approximate)	Voltage Breakdown Rating in Volts per Mil (V/0.001 in.)
Vacuum	1.	75
Air	1.0006	75–80
Teflon (plastic)	2.	1500
Paper (wax)	2.5	1200–1800
Rubber	3.	700
Oil (transformer oil)	4.	400
Mica	5.	5000
Porcelain (ceramic)	6.	200–700
Bakelite	7.	400–500
Glass	6 to 8.	3000
Water	80.	—
Barium-strontium Titanate (ceramic)	7500.	75

Refer to End-of-Chapter Problems 13-9 and 13-10

Voltage Breakdown Rating of Dielectric. When a capacitor has become charged, the insulation (the dielectric) between the plates is subjected to the stress of having its electron orbits distorted, as discussed in the preceding section. If sufficient voltage is applied to the capacitor plates, electrons will be forced off the atoms of the dielectric and it acts like a conductor. The capacitor is said to be *shorted* and no longer acts as a capacitor, with arcing occurring between the plates. The voltage that caused this is said to exceed the voltage *breakdown rating* of the capacitor dielectric. If the dielectric material is a solid such as mica, paper, ceramics, and the like, it is ruined and the capacitor must be replaced. An ohmmeter check indicates such a *short* or *leakage* condition. However, when the dielectric is air, vacuum, or oil, it is "self-healing." Once the applied voltage is either removed or reduced to below the breakdown value, the "short" or "leakage" disappears and the capacitor may be used again. Table 13-1 lists approximate *voltage breakdown* or *dielectric strength* values of several insulating materials.

Refer to End-of-Chapter Problems 13-11 and 13-12

Types of Capacitors. Capacitors are either the *variable* type or of a fixed value.

Variable Tuning Capacitor. One variable type is shown in Figs. 13-2 and 13-3. This type uses air as the dielectric, and changes the plate areas that face one another by having the plates brought into mesh or out of mesh, as discussed in the previous section, Area of Plates. Often, there are two or three separate variable capacitors *ganged* together. This means that the rotors of these capacitors are mounted on a common shaft, or axle. The tuning knob, or control, is rigidly attached to this shaft. As the shaft is turned, all rotor plates move together. This type is usually used in the tuner portion of a radio, where one section of the variable capacitor tunes or resonates with a coil (inductance) at the frequency of the desired station; a second section of the variable capacitor resonates with another coil at a frequency higher than that of the desired station. This is the *oscillator* section of the radio receiver. A large variable capacitor of this kind often has a minimum value (plates out of mesh) of a few picofarads, about 5, and a maximum value (plates turned in mesh) of a few hundred picofarads, about 300.

Variable Trimmer Capacitor. Another, but much smaller, variable capacitor usually called a *trimmer* or *padder* is adjusted by a screwdriver. This type consists of two small metal plates with a flat piece of mica between them. The plates are so constructed that their spring tension keeps them apart. By tightening, a screw, the spacing between the metal plates can be changed, resulting in a different capacitor value. A similar variable mica capacitor, also screwdriver adjusted, permits a single half-moon-shaped metal plate to be rotated in or out of mesh with a similar but stationary plate, thus varying the areas that face adjacent to each other, as shown in Fig. 13-3. A typical value of these small variable capacitors is from a minimum of about 2 pF to a maximum of about 30 pF.

Fixed Capacitors and Color Code. This type, the values of which cannot be varied, come in a wide assortment of shapes and sizes. The smaller values are often either the flat rectangular shapes (about 1/2 in. by 3/4 in. and 1/8 in. thick) often called "postage stamp" capacitors, or the small flat, round ceramic disc, resembling a button. Another common small value capacitor has a hollow, tubular shape, and is called the tubular ceramic. The ceramic types have copper or silver coatings on opposite sides of a thin ceramic disc (about the size of a small button), or on opposite faces of a thin-walled ceramic tube. The square or rectangular-shaped "postage stamp" capacitors consist of a thin mica wafer sandwiched between two silver-plated copper wafers, as drawn in Fig. 13-1*a*. The larger mica capacitors have several metal plates with mica between each; alternate plates are tied together to form one large plate, with the other similarly connected to form the other plate.

Ceramic capacitors (disc and tubular) are generally between 1 pF and up to about 50,000 pF (or 0.05 μF). The disc of button type usually have the value of the capacitor printed on it, but like the ceramic tubular type may employ the color code markings with color dots. The tubular often has color bands. Some of these are shown in the Color Code Chart of Appendix VI.

The mica capacitor values similarly start with a few picofarads and go up to about 50,000 pF. Most of the smaller ones use the color coded dots or stripes to signify their values, as shown in the chart of Appendix VI.

The colors listed are the same ones discussed previously in the chapter on resistors (Chapter 8). Each color denotes a numerical value as listed in the chart. For the actual value of the capacitor, usually rated as a two-digit number followed by one or more zeros (or none), the capacitor *value* is given by a group of three colors in sequence. The first and second colors of this group denote the first and second digits, while the *third* color signifies the *number of zeros* (if any) to be added or tacked on behind the first two numbers. As an example, the colors brown (1), orange (3), and black (0), signify 13 and no zeros, or 13 pF. The colors yellow (4), violet (7), and brown (1), signify 47 and 1 zero, or 470 pF. If the third color in the above illustration were red (2), the value would be 47 and 2 zeros, or 4700 pF.

As an aid in remembering the sequence of colors listed in the Color Code chart of Appendix VI, many students use the following sentence, noting that the first letter of each word is a clue to the actual color: *B*ad *B*oys *R*ob *O*ur *Y*oung *G*irls *B*ut *V*iolet *G*ives *W*illingly *S*ilver and *G*old.

Example 13-1

A 5-color band ceramic tubular capacitor with axial leads has the following colors reading from left to right: brown, red, green, red, and white. What information is included in these colors?

Solution

Referring to the color code table and example drawings of Appendix VI:

First color denotes *temperature coefficient*; *brown* is -30.
Second color denotes *first digit*; *red* is 2.
Third color denotes *second digit*; *green* is 5.
Fourth color denotes *number of zeroes to be added*; *red* means add two zeros, 00.
Fifth color denotes *tolerance*; *white* is 10 %.

Therefore, this capacitor is a 2500-pF capacitor with a temperature coefficient of -30 and a 10% tolerance.

NOTE

The electronic technician usually is primarily interested in the value of the capacitor and its voltage rating, and seldom in its other characteristics such as its temperature coefficient.

Example 13-2

A flat rectangular molded mica capacitor has six color dots. Reading in the direction of the arrow on the capacitor, from left to right on the upper row, and then continuing in a clockwise direction (from right to left) on the lower row, the colors are: white, red, red, orange, silver and violet. What information is denoted?

Solution

Referring to the table and example of Appendix VI, the reader should find that this capacitor has a 22,000-pF value, with a 10% tolerance and a temperature coefficient of −750.

For Similar Problems Refer to End-of-Chapter Problems 13-13 to 13-17

Electrolytic Capacitors. The very large capacitors, about five microfarads and up to several hundred microfarads are called *electrolytics.* These capacitors owe their large values to the extreme closeness of the plates due to the almost molecular thinness of the dielectric material. Two aluminum strips rolled up with strips of gauze cloth between the metal comprise the elements of this capacitor. The gauze has been soaked in an electrolyte such as borax. In the manufacturing process, a dc voltage is applied across the two metal plates. The current flow causes a chemical action to take place, and an extremely thin coating of aluminum oxide forms on the positive plate. The oxide acts as the dielectric or insulator between the two plates. One capacitor plate is the positive aluminum strip, while the other is the electrolyte-soaked gauze and negative aluminum strip.

Since this capacitor has been formed by the electrolytic chemical action due to current flowing through it, it is very important to observe this negative and positive polarity when connecting an *electrolytic* capacitor in a dc circuit. The polarity of this type of capacitor is always indicated on the container, and the positive lead must be tied to the more positive end of the part of the circuit to which it is connected, while the negative lead goes to the less positive (or negative) end of this part of the circuit.

Failure to connect the electrolytic capacitor with its proper polarity in a dc circuit would cause the current to reverse the chemical action, producing a gas that often results in an exploded capacitor. Polarity need not be followed in other capacitor types (mica, paper, ceramic, etc.), since either plate may act as the positive or negative one.

Where a current flows first in one direction, and then in another (ac), and where a large value capacitor is required, a special electrolytic capacitor is available which requires that no polarity be followed. It is in reality two electrolytics connected in series, but with their polarities opposing. That is, the positive of one is connected to the positive of the other, with the negative ends of each becoming the two available leads.

Ohmmeter Check of an Electrolytic Capacitor. An electrolytic capacitor, unlike ones that are mica, paper, and ceramic, has some leakage current through the chemical oxide dielectric. Checking an *uncharged* electrolytic capacitor using a high-range ohmmeter (about $R \times 100$ K or $R \times 1$ M) results in readings of about 200 kΩ or larger. A similar test of an uncharged mica, paper, or ceramic capacitor indicates an *infinity* reading on the ohmmeter. When the high-range ohmmeter is first connected across the leads of the *uncharged* electrolytic capacitor, the ohmmeter pointer swings immediately toward zero, and then slowly starts climbing toward progressively higher and higher values. After about 10 or 20 seconds, with the ohmmeter still connected across the electrolytic, the pointer becomes almost stationary and indicates a reading of about 200 kΩ or higher. An *open* electrolytic capacitor produces no movement of the ohmmeter pointer, with the needle remaining motionless at *infinity*. A *shorted* or excessively *leaky* electrolytic results in the ohmmeter needle indicating *zero* ohms or up to several thousand ohms.

Two precautions should be observed in checking a capacitor with an ohmmeter: (a) the capacitor must be completely discharged first, before the ohmmeter is connected to it, and (b) one lead of the capacitor must be disconnected from the circuit so that there is noting in parallel with the capacitor.

Refer to End-of-Chapter Problems 13-18 to 13-25

13-2. Capacitors, Series and Parallel

Capacitors, like resistors and inductors, may be connected in parallel and also in series. When capacitors are connected in *parallel*, as shown in Fig. 13-5*a*,

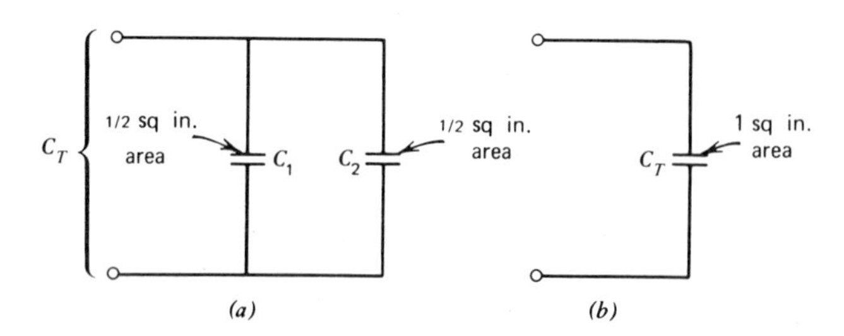

$C_T = C_1 + C_2$

Figure 13-5. Capacitors in parallel. (*a*) Circuit. (*b*) Equivalent.

the upper plates of C_1 and C_2 result in a larger area plate. If the plate of C_1 has an area 1/2 sq in., with the plate of C_2 having the same area, the total upper plate area is now the sum of the two, or 1 sq in., as shown in Fig. 13-5*b*. The same holds true for the lower plates. As a result, C_1 and C_2 in parallel form a larger capacitor. The value of this combination, forming the total capacitor C_T, is

$$C_{T\ \text{parallel}} = C_1 + C_2 \tag{13-1}$$

If there were three or more parallel capacitors, C_{total} would be the sum, or $C_T = C_1 + C_2 + C_3 + etc.$

The following examples illustrate the above.

Example 13-3

A 4-μF capacitor is connected in parallel with a 6-μF one. What is the total capacitance?

Solution

$$C_{T\ \text{parallel}} = C_1 + C_2 \tag{13-1}$$

$$C_{T\ \text{parallel}} = 4 + 6$$

$$C_{T\ \text{parallel}} = 10\ \mu\text{F}$$

Example 13-4

Three capacitors are connected in parallel. What is the total capacitor formed by this combination if the capacitors are 30 pF, 50 pF, and 70 pF?

Solution

The reader should find that C_T is 150 pF.

Example 13-5

A 250-pF capacitor is connected in parallel with a 0.005 μF capacitor. What is the total capacitance of this combination?

Solution

The reader should first change the *microfarads* into *picofarads*, and should then find that C_T is 5250 pF.

For Similar Problems Refer to End-of-Chapter Problems 13-26 to 13-29

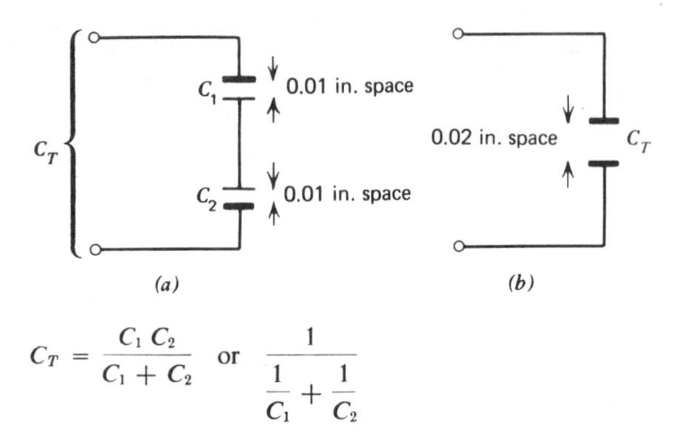

Figure 13-6. Capacitors in series. (*a*) Circuit. (*b*) Equivalent.

Capacitors in Series. When two capacitors are connected in *series*, as is shown in Fig. 13-6*a*, the total capacitance of this combination becomes *less* than the smaller of the capacitors. The reason for this may be seen from the following. Since the lower plate of C_1, in Fig. 13-6*a*, is connected to the upper plate of C_2, these center common plates may be removed without altering the total capacitance C_T of the circuit. Then C_T actually consists of the remaining plates, *upper* plate of C_1 and *lower* plate of C_2, shown as the thicker-line plates. If the original spacing between C_1 plates is 0.01 in., as is shown in Fig. 13-6*a*, with C_2 plates having the same spacing, then the combination or C_T *spacing* is the sum of the two or 0.02 in., as depicted in Fig. 13-6*b*. When plates are further apart, the capacitor value is less.

The total capacitance C_T of *two* series-connected capacitors is their *product divided by* their *sum*:

$$C_{T\,\text{series}} = \frac{C_1 C_2}{C_1 + C_2} \tag{13-2}$$

When *two or more* capacitors are connected in series, the following equation is used to find C_T. (It is called the *reciprocal* equation.)

$$C_{T\,\text{series}} = \frac{1}{\frac{1}{C_1} + \frac{1}{C_2} + \text{etc.}} \tag{13-3}$$

The following examples illustrate the use of the above equations in solving for C_T where capacitors are connected in series.

Example 13-6

A 10-pF capacitor is connected in series with one of 15 pF value. Find the total capacitance C_T of this combination.

Solution

One method is to use the *product* divided by the sum.

$$C_{T\text{ series}} = \frac{C_1\,C_2}{C_1 + C_2} \tag{13-2}$$

$$C_{T\text{ series}} = \frac{(10)\,(15)}{10 + 15}$$

$$C_{T\text{ series}} = \frac{150}{25}$$

$$C_{T\text{ series}} = 6\text{ pF}$$

A second method is to use the *reciprocal* equation. The reader should solve for C_T by this method and should find that C_T is 6 pF.

Example 13-7

Three capacitors are connected in series. If their values are 30 μF, 20 μF, and 6 μF, find the total capacitance C_T.

Solution

The reader, using the reciproval equation, should find that C_T is 4 μF.

For Similar Problems Refer to End-of-Chapter Problems 13-30 and 13-31

Capacitors, Combination of Series and Parallel. In Fig. 13-7*a*, three capacitors are shown. C_2 and C_3 are in *parallel*, and this combination is in

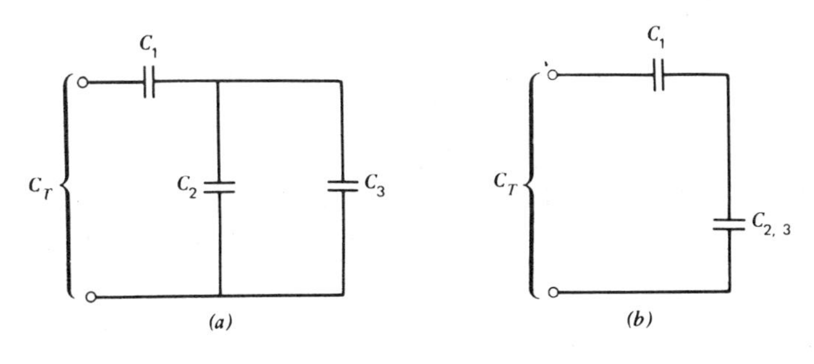

$$C_T = \frac{(C_1)\,(C_2 + C_3)}{C_1 + (C_2 + C_3)}$$

Figure 13-7. Capacitors, combination of series and parallel. (*a*) Circuit. (*b*) Equivalent.

series with C_1. The total capacitance C_T of this circuit may be found by first finding the *equivalent* capacitance of the parallel C_2 and C_3 by getting their sum $(C_2 + C_3)$. Then this resultant is combined with *series* C_1 to produce the total capacitance as shown in Fig. 13-7*b*, using the *product divided* by the *sum* method (equation 13-2), or by employing the *reciprocal* formula (equation 13-3).

The following example illustrates this.

Example 13-8

Find the total capacitance C_T of the circuit of Fig. 13-7*a*, if $C_1 = 6\ \mu F$, $C_2 = 9\ \mu F$, and $C_3 = 3\ \mu F$.

Solution

Solve for C_T, first finding the equivalent of the parallel C_2 and C_3. As a guide, $C_{2,3}$ should be found to be 12 μF, and C_T should be found to be 4 μF.

In Fig. 13-8*a*, four capacitors are shown in a series-parallel arrangement. Note that C_2 *and* C_3 are in *parallel*, and that this combination is in *series* with C_1 and C_4. To find the total capacitance C_T, first find the equivalent of the parallel C_2 and C_3 by getting their sum (equation 13-1). Then the three *series* capacitors (C_1, $C_{2,3}$ and C_4) may be combined to find C_T as shown in Fig. 13-8*b* by using the *reciprocal* method (equation 13-3). This is shown by the following example.

Example 13-9

Find the total capacitance C_T of Fig. 13-8 if $C_1 = 5\ \mu F$, $C_2 = 13\ \mu F$, $C_3 = 7\ \mu F$, and $C_4 = 12\ \mu F$.

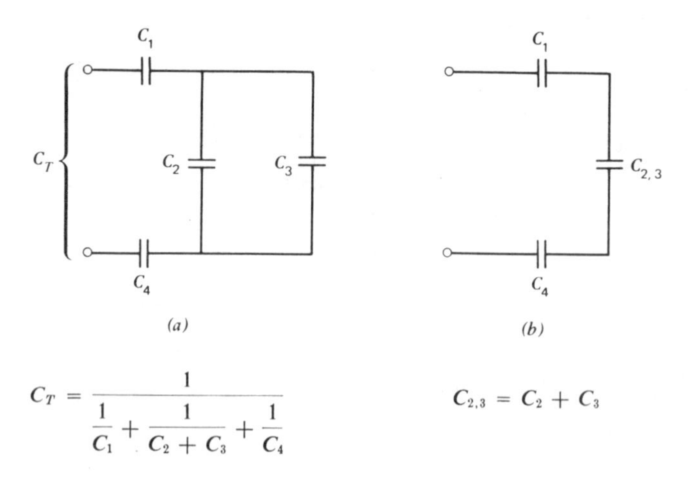

Figure 13-8. Series-paralel capacitors. (*a*) Circuit. (*b*) Equivalent.

Solution

The reader should find that the equivalent of C_2 and C_3 in parallel should be 20 μF, and that C_T should be found to be 3 μF.

13-3. Quantity (Q) of Charge on a Capacitor

When a capacitor has become charged, the *quantity* of electrons that have accumulated on its negative plate is described in units called *coulombs.*A coulomb is a huge quantity of electrons, consisting approximately of 6.24×10^{18} electrons. The *quantity* of electrons, called Q, on a charged capacitor depends on the *size of the capacitor* and the *voltage* to which it is charged. This relationship is shown in the following.

$$\text{quantity} = (\text{capacitor})\,(\text{voltage}) \qquad (13\text{-}4)$$

or

$$Q = CE$$

where Q is the *quantity* in *coulombs*, C is the value of the *capacitor* in *farads*, and E is the voltage to which the capacitor is charged.

A *farad* may therefore be defined as that value of *capacitance* that produces *one coulomb* (or 6.24×10^{18} electrons) of charge when *one volt* is applied to the capacitor.

The following examples illustrate the use of the Q equation.

Example 13-10

A 500-pF capacitor, in the high-voltage section of a television receiver, is charged to 15,000 V. Find the quantity of electrons in coulombs stored on this capacitor.

Solution

$$Q = CE \qquad (13\text{-}4)$$

$$Q = (500 \text{ pF})\,(15{,}000 \text{ V})$$

$$Q = (500 \times 10^{-12} \text{ F})\,(15 \times 10^{3} \text{ V})$$

$$Q = (5 \times 10^{-10})\,(15 \times 10^{3})$$

$$Q = 75 \times 10^{-7} \text{ C} \quad \text{or} \quad 7.5 \times 10^{-6} \text{ C} \quad \text{or} \quad 7.5\ \mu\text{C}$$

Example 13-11

A large 50-μF capacitor, in the low-voltage section of a television receiver, is charged to 300 V. Find the quantity of electrons in coulombs stored on this capacitor.

Solution

The reader should solve this example and should find that Q is 15×10^{-3} C, or 15,000 μC.

An interesting observation could be made from the previous two examples. Note that the small *500 pF* capacitor charged to *15,000 V* only holds *7.5 μC*, while the large *50-μF* capacitor only charged to *300 V* is holding *15,000 μC*. In other words, the *low voltage circuit large capacitor* holds *15,000 μC*, which is about 2000 times as much as that held by the *high-voltage*-circuit *small capacitor* of *7.5 μC*. This explains why the low-voltage section of a television reciever could give much more of a jolt to a careless technician than could the high-voltage section.

For Similar Problems Refer to End-of-Chapter Problems 13-32 to 13-34

Quantity (Q) of Charge on Capacitors in Series. When capacitors are connected in series, as shown in Fig. 13-9*a* and a dc voltage is applied, the same charging current flows onto each capacitor. This means that when each capacitor has fully charged, there will be an equal quantity of electrons on each

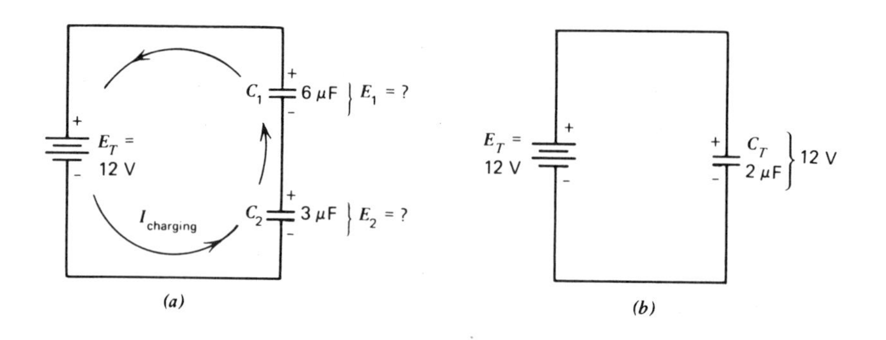

$$C_T = \frac{C_1\,C_2}{C_1 + C_2} = \frac{(6)\,(3)}{6 + 3} = \frac{18}{2} = 2\ \mu\text{F}$$

$$Q_T = Q_1 = Q_2 = 24 \times 10^{-6}\ \text{C}$$

$$Q_1 = C_1\,E_1$$

$$24 \times 10^{-6} = (6 \times 10^{-6})\,(E_1)$$

$$4\ \text{V} = E_1$$

$$Q_T = C_T\,E_T = (2 \times 10^{-6})\,(12) = 24 \times 10^{-6}\ \text{C}$$

$$Q_2 = C_2\,E_2$$

$$24 \times 10^{-6} = (3 \times 10^{-6})\,(E_2)$$

$$8\ \text{V} = E_2$$

Figure 13-9. Quantity of charge (Q) on series capacitors. (*a*) Circuit. (*b*) Equivalent.

capacitor. The *equivalent* or *total* capacitance, C_T, will also have this same equal quantity of charge or quantity of electrons. In a *series* capacitor circuit then:

$$Q_{T\text{ series}} = Q_1 = Q_2 \qquad (13\text{-}5)$$

where Q_T is the quantity of electrons, in coulombs, stored on the *equivalent* or *total* capacitance C_T, and Q_1 and Q_2 are the quantities on each capacitor C_1 and C_2.

Capacitors are sometimes connected in series, acting as a *capacitor voltage divider*. By solving for C_T, Q_T may then be found as shown in Fig. 13-9*b*. Then, knowing Q_1 and Q_2, as well as C_1 and C_2, voltage across each capacitor E_1 and E_2 may be found.

These examples illustrate this:

Example 13-12

In the series capacitor circuit of Fig. 13-9*a*, find (a) C_T, (b) Q_T, (c) Q_1, (d) Q_2, (e) E_1, and (f) E_2.

Solution

The reader should follow the solutions shown in the text of Fig. 13-9. Note that the sum of E_1 (4 V) and E_2 (8 V) add up to E_T (12 V). Note also that in *series*, the *larger* capacitor ($C_1 = 6\ \mu F$) has less voltage across it than the *smaller* capacitor ($C_2 = 3\ \mu F$).

Example 13-13

As shown in Fig. 13-10*a*, three capacitors ($C_3 = 12\ \mu F$, $C_4 = 15\ \mu F$, and $C_5 = 10\ \mu F$) are connected in series, with 120 V applied dc voltage. Find: (a) C_T, (b) Q_T, (c) Q_3, (d) Q_4, (e) Q_5, (f) E_3, (g) E_4, and (h) E_5.

Solution

Solve this example by using the text of Fig. 13-10 as a guide.

Note that the sum of the capacitor voltages ($E_3 = 40$ V, $E_4 = 32$ V, and $E_5 = 48$ V) is equal to the total applied voltage ($E_T = 120$ V). Also note that in *series*, the *largest capacitor* ($C_4 = 15\ \mu F$) has the *smallest voltage*.

For Similar Problems Refer to End-of-Chapter Problems 13-35 to 13-37

Quantity (Q) of Charge on Parallel Capacitors. In the circuit of Fig. 13-11*a*, capacitor C_1 (10 μF) is in parallel with C_2 (5 μF), with an applied dc voltage of 30 V. Each capacitor will charge to the 30 V with separate charging

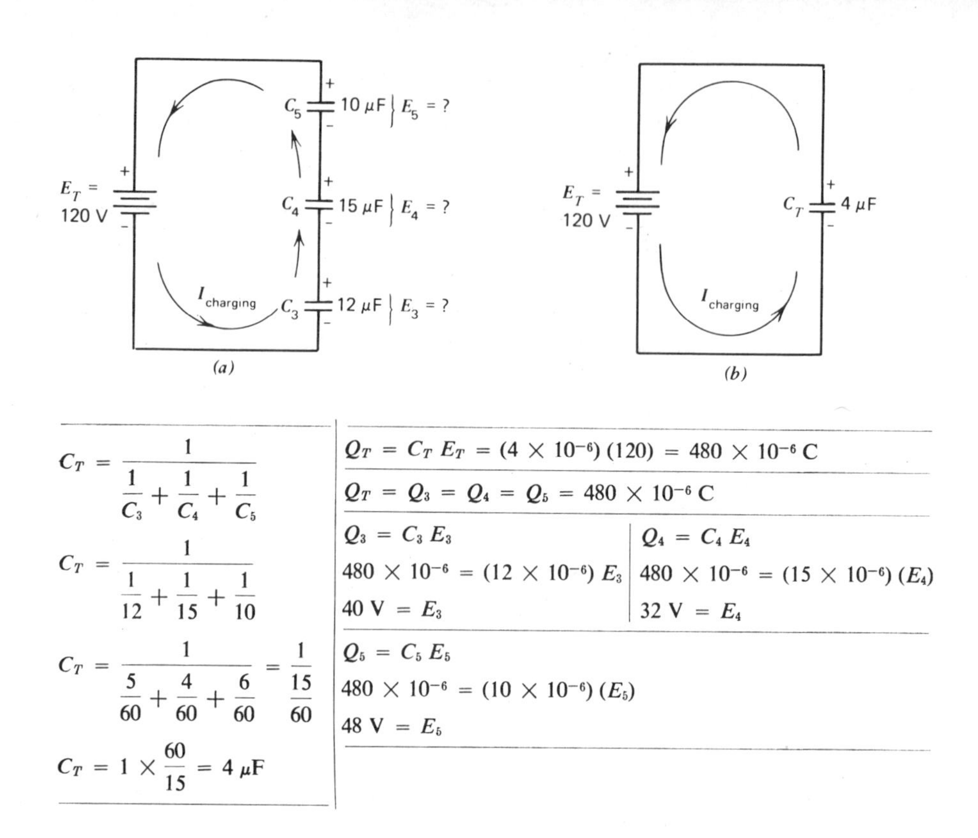

$$C_T = \frac{1}{\frac{1}{C_3} + \frac{1}{C_4} + \frac{1}{C_5}}$$

$$C_T = \frac{1}{\frac{1}{12} + \frac{1}{15} + \frac{1}{10}}$$

$$C_T = \frac{1}{\frac{5}{60} + \frac{4}{60} + \frac{6}{60}} = \frac{1}{\frac{15}{60}}$$

$$C_T = 1 \times \frac{60}{15} = 4\ \mu\text{F}$$

$$Q_T = C_T\, E_T = (4 \times 10^{-6})\,(120) = 480 \times 10^{-6}\ \text{C}$$

$$Q_T = Q_3 = Q_4 = Q_5 = 480 \times 10^{-6}\ \text{C}$$

$Q_3 = C_3\, E_3$	$Q_4 = C_4\, E_4$
$480 \times 10^{-6} = (12 \times 10^{-6})\, E_3$	$480 \times 10^{-6} = (15 \times 10^{-6})\,(E_4)$
$40\ \text{V} = E_3$	$32\ \text{V} = E_4$

$$Q_5 = C_5\, E_5$$

$$480 \times 10^{-6} = (10 \times 10^{-6})\,(E_5)$$

$$48\ \text{V} = E_5$$

Figure 13-10. Quantity of charge (Q) on series capacitors. (*a*) Circuit. (*b*) Equivalent.

currents. As shown, C_1 charging current will take the path shown by the solid-line arrows. C_2 charging current will take the path shown by the dashed-line arrows. The quantity of charge or quantity of electrons Q stored on each capacitor will depend on the size of each capacitor and, unlike the *series* circuit where the *same* current flows onto each capacitor, the quantity on each parallel capacitor may be different. The quantity of electrons on parallel capacitors will be equal only for equal-value capacitors.

The parallel capacitors are equivalent to the larger capacitor C_T (where $C_T = C_1 + C_2$, equation 13-1) (Fig. 13-11*b*). The quantity of electrons stored on C_T (or $Q_T = C_T\, E_T$, equation 13-4) is equal to the *sum* of the *quantity on each parallel capacitor*, or

$$Q_{T\ \text{parallel}} = Q_1 + Q_2 \qquad (13\text{-}6)$$

The following examples illustrate this.

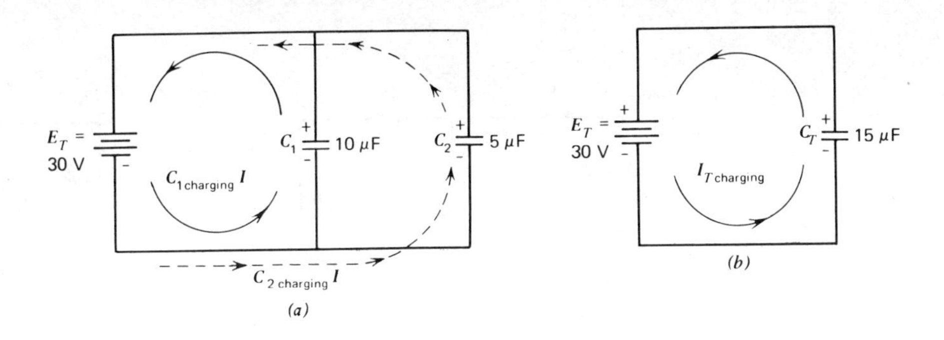

$C_T = C_1 + C_2$	$Q_1 = C_1\, E_1$	$Q_2 = C_2\, E_2$	$Q_T = C_T\, E_T$
$C_T = 10 + 5$	$Q_1 = (10\ \mu\text{F})\,(30\ \text{V})$	$Q_2 = (5\ \mu\text{F})\,(30)$	$Q_T = (15\ \mu\text{F})\,(30\ \text{V})$
$C_T = 15\ \mu\text{F}$	$Q_1 = 300 \times 10^{-6}\ \text{C}$	$Q_2 = 150 \times 10^{-6}\ \text{C}$	$Q_T = 450 \times 10^{-6}\ \text{C}$

$Q_T = Q_1 + Q_2$	
$Q_T = (300 \times 10^{-6}) + (150 \times 10^{-6})$	
$Q_T = 450 \times 10^{-6}\ \text{C}$	

Figure 13-11. Quantity of charge (Q) on parallel capacitors. (*a*) Circuit. (*b*) Equivalent.

Example 13-14

In the parallel capacitor circuit of Fig. 13-11*a*, C_1 (10 μF) and C_2 (5 μF) are in parallel, with 30 V dc applied. Find: (a) Q_1, (b) Q_2, (c) C_T, and (d) Q_T.

Solution

Follow the solutions shown in the text of Fig. 13-11.

Example 13-15

Three capacitors, $C_3 = 20\ \mu$F, $C_4 = 20\ \mu$F, and $C_5 = 10\ \mu$F, are connected in parallel, with an applied dc voltage $E_T = 3$ V. Find: (a) Q_3, (b) Q_4, (c) Q_5, (d) C_T, and (e) Q_T.

Solution

The reader should solve this example by using the method shown previously and should find the following results: $Q_3 = 60 \times 10^{-6}$ C, $Q_4 = 60 \times 10^{-6}$ C, $Q_5 = 30 \times 10^{-6}$ C, $Q_T = 150 \times 10^{-6}$ C, $C_T = 50\ \mu$F.

For Similar Problems See End-of-Chapter Problems 13-38 to 13-41

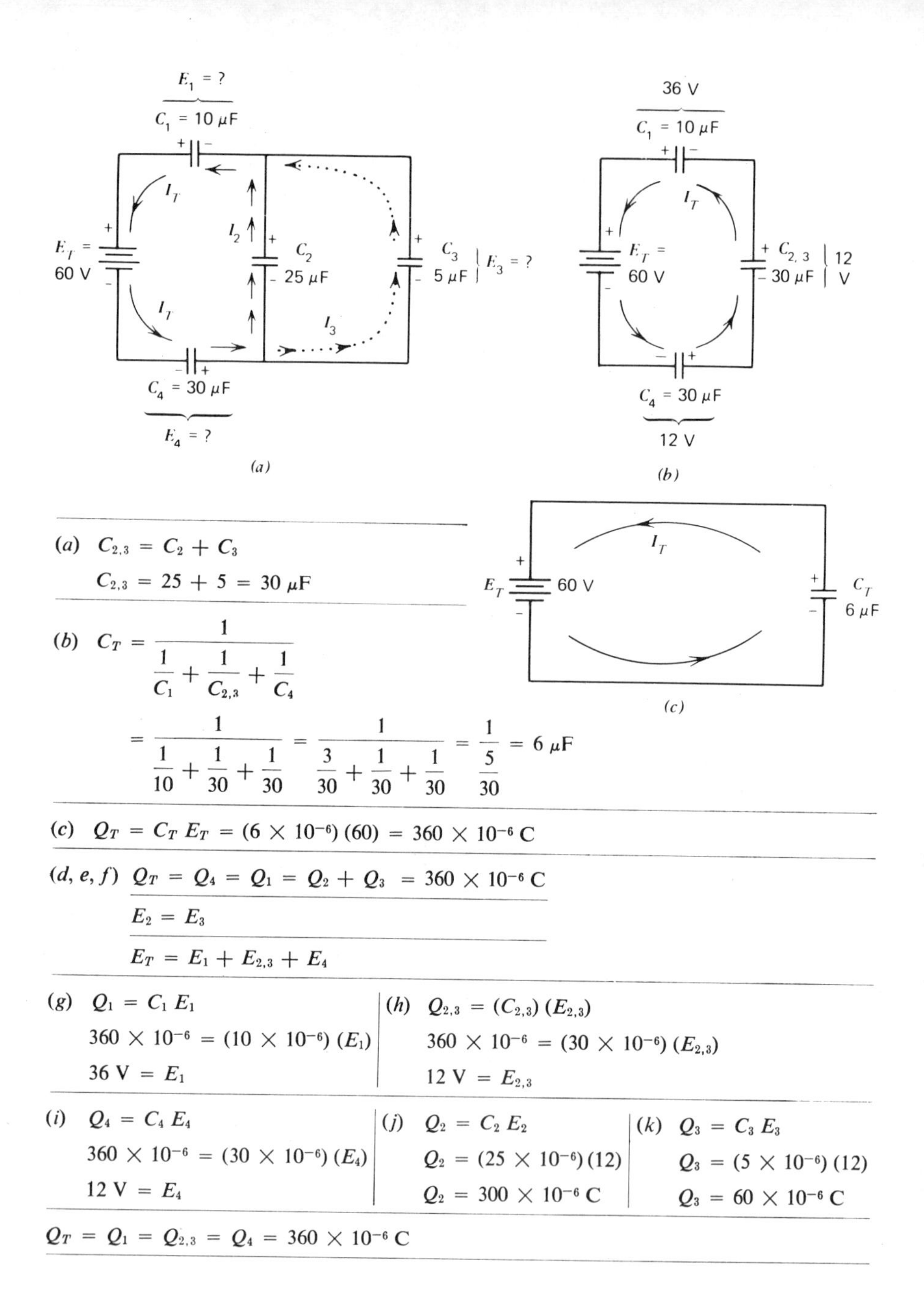

(a) $C_{2,3} = C_2 + C_3$

$C_{2,3} = 25 + 5 = 30\ \mu F$

(b) $$C_T = \frac{1}{\frac{1}{C_1} + \frac{1}{C_{2,3}} + \frac{1}{C_4}}$$

$$= \frac{1}{\frac{1}{10} + \frac{1}{30} + \frac{1}{30}} = \frac{1}{\frac{3}{30} + \frac{1}{30} + \frac{1}{30}} = \frac{1}{\frac{5}{30}} = 6\ \mu F$$

(c) $Q_T = C_T E_T = (6 \times 10^{-6})(60) = 360 \times 10^{-6}$ C

(d, e, f) $Q_T = Q_4 = Q_1 = Q_2 + Q_3 = 360 \times 10^{-6}$ C

$E_2 = E_3$

$E_T = E_1 + E_{2,3} + E_4$

(g) $Q_1 = C_1 E_1$

$360 \times 10^{-6} = (10 \times 10^{-6})(E_1)$

$36\ V = E_1$

(h) $Q_{2,3} = (C_{2,3})(E_{2,3})$

$360 \times 10^{-6} = (30 \times 10^{-6})(E_{2,3})$

$12\ V = E_{2,3}$

(i) $Q_4 = C_4 E_4$

$360 \times 10^{-6} = (30 \times 10^{-6})(E_4)$

$12\ V = E_4$

(j) $Q_2 = C_2 E_2$

$Q_2 = (25 \times 10^{-6})(12)$

$Q_2 = 300 \times 10^{-6}$ C

(k) $Q_3 = C_3 E_3$

$Q_3 = (5 \times 10^{-6})(12)$

$Q_3 = 60 \times 10^{-6}$ C

$Q_T = Q_1 = Q_{2,3} = Q_4 = 360 \times 10^{-6}$ C

Figure 13-12. Quantity of charge (Q) on series-parallel capacitors, Example 13-16. (*a*) Circuit. (*b*) Simplified equivalent circuit. (*c*) More simplified equivalent circuit.

Quantity (Q) of Charge on Series-Parallel Capacitors. In the series-parallel capacitor circuit of Fig. 13-12*a*, total charging current I_T flows onto C_4 and C_1, but this current splits up with some flowing onto C_2 (shown as dashed-line arrows I_2) and some onto C_3 (shown as dotted-line arrows I_3. Capacitors C_4 and C_1 are in *series* with each other and also in *series* with the *parallel combination* of C_2 *and* C_3. $I_T = I_2 + I_3$, and $Q_T = Q_2 + Q_3$ (equation 13-6).

In Fig. 13-12*b*, C_2 and C_3 have been combined into its equivalent, $C_{2,3}$. Figure 13-12*b* is a series capacitor circuit, with the same I_T flowing onto C_4, $C_{2,3}$ and C_1. $Q_T = Q_4 = Q_{2,3} = Q_1$ (equation 13-5).

Finally in Fig. 13-12*c*, C_1, $C_{2,3}$ and C_4 have been combined into its equivalent C_T. Here, $Q_T = C_T E_T$ (equation 13-4).

The following example shows how to solve the series-parallel capacitor circuit of Fig. 13-12*a*.

Example 13-16

In the circuit of Fig. 13-12*a*, find: (a) $C_{2,3}$, (b) C_T, (c) Q_T, (d) Q_1, (e) $Q_{2,3}$, (f) Q_4, (g) E_1, (h) $E_{2,3}$, (i) E_4, (j) Q_2, and (k) Q_3.

Solution

Solve this example by using the previous methods shown, and refer to the text of Fig. 13-12 as a guide.

Figure 13-13 is a series-parallel capacitor circuit used to illustrate Problem 13-43.

For Similar Problems Refer to End-of-Chapter Problems 13-42 and 13-43

13-4. RC Time Constant, Capacitor Charging

When a dc voltage is applied to a capacitor alone, the capacitor becomes charged to the applied voltage almost immediately. When a resistor is added in series with the capacitor, it takes some time before the capacitor can become fully charged to the applied voltage. The *time* depends on the values of the *resistor* and the *capacitor*. The *product* of the *resistor value in ohms* and the *capacitor value* in *farads* gives a *time in seconds* called the *time constant.*

$$\text{(resistor in ohms) (capacitor in farads)} = \text{one time constant (in seconds)} \qquad (13\text{-}7)$$

The following example illustrates this.

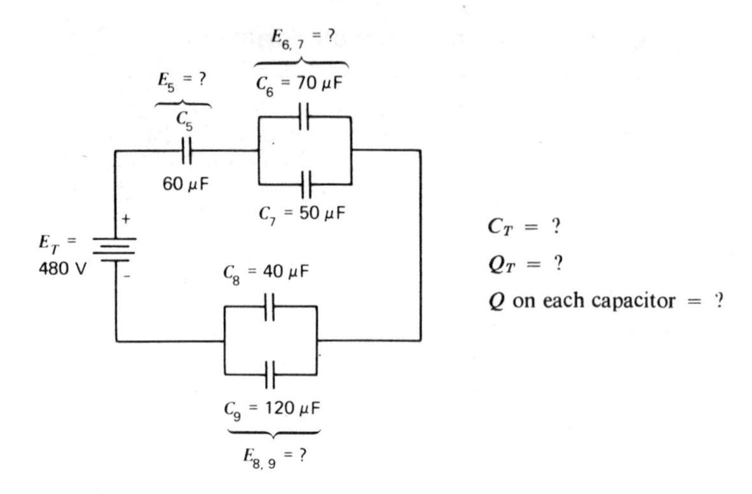

Figure 13-13. Quantity of charge (Q) on series-parallel capacitors, Problem 13-43.

Example 13-17

What is the *RC* time constant in the circuit of Fig. 13-14 where $C = 0.1$ μF and $R = 100$ kΩ?

Solution

$$\text{time constant} = (R_{\text{in ohms}})(C_{\text{in farads}}) \tag{13-7}$$

$$\text{time constant} = (100 \text{ K})(0.1\ \mu\text{F})$$

$$\text{time constant} = (1 \times 10^5)(0.1 \times 10^{-6})$$

$$\text{time constant} = 0.1 \times 10^{-1} \text{ sec} \quad \text{or} \quad 0.01 \text{ sec}$$

Note that *each time constant* for the values in the circuit of Fig. 13-14 is 0.01 sec. Two time constants is, therefore: (2) (0.01) = 0.02 sec; three time constants is (3) (0.01) 0.03 sec, and so on.

In the period of the first time constant, the capacitor voltage rises from zero to 63% of the applied dc voltage. When the switch in Fig. 13-14 is placed in the ON position, a current starts flowing and the capacitor starts *charging* up toward the applied 100 V. This charging current is depicted as the solid-line arrows in Fig. 13-14. Voltage across the resistor, E_R or V_R, has a polarity with the upper end of R positive.

At the instant that the switch is placed to the ON position, applying 100 V, a current starts flowing, but the capacitor has not had time to charge yet.

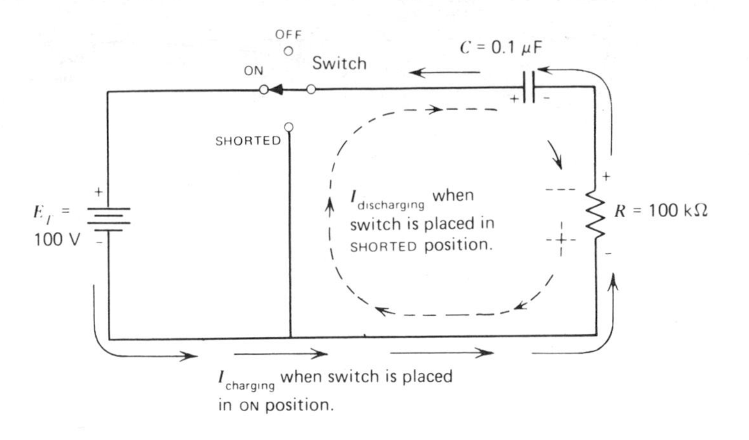

$$\underset{\text{(in seconds)}}{\text{time constant}} = \underset{\text{(in ohms)}}{(R)}\ \underset{\text{(in farads)}}{(C)}$$

$$\text{time constant} = (100\ \text{k}\Omega)\ (0.1\ \mu\text{F})$$

$$\text{time constant} = (1 \times 10^{5})\ (0.1 \times 10^{-6})$$

$$\text{time constant} = 0.1 \times 10^{-1}\ \text{sec}$$

$$\text{time constant} = 0.01\ \text{sec}$$

Figure 13-14. Charging and discharging paths of a capacitor, and *RC* time constant.

Capacitor voltage E_C or V_C is still zero, while resistor voltage E_R is 100 V. From Kirchhoff's voltage law for series circuits $E_R + E_C = E_{\text{total}}$. The *RC* time-constant graphs or curves of Fig. 13-15 show that at this first instant (time = 0), E_R and I are at maximum value, while E_C is zero. E_R is equal to the full applied E_T of 100 V. Current *maximum* may be found from Ohm's law where $I = E_R/R$, $I = 100\ \text{V}/100\ \text{k}\Omega$, $I = 0.001$ A or 1 mA.

At the end of the first time period (0.01 sec in Fig. 13-14), E_C has risen to 63% of E_T (from the graph of Fig. 13-15). E_C at *first time constant* = 63% of 100 V = 63 V. I and E_R have each decreased to 37% of their former maximum values (graph of Fig. 13-15). I is now at 37% of its former maximum value of 1 mA, or *I at first time constant* = (0.37) (1 mA) = 0.37 mA, and *E_R at first time constant* = (0.37) 100 = 37 V. Again note that E_C (63 V) + E_R (37 V) = E_T (100 V).

At the end of the second time constant period, E_C has risen to 86% of maximum, while I and E_R have each decreased to 14% of their maximum values, as shown in the graphs of Fig. 13-15. Observe from these graphs that at the end of the *fifth time constant*, E_C has become practically fully charged, while E_R and I are just about zero. This is an approximation.

The following examples illustrate the preceding discussion.

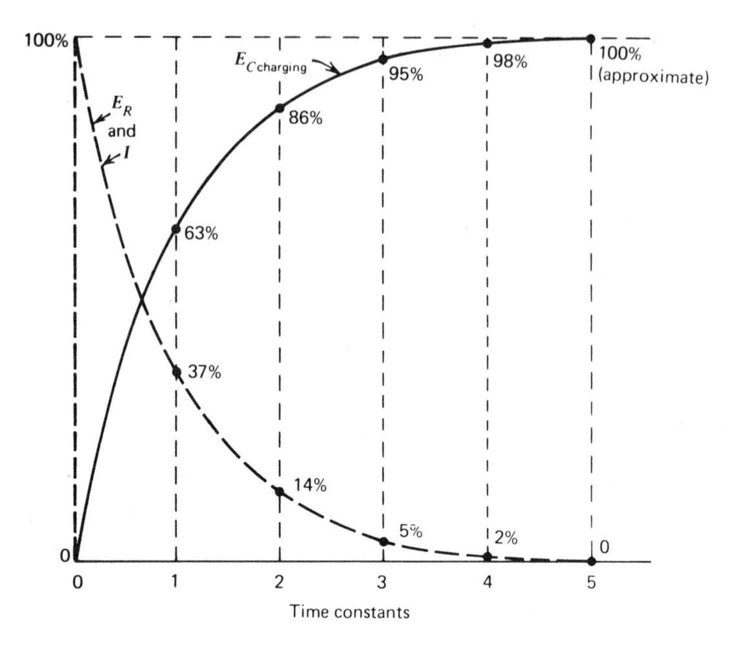

$$1\,RC \text{ time constant (in seconds)} = (R_{\text{in ohms}})\,(C_{\text{in farads}})$$

Figure 13-15. *RC* time constant graphs during capacitor charging period.

Example 13-18

A 0.5-μF capacitor is connected in series with a 1 MΩ resistor, with 120 V dc voltage applied. What is the time of one *RC* time constant?

Solution

$$\text{time constant} = (R)\,(C) \qquad (13\text{-}7)$$

$$\text{time constant} = (1\ \text{M}\Omega)\,(0.5\ \text{mF})$$

$$\text{time constant} = (1 \times 10^{6})\,(0.5 \times 10^{-6})$$

$$\text{time constant} = 0.5\ \text{sec}$$

Example 13-19

How long would it take the capacitor in the previous example to reach approximately full charge?

Solution

The capacitor will charge up to the full applied 120 V dc in about 5 *RC* time constants. Therefore, 5 times 0.5 sec is 2.5 sec.

Example 13-20

In Exəmple 13-18, what is the value of:

(a) I at the start, when current just begins?
(b) E_R at the beginning?
(c) E_C at the beginning?

Solution

Solve this example by referring to the graph of Fig. 13-15. The results should be: I is maximum and is 120 μA (from Ohm's law); E_R is 120 V, and E_C is still zero.

Example 13-21

In Example 13-18, find (a) the period fo three time constants, (b) E_C at the end of three time constants, (c) E_R at the end of this period, and (d) I at the end of this period.

Solution

Solve this example by referring to the graphs of Fig. 13-15 and to the previous three examples. The following results should be found: (a) 1.5 sec; (b) $E_C = 114$ V, or 95% of 120; (c) $E_R = 6$ V or 5% of 120; (d) $I = 6$ μA, or 5% of 120 μA.

For Similar Problems Refer to End-of-Chapter Problems 13-44 to 13-47

Capacitor Discharging. When the switch in the circuit of Fig. 13-14 is placed in the ON position, the 0.1-μF capacitor starts *charging*, and *charging* current flows up through the 100 kΩ resistor. E_R has a polarity with *positive* at the upper end as shown by the solid-line polarity. If the switch is kept ON a sufficiently long time (at least five time constants), the capacitor will charge fully to the applied E_T of 100 V. With $E_C = 100$ V, current stops, and E_R becomes zero.

If the switch in Fig. 13-14 is now moved to the SHORTED position, the applied voltage is disconnected, and a path is now available through which the capacitor will discharge. The 100 V on the capacitor will now cause a *discharge* current to flow *down* through the resistor as shown by the dotted-line arrows in Fig. 13-14. E_R now has a polarity with *negative* at the upper end as shown by the dotted-line polarity. Note that this is the opposite of the polarity of E_R which occurred during charging time.

The graphs of Fig. 13-16 show E_C, E_R, and I during the charging time of the capacitor, and also during the discharging time of the capacitor. The left half of Fig. 13-16 duplicates Fig. 13-15, showing E_C, E_R, and I during charging time.

When the switch of Fig. 13-14 is *first* moved to the SHORTED position, the capacitor starts discharging, and E_C starts decreasing toward zero. As is shown in the right half of the Fig. 13-16 graphs, E_C takes about five time constants, after the switch is moved, to decay to zero. At the first instant that the switch is moved, E_C is still at a full charge of 100 V. It acts as an applied voltage, causing a current to flow *down* through R in the circuit of Fig. 13-14. E_R, as in any series circuit, is equal to the applied voltage (which is really E_C here), or $E_R = E_C$.

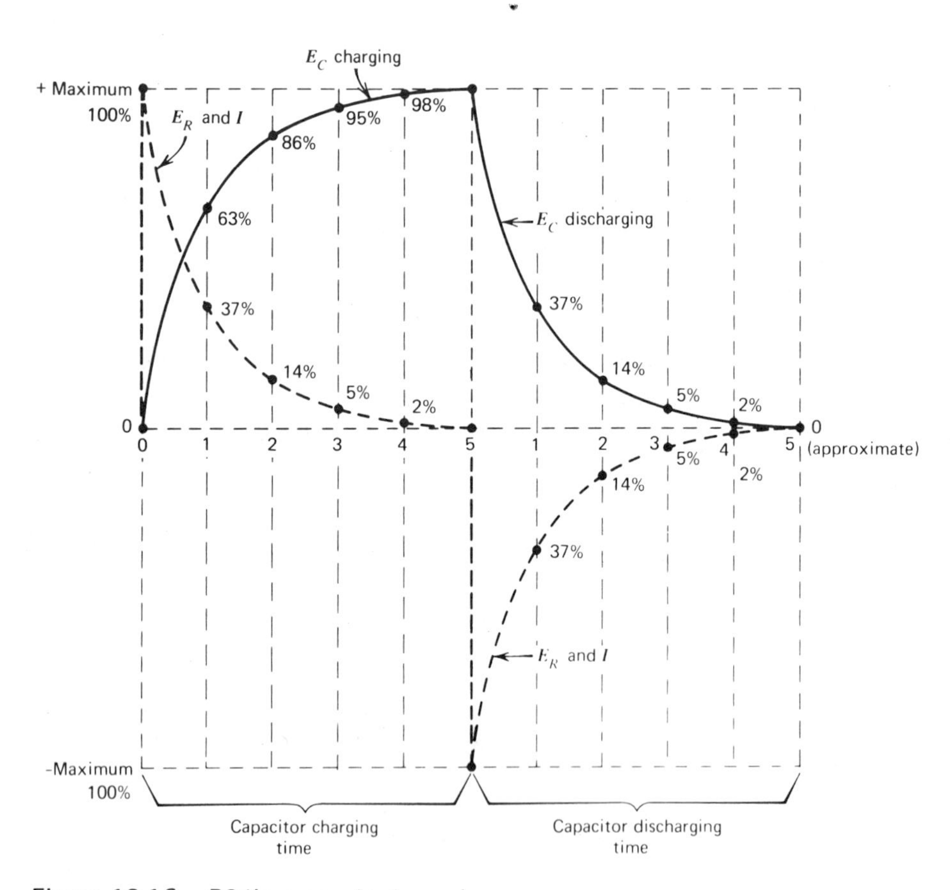

Figure 13-16. *RC* time constant graphs.

One time constant after the switch has been moved to the SHORTED position, E_C has lost 63% of its charge and is down to 37% of its maximum voltage, as shown in the *capacitor discharging* portion of the Fig. 13-16 graphs. At this instant, E_R (which is equal to the "applied voltage" of E_C) has also decreased to 37% of its former maximum value. Note that the E_R graph of Fig. 13-16 during *capacitor charge* time is shown *above* the zero axis, while the E_R graph during *capacitor discharge* time is shown *below* the zero axis. This simply shows in graph form that the polarity of E_R has reversed.

The following examples illustrate the use of the time constant graphs of Fig. 13-16.

Example 13-22

The capacitor of Fig. 13-14 has been charged fully to 100 V. When the switch is moved to the SHORTED position, what is (a) E_C, (b) E_R, and (c) I, two time constants after the switch position has been set to SHORTED?

Solution

(a) Moving the switch to its SHORTED position, permits the capacitor to start discharging. Referring to the discharge period of the graphs of Fig. 13-16, it can be seen that after two time constants, E_C has decreased to 14% of its previous charge. Therefore, $E_C = 14\%$ of 100 V $= (0.14)(100) = 14$ V.

Solve the remainder of this example by using the following as a guide: (b) $E_R = 14$ V, or 14% of 100; (c) $I = 0.14$ mA, or 14% of I_{max}.

Example 13-23

A capacitor has been charged to 300 V. If it is now allowed to start discharging through a 25-kΩ resistor, find : (a) E_R at the instant the capacitor starts discharging, (b) I_{max} at this first instant, (c) E_R, one time constant later, (d) E_C at the end of three time constants, and (e) E_R at the end of three time constants.

Solution

Solve this example by using the previous methods shown and the following results as a guide: (a) $E_R = 300$ V, (b) 12 mA from Ohm's law, (c) $E_R = 111$ V, or 37% of 300, (d) $E_C = 15$ V, or 5% of 300 V, and (e) $E_R = E_C = 15$ V.

Example 13-24

Find E_C of a capacitor at an instant during its discharge if 1.5 mA of discharge current flows through the discharge path resistor of 2500 Ω.

Solution

From Ohm's law, voltage across the resistor is: $E_R = IR = (1.5\text{ mA})(2500\ \Omega) = (1.5 \times 10^{-3})(2.5 \times 10^3) = 3.75$ V. *During discharge*, $E_C = E_R$. Therefore, E_C is also 3.75 V.

For Similar Problems Refer to End-of-Chapter Problems 13-48 to 13-51

13-5. Capacitor with AC Applied Voltage

An ac voltage first has one polarity and then it has the reverse polarity. When ac voltage is applied to a capacitor, it reacts much differently than when a dc voltage is applied. Applying 100 V dc to a capacitor in series with a resistor, as shown in Fig. 13-14, causes a current flow that starts charging up the capacitor. After about five *RC* time constants, the capacitor has been charged fully. That is E_C (or V_C) has become equal to the 100 V applied. When this happens, *no further current flows* and $E_C = 100$ V while E_R and I = zero. The capactior is said, therefore, to *block* direct current (that is, after the capacitor has become fully charged).

If the 100 V were only applied for a short time, say about the time of 1 *RC* time constant, then E_C has only become 63% of 100 V (from the graph of Fig. 13-15) or 63 V. At this instant, $E_R = 37$ V and *current is still flowing*. Charging current will flow as long as the capacitor has not become completely charged. That is, as long as E_C is less than the applied dc voltage.

Similarly, when the SWITCH in Fig. 13-14 is moved to the SHORTED position, the capacitor starts discharging and a discharge current now flows in the opposite direction from the direction of the previous charging current. If sufficient time is given the capacitor (5 *RC* time constants or longer), then E_C will decrease to zero, and current becomes zero. If, however, too little time is given the capacitor (less than 5 time constants), then E_C will not have fully discharged, and a discharge current continues to flow.

This lack of sufficient time, either not giving the capacitor enough time to fully charge or to fully discharge, results in a continuance of the current flow. That is, the capacitor will not block the current. This is what occurs when an *ac voltage* is applied to a capacitor in series with a resistor. Figure 13-17*a* shows a 5000 Hz (5000 cps) ac voltage having a peak or maximum value of 100 V. At a frequency of 5000 Hz, the time of *one complete cycle* (from instant A to instant E) is 1/5000 sec or 0.0002 sec, or 200 μsec, as shown. The time of a *half cycle* or *alternation* (from instant A to instant C) is one half of 200, or 100 μsec, as shown.

As is shown in Fig. 13-17*a*, the ac voltage rises from zero to its peak of 100 V in one fourth of a cycle, from instant A to instant B, or 1/4 of 200, or 50 μsec. The ac applied voltage then decreases from 100 V to zero in the next 1/4 cycle (instant B to instant C), or the next 50 μsec.

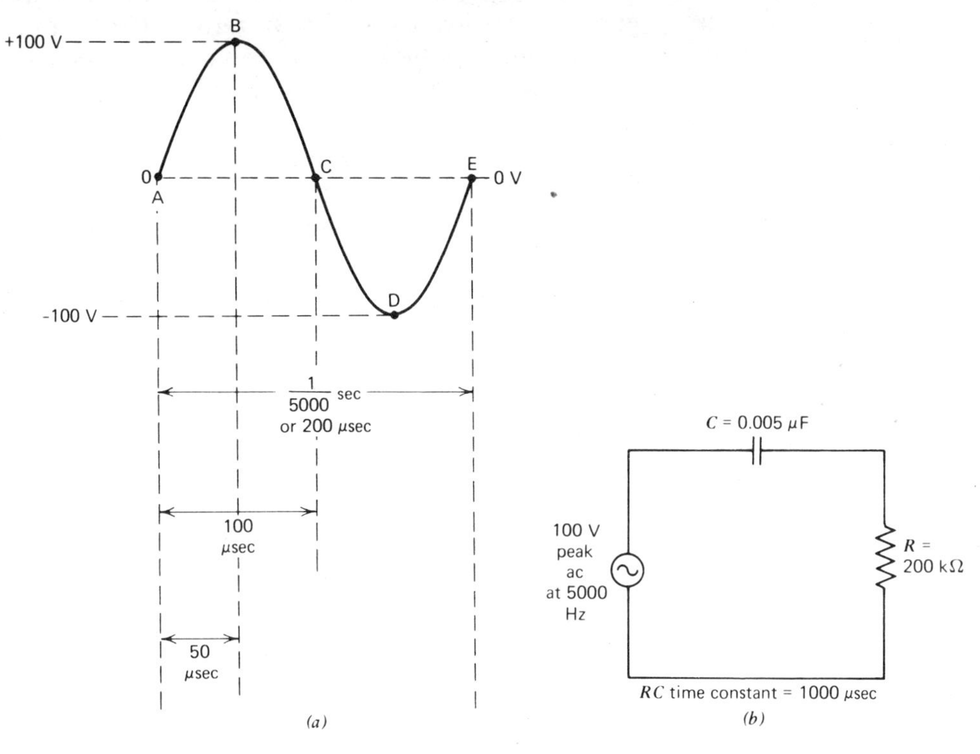

Figure 13-17. An ac voltage applied to series *RC* circuit. (*a*) A 500-Hz ac sine wave. (*b*) An ac series *RC* circuit.

In Fig. 13-17*b*, this ac voltage is shown applied to a 0.005-μF capacitor in series with a 200-kΩ resistor. The *RC* time constant is: $(R)\ (C) = (200\ \text{K})\ (0.005\ \mu\text{F}) = (200 \times 10^3)\ (5 \times 10^{-9}) = 1000 \times 10^{-6}$ sec, or 1000 μsec.

Note that the time during which the applied voltage rises from zero to 100 V (instant A to instant B) in Fig. 13-17*a*, is only 50 μsec, while one *RC* time constant of the circuit is 1000 μsec. This means that the applied voltage rises from zero to 100 V in only 1/20 time constant. Since it requires about five time constants for the capacitor to become fully charged, then in only 1/20 time constant, the capacitor will only become slightly charged. As a result, current will not stop, as it would have if the capacitor had become fully charged. The smaller the capacitor voltage, the greater remains the current. The larger the capacitor voltage, the smaller becomes the current.

During the period (instant B to instant C) of the next 50 μsec, the applied ac voltage decreases to zero. In this short period the capacitor starts discharging but doesn't have sufficient time to completely discharge. As a result, discharge current keeps flowing in this opposite direction and does not stop, as it would if the capacitor had completely discharged.

In the time between instant C and instant D (of Fig. 13-17*a*), the ac voltage has reversed its polarity, and rises from zero to 100 V in the next 50 μsec. Again, in this short time, the capacitor does not have sufficient time to charge up to this reverse-polarity 100 V, and current keeps flowing.

Refer to End-of-Chapter Problems 13-52 to 13-54

Capacitive Reactance. The capacitor offers, to an ac applied voltage, some *opposition* to its current flow. This opposition is called *capacitive reactance* or X_C. X_C depends on the *frequency* of the alternating current and also on the *value of the capacitor.* The *higher* the frequency, the *less time* the capacitor is given to charge. This smaller capacitor voltage then permits a greater current flow, or the capacitor offers less opposition (X_C). Therefore, X_C is *smaller for higher frequencies.*

Similarly, X_C *is smaller for larger capacitors.* The larger the capacitor, the larger is the RC time constant, requiring a longer time for the capacitor to charge. In a given short time, a larger capacitor therefore charges more slowly, only reaching a smaller voltage. This smaller voltage on the capacitor permits a greater current flow, meaning that the capacitor offers less opposition (X_C).

The value of *capacitive reactance*, X_C is

$$X_C = \frac{1}{2\pi f C} \tag{13-8}$$

where X_C is in ohms, $\pi = 3.14$ and $2\pi = 6.28$, f is the *frequency* of the ac applied voltage in hertz (or cycles per second), and C is the *value of the capacitor* in *farads.*

The following examples illustrate the use of the X_C equation.

Example 13-25

Find the opposition (X_C) of a 0.05-μF capacitor where the frequency of the ac voltage is 0.2 MHz.

Solution

$$X_C = \frac{1}{2\pi f C} \tag{13-8}$$

$$X_C = \frac{1}{(2)\,(3.14)\,(0.2\ \text{MHz})\,(0.05\ \mu\text{F})}$$

$$X_C = \frac{1}{(6.28)\,(0.2 \times 10^{6})\,(0.05 \times 10^{-6})}$$

$$X_C = \frac{1}{(6.28)\,(0.01)}$$

$$X_C = \frac{1}{6.28 \times 10^{-2}}$$

$$X_C = 0.159 \times 10^2$$

$$X_C = 15.9\ \Omega$$

Example 13-26

Find the capacitive reactance of a 150-pF capacitor at a frequency of 250 kHz.

Solution

By using equation 13-8, find X_C as shown previously. The 150 pF should be written as 150×10^{-12}, and the 250 kHz becomes 250×10^3. X_C should be found to be 4250 Ω.

Sometimes, X_C is known along with the frequency of the ac voltage, but the required value of the capacitor must be found. These following examples illustrate this.

Example 13-27

Find the value of a capacitor if its capacitive reactance (X_C) must be 1/10 that of a 500-Ω resistor, and where the ac voltage has a frequency of 30 Hz. (This example is a practical one where the capacitor is in the cathode (or emitter) circuit of an audio amplifier and where the lowest frequency audio signal is 30 Hz.)

Solution

X_C must be 1/10 of 500 Ω, or 50 Ω.

and

$$X_C = \frac{1}{2\pi f C} \tag{13-8}$$

Solving for C by *multiplying each side of equation by C and also dividing by X_C*, and then canceling the X_C's at the left side, and canceling the C's at the right side yields

$$C = \frac{1}{2\pi f X_C}$$

Then

$$C = \frac{1}{(6.28)\ (30\ \text{Hz})\ (50\ \Omega)}$$

$$C = \frac{1}{9420}$$

$$C = \frac{1}{9.42 \times 10^3}$$

$$C = 0.106 \times 10^{-3}\ \text{F}$$

$$C = 0.000106\ \text{F} \qquad \text{or} \qquad 106.\ \mu\text{F}$$

Example 13-28

The screen grid of a pentode radio-frequency amplifier electron tube requires a capacitor having an X_C 1/20 that of a 10-kΩ resistor. Find the value of this capacitor if the frequency of the ac signal voltage is 7.5 MHz.

Solution

Solve this example by using the equation in the previous example. As a guide, the results should be: $X_C = 0.5$ kΩ, frequency $= 7.5 \times 10^6$ Hz, and c should be 42.5 pF.

For Similar Problems Refer to End-of-Chapter Problems 13-55 to 13-59

13-6. Capacitor and Resistor in Series with AC Voltage Applied

When an ac sine-wave voltage is applied to a capacitor and a resistor connected in series, as in Fig. 13-18*b*, current flows first in one direction and then in the reverse direction. As this alternating current flows, the capacitor attempts to charge, but because of the short time that the applied voltage gives the capacitor, it cannot charge very much. The *voltage* rise on the capacitor takes time, and is said to be delayed behind the current, or to *lag* the current.

Phase Relationships, Sine Waves. As shown in Fig. 13-18*a*, a sine-wave alternating current flowing in the capacitor and resistor circuit of Fig. 13-18*b*, produces a sine-wave voltage across the *resistor*, E_R (or V_R), which is *in phase* with the current. The term *in phase*, using sine waves, means that both sine waves start at zero together and rise in the same direction together, reaching their maximums at the same instant. Note that in Fig. 13-18*a*, I and E_R are *in phase.*

This same alternating current flowing in and out of the capacitor of Fig. 13-18*b*, produces a voltage across the *capacitor*, E_C (or V_C), which *lags* the current by 90° as shown in the sine waves of Fig. 13-18*a*. The term *lagging by* 90 *degrees*, using sine waves, means that one sine wave starts 90° (1/4 of a cycle) behind the other. Observe in Fig. 13-18*a* that the capacitor voltage E_C is shown starting at *instant A* which is 1/4 of a cycle after the current, I, sine wave had started.

The sum of the E_R *sine wave* (in phase with I) and the E_C *sine wave* (lagging I by 90°) is the ac *applied voltage* E_T. This E_T sine wave (Fig. 13-18*a*) is actually lagging the current sine wave by 45°. It may be seen that when E_T sine wave is at *zero for the first time* (1/8 cycle before instant A), the I sine wave has already increased above zero for 1/8 of its cycle. Since a complete cycle is 360°, then 1/8 cycle is 45°. As a result E_T is said to lag I by 45°.

By adding the E_C and E_R sine waves at several instants, their sum (E_T) may be drawn. This is shown in the following discussion of Fig. 13-18*a*.

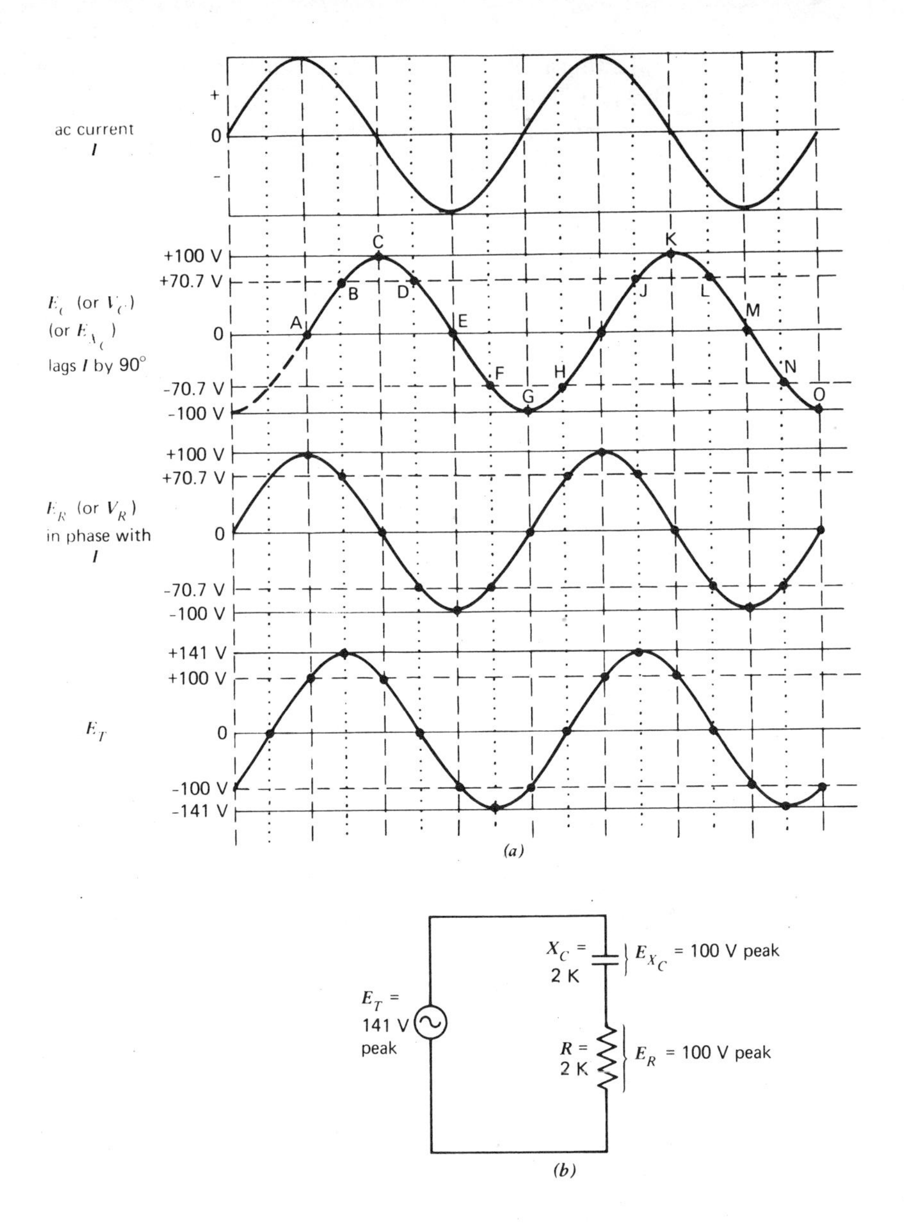

Figure 13-18. A sine-wave phase relationship in *RC* series circuit. (*a*) The ac sine wave voltages and current in an *RC* series circuit. (*b*) The circuit.

Starting at *instant A*, E_C is zero and E_R is at its positive maximum of +100 V. The sum, E_T, is +100 V.

At *instant B*, E_C is +70.7 V, while E_R is also +70.7 V. Their sum, E_T, is now at its positive maximum of +141 V, as shown.

At *instant C*, E_C is at its positive maximum of +100 V, while E_R is zero. Their sum, E_T, is now +100 V, as shown.

At *instant D*, E_C is +70.7 V, while E_R is negative, −70.7 V. Their sum, E_T, is now zero, as shown.

At *instant E*, E_C is zero, while E_R is −100 V. Their sum, E_T, is now −100 V, as shown.

At *instant F*, E_C is *negative*, −70.7 V, while E_R is also −70.7 V. Their sum, E_T, is now at its negative maximum of −141 V, as shown.

At *instant G*, E_C is −100 V, while E_R is zero. Their sum, E_T, is now −100 V, as shown.

At *instant H*, E_C is −70.7 V, while E_R is +70.7 V. Their sum, E_T, is now zero, as shown.

At *instant I*, E_C is zero, while E_R is +100 V. Their sum, E_T, is now +100 V, as shown.

Instants A to I completes *one cycle of the E_C sine wave.* That portion of the *next* cycle continuing with instants J, K, and so on, and the previous part of the E_C cycle *before* instant A, is left for the reader to complete by *referring to end-of-chapter Problem 13-60.*

Phase Relationships, Phasors and Vectors. The 90° difference between the E_C and E_R sine waves of Fig. 13-18, and the 45° difference between E_T and I are not too easily seen when sine waves are used. Another method of indicating phase relationships is to use *vectors or phasors.* A discussion of these is given in the previous chapter, Section 12-3, and also in the Appendix, Mathematics for Electronics. For students who require it at this time, it is suggested that they read the Phasors and Vectors portion of Section 12-3 before continuing with the present chapter. A very brief discussion of this topic is also presented below.

A *vector* is a line, the *length* of which signifies *magnitude.* The *arrowhead* on the vector signifies *direction.* For example, an arrow-line four in. long pointing horizontally to the right could denote a swimmer moving east at a speed of 4 mph.

A *rotating vector* could also represent an ac sine-wave voltage or current. The *frequency of rotation* of the vector would be the same as the *frequency* of the alternating current, while the *length* of the vector would represent the *amount of amplitude* of the voltage, either in peak (maximum) or effective (rms) volts. The *direction* of the rotating vector is normally assumed to be *counterclockwise*, The *rotating vector*, called a *phasor*, is always assumed to be viewed under a stroboscopic or flashing light. The frequency of this imaginary

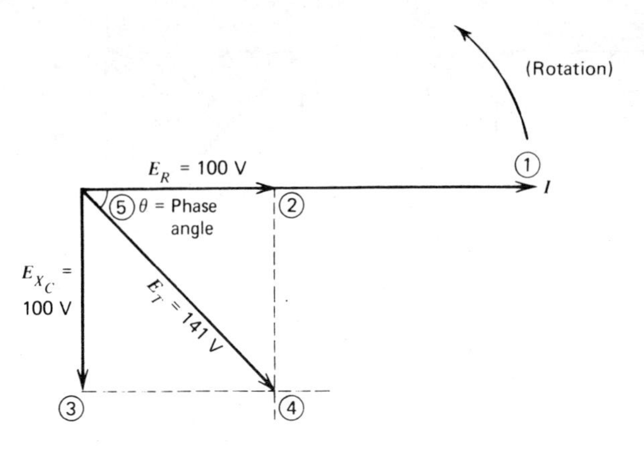

Figure 13-19. Phase relationship using phasors.

flashing light is the same as the phasor frequency of rotation, allowing the phasor to appear in a seemingly stationary position.

The phasor diagram of Fig. 13-19 shows four different phasors with the phase relationship between each being much more apparent than the sine-wave diagram of Fig. 13-18*a*.

The phasors of Fig. 13-19 show the current and voltages in the ac series circuit of X_C and R of Fig. 13-18*b*. The following discussion refers to the series circuit and to the phasor diagram, and each numbered step refers to the phasor having this same number.

1. The same current, I, flows through R and X_C in the series circuit. Since I is common to both R and X_C, this I phasor is used as a reference.

2. Voltage across the resistor, E_R, is *in phase* with I, and the E_R phasor is shown pointing in the same direction as the I phasor.

3. Voltage across the capacitive reactance, E_{X_C}, always *lags* I by 90°. Since phasors are normally assumed to be rotating counterclockwise, then the E_C phasor is shown behind I by 90°.

4. Total voltage, E_T, is the sum of E_R and E_{X_C}. Since the E_R and E_C phasors can be seen to be 90° out of phase with each other (Fig. 13-19), then their sum, called the *vector sum* or the *phasor sum*, is added vectorially, and is shown as the E_T phasor.

5. The angle between E_T and I is called the *phase angle*. The Greek letter θ (pronounced "theta") is usually used to denote the phase angle.

By comparing the phase relationships of the sine waves of Fig. 13-18*a* with those of the phasors, it is apparent that the phasors more clearly indicate the phase between the current and the voltages.

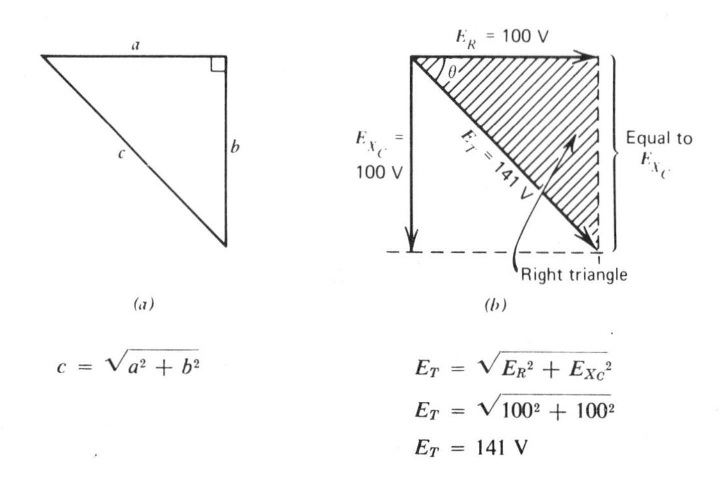

Figure 13-20. (*a*) Right triangle. (*b*) The addition of E_R and E_{XC} phasors.

E_R and E_{X_C} Addition of Phasors. To add two phasors that are out of phase by 90°, as are E_R and E_{X_C} of Fig. 13-19, and as are shown repeated in Fig. 13-20*b*, the sum E_T may be found by either a construction method or a simple mathematical equation.

The *construction method* consists of the following.

If E_R and E_{X_C} are drawn to scale, that is, if each is 100 V, and are each drawn 1 in. long (1 in. then represents 100 V), then at the arrowhead of *each* phasor, draw a line (shown as the dashed lines of Fig. 13-20*b*), *parallel* to the other phasor. A rectangle (or a square) is then formed. The *diagonal of the rectangle*, from the junction of the E_R and E_{X_C} phasors to the junction of the dashed lines, is the *phasor sum* or E_T. If this diagonal is then measured, it will be found to be 1.4 in. approximately. Since the E_R and E_{X_C} phasors were drawn to a 1 in. = 100 V scale, then the 1.4 in. length of E_T means that E_T is 140 V. E_T will be larger than E_R or E_{X_C}, but less than their arithmetical sum.

The mathematical equation method of adding two 90° phasors is explained in the following.

In Fig. 13-20*a*, a right triangle is shown. This is a triangle containing a 90° or right angle. A rule called the *Pythagoras theorem* states that the longest side of a right triangle (side *c* in Fig. 13-20*a*) is equal to the square root of the sum of the squares of the other two sides. This has been given in the previous chapter, and is restated here.

$$c = \sqrt{a^2 + b^2} \qquad (12\text{-}3)$$

In Fig. 13-20*b*, note that when the dashed lines are drawn at each phasor arrowhead parallel to the other phasor, a square or a rectangle is formed. With the diagonal, the phasor sum E_T, there now is formed a right triangle (the shaded portion of Fig. 13-20*b*). From the *Pythagoras theorem*, the longest side E_T is equal to the square root of the sum of the squares of the other two sides, E_R and E_{X_C}, or

$$E_T = \sqrt{E_R^2 + E_{X_C}^2} \qquad (13\text{-}9)$$

The following examples illustrate the use of this equation.

Example 13-29

In the circuit diagram of Fig. 13-18*b*, and in the phasor diagram of Fig. 13-20*b*, find the sum E_T where E_R is 100 V and E_{X_C} is also 100 V

Solution

$$E_T = \sqrt{E_R^2 + E_{X_C}^2} \qquad (13\text{-}9)$$

$$E_T = \sqrt{100^2 + 100^2}$$

$$E_T = \sqrt{(1 \times 10^2)^2 + (1 \times 10^2)^2}$$

$$E_T = \sqrt{1 \times 10^4 + 1 \times 10^4}$$

$$E_T = \sqrt{2 \times 10^4}$$

$$E_T = 1.41 \times 10^2 \text{ V} \qquad \text{or} \qquad 141 \text{ V}$$

Note that E_T (141 V) is, of course, more than E_R (100 V) or E_{X_C} (100 V), but *less* than their arithmetical sum (100 + 100 = 200).

Example 13-30

A resistor and a capacitor are connected in series to an applied ac voltage. If E_R = 300 V, and E_{X_C} = 400 V, find E_{total}.

Solution

$$E_T = \sqrt{E_R^2 + E_{X_C}^2} \qquad (13\text{-}9)$$

$$E_T = \sqrt{300^2 + 400^2}$$

$$E_T = \sqrt{(3 \times 10^2)^2 + (4 \times 10^2)^2}$$

$$E_T = \sqrt{9 \times 10^4 + 16 \times 10^4}$$

$$E_T = \sqrt{25 \times 10^4}$$

$$E_T = 5 \times 10^2 \text{ V} \qquad \text{or} \qquad 500 \text{ V}$$

Again, note that E_T is, as expected, more than either E_R (300 V) or E_{X_C} (400 V), but *less* than their arithmetical sum (300 + 400 = 700).

If, in the right triangle phasor diagram of Fig. 13-20*b*, the longest side E_T is known, as well as one side E_R, then the third side E_{Xc} can be found as follows.

$$E_T = \sqrt{E_R^2 + E_{Xc}^2} \qquad (13\text{-}9)$$

Squaring both sides of equation gives

$$E_T^2 = \left(\sqrt{(E_R^2 + E_{Xc}^2)}\right)^2$$

This becomes

$$E_T^2 = E_R^2 + E_{Xc}^2$$

Solving for E_{Xc}^2, transpose the E_R^2 term to the left side of the equation:

$$E_T^2 - E_R^2 = E_{Xc}^2$$

And taking the square root of both sides of equation yields

$$\sqrt{E_T^2 - E_R^2} = \sqrt{E_{Xc}^2}$$

and, finally,

$$\sqrt{E_T^2 - E_R^2} = E_{Xc} \qquad (13\text{-}10)$$

If E_R were the unknown, the equation would be

$$\sqrt{E_T^2 - E_{Xc}^2} = E_R \qquad (13\text{-}11)$$

The following examples illustrate the use of the last two equations.

Example 13-31

If 150 V ac were applied to a resistor in series with a capacitor, find E_{Xc} if E_R is 90 V.

Solution

$$E_{Xc} = \sqrt{E_T^2 - E_R^2} \qquad (13\text{-}10)$$

$$E_{Xc} = \sqrt{150^2 - 90^2}$$

$$E_{Xc} = \sqrt{(1.5 \times 10^2)^2 - (0.9 \times 10^2)^2}$$

$$E_{Xc} = \sqrt{2.25 \times 10^4 - 0.81 \times 10^4}$$

$$E_{Xc} = \sqrt{1.44 \times 10^4}$$

$$E_{Xc} = 1.2 \times 10^2 \text{ V} \quad \text{or} \quad 120 \text{ V}$$

Example 13-32

Find the voltage across a resistor, E_R, which is connected in series with a capacitor, if E_{Xc} is 200 V and the total applied ac voltage E_T is 250 V.

Solution

Using equation 13-11, the reader should solve for E_R and should find it to be 150 V.

For Similar Problems Refer to End-of-Chapter Problems 13-61 to 13-64

Impedance (Z), or Total Opposition. A resistor in series with a capacitor in an ac circuit offer a *total opposition* called *impedance* (Z). This impedance is the phasor sum of R and X_C. In the circuit of Fig. 13-21*a*, the *voltages* E_R and E_{X_C} are shown in the previous phasor diagram of Fig. 13-20*b*. Since the same current I flows through R and X_C, and from Ohm's law, $E_R = IR$ and $E_{X_C} = IX_C$, then the phasors E_R and E_{X_C} may be called R and X_C, respectively, as shown in Fig. 13-21*b*. The R and X_C are 90° out of phase. The sum of these phasors is the total opposition called the *impedance*, Z. To add R and X_C, lines are drawn at each phasor arrowhead parallel to the other phasor. These lines are shown as the dashed lines in Fig. 13-21*b*, forming the rectangle (or square, where $R = X_C$). The diagonal of the rectangle is the impedance Z. Note that the shaded area forms a right triangle with R and X_C

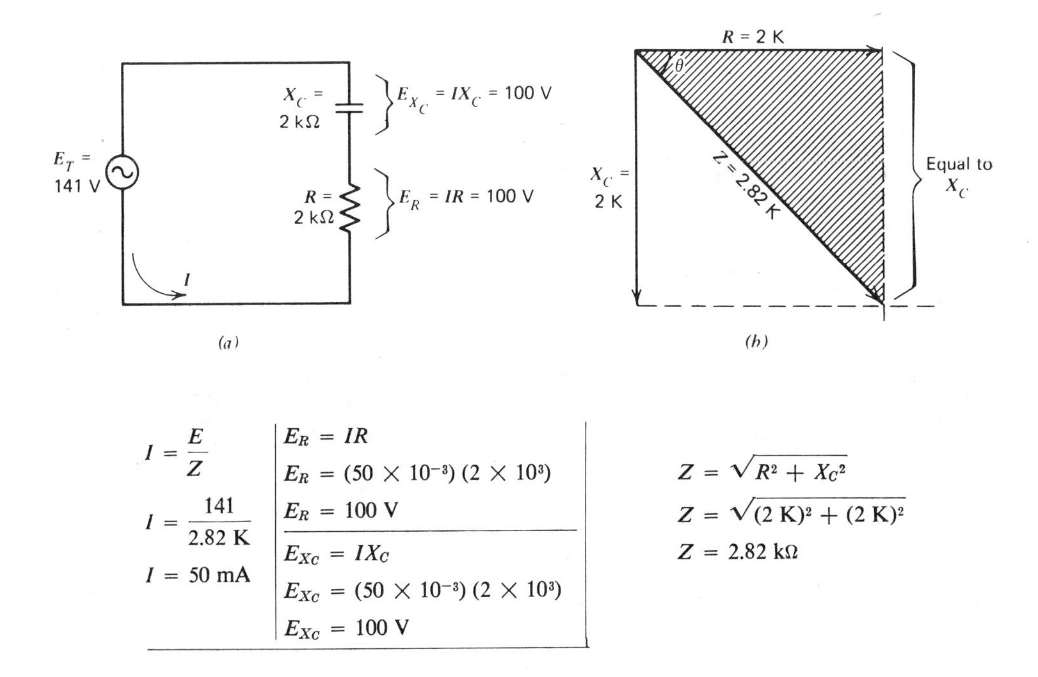

Figure 13-21. An *RC* series ac circuit. (*b*) Phasor addition of X_C and R.

as the two smaller sides and with Z the longest side. From the Pythagoras theorem,

$$Z = \sqrt{R^2 + X_C^2} \qquad (13\text{-}12)$$

Using the values of $R = 2\ \text{k}\Omega$ and $X_C = 2\ \text{k}\Omega$, as shown in the circuit of Fig. 13-21*a*, the following example illustrates the use of the last equation and the solution of the circuit.

Example 13-33

Find the total opposition Z in the circuit of Fig. 13-21*a*.

Solution

Solve for Z by using equation 13-12, and the solution shown in Fig. 13-21*b* as a guide.

Example 13-34

In the circuit of Fig. 13-21*a*, where $R = 2$ K, $X_C = 2$ K, $Z = 2.82$ K, and $E_T = 141$ V, find: (a) I, (b) E_R, (c) E_{X_C}, and (d) vectorially add E_R (100 V) and E_{X_C} (100 V) to prove that their sum is equal to E_T (141 V).

Solution

The reader should solve this example by using the texts in Figs. 13-21*a* and 13-20*b* as a guide.

Example 13-35

An ac voltage of 30 V at a frequency of 20 MHz is applied to a 15-pF capacitor in series with a 250-Ω resistor. Find: (a) X_C, (b) Z, (c) I, (d) E_R, (e) E_{X_C}, and (f) by using the values of E_R and E_{X_C} found, add them vectorially to prove that their sum is equal to E_T.

Solution

Again solve this example by using procedures shown previously and the following results as a guide: $X_C = 530\ \Omega$, $Z = 586\ \Omega$, $I = 51.2$ mA, $E_R = 12.8$ V, and $E_{X_C} = 27.1$ V.

For Similar Problems Refer to End-of-Chapter Problems 13-65 to 13-67

Resistors and Capacitors in AC Series Circuits. When two or more resistors are connected in series with two or more capacitors in an ac circuit, as shown in Fig. 13-22*a*, the resistor values must be added to get the total resistance R_T. Total capacitance is found by using either: $C_T = C_1 C_2 / C_1 + C_2$ (equation 13-2) or $C_T = 1/(1/C_1 + 1/C_2 + \text{etc.})$ (equation 13-3). If the *reactance* of each capacitor is known, then the total capacitive reactance, X_{C_T}, is simply the sum of the series X_{C_1}, X_{C_2}, and so on. Current and all voltages are found as shown in the following discussion.

In the circuit of Fig. 13-22*a*, $R_1 = 70\ \Omega$ and $R_2 = 80\ \Omega$. R_T is then $70 + 80$ or $R_T = 150\ \Omega$. This is shown in the simplified equivalent circuit of Fig. 13-22*b*, and also depicted in the phasors of Fig. 13-22*c*.

In the circuit of Fig. 13-22*a*, $C_1 = 5\ \mu\text{F}$ and $C_2 = 20\ \mu\text{F}$. X_C of each at the frequency of 100 Hz is

$$X_{C_1} = \frac{1}{2\pi f C_1} \tag{13-8}$$

$$X_{C_1} = \frac{1}{(6.28)\ (100\ \text{Hz})\ (5\ \mu\text{F})}$$

$$X_{C_1} = \frac{1}{(6.28)\ (1 \times 10^{2})\ (5 \times 10^{-6})}$$

$$X_{C_1} = \frac{1}{31.4 \times 10^{-4}}$$

$$X_{C_1} = \frac{1}{3.14 \times 10^{-3}}$$

$$X_{C_1} = 0.318 \times 10^{3}\ \Omega \qquad \text{or} \qquad 318\ \Omega$$

This is shown in the circuit of Fig. 13-22*a* and also in the phasor diagram of Fig. 13-22*c*.

In a similar manner, X_{C_2} is found to be 79.6 Ω which the reader should also find. X_{C_2} is shown in Fig. 13-22*a* and *c*.

Total capacitance is found as follows

$$C_T = \frac{C_1\ C_2}{C_1 + C_2} \tag{13-2}$$

$$C_T = \frac{(5)\ (20)}{5 + 20}$$

$$C_T = \frac{100}{25}$$

$$C_T = 4\ \mu\text{F}$$

This is shown in the simplified equivalent circuit of Fig. 13-22*b*.

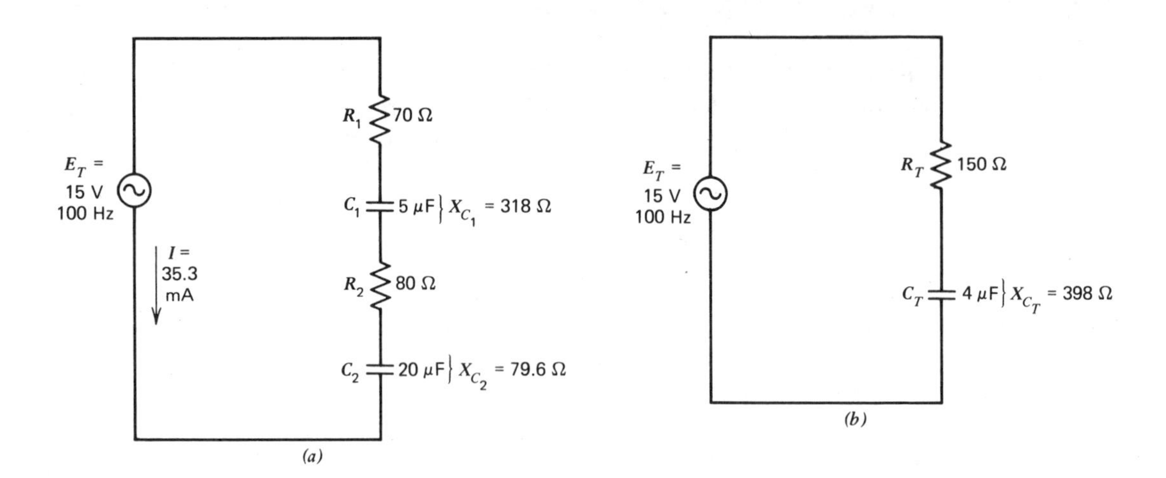

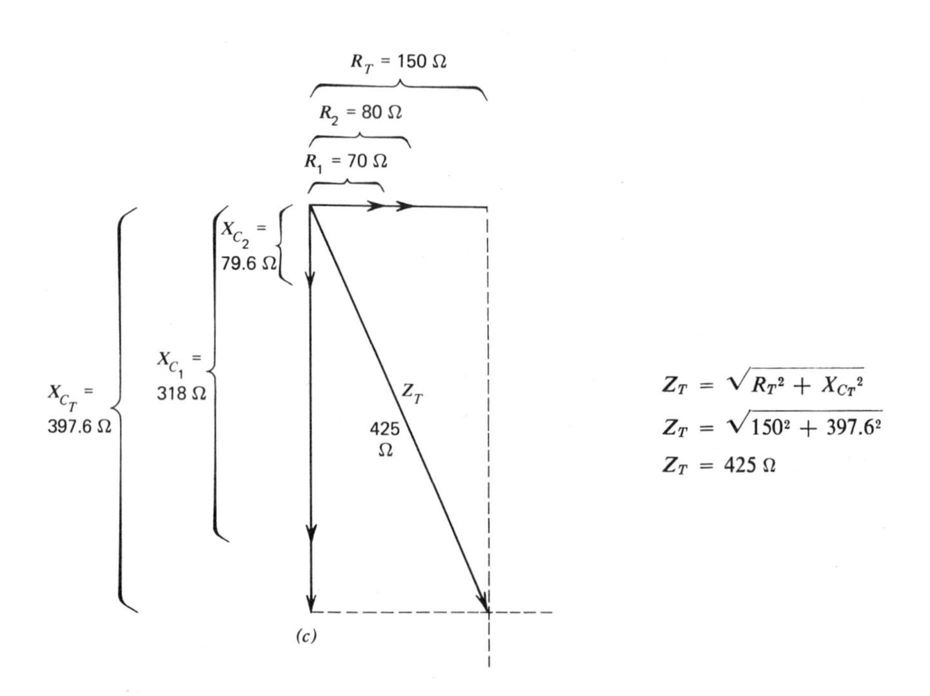

Figure 13-22. Resistors and capacitors in series ac circuit. (*a*) Circuit. (*b*) Simplified equivalent circuit. (*c*) Phasors.

Total capacitive reactance, X_{C_T}, is

$$X_{C_T} = \frac{1}{2\pi f C_T} \qquad (13\text{-}8)$$

$$X_{C_T} = \frac{1}{(6.28)\,(100\text{ Hz})\,(4\ \mu\text{F})}$$

$$X_{C_T} = \frac{1}{(6.28)\,(1 \times 10^2)\,(4 \times 10^{-6})}$$

$$X_{C_T} = \frac{1}{25.1 \times 10^{-4}}$$

$$X_{C_T} = \frac{1}{2.51 \times 10^{-3}}$$

$$X_{C_T} = 0.398 \times 10^3\ \Omega \qquad \text{or} \qquad 398\ \Omega$$

This is shown in the simplified equivalent circuit of Fig. 13-22*b*, and also in the phasors of Fig. 13-22*c*. Note that this value of X_{C_T} is approximately equal to the sum of X_{C_1} and X_{C_2}.

Total impedance, Z_T, is the vector sum of R_T (150 Ω) and X_{C_T} (398 Ω), as shown in the phasors of Fig. 13-22*c*.

$$Z_T = \sqrt{R_T^2 + X_{C_T}^2} \qquad (13\text{-}12)$$

$$Z_T = \sqrt{150^2 + 398^2}$$

$$Z_T = \sqrt{(1.5 \times 10^2)^2 + (3.98 \times 10^2)^2}$$

$$Z_T = \sqrt{2.25 \times 10^4 + 15.8 \times 10^4}$$

$$Z_T = \sqrt{18.05 \times 10^4}$$

$$Z_T = 4.25 \times 10^2\ \Omega \qquad \text{or} \qquad 425\ \Omega$$

This is shown in the phasor diagram of Fig. 13-22*c*.

Current and each voltage may now be found by using Ohm's law. $I = E_T/Z_T = 15/425 = 35.3$ mA. $E_{R_1} = IR_1 = (35.3 \times 10^{-3})\,(70) = 2.47$ V. $E_{R_2} = IR_2 = (35.3 \times 10^{-3})\,(80) = 2.82$ V. $E_{X_{C1}} = IX_{C_1} = (35.3 \times 10^{-3})\,(318) = 11.2$ V. $E_{X_{C2}} = IX_{C_2} = (35.3 \times 10^{-3})\,(79.6) = 2.81$ V.

A check for proving that all the voltages found are correct is the following. The vector sum of all must be equal to the total applied voltage, $E_T = 15$ V. E_{R_T} is the sum of E_{R_1} and E_{R_2}, 2.47 + 2.82, or $E_{R_T} = 5.29\ V$. Similarly, $E_{X_{CT}}$ is the sum of $E_{X_{C1}}$ and $E_{X_{C2}}$, or 11.2 + 2.81, or $E_{X_{CT}} = 14.01\ V$. Therefore

$$E_T = \sqrt{E_{R_T}^2 + E_{X_{CT}}^2} \qquad (13\text{-}9)$$

$$E_T = \sqrt{5.29^2 + 14.01^2}$$

$$E_T = \sqrt{28. + 196.}$$

$$E_T = \sqrt{224.}$$

$$E_T = 14.97\text{ V}$$

Note that this is approximately equal to the 15 volts E_T.

The following example illustrates the preceding discussion.

Example 13-37

In the circuit of Fig. 13-23a, two resistors ($R_1 = 8$ K and $R_2 = 2$ K) are connected in series with three capacitors ($C_1 = 300$ pF, $C_2 = 200$ pF, and $C_3 = 40$ pF). The applied ac voltage is 50 V at a frequency of 200 kHz. Find:

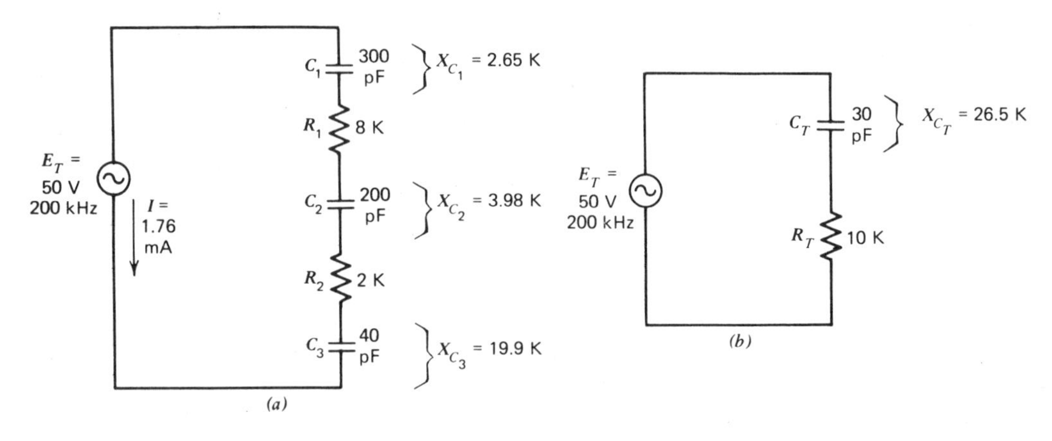

$$R_T = R_1 + R_2 = 8\text{ K} + 2\text{ K} = 10\text{ K}$$

$$X_{C_T} = X_{C_1} + X_{C_2} + X_{C_3} = 2.65\text{ K} + 3.98\text{ K} + 19.9\text{ K} = 26.53\text{ K}$$

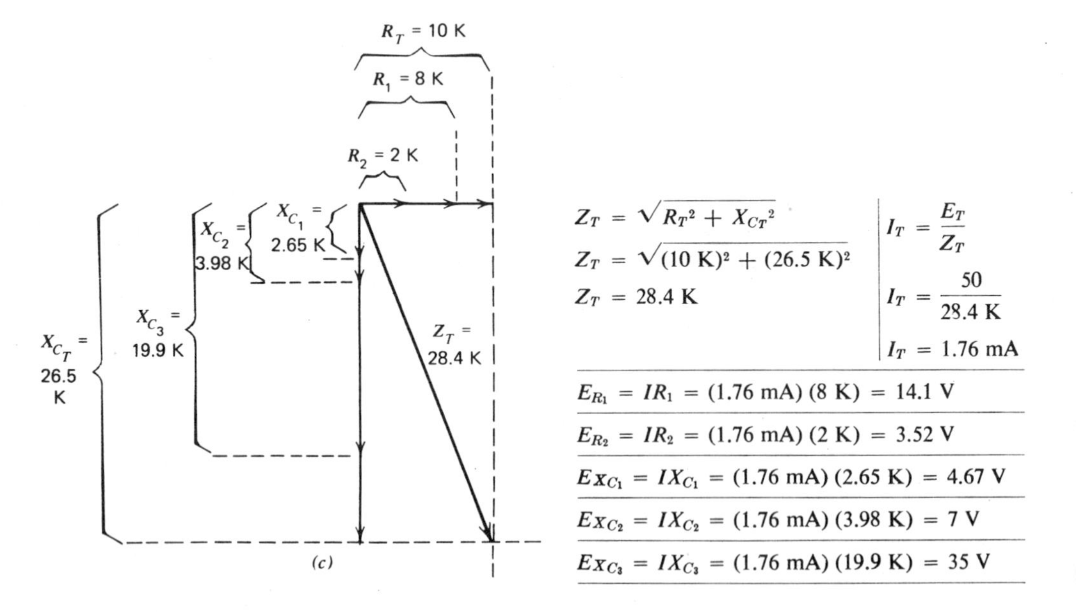

Figure 13-23. Example 13-37. (*a*) Circuit. (*b*) Simplified equivalent circuit.

(a) X_{C_1}, (b) X_{C_2}, (c) X_{C_3}, (d) C_T, (e) X_{CT}, (f) R_T, (g) Z_T, (h) I, (i) E_{R_1}, (j) E_{R_2}, (k) $E_{X_{C1}}$, (l) $E_{X_{C2}}$, (m) $E_{X_{C3}}$.

Solution

Solve this example by using the previous procedure and the results shown in Fig. 13-23 as a guide.

For Similar Problems Refer to End-of-Chapter Problems 13-68 and 13-69

Power in Resistor-Capacitor Circuits. A pure capacitor, similar to an inductor, consumes no power. Energy is stored in the capacitor during its charge, but it is then returned to the circuit during the capacitor discharge. Only the resistance in the circuit dissipates power. The power equation, $P_T = E_T I_T$, is only true for a pure resistor circuit. It is not true when a reactance is also present. The equation, $P_T = E_T I_T$, where X_C and R are both present, therefore does not give the actual or true power. What it does give is called the *apparent* power.

True power in a series ac circuit may be found by using the other power equations which concern only the resistance in the circuit. These are: $P_{true} = I^2R$, or $E_R I$, or $E_R{}^2/R$.
True power is the actual power dissipated as read on a wattmeter.

Refer back to the circuit of Fig. 13-21*a*, where E_T is 141 V, $X_C = 2$ K with 100 V across it, $R = 2$ K with 100 V across it, and $I = 50$ mA. The apparent power is then

$$P_{apparent} = E_T I_T$$

$$P_{apparent} = (141 \text{ V}) (50 \text{ mA})$$

$$P_{apparent} = (1.41 \times 10^2) (50 \times 10^{-3})$$

$$P_{apparent} = 70.5 \times 10^{-1} \text{ W} \quad \text{or} \quad \text{VA} \quad \text{or}$$
$$7.05 \text{ W} \quad \text{or} \quad \text{VA}$$

The actual or *true power* dissipated is *less* than the *apparent* power. True power may be found from: $P_{true} = I^2R = (50 \text{ mA})^2 (2 \text{ K}) = (50 \times 10^{-3})^2 (2 \times 10^3) = (2500 \times 10^{-6}) (2 \times 10^3) = 5$ W.

The following examples further illustrate this.

Example 13-38

In the circuit of Fig. 13-22*a*, two capacitive reactances are connected in series with two resistors. $X_{C_1} = 318\ \Omega$, $X_{C_2} = 79.6\ \Omega$, $R_1 = 70\ \Omega$, $R_2 = 80\ \Omega$, $Z_T = 425\ \Omega$, $E_T = 15$ V, and $I = 35.3$ mA. Find: (a) $P_{apparent}$, and (b) P_{true}.

Solution

The reader should solve this example by using the method shown previously. The following results should be used as a guide: $P_{apparent} = 530$ mVA, $P_{true} = 186$ mW.

Example 13-39

In Fig. 13-23*a*, three capacitors are connected in series with two resistors, $R_1 = 8$ K and $R_2 = 2$ K. $E_T = 50$ V and $I_T = 1.76$ mA. Find: (a) $P_{apparent}$ and P_{true}.

Solution

Solve this example by using the method shown previously and by using the following results as a guide: $P_{apparent} = 88$ mVA, $P_{true} = 31$ mW.

For Similar Problems Refer to End-of-Chapter Problems 13-70 and 13-71

Power Factor. In the previous discussion of power in *RC* circuits, it was shown that the *true* power in a circuit containing reactance and resistance is always *less* than the apparent power. The ratio between these powers is called the *power factor*. This was shown in the previous chapter on inductors and resistors.

$$\text{power factor (or PF)} = \frac{P_{true}}{P_{apparent}} \tag{12-7}$$

Since the true power is always less than the apparent power, then the PF is always *less* than one.

In the previous section on power in *RC* circuits the true power of Fig. 13-21*a* is found to be 5 W, while the apparent power is 7.05 W or VA. In this case, the power factor would be

$$\text{power factor} = \frac{P_{true}}{P_{apparent}} \tag{12-7}$$

$$\text{power factor} = \frac{5}{7.05}$$

$$\text{power factor} = 0.71$$

The following examples illustrate this further.

Example 13-40

In Example 13-38, true power is 186 mW, while apparent power is 530 mW or mVA. Find the power factor.

Solution

By using equation 12-7, the reader should find that PF is 0.351.

Example 13-41

In Example 13-39, true power is 31 mW, and apparent power is 88 mW or mVA. Find the power factor.

Solution

The reader should find that PF here is 0.352.

In equation 12-7, solving for the true power is shown in the following.

$$\text{PF} = \frac{P_{\text{true}}}{P_{\text{apparent}}} \tag{12-7}$$

$$(P_{\text{apparent}})\,(\text{PF}) = P_{\text{true}}$$

Note that the *true* power is the *product* of *apparent* power and *power factor*.

Power factor may be defined as the factor that, when multiplied with the apparent power, gives the actual or true power. In a purely resistive circuit, that is, where there is no reactance, true and apparent power are the same. Here, power factor is 1.

The following example illustrates further use of the power factor.

Example 13-42

An alternating applied voltage of 100 V causes 20 mA of current flow through a resistor and a series capacitive reactance. If the power factor is 0.8, find (a) P_{apparent}, and (b) P_{true}.

Solution

(a) $P_{\text{apparent}} = E_T\,I_T = (100)\,(20 \text{ mA}) = (100)\,(20 \times 10^{-3}) = 2$ W or VA.
(b) $P_{\text{true}} = (P_{\text{apparent}})\,(\text{PF}) = (2)\,(0.8) = 1.6$ W.

For Similar Problems Refer to End-of-Chapter Problems 13-72 to 13-75

13-7. Capacitor and Resistor in Parallel, Phase Relationship

In Fig. 13-24*a*, a resistor and a capacitor are shown connected in parallel, each across the applied ac voltage. The current phasors are shown in Fig. 13-24*b*

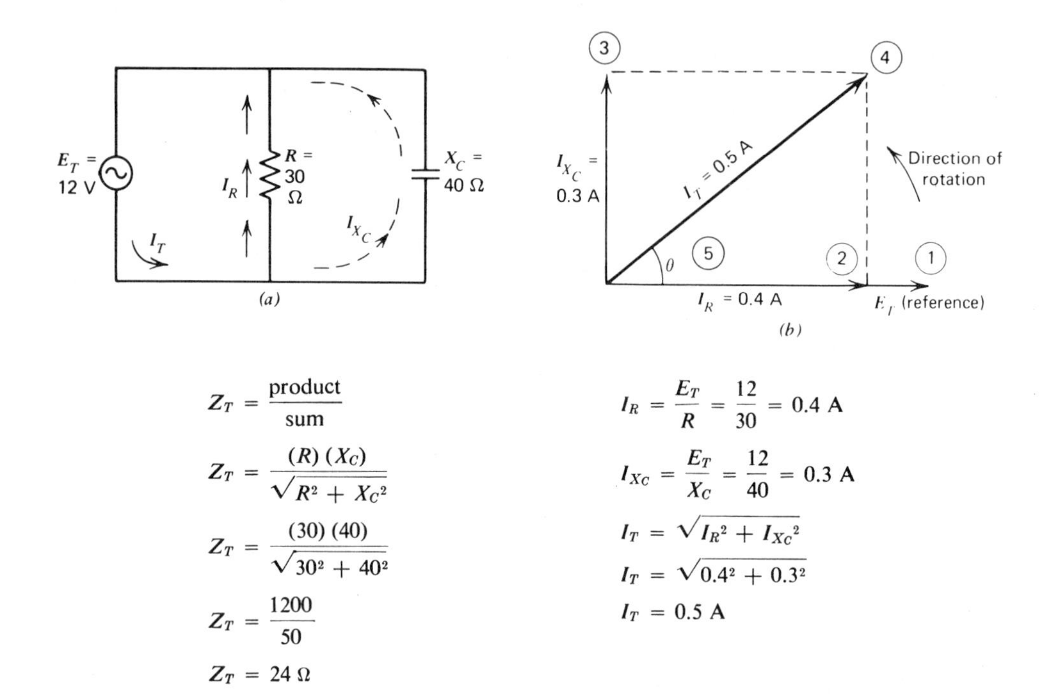

Figure 13-24. Capacitor and resistor in parallel. (*a*) Circuit. (*b*) Current phasors.

The following steps explain the positions of each phasor. The number at each phasor tip corresponds to the number in the steps below.

1. Since the applied voltage E_T is across R and also across X_C, or is common to both, the E_T phasor is used as a reference, and is shown pointing horizontally to the right, denoting 0°.

2. Current through the resistor, I_R, is always in phase with voltage across the resistor. The I_R phasor is shown alongside of, and parallel to, the E_T phasor.

3. Current through X_C is shown pointing vertically upward, denoting the 90° point in the normal counterclockwise rotation of phasors. Note that E_T (the voltage across the capacitor) is *lagging* the *current* I_{X_C} by 90°. This is true of any X_C.

4. Total current I_T is the phasor sum of I_R and I_{X_C}. By drawing a line (shown as the dashed lines of Fig. 13-24*b*) at each phasor arrowhead parallel to the other phasor, a rectangle is formed. The diagonal of the rectangle, going from the junction of the two phasors, I_R and I_{X_C}, to the junction of the two dashed lines, is the phasor sum I_T.

5. The phase angle θ is the angle between I_T and E_T. Note that E_T *lags* I_T by the angle θ.

Impedance. In the circuit of Fig. 13-24*a*, R and X_C are in parallel. The *total* opposition of this combination, Z_T, may be found by using the method of *the product divided by the sum.* This is similar to two parallel resistors (Chapter 4, equation 4-3), and also to R and X_L in parallel (Chapter 12, equation 12-9). Z_T is then

$$Z_T = \frac{\text{product}}{\text{sum}}$$

$$Z_T = \frac{(R)\,(X_C)}{\sqrt{R^2 + X_C^2}} \qquad (13\text{-}13)$$

Note that the *sum* is actually the *phasor sum* shown in the denominator of equation 13-13. Then

$$Z_T = \frac{(30)\,(40)}{\sqrt{30^2 + 40^2}}$$

$$Z_T = \frac{1200}{\sqrt{900 + 1600}}$$

$$Z_T = \frac{12 \times 10^2}{\sqrt{9 \times 10^2 + 16 \times 10^2}}$$

$$Z_T = \frac{12 \times 10^2}{\sqrt{25 \times 10^2}}$$

$$Z_T = \frac{12 \times 10^2}{5 \times 10^1}$$

$$Z_T = 2.4 \times 10^1\ \Omega \qquad \text{or} \qquad 24\ \Omega$$

Current in each branch, I_R and I_{X_C}, may be found from Ohm's law. $I_R = E_T/R = 12/30 = 0.4$ A. $I_{X_C} = E_T/X_C = 12/40 = 0.3$ A.

As shown in the phasors of Fig. 13-24*b*, total current I_T is the vector sum of $I_R = 0.4$ A and $I_{X_C} = 0.3$ A which are 90° out of phase with each other. These currents must be added by using the Pythagoras theorem.

$$I_T = \sqrt{I_R^2 + I_{X_C}^2} \qquad (13\text{-}14)$$

Therefore, I_T is

$$I_T = \sqrt{0.4^2 + 0.3^2}$$

$$I_T = \sqrt{(4 \times 10^{-1})^2 + (3 \times 10^{-1})^2}$$

$$I_T = \sqrt{16 \times 10^{-2} + 9 \times 10^{-2}}$$

$$I_T = \sqrt{25 \times 10^{-2}}$$

$$I_T = 5 \times 10^{-1}\ \text{A} \qquad \text{or} \qquad 0.5\ \text{A}$$

Knowing that $I_T = 0.5$ A and that $E_T = 12$ V, total opposition Z_T could be found by using Ohm's law. $Z_T = E_T/I_T = 12/0.5 = 24\ \Omega$.

Note that this value of $Z_T = 24\ \Omega$ agrees with that found previously using the product divided by the sum method of equation 13-13.

The following example further illustrates this.

Example 13-43

A voltage E_T of 75 V is applied across a 2500-Ω resistor in parallel with an X_C of 3000 Ω, in the same type of circuit as shown in Fig. 13-24*a*. Find (a) Z_T, using the product divided by the sum method, (b) I_R, (c) I_{X_C}, (d) I_T, using the vector sum of I_R and I_{X_C}, and (e) Z_T, using Ohm's law.

Solution

Solve this example by using the procedure shown previously and the following results as a guide: $Z_T = 1920\ \Omega$, $I_R = 30$ mA, $I_{X_C} = 25$ mA, $I_T = 39$ mA.

For Similar Problems Refer to End-of-Chapter Problem 13-76

Capacitors and Resistors in Parallel, Phase Relationship. When more than one resistor and more than one capacitor are connected in parallel, one method of solution is that the *total* resistance R_T, and the total reactance X_{C_T} must first be found. Then the total impedance Z_T may be found. The circuit of Fig. 13-25*a* has two resistors R_1 and R_2 in parallel, and also two X_C's (X_{C_1} and X_{C_2}) in parallel. Total resistance R_T and total reactance X_{C_T} are first found, and the simplified equivalent circuit is shown in Fig. 13-25*b*. A second solution, to be discussed later, is to vectorially add all branch currents.

Total resistance R_T may be found using the product divided by the sum method.

$$R_T = \frac{R_1\ R_2}{R_1 + R_2} \qquad (4\text{-}3)$$

$$R_T = \frac{(30)\ (20)}{30 + 20}$$

$$R_T = \frac{600}{50}$$

$$R_T = 12\ \Omega$$

The R_T of 12 Ω is shown in Fig. 13-25*b*.

Total reactance X_{C_T} may also be found using the product divided by the sum method.

$$X_{C_T} = \frac{\text{product}}{\text{sum}} \qquad (13\text{-}15)$$

$$X_{C_T} = \frac{(X_{C_1})(X_{C_2})}{X_{C_1} + X_{C_2}}$$

Note that the vector sum of the reactances X_{C_1} and X_{C_2}, in the denominator of equation 13-15, is simply the arithmetic sum, since there is no phase angle between them. Therefore, X_{C_T} is

$$X_{C_T} = \frac{(5)(20)}{5 + 20}$$

$$X_{C_T} = \frac{100}{25}$$

$$X_{C_T} = 4\ \Omega$$

X_{C_T} is shown in the simplified equivalent circuit of Fig. 13-25*b*.

Total impedance, consisting of R_T and X_{C_T} in parallel, may now be found.

$$Z_T = \frac{(R_T)(X_{C_T})}{\sqrt{R_T^2 + X_{C_T}^2}} \qquad (13\text{-}13)$$

$$Z_T = \frac{(12)(4)}{\sqrt{12^2 + 4^2}}$$

$$Z_T = \frac{48}{\sqrt{144 + 16}}$$

$$Z_T = \frac{48}{\sqrt{160}}$$

$$Z_T = \frac{48}{\sqrt{1.6 \times 10^2}}$$

$$Z_T = \frac{48}{1.26 \times 10^1}$$

$$Z_T = 38.1 \times 10^{-1}\ \Omega \qquad \text{or} \qquad 3.81\ \Omega$$

A second method to find Z_T is to first find the current flowing in each branch of the circuit of Fig. 13-25*a*, that is I_{R_1}, $I_{X_{C1}}$, I_{R_2}, and $I_{X_{C2}}$.

Each branch current is found using Ohm's law. $I_{R_1} = E_T/R_1 = 60/30 = 2$ A. The I_{R_1} phasor (number 1) is shown in phase with E_T in Fig. 13-25*c*. $I_{R_2} = E_T/R_2 = 60/20 = 3$ A. I_{R_2} phasor (number 2) is also shown in phase with E_T in the phasor diagram.

Since both resistor currents are in phase with each other, as shown in Fig. 13-25*c*, then total resistance current I_{R_T} is simply the sum of each. $I_{R_T} = I_{R_1} +$

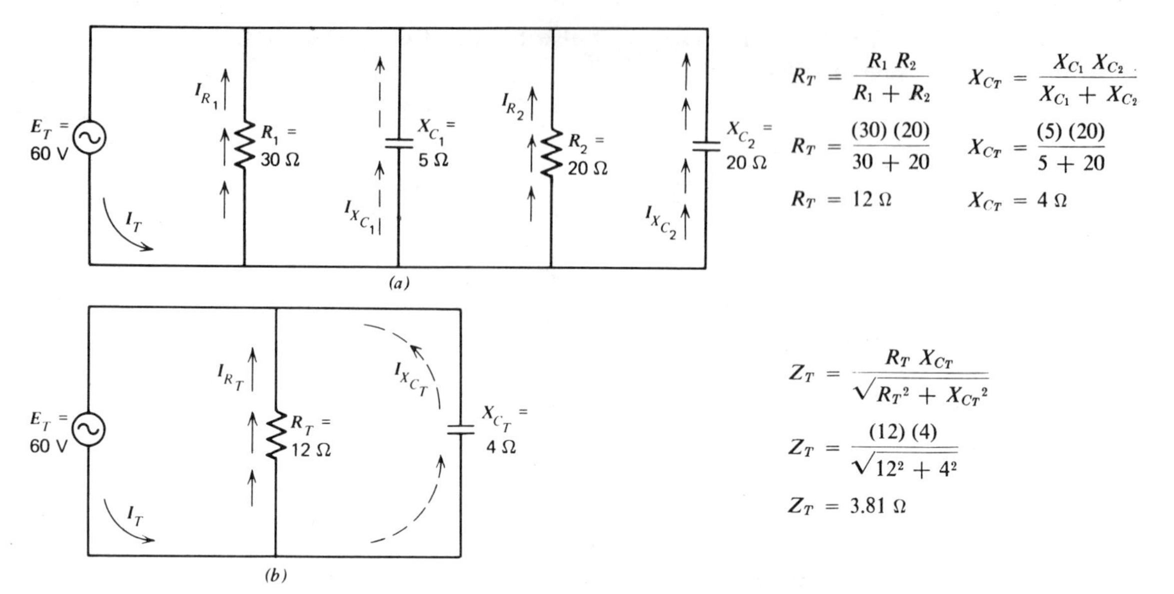

Figure 13-25. Capacitor and resistors in parallel. (*a*) Circuit. (*b*) Simplified equivalent circuit.

$I_{R_2} = 2 + 3 = 5$ A. The I_{R_T} phasor (number 3) is shown in phase with E_T in the phasor diagram of Fig. 13-25*c*.

Current in each reactance is: $I_{X_{C_1}} = E_T/X_{C_1} = 60/5 = 12$ A. The $I_{X_{C_1}}$ phasor (number 4) is shown in the phasor diagram of Fig. 13-25*c*. Note that the E_T phasor is lagging behind $I_{X_{C_1}}$ by 90°. This is true of all capacitors. $I_{X_{C_2}} = E_T/X_{C_2} = 60/20 = 3$ A. The $I_{X_{C_2}}$ phasor (number 5) is also shown in Fig. 13-25*c*.

Total reactance current $I_{X_{CT}}$ is simply the sum of $I_{X_{C_1}}$ and $I_{X_{C_2}}$, since they are in phase with each other. $I_{X_{CT}} = I_{X_{C_1}} + I_{X_{C_2}} = 12 + 3 = 15$ A. The $I_{X_{CT}}$ phasor (number 6) is shown in the phasor diagram of Fig. 13-25*c*.

Total current I_T (number 7) is the vector sum of I_{R_T} (number 3) and $I_{X_{CT}}$ (number 6) shown in the phasor diagram. Since I_{R_T} and $I_{X_{CT}}$ are 90° out of phase, they must be added by using Pythagoras' theorem.

$$I_T = \sqrt{I_{R_T}{}^2 + I_{X_{CT}}{}^2} \qquad (13\text{-}14)$$

$$I_T = \sqrt{5^2 + 15^2}$$

$$I_T = \sqrt{25 + 225}$$

$$I_T = \sqrt{250}$$

$$I_T = \sqrt{2.5 \times 10^2}$$

$$I_T = 1.58 \times 10^1 \text{ A} \qquad \text{or} \qquad 15.8 \text{ A}$$

Note that E_T lags I_T by the angle θ (theta) in Fig. 13-25*c*.

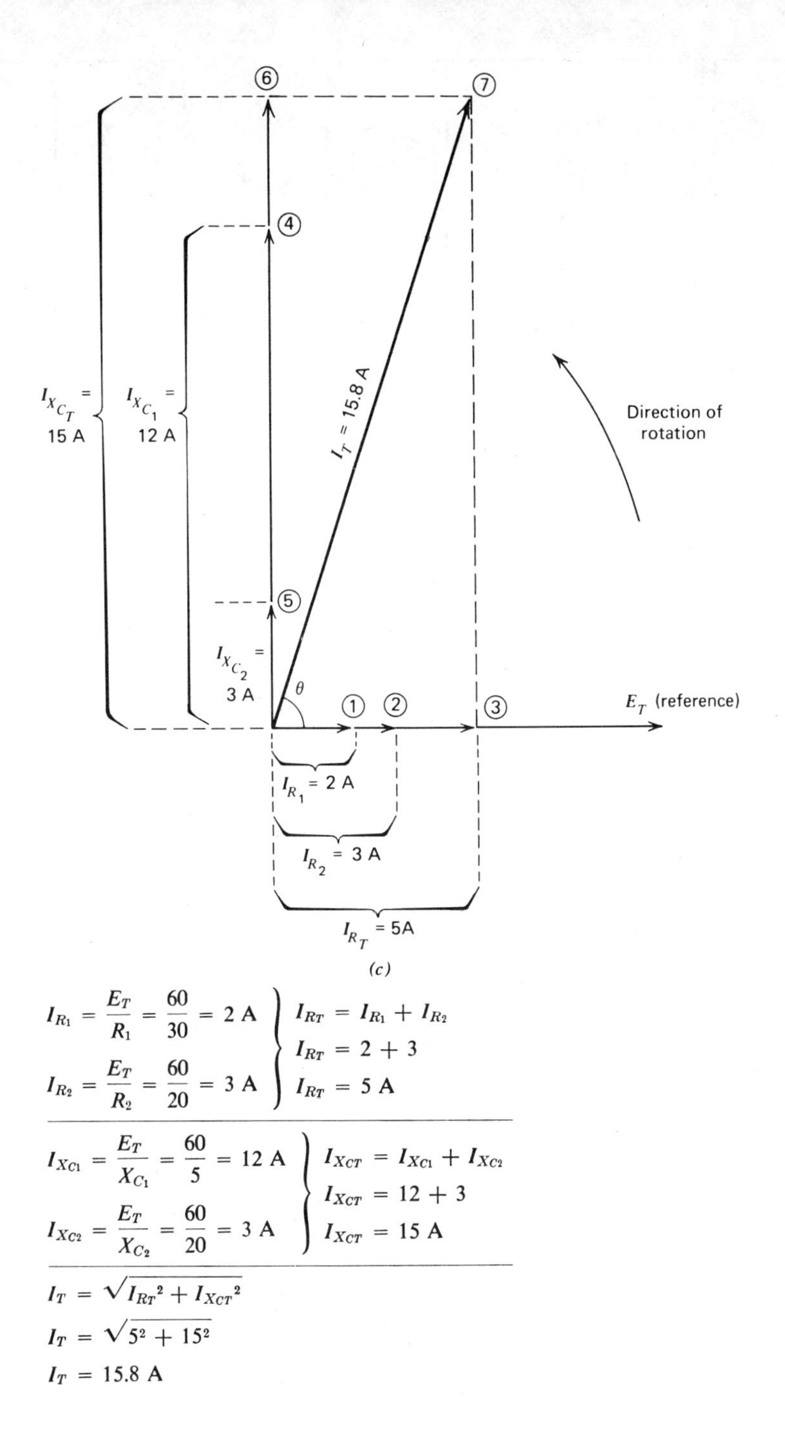

$$I_{R_1} = \frac{E_T}{R_1} = \frac{60}{30} = 2\text{ A} \qquad I_{R_2} = \frac{E_T}{R_2} = \frac{60}{20} = 3\text{ A}$$

$$I_{RT} = I_{R_1} + I_{R_2} \qquad I_{RT} = 2 + 3 \qquad I_{RT} = 5\text{ A}$$

$$I_{X_{C1}} = \frac{E_T}{X_{C_1}} = \frac{60}{5} = 12\text{ A} \qquad I_{X_{C2}} = \frac{E_T}{X_{C_2}} = \frac{60}{20} = 3\text{ A}$$

$$I_{XCT} = I_{X_{C1}} + I_{X_{C2}} \qquad I_{XCT} = 12 + 3 \qquad I_{XCT} = 15\text{ A}$$

$$I_T = \sqrt{I_{RT}^2 + I_{XCT}^2}$$

$$I_T = \sqrt{5^2 + 15^2}$$

$$I_T = 15.8\text{ A}$$

Figure 13-25 (*cont.*) (c) Current phasors.

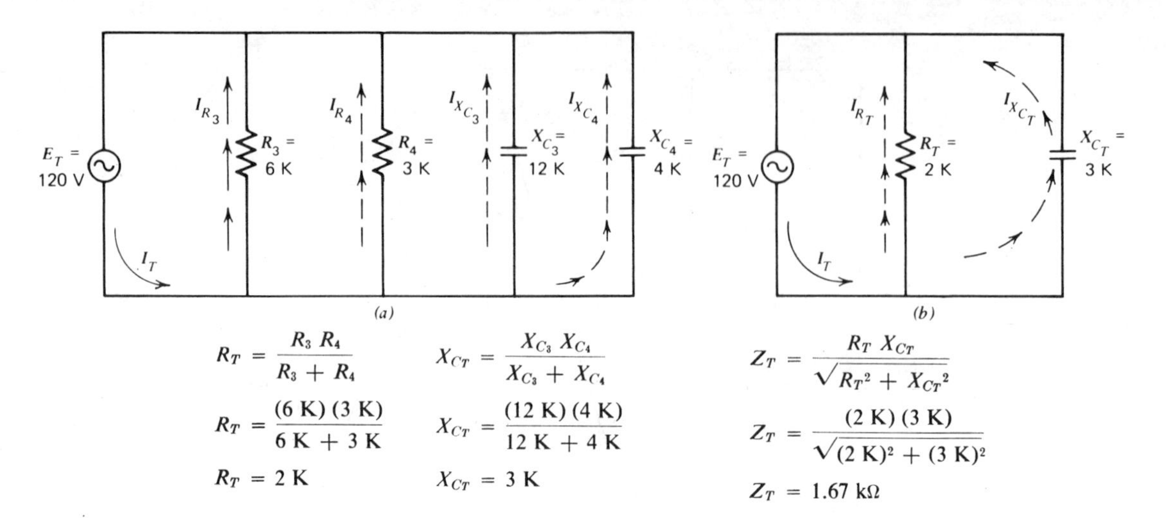

Figure 13-26. Examples 13-44 and 13-45. (*a*) Circuits. (*b*) Simplified equivalent circuit.

Finally, knowing E_T and I_T, total impedance Z_T may be found by using Ohm's law. $Z_T = E_T/I_T = 60/15.8 = 3.8\ \Omega$. Note that this value of Z_T agrees with the value found previously.

The following examples illustrate the previous discussion.

Example 13-44

In the circuit for this example shown in Fig. 13-26*a*, two resistors (R_3 = 6 K and R_4 = 3 K) are in parallel. Also in parallel are two capacitive reactances (X_{C_3} = 12 K and X_{C_4} = 4 K). If the applied voltage E_T is 120 V, find: (a) R_T, (b) X_{C_T}, and (c) Z_T.

Solution

Solve this example by using the method shown previously and the results shown in the text of Fig. 13-26*a* and *b* as a guide, R_T is shown as 2 K, X_{C_T} as 3 K, and Z_T as 1.67 K.

Example 13-45

In the circuit shown in Fig. 13-26*a*, find: (a) I_{R_3}, (b) I_{R_4}, (c) I_{R_T}, (d) $I_{X_{C_3}}$, (e) $I_{X_{C_4}}$, (f) $I_{X_{C_T}}$, (g) I_T, and (h) Z_T.

Solution

Solve this example in the method shown previously. As a guide, the results are worked out in the text of Fig. 13-26*c*. The phasors are shown for each of the

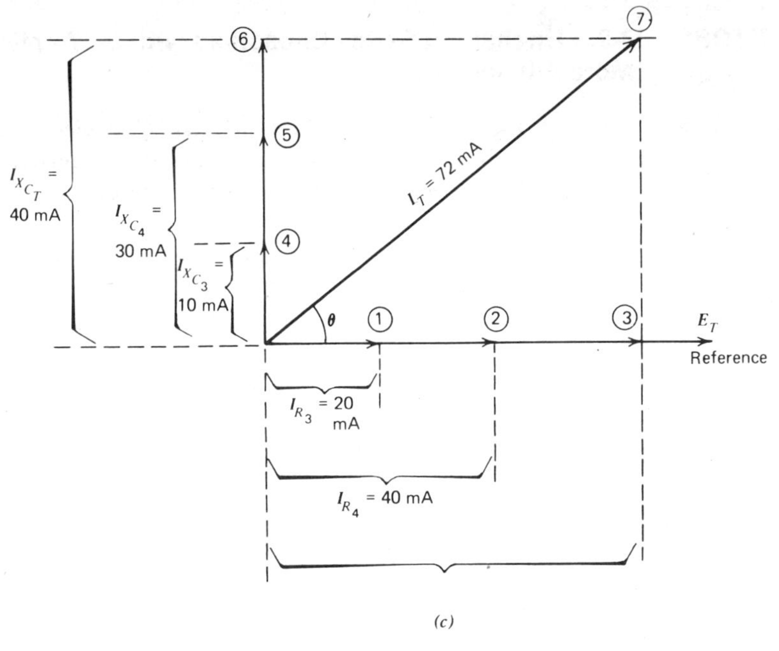

(c)

$$I_{R_3} = \frac{E_T}{R_3} = \frac{120}{6\text{ K}} = 20\text{ mA}$$

$$I_{R_4} = \frac{E_T}{R_4} = \frac{120}{3\text{ K}} = 40\text{ mA}$$

$$I_{RT} = I_{R_3} + I_{R_4}$$

$$I_{RT} = 20 + 40$$

$$I_{RT} = 60\text{ mA}$$

$$I_{XC_3} = \frac{E_T}{X_{C_3}} = \frac{120}{12\text{ K}} = 10\text{ mA}$$

$$I_{XC_4} = \frac{E_T}{X_{C_4}} = \frac{120}{4\text{ K}} = 30\text{ mA}$$

$$I_{XCT} = I_{XC_3} + I_{XC_4}$$

$$I_{XCT} = 10 + 30$$

$$I_{XCT} = 40\text{ mA}$$

$$I_T = \sqrt{I_{RT}^2 + I_{XCT}^2}$$

$$I_T = \sqrt{(60\text{ mA})^2 + (40\text{ mA})^2}$$

$$I_T = 72\text{ mA}$$

$$Z_T = \frac{E_T}{I_T} = \frac{120}{72\text{ mA}} = 1.67\text{ k}\Omega$$

Figure 13-26 (cont.) (c) Current phasors.

following: $I_{R_3} = 20$ mA, phasor number 1; $I_{R_4} = 40$ mA, phasor number 2; $I_{RT} = 60$ mA, phasor 3; $I_{XC_3} = 10$ mA, phasor 4; $I_{XC_4} = 30$ mA, phasor 5; $I_{XCT} = 40$ mA, phasor 6; $I_T = 72$ mA, phasor 7; $Z_T = 1.67$ K.

For Similar Problems Refer to End-of-Chapter Problems 13-77 and 13-78

13-8. Uncharged Series Capacitors with DC Applied Voltage (More Advanced)

In Section 13-3, a discussion of the *quantity* (Q) of electrons stored on a charged capacitor is given. Also, using this Q, the voltage across a capacitor, which is one of several in series, is also solved. Another method is explained in this present, more advanced, section.

Figure 13-27 shows two series resistors ($R_1 = 10\Omega$ and $R_2 = 20\ \Omega$) connected to an applied voltage, $E_T = 90$ V. One method of finding the voltage across each resistor (E_{R_1} or V_{R_1}, and E_{R_2} or V_{R_2}) is by the use of an equation from Chapter 3, equation 3-1: $V_{R_1} = (R_1/R_T)\ E_T$, and $V_{R_2} = (R_2/R_T)\ E_T$. Note that the voltage across each resistor is a *fraction* of the total applied voltage.

Voltage across R_1 is: $V_{R_1} = (R_1/R_T)\ E_T = (10\ \text{K}/30\ \text{K})\ 90 = 30\text{V}$; and across R_2, $V_{R_2} = (R_2/R_T)\ E_T = (20\ \text{K}/30\ \text{K})\ 90 = 60$ V.

A similar method may be used for series capacitors. In Fig. 13-28, two capacitors ($C_1 = 20\ \mu$F and $C_2 = 30\ \mu$F) are connected in series with an applied voltage, $E_T = 90$ V, when the switch is closed. As in the series resistors of Fig. 13-27, voltage across each capacitor becomes a *fraction* of the total voltage as a result of current flow that charges the two capacitors. The fraction is a proper fraction, having a value less than 1. The numerator must be less than the denominator. With resistors in series, the value of one resistor is, of course, less than the total resistance. The *fraction* is, therefore, R_1/R_T.

With *series capacitors*, the opposite is true. That is, C_{total} is *less* than any one capacitor. Therefore, the fraction is C_T/C_1

In Fig. 13-28, C_T is: $C_T = C_1\ C_2/(C_1 + C_2) = (20)\ (30)/(20 + 30) = 600/50 = 12\ \mu$F.

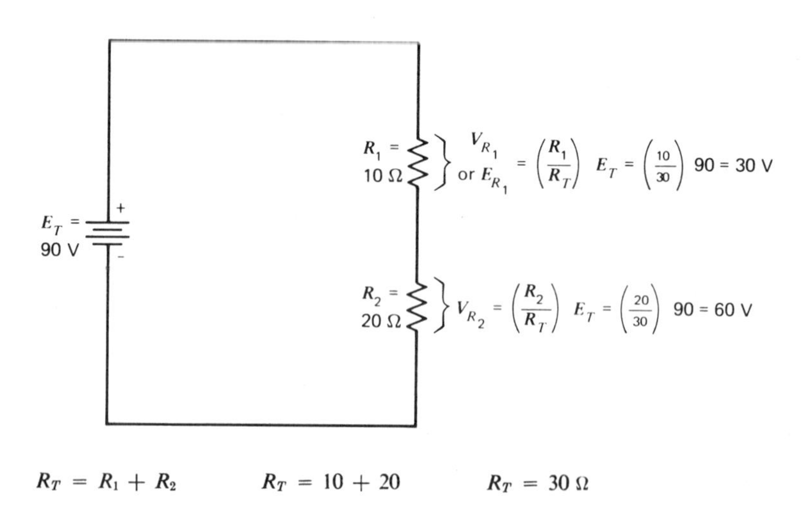

Figure 13-27. Voltages across series resistors.

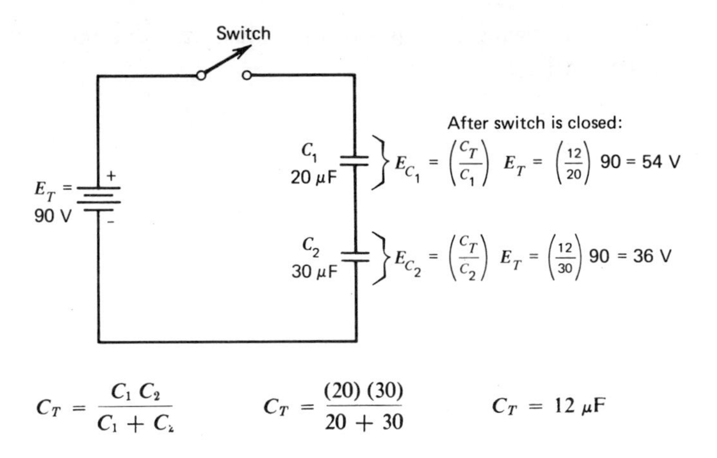

Figure 13-28. Voltages across series capacitors.

Voltage across C_1 in Fig. 13-28 is, therefore, a fraction of E_T:

$$V_{C_1} \text{ or } E_{C_1} = \left(\frac{C_T}{C_1}\right) E_T \tag{13-16}$$

$$V_{C_1} \text{ or } E_{C_1} = \left(\frac{12}{20}\right) 90$$

$$V_{C_1} \text{ or } E_{C_1} = 54 \text{ V}$$

Voltage on C_2 is

$$V_{C_2} \text{ or } E_{C_2} = \left(\frac{C_T}{C_2}\right) E_T \tag{13-16}$$

$$V_{C_2} \text{ or } E_{C_2} = \left(\frac{12}{30}\right) 90$$

$$V_{C_2} \text{ or } E_{C_2} = 36 \text{ V}$$

Note that the sum of E_{C_1} (54 V) and E_{C_2} (36 V) is equal to E_T (90 V), and also that the smaller capacitor ($C_1 = 20\ \mu F$) has the larger voltage (54 V). After these capacitors have become charged, current stops.

As proof of the above series capacitor voltages, the quantity of electrons (Q) on each capacitor must be equal (since the same current flows at all points in a series circuit) and also must equal the total Q on the equivalent or total C. This is shown in the following. $Q_T = C_T E_T = (12 \times 10^{-6})(90) = 1080 \times 10^{-6}$ coul. Similarly, $Q_1 = C_1 E_1$ (where E_1 is E_{C_1} or V_{C_1}) $= (20 \times 10^{-6})(54) = 1080 \times 10^{-6}$ coul. $Q_2 = C_2 E_2 = (30 \times 10^{-6})\ 36 = 1080 \times 10^{-6}$ C.

The following example further illustrates this method of finding voltages across series capacitors.

Example 13-46

When the switch in Fig. 13-29 is closed, find (a) C_T, (b) E_{C_3}, (c) E_{C_4}, and (d) E_{C_5}.

Solution

Solve this example by using the methods shown previously and the results shown in Fig. 13-29 as a guide. Note that the sum of the capacitor voltages (9 + 36 + 15) is equal to the total applied voltage E_T of 60 V. Current stops after the capacitors have charged up to the voltages listed above.

For Similar Problems Refer to End-of-Chapter Problems 13-79 and 13-80

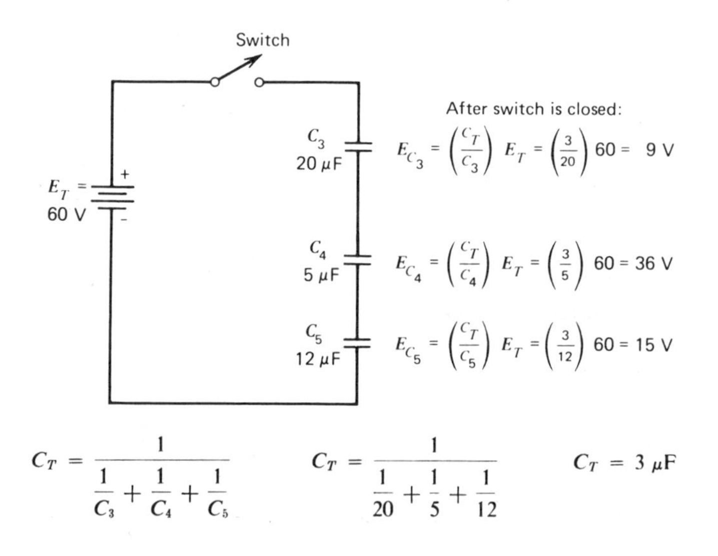

Figure 13-29. Example 13-46.

Previously Charged Series Capacitors with DC Applied Voltage. When a dc voltage is applied to series capacitors, one or more of which have been previously charged, the capacitors undergo a *change of voltage*, Δe_c, becoming charged to the total applied voltage. In Fig. 13-30*a*, capacitor C_1 has been previously charged to 40 V with the polarity shown. C_2 has not been charged yet.

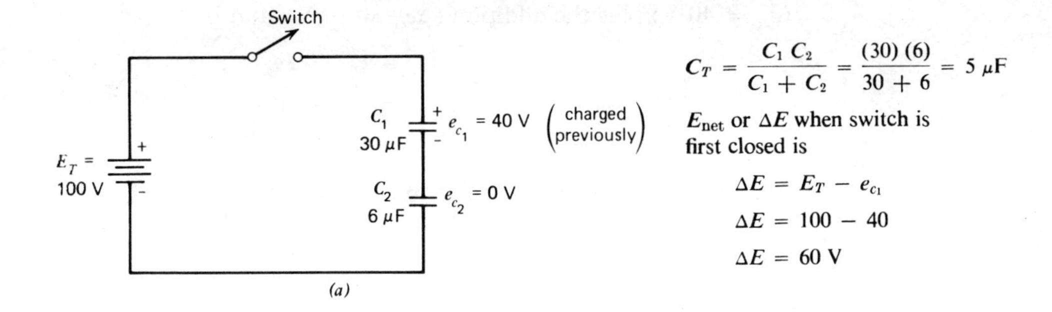

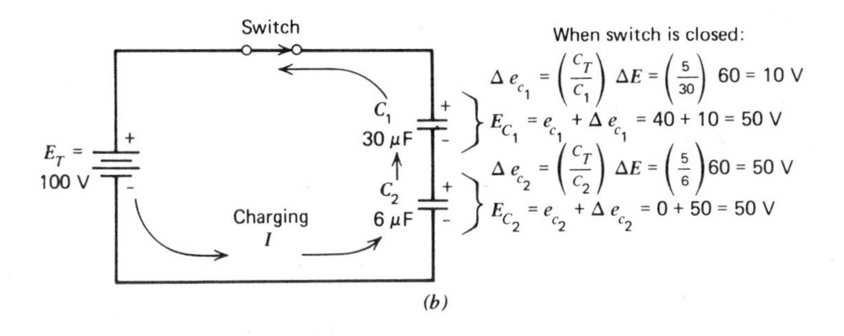

Figure 13-30. (*a*) Switch open. (*b*) Switch closed.

If the switch were to be closed, the 100 V E_T would be opposed by the 40 V (e_{c_1}) on C_1, shown in Fig. 13-30*a*. With the switch in the closed position, Fig. 13-30*b*, a charging current, I, flows as shown. This current flows due to a *net voltage* ΔE, which is the difference between E_T and e_{c_1} (the original charge on C_1). Net voltage, ΔE, is then: $E_T - e_{c_1} = 100 - 40 = 60$ V.

Each series capacitor will get a *fraction* of this *net voltage*. C_1 gets Δe_{c_1}, while C_2 gets Δe_{c_2}. The amount that C_1 receives is

$$\Delta e_{c_1} = \left(\frac{C_T}{C_1}\right) \Delta E \qquad \text{(from equation 13-16)}$$

where $C_T = C_1 C_2/(C_1 + C_2) = (30)(6)/(30 + 6) = 180/36 = 5\ \mu F$. Therefore

$$\Delta e_{c_1} = \left(\frac{5}{30}\right) 60$$

$$\Delta e_{c_1} = 10 \text{ V}$$

Since the current flow from E_T is in the direction shown in Fig. 13-30*b*, C_1 will gain voltage ($\Delta e_{c_1} = 10$ V). Final voltage on C_1, E_{C_1} is its original voltage

($e_{c_1} = 40$ V) plus the addition ($\Delta e_{c_1} = 10$ V), and is

$$E_{C_1} = e_{c_1} + \Delta e_{c_1} \qquad (13\text{-}17)$$

$$E_{C_1} = 40 + 10$$

$$E_{C_1} = 50 \text{ V}$$

Capacitor C_2 receives its fraction of the net voltage ($\Delta E = 60$ V) which is

$$\Delta e_{c_2} = \left(\frac{C_T}{C_2}\right) \Delta E \qquad \text{(from equation 13-16)}$$

$$\Delta e_{c_2} = \left(\frac{5}{6}\right) 60$$

$$\Delta e_{c_2} = 50 \text{ V}$$

Since C_2 started off uncharged, its gain of voltage ($\Delta e_{c_2} = 50$ V) is added to its starting zero voltage, resulting in

$$E_{C_2} = e_{c_2} + \Delta e_{c_2} \qquad (13\text{-}17)$$

$$E_{C_2} = 0 + 50$$

$$E_{C_2} = 50 \text{ V}$$

Note that the sum of E_{C_1} and E_{C_2} is equal to E_{total}. Current stops when this has occurred. The fact that E_{C_1} and E_{C_2} are equal is simply a coincidence.

In Fig. 13-31*a*, a more complex situation is shown. Here, two capacitors, C_3 and C_4, have been previously charged to 10 V and 6 V, respectively, as shown, but with opposing voltages. If the switch were to be closed, the applied voltage ($E_T = 40$ V) would be connected series *aiding* with the voltage previously placed across C_4 ($e_{c_4} = 6$ V), and series *opposing* with the voltage previously placed across C_3 ($e_{c_3} = 10$ V). *Net voltage*, ΔE, the instant that the switch were closed would be the *sum* of the *aiding* voltages (E_T and e_{c_4}), *subtracted* by the *opposing* voltage of e_{c_3}. Net voltage ΔE, which causes a current flow, is then

$$\Delta E = E_T + e_{c_4} - e_{c_3}$$

$$\Delta E = 40 + 6 - 10$$

$$\Delta E = 36 \text{ V}$$

Each capacitor receives a fraction of this ΔE of 36 V. C_3 gets the following change:

$$\Delta e_{c_3} = \left(\frac{C_T}{C_3}\right) \Delta E \qquad \text{(from equation 13-16)}$$

where $C_T = C_3 C_4/(C_3 + C_4) = (12)(4)/(12 + 4) = 48/16 = 3\ \mu\text{F}$. Therefore

$$\Delta e_{c_3} = \left(\frac{3}{12}\right) 36$$

$$\Delta e_{c_3} = 9 \text{ V}$$

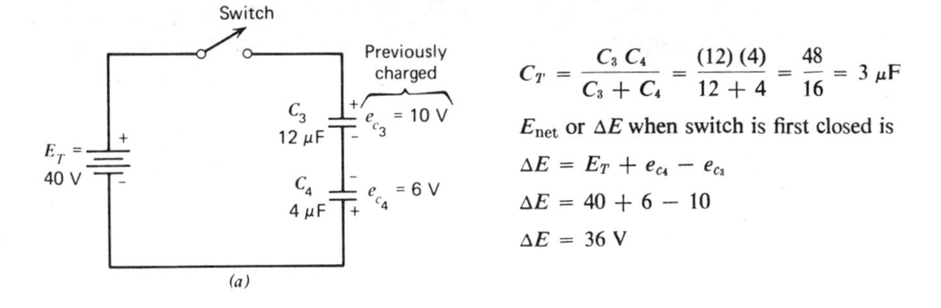

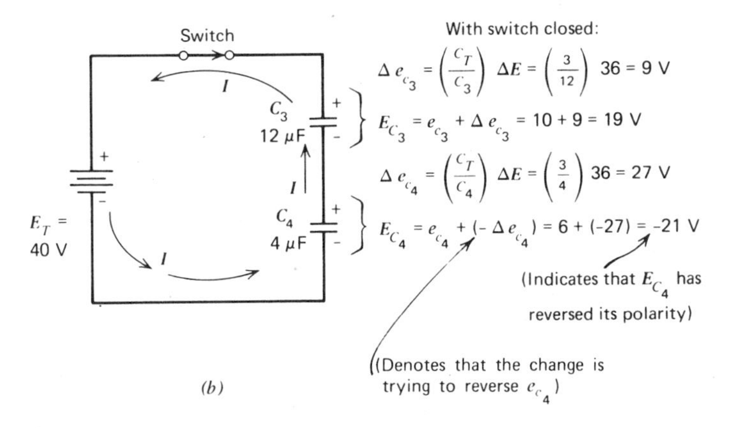

Figure 13-31. (*a*) Switch open. (*b*) Switch closed.

In Fig. 13-31*b*, with the switch closed, the direction of current is shown flowing up onto the lower plate of C_3, making this plate more negative. C_3 is therefore being charged to a larger voltage, and the change ($\Delta e_{C_3} = 9$ V) adds to its original voltage. As a result, voltage across C_3 becomes

$$E_{C_3} = e_{C_3} + \Delta e_{C_3} \qquad (13\text{-}17)$$

$$E_{C_3} = 10 + 9$$

$$E_{C_3} = 19 \text{ V}$$

Capacitor C_4 also receives a fraction of the net voltage ($\Delta E = 36$ V) when the switch of Fig. 13-31*a* is closed. C_4 receives the following change:

$$\Delta e_{C_4} = \left(\frac{C_T}{C_4}\right) \Delta E \qquad \text{(from equation 13-16)}$$

$$\Delta e_{C_4} = \left(\frac{3}{4}\right) 36$$

$$\Delta e_{C_4} = 27 \text{ V}$$

In Fig. 13-31*b*, with the switch closed, current direction is shown flowing up onto the lower plate of C_4, attempting to charge this plate *negatively*. The original polarity, shown in Fig. 13-31*a* with the switch open, is *positive* at the lower plate. This means that the current flow is attempting to reverse the polarity on C_4. If the change of voltage that C_4 receives ($\Delta e_{c_4} = 27$ V) is greater than the *original opposite* polarity voltage ($e_{c_4} = 6$ V), then C_4 will have its original voltage reversed. This is shown on C_4 when Fig. 13-31*a* and *b* are compared. Voltage on C_4 is therefore its orginal voltage ($e_{c_4} = 6$ V) *subtracted* from the change ($\Delta e_{c_4} = 27$ V), or E_{C_4} becomes

$$E_{C_4} = e_{c_4} + (-\Delta e_{c_4}) \qquad (13\text{-}17)$$

The *negative* Δe_{c_4} denotes that the *change* is trying to reverse the original polarity.

$$E_{C_4} = 6 + (-27)$$

$$E_{C_4} = 6 - 27$$

$$E_{C_4} = -21 \text{ V}$$

The *negative sign* before the 21 V indicates that E_{C_4} has actually reversed its polarity from that shown in part (a) of Fig. 13-31 to that shown in part (b). Note also that the *sum* of E_{C_3} (= 19 V) and E_{C_4} (= 21 V), with their polarities depicted in part (b), equals E_T (= 40 V). Current stops when this has occurred.

The following examples further illustrate the discussions of this section.

Example 13-47

As shown in Fig. 13-32, capacitor $C_1 = 20\ \mu$F and has been previously charged to 30 V with polarity as indicated. An applied voltage, $E_T = 90$ V, is applied through the switch along with series capacitor $C_2 = 30\ \mu$F. With the switch closed, find: (a) C_{total}, (b) *net voltage*, ΔE, which causes a current flow when the switch is *first* closed, (c) the direction of I when the switch is *first* closed, (d) *change* of C_1 voltage, Δe_{c_1}, (e) voltage across C_1, E_{C_1}, (f) *change* of C_2 voltage, Δe_{c_2}, and (g) voltage across C_2, E_{C_2}.

Solution

Solve this example by using the method shown previously and by using the results shown in the text of Fig. 13-32 as a guide.

The polarity of E_{C_2} is the same as that of E_{C_1}, that is, the lower plate is the negative one. Note that the *sum* of E_{C_1} (= 66 V) and E_{C_2} (= 24 V) is equal to E_T (= 90 V).

Example 13-48

The circuit for this example is shown in Fig. 13-33. Two capacitors, $C_5 = 30\ \mu$F and $C_6 = 60\ \mu$F, have been charged *previously* to $e_{c_5} = 100$ V and $e_{c_6} = 120$

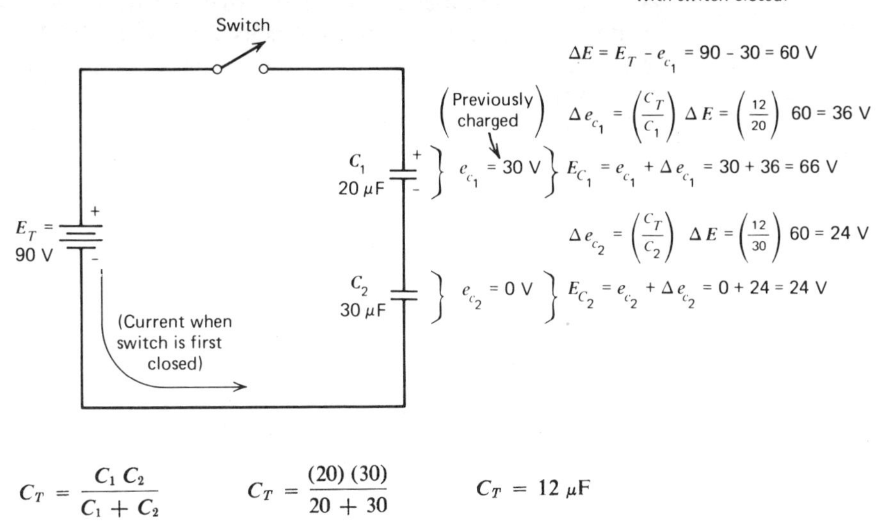

Figure 13-32. Example 13-47.

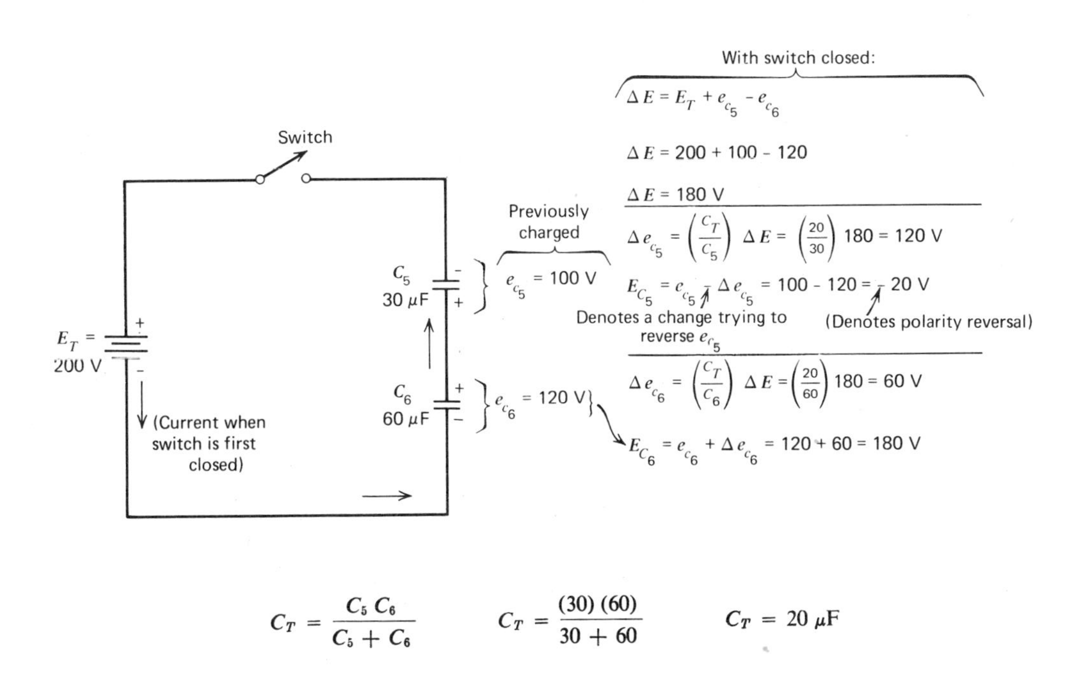

Figure 13-33. Example 13-48.

V with the polarities shown. Note that the lower plate of C_5 is its positive plate, while the lower plate of C_6 is its negative one. When the switch is closed, an applied dc voltage, $E_T = 200$ V, is connected in series with the capacitors. With the switch closed, find: (a) C_{total}, (b) *net voltage*, ΔE, which causes a current flow when switch is *first* closed, (c) direction of this current, (d) *change* of C_5 voltage, Δe_{c_5}, (e) voltage and polarity across C_5, E_{C_5}, (f) *change* of C_6 voltage, Δe_{c_6}, and (g) voltage and polarity across C_6, E_{C_6}.

Solution

Solve this example from the method shown previously. As a guide, the results shown in the text of Fig. 13-33 should be used.

Polarity of E_{C_6} is the same as it was originally when e_{c_6} was 120 V. Since E_{C_5} (= 20 V and has reversed its polarity, and E_{C_6}, (= 180 V), each have the same polarity as that shown across C_6, that is, the *lower* plates are each negative, then the capacitors have charged to the total applied voltage E_T (= 200 V). Current stops now.

For Similar Problems Refer to End-of-Chapter Problems 13-81 and 13-82

Charged Capacitors Connected to Others. When a charged capacitor is connected to an uncharged one, the charged capacitor starts discharging. In so doing, the uncharged capacitor starts charging. Current flows due to the voltage on the charged capacitor which acts as an applied voltage. Current continues to flow, discharging one capacitor and charging the other, until the voltages across the capacitors are equal.

In Fig. 13-34*a*, two capacitors ($C_1 = 30\ \mu$F and $C_2 = 20\ \mu$F) are shown with the switch open. C_1 has been previously charged to 60 V with the polarity shown, negative at the lower plate. C_2 is uncharged. When the switch is first closed, the two capacitors are connected in series as far as the voltage and current are concerned. Total capacitance is then: $C_T = C_1 C_2/(C_1 + C_2) = (30)(20)/(30 + 20) = 600/50 = 12\ \mu$F.

As shown in Fig. 13-34*b*, when the switch is *first* closed, the *net* voltage ΔE, which causes a current flow, is simply the voltage across C_1, or $\Delta E = 60$ V. Electron current flows counterclockwise as shown, leaving the negative plate of C_1 and flowing onto the lower plate of C_2. This causes C_1 to *lose* charge, while C_2 *gains* charge. Change of voltage on C_1 is

$$\Delta e_{C_1} = \left(\frac{C_T}{C_1}\right) \Delta E \qquad \text{(from equation 13-16)}$$

$$\Delta e_{C_1} = \left(\frac{12}{30}\right) 60$$

$$\Delta e_{C_1} = 24 \text{ V}$$

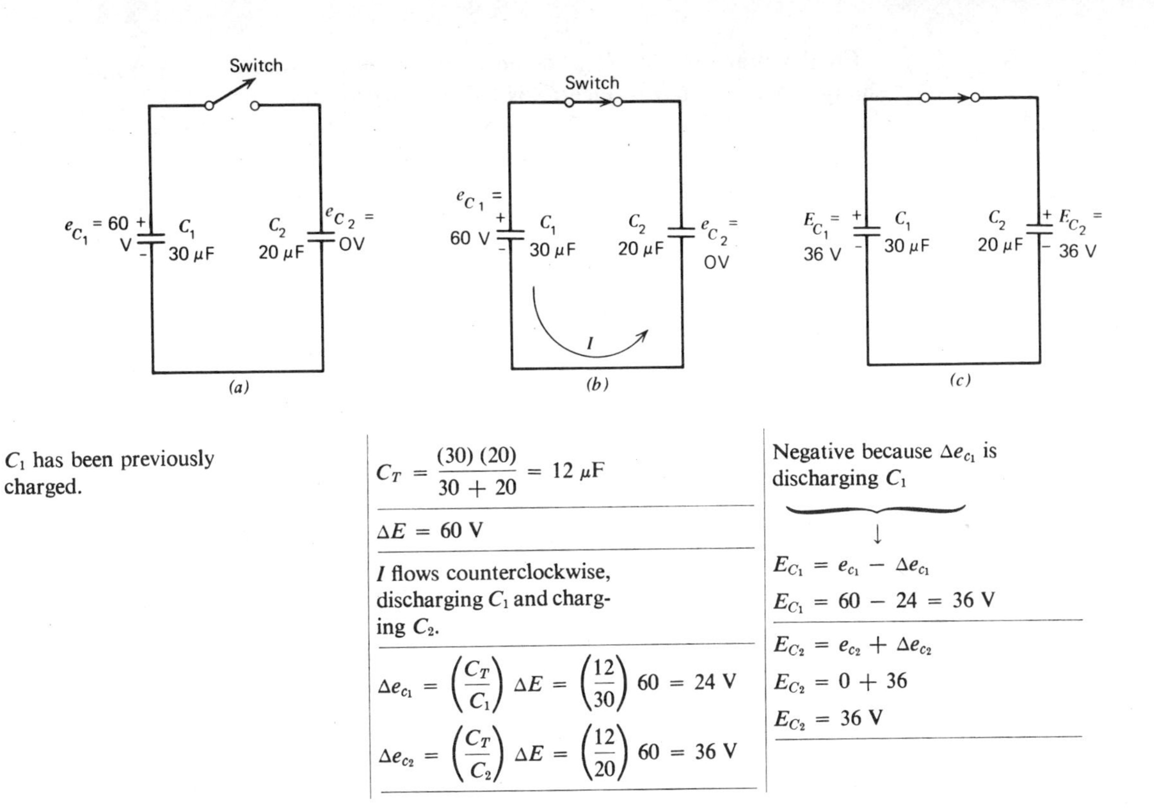

Figure 13-34. Charged capacitor connected to uncharged capacitor. (a) Switch open. (b) Switch first closed. (c) Switch remains closed.

Change of voltage on C_2 is

$$\Delta e_{C_2} = \left(\frac{C_T}{C_1}\right) \Delta E \qquad \text{(from equation 13-16)}$$

$$\Delta e_{C_2} = \left(\frac{12}{20}\right) 60$$

$$\Delta e_{C_2} = 36 \text{ V}$$

With the switch remaining closed, Fig. 13-34c, C_1 *loses* voltage, and C_2 *gains* voltage. Final voltage on C_1, E_{C_1}, therefore becomes its starting voltage ($e_{C_1} = 60$ V) *less* the change ($\Delta e_{C_1} = 24$ V), since C_1 is being discharged, or

$$E_{C_1} = e_{C_1} - \Delta e_{C_1}$$

$$E_{C_1} = 60 - 24$$

$$E_{C_1} = 36 \text{ V}$$

E_{C_1} is shown in Fig. 13-34c with its polarity.

Final voltage on C_2, E_{C_2}, becomes its starting voltage ($e_{C_2} = 0$ V) *plus* the change ($\Delta e_{C_2} = 36$ V), since C_2 is being charged, or

$$E_{C_2} = e_{C_2} + \Delta e_{C_2} \tag{13-17}$$

$$E_{C_2} = 0 + 36$$

$$E_{C_2} = 36 \text{ V}$$

E_{C_2} is shown in Fig. 13-34*c* with its polarity. Note that E_{C_1} is equal to E_{C_2} (each is 36 V). These voltages are equal and are opposing each other, and current therefore stops.

Another circuit is shown in Fig. 13-35*a*. Here, two capacitors such as C_3 and C_4 have been previously charged to different voltages and then connected

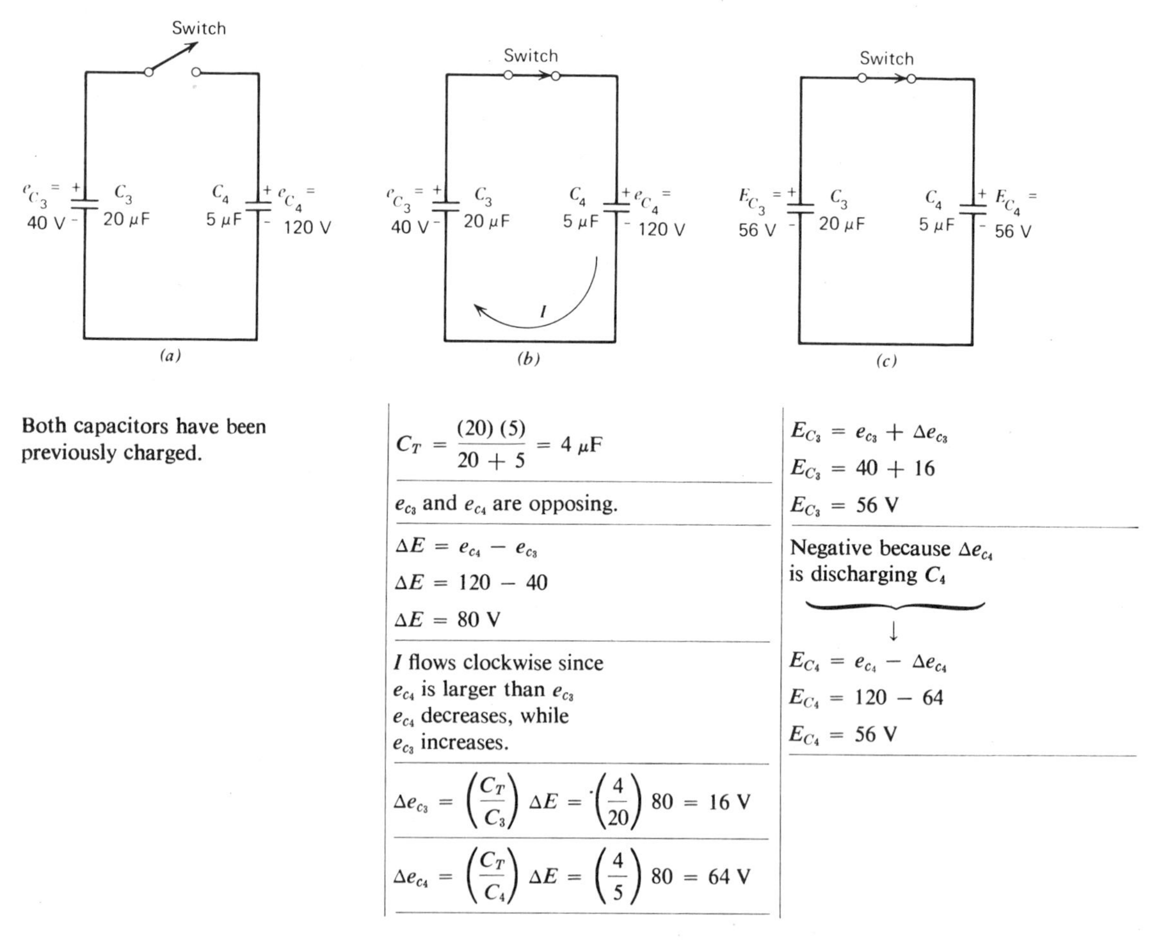

Figure 13-35. Charged capacitors connected together (voltages opposing). (*a*) Switch open. (*b*) Switch first closed. (*c*) Switch remains closed.

together. Current flows until the voltages across the capacitors are equal. The original voltages on the capacitors (e_{C_3} and e_{C_4}) of Fig. 13-35*b* are *opposing* when the switch is first closed. This may be seen, since e_{C_3} will attempt to push electrons *counterclockwise*, while e_{C_4} tries to produce a *clockwise* current.

The net voltage, ΔE, is the difference (or algebraic sum) of these voltages. This net voltage ΔE is the force that causes a current flow. Since e_{C_4} (120 V) is larger than e_{C_3} (40 V), current flows clockwise due to C_4 *discharging* somewhat. This current causes e_{C_4} to *decrease*, but *increases* e_{C_3}.

The following discussion shows how to find the final voltages E_{C_3} and E_{C_4} when the switch remains closed.

Total capacitance of C_3 (20 μF) in series with C_4 (5 μF) is: $C_T = C_3 C_4/(C_3 + C_4) = (30)(5)/(20 + 5) = 100/25 = 4\ \mu$F.

When the switch is first closed in Fig. 13-34*b*, *net* voltage ΔE causing a current flow is the sum of the opposing e_{C_4} and e_{C_3} (actually, it is the difference, since they are bucking or opposing), or

$$\Delta E = e_{C_4} - e_{C_3}$$

$$\Delta E = 120 - 40$$

$$\Delta E = 80 \text{ V}$$

C_3 and C_4 will each *change* by a fraction of this 80 V. Change of C_3 voltage, Δe_{C_3}, is

$$\Delta e_{C_3} = \left(\frac{C_T}{C_3}\right) \Delta E \qquad \text{(from equation 13-16)}$$

$$\Delta e_{C_3} = \left(\frac{4}{20}\right) 80$$

$$\Delta e_{C_3} = 16 \text{ V}$$

Change of C_4 voltage, Δe_{C_4}, is

$$\Delta e_{C_4} = \left(\frac{C_T}{C_4}\right) \Delta E \qquad \text{(from equation 13-16)}$$

$$\Delta e_{C_4} = \left(\frac{4}{5}\right) 80$$

$$\Delta e_{C_4} = 64 \text{ V}$$

With the switch remaining closed, Fig. 13-35*c*, C_3 *gains* voltage, while C_4 *decreases* voltage due to the clockwise current shown in part (b) of the diagram. As a result, C_3 final voltage is

$$E_{C_3} = e_{C_3} + \Delta e_{C_3} \qquad (13\text{-}17)$$

$$E_{C_3} = 40 + 16$$

$$E_{C_3} = 56 \text{ V}$$

C_4, due to its discharging, loses voltage. Its final voltage is its starting voltage ($e_{C_4} = 120$ V) *less* its change ($\Delta e_{C_4} = 64$ V), or

$$E_{C_4} = e_{C_4} - \Delta e_{C_4} \qquad (13\text{-}17)$$

$$E_{C_4} = 120 - 64$$

$$E_{C_4} = 56 \text{ V}$$

Note that E_{C_3} and E_{C_4} are each 56 V with the same polarity, as shown in Fig. 13-35*c*, and current stops.

A similar circuit is shown in Fig. 13-36*a*. Here, capacitors C_5 and C_6 have each been previously charged, but with voltage polarities such as to make their voltages *aid* or add. When the switch is first closed, Fig. 13-36*b*, e_{C_5} and e_{C_6} each attempts to push electrons counterclockwise as shown. The *net* voltage ΔE which causes this current flow is therefore the sum of the aiding voltages e_{C_5} and e_{C_6}. Note that the counterclockwise current means that electrons are leaving the lower negative plate of C_5, making that plate less negative. This current adds electrons onto the lower positive plate of C_6, making this plate less positive. As a result, e_{C_5} and e_{C_6} both decrease.

When the switch is first closed in Fig. 13-36*b*, total capacitance is $C_T = C_5 C_6/(C_5 + C_6) = (6)(30)/(6 + 30) = 180/36 = 5\ \mu$F.

Net voltage ΔE is the sum of the aiding voltages e_{C_5} and e_{C_6}, or

$$\Delta E = e_{C_5} + e_{C_6}$$

$$\Delta E = 10 + 20$$

$$\Delta E = 30 \text{ V}$$

Each capacitor changes by a fraction of this 30 V. The change of voltage, Δe_{C_5}, that C_5 undergoes is

$$\Delta e_{C_5} = \left(\frac{C_T}{C_5}\right) \Delta E \qquad \text{(from equation 13-16)}$$

$$\Delta e_{C_5} = \left(\frac{5}{6}\right) 30$$

$$\Delta e_{C_5} = 25 \text{ V}$$

The change of voltage, Δe_{C_6}, that C_6 undergoes is

$$\Delta e_{C_6} = \left(\frac{C_T}{C_6}\right) \Delta E \qquad \text{(from equation 13-16)}$$

$$\Delta e_{C_6} = \left(\frac{5}{30}\right) 30$$

$$\Delta e_{C_6} = 5 \text{ V}$$

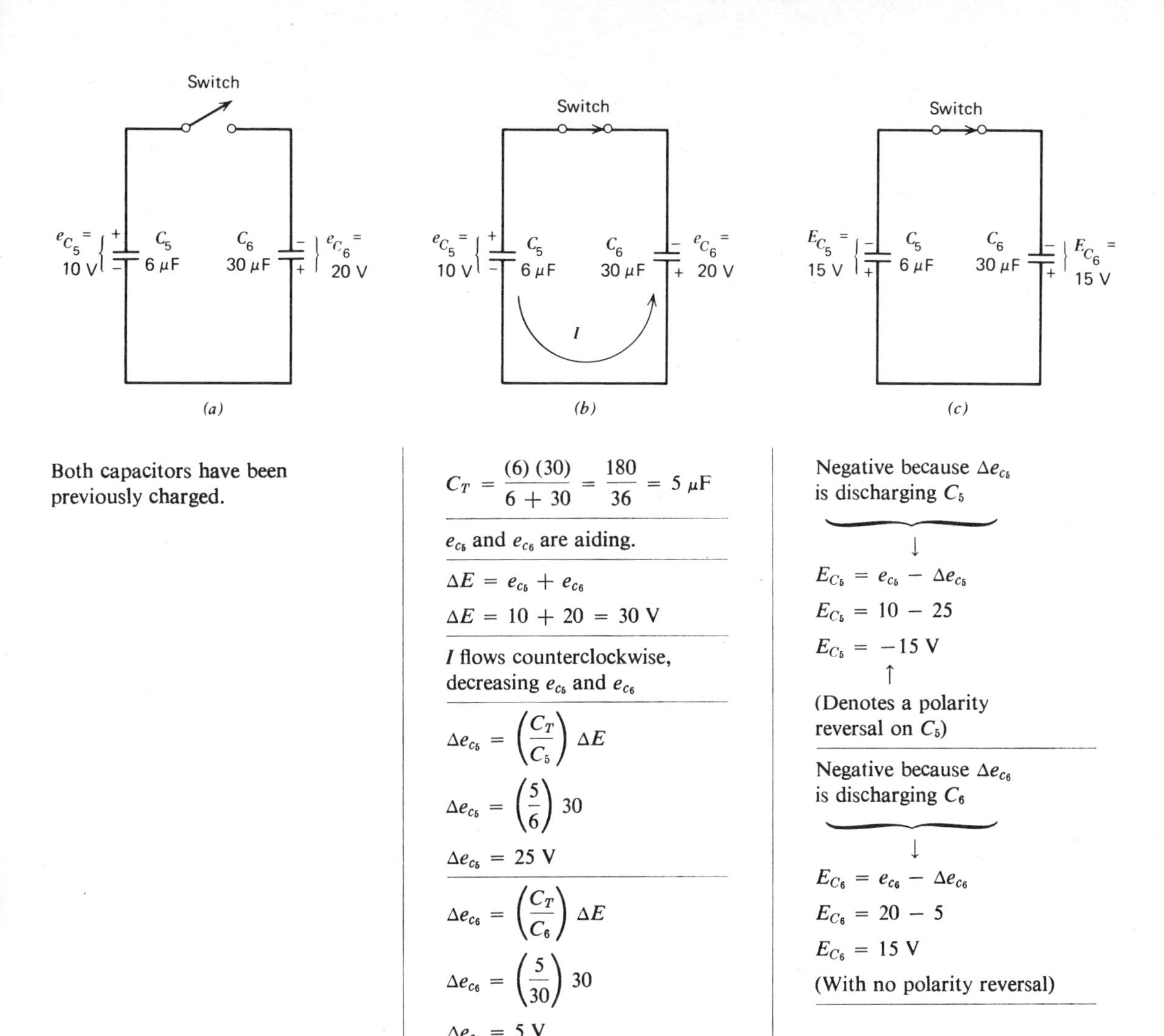

Figure 13-36. Charged capacitors connected together (voltages aiding). (a) Switch open. (b) Switch first closed. (c) Switch remains closed.

With the switch remaining closed in Fig. 13-36c, each capacitor finishes up with its original voltage e_C *less* the change Δe_C, since the counterclockwise current is *decreasing* the voltage on each capacitor. Final voltage on C_5 is

$$E_{C_5} = e_{C_5} - \Delta e_{C_5} \tag{13-17}$$

$$E_{C_5} = 10 - 25$$

$$E_{C_5} = -15 \text{ V}$$

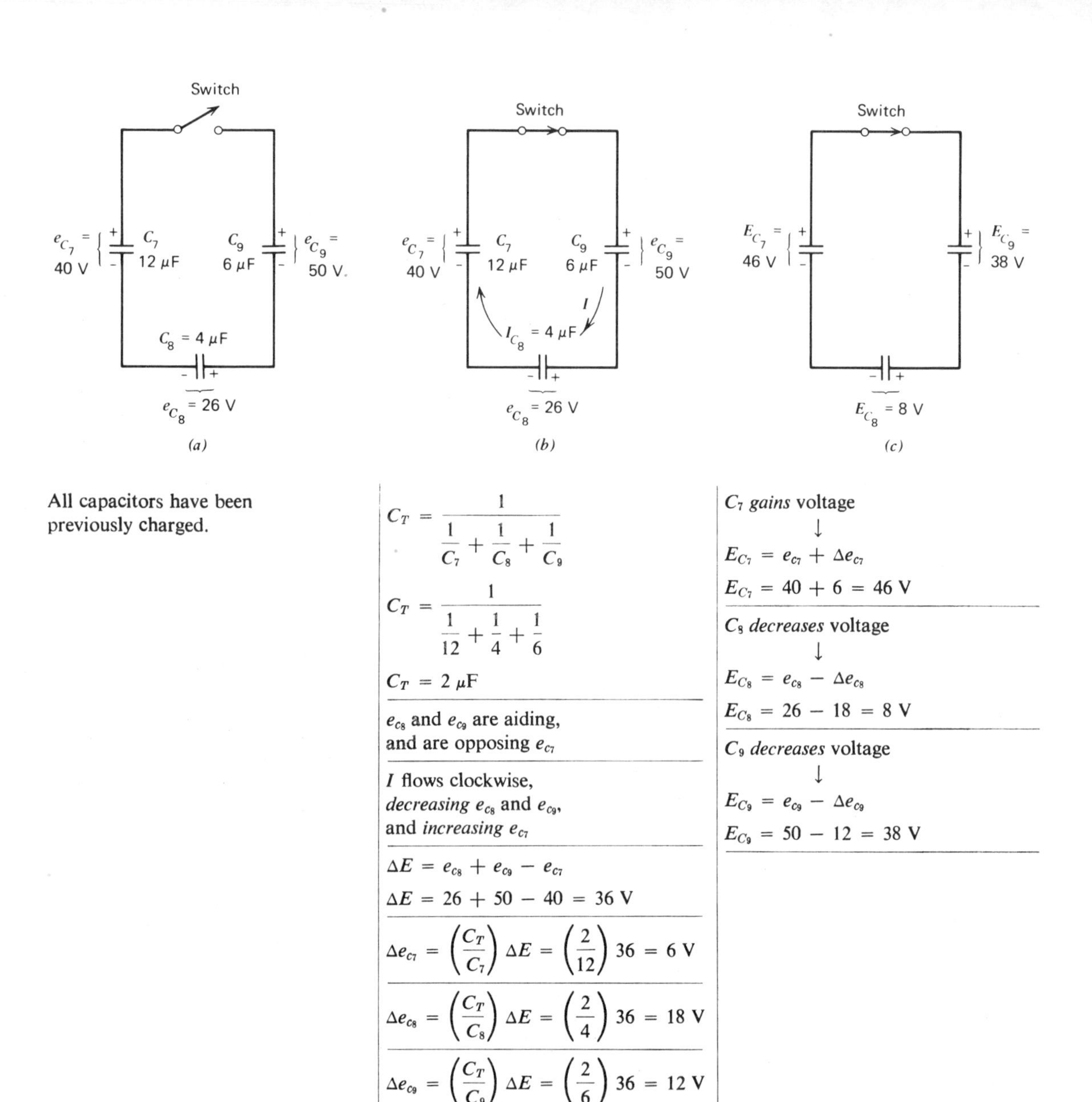

Figure 13-37. Example 13-49. (*a*) Switch open. (*b*) Switch first closed. (*c*) Switch remains closed.

The negative sign means that the voltage on C_5 has reversed its polarity as shown in Fig. 13-36*c*. Note the polarity on C_5, comparing parts (b) and (c) of Fig. 13-36.

Capacitor C_6 final voltage is

$$E_{C_6} = e_{C_6} - \Delta e_{C_6} \tag{13-17}$$

$$E_{C_6} = 20 - 5$$

$$E_{C_6} = 15 \text{ V}$$

Note that C_6 finishes with 15 V and its original polarity, as shown in Fig. 13-36*c*. Note that C_5 and C_6 finish with 15 V each with the same polarities and as a result, current stops.

The following example illustrates the methods discussed previously in this section combining the *opposing* capacitor voltages of Fig. 13-35*a*, and the *aiding* capacitor voltages of Fig. 13-36*a*.

Example 13-49

In Fig. 13-37*a*, three capacitors ($C_7 = 12\ \mu\text{F}$, $C_8 = 4\ \mu\text{F}$, and $C_9 = 6\ \mu\text{F}$) have been previously charged with the polarities shown. The voltages are $e_{C_7} = 40$ V, $e_{C_8} = 26$ V, and $e_{C_9} = 50$ V. When the switch is first closed (Fig. 13-37*b*), and then remains closed (Fig. 13-37*c*), find: (a) C_{total}, (b) net voltage that causes a current flow, (c) direction of this current, (d) which capacitors are increasing or decreasing voltage, (e) *change* of voltage, Δe_{C_7}, on C_7, (f) *change* on C_8, Δe_{C_8}, (g) *change* on C_9, Δe_{C_9}, (h) final voltage on C_7, E_{C_7}, (i) final voltage on C_8, E_{C_8}, and (j) final voltage on C_9, E_{C_9}.

Solution

Solve this example by using the methods shown previously. As a guide, refer to the text of Fig. 13-37.

For Similar Problems Refer to End-of-Chapter Problems 13-83 to 13-86

13-9. RC Time Constant (More Advanced)

In Section 13-4 an elementary discussion of *RC* time constants is given. The reader should first understand that section which covers both capacitor charging and capacitor discharging before studying this present more advanced section.

An *RC time constant* (τ) is a period of time during which a capacitor charges up, or discharges, a specific amount of voltage. By referring back to the circuit and to the *RC* time constant graphs of Figs. 13-14 to 13-16, it is seen that in the first time constant, a capacitor will charge to approximately 63% of the applied dc voltage, while voltage across the series resistor E_R, and the current *I*, decrease down to 37% of their former maximum values. In the second time constant, E_C rises to about 86%. In this second time-constant period, the capacitor has repeated its action of the previous period. That is, the capacitor has charged up 63% of the *remaining* voltage. As an example, if the applied voltage is 100 V, then, at the end of the first time constant, the capacitor has reached 63 V. This is 63% of the applied 100 V. The capacitor still has to rise

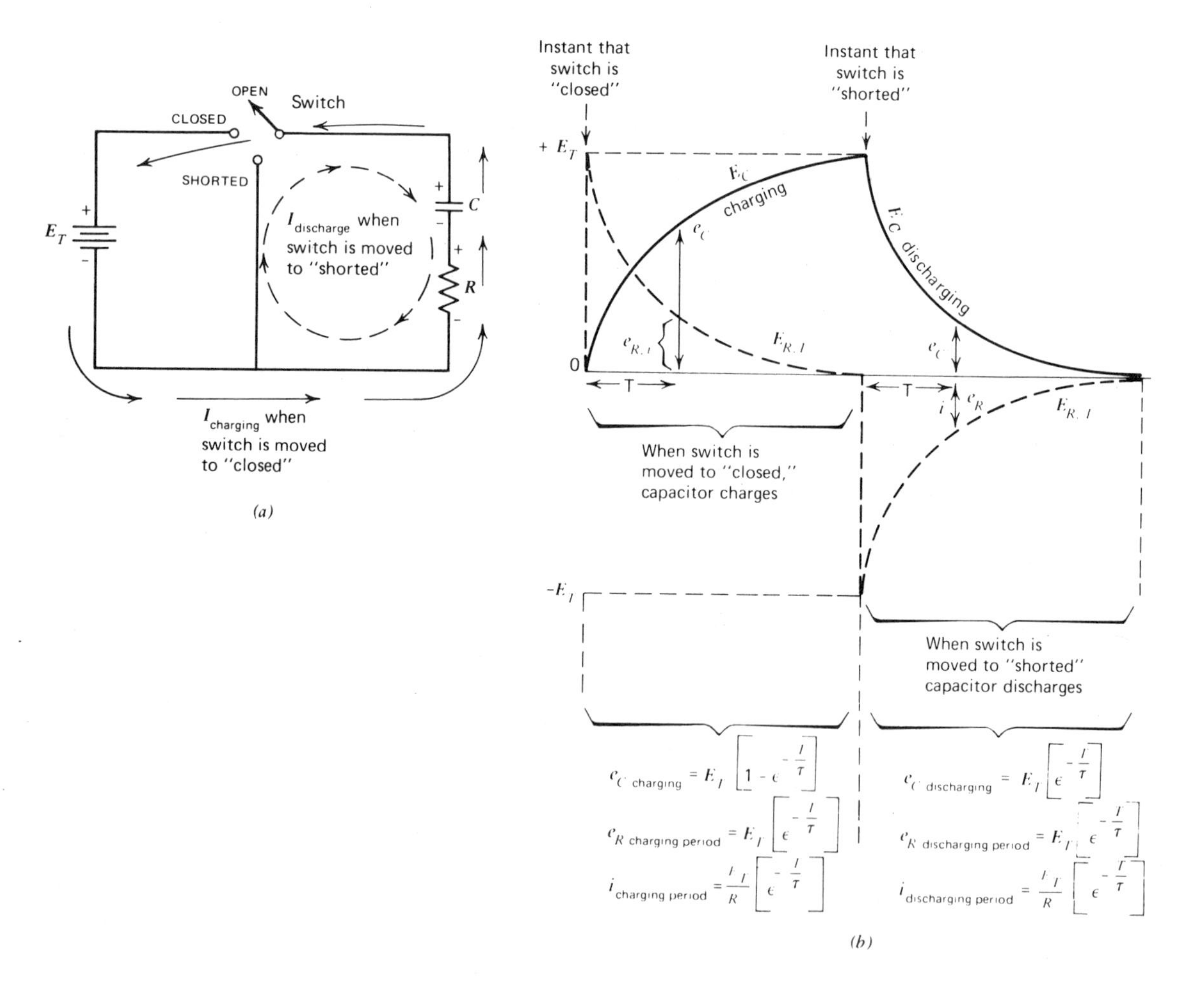

Figure 13-38. *RC* time constant. (*a*) The *RC* series circuit. (*b*) The *RC* time constant (τ) graphs and equations.

37 V more to reach 100 V. In the second time constant, the capacitor will gain 63% of 37 V, or about 23 V. This 23 V, added to the previous capacitor voltage of 63, brings the capacitor voltage to 86. This same procedure is repeated in every time constant period.

At the end of the second period, with 86 V across the capacitor, and with 100 V applied, the capacitor still has 14 V to go to reach full charge. In the third time constant period, the capacitor gains 63% of this 14 V, or approximately 9 V. This 9 V added to the previous capacitor voltage of 86, brings the capacitor voltage to 95. The reader should now see that the capacitor never actually becomes fully charged, always charging up 63% of the *remaining* voltage in the next time constant. However, after about five or six time-constant periods, E_C is so close to 100% that it is regarded as having completely charged.

The graphs of Figs 13-15 and 13-16 give the approximate values of E_C, E_R, and I. The following discussion shows how more accurate results are obtained.

Capacitor Charging. When the switch in Fig. 13-38*a* is moved to the "closed" position, charging current (electron drift) flows from the negative end of E_T, up through the resistor, onto the lower plate of C, off the upper plate of C, and into the positive end of E_T. Voltage polarity across C is shown, as well as the polarity across R (upper end is positive). Values of the voltage across C, e_C, at any instant, also across R, e_R, at any instant, and the current, i, may be found by using the following equations that are also shown in the left half of the graphs of Fig. 13-38*b*:

$$e_{C\text{ charging}} = E_T\,[1 - \epsilon^{-(T/\tau)}] \qquad (13\text{-}18)$$

$$e_{R\text{ (charge period)}} = E_T\,[\epsilon^{-(T/\tau)}] \qquad (13\text{-}19)$$

$$i_{\text{(charge period}} = \frac{E_T}{R}\,[\epsilon^{-(T/\tau)}] \qquad (13\text{-}20)$$

In the above equations, T is the *time involved*, τ (Greek letter, pronounced "tau") is the *time constant*, and the value of ϵ (Greek letter, pronounced "epsilon" to its *negative exponent* is given in Appendix V, "Values and Common Logarithms of Exponential Functions."

The following examples illustrate the use of these equations.

Example 13-50

In the diagram of Fig. 13-38*a*, if $C = 0.005\ \mu$F, $R = 2$ kΩ, and $E_T = 50$ V, find the values of (a) time constant, τ, (b) capacitor voltage, e_C, (c) resistor voltage, e_R, and (d) current, i, at an instant 15 μsec after the switch has been moved to "closed" position ($T = 15\ \mu$sec).

Solution

(a) $$\text{time constant, } \tau = (R_{\text{in ohms}})\,(C_{\text{in farad}}) \tag{13-17}$$

$$\text{time constant, } \tau = (2\text{ K})\,(0.005\ \mu\text{F})$$

$$\text{time constant, } \tau = (2 \times 10^3)\,(0.005 \times 10^{-6}$$

$$\text{time constant, } \tau = (2 \times 10^3)\,(5 \times 10^{-9})$$

$$\text{time constant, } \tau = 10 \times 10^{-6}\text{ sec} \quad \text{or} \quad 10\ \mu\text{sec}$$

(b) $$e_{C\,\text{charging}} = E_T\,[1 - \epsilon^{-(T/\tau)}] \tag{13-18}$$

$$e_{C\,\text{charging}} = 50\,(1 - \epsilon^{-(15\ \mu\text{sec}/10\ \mu\text{sec})})$$

$$e_{C\,\text{charging}} = 50\,(1 - \epsilon^{-1.5})$$

The value of $\epsilon^{-1.5}$ may be found in the table of Appendix V, under the column of ϵ^{-x}. By referring to the table of Values and Common Logarithms of Exponential Functions in Appendix V, find in the left column headed "x" the figure 1.5; then move horizontally across to the column headed "ϵ^{-x} Value," and read the number 0.22313. Then

$$e_{C\,\text{charging}} = 50\,(1 - 0.22313)$$

$$e_{C\,\text{charging}} = 50\,(0.77687)$$

$$e_{C\,\text{charging}} = 38.8\text{ V}$$

(c) $$e_{R\,(\text{charge period})} = E_T\,[\epsilon^{-(T/\tau)}] \tag{13-19}$$

$$e_{R\,(\text{charge period})} = 50\,(\epsilon^{-(15\ \mu\text{sec}/10\ \mu\text{sec})})$$

$$e_{R\,(\text{charge period})} = 50\,(\epsilon^{-1.5})$$

From the table, as in the previous part (b), $\epsilon^{-1.5}$ is 0.22313,

$$e_R = 50\,(0.22313)$$

$$e_R = 11.15\text{ V} \quad \text{or} \quad 11.2\text{ V (approximate)}$$

Note that the sum of e_C (38.8) and e_R (11.2) must equal E_T (50 V). Therefore, once e_C has been found, it may be subtracted from E_T in order to find e_R. This is a simpler method than using equation 13-19.

(d) $$i_{(\text{charge period})} = \frac{E_T}{R}\,[\epsilon^{-(T/\tau)}] \tag{13-20}$$

$$i_{(\text{charge period})} = \frac{50}{2000}\,(\epsilon^{-(15\ \mu\text{sec}/10\ \mu\text{sec})})$$

$$i_{(\text{charge period})} = \frac{50}{2 \times 10^3}\,(\epsilon^{-1.5})$$

$$i_{(\text{charge period})} = 25 \times 10^{-3}\,(\epsilon^{-1.5})$$

As in part (b) of this example, $\epsilon^{-1.5}$ has a value of 0.22313, and

$$i = (25 \times 10^{-3})\,(0.22313)$$

$$i = 5.58 \times 10^{-3}\ \text{A} \qquad \text{or} \qquad 5.58\ \text{mA}$$

Note that this current i could have been found using Ohm's law, once the resistor voltage has been found [part (c)].

$$i = \frac{e_R}{R}$$

$$i = \frac{11.15}{2 \times 10^3}$$

$$i = 5.58 \times 10^{-3}\ \text{A}$$

Example 13-51

If 300 V dc (E_T) is applied to a 0.1-μF capacitor in series with a 30-kΩ resistor, find: (a) time-constant τ, (b) capacitor voltage, e_C, and (c) resistor voltage, e_R, and (d) current, i, after E_T has been applied for 2 msec.

Solution

Solve this example by using the method shown previously and the following results as a guide: $\tau = 3$ msec, $\epsilon^{-0.667} = 0.51171$, e_C after 2 msec $= 146$ V, $e_R = 154$ V, $i = 5.12$ mA.

For Similar Problems Refer to End-of-Chapter Problems 13-87 and 13-88

Time Between Capacitor or Resistor Voltages. When a capacitor is charging, the time T between the capacitor voltage e_{C_1} at an instant t_1 and at a later instant t_2 may be found if E_T is known and the values of R and C are known. This is depicted in Fig. 13-39. The solution involves an equation similar to that of equation 13-19, repeated here:

$$e_R = E_T\,[\epsilon^{-(T/\tau)}] \qquad (13\text{-}19)$$

Note that the larger voltage E_T is on the right side of the equation. This will hold true for the equation to be used to find time period T. In Fig. 13-39 it is desired to find the length of time T for the capacitor voltage to rise from $e_{C_1} = 63$ V to $e_{C_2} = 95$ V, where the total applied dc is 100 V. Note that in equations 13-18 to 13-20, the voltage E_T appears. When the values of capacitor voltages e_{C_1} and e_{C_2} are known, the *differences* between E_T and each must be used. When e_{C_1} is 63 V, the difference between it and $E_T = 100$ V is 37 V. Note that this is the voltage across the resistor, or e_{R_1} is 37 V, as shown. Similarly, the difference between E_T (100 V) and the later value of the capacitor voltage e_{C_2} (95 V) is 5 V. This is shown in Fig. 13-39 as $e_{R_2} = 5$ V.

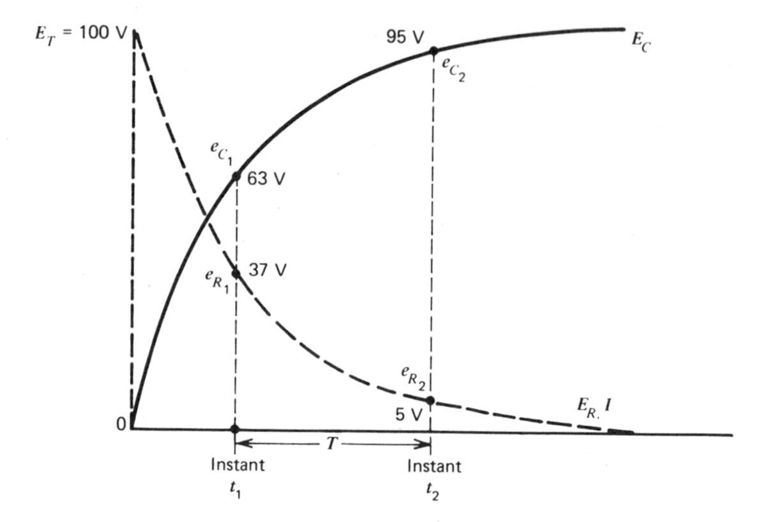

Figure 13-39.

The equation used to find the time period T is then

$$E_T - e_{C_2} = (E_T - e_{C_1}) [\epsilon^{-(T/\tau)}] \tag{13-21}$$

Since $E_T - e_{C_2}$ (100 − 95) is actually e_{R_2} (5 V), then the left side of the equation is e_{R_2}. The expression on the right side of the equation, $E_T - e_{C_1}$ (100 − 63) is actually e_{R_1} (37 V). The equation may then be rewritten as

$$e_{R_2} = E_{R_1} (\epsilon^{-(T/\tau)}) \tag{13-22}$$

Note that e_{R_2} (5 V in this example) at the left side of the equation will always be smaller than the voltage on the right side, e_{R_1} (37 V in this example).

If the time constant τ is 10 μsec, then the time T required for the capacitor voltage to rise from 63 V (e_{C_1}) to 95 V (e_{C_2}) is

$$E_T - e_{C_2} = (E_T - e_{C_1}) [\epsilon^{-(T/\tau)}] \tag{13-21}$$

$$100 - 95 = (100 - 63) (\epsilon^{-(T/10\ \mu\text{sec})})$$

or

$$e_{R_2} = e_{R_1} (\epsilon^{-(T/10\ \mu\text{sec})}) \tag{13-22}$$

$$5 = 37 (\epsilon^{-[T/(10 \times 10^{-6})]})$$

$$\frac{5}{37} = \epsilon^{-[T/(10 \times 10^{-6})]}$$

$$0.135 = \epsilon^{-[T/(10 \times 10^{-6})]}$$

From the table, Values and Common Logarithms of Exponential Functions, in Appendix V, it can be seen that if the value of ϵ^{-x} is 0.13534 (right-hand column), then the exponent x is 2 (left-hand column). Therefore, if

$$0.13534 = \epsilon^{-x}$$

then

$$2 = x$$

And, repeating the previous equation,

$$0.135 = \epsilon^{-[T/(10 \times 10^{-6})]}$$

then (approximately)

$$2 = \frac{T}{10 \times 10^{-6}}$$

$$(2)(10 \times 10^{-6}) = T$$

$$20 \times 10^{-6} \text{ sec} = T$$

or

$$20\ \mu\text{sec} = T$$

Note that this time = 20 μsec, or 2 *time constants*, agrees with the period of time shown in the *RC* time constant graphs of Fig. 13-15. Here, the capacitor voltage rose from 63% of E_T at the *first* time constant, to 95% of E_T at the *third* time constant, or a time period equal to *two time constants.*

The following example illustrates the use of equation 13-22 to find the period of time T.

Example 13-52

If an applied dc of 50 V is connected to a 0.005-μF capacitor connected in series with a 2-kΩ resistor, find the period of time T it requires for the capacitor voltage to rise from 0.1 to 38.8 V.

Solution

Solve this example by using the method shown previously, and by using the following results as a guide: $\tau = 10\ \mu\text{sec}$, $e_{R_1} = 50 - 0.1 = 49.9$ V, $e_{R_2} = 50 - 38.8 = 11.2$ V, and $T = 15\ \mu\text{sec}$.

Time Constant Required. When it is desired to change a capacitor voltage from one known value e_{C_1} to another e_{C_2}, in a known period of time, T, the previous equation 13-22 may be employed to solve for the required time constant τ. The following example illustrates this.

Example 13-53

What value of time constant, τ, will permit a capacitor voltage to rise from 0.2 V (e_{C_1}) to 146 V (e_{C_2}) in a 2-msec time (T), if the applied dc is 300 V?

Solution

$$e_{R_2} = e_{R_1}\,[\epsilon^{-(T/\tau)}] \qquad (13\text{-}22)$$

or

$$(E_T - e_{C_2}) = (E_T - e_{C_1})\,[\epsilon^{-(T/\tau)}]$$

$$300 - 146 = (300 - 0.2)\,[\epsilon^{-(2\text{ msec}/\tau)}]$$

$$154 = (299.8)\,[\epsilon^{-[(2\times 10^{-3})/\tau]}]$$

$$\frac{154}{299.8} = \epsilon^{-[(2\times 10^{-3})/\tau]}$$

$$0.514 = \epsilon^{-[(2\times 10^{-3})/\tau]}$$

From the table, Values and Common Logs of Exponential Functions, Appendix V, it can be seen that when "ϵ^{-x} value" is 0.514 (in the right-hand column), the value of x (in the left-hand column) is between 0.66 and 0.67, or about 0.665. Therefore, repeating

$$0.514 = \epsilon^{-[(2\times 10^{-3})/\tau]}$$

$$0.665 \quad \frac{2\times 10^{-3}}{\tau}$$

$$0.665\,\tau = 2\times 10^{-3}$$

$$\tau = \frac{2\times 10^{-3}}{0.665}$$

$$\tau = 3.01\times 10^{-3}\text{ sec} \qquad \text{or} \qquad 3.01\text{ msec}$$

For Similar Problems Refer to End-of-Chapter Problems 13-89 and 13-90

Capacitor Discharging. When the applied dc voltage is removed from a charged capacitor, and a path is provided through which current could flow, the capacitor discharges. In Fig. 13-38*a*, while the switch has been in the "closed" position, the capacitor has charged up. Charging current, as shown, flows *up* through the resistor, making the upper end of R its positive end, as shown. If the switch is kept in its "closed" position for a sufficient time (5 RC time constants or longer) C will have charged completely to the value of E_T. If the switch remains closed, after C has charged completely, then no further current flows, and voltage across R becomes zero, while E_C remains equal to E_T. A capacitor is said to block direct current, that is, *after* C has become charged.

When the switch in Fig. 13-38*a* is moved to the "shorted" position, E_C now acts as if it were an applied dc voltage to the resistor. A current now flows from the lower, negative plate of C *down* through R, through the switch, and onto the upper positive plate of C. This current is in the opposite direction through R, as shown by the dotted-line arrows. The polarity of the voltage across R is opposite to the previous voltage, during charging time, and is now negative at the upper end of R.

Electrons leaving the lower, negative plate of C, make this plate less negative, while electrons moving onto the upper, positive plate of C, make this plate less positive. In other words, the difference of charge between these plates, or voltage across the capacitor decreases. The capacitor is *discharging*. The right half of the graphs of Figs. 13-16 and 13-38*b* show E_C decreasing or discharging, requiring approximately 5 time constant to reach zero volts.

Note that at the instant that C starts discharging, resistor voltage, E_R, becomes a maximum value (equal to the voltage on C, which now acts as the applied voltage). This E_R is also opposite in polarity from the E_R during charging time, since the discharge current is now flowing in the opposite direction. E_R is shown as a negative graph (below the zero axis) during the discharge time.

To find the instantaneous values of capacitor voltage e_C, resistor voltage e_R, or current i, during discharge time, the following equations are employed.

$$e_{C\text{ discharging}} = E_T\,[\epsilon^{-(T/\tau)}] \qquad (13\text{-}23)$$

During discharge, e_R is equal to e_{C_1} and is the same equation.

$$E_{R\text{ (discharging period)}} = E_T\,[\epsilon^{-(T/\tau)}] \qquad (13\text{-}24)$$

$$i_{\text{(discharging period)}} = \frac{E_T}{R}\,[\epsilon^{-(T/\tau)}] \qquad (13\text{-}25)$$

Note that the e_R and i equations are exactly the same as those employed during charging time (equations 13-19 and 13-30) since e_R and i decrease during both charge and discharge periods.

The following example illustrates the use of equations 13-23, 13-24 and 13-25.

Example 13-54

A 0.05-μF capacitor has been charged to 6 V. If it is now allowed to discharge through a 40-kΩ resistor, find: (a) time constant τ, (b) capacitor voltage, e_C, 1 msec after C starts discharging, (c) resistor voltage, e_R, at this instant, and (d) current, i, at this same instant.

Solution

Solve this example by using the method shown previously, and by using the following results as a guide: $\tau = 2$ msec, $\epsilon^{-0.5} = 0.60653$, $e_C = 3.64$ V, $e_R = 3.64$ V, and $i = 91$ μA.

For a Similar Problem Refer to End-of-Chapter Problem 13-91

13-10. Capacitive Reactance and Resistance in Series AC Circuits (More Advanced)

In Sections 13-5 and 13-6, capacitive reactance and resistance in series are discussed. A full understanding of these sections should be attained before this present section is studied. The reader should also be familiar with the trigonometric functions and complex numbers that are presented in Section 12-6 of the preceding chapter, and in Appendix I, on mathematics.

In Fig. 13-40*a*, an X_C of 4 kΩ is in series with an R of 3 kΩ, with an applied 120 V. The phasors of Fig. 13-40*b* show the current phasor I pointing horizontally to the right. Since the current flows through both R and X_C, then I is used as the reference phasor at zero degrees. E_R is in phase with I, while E_{X_C} *lags* behind I by 90°. Since phasors are rotating vectors, rotating counterclockwise, then in order to show that E_{X_C} is lagging behind the horizontal I by 90°, E_{X_C} must be shown pointing vertically down. This is also referred to as − 90°, and also as a *negative j* term. Note that in the *rectangular form* of the total voltage E_T in Fig. 13-40*b*, it is shown to be equal to $E_R - jE_{X_C}$, where the first term E_R describes the *right horizontal* component, and the $-jE_{X_C}$ signifies a *downward vertical* component.

In Fig. 13-40*c*, the R phasor replaces the E_R phasor of part (b), while the X_C is substituted for the E_{X_C} phasor. The total opposition or impedance Z is then written in the *rectangular form* of a *complex number* as

$$Z = R - jX_C$$

Note that the shaded area of Fig. 13-40*c* phasors forms a right triangle with the hypotenuse (longest side) being Z, and the sides being R (horizontal) and the vertical being the same as X_C. Solving for Z by using the Pythagoras theorem gives

$$Z = \sqrt{R^2 + X_C{}^2} \qquad (13\text{-}12)$$

$$Z = \sqrt{3000^2 + 4000^2}$$

$$Z = \sqrt{(3 \times 10^3)^2 + (4 \times 10^3)^2}$$

$$Z = \sqrt{9 \times 10^6 + 16 \times 10^6}$$

$$Z = \sqrt{25 \times 10^6}$$

$$Z = 5 \times 10^3\ \Omega \qquad \text{or} \qquad 5000\ \Omega$$

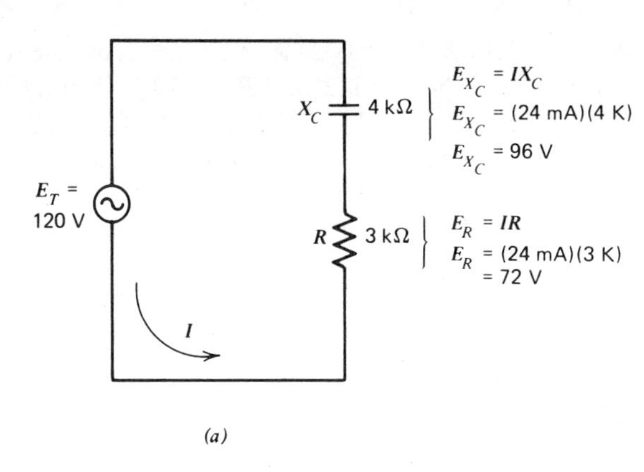

(a)

E_R = 72 V

I = 24 mA

0°

θ = -53°

(Reference)

E_{X_C} =
96 V

E_T =
120 V

(Same as
E_{X_C})

-90°

I

(b)

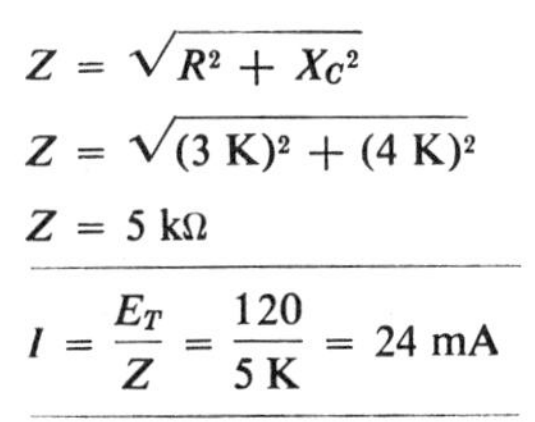

$$Z = \sqrt{R^2 + X_C^2}$$

$$Z = \sqrt{(3\text{ K})^2 + (4\text{ K})^2}$$

$$Z = 5\text{ k}\Omega$$

$$I = \frac{E_T}{Z} = \frac{120}{5\text{ K}} = 24\text{ mA}$$

Rectangular Form

$$E_T = E_R - jE_{X_C}$$

$$E_T = \sqrt{E_R^2 + E_{X_C}^2}$$

$$E_T = \sqrt{72^2 + 96^2}$$

$$E_T = 120\text{ V}$$

$$\tan\theta = \frac{E_{X_C}}{E_R}$$

$$\tan\theta = \frac{96}{72}$$

$$\tan\theta = 1.33$$

$$\theta = -53°$$

Polar Form

$$E_T = 120\,\underline{|-53°}$$

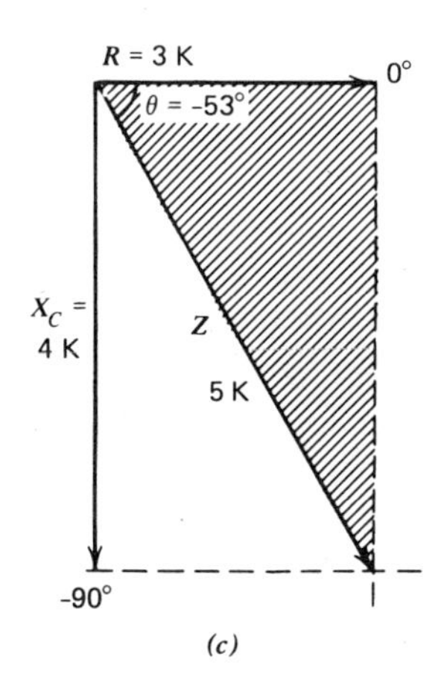

(c)

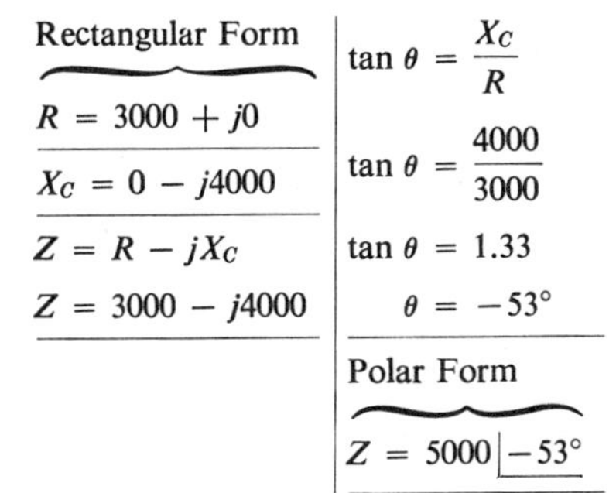

Rectangular Form

$$R = 3000 + j0$$

$$X_C = 0 - j4000$$

$$Z = R - jX_C$$

$$Z = 3000 - j4000$$

$$\tan\theta = \frac{X_C}{R}$$

$$\tan\theta = \frac{4000}{3000}$$

$$\tan\theta = 1.33$$

$$\theta = -53°$$

Polar Form

$$Z = 5000\,\underline{|-53°}$$

Figure 13-40. Series R and X_c. (a) X_c and R in series. (b) Current and voltage phasors. (c) R, X_c, and Z phasors.

Current is then: $I = E_T/Z = 120/5000 = 0.024$ A. Voltage across the resistor, E_R or V_R is then: $E_R = IR = (24 \times 10^{-3})(3 \times 10^3) = 72$ V. Voltage across the capacitor, E_{X_C} or V_{X_C} is then: $E_{X_C} = IX_C = (24 \times 10^{-3})(4 \times 10^3) = 96$ V. These voltage values may be checked by finding their vector sum (which should equal the E_T of 100 V), as shown in the text of Fig. 13-40*b*. The shaded portion of the phasor diagram of Fig. 13-40*b* forms a right triangle. The angle between E_T and I is called the *phase angle* or θ. Since all three sides of the triangle are known ($E_T = 120$, $E_R = 72$, and $E_{X_C} = 96$), then the angle may be found by using any of the trigonometric functions, sine, cosine or tangent. Using the tangent, gives

$$\tan\theta = \frac{\text{opposite side}}{\text{adjacent side}}$$

$$\tan\theta = \frac{E_{X_C}}{E_R}$$

$$\tan\theta = \frac{96}{72}$$

$$\tan\theta = 1.33$$

$$\theta = 53^\circ$$ (approximate, from the trigonometric table of Appendix IV)

Since the E_T phasor is shown in Fig. 13-40*b* as being between I (zero degrees) and E_{X_C} (-90°), then the 53° for θ is called -53°.

Another form of a *complex number* is called the *polar form*. E_T in this form is

$$E_T = 120 \text{ V}\underline{|-53^\circ}$$

This is read as "*120 volts at an angle of −53 degrees.*"

As shown in Fig. 13-40*c*, the phase angle θ between R and Z could also be found by using the shaded area right triangle, and any of the trigonometric functions. Using the tangent yields

$$\tan\theta = \frac{\text{opposite side}}{\text{adjacent side}}$$

$$\tan\theta = \frac{X_C}{R}$$

$$\tan\theta = \frac{4000}{3000}$$

$$\tan\theta = 1.33$$

$$\theta = -53^\circ$$ (approximate, using Appendix IV trigonometric table)

Again, this angle is called *negative*, since the Z phasor is between R (zero degrees) and X_C (-90°).

The *polar form* of Z is then

$$Z = 5000\ \Omega \underline{|-53^\circ}$$

The following example illustrates the above discussion.

Example 13-55

A 25-kΩ resistor is connected in series with a 40 kΩ X_C. If the applied ac voltage is 30 V, find: (a) impedance Z in its rectangular form, (b) phase angle θ, (c) impedance Z in its polar form, (d) I, (e) E_R, (f) E_{X_C}, (g) E_T in its rectangular form, and (h) E_T in its polar form.

Solution

Solve this example by using the method shown previously, and by using these results as a guide: $Z = R - jX_C = 25 \times 10^3 - j40 \times 10^{-3}$; $\tan\theta = X_C/R = 40\ \text{K}/25\ \text{K} = 1.6$, $\theta = 58^\circ$; $Z = 47.2\ \text{K}\underline{|-58^\circ}$; $I = 636\ \mu\text{A}$; $E_R = 15.9\ \text{V}$; $E_{X_C} = 25.4\ \text{V}$; $E_T = E_R - jE_{X_C} = 15.9 - j25.4$; $E_T = 30\underline{|-58^\circ}$.

For Similar Problems Refer to End-of-Chapter Problems 13-92 to 13-94

13-11. Capacitive Reactance and Resistance in Parallel AC Circuits (More Advanced)

The reader should fully understand the parallel ac circuit discussion given in Section 13-6, as well as the complex numbers presented in Chapter 12, Section 12-6, before continuing with this advanced section.

Figure 13-41 shows a two-branch parallel ac circuit consisting of the left branch Z_1 and the right branch Z_2. Z_1 is made up of X_{C_1} and R_1 in series, while Z_2 is X_{C_2} and R_2 in series. As shown in the text portion of Fig. 13-41, total impedance Z_T and total current I_T will be found.

One method of finding Z_T is by using the *product* of the two parallel branches *divided by their sum*. To do this, the two branches Z_1 and Z_2 must be found in their *rectangular forms* and also in their polar forms. The polar forms are easily multiplied and divided, but cannot be added. The rectangular forms are easily added. All steps are shown in Fig. 13-41 and will be referred to in the following.

The rectangular form of Z_1 is: $R_1 - jX_{C_1} = 60 - j80$. The actual value of Z_1 may be found by using the Pythagoras theorem. $Z_1 = 100\ \Omega$. The phase angle θ_1 may be found by using any one of the trigonometric functions. Using the tangent function, $\theta_1 = -53^\circ$. The *negative* sign is added in front of the 53° since Z_1 (similar to Z in Fig. 13-40c is between R (which is zero degrees) and

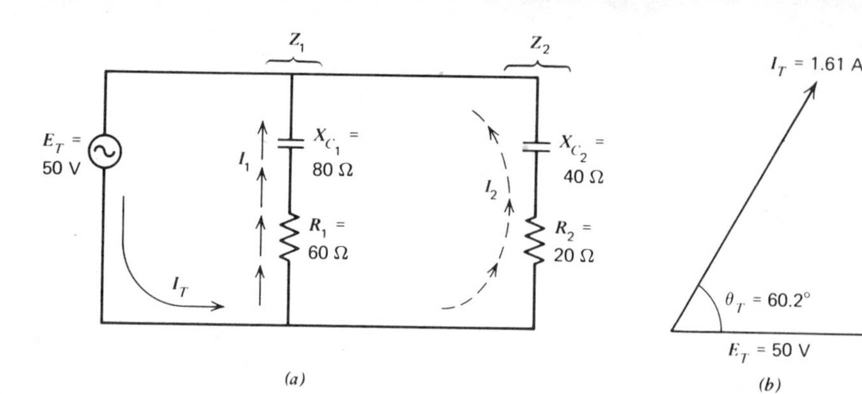

Left Branch

$Z_1 = R_1 - jX_{C_1}$ (rectangular form)

$Z_1 = 60 - j80$

$Z_1 = \sqrt{60^2 + 80^2}$

$Z_1 = 100\ \Omega$

$\tan\theta_1 = \dfrac{X_{C_1}}{R_1} = \dfrac{80}{60} = 1.33$

$\theta_1 = 53^\circ$

$Z_1 = 100\angle{-53^\circ}\ \Omega$ (polar form)

$I_1 = \dfrac{E_T}{Z_1} = \dfrac{50\angle 0^\circ}{100\angle{-53^\circ}} = 0.5\angle 53^\circ\ \text{A}$ (polar form)

Right Branch

$Z_2 = R_2 - jX_{C_2}$ (polar form)

$Z_2 = 20 - j40$

$Z_2 = \sqrt{20^2 + 40^2}$

$Z_2 = 44.7\ \Omega$

$\tan\theta_2 = \dfrac{X_{C_2}}{R_2} = \dfrac{40}{20} = 2$

$\theta_2 = 63.5^\circ$

$Z_2 = 44.7\angle{-63.5^\circ}\ \Omega$ (polar form)

$I_2 = \dfrac{E_T}{Z_2} = \dfrac{50\angle 0^\circ}{44.7\angle{-63.5^\circ}} =$

$I_2 = 1.12\angle 63.5^\circ\ \text{A}$ (polar form)

Total

$Z_T = \dfrac{(Z_1)(Z_2)}{Z_1 + Z_2}$

$Z_T = \dfrac{(100\angle{-53^\circ})(44.7\angle{-63.5^\circ})}{60 - j80 + 20 - j40}$ (polar form) (rectangular form)

$Z_T = \dfrac{4470\angle{-116.5^\circ}}{80 - j120}$

$Z_T = \dfrac{4470\angle{-116.5^\circ}}{\sqrt{80^2 + 120^2};\ \tan\theta_1 = \dfrac{120}{80} = 1.5}$

$Z_T = \dfrac{4470\angle{-116.5^\circ}}{144\angle{-56.3^\circ}}$

$Z_T = 31\angle{-60.2^\circ}\ \Omega$ (polar form)

$I_T = \dfrac{E_T}{Z_T} = \dfrac{50\angle 0^\circ}{31\angle{-60.2^\circ}} = 1.61\angle 60.2^\circ\ \text{A}$ (polar form)

$I_T = I_1 + I_2$

$I_T = 0.5\angle 53^\circ + 1.12\angle 63.5^\circ$ (polar forms)

$I_T = 0.5(\cos 53^\circ + j\sin 53^\circ) + 1.12(\cos 63.5^\circ + j\sin 63.5^\circ)$ (trigonometric forms)

$I_T = 0.5(0.602 + j0.799) + 1.12(0.446 + j0.895)$

$I_T = (0.301 + j0.4) + (0.5 + j1)$

$I_T = 0.801 + j1.4$ (rectangular form)

$I_T = \sqrt{0.801^2 + 1.4^2};\ \tan\theta_T = \dfrac{1.4}{0.801} = 1.75;\ \theta_T = 60.2^\circ$

$I_T = 1.615\angle 60.2^\circ$ (polar form)

Figure 13-41. Parallel ac circuit solution. (*a*) Parallel circuit. (*b*) E_T, I_T phasors total.

X_C (which is $-90°$). The polar form of Z_1, which describes its magnitude and its phase angle, is then

$$Z_1 = 100\underline{|-53°}$$

The *polar form* of current I_1 may be found by using Ohm's law and is $0.5\underline{|53°}$ A.

The *rectangular form* of Z_2 is: $R_2 - jX_{C_2} = 20 - J40$. The actual value of Z_2 may be found by using Pythagoras' theorem and is 44.7 Ω. Phase angle θ_2 may be found by referring to the similar right triangle of Fig. 13-40*c*. Using the tangent, $\theta_2 = -63.5°$. The *negative* sign is added in front of the 63.5°, since Z_2 is between R_2 (0°) and X_{C_2} ($-90°$), similar to Z of Fig. 13-40*c*.

The *polar form* of Z_2, describing its value and its phase angle is then:

$$Z_2 = 44.7\underline{|-63.5°}\ \Omega$$

The *polar form* of I_2 may be found from Ohm's law and is $1.12\underline{|63.5°}$ A.

Total impedance Z_T may be found by using the product/sum method. For the *product*, the *polar forms* of Z_1 and Z_2 will be used. For the *sum*, their *rectangular forms* are used, as shown in Fig. 13-41. $Z_T = 31\underline{|-60.2°}$ Ω.

The *polar form* of I_T may be found using Ohm's law and is $1.61\underline{|60.2°}$ A.

This polar form of I_T could also be found by adding the values of I_1 and I_2. Since the *polar forms* of I_1 and I_2 are known but cannot be added, they must each be changed to their *rectangular forms*, which can be added. To convert from *polar* to *rectangular*, the *trigonometric form* must be found. The following illustrates this.

$$I_T = I_1 + I_2$$

$$I_T = 0.5\underline{|53°} + 1.12\underline{|63.5°} \qquad \text{(polar forms)}$$

$$I_T = 0.5\,(\cos 53° + j \sin 53°) + 1.12\,(\cos 63.5° + j \sin 63.5°) \qquad \text{(trigonometric forms)}$$

$$I_T = 0.5\,(0.602 + j0.799) + 1.12\,(0.446 + j0.895)$$

$$I_T = 0.301 + j0.4 + 0.5 + j1$$

$$I_T = 0.801 + j1.4 \qquad \text{(rectangular form)}$$

$$I_T = \sqrt{0.801^2 + 1.4^2}; \quad \tan \theta_T = \frac{1.4}{0.801}$$

$$I_T = \sqrt{0.64 + 1.96}; \quad \tan \theta_T = 1.75$$

$$I_T = \sqrt{2.6}; \quad \theta_T = 60.2°$$

$$I_T = 1.615\underline{|60.2°}\ \text{A} \qquad \text{(polar form)}$$

Note that this value of I_T, which is the result of adding I_1 and I_2, agrees with the previous value of I_T found by E_T/Z_T.

For Similar Problems Refer to End-of-Chapter Problems 13-95 and 13-96

13-12. Power and Power Factor in AC *R* and X_C Circuits

In Section 13-6, a discussion of power and power factor is given. In an ac circuit containing resistors only, the total power dissipated may be calculated by using any one of the power equations given in Chapter 2, Section 2-9. These are: $E_T I_T$, or $I_T^2 R_T$, or E_T^2/R_T. Power is consumed in the resistors only. In an ac circuit containing only a pure capacitive reactance (practically no resistance), no power is consumed. Energy is stored in the electrostatic field between the capacitor plates during the charge of the capacitor, but is returned when the capacitor discharges. This is similar to the inductive reactance, which stores the energy in its magnetic field, but which returns the energy to the circuit when the magnetic field collapses.

Where both reactance and resistance are present in an ac circuit, the product $E_T I_T$ does not give the actual or true power dissipated, but results in a larger value called *apparent power*. This may be stated in *watts*, but since it is not the true power, the term *volt-amperes* is more desirable. The *true power* may be found by either using the resistor component alone, such as E_R^2/R, $E_R I$, or I^2R. Another method is to make use of the *power factor*. This was shown in Section 13-6, Power Factor, and also in the preceding chapter as equation 12-8, $P_{\text{true}} = (P_{\text{apparent}})$ (power factor). The power factor is due to the phase angle θ between E_T and I_T, and is actually the cosine of this angle, or

$$\text{power factor} = \cos\theta \qquad (13\text{-}26)$$

As shown previously in the trigonometric functions of Fig. 12-14 and elsewhere, the *cosine* function is the *adjacent side* divided by the *hypotenuse* of the right triangle. In the phasor diagram of Fig. 13-40*c*, the R, X_C, and Z phasors are shown. The shaded area is a right triangle, with the phase angle θ between R and Z, or as shown in Fig. 13-40*b*, θ is between E_T and I_T. In part (c), the adjacent side to θ is R, while the hypotenuse (the longest side) is Z. Therefore, the cosine of θ is

$$\cos\theta = \frac{R}{Z}$$

In the Fig. 13-40*a* circuit, an X_C of 4 kΩ is connected in series with an R of 3 kΩ. E_T is shown as 120 V. Z, in the text of part (a), is 5 kΩ, and I is 24 mA.

The apparent power dissipated in the circuit of Fig. 13-40*a* is

$$P_{\text{apparent}} = E_T I_T = (120\text{ V})(24\text{ mA}) = 2.88\text{ W} \quad \text{or} \quad \text{volt-amperes.}$$

The *power factor* is (equation 13-26), PF $= \cos\theta = R/Z = 3000/5000 = 0.6$. The actual or *true* power dissipated in this circuit is: (equation 12-8), $P_{\text{true}} = (P_{\text{apparent}})(\text{PF}) = (2.88)(0.6) = 1.73$ W.

True power could also be found by using the voltage across the resistor, not the total voltage, as shown in the following. E_R is shown in Fig. 13-40*a* and *b* as 72 V.

$$P_{\text{true}} = E_R\, I = (72)\,(24 \text{ mA})\; 1.73 \text{ W}$$

This true power could also be found using by $I_T^2\, R$ or E_R^2/R.

VAR. There is no actual power dissipated in a pure reactance, but the product of the voltage across a reactance and the current through it is called the *reactive power*, or the *volt-ampere reactive*, or *VAR*. In the circuit of Fig. 13-40*a* there is 96 V across X_C while 24 mA flows through it. The product of the two is the reactive volt-amperes or VAR.

$$\text{VAR} = (E_{X_C})\,(I) = (96)\,(24 \text{ mA}) = 2.3$$

VAR is the *volt-ampere* at a 90° angle. This reactive volt-ampere may also be found by using the sine function of the phase angle. In the phasors of Fig. 13-40*b*, the vertical phasor E_{X_C} is the *opposite side* with respect to phase angle θ. The *sine* function of this angle is the ratio of opposite side to hypotenuse. To find the *vertical* or *reactive* part of the apparent power (volt-amperes) the sine function is used. Therefore, in Fig. 13-40*a* circuit,

$$\text{VAR} = E_T I_T \sin\theta$$

$$\text{VAR} = (120)\,(24 \text{ mA})\,(\sin 53°)$$

$$\text{VAR} = (1.2 \times 10^2)\,(24 \times 10^{-3})\,(0.8)$$

$$\text{VAR} = 23 \times 10^{-1} \qquad \text{or} \qquad 2.3$$

When a reactive device such as an electric motor is connected to the ac line voltage from the electric power company, the inductive reactance of the motor results in a power factor of less than one. This requires a greater current in the line than would be required if the load (the motor) were not reactive. This greater line current produces greater power losses in the line. To reduce this line current without decreasing the true power in the load, the power factor must be brought to a value of one or *unity*. Adding a capacitive reactance in parallel with the inductive load (the motor), where the capacitor reactive power (VAR) is equal to that of the load, results in the power factor's being *corrected* to a value of one.

The following example illustrates the previous discussion in the *parallel* ac circuit of Fig. 13-41*a*.

Example 13-56

In the parallel circuit of Fig. 13-41*a*, current in the left branch I_1 is 0.5 A through a resistor $R_1 = 60\Omega$, while current in the right branch I_2 is 1.12 A through a resistor $R_2 = 20\ \Omega$. Total applied voltage is 50 V, and as shown in the phasor diagram of Fig. 13-41*b*, $I_{\text{total}} = 1.61$ A, and total phase angle between I_T and E_T is 60.2°. Find: (a) total true power, $P_{T\,\text{true}}$, using the power factor, (b) true power in the left branch, $P_{1\ true}$, (c) true power in the right branch, $P_{2\ true}$, and (d) total true power, $P_{T\ true}$, by adding $P_{1\ true}$ and $P_{2\ true}$.

Solution

(a) Total true power is: (equation 12-8), $P_{T\ true} = (P_{apparent})\,(PF) = (E_T\, I_T)$ $(PF) = (50)\,(1.61)\,(\cos 60.2°) = (80.5)\,(0.496) = 40$ W.
(b) True power in left branch is: $I_1^2\, R_1 = (0.5)^2\,(60) = 15$ W.
(c) True power in right branch is: $I_2^2\, R_2 = (1.12)^2\,(20) = 25$ W.
(d) Total true power is also: $P_{1\ true} + P_{2\ true} = 15 + 25 = 40$ W.

For Similar Problems Refer to End-of-Chapter Problems 13-97 and 13-98

Energy Stored in a Capacitor. Energy is the ability to do work. The voltage difference between the plates of a charged capacitor has the potential of performing work, for example, when the capacitor is discharging. The potential energy stored in the charged capacitor is

$$\text{energy} = \tfrac{1}{2}\, CE^2 \qquad (13\text{-}27)$$

where *energy* is in *joules* (1 W of power is 1 J of work performed in 1 sec), C is in *farads*, and E is in *volts*.

As an illustration, in the low-voltage power supply of a television receiver a 25-μF capacitor which has been charged to 300 V has a potential energy of: (equation 13-27) $= \tfrac{1}{2}\, CE^2 = (\tfrac{1}{2})\,(25 \times 10^{-6})\,(300^2) = 1.125$ J. This energy is stored in the electrostatic field between the plates.

Compare the energy stored in this large capacitor low-voltage circuit with that of a small capacitor in the following high-voltage example.

Example 13-57

What energy is stored in the high-voltage capacitor of a television receiver if the capacitor is 500 pF, and it is charged to 20 kV?

Solution

The reader should solve this example by using the previous procedure and should find the result to be 0.1 J.

For a Similar Problem Refer to End-of-Chapter Problem 13-99

Capacitor Electrostatic Field. Between the charged plates of a capacitor there exists an electrostatic field. This consists of electrostatic lines (Fig. 13-42). The *direction* of these lines depends on whether a positively charged

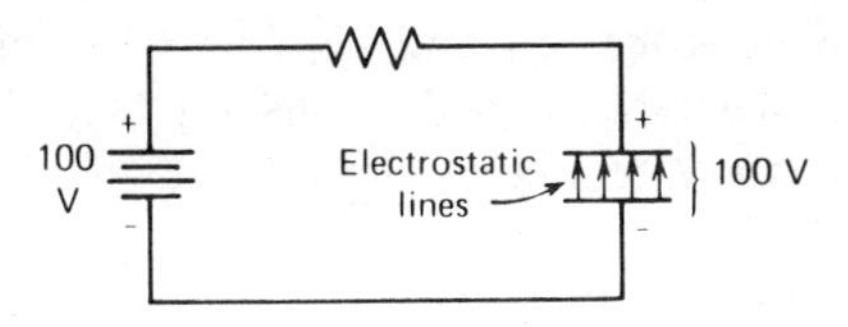

Figure 13-42. Electrostatic field between plates of a charged capacitor.

particle or a negatively charged particle is assumed to be placed at some point between the plates. Since, in this book, current has been taken as the drift or movement of *electrons*, then it will be assumed that a negatively charged particle (an electron) is placed between the charged capacitor plates. The electron will be repelled by the negative plate, and will be attrated by the positive plate, moving in the direction of the arrows (the electrostatic lines) in Fig. 13-42.

Figure 13-43*a* and *b* shows the electrostatic field present between two conductors. Part (a) shows a pair of parallel wires such as the familiar twin-lead ribbon-type transmission line from a television antenna to the receiver. Part (b) shows a coaxial cable (an inner conductor held in place at the center of a hollow, tubular outer conductor). The outer conductor is usually grounded like, for example, the cable from a car antenna to the radio receiver.

The electrostatic lines of Fig. 13-42 indicate the direction of force exerted on an electron that is placed in the electrostatic field. The amount of this force depends on the quantity of charge on each capacitor plate at a point directly above, and directly below, the electron, as well as the distance between the plates. This relationship is given by Coulomb's law:

$$\text{force} = \left[\frac{(Q_1)\,(Q_2)}{\text{distance}^2}\right] \text{(constant conversion factor)} \qquad (13\text{-}28)$$

$$F = \left(\frac{Q_1\; Q_2}{d^2}\right)(K) \qquad (13\text{-}28)$$

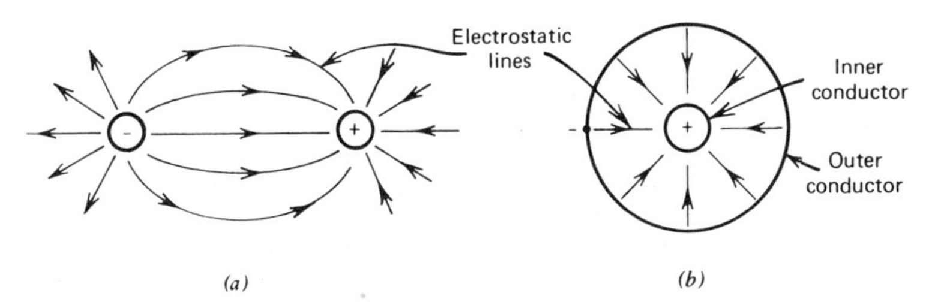

Figure 13-43. Electrostatic field between conductors. (*a*) Parallel wires such as television twin-lead. (*b*) Concentric cable such as coaxial shielded cable.

where the force is in newtons, Q_1 and Q_2 (the two point charges) are in coulombs, distance is in *meters*, and the constant, K ($= 8.99 \times 10^9$) is the conversion factor necessary to obtain the answer for the force in newtons (the mks system) in air.

The following example illustrates this.

Example 13-58

Find the force in *newtons* between two 5-μC charges separated in air at a distance of 3 mm.

Solution

$$F = \left(\frac{Q_1 \, Q_2}{d^2}\right)(K)$$

$$F = \left(\frac{(5\mu\text{C})\,(5\mu\text{C})}{3\ \text{mm}^2}\right)(8.99 \times 10^9) \qquad (13\text{-}28)$$

$$F = \left(\frac{(5 \times 10^{-6})\,(5 \times 10^{-6})}{(3 \times 10^{-3})^2}\right)(8.99 \times 10^9)$$

$$F = \left(\frac{25 \times 10^{-12}}{9 \times 10^{-6}}\right)(8.99 \times 10^9)$$

$$F = 25 \times 10^3\ \text{N}$$

For a Similar Problem Refer to End-of-Chapter Problem 13-100

13-13. Capacitor Value

At the beginning of this chapter, in Section 13-1, it was shown that the size or value of a capacitor depended on the area of the plates, the spacing between them, and the type of material used as the insulator or dielectric. The equation to determine the value of a parallel-plate capacitor is

$$\text{capacitor (in farads)} = \left[\frac{\text{area of a plate (in square meters)}}{\text{spacing between plates (in meters)}}\right]$$

$$[(\text{dielectric constant})\ (\text{permittivity of air})] \qquad (13\text{-}29)$$

$$C = \frac{(A)\,(K)\,(8.85 \times 10^{-12})}{s}$$

The term *permittivity* is the ability of an insulator to *permit* the establishing of electrostatic lines. The permittivity of air or free space is the number shown in equation 13-29, *8.85 × 10⁻¹²*. The term *dielectric constant* (K in the equation) is actually the permittivity of an insulator (the dielectric) compared to that of air. Dielectric constant is actually the relative permittivity. In Table 13-1, the dielectric constants of several insulator materials are listed. Mica for example has a constant (K) of about 5. This means that mica has five times more permittivity than air, or simply that mica will permit electrostatic lines to be set up five times more easily than air, or five times more numerous than air.

The following example illustrates the use of equation 13-29.

Example 13-59

A capacitor consists of a pair of silver-plated copper plates each 20 mm wide and 30 mm long, with a strip of mica between them 0.25 mm thick. Find the value of this capacitor [Note that from Table 13-1, mica is listed as having a dielectric constant (K) = 5.]

Solution

The area of a plate is

$$\text{area} = (20\text{ mm})\,(30\text{ mm})$$

$$\text{area} = (20 \times 10^{-3})\,(30 \times 10^{-3})$$

$$\text{area} = 600 \times 10^{-6}\text{ m}^2$$

The capacitor value is then

$$C = \frac{(\text{area})\,(K)\,(8.85 \times 10^{-12})}{\text{spacing}} \qquad (13\text{-}29)$$

$$C = \frac{(600 \times 10^{-6})\,(5)\,(8.85 \times 10^{-12})}{(0.25 \times 10^{-3})}$$

$$C = \frac{(6 \times 10^{-4})\,(5)\,(8.85 \times 10^{-12})}{2.5 \times 10^{-4}}$$

$$C = \frac{265 \times 10^{-16}}{2.5 \times 10^{-4}}$$

$$C = 106 \times 10^{-12}\text{ F} \qquad \text{or} \qquad 106\text{ pF}$$

For a Similar Problem See End-of-Chapter Problem 13-101

13-14. Capacitor Current

When a current flows onto one plate of a capacitor, it makes that plate negative. This repels electrons from the other plate, leaving it positive. As a result, a difference of charge or a voltage appears between the two plates, and the capacitor is said to be "charged." The length of *time* that this charging current flows, the amount of this *current*, and the size or value of the *capacitor* determine the voltage to which the capacitor becomes charged.

The quantity (Q) of electrons (in coulombs) that accumulate on the negative plate of the charged capacitor depends on the amount of this *current* and the length of *time* that it flows. This is shown in the following relationship.

$$Q \text{ (in coulombs)} = I \text{ (in amperes) time (in seconds)} \qquad (13\text{-}30)$$

or

$$Q = It$$

Q (for quantity of electrons stored on a capacitor plate) is discussed in this chapter in Section 13-3.

From equation 13-20, it may be seen that the longer the current flows, or the larger the current, the greater is the quantity of electrons piled up on the capacitor plate. These examples show this relationship:

Example 13-60

A steady 5 mA flows for 3 μsec into a 0.05-μF capacitor. Find (a) the quantity (Q) of electrons accumulated on this capacitor, and (b) the voltage across the capacitor.

Solution

(a) $Q = It$

$$Q = (5 \text{ mA})(3\ \mu\text{sec})$$

$$Q = (5 \times 10^{-3})(3 \times 10^{-6})$$

$$Q = 15 \times 10^{-9} \text{ C}$$

(b) Knowing the Q, (15×10^{-9} C), and the value of the capacitor (C = 0.05 μF), voltage may be found by using the relationship: $Q = CE$ (equation 13-4).

$$Q = CE \qquad (13\text{-}4)$$

$$15 \times 10^{-9} = (0.05 \times 10^{-6})(E)$$

Solving for E by dividing both sides of equation by (0.05×10^{-6}) gives

$$\frac{15 \times 10^{-9}}{0.05 \times 10^{-6}} = E$$

$$\frac{15 \times 10^{-9}}{5 \times 10^{-8}} = E$$

$$3 \times 10^{-1}\ \text{V} = E$$

$$0.3\ \text{V} = E$$

The following example uses the same capacitor, 0.05 μF, but a larger current and a longer time.

Example 13-61

If a constant 150 mA flows for 200 μsec onto a 0.05-μF capacitor, find (a) the quantity (Q) of electrons accumulated on this capacitor, and (b) the voltage across the capacitor.

Solution

Solve this example from the procedure shown previously. As a guide, the following results should be found: $Q = 3 \times 10^{-5}$ C, $E_C = 600$ V.

For a Similar Problem Refer to End-of-Chapter Problem 13-102

Capacitor Voltage Lags Current in AC Circuit. In Sections 13-5 and 13-6 it is shown that in an ac circuit, *voltage* across the *capacitor lags* the *current* 90°. There we simply state that the capacitor takes time to charge up. Voltage across C, therefore, slowly rises, while current flows immediately. A more rigorous explanation is given in the following disscussion.

In Section 11-10, it is shown that the voltage induced in a coil when the current through it changes is: $e_L = L\,(dI/dT)$ (equation 11-9), where dI/dT is a change of I for a change of time. A similar expression for a capacitor states that the current (in amperes) flowing in a capacitor at some instant is the value of the capacitor (in farads) multiplied by the *change* of capacitor voltage for a change of time (in seconds), or

$$i_c = C\left(\frac{de_c}{dt}\right) \qquad (13\text{-}31)$$

When a sine-wave ac voltage is applied to a capacitor or to a capacitor-resistor combination (series or parallel), voltage across the capacitor, and current flow "through" it are both sinusoidal (sine wave shape), with E_C lagging I_C by 90° as shown in Fig. 13-44. Values of I may be found using equation 13-31, as shown in the following.

The capacitor value is 1 μF, and each time interval, $dt_1 = dt_2 = dt_3 =$ etc. $=$ 1 μsec.

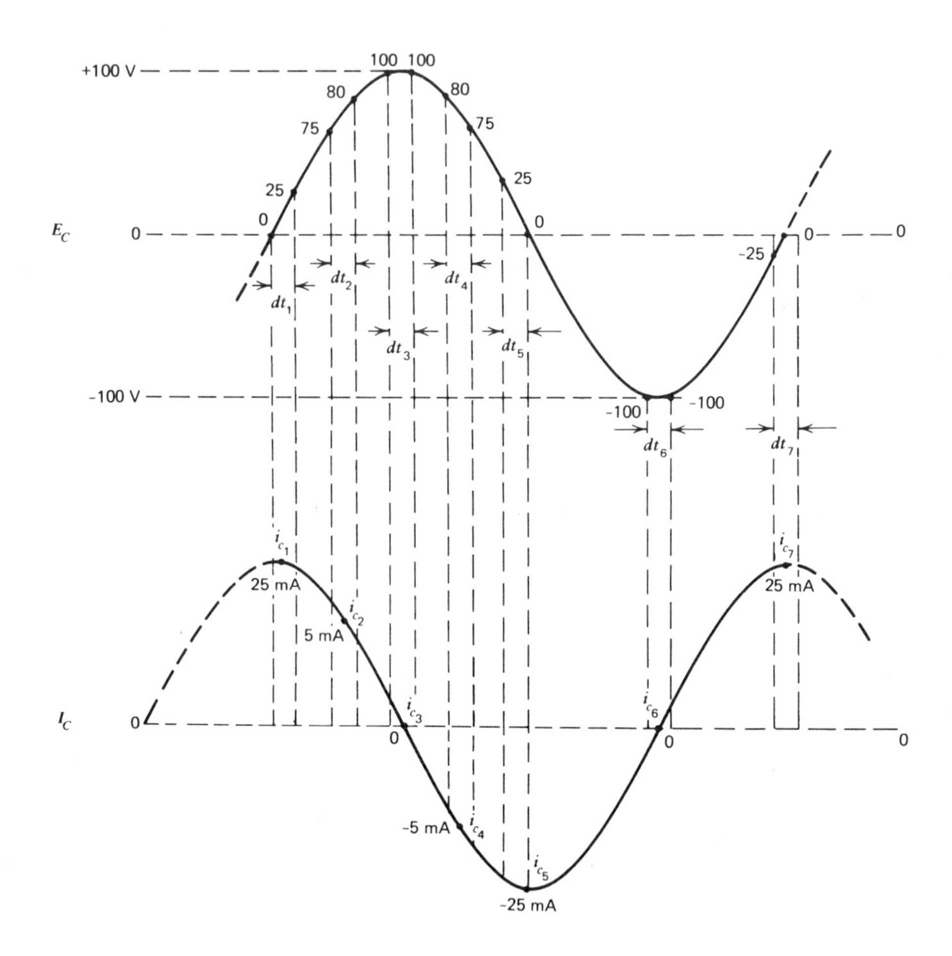

Each time period, $dt_1 = dt_2 = dt_3 =$ etc. $=$ 1 msec

$C = 1\ \mu\text{F}$

$$i_c = C\left(\frac{de_c}{dt}\right)$$

Figure 13-44.

At the first time interval, dt_1, e_C rises from 0 to 25 V, and i_{c_1} is

$$i_{c_1} = C\left(\frac{de_{c_1}}{dt_1}\right) \tag{13-31}$$

$$i_{c_1} = 1\ \mu\text{F}\left(\frac{25\ \text{V}}{1\ \text{msec}}\right)$$

$$i_{c_1} = (1 \times 10^{-6})\left(\frac{25}{1 \times 10^{-3}}\right)$$

$$i_{c_1} = 25 \times 10^{-3}\ \text{A} \qquad \text{or} \qquad 25\ \text{mA}$$

as shown on the I_C curve as i_{c_1} in Fig. 13-44.

At the second time interval, dt_2, e_C rises from 75 to 80 V, ($de_{c_2} = 5$ V) and i_{c_2} is: $(1 \times 10^{-6})\ (5/1 \times 10^{-3}) = 5$ mA, as shown in Fig. 13-44.

At the third time interval, dt_3, e_C does not change, bur remains at 100 V, ($de_{c_3} = 0$ V), and i_{c_3} is

$$(1 \times 10^{-6})\ (0/1 \times 10^{-3}) = 0\ \text{mA}$$

as shown.

At the fourth time interval, dt_4, e_C *decreases* from 80 to 75 V, ($de_{C_4} = -5$ V, the *negative* denoting a decrease). The decreasing e_C means that the current has reversed its direction. This reversal is shown in the I_C wave shape at i_{c_4} which is below the zero axis, and i_{c_4} is

$$(1 \times 10^{-6})\ [-5/(1 \times 10^{-3})] = -5\ \text{mA}$$

as shown.

Note that time interval dt_5 is a repeat of dt_1 except that the current is reversed as shown for i_{c_5}. Also note that at time interval dt_6, there is no change in E_C, and i_{c_6} is again zero. Finally, time period dt_7 is an exact duplicate of dt_1.

The E_C wave shape of Fig. 13-44 can now be seen as lagging the I_C wave shape by 90°. That is, I_C has already reached its maximum value at dt_1, while E_C is only just starting to increase from zero.

The following example further illustrates use of equation 13-31.

Example 13-62

In Fig. 13-45 voltage changes across a 0.005-μF capacitor are as shown. Find: (a) value of current i_{c_1} during the time interval dt_1 (50 μsec) when E_C goes from 0 to 100 V, (b) value of current i_{c_2} during time period dt_2 (50 μsec) when E_C remains at 100 V, (c) value of current i_{c_3} during time period dt_3 (50 μsec) when E_C rises from 100 to 150 V, and (d) value of current i_{c_4} during time period dt_4 (10 μsec when E_C *falls* from 150 to 0 V).

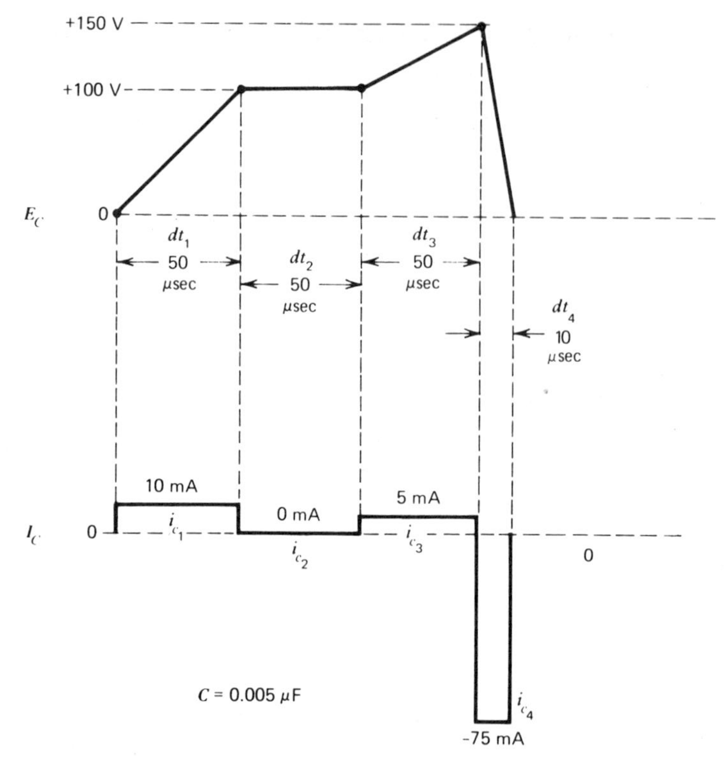

Figure 13-45. Example 13-62.

Solution

Solve this example from the previous procedure shown and, using the results given in Fig. 13-45 as a guide: $i_{c_1} = 10$ mA, $i_{c_2} = 0$, $i_{c_3} = 5$ mA, $i_{c_4} = -75$ mA.

For a Similar Problem See End-of-Chapter Problem 13-103

13-15. Voltage Square Waves and RC Circuits

A series capacitor and resistor, as shown in Fig. 13-46*a*, is used very often between one amplifier and the next. If the voltage input to the *RC* circuit is a square wave, or a pulse as shown in Fig. 13-46*b*, then the output voltage E_R may be a square wave also, as shown in Fig. 13-46*c*. A large or long *RC* time constant (that is, large compared to the time period or duration, *T*, of the input pulse) results in an output voltage across *R* that practically duplicates the input. In special purpose circuits the output across *R* is purposely changed

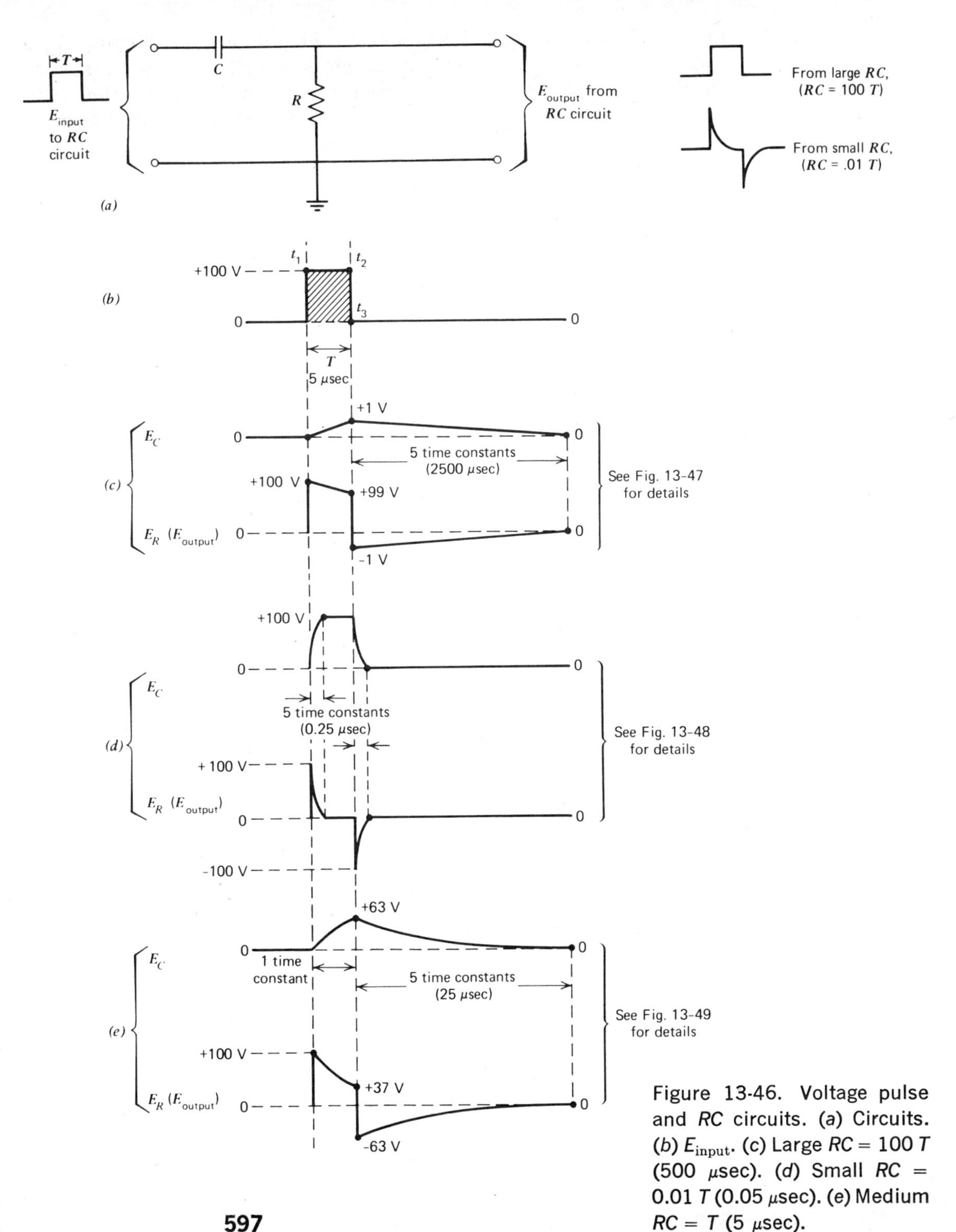

Figure 13-46. Voltage pulse and RC circuits. (a) Circuits. (b) E_{input}. (c) Large $RC = 100\,T$ (500 μsec). (d) Small $RC = 0.01\,T$ (0.05 μsec). (e) Medium $RC = T$ (5 μsec).

to either narrow voltage spikes or a pulse with a sloping top. Narrow, sharp spikes of voltage, as shown in part (d), are produced from a short or small *RC* time constant. That is, the *RC* time is very much less than the time duration T of the input voltage pulse. An output having sloping or tilted tops and bottoms, as shown in Fig. 13-46*e*, results when the *RC* time constant is neither long nor short but is, instead, medium. That is, the *RC* time equals the time duration T of the input pulse.

The following discussions explain the action of the large, small, and medium *RC* time constants with the input voltage pulse of Fig. 13-46*b*.

Large RC Time Constant. The 100-V input pulse of Fig. 13-46*b* has a time duration of 5 μsec. This, for example, would be the time of a half cycle of a 100-kHz square wave. The time of one complete square-wave cycle would be the *reciprocal* of the frequency (equation 11-20), or: time = 1/frequency = 1/100 kHz = $1/100 \times 10^3 = 1/1 \times 10^5 = 1 \times 10^{-5}$ sec = 0.00001 sec = 10 μsec. Therefore, the time of a half cycle is 5 μsec.

Applying this 100-V pulse with a 5-μsec time duration into a series R and C causes C to start charging. Whether C could eharge up depends on the *RC* time constant compared to the 5-μsec time. If, as shown in the circuit of Fig. 13-47*b*, R = 200 kΩ and C = 0.0025 μF, then the *RC* time constant is: RC = (200 K) (0.0025 μF) = $(200 \times 10^3)(0.0025 \times 10^{-6} = (2 \times 10^5)(2.5 \times 10^{-9}) = 5 \times 10^{-4}$ sec = 0.0005 sec = 500 μsec. Since RC = 500 μsec and time T = 5 μsec, then this is called a long *RC* time constant.

With an RC = 500 μsec, the capacitor would not charge very much in only a time period of 5 μsec. When the input pulse of Fig. 13-47*a* rises to +100 V at instant t_1, E_R immediately rises from 0 to +100 V, as is shown in Fig. 13-47*b*, and C starts charging. At instant t_1, E_C is still zero.

At instant t_2, 5 μsec later, voltage across C becomes, as is shown in Section 13-9,

$$e_{C\,\text{charge}} = E_T\,[1 - \epsilon^{-(T/\tau)}] \qquad (13\text{-}18)$$

where T is the time, 5 μsec, and τ is the RC, 500 μsec.

$$e_C = 100\,(1 - \epsilon^{-[(5 \times 10^{-6})/(500 \times 10^{-6})]})$$

$$e_C = 100\,(1 - \epsilon^{-0.01})$$

From the table of Values and Common Logarithms of Exponential Functions in Appendix V, the value of $\epsilon^{-0.01}$ is 0.99 (approximate). Then

$$e_C = 100\,(1 - 0.99)$$

$$e_C = 100\,(0.01)$$

$$e_C = 1 \text{ V}$$

As shown in the circuit of Fig. 13-47*c*, E_C is now 1 V. E_{input} is still 100 V, and E_R is 99 V, having decreased from its previous value of 100 V, as C charged up 1 V.

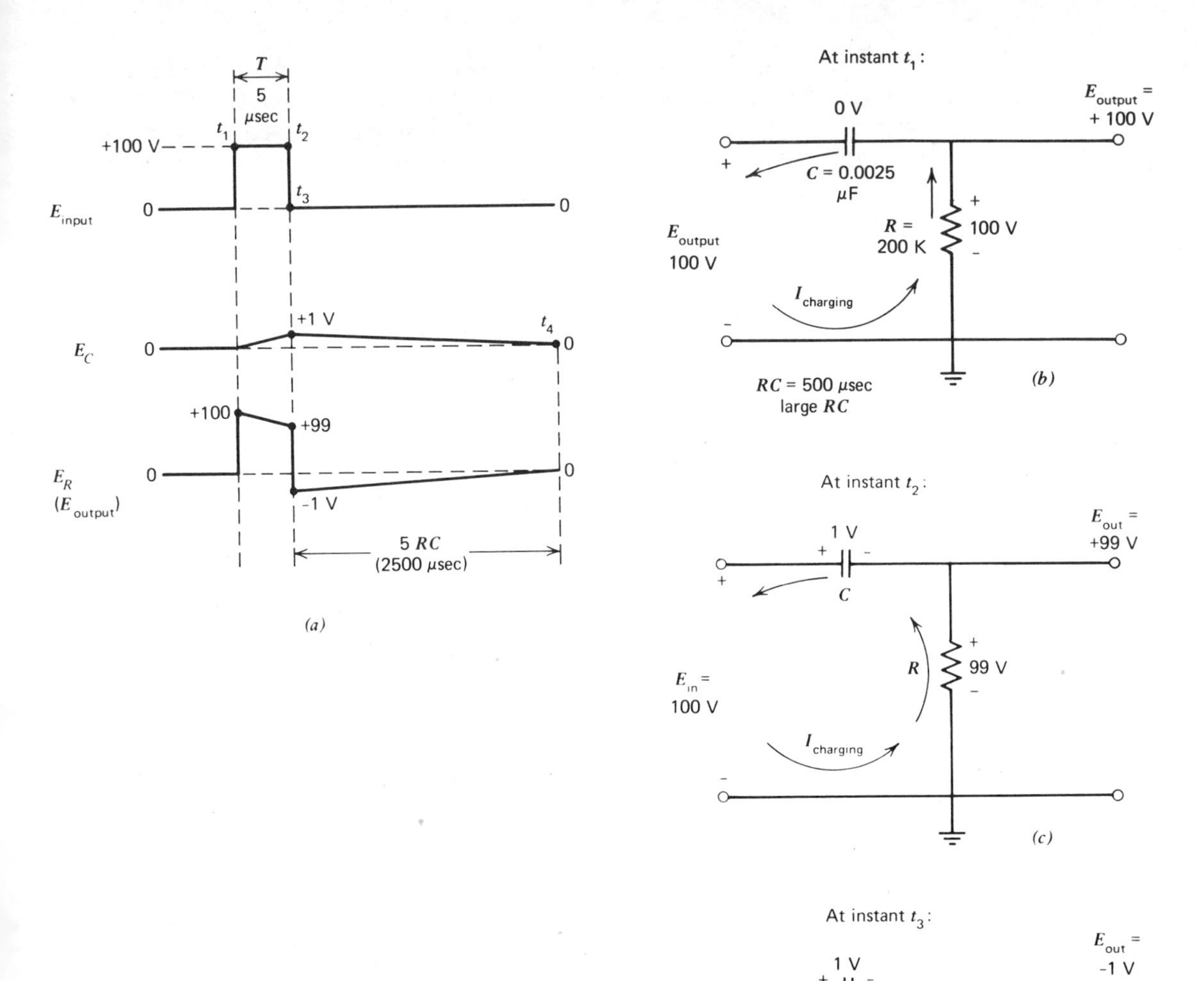

Figure 13-47. Voltage pulse and large RC (RC = 500 μsec) T = 5 μsec. (a) E_{input} and E_{output}. (b) At the start, E_{in} = 100 V. (c) 5 μsec after (b), E_{in} is still 100. (d) E_{input} abruptly falls to zero.

At instant t_3, E_{input} immediately drops to zero, as the pulse of Fig. 13-47*a* shows, and C starts discharging. This is shown in the circuit of Fig. 13-47*d*. Since E_{input} is zero, then the 1 V across C, causing a discharge current to flow down through R, acts as an applied voltage, producing 1 V across R. Note that the top of R is negative, and that E_R or E_{output} is -1 V.

It will take C about 5 RC time constants (5×500 μsec $= 2500$ μsec) to discharge from 1 V down to zero. At instant t_4 (2500 μsec after t_3), $E_C = 0$, and E_R, or E_{output}, $= 0$ V. Note that E_{output}, as shown in Fig. 13-47*a*, and also in Fig. 13-46*c*, is practically a duplicate of E_{input}. This is true for large RC time constants. When E_C is used as the output signal in a large time constant, the circuit is called an *integrator*.

Short RC Time Constant. If a small or short RC time constant is used, such as when $R = 5$ kΩ and $C = 10$ pF, as shown in the circuit of Fig. 13-48*b*, the RC time constant is: $RC = (5 \text{ k}\Omega)(10 \text{ pF}) = (5 \times 10^3)(10 \times 10^{-12}) = 50 \times 10^{-9}$ sec $= 0.05$ μsec, then E_{output} is not a duplicate of E_{input}. Since the RC time $= 0.05$ μsec, while time $T = 5$ μsec, then this called a short or small RC, or a *differentiator*.

At instant t_1, when E_{input} changes abruptly from 0 to $+100$ V, the capacitor starts charging. At this first instant, C has not charged yet, and E_R is 100 V. This is shown in the circuit of Fig. 13-48*b*, and also for E_R at t_1 of Fig. 13-48*a*.

At instant t_2, 5 μsec later, the capacitor voltage rises to, as shown in Section 13-9, equation 13-18,

$$e_{C\,\text{charge}} = E_T\,[1 - \epsilon^{-(T/\tau)}] \tag{13-18}$$

where $T = 5$ μsec, and $\tau = 0.05$ μsec.

$$e_C = 100\,(1 - \epsilon^{-(5/0.05)})$$

$$e_C = 100\,(1 - \epsilon^{-100})$$

from the Appendix V, table of Values and Common Logs of Exponential Functions, the value of ϵ^{-100} is 0 (approximate). Then

$$e_C = 100\,(1 - 0)$$

$$e_C = 100\,(1)$$

$$e_C = 100 \text{ V}$$

As shown in the circuit of Fig. 13-48*c*, $E_C = 100$ V. C has completely charged in the 5-μsec period of the input pulse, since C will charge up in five RC time constants (5×0.05 μsec $= 0.25$ μsec). With C completely charged, no further current flows, and $E_R = 0$. Note that in the E_R or E_{output} voltage wave shape of Fig. 13-48*a*, E_R rises from 0 to $+100$ V at instant t_1 and then quickly (in 5 RC time constants $= 0.25$ μsec) decreases back to zero, remaining at zero for the rest of the 5 μsec time period. E_{output} is, therefore, a narrow, sharp, positive-going 100-V spike.

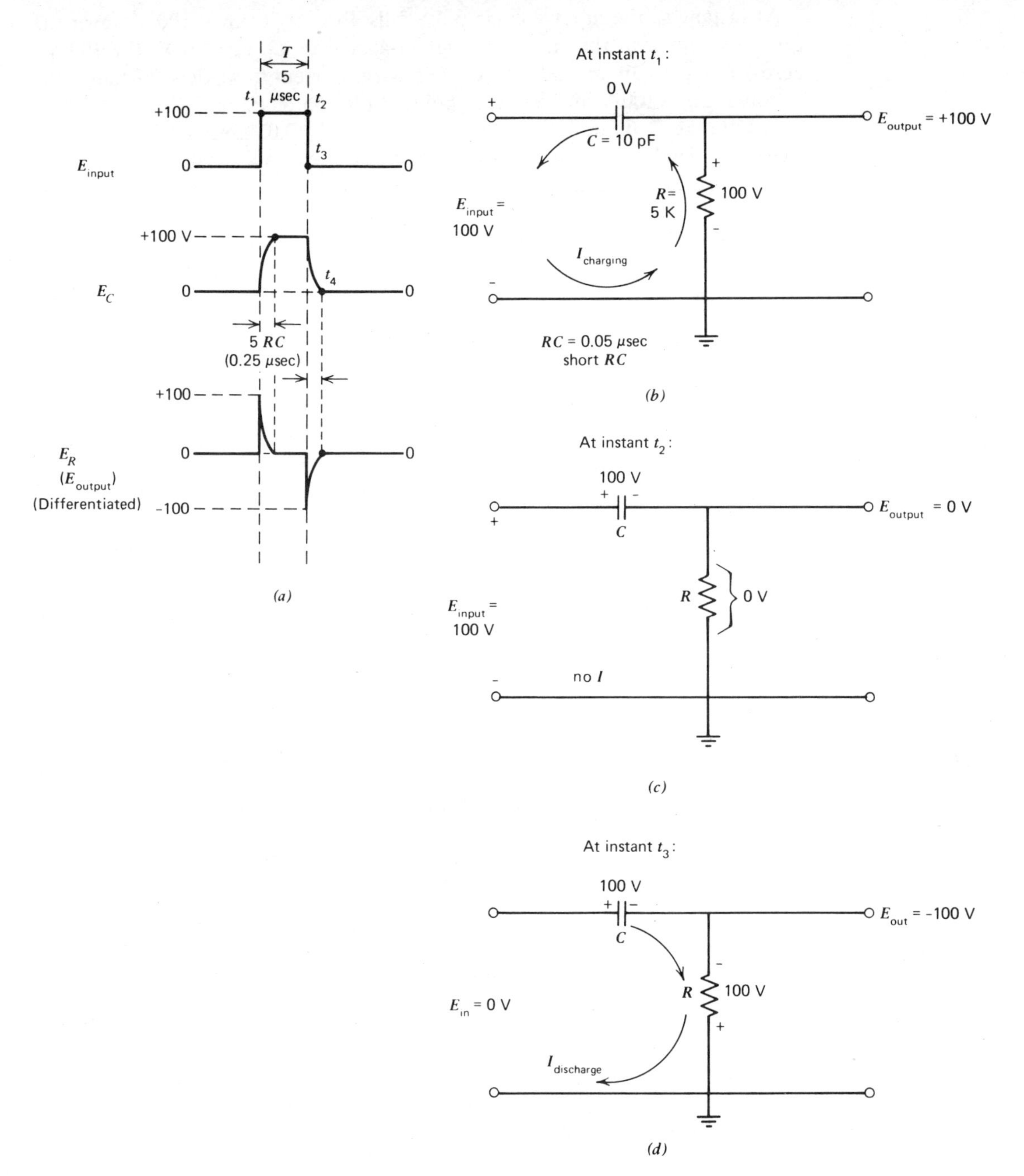

Figure 13-48. Voltage pulse and short RC (RC = 0.05 μsec; T = 5 μsec). (a) E_{input} and E_{output}. (b) At the start, E_{in} = 100 V. (c) 5 μsec after (b), E_{in} is still 100 V. (d) E_{input} abruptly falls to zero.

At instant t_3, the input voltage pulse falls abruptly from +100 V down to zero. As shown in the circuit of Fig. 13-48*d*, with $E_{\text{input}} = 0$ V, the 100 V across C acts as an applied voltage. Discharge current flows down through R, producing a −100 V at the output point at the top of R.

It will take C about five RC time constants ($5 \times 0.05\ \mu\text{sec} = 0.25\ \mu\text{sec}$) for C to completely discharge from its 100 V to zero. As a result, when E_C becomes zero, at instant t_4, E_R (or E_{output}) likewise decreases to zero. This is shown as the sharp, narrow, negative-going spike in the E_R wave shape of Fig. 13-48*a*, and also in Fig. 13-46*d*. The sharp positive and negative-going output voltage spikes taken across R, resulting from the square pulse input, are called *differentiated* waves.

Medium RC Time Constant. A medium time constant is neither much larger nor much smaller than the time duration T. In the circuit of Fig. 13-49*b*, $R = 50\ \text{k}\Omega$ and $C = 100$ pF. The RC time constant is: $RC = (50\text{k}\Omega)(100\ \text{pF}) = (50 \times 10^3)(100 \times 10^{-12}) = (5 \times 10^4)(1 \times 10^{-10}) = 5 \times 10^{-6}\ \text{sec} = 5\ \mu\text{sec}$. This RC time constant is equal to the time duration ($T = 5$ usec) of the input pulse of Fig. 13-49*a*. The RC could be called medium.

At instant t_1 E_{input} goes abruptly to 100 V. C cannot charge in this short instant, and E_C is still zero. E_R then rises immediately to 100 V, as shown in the circuit of Fig. 13-49*b*.

At instant t_2, E_{input} remains at +100 V for 5 μsec, and C charges. E_C rises to

$$e_{C\,\text{charge}} = E_T\,(1 - \epsilon^{-(T/\tau)}) \tag{13-18}$$

where T and τ are each 5 μsec.

$$e_C = 200\,(1 - \epsilon^{-[(5 \times 10^{-6})/(5 \times 10^{-6})]})$$

$$e_C = 100\,(1 - \epsilon^{-1})$$

from the table of Values and Common Logs of Exponential Functions of Appendix V, the value of ϵ^{-1} is 0.37 (approximate). Then

$$e_C = 100\,(1 - 0.37)$$

$$e_C = 100\,(0.63)$$

$$e_C = 63\ \text{V}$$

As shown in the circuit of Fig. 13-49*c*, $E_C = 63$ V. With 100 V E_{input}, E_R is then 37 V, as shown. Note in the E_R wave shape of Fig. 13-49*a*, E_R has tilted downward from +100 V to +37 V.

At instant t_3, E_{input} has abruptly changed from 100 V to zero. As shown in the circuit of Fig. 13-49*d*, the 63 V across C acts as if it were the applied voltage. Discharge current flows down through R, making $E_R = 63$ V with the top end negative. E_{output} is then −63 V.

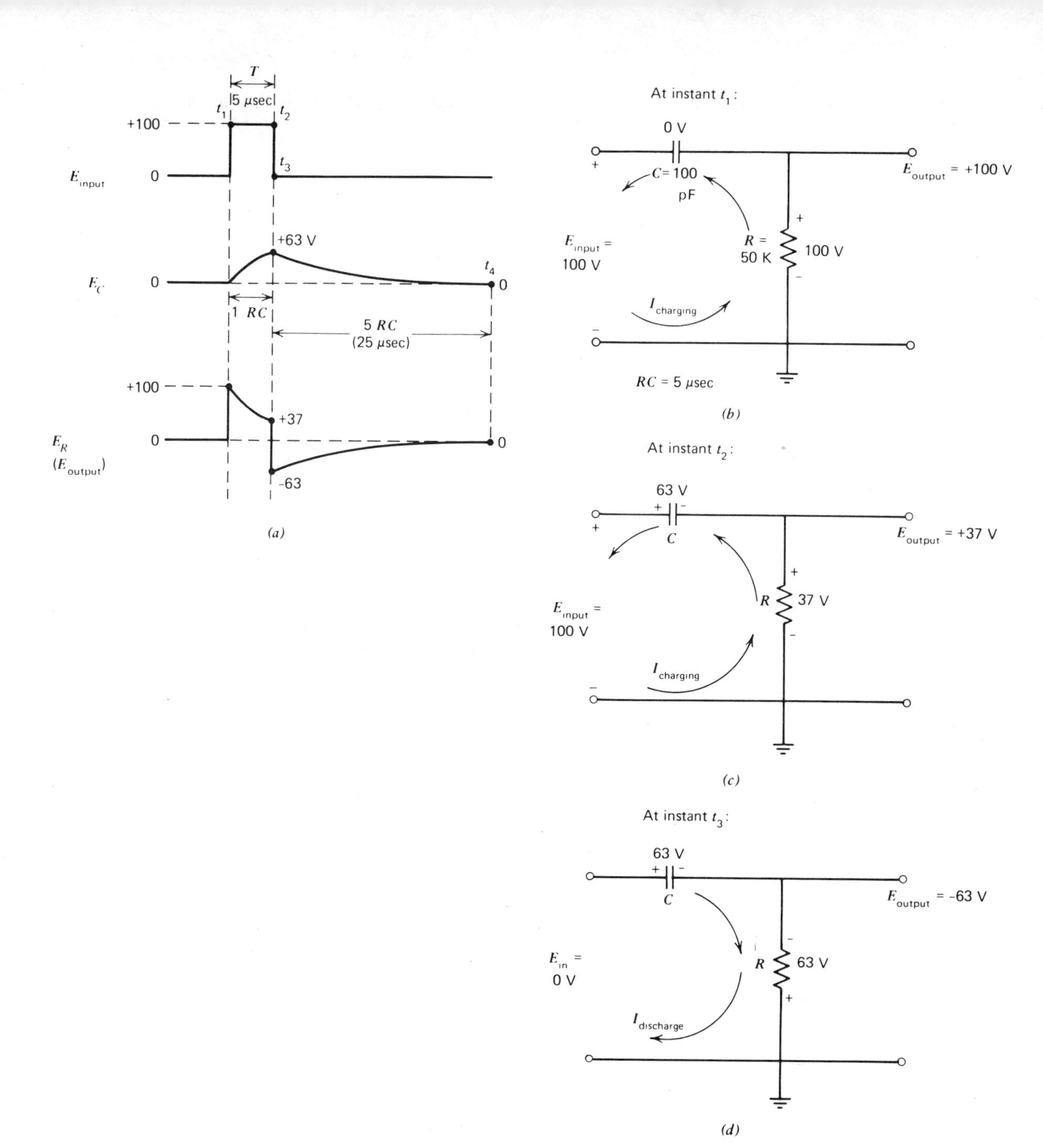

Figure 13-49. Voltage pulse and medium *RC* ($RC = T = 5$ μsec). (*a*) E_{input} and E_{output}. (*b*) At the start, $E_{input} = 100$ V. (*c*) 5 μsec after (*b*), E_{input} is still 100 V. (*d*) E_{input} abruptly falls to zero.

With E_{input} remaining at zero (after t_3), C will discharge in about five RC time constants, or (5) (5 μsec) = 25 μsec at instant f_4. E_C and E_R then each decrease to zero at f_4.

For Similar Problems Refer to End-of-Chapter Problems 13-104 to 13-106

PROBLEMS

See section 13-1, The Capacitor, for discussion covered by the following.

13-1. What parts make up a capacitor?
13-2. What is the dielectric?
13-3. Briefly describe the charging of a capacitor.

See section 13-1, Factors Affecting Capacitor Values, for discussion covered by the following.

13-4. What are the practical units of measurement of capacitors?
13-5. What three things determine the value of a capacitor?

See section 13-1, Area of Plates, for discussion covered by the following.

13-6. In a variable capacitor containing a rotor plate and a stator plate, in which position of the rotor, in mesh or out of mesh, would (a) minimum capacitance exist, and (b) maximum capacitance exist?

See section 13-1, Distance Between Plates, for discussion covered by the following.

13-7. In a variable capacitor where the spacing between the plates can be changed, when does maximum capacitance exist?
13-8. Briefly describe why the distance between plates affects the value of a capacitor.

See section 13-1, Dielectric Between Plates, for discussion covered by the following.

13-9. If a capacitor has air between its two plates and its value is 300 pF, what is its value if a sheet of mica is slipped between the plates without changing the spacing between the plates. (Use the dielectric constant of 5 shown in Table 13-1.)

13-10. Briefly describe how the dielectric material increases the value of the capacitor compared to an air or vacuum dielectric.

See section 13-1, Voltage Breakdown Rating of Dielectric, for discussion covered by the following.

13-11. What does the voltage rating of a capacitor mean?

13-12. What simple test could be used to check for a damaged capacitor dielectric?

See section 13-1, Types of Capacitors, Variable, Fixed, and Color Code, for discussion covered by the following.

13-13. What two methods are used to change the value of a variable capacitor?

What values in picofarads are indicated by the following three-color groupings:

13-14. Green, red, orange?

13-15. Yellow, brown, brown?

13-16. Violet, yellow, black?

13-17. Orange, black, red?

See section 13-1, Electrolytic Capacitors, for discussion covered by the following.

13-18. A ceramic capacitor is connected between the terminals of a dc voltage without regard to polarities. What could occur?

13-19. A single electrolytic capacitor is connected across a dc voltage with the + capacitor lead connected to the + voltage terminal. What could occur?

13-20. A single electrolytic capacitor is connected across a dc voltage with the − capacitor lead connected to the + voltage terminal. What could occur?

13-21. Describe the action of the ohmmeter dial pointer when the ohmmeter, operating on $R \times 100$ K range, is connected across a good electrolytic capacitor (uncharged).

13-22. Describe the action of the ohmmeter dial pointer when the ohmmeter, operating on $R \times 100$ K range, is connected across a shorted or leaky capacitor.

13-23. Describe the action of the ohmmeter dial pointer when the ohmmeter, operating on $R \times 100$ K range, is connected across a good small value ceramic capacitor (uncharged).

13-24. Same as previous question but for an open capacitor.
13-25. Same as previous question but for a good mica capacitor (uncharged) in parallel with a 5-kΩ resistor.

See section 13-2, Capacitors, Series and Parallel, for discussion covered by the following.

13-26. Why is the total capacitance larger when capacitors are connected in parallel?
13-27. A 350-pF capacitor is in parallel with a 150-pF capacitor. What is the total capacitance of this combination?
13-28. If a third capacitor, 200 pF, is placed in parallel with those of the previous problem, what total capacitance is now present?
13-29. A 0.01 μF capacitor is connected in parallel with a 4700-pF capacitor. What is the total capacitance in (a) picofarads, (b) microfarads?

See section 13-2, Capacitors in Series, for discussion covered by the following.

13-30. A 30-μF and a 6-μF capacitor are connected in series. What is the value of the total resultant capacitor?
13-31. Three capacitors are connected in series. If they are respectively: 10 μF, 8 μF, and 40 μF, what is the value of the total resultant capacitor?

See section 13-3, Quantity of Charge on a Capacitor, for discussion covered by the following.

13-32. Explain how the high voltage section of a television receiver may actually cause less of a shock than the low voltage section.
13-33. Find the quantity of electrons, in coulombs, stored on a small 200-pF capacitor charged to 10 kV.
13-34. A large 40-μF capacitor has been charged up to 150 V. What quantity of electrons, in coulombs, has been stored on this capacitor?

See section 13-3, Quantity of Charge on Capacitors in Series, for discussion covered by the following.

13-35. Why do capacitors connected in series have the same quantity of charge if they each start out uncharged, and then charge up?
13-36. Two capacitors, $C_6 = 100$ pF, $C_7 = 25$ pF, are connected in series with 10 V dc applied. Find: (a) C_T, (b) Q_T, (c) Q_6, (d) Q_7, (e) E_6, and (f) E_7, after the capacitors have charged.

13-37. If a third capacitor, $C_8 = 5$ pF, were placed in series with the other two of the previous problem, ($C_6 = 100$ pF and $C_7 = 25$ pF, with 10 V dc applied, and all capacitors started off from an uncharged state, find: (a) C_T, (b) Q_T, (c) Q_6, (d) Q_7, (e) Q_8, (f) E_6, (g) E_7, and (h) E_8, after the capacitors have charged.

See section 13-3, Quantity of Charge (Q) on Parallel Capacitors, for discussion covered by the following.

13-38. Explain how parallel capacitors may have a different quantity of electrons on each.
13-39. Explain how parallel capacitors may have an equal quantity of electrons on each.
13-40. Two parallel capacitors ($C_1 = 50$ μF and $C_2 = 10$ μF) are connected to an applied dc voltage source, $E_T = 12$ V. Find: (a) Q_1, (b) Q_2, and (c) Q_T.
13-41. If a third capacitor ($C_3 = 25$ μF) were connected in parallel with the two parallel capacitors ($C_1 = 50$ μF and $C_2 = 10$ μF) of the previous problem, find: (a) Q_3 and (b) new Q_T.

See section 13-3, Quantity (Q) of Charge on Series-Parallel Capacitors, for discussion covered by the following.

13-42. In the series-parallel capacitor circuit of Fig. 13-12*a*, if the values of the capacitors were: $C_1 = 30$ μF, $C_2 = 25$ μF, $C_3 = 35$ μF, and $C_4 = 20$ μF, with E_T still 60 V, find: (a) $C_{2,3}$, (b) C_T, (c) Q_T, (d) Q_1, (e) $Q_{2,3}$, (f) Q_4, (g) E_1, (h) $E_{2,3}$, (i) E_4, (j) Q_2, and (k) Q_3.
13-43. In the series-parallel capacitor circuit of Fig. 13-13, C_5 is in *series* with the *parallel combination* of C_6 and C_7, and also in *series* with the *parallel combination* of C_8 and C_9. Find the following: (a) $C_{6,7}$, (b) $C_{8,9}$, (c) C_T, (d) Q_T, (e) Q_5, (f) $Q_{6,7}$, (g) $Q_{8,9}$, (h) E_5, (i) $E_{6,7}$, (j) $E_{8,9}$, (k) Q_6, (l) Q_7, (m) Q_8, and (n) Q_9.

See section 13-4, RC Time Constant, Capacitor Charging, and Capacitor Discharging, for discussion covered by the following.

13-44. What is the period of time for one time constant if a 250-pF capacitor is connected in series with a 20-kΩ resistor?
13-45. What is the period of time of three time constants if one time constant is 5 μsec?
13-46. If 30 V dc is applied to the circuit of Problem 13-44, find: (a) E_C at the first instant, (b) E_R at the first instant, and (c) I at the first instant.

13-47. If 30 V dc is applied to the circuit of Problem 13-44, find: (a) E_C at the end of two time constants, (b) E_R at the end of two time constants, and (c) I at this same instant.

13-48. A 0.005 μF capacitor has been charged to 200 V. If it is permitted to discharge through a 50 -kΩ resistor, find: (a) E_C, (b) E_R, and (c) I, at the instant the capacitor starts discharging.

13-49. In the previous problem, find: (a) E_C one time constant after the capacitor started its discharge, (b) E_R at this same instant, and (c) I at this same instant.

13-50. In Problem 13-48, find: (a) E_C two time constants after the capacitor started its discharge, (b) E_R at this same instant, and (c) I at this same instant.

13-51. If 0.5 mA of discharge current flows through a 2 MΩ resistor path, find the voltage across the capacitor, E_C, at this instant.

See section 13-5, Capacitor with AC Applied Voltage, for discussion covered by the following.

13-52. When a dc voltage is applied to a capacitor, allowing the capacitor to completely charge, what happens to the current?

13-53. How does a capacitor block direct current?

13-54. Why does not a capacitor block alternating current?

See section 13-5, Capacitive Reactance, for discussion covered by the following.

13-55. What is the name of the opposition that a capacitor offers to alternating current?

13-56. How much capacitive reactance (X_C) is offered by a 0.01 μF capacitor to an alternating current where the frequency is 5000 kHz?

13-57. An ac voltage at a frequency of 6 MHz is applied to a circuit containing a 0.00025-μF capacitor. Find X_C.

13-58. A capacitor should have an X_C of 100 Ω to an ac voltage the frequency of which is 200 Hz. Find the value of the capacitor.

13-59. What value capacitor is required if X_C must be 5 Ω and the frequency of the ac voltage is 20 Kilohertz?

See section 13-6, Phase Relationships, for discussion covered by the following.

13-60. Check the values of E_T in Fig. 13-18 at instants J, K, L, M, N, and O by adding the values of E_C and E_R at those instants.

See section 13-6, E_R and E_{X_C} Addition of Phasors, for discussion covered by the following.

13-61. Find the total applied ac voltage for a capacitor in series with a resistor if E_{X_C} is 250 V and E_R is 150 V.
13-62. If E_R is 60 V and E_{X_C} is 40 V where a resistor and a capacitor are connected in series, find the total applied ac voltage.
13-63. A resistor and a capacitor are connected in series. If E_T the ac applied voltage is 200 V and E_R is 50 V, find E_{X_C}.
13-64. Find E_R where a resistor and capacitor are connected in series to a total applied ac voltage of 30 V and where E_{X_C} is 10 V.

See section 13-6, Impedance (Z) or Total Opposition, for discussion covered by the following.

13-65. A 300-Ω resistor is connected in series with a 400-Ω capacitive reactance in an ac circuit. Find the total opposition or impedance, Z.
13-66. A 200-Ω resistor is connected in series with a 500-Ω capacitive reactance, with an applied 30 V ac. Find: (a) Z, (b) I, (c) E_R, and (d) E_{X_C}.
13-67. An ac voltage of 50 V at a frequency of 1 MHz is applied to a 2-kΩ resistor in series with a 150-pF capacitor. Find (a) X_C, (b) Z, (c) I, (d) E_R and (e) E_{X_C}.

See section 13-6, Resistors and Capacitors in AC Series Circuit, for discussion covered by the following.

13-68. Two resistors (R_1 = 20 Ω, and R_2 = 10 Ω) are connected in series with two capacitors (X_{C_1} = 25 Ω, and X_{C_2} = 15 Ω). If the applied ac voltage is 10 V, find: (a) R_T, (b) X_{C_T}, (c) Z_T, (d) I, (e) E_{R_1}, (f) E_{R_2}, (g) $E_{X_{C_1}}$, (h) $E_{X_{C_2}}$, and (i) vector sum of all resistor and capacitor voltages.
13-69. In the diagram for this problem, find: (a) X_{C_3}, (b) X_{C_4}, (c) C_T, (d) X_{C_T}, (e) R_T, (f) Z_T, (g) I, (h) E_{R_3}, (i) E_{R_4}, (j) $E_{X_{C_3}}$, (k) $E_{X_{C_4}}$, and (l) vector sum of all capacitor and resistor voltages.

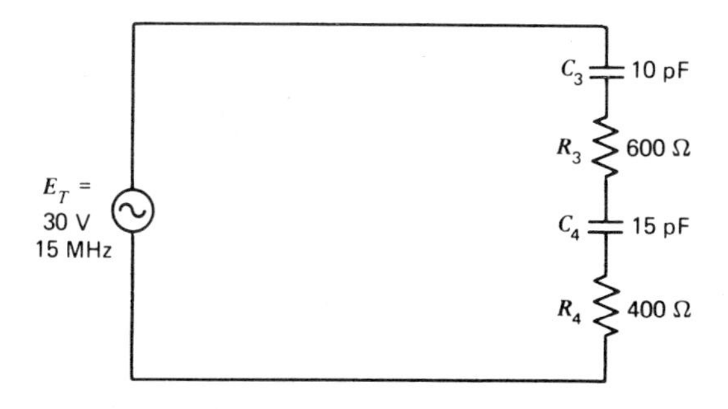

Problem 13-69.

See section 13-6, Power in Resistor-Capacitor Circuits, for discussion covered by the following.

13-70. If 30 V ac is applied to a 1 kΩ resistor in series with an X_C of 1770 Ω, with a current of 14.8 mA, find (a) P_{apparent}, and (b) P_{true}.
13-71. In a series ac circuit, an applied voltage of 10 V causes a current of 0.2 A to flow through a resistor of 10 Ω, another resistor of 20 Ω, an X_C of 5 Ω, and another X_C of 35 Ω. Find (a) P_{apparent}, and (b) P_{true}.

See section 13-6, Power Factor, for discussion covered by the following.

13-72. Find the power factor of a circuit if the true power is 150 W, while the apparent power is 300 W.
13-73. Find the power factor of a circuit where an applied alternating voltage of 120 V causes a current of 300 mA, and where the true power read by a wattmeter is 10.8 W.
13-74. The apparent power of a circuit is 50 W. If the power factor is 0.6, find the true power.
13-75. An ac circuit has an $E_T = 75$ V and an $I_T = 400$ mA. If the power factor is 0.8, find: (a) P_{apparent}, and (b) P_{true}.

See section 13-7, Capacitor and Resistor in Parallel, and Impedance, for discussion covered by the following.

13-76. An E_T of 60 V is applied to a 2-kΩ resistor in parallel with a 5-kΩ X_C. Find: (a) Z_T, using the product divided by the sum method, (b) I_R, (c) I_{X_C}, (d) I_T, by vectorially adding I_R and I_{X_C}, and (e) Z_T, using Ohm's law.

See section 13-7, Capacitors and Resistors in Parallel, for discussion covered by the following.

13-77. Refer to the four-branch parallel circuit of Fig. 13-25*a*, but change all values to the following: $E_T = 60$ V, $R_1 = 30$ K, $R_2 = 6$ K, $X_{C_1} = 10$ K, $X_{C_2} = 15$ K. Find: (a) R_T, (b) X_{C_T}, and (c) Z_T. (This should agree with Z_T in next problem.)
13-78. In the circuit of the previous problem, find: (a) I_{R_1}, (b) I_{R_2}, (c) I_{R_T}, (d) $I_{X_{C1}}$, (e) $I_{X_{C2}}$, (f) $I_{X_{CT}}$, (g) I_T, and (h) Z_T. (This should agree with Z_T of previous problem.)

See section 13-8, Uncharged Series Capacitors with DC Applied Voltage, for discussion covered by the following.

13-79. Two capacitors are connected in series with an applied direct voltage of 18 V. If the capacitors are 30 μF and 6 μF, find: (a) C_T, (b) E across the larger capacitor, and (c) E across the smaller capacitor.
13-80. Three capacitors, 6, 12, and 4 μF, are connected in series with an applied dc voltage of 36 V. Find: (a) C_T, (b) E across the smallest capacitor, (c) E across the largest capacitor, and (d) E across the 6 μF capacitor.

See section 13-8, Previously Charged Series Capacitors with DC Applied Voltage, for discussion covered by the following.

13-81. Refer to Fig. 13-30*a*, but change all values to the following: E_T = 145 V, C_1 = 60 μF, and C_2 = 40 μF. Capacitor C_1 has been charged previously to 25 V with the polarity shown on C_1 in Fig. 13-30*a*, that is, the lower plate is the negative plate. When the switch is closed, find: (a) C_T, (b) net voltage, ΔE, which causes a current flow when the switch is first closed, (c) change of voltage, Δe_{C_1}, that C_1 undergoes, (d) final voltage across C_1, E_{C_1}, (e) change of voltage, Δe_{C_2}, that C_2 undergoes, and (f) final voltage across C_2, E_{C_2}.
13-82. Refer to Fig. 13-31*a* but change all values to the following: E_T = 90 V, C_3 = 5 μF, and C_4 = 20 μF. Both capacitors have been previously charged with the polarities shown in Fig. 13-31*a*, that is, e_{C_3} = 40 V with its *lower* plate *negative*, while e_{C_4} = 30 V with its *lower* plate *positive*. When the switch is closed, find: (a) C_T, (b) net voltage, ΔE, which causes a current flow when the switch is first closed, (c) change of voltage, Δe_{C_3}, that C_3 undergoes, (d) final voltage and polarity across C_3, E_{C_3}, (e) change of voltage, Δe_{C_4}, that C_4 undergoes, and (f) final voltage and polarity across C_4, E_{C_4}.

See section 13-8, Charged Capacitors Connected to Others, for discussion covered by the following.

13-83. Redraw Fig. 13-34*a* keeping the values of C_1 and C_2 the same as shown, but exchange the starting voltages, where e_{C_1} = 0 and e_{C_2} = 60 V, negative at the lower plate. When the switch is closed and remains closed as shown in parts (b) and (c) of the diagram, find: (a) C_{total}, (b) ΔE, the voltage causing a current flow, (c) direction of this current, (d) change of C_1 voltage, Δe_{C_1}, (e) change of C_2 voltage, Δe_{C_2}, (f) final voltage on C_1, E_{C_1}, and (g) final voltage on C_2, E_{C_2}.
13-84. Redraw Fig. 13-35*a* keeping the values of C_3 and C_4 the same as shown, but exchange the starting voltages on each capacitor. Capacitor C_3 has been previously charged to 120 V, negative at the lower plate, while C_4 has

been previously charged to 40 V, also negative at the lower plate. When the switch is closed and remains closed as in Fig. 13-35*b* and *c*, find: (a) C_{total}, (b) ΔE, the voltage causing a current flow, (c) direction of this current, (d) change of C_3 voltage Δe_{C_3}, (e) change of C_4 voltage, Δe_{C_4}, (f) final voltage on C_3, E_{C_3}, and (g) final voltage on C_4, E_{C_4}.

13-85. Redraw Fig. 13-36*a* keeping the values of C_5 and C_6 the same as shown, but exchange the starting voltages on each capacitor. Make $e_{C_5} = 20$ V, negative at the upper plate, while $e_{C_6} = 10$ V, negative at the lower plate. When the switch is closed and remains closed as in Fig. 13-36*b* and *c*, find: (a) C_{total}, (b) ΔE, the voltage producing a current flow, (c) direction of this current, (d) change of C_5 voltage, Δe_{C_5}, (e) change of C_6 voltage, Δe_{C_6}, (f) final voltage and polarity on C_5, E_{C_5}, (g) final voltage and polarity on C_6, E_{C_6}.

13-86. Refer to the diagram shown for this problem. When the switch is closed and remains closed, find: (a) C_{total}, (b) ΔE, the voltage producing a current, (c) direction of this current, (d) change of C_7 voltage, Δe_{C_7}, (e) change of C_8 voltage, Δe_{C_8}, (f) change of C_9 voltage, Δe_{C_9}, (g) final voltage and polarity on C_7, E_{C_7}, (h) final voltage and polarity on C_8, E_{C_8}, and (i) final voltage and polarity on C_9, E_{C_9}.

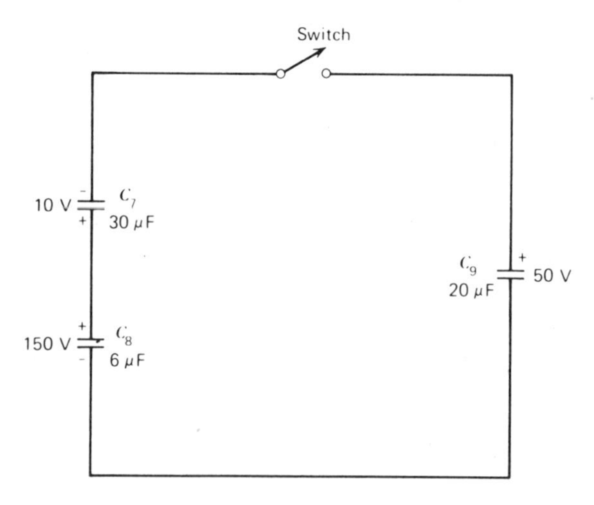

Problem 13-86. (All capacitors have been previously charged.)

See section 13-9, Capacitor Charging, for discussion covered by the following.

13-87. In the series *RC* circuit of Fig. 13-38*a*, if $R = 0.02$ MΩ, $C = 0.05$ μF, and $E_T = 60$ V, find: (a) time constant, (b) e_C, 2.5 msec after the switch is closed, (d) e_R at this same instant, and (d) i, at this instant.

13-88. 120 V dc is applied through a switch to a 500-pF capacitor in series with a 250-kΩ resistor. Find: (a) time constant, (c) e_C, 25 μsec after the switch is closed, (c) e_R at this same instant, and (d) i, at this instant.

See section 13-9, Time Between Capacitor or Resistor Voltages, and also, Time Constant Required, for discussion covered by the following.

13-89. An applied 60 V dc is applied to a 0.05 μF capacitor in series with a 0.02 MΩ resistor. Find the time required for the capacitor voltage to rise from 0.02 to 55.03 V.

13-90. Find the time constant of a series RC that will cause a capacitor voltage to rise from 0.3 to 22 V in 25 μsec, where the applied dc is 120 V.

See section 13-9, Capacitor Discharging, for discussion covered by the following.

13-91. A 0.2-μF capacitor has been charged to 15 V. If it is allowed to discharge for 2 msec through a 25-kΩ resistor, find: (a) time constant τ, (b) capacitor voltage e_C after 2 msec, (c) resistor voltage e_R at this instant, and (d) current i at this same instant.

See section 13-10 for discussion covered by the following.

13-92. A 150-Ω resistor is connected in series with a 0.025-μF capacitor. At the frequency of the applied 50 V ac, X_C is 300 Ω. What is the rectangular form of the impedance Z?

13-93. What is the phase angle θ in the previous problem?

13-94. What is the polar form of Z in Problem 13-92?

See section 13-11 for discussion covered by the following.

13-95. An ac voltage of 50 V is applied across a two-branch parallel circuit consisting of Z_3 and Z_4. Z_3 is made up of $X_{C_3} = 400\ \Omega$ in series with $R_3 = 300\ \Omega$. Z_4 consists of $X_{C_4} = 600\ \Omega$ in series with $R_4 = 200\ \Omega$. Find: (a) Z_3 in its rectangular form, (b) Z_3 in its polar form, (c) I_3 in its polar form, (d) Z_4 in its rectangular form, (e) Z_4 in its polar form, (f) I_4 in its polar form, (g) Z_T in its polar form, and (h) I_T in its polar form.

13-96. Z_5 consists of $X_{C_5} = 30\ \Omega$ in series with $R_5 = 10\ \Omega$. This combination is in parallel with Z_6, which contains $X_{C_6} = 40\ \Omega$ in series with $R_6 = 20\ \Omega$. An applied 50 V ac is connected across the two parallel branches Z_5 and Z_6. Find the values of the following in their polar forms: (a) Z_5, (b) Z_6, (c) Z_T, and (d) I_T.

See section 13-12 for discussion covered by the following.

13-97. Example 13-55 consists of 30 V E_T applied to an $R = 25$ kΩ in series with an $X_C = 40$ kΩ. Information found in this example is: $Z_T = 47.2$ kΩ, phase angle $\theta = 58$ °, $E_R = 15.9$ V, and $I = 636$ μA. Find: (a) apparent power, (b) power factor, (c) true power using answers to parts (a) and (b), and (d) true power using voltage across the resistor.

13-98. An ac voltage $E_T = 50$ V is applied across Z_3 in parallel with Z_4. Z_3 consists of $R_3 = 300$ Ω in series with $X_{C_3} = 400$ Ω. Z_4 consists of $R_4 = 200$ Ω in series with $X_{C_4} = 600$ Ω. If $I_T = 0.176$ A, $I_3 = 0.1$ A, $I_4 = 0.079$ A and $\theta_T = 61°$, find: (a) total apparent power, (b) power factor, (c) total true power by using the answers of parts (a) and (b), (d) true power in Z_3, (e) true power in Z_4, and (f) true power by adding the answers of parts (d) and (e). Note that the answers to parts (c) and (f) should agree.

See section 13-12, Energy Stored on a Capacitor, for discussion covered by the following.

13-99. Find the energy in joules stored on a 0.5-μF capacitor that is charged to 600 V.

See section 13-12, Capacitor Electrostatic Field, for discussion covered by the following.

13-100. Find the force in newtons acting between two 20-μC charges separated by 10 cm of air.

See section 13-13, Capacitor Value, for discussion covered by the following.

13-101. What is the capacitance value of a flat, round ceramic disk made of barium-strontium-titanate which is silver plated on the upper and lower flat surfaces if the disk is 2 cm in diameter and 1 mm thick. [*Note*: from Table 13-1, the dielectric constant K of barium-strontium-titanate is 7500. The area of a circle is: (π) (diam/2)^2.]

See section 13-14, Capacitor Current, for discussion covered by the following.

13-102. If a constant 4 A flows for 3 msec into a 50 μF capacitor, find: (a) the quantity of electrons (in coulombs) on the capacitor, and (b) voltage across the capacitor.

See section 13-14, E_C Lags I in Capacitive AC Circuit, for discussion covered by the following.

13-103. Refer to the E_C wave shape of Fig. 13-45, but change the voltages to 10 and 15 V (instead of the 100 and 150 V shown. The capacitor is changed to 0.2 μF. Also double all time periods so that dt_1, dt_2, and dt_3 are each 100 μsec, while dt_4 is 20 μsec. Find: (a) i_{C_1} during dt_1, (b) i_{C_2} during dt_2, (c) i_{C_3} during dt_3, and (d) i_{C_4} during dt_4.

See section 13-15 for discussions covered by the following.

13-104. A square-shaped pulse of voltage is 50 μsec wide, and rises from 0 to +20 V, similar to the shape of the input pulse of Fig. 13-47*a*. If the pulse is applied across a series *RC* circuit, where $R = 100$ kΩ and $C = 0.005$ μF, find: (a) E_C at instant t_1, (b) E_C at instant t_2, (c) E_C at instant t_3, (d) E_R at t_1, (e) E_R at t_2, and (f) E_R at t_3. (E_C and E_R will be similar in shape to those of Fig. 13-47*a*).

13-105. If the 50 μsec, +20-V pulse of the previous problem were applied across a series *RC* circuit, where $R = 25$ kΩ and $C = 200$ pF, find: (a) E_C at instant t_1, (b) E_C at t_2, (c) E_C at t_3, (d) E_R at t_1, (e) E_R at t_2, and (f) E_R at t_3. (E_C and E_R will be similar in shape to those of Fig. 13-48*a*).

13-106. If the 50 μsec, +20-V pulse of Problem 13-104 were applied across a series *RC* circuit, where $R = 10$ kΩ and $C = 0.0025$ μF, find: (a) E_C at instant t_1, (b) E_C at t_2, (c) E_C at t_3, (d) E_R at t_1, (e) E_R at t_2, and (f) E_R at t_3. (E_C and E_R will be similar in shape to those of Fig. 13-49*a*).

chapter

14

Inductance, Capacitance, and Resistance in AC Circuits

Inductors and resistors in ac circuits are discussed in Chapter 12, and then capacitors and resistors in ac circuits are considered in Chapter 13. In this chapter, circuits containing all three, inductance, capacitance, and resistance are discussed.

14-1. *L*, C, and *R* in Series Circuits

In Figure 14-1*a*, a series ac circuit is shown consisting of an $X_L = 500\ \Omega$, an $R = 400\ \Omega$, and an $X_C = 200\ \Omega$. The voltage phasor diagram is shown in Fig. 14-1*b*, and is explained in the following step by step. (The number of each step corresponds to the number shown on each phasor.) Note the similarity to previous phasor diagrams of Figures 12-6 and 13-19, and 13-21.

1. Since the same current I flows through each part of a series circuit, then the I phasor is common to X_L, R, and X_C and this phasor is shown horizontally to the right, or at zero degrees (reference).

2. Voltage across the resistor E_R is always in phase with the current through it, as shown in Fig. 14-1*b*.

3. Voltage across a capacitive reactance E_{X_C} always *lags* 90° behind the current through it. Since phasors rotate counterclockwise, then the E_{X_C} phasor

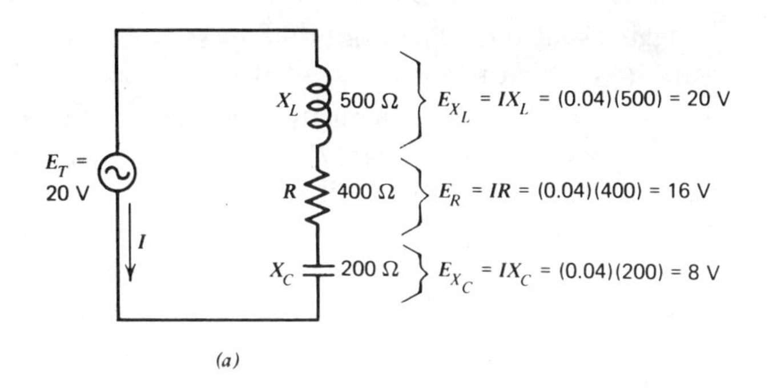

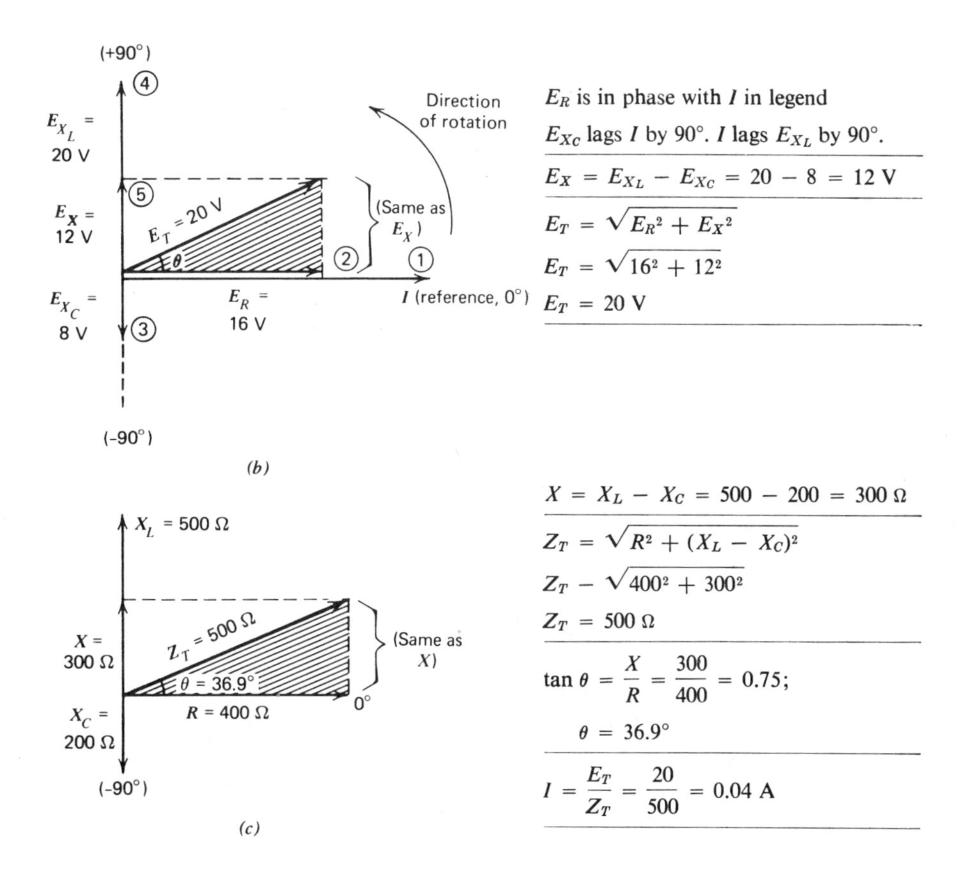

E_R is in phase with I in legend

E_{X_C} lags I by 90°. I lags E_{X_L} by 90°.

$$E_X = E_{X_L} - E_{X_C} = 20 - 8 = 12 \text{ V}$$

$$E_T = \sqrt{E_R^2 + E_X^2}$$

$$E_T = \sqrt{16^2 + 12^2}$$

$$E_T = 20 \text{ V}$$

$$X = X_L - X_C = 500 - 200 = 300 \ \Omega$$

$$Z_T = \sqrt{R^2 + (X_L - X_C)^2}$$

$$Z_T - \sqrt{400^2 + 300^2}$$

$$Z_T = 500 \ \Omega$$

$$\tan\theta = \frac{X}{R} = \frac{300}{400} = 0.75;$$

$$\theta = 36.9^\circ$$

$$I = \frac{E_T}{Z_T} = \frac{20}{500} = 0.04 \text{ A}$$

Figure 14-1. L, C, R in series. (*a*) Circuit. (*b*) Voltage phasors. E_R is in phase with I. E_{X_C} lags by 90°. I lags E_{X_L} by 90°. (c) Opposition (R, X, Z) phasors.

is lagging behind the I phasor by 90° by showing E_{X_C} pointing vertically downward. (Note that this is called $-90°$).

4. Voltage across the inductive reactance E_{X_L} always leads the current or, as it is usually stated, *I lags* E_{X_L}. E_{X_L} is shown pointing vertically upward. (Note that this is called $+90°$.)

5. E_{X_C} (phasor 3) and E_{X_L} (phasor 4) are pointing in opposite directions. This is called *180° out of phase*. The sum of these two phasors is actually their arithmetic difference, or $E_{X_L} - E_{X_C}$ is E_X, where X is the *net reactance*. E_X will be in the same direction as the larger reactive voltage, either E_{X_L} or E_{X_C}. In this drawing, E_{X_L} is larger than E_{X_C}. Therefore, E_X is shown pointing vertically upward (the same as the larger E_{X_L}). *E_X is simply the smaller one subtracted from the larger one.*

6. The total voltage E_T phasor, as shown in Fig. 14-1*b* is the hypotenuse (longest side) of the right triangle (the shaded area). From the Pythagoras theorem, the hypotenuse $E_T = \sqrt{E_R^2 + E_X^2}$ (from equations 12-4 and 13-9). The angle between the E_T and I phasors is called the *phase angle*, or θ (Greek letter, theta).

The opposition (R, X_L, X_C, and Z) phasors shown in Fig. 14-1*c* are similar to the voltage phasors of Fig. 14-1*b*. Note that the phasor, R, like the E_R phasor, points horizontally (0°). The X_L phasor, like the E_{X_L} one, points vertically upward ($+90°$), while the X_C phasor, like E_{X_C}, points vertically downward ($-90°$). Impedance Z_T is the phasor sum of R and X (the net reactance or difference between X_L and X_C), and is the hypotenuse of the shaded area right triangle. $Z = \sqrt{R^2 + X^2}$ (from equations 12-5 and 13-12).

In the circuit of Fig. 14-1*a*, the net reactance $X = X_L - X_C = 500 - 200$, or $X = 300\ \Omega$. *X is simply the smaller one subtracted from the larger one.* Total impedance Z_T is: $Z_T = \sqrt{R^2 + (X_L - X_C)^2}$ (equation 14-1) $Z_T = \sqrt{400^2 + (500 - 200)^2} = 500\ \Omega$.

Current, $I = E_T/Z_T = 20/500 = 0.04$ A. Voltage across each component of opposition is: $E_R = IR = (0.04)(400) = 16$ V, as shown in Fig. 14-1*a* and*b*. $E_{X_C} = IX_C = (0.04)(200) = 8$ V as shown. $E_{X_L} = IX_L = (0.04)(500) = 20$ V, as shown.

Note that the sum of these voltages ($E_R = 16$ V, $E_{X_C} = 8$ V, and $E_{X_L} = 20$ V) would exceed the total applied voltage, $E_T = 20$ V, if added arithmetically. However, as shown in the phasors of Fig. 14-1*b*, they are added *vectorially* (the phasor sum): $E_T = \sqrt{E_R^2 + (E_{X_L} - E_{X_C})^2}$ (equation 14-2); $E_T = \sqrt{16^2 + (20 - 8)^2} = 20$ V.

Phase angle θ may be found by using the tangent function as shown in Fig. 14-1*c*, $\theta = 36.9°$. The following example further illustrates the previous discussion.

Example 14-1

An ac voltage is applied to a series circuit consisting of $X_L = 90\ \Omega$, $R = 50\ \Omega$, and $X_C = 30\ \Omega$. If current $I = 0.05$ A, find: (a) Z_T, (b) E_{X_L}, (c) E_R, (d) E_{X_C}, (e) E_T, and (f) θ.

Solution

Solve this example by using the method shown previously. As a guide the following results are given: $Z_T = 78.1\ \Omega$, $E_{X_L} = 4.5$ V, $E_R = 2.5$ V, $E_{X_C} = 1.5$ V, $E_T = 3.91$ V, and $\theta = 50.2°$.

Power. The dissipation of power in the circuit of Fig. 14-1*a* occurs only in the resistor. X_L and X_C do not produce power dissipation. The *actual* or *true power* may then be found by using either: I^2R (equation 2-9), or $E_R\ I$ (equation 12-6), or E_R^2/R. A wattmeter reads the true power.

In the Fig. 14-1*a* circuit then, true power then is: *Power true* $= I^2\ R = (0.04)^2\ (400) = (4 \times 10^{-2})^2\ (4 \times 10^2) = (16 \times 10^{-4})\ (4 \times 10^2) = 64 \times 10^{-2}$ W, or 0.64 W.

A second method is: *power true* $= E_R\ I = (16)\ (0.04) = 0.64$ W. The third alternative is: *power true* $= E_R^2/R = 16^2/400 = 256/400 = 0.64$ W.

The apparent power which is not the actual or true power in a reactive circuit is: *power apparent* $= E_T\ I = (20)\ (0.04) = 0.8$ W or *volt-ampere*. *True* power, from eq. 12-8, $= (P_{\text{apparent}})\ (\text{PF}) = (0.8)\ (\cos\theta) = (0.8)\ (\cos 36.9°) = (0.8)\ (0.8) = 0.64$ W.

For Similar Problems Refer to End-of-Chapter Problems 14-1 to 14-3

14-2. *L*, *C*, and *R* in Parallel Circuits

Figure 14-2*a* shows a three-branch parallel circuit consisting of an $X_L = 60\ \Omega$, an $X_C = 40\ \Omega$, and an $R = 30\ \Omega$, with a total voltage $E_T = 120$ V applied across each. The current in each branch, and the total current are out of phase with each other. This may be seen in the phasor diagram of Fig. 14-2*b* and from the following step-by-step numbered explanation. (Each step number corresponds to the number shown on each phasor.) Note the similarity to the previous current phasors of Figs. 12-10 and 13-24.

1. The total voltage E_T is applied across *each* branch. Since E_T is common to all, it is used as the reference phasor, E_T phasor in Fig. 14-2*b* is shown pointing horizontally to the right at zero degrees.
2. Current through a resistor I_R is always in phase with voltage across the resistor. Phasor I_R is then shown pointing horizontally to the right, or also at zero degrees.

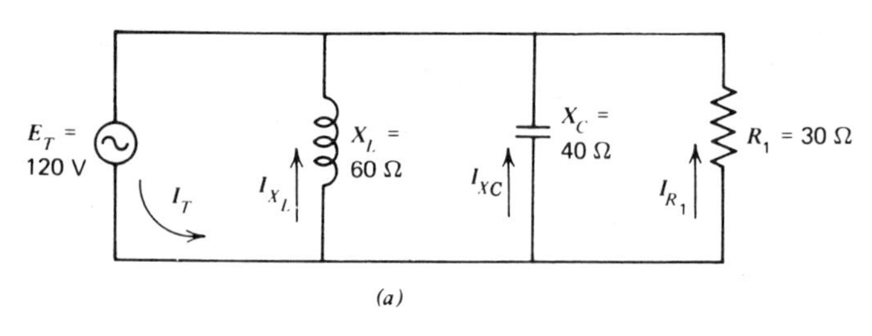

(a)

$$I_{X_L} = \frac{E_T}{X_L} = \frac{120}{60} = 2 \text{ A}$$

$$I_{X_C} = \frac{E_T}{X_C} = \frac{120}{40} = 3 \text{ A}$$

$$I_{R_1} = \frac{E_T}{R_1} = \frac{120}{30} = 4 \text{ A}$$

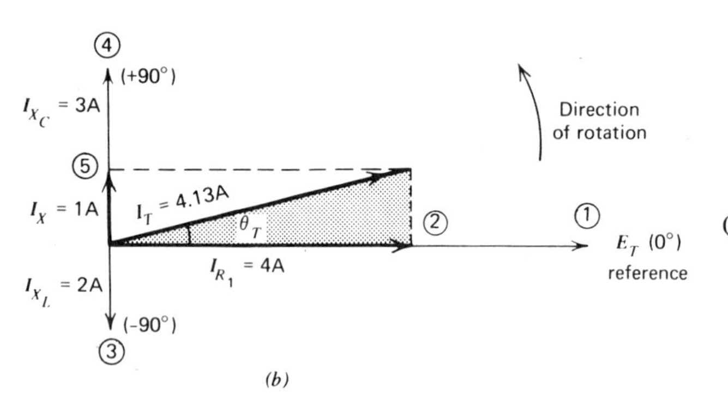

(b)

I_{R_1} is in phase with E_T

I_{X_L} lags E_T by 90°

E_T lags I_{XC} by 90°

$$I_X = I_{XC} - I_{X_L} = 3 - 2 = 1 \text{ A}$$

$$I_T = \sqrt{I_{R_1}{}^2 + (I_{XC} - I_{X_L})^2}$$

$$I_T = \sqrt{4^2 + 1^2}$$

$$I_T = 4.13 \text{ A}$$

$$\tan \theta_T = \frac{I_X}{I_{R_1}} = \frac{1}{4} = 0.25; \ \theta_T = 14.1°$$

$$I_T = 4.13\angle 14.1° \text{ A} \quad \text{(polar form)}$$

$$Z_T = \frac{E_T}{I_T} = \frac{120\angle 0°}{4.13\angle 14.1°} = 29.1\angle -14.1° \quad \text{(polar form)}$$

$$P_{\text{true}} = (I_{R_1}{}^2)(R_1) = (4^2)(30) = 480 \text{ W}$$

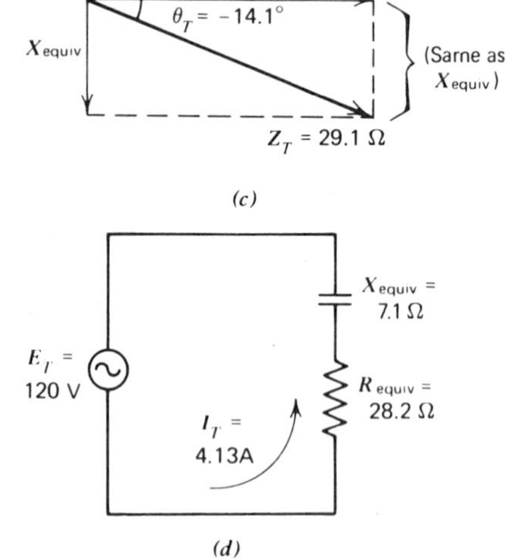

(c)

$\cos \theta_T = \dfrac{R_{\text{equiv}}}{Z_T}$	$\sin \theta_T = \dfrac{X_{\text{equiv}}}{Z_T}$
$Z_T (\cos \theta_T) = R_{\text{equiv}}$	$Z_T (\sin \theta_T) = X_{\text{equiv}}$
$29.1 (\cos - 14.1°) = R_{\text{equiv}}$	$29.1 (\sin -14.1°) = X_{\text{equiv}}$
$29.1 (0.968) = R_{\text{equiv}}$	$29.1 (-0.244) = X_{\text{equiv}}$
$28.2\ \Omega = R_{\text{equiv}}$	$-7.1\ \Omega = X_{\text{equiv}}$

(d)

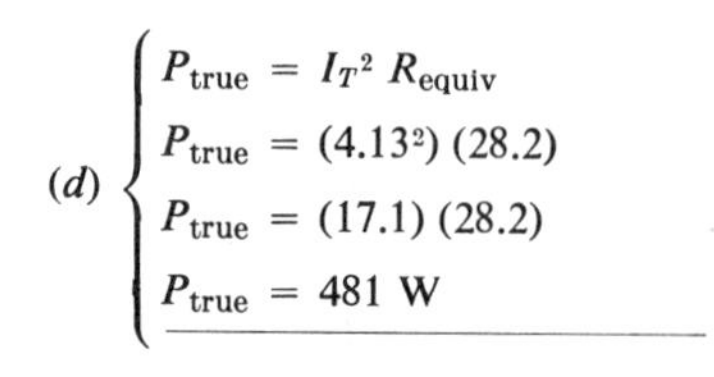

$$P_{\text{true}} = I_T{}^2 R_{\text{equiv}}$$

$$P_{\text{true}} = (4.13^2)(28.2)$$

$$P_{\text{true}} = (17.1)(28.2)$$

$$P_{\text{true}} = 481 \text{ W}$$

Figure 14-2. *L*, *C*, and *R* in parallel. (*a*) Parallel circuit. (*b*) Current phasors. (*c*) Z_T, R_{equiv}, X_{equiv} phasors. (*d*) Series equivalent circuit.

3. Current through the inductive reactance branch I_{X_L} always *lags* behind the voltage across X_L by 90°. With phasors rotating counterclockwise, I_{X_L} is shown pointing vertically downward (at the −90° point), or lagging behind E_T by 90°.

4. *Current* "through" the capacitive reactance I_{X_C} always *leads voltage* across X_C (or, as usually stated, *voltage* across X_C always lags its current by 90°). Phasor I_{X_C} is shown in Fig. 14-2*b* pointing vertically upward (at the +90° point), with E_T therefore lagging I_{X_C} by 90°.

5. The phasor sum of I_{X_C} and I_{X_L}, or the *net reactance* current I_X is actually the arithmetic difference of I_{X_C} and I_{X_L} since these two phasors are pointing in opposite directions, one upward and one downward. This is called *180° out of phase.* I_X is in the same direction as the larger of the two. *I_X is simply the smaller one subtracted from the larger one.*

6. Total current I_T, as shown by the phasors of Fig. 14-2*b*, is the phasor sum of I_R and I_X. Note that I_T is the hypotenuse (longest side) of the right triangle (the shaded area). From the Pythagoras theorem, $I_T = \sqrt{I_R^2 + I_X^2}$ (from equations 12-10 and 13-14). The angle between I_T and E_T is called the phase angle θ.

The value of each branch current may be found by using Ohm's law. $I_{X_L} = E_T/X_L = 120/60 = 2$ A, as shown in the phasors of Fig. 14-2*b*, $I_{X_C} = E_T/X_C = 120/40 = 3$ A. $I_{R_1} = E_T/R_1 = 120/30 = 4$ A as shown.

Net reactive current I_X is simply the difference between the larger and smaller reactive branch currents, or: $I_X = I_{X_C} - I_{X_L} = 4 - 3 = 1$ A, as shown.

Total current is: $I_T = \sqrt{I_{R_1}^2 + (I_{X_C} - I_{X_L})^2}$ (equation 14-3); $I_T = \sqrt{4^2 + (3-2)^2} = 4.13$ A, as shown in Fig. 14.2*b*.

Total impedance Z_T may be found using Ohm's law. $Z_T = E_T/I_T = 120/4.13 = 29.1\ \Omega$.

Since actual or *true* power is only dissipated by the resistor, then true power is: $P_{\text{true}} = E_R I_R = (120)(4) = 480$ W. The *apparent* power is: power apparent $= E_T I_T = (120)(4.13) = 495$.

Voltamperes. The polar forms of I_T and Z_T are, as shown in the text of Fig. 14-2*b*, $I_T = 4.13\underline{/14.1°}$ A, $Z_T = 29.1\underline{/-14.1°}\ \Omega$. In the phasor diagram of Z_T in Fig. 14-2*c*, it can be seen that the *horizontal* component of Z_T is R_{equiv}, while the *vertical* component is X_{equiv}. As shown in the text of Fig. 14-2*c*, using the cosine of θ, R_{equiv}, is found to be 28.2Ω. By using the sine of θ, X_{equiv} is found to be −7.1 Ω. (The negative sign means that X_{equiv} is capacitive.) The *parallel* X_L, X_C, and R_1 circuit of Fig. 14-2*a* is equivalent to the *series* circuit of X_{equiv} and R_{equiv} of Fig. 14-2*d*.

True power could also be found from this series equivalent circuit; $P_{\text{true}} = I_T^2 R_{\text{equiv}} = (4.13)^2 (28.2) = 481$ W.

The following example illustrates the previous discussion.

Example 14-2

An ac voltage $E_T = 30$ V is applied across a three-branch parallel circuit consisting of an $X_L = 3$ kΩ, an $X_C = 6$ kΩ, and an $R = 5$ kΩ. Find: (a) I_{X_L}, (b) I_{X_C}, (c) I_R, (d) I_T in polar form, (e) Z_T in polar form, (f) true power, (g) apparent power, (h) R_{equiv}, and (i) X_{equiv}.

Solution

Solve this example by using the method shown previously and by referring to the phasors and text of Fig. 14-2. As a guide, the following results are given: (a) $I_{X_L} = 10$ mA, (b) $I_{X_C} = 5$ mA, (c) $I_R = 6$ mA, (d) $I_T = 7.81\underline{/-39.8^\circ}$ mA, (e) $Z_T = 3.84\underline{/39.8^\circ}$ kΩ, (f) $P_{\text{true}} = 0.18$ W, (g) $P_{\text{apparent}} = 0.234$ VA, (h) $R_{\text{equiv}} =$ 2.95 kΩ, and (i) $X_{\text{equiv}} = 2.46$ kΩ.

For a Similar Problem See End-of-Chapter Problem 14-4

14-3. *L*, *C*, and *R* in Parallel-Series Circuits

A *parallel-series* circuit is basically a parallel circuit, but one or more branches contain *series* components. As shown in the circuit of Fig. 14-3*a*, two parallel branches Z_1 and Z_2 are each connected across the voltage source $E_T = 25$ V. The left branch Z_1 consists of X_L and R in series with each other, while the right branch Z_2 is only X_C.

The method used to find total current I_T and total impedance Z_T in the previous section cannot be employed here because the two branch currents are *not* at a right angle (90°) with each other. Therefore, equation 14-3, $I_T = \sqrt{I_R^2 + I_X^2}$ cannot be used. *Current* (I_1) through the left branch, X_L and R, *lags* E_T by some angle between 0 and 90°, since Z_1 is partially inductive and partially resistive, as shown in the $I_{X_{L,R}}$ phasor of Fig. 14-3*c*. If R and X_L were equal, the current in Z_1 would lag E_T by 45°. Since $X_L = 120$ Ω and $R = 30$ Ω, as shown in Fig. 14-3, then with X_L larger than R, this current lags E_T by more than 45°, as shown by the phasors.

Current (I_2) in the right branch, I_{X_C}, is 90° out of phase with E_T, with E_T *lagging* I_{X_C} *by 90°* (always true of any pure X_C), as shown by the I_{X_C} phasor of Fig. 14.3*c*. Therefore, by using E_T as a reference phasor, it can be seen in the phasor diagram that I_{X_C} and $I_{X_{L,R}}$ are neither 180° nor 90° out of phase with each other. As a result, they cannot easily be combined into total current I_T by the method previously used. Total impedance Z_T, and then I_T, will be found by using the following methods shown in the text of Fig. 14-3. The reader should follow each step shown in Fig. 14-3.

Z_1 in the circuit of Fig. 14-3*a* consists of $R = 30\ \Omega$ in series with $X_L = 120\ \Omega$. Z_1 is: $R + jX_L = 30 + j120$ (rectangular form) $= \sqrt{30^2 = 120^2} = 123.8\ \Omega$; $\tan \theta_1 = X_L/R = 120/30 = 4$; $\theta_1 = 76°$; $Z_1 = 123.8\angle 76°$ (polar form). Z_2 consists only of $X_C = 80\ \Omega$, $= 0 - j80$ (rectangular form), $= 80\angle -90°$ (polar form). Z_T, as shown in the text of Fig. 14-3, is: $198\angle -67.2°\ \Omega$ (polar form), and $76.8 - j183\ \Omega$ (rectangular form).

The rectangular form of Z_T means that the actual circuit of Fig. 14-3*a* is equivalent to the simple circuit of part (d), having an $R_{\text{equiv}} = 76.8\ \Omega$ in *series* with an $X_{\text{equiv}} = 183\ \Omega$. Notice that since the rectangular form $(76.8 - j183)$ of Z_T has a $-j$ term, then X_{equiv} is *capacitive*, as shown in Fig. 14-3*d*.

I_T is then $= E_T/Z_T = 25\angle 0°/198\angle -67.2° = 0.126\angle 67.2°$.

Another method of solution for the circuit of Fig. 14-3*a* is also shown in the text of this figure. It consists of finding the current in each branch (I_1 and I_2), and then adding them *vectorially* to find I_T. As shown, $I_1 = 0.202\angle -76°$ A (polar form), and $0.0505 - j0.196$ (rectangular form). I_2 is: $0.313\angle 90°$ A (polar form), and $0 + j0.313$ (rectangular form). Total current I_T is: $0.0505 + j0.117$ (rectangular form), and $0.127\angle 66.7°$ A (polar form). Note that this agrees with value of I_T found previously, within slide-rule tolerances. Finally $Z_T = 197\angle -66.7°\ \Omega$, from Ohm's law, again agreeing with the value found previously.

Figure 14-4*a* is similar to Fig. 14-3*a* except that the right branch now contains a resistor besides X_C. Z_1 consists of $X_L = 4\ \text{k}\Omega$ and a series resistor $R_1 = 2\ \text{k}\Omega$. Z_1 is partially inductive and partially resistive. As shown in the phasor diagram of Fig. 14-5, current through this branch I_1 lags behind E_T by the angle θ_1. Since X_L is larger than R_1, this angle will be nearer to 90° than to zero degrees.

Z_2 consists of an $X_C = 5\ \text{k}\Omega$ and a series resistor $R_2 = 1\ \text{k}\Omega$. Z_2 is partially capacitive and partially resistive. As shown in the phasor diagram of Fig. 14-5, E_T lags behind current I_1 in Z_2 branch by the angle θ_2. Since X_C is larger than R_2, this angle θ_2 will be nearer to 90° than to zero degrees. Since I_1 and I_2 are neither 180 nor 90° out of phase with each other, these currents cannot be added by the method discussed previously to find total current I_T. As shown in the text of Fig. 14-4, Z_T is found using equation 14-4 (product/sum). Then, from Ohm's law, I_T is found. Later, in Example 14-3, I_T is found first by vectorially adding I_1 and I_2. Then, from Ohm's law, Z_T is found.

In the circuit of Fig. 14-4*a*, Z_1 (the left branch) and θ_1 are found from the phasor diagram of Fig. 14-4*b*. As shown in the text of the figure, $Z_1 = 4.47\ \text{K}\angle 63.4°$ (polar form). From the phasor diagram of part (c), the right branch Z_2 is found to be $5.1\ \text{K}\angle -78.7°$ (polar form.) By using equation 14-4, Z_T is found to be $7.23\ \text{K}\angle 3.1°$ (polar form), and $7.2\ \text{K} + j0.39\ \text{K}$ (rectangular form). This rectangular form of Z_T shows that the horizontal component or $R_{\text{equiv}} = 7.2$ K, and that the vertical component or $X_{\text{equiv}} = 0.39$ K. The *series equivalent circuit* of the parallel circuit of Fig. 14-4*a* is shown in part (d). Total current I_T, found from Ohm's law, is shown to be $6.93\angle -3.1°$ mA.

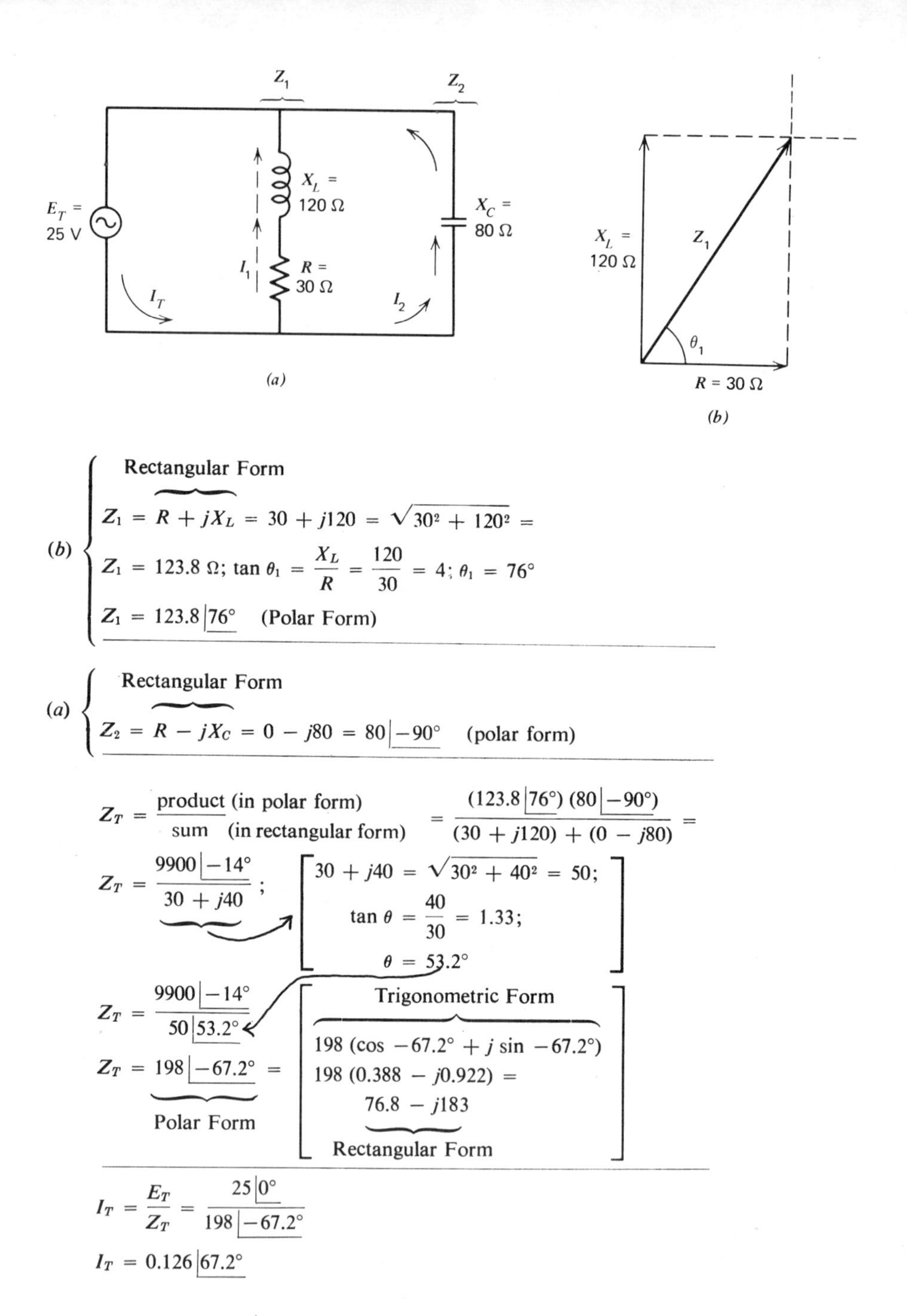

Figure 14-3. *L*, *C*, and *R* in parallel-series circuit. (*a*) Circuit. (*b*) *R*, X_L, and Z_1 phasors.

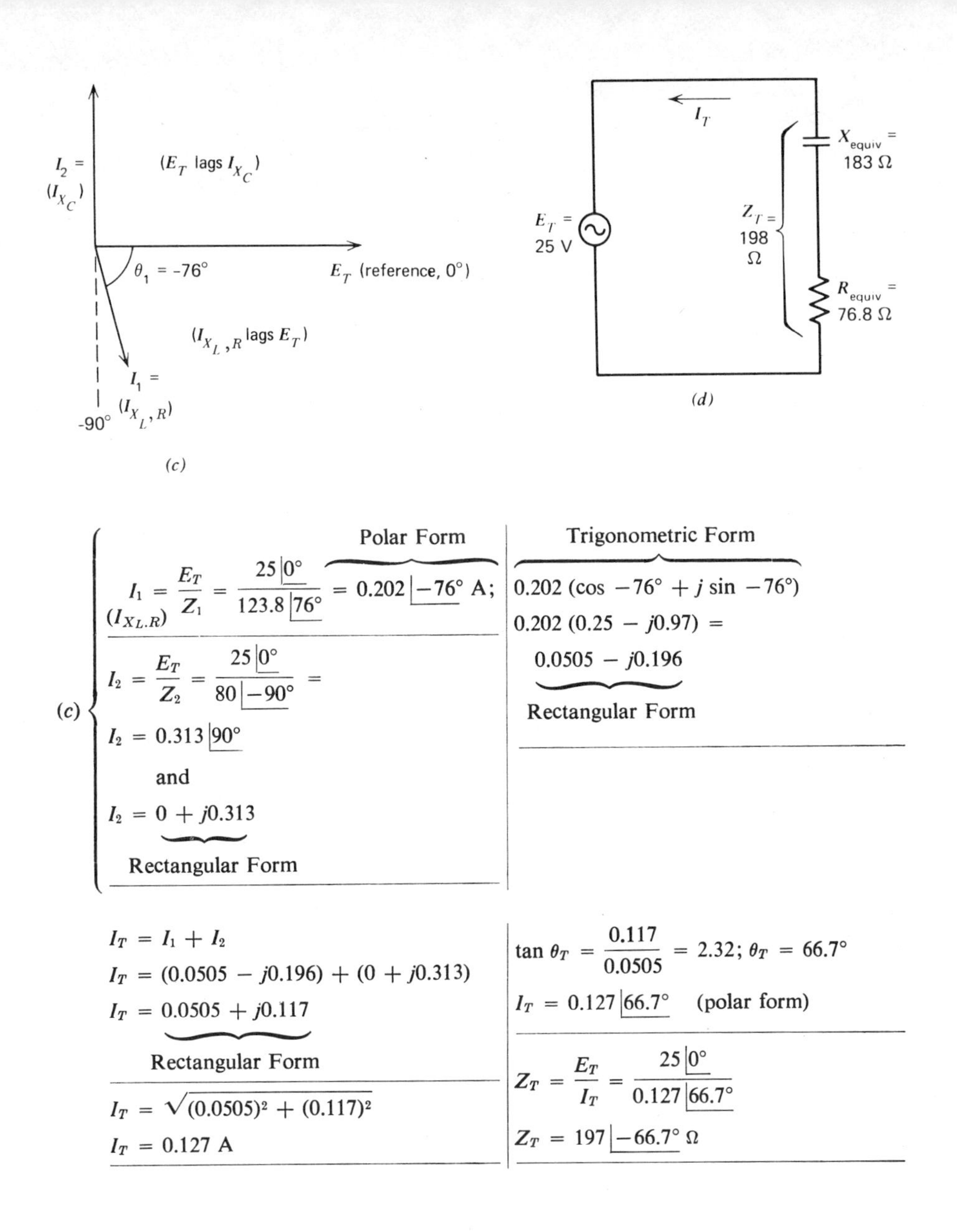

Figure 14-3 (*cont.*) (c) Current phasors. (*d*) Series equivalent circuit.

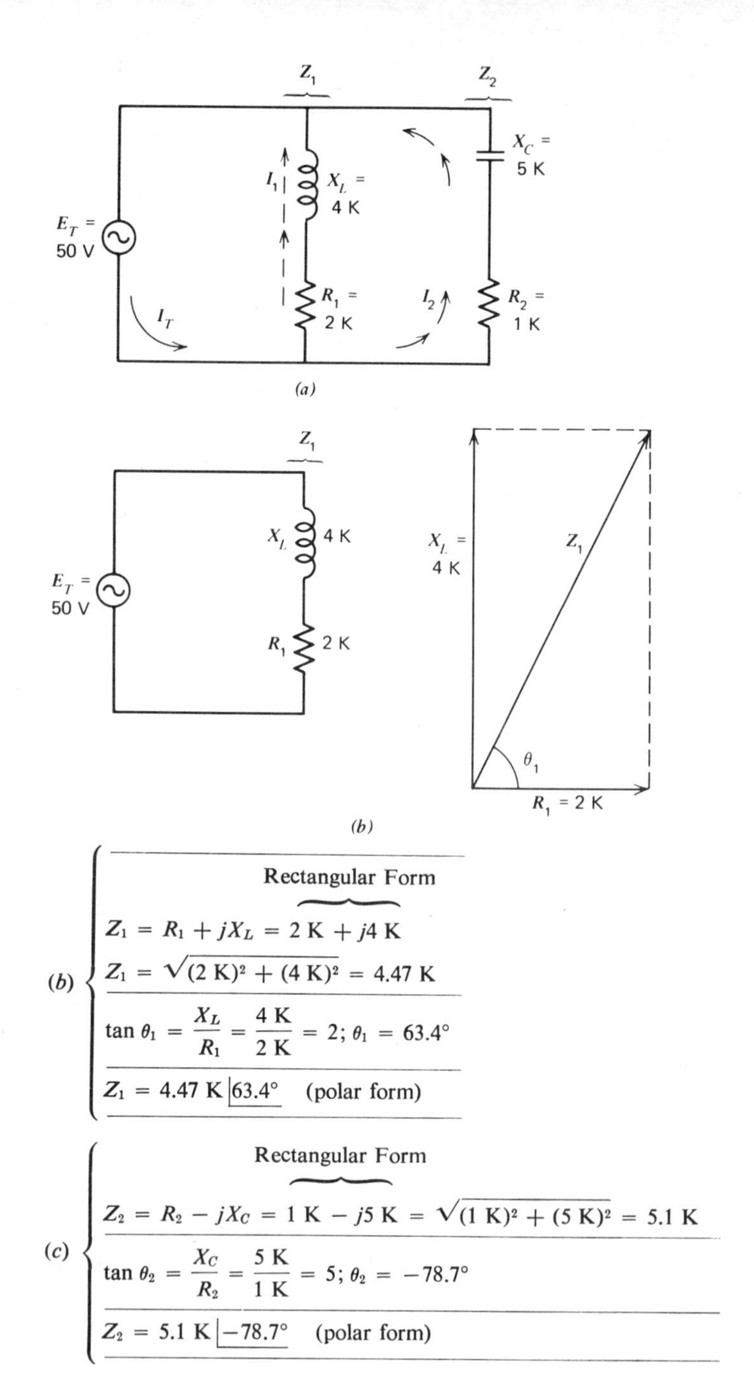

Figure 14-4. *L*, *C*, and *R* in parallel-series circuit. (*a*) Circuit. (*b*) Z_1 circuit and phasors.

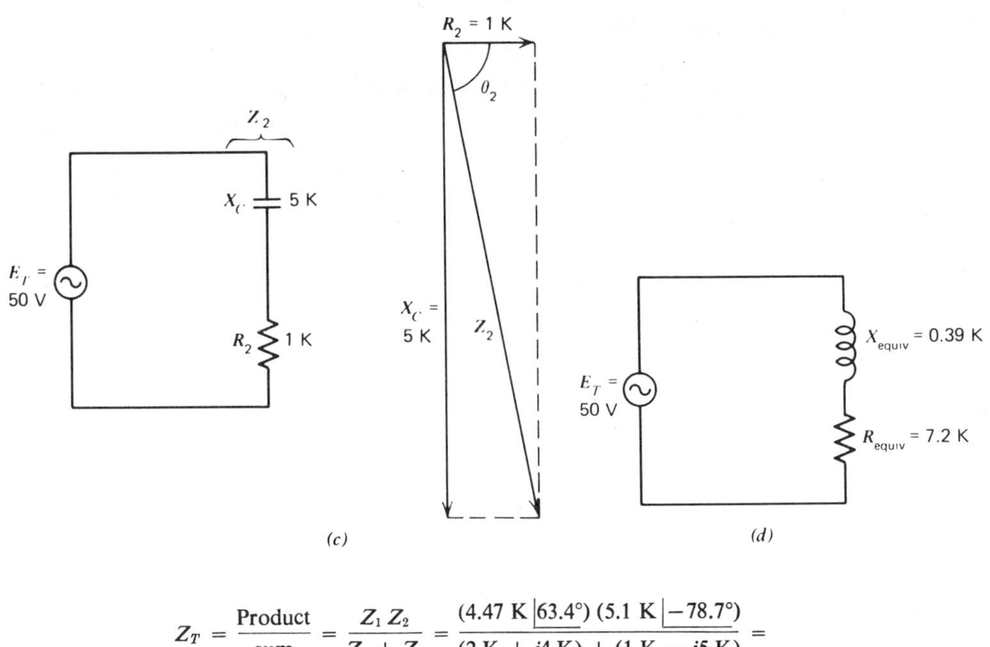

$$Z_T = \frac{\text{Product}}{\text{sum}} = \frac{Z_1 Z_2}{Z_1 + Z_2} = \frac{(4.47\text{ K}\angle 63.4^\circ)(5.1\text{ K}\angle -78.7^\circ)}{(2\text{ K} + j4\text{ K}) + (1\text{ K} - j5\text{ K})} =$$

$$Z_T = \frac{22.8 \times 10^6\angle -15.3^\circ}{3\text{ K} - j1\text{ K}}; \quad \sqrt{(3\text{ K})^2 + (1\text{ K})^2} = 3.16\text{ K}$$

$$\tan\theta = \frac{1\text{ K}}{3\text{ K}} = 0.333$$

$$Z_T = \frac{22.8 \times 10^6\angle -15.3^\circ}{3.16 \times 10^3\angle -18.4^\circ} \leftarrow \theta = -18.4^\circ$$

$$Z_T = 7.23\text{ K}\angle 3.1^\circ \quad \text{(polar form)}$$

$$I_T = \frac{E_T}{Z_T} = \frac{50\angle 0^\circ}{7.23\text{ K}\angle 3.1^\circ}$$

$$I_T = 6.93\text{ mA}\angle -3.1^\circ$$

(d)

Polar Form		Trigonometric Form
$Z_T = 7.23\text{ K}\angle 3.1^\circ$	=	$7.23\text{ K}(\cos 3.1^\circ + j\sin 3.1^\circ)$
	=	$7.23\text{ K}(0.999 + j0.054)$
	=	$7.2\text{ K} + j0.39\text{ K}$ (rectangular form)

Figure 14-4 (*cont.*) (c) Z_2 and phasors. (*d*) Series equivalent circuit.

Example 14-3

In the circuit of Fig. 14-4*a*, find: (a) I_{total} in polar form by vectorially adding I_1 and I_2, (b) Z_{total} in polar form, from Ohm's law, (c) $P_{\text{total true}}$ by using current in each branch resistor, and (d) $P_{\text{total true}}$ by using I_T and R_{equiv}.

Solution

Solve this example by following the step-by-step discussion shown in the text of Fig. 14-5, and by comparing the results with I_T and Z_T shown previously in Fig. 14-4: (a) $I_T = 6.93\angle{-3.1^\circ}$ mA, (b) $Z_T = 7.23\text{ K}\angle{3.1^\circ}$, (c) 344 mW, (d) 346 mW.

For a Similar Problem Refer to End-of-Chapter Problem 14-5

14-4. Conductance, Susceptance, and Admittance

In the previous two sections (14-2 and 14-3), a parallel circuit and a parallel-series circuit is solved using two methods. In one procedure, the total impedance is found by using the product of the two branch impedances divided by their sum. In the second method, the individual branch currents are added vectorially to find total current, and then from Ohm's law, total impedance is found.

A third method to solve a parallel or parallel-series ac circuit involves the use of *reciprocal* terms. This method is not being demonstrated here, but is being simply introduced. *Resistance* is the *opposition* to current by a pure resistor, R. The reciprocal is called *conductance*, G, and is the ability of R to *allow* current flow. $G = 1/R$ and is measured in *mhos* (℧).

Similarly, *reactance*, X, is the *opposition* by a pure L or a pure C to current flow. The reciprocal is called *susceptance*, B, and is the ability of a pure L or C to *allow* current flow. Susceptance $B = 1/X$, and is measured in *mhos*. Susceptance B, as reactance X, may be either inductive, B_L, or capacitive, B_C. As in the individual *branch currents* of a parallel circuit such as in Fig. 14-3*a*, capacitive current leads the applied voltage (phasors of Fig. 14-3*c*). As a result, susceptance of the capacitive branch is written as $+jB_C$, while that of the inductive branch is written as $-jB_L$.

Impedance is the total *opposition* of an ac circuit, while its reciprocal, called *admittance*, Y, is the ability of the circuit to *allow* current to flow. Admittance $Y = 1/Z$ and is measured in *mhos*. Similar to Z, $Y = \sqrt{G^2 + B_{\text{net}}^2}$, where B_{net} is the difference between B_L and B_C. Similarly, admittance $Y = G - jB_L + jB_C$, and $\tan\theta = B/G$.

The horizontal and vertical components of Z_T are the *series* circuit R_{equiv} and X_{equiv}. These components of Y_T are the *parallel* circuit G_{equiv} and B_{equiv}.

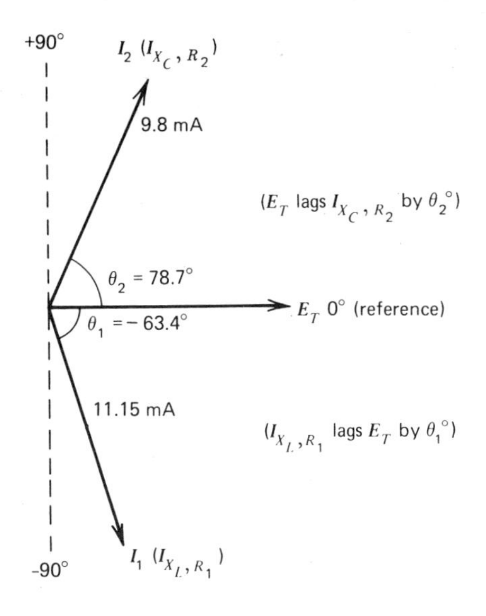

I Phasors of Circuit of Fig. 14-4*a*

$$I_1 = \frac{E_T}{Z_1} = \frac{50\underline{|0^\circ}}{4.47\ \text{K}\underline{|63.4^\circ}} = \underbrace{11.15\ \text{mA}\underline{|-63.4^\circ}}_{\text{Polar Form}} = \underbrace{11.15\,(\cos -63.4^\circ + j\sin -63.4^\circ)}_{\text{Trigonometric Form}}$$

$$11.15\,(0.447 - j0.895) = \underbrace{4.99 - j9.97}_{\text{Rectangular Form}}$$

$$I_2 = \frac{E_T}{Z_2} = \frac{50\underline{|0^\circ}}{5.1\ \text{K}\underline{|-78.7^\circ}} = \underbrace{9.8\ \text{mA}\underline{|78.7^\circ}}_{\text{Polar Form}} = \underbrace{9.8\,(\cos 78.7^\circ + j\sin 78.7^\circ)}_{\text{Trigonometric Form}}$$

$$9.8\,(0.196 + j0.98) = \underbrace{1.92 + j9.6}_{\text{Rectangular Form}}$$

$$I_T = I_1 + I_2 = (4.99 - j9.97) + (1.92 + j9.6) =$$

$$I_T = 6.91 - j0.37 = \sqrt{6.91^2 + 0.37^2} =$$

$$I_T = 6.93\ \text{mA};\ \tan\theta_T = \frac{0.37}{6.91} = 0.534;\ \theta_T = 3.1^\circ$$

$$I_T = 6.93\ \text{mA}\underline{|-3.1^\circ}\ \text{(polar form)}$$

$$Z_T = \frac{E_T}{I_T} = \frac{50\underline{|0^\circ}}{6.93\ \text{mA}\underline{|-3.1^\circ}} = 7.23\ \text{K}\underline{|3.1^\circ}$$

$$P_{T\,\text{true}} = P_{1\,\text{true}} + P_{2\,\text{true}}$$

$$P_{T\,\text{true}} = I_1^2 R_1 + I_2^2 R_2$$

$$P_{T\,\text{true}} = (11.15\ \text{mA})^2\,(2\ \text{K}) + (9.8\ \text{mA})^2\,(1\ \text{K})$$

$$P_{T\,\text{true}} = (11.15 \times 10^{-3})\,(2 \times 10^3) + (9.8 \times 10^{-3})^2\,(1 \times 10^3)$$

$$P_{T\,\text{true}} = (124 \times 10^{-6})\,(2 \times 10^3) + (96 \times 10^{-6})\,(1 \times 10^3)$$

$$P_{T\,\text{true}} = 248 \times 10^{-3} + 96 \times 10^{-3}$$

$$P_{T\,\text{true}} = 344\ \text{mW}$$

or

$$P_{T\,\text{true}} = I_T^2 R_{\text{equiv}}$$

$$P_{T\,\text{true}} = (6.93\ \text{mA})^2\,(7.2 \times 10^3)$$

$$P_{T\,\text{true}} = 346\ \text{mW}$$

Figure 14-5. Example 14-3.

PROBLEMS

See section 14-1 for discussion covered by the following.

14-1. An $X_L = 20\ \Omega$, an $R = 10\ \Omega$, and an $X_C = 60\ \Omega$ are connected in series. Find total impedance Z_T.

14-2. If 100 mA flows through the series circuit of the previous problem, find: (a) E_R, (b) E_{X_L}, (c) E_{X_C}, and (d) E_T. Find E_T by using two methods: Ohm's law and phasor sum of E_R, E_{X_L}, and E_{X_C}.

14-3. If, in the circuit of Problems 14-1, a current of 2 A flows resulting in $E_{X_L} = 40$ V, $E_R = 20$ V, $E_{X_C} = 120$ V, then find: (a) E_T, (b) apparent power, and (c) true power.

See section 14-2 for discussion covered by the following.

14-4. In a circuit similar to that of Fig. 14-2, an ac voltage $E_T = 60$ V is applied across a three-branch parallel circuit consisting of an $X_L = 1\ \text{k}\Omega$, an $X_C = 2\ \text{k}\Omega$, and an $R = 3\ \text{k}\Omega$. Find: (a) I_{X_L}, (b) I_{X_C}, (c) I_R, (d) I_T, (e) Z_T, (f) true power, and (g) apparent power.

See section 14-3 for discussion covered by the following.

14-5. In a parallel-series ac circuit similar to the circuit of Fig. 14-3*a*, 300 V E_T is applied across an $X_L = 500\ \Omega$ in series with an $R = 100\ \Omega$. E_T is also applied across a parallel branch $X_C = 200\ \Omega$. If the series branch (R and X_L) is called Z_1, while the other branch (X_C) is called Z_2, as shown in Fig. 14-3*a*, find: (a) Z_1 in polar form, (b) Z_2 in polar form, (c) Z_T in polar form, (d) I_1 in polar form, (e) I_2 in polar form, (f) I_T in polar form by adding vectorially I_1 and I_2, (g) R_{equiv}, and (h) X_{equiv}.

chapter

15

Resonance and Filters

In any ac circuit containing L, C, and R, X_L and X_C depend on the *frequency* of the applied voltage. At some particular frequency, called the *resonant* frequency, F_r, $X_L = X_C$. The circuit is called a resonant or *tuned* circuit. At this F_r, the circuit acts as a tuning or selective device, resulting in amplification of the ac voltage at the resonant frequency and a narrow band of adjacent frequencies, but in attenuation (reduction) of all other ac voltages. The tuner of any radio or television receiver, therefore, must contain one or more resonant circuits to select the desired station (transmitter) and to reject all others.

15-1. Series Resonant Circuits

A series L, C, and R is shown in Fig. 15-1 with $L = 50.7\ \mu\text{H}$, $R = 63.5\ \Omega$, and $C = 500$ pF. Total voltage applied is 10 mV. Table 15-1 lists the results when the *frequency* of the applied voltage is varied over a wide range, both above and below the resonant frequency. Note that at the *low* frequency of 100 kHz, X_L is small (31.8 Ω), while X_C is large (3180 Ω). As the *frequency* is *increased*, X_L *increases* but X_C *decreases*. At the resonant frequency, F_r, (1000 kHz) $X_L = X_C = 318\ \Omega$. At the highest frequency shown (4000 kHz), X_L is up to 1272 Ω while X_C is down to 79.5 Ω.

Note also that total impedance Z is at its smallest value, being equal to R alone (63.5 Ω), at the resonant frequency. This occurs because X_L and X_C effectively cancel each other so that the net reactance X_{net} is zero. With Z at its minimum value at F_r, *current* becomes *maximum* (157 μA). This results in

TABLE 15-1 (For Series L, R, and C of Fig. 15-1)

	Frequency	X_L ($2\pi FL$)	X_C ($1/2\pi FC$)	X_{net} ($X_L - X_C$)	Z ($\sqrt{R^2 + X^2}$)	I (E/Z) μA	E_R (IR) mV	E_{X_L} (IX_L) mV	E_{X_C} (IX_C) mV
	100 kHz	31.8 Ω	3180 Ω	3148 Ω	3148 Ω	3.18 μA	0.202 mV	0.101 mV	10.1 mV
	200 kHz	63.6 Ω	1590 Ω	1526 Ω	1526 Ω	6.56 μA	0.417 mV	0.417 mV	10.43 mV
	300 kHz	95.4 Ω	1060 Ω	965 Ω	967 Ω	10.35 μA	0.657 mV	0.988 mV	11 mV
	400 kHz	127.2 Ω	795 Ω	668 Ω	670 Ω	14.9 μA	0.947 mV	1.89 mV	11.85 mV
	500 kHz	159 Ω	636 Ω	477 Ω	482 Ω	20.7 μA	1.315 mV	3.29 mV	13.2 mV
	600 kHz	191 Ω	530 Ω	339 Ω	345 Ω	29 μA	1.84 mV	5.54 mV	15.4 mV
	700 kHz	223 Ω	455 Ω	232 Ω	240 Ω	41.7 μA	2.65 mV	9.3 mV	19 mV
	800 kHz	255 Ω	398 Ω	143 Ω	156 Ω	64 μA	4.06 mV	16.3 mV	25.5 mV
	900 kHz	286 Ω	354 Ω	68 Ω	93 Ω	107 μA	6.8 mV	30.6 mV	37.8 mV
F_r →	1000 kHz	318 Ω	318 Ω	0 Ω	63.5 Ω	157 μA	10 mV	50 mV	50 mV
	1100 kHz	350 Ω	290 Ω	60 Ω	87.4 Ω	114 μA	7.25 mV	39.9 mV	33.1 mV
	1200 kHz	382 Ω	266 Ω	116 Ω	132 Ω	75.8 μA	4.81 mV	29 mV	20.2 mV
	1300 kHz	413 Ω	246 Ω	167 Ω	178.5 Ω	56 μA	3.56 mV	23.2 mV	13.7 mV
	1400 kHz	445 Ω	228 Ω	217 Ω	226 Ω	44.3 μA	2.81 mV	19.8 mV	10.1 mV
	1500 kHz	477 Ω	212 Ω	265 Ω	268 Ω	37.3 μA	2.37 mV	17.8 mV	7.9 mV
	2000 kHz	636 Ω	159 Ω	477 Ω	482 Ω	20.7 μA	1.315 mV	13.1 mV	3.29 mV
	4000 kHz	1272 Ω	79.5 Ω	1193 Ω	1193 Ω	8.4 μA	0.534 mV	10.7 mV	0.667 mV

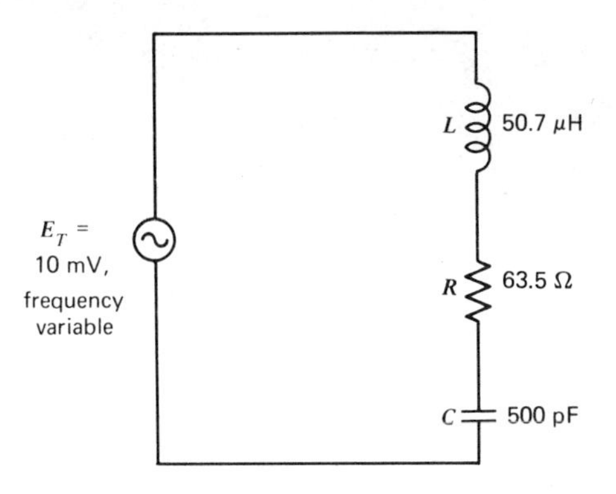

Figure 15-1. Series *L*, *R*, and *C*.

maximum voltages across X_L and X_C at F_r. ($E_{X_L} = E_{X_C} = 50$ mV). Note that this is much larger than the applied E_T of 10 mV, and is called a *resonant rise of voltage.*

The following example shows the calculations for the above results at F_r.

Example 15-1

In the series *L*,*C*,*R* circuit of Fig. 15-1, find each of the following at a frequency of 1000 kHz: (a) X_L, (b) X_C, (c) Z, (d) I, (e) E_{X_L} (or V_{X_L}), and (f) E_{X_C} (or V_{X_C}).

Solution

(a) $X_L = 2\pi FL$ (12-2)

$$X_L = (6.28)\,(1000\text{ kHz})\,(50.7\ \mu\text{H})$$

$$X_L = (6.28)\,(1000 \times 10^3)\,(50.7 \times 10^{-6})$$

$$X_L = 318\ \Omega$$

(b) $X_C = \dfrac{1}{2\pi FC}$ (13-19)

$$X_C = \frac{1}{(6.28)\,(1000\text{ kHz})\,(500\text{ pF})}$$

$$X_C = \frac{1}{(6.28)\,(1000 \times 10^3)\,(500 \times 10^{-12})}$$

$$X_C = 318\ \Omega$$

(c) $Z = \sqrt{R^2 + (X_L - X_C)^2}$

$Z = \sqrt{63.5^2 + (318 - 318)^2}$

$Z = \sqrt{63.5^2 + 0^2}$

$Z = 63.5\ \Omega$ (or same as R)

(d) $I = \dfrac{E_T}{Z}$

$I = \dfrac{10\ \text{mV}}{63.5\ \Omega}$

$I = \dfrac{10 \times 10^{-3}}{63.5}$

$I = 0.157 \times 10^{-3}$ A or 157 μA

(e) $E_{X_L} = IX_L$

$E_{X_L} = (157 \times 10^{-6})\ (318)$

$E_{X_L} = 0.05$ V or 50 mV

(f) $E_{X_C} = IX_C$

$E_{X_C} = (157 \times 10^{-6})\ (318)$

$E_{X_C} = 50$ mV

At frequencies above and below the resonant frequency of 1000 kHz, X_L and X_C are not equal, and the net reactance X_{net} is not zero. As a result, total impedance Z becomes larger than it is at resonance, and current decreases from its maximum value. This is shown in Table 15-1. As an example, refer to this table at a frequency of 600 kHz. Solve for the results shown listed for this frequency. X_L is 191 Ω, Z_C is 530 Ω, X_{net} is 339, Z is 345 Ω, and I is now down to 29 μA (compared to $I = 157\ \mu$A at resonance).

Figure 15-2 shows in graphical form the results of X_L, X_C, and I at various frequencies. Note that in Fig. 15-2*a*, X_L *increases linearly* (along a straight line) as the frequency is increased. In part (b), X_C is shown *decreasing nonlinearly* (along a curved line) as the frequency is increased. Note that at the resonant frequency (F_r) of 1000 kHz, the X_L and X_C graphs intersect or cross when their values are each 318 Ω.

Graph (c) depicts the values of I at each of the frequencies. Note that this curve shows an increasing value of I as the frequency is increased from below resonance towards F_r. I becomes maximum at F_r, then I decreases at frequencies going higher above resonance. The I graph is often called the *frequency response curve*.

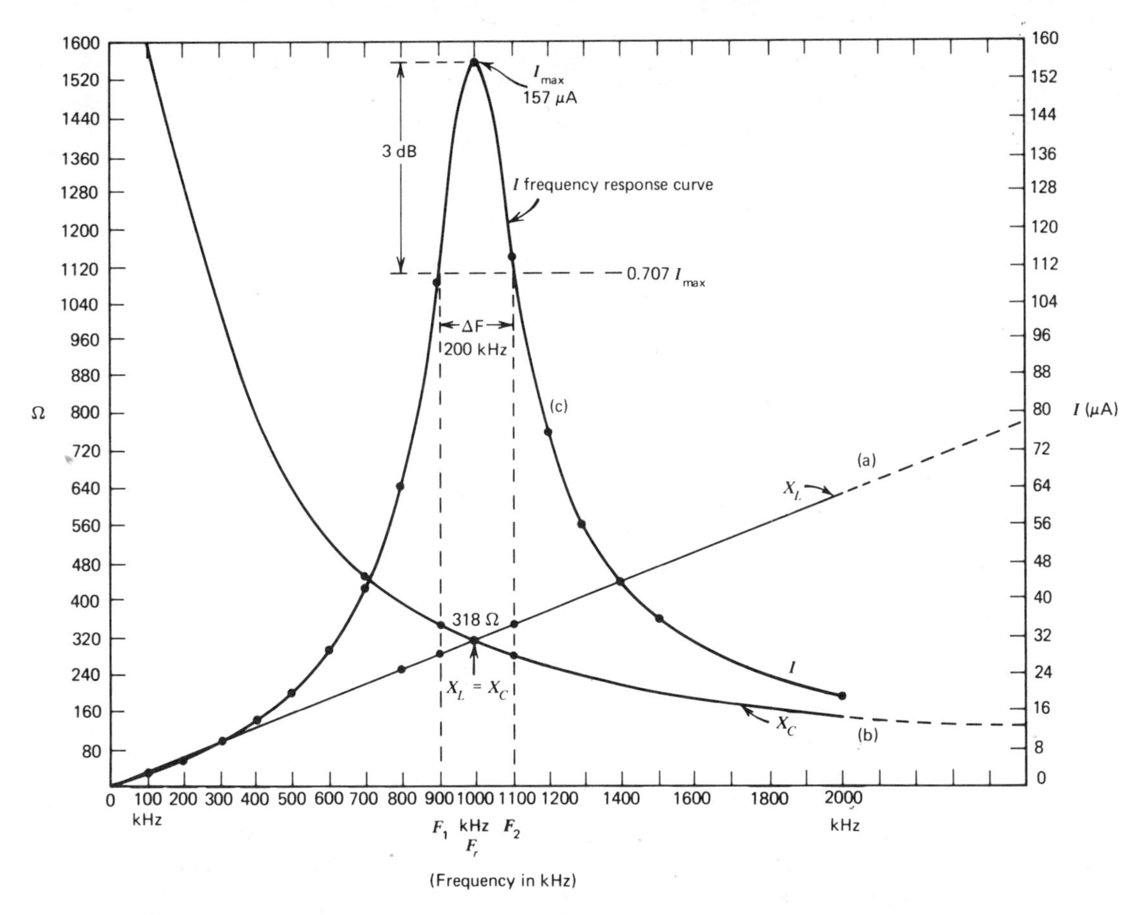

Figure 15-2. (a) X_L versus frequency. (b) X_c versus frequency. (c) I versus frequency of circuit of Fig. 15-1.

The impedance Z curve of Fig. 15-3 depicts graphically the results of Table 15-1. Note that Z is *minimum* at the resonant frequency, when I is maximum.

For a Similar Problem Refer to End-of-Chapter Problem 15-1

Quality (Q) of Tuned Circuit. In Chapter 12, Inductors and Resistors, we show that the ratio of the inductive reactance to the resistance of a coil is called the *quality* (Q) of a coil, or $Q = X_L/R$ (equation 12-23). In a series resonant circuit, the Q of the coil is practically the same as the Q of the circuit unless some additional R has been added. The *quality* (Q) of the resonant circuit has a great effect on the current graph shown in Fig. 15-2*c*. As shown, I reaches its maximum value at F_r, and decreases at frequencies below and above F_r, just the opposite of the Z graph of Fig. 15-3. At a point 0.707 of I_{max}

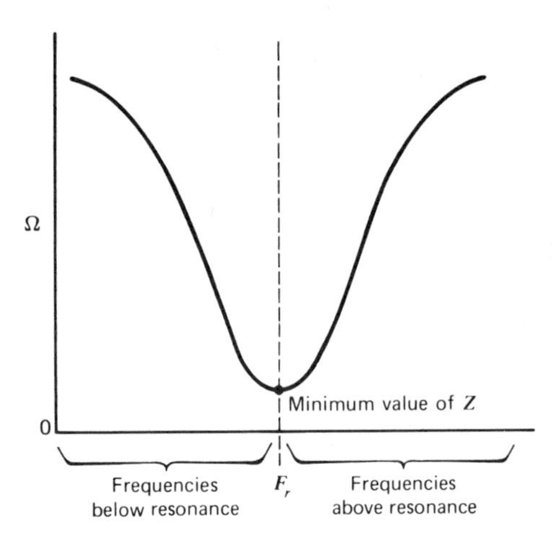

Figure 15-3. Graph of Z of series L, C, R circuit.

in Fig. 15-2*c*, on each side of F_r, are frequencies F_1 and F_2. These are shown as 900 kHz and 1100 kHz on the I graph, with 1000 kHz as F_r. Current values at frequencies between 900 kHz (F_1) and 1100 kHz (F_2) are sufficiently large as to be usable, and are said to be accepted by the resonant circuit. These are called the *band-pass or bandwidth frequencies* (Δf). Current values below F_1 and above F_2 are too small to be usable, and are said to be rejected by the resonant circuit. The term *decibel* refers to the ratio of two ac voltages, currents or powers. In Fig. 15-2, I at resonance is shown at maximum value. When I has decreased (at frequencies F_1 and F_2) to 0.707 of maximum, I is said to have decreased 3 decibels (dB).

The Q of the circuit at resonance is the ratio of the resonant frequency (F_r) to the band-pass or bandwidth frequencies (Δf) or,

$$Q_{\text{at resonance}} = \frac{F_r}{\Delta f} \tag{15-1}$$

From the I frequency response graph of Fig. 15-2*c*, $F_r = 1000$ kHz, and Δf is 200 kHz (difference between 1100 kHz, F_2, and 900 kHz, F_1). Q is then $F_r/\Delta f = 1000$ kHz/200 KHz $= 5$. This is also referred to as the *sharpness of tuning*. Note that this is also the Q of the coil at resonance, where from Table 15-1, at F_r, $X_L = 318\ \Omega$, and $R = 63.5\ \Omega$, and $Q = X_L/R$ (equation 12-23), $= 318/63.5 = 5$.

Example 15-2

At a resonant frequency of 3 MHz, $X_L = X_C = 1500\ \Omega$, and $R = 100\ \Omega$. Find: (a) Q of coil, (b) band-pass frequencies, Δf, and (c) lowest frequency F_1, and highest frequency F_2, passed.

Solution

(a) $Q_{\text{of coil}} = X_L/R = 1500/100 = 15$

(b) $$Q_{\text{of circuit}} = \frac{F_r}{\Delta f} \quad (15\text{-}1)$$

$$15 = \frac{3 \times 10^6}{\Delta f}$$

$$\Delta f = \frac{3 \times 10^6}{15}$$

$$\Delta f = 0.2 \times 10^6 \text{ Hz} \qquad \text{or} \qquad 200{,}000 \text{ Hz}$$

(c) With $\Delta f = 0.2 \times 10^6$, the lowest frequency passed, F_1, is 100,000 Hz (1/2 of 200,000) or 0.1 MHz *below* the F_r of 3 MHz, or $F_1 = 2.9$ MHz. The highest frequency F_2 is 100,000 Hz or 0.1 MHz *above* the F_r of 3 MHz, or $F_2 = 3.1$ MHz.

For a Similar Problem See End-of-Chapter Problem 15-2

High and Low Q Resonant Circuits. If the *resistance* in a series resonant circuit is *increased,* Q will be *decreased.* From $Q = X_L/R$ (equation 12-23), note that Q varies inversely as R. From equation 15-1, $Q = F_r/\Delta f$, then $\Delta f = F_r/Q$. Therefore, with a *lower* Q, Δf *increases.* The following illustrates this. In the circuit of Fig. 15-4*b*, $X_L = 318\ \Omega$ and $R_1 = 63.5\ \Omega$. This is the same as the circuit of Fig. 15-1 at resonance. As shown previously, with a $Q_1 = X_L/R_1 = 318/63.5 = 5$, the bandwidth frequencies, $\Delta f_1 = F_r/Q_1 = 1000$ kHz/5 = 200 kHz. This is 100 kHz *below,* and 100 kHz *above* the F_r of 1000 kHz, or from 900 kHz to 1100 kHz, as shown in Fig. 15-2*c*, and also in Fig. 15-4*a*.

If R is increased in value, as shown in Fig. 15-4*b*, where R_2 is 127 Ω, then Q_2 becomes $= X_L/R_2 = 318/127 = 2.5$. Band-pass or bandwidth frequencies Δf_2 is now $= F_r/Q_2 = 1000$ kHz/2.5 = 400 kHz. This is much wider than the former Δf_1 of only 200 kHz. As shown in the dashed-line I graph of Fig. 15-4*a*, the band-pass (Δf_2) is from F_A (800 kHz) to F_B (1200 kHz), which is wider than the previous Δf (of the higher Q), which passed frequencies from F_1 (900 kHz) to F_2 (1100 kHz). The solid-line I curve of the $Q_1 = 5$ is said to be a sharper or steeper curve than the more broadly tuned dashed-line curve for $Q_2 = 2.5$. Various tuned circuits must pass different bandwidths. For example, the resonant circuit of the intermediate frequency (IF) amplifier of an FM radio receiver has an F_r of 10.7 MHz and a required bandwidth (F_1 to F_2) of 0.15 MHz. This requires a $Q = F_r/\Delta f = 10.7/0.15 = 71.3$. The IF tuned circuit of an AM radio receiver may have an F_r of 465 kHz and a desired bandwidth of 20 kHz. This $Q = F_r/\Delta f = 465/20 = 23.25$. To attain

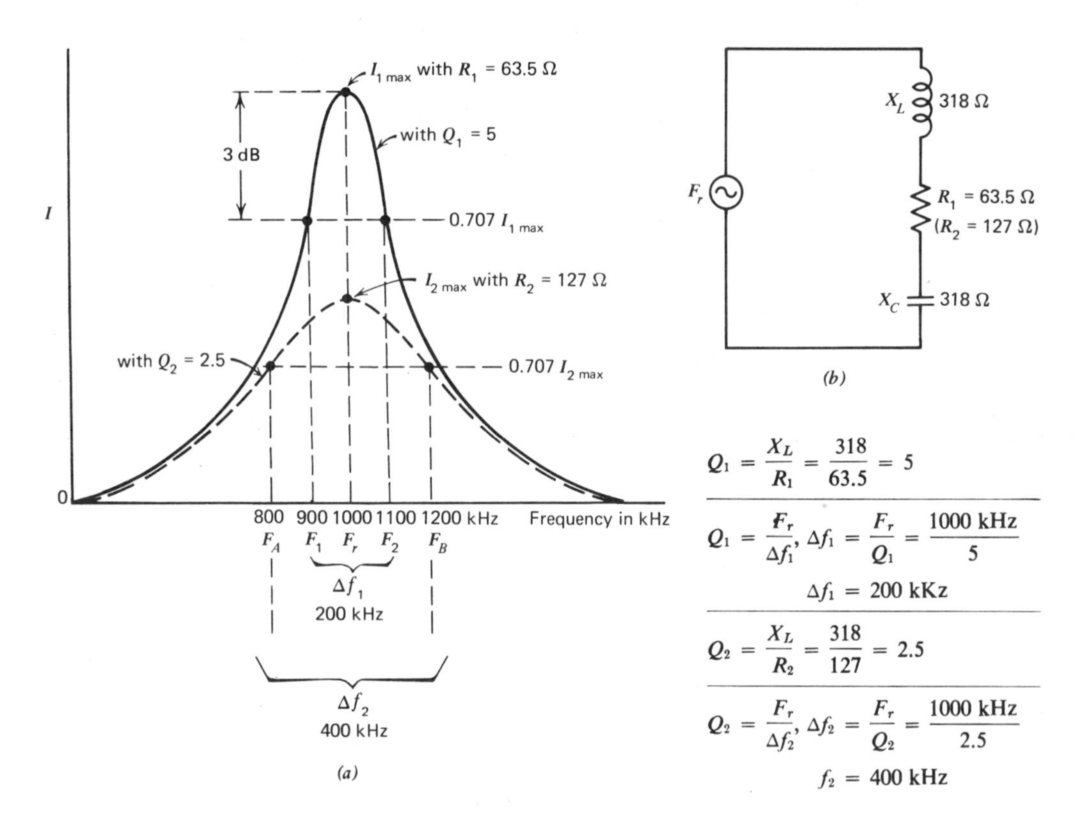

Figure 15-4. (*a*) *I* graphs for $Q_1 = 5$, and $Q_2 = 2.5$. (*b*) Series *L*, *R*, *C* circuit.

a desired Q, resistance may be added to the resonant circuit, or resistance may be decreased by selecting a coil having less R.

Q Rise of Voltage at F_r. As shown in Table 15-1, voltages across X_L and also across X_C rise to maximum values (50 mV) at F_r. This is larger than the 10 mV E_T applied. At resonance, E_{X_L} and E_{X_C} are each larger than E_T by the factor Q, or

$$E_{X_C} \text{ (or } E_{X_L}) \quad \text{at} \quad F_r = Q\,E_T \qquad (15\text{-}2)$$

Q of the circuit of Fig. 15-1 at F_r is: $X_L/R = 318/63.5 = 5$. $E_{X_C} = QE_T = (5)(10 \text{ mV}) = 50$ mV.

Example 15-3

In Example 15-2, $F_r = 3$ MHz, $X_L = X_C = 1500\ \Omega$, $R = 100\ \Omega$, and Q was found to be 15. If the applied voltage E_T is 30 μV, find E_{X_C}.

Solution

$E_{X_C} = QE_T$ (equation 15-2), $= (15)\,(30\ \mu V) = 450\ \mu V$.

For a Similar Problem Refer to End-of-Chapter Problem 15-3

Resonant Frequency. The *frequency* of the ac voltage where $X_L = X_C$, called the *resonant frequency* F_r, depends on the values of L and C, and may be calculated from the following equation.

$$F_r = \frac{1}{2\pi\sqrt{LC}} \qquad (15\text{-}3)$$

This equation is derived from the relationship:

$$X_L = X_C$$

$$2\pi F_r L = \frac{1}{2\pi F_r C}$$

multiplying each side by F_r, and also dividing each side by 2π gives

$$\frac{(2\pi F_r L)\,(F_r)}{2\pi} = \left(\frac{1}{2\pi F_r C}\right)\left(\frac{F_r}{2\pi}\right)$$

the 2π's cancel at the left side, and the F_r's cancel at the right, giving

$$F_r^2 L = \frac{1}{4\pi^2 C}$$

solving for F_r^2 (equation 15-4)

$$F_r^2 = \frac{1}{4\pi^2 LC}$$

taking the square root of both sides yields

$$F_r = \frac{1}{2\pi\sqrt{LC}} \qquad (15\text{-}3)$$

From equation 15-4, L or C may be found in terms of one or the other, and in terms of F. This is shown in the following (equation 15-4):

$$F_r^2 = \frac{1}{4\pi^2 LC}$$

solving for L yields

$$L = \frac{1}{4\pi^2 F_r^2 C} \qquad (15\text{-}5)$$

or solving for C gives

$$C = \frac{1}{4\pi^2 F_r^2 L} \tag{15-6}$$

From the circuit of Fig. 15-1, where $L = 50.7\ \mu\text{H}$ and $C = 500$ pF, find the resonant frequency F_r, using equation 15-3. $F_r = 1/(2\pi\sqrt{LC}) = 1/[6.28\sqrt{(50.7 \times 10^{-6})(500 \times 10^{-12})}] = 1/(6.28\sqrt{2.535 \times 10^{-14}}) = 1/[6.28(1.59 \times 10^{-7})] = 1/(10 \times 10^{-7}) = 0.1 \times 10^7 = 1000$ kHz. Note that this agrees with the F_r shown in Table 15-1.

The following example illustrates the use of equation 15-5 in finding the required value of L to produce a desired F_r with a known value of C.

Example 15-4

What value of L will produce resonance at a frequency of 1000 kHz with a 500-pF capacitor? (This is F_r and C of the circuit of Fig. 15-1.)

Solution

$$L = \frac{1}{4\pi^2 F_r^2 C} \tag{15-5}$$

$$L = \frac{1}{(39.5)(1 \times 10^6)^2 (500 \times 10^{-12})}$$

$$L = \frac{1}{(3.95 \times 10^1)(1 \times 10^{12})(5 \times 10^{-10})}$$

$$L = \frac{1}{19.75 \times 10^3}$$

$$L = 0.0507 \times 10^{-3}\ \text{H} \qquad \text{or} \qquad 50.7\ \mu\text{H}$$

Note that this agrees with the circuit inductance of Fig. 15-1.

For Similar Problems Refer to End-of-Chapter Problems 15-4 to 15-5

L/C Ratio. The resonant frequency, $F_r = 1/(2\pi\sqrt{LC})$ (equation 15-3), depends on the *product* of L and C. Many different values of L and C produce the same product. As a simple example, 8 H and 3 μF result in a product of 24×10^{-6}, but so also do 6 H and 4 μF, and 12 H and 2 μF, and so on. The *ratio* of L to C determines the sharpness or steepness of the current graph or frequency response curve.

Figure 15-6 shows curves *a* and *b*. Part (a) is the I graph for the L, C, R series circuit of Fig. 15-1. This I curve is also shown in greater detail in Fig.

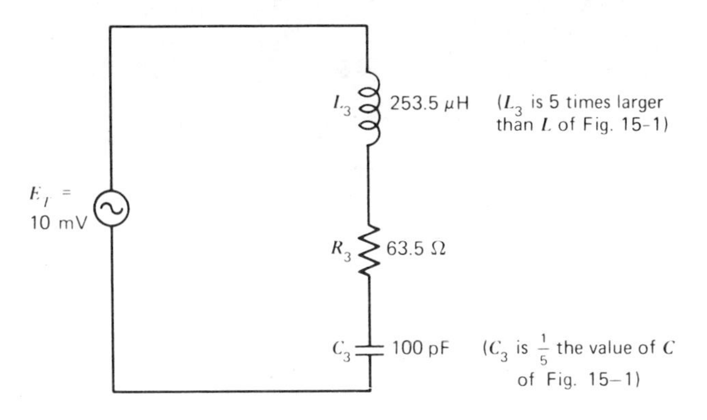

Figure 15-5. The L_3 C_3 ratio 25 times larger than the L/C ratio of Fig. 15-1.

15-2*c*. Curve *b* of Fig. 15-6 (the dashed-line graph) is much narrower, resulting in a higher *Q*. This sharper or steeper-sided curve is the result of *increasing* $L = 50.7\ \mu H$ (of Fig. 15-1) five times, where L_3 is now 253.5 μH (Fig. 15-5), and *decreasing* *C* by the same proportion. The original *C* (of Fig. 15-1) = 500 pF. In Fig. 15-5, it has been decreased down to 1/5, or $C_3 = 100$ pF. Note that the *product* of *L* and *C* of Fig. 15-1, $(50.7 \times 10^{-6})(500 \times 10^{-12}) = 2.535 \times 10^{-14}$, is equal to the product of L_3 and C_3 of Fig. 15-5, $(253.5 \times 10^{-6})(100 \times 10^{-12}) = 2.535 \times 10^{-14}$. Therefore, F_r is the same for both circuits. The *L/C ratio* has been increased 25 times.

In the short Table 15-2, the values of X_L, X_C, X_{net}, Z, and I are listed for several frequencies from 900 kHz to 1100 kHz. The *I* values are graphed as the dashed-line curve *b* of Fig. 15-6. Note that the bandwidth Δf of the original *I* curve (200 kHz) is much wider or broader than Δf_3 of the high ratio

TABLE 15-2 (For Circuit of Fig. 15-5) High L/C Ratio

Frequency (kHz)	X_{L_3} $(2\pi FL_3)$	X_{C_3} $\left(\frac{1}{2\pi FC_3}\right)$	X_{net} $(X_{L_3} - X_{C_3})$	Z $(\sqrt{R^2 + X_{net}^2})$	I (E_T/Z)
900 kHz	1430 Ω	1770 Ω	340 Ω	346 Ω	28.9 μA
980 kHz	1560 Ω	1625 Ω	65 Ω	91 Ω	110 μA
$F_r \rightarrow$ 1000 kHz	1590 Ω	1590 Ω	0 Ω	63.5 Ω	157 μA
1020 kHz	1625 Ω	1560 Ω	65 Ω	91 Ω	110 μA
1100 kHz	1750 Ω	1450 Ω	300 Ω	307 Ω	32.6 μA

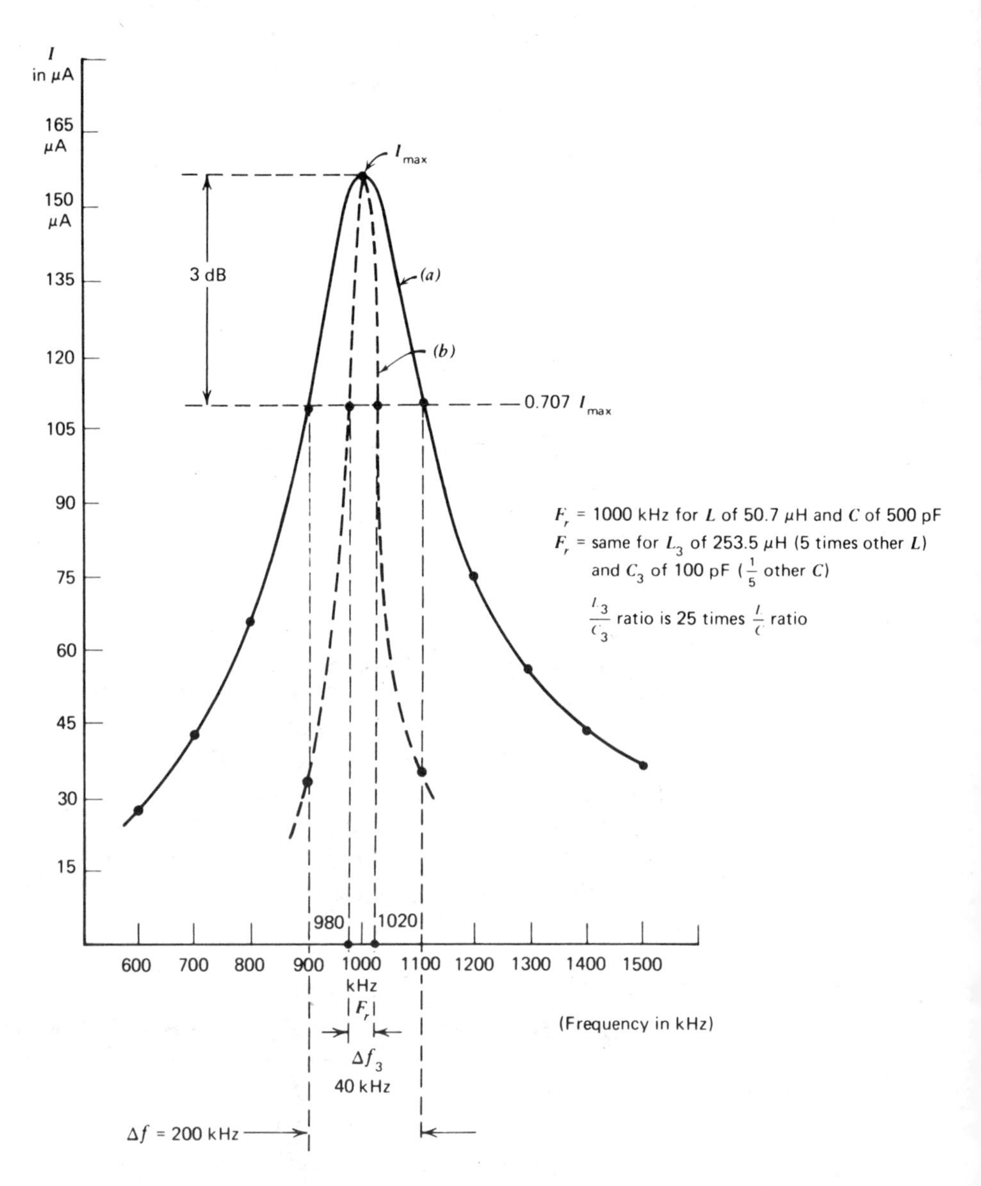

Figure 15-6. The *L/C* ratio, bandwidth (Δf), and *I* frequency-response curves. (*a*) *I* curve for $L = 50.7\ \mu\text{H}$ and $C = 500$ pF (circuit of Fig. 15-1). (*b*) *I* curve for $L_3 = 253.5\ \mu\text{H}$ and $C_3 = 100$ pF (higher *LC* ratio) (circuit of Fig. 15-5).

L_3/C_3 dashed-line curve, $\Delta f_3 = 40$ kHz. Note also that Q_3 at F_r, from Table 15-2 (with the larger value of L_3) $= X_{L_3}/R_3 = 1590/63.5 = 25$. This is five times larger than the Q of the original circuit of Fig. 15-1. This also agrees with Q_3 from the b curve of Fig. 15-6, where $Q_3 = F_r/\Delta f_3 = 1000$ kHz/40 kHz = 25.

In other words, increasing the L/C ratio (without changing R), produces a higher Q circuit with a narrower bandwidth.

Example 15-5

A 2 μH coil resonates with a 5-pF capacitor at approximately 50 MHz. Q of the coil, and the circuit, at F_r is 100. If the coil is increased 10 times to 20 μH, with no change in R, but C is decreased to 1/10 of its previous value, or 0.5 pF, find: (a) new Q, and (b) newbandwidth, Δf.

Solution

(a) Since L is 10 times larger, then X_L is 10 times larger. With no change of R, then Q is 10 times larger ($Q = X_L/R$), or Q is now (10) (100) = 1000.

(b) $\Delta f = F_r/Q = 50$ MHz/1000 = 0.05 MHz, or 50 kHz.

For a Similar Problem Refer to End-of-Chapter Problem 15-6

15-2. Parallel Resonant Circuits

A parallel resonant circuit, also called a *tank* circuit, is shown in Fig. 15-7. The inductance L is 50.7 μH, having a resistance R of 63.5 Ω. This series combination comprises the left branch Z_1. The right branch Z_2 is the 500-pF capacitor. These values of L, C, and R are identical to those of the series resonant circuit of Fig. 15-1. The resonant frequency F_r of the *series* resonant circuit, where $X_L = X_C$, is 1000 kHz. Resonance in the *parallel* circuit of Fig. 15-7 is more complex than in the series circuit. When $X_L = X_C$ in the parallel circuit, Z_1 is not equal to Z_2, since Z_1 has X_L in series with R, making Z_1 larger than Z_2. Parallel resonance is most accurately defined as the frequency where $Z_1 = Z_2$. Then, branch currents I_1 and I_2 are equal, resulting in equal reactive

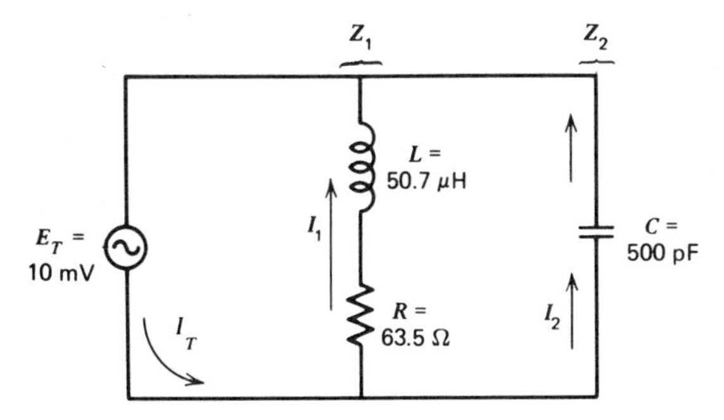

Figure 15-7. Parallel resonant circuit. $F_r = 1000$ kHz (approximate).

power in X_L and X_C. The power factor of the tank is then *unity* or 1, and the circuit acts as a pure R. I_T (also called I_{line}) is then in phase with the applied voltage E_T.

Parallel Resonant Frequency. To derive an equation for the resonant frequency $F_{r\ parallel}$ of a parallel circuit such as in Fig. 15-7, the *series* portion Z_1 consisting of X_L and R must be converted to an *equivalent parallel* circuit. This is shown in Fig. 15-8. The equivalent of X_L and R in series consist of an $X_{L\ parallel}$ and an $R_{parallel}$ which are

$$X_{L\ \text{parallel}} = \frac{X_L^2 + R^2}{X_L} \tag{15-7}$$

where X_L and R are the *series* components of Z_1 in Fig. 15-7.

$$R_{\text{parallel}} = \frac{X_L^2 + R^2}{R} \tag{15-8}$$

At the *parallel resonant frequency*, X_C should equal $X_{L\ \text{parallel}}$ in Fig. 15-8.

$$X_C = X_{L\ \text{parallel}}$$

$$\frac{1}{2\pi FC} = \frac{X_L^2 + R^2}{X_L}$$

$$\frac{1}{2\pi FC} = \frac{(2\pi FL)^2 + R^2}{2\pi FL} \qquad \text{(multiply both sides of equation by } 2\pi FL\text{)}$$

$$\frac{L}{C} = (2\pi FL)^2 + R^2$$

$$\frac{L}{C} = 4\pi^2F^2L^2 + R^2 \qquad \text{(transpose the } + R^2\text{)}$$

$$\frac{L}{C} - R^2 = 4\pi^2F^2L^2 \qquad \text{(combine left side into one fraction)}$$

$$\frac{L - CR^2}{C} = 4\pi^2F^2L^2 \qquad \text{(transpose the } 4\pi^2L^2\text{)}$$

$$\frac{L - CR^2}{4\pi^2L^2C} = F^2 \qquad \text{(take square root of both sides)}$$

$$\frac{1}{2\pi}\sqrt{\frac{L - CR^2}{L^2C}} = F \qquad \text{(divide numerator and denominator by } L\text{)}$$

$$\frac{1}{2\pi}\sqrt{\frac{(L/L) - (CR^2/L)}{L^2C/L}} = F$$

$$\frac{1}{2\pi}\sqrt{\frac{1 - (CR^2/L)}{LC}} = F$$

$$\frac{1}{2\pi\sqrt{LC}}\left(\sqrt{1-\frac{CR^2}{L}}\right) = F_{r\,\text{parallel}} \tag{15-9}$$

Note that the first part, $1/2\pi\sqrt{LC}$, is the resonant frequency of a *series* resonant circuit, and is the approximate parallel resonant frequency. By multiplying by the factor $[\sqrt{1-(CR^2/L)}]$, the result is the resonant frequency of the *parallel* circuit of Fig. 15-7, the Z_1 and Z_2 branches. If CR^2/L in a *parallel* circuit is larger than 1, then the number under the radical sign (the square root) will be a *negative* quantity. Since there is no answer to this, then such a circuit has no parallel resonant frequency.

An alternate equation to find the resonant frequency of a parallel circuit may be derived from the following. Again referring to the parallel *equivalent* circuit and X_C of Fig. 15-8, $X_C = X_{L\,\text{parallel}}$. This becomes $1/(2\pi FC) = (R^2 + X_L{}^2)/X_L$. Since $Q = X_L/R$, then $R = X_L/Q$. Substituting for R^2, the reader should solve for F. The result should be

$$F_{r\,\text{parallel}} = \frac{1}{2\pi\sqrt{LC}}\left(\sqrt{\frac{Q^2}{1+Q^2}}\right) \tag{15-10}$$

Note that, as in equation 15-9, the first part $(1/2\pi\sqrt{LC})$ is the resonant frequency of a *series* circuit. The factor $[\sqrt{Q^2/(1+Q^2)}]$ multiplied by the series resonant frequency gives the parallel resonant frequency. If Q is about 10 or larger, then the factor $[\sqrt{Q^2/(1+Q^2)}]$ is just about 1, and the parallel *resonant frequency* is about the same as the series resonant frequency $(1/2\pi\sqrt{LC})$.

The following examples illustrate the use of equations 15-9 and 15-10.

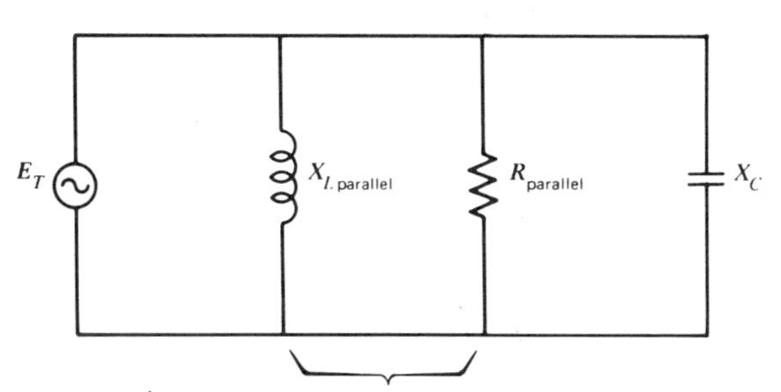

Parallel equivalents of X_L and R in series (Z_1) of Fig. 15–7

$$R_{\text{parallel}} = \frac{X_L{}^2 + R^2}{R}$$

(where X_L and R are the *series* components of Fig. 15-7)

$$X_{L\,\text{parallel}} = \frac{X_L{}^2 + R^2}{X_L}$$

Figure 15-8.

Example 15-6

In the parallel resonant circuit of Fig. 15-7 where $L = 50.7\ \mu H$, $R = 63.5\ \Omega$, and $C = 500$, find the *parallel resonant frequency* using equation 15-9.

Solution

$$F_{r\ \text{parallel}} = \frac{1}{2\pi\sqrt{LC}}\sqrt{1 - \frac{CR^2}{L}} \qquad (15\text{-}9)$$

$$F_{r\ \text{parallel}} = \left(\frac{1}{(6.28)\sqrt{(50.7 \times 10^{-6})\,(500 \times 10^{-12})}}\right)$$

$$\left(\sqrt{1 - \frac{(500 \times 10^{-12})\,(63.5)^2}{50.7 \times 10^{-6}}}\right)$$

The reader should solve this example, and should find that

$$F_{r\ \text{parallel}} = 1000\ \text{kHz}\ (0.98)$$

$$F_{r\ \text{parallel}} = 980\ \text{KHz}$$

Example 15-7

Find the parallel resonant frequency of the circuit of Fig. 15-7, using equation 15-10, and compare it with the previous example.

Solution

$$F_{r\ \text{parallel}} = \frac{1}{2\pi\sqrt{LC}}\sqrt{\frac{Q^2}{1 + Q^2}} \qquad (15\text{-}10)$$

First the value of Q must be found. By using the approximate resonant frequency $(1/2\pi\sqrt{LC})$, $F_r = 1000$ kHz, then $Q = X_L/R = 2\pi FL/R = (6.28)(1000 \times 10^3)\,(50.7 \times 10^{-6})/63.5 = 318/63.5 = 5$.

Therefore,

$$F_{r\ \text{parallel}} = \frac{1}{2\pi\sqrt{LC}}\sqrt{\frac{Q^2}{1 + Q^2}}$$

$$F_{r\ \text{parallel}} = \left(\frac{1}{(6.28)\sqrt{(50.7 \times 10^{-6})\,(500 \times 10^{-12})}}\right)\sqrt{\frac{5^2}{1 + 5^2}}$$

$$F_{r\ \text{parallel}} = (1000\ \text{kHz})\sqrt{\frac{25}{26}}$$

$$F_{r\ \text{parallel}} = (1000\ \text{kHz})\,(0.98)$$

$$F_{r\ \text{parallel}} = 980\ \text{kHz}$$

Note that this is the same as in the previous example. If R were much smaller, resulting in a much higher Q, then the parallel resonant frequency would be the same as series resonance $(1/2\pi\sqrt{LC})$, or 1000 kHz.

For Similar Problems Refer to End-of-Chapter Problems 15-8 and 15-9

In a parallel resonant circuit, there are three slightly different frequencies that may be regarded as resonance.

1. The frequency where $X_L = X_C$ (same as series resonance).

2. The frequency where Z_T is maximum, and I_T (or I_{line}) is minimum.

3. The frequency where the circuit acts as a pure R; $I_1 = I_2$; or the reactive power of L and C are equal; power factor = 1.

When the Q of the circuit is about 10 or larger, the above three frequencies are just about the same as the *series* resonant frequency of $1/(2\pi\sqrt{LC})$.

Table 15-3 lists the values of X_C, X_L, Z_1, Z_2, Z_T, and the branch currents I_1 and I_2 of the parallel resonant circuit of Fig. 15-7 for frequencies below, at, and above the *parallel resonant frequency* of 980 kHz. Note that the total impedance Z_T *is maximum* at $F_{r\text{ parallel}}$, resulting in minimum *line current* I_T at this frequency. At a frequency *below* $F_{r\text{ parallel}}$ (500 kHz), the circuit is *inductive* since X_L (and Z_1) is less than X_C (or Z_2), and more current flows through the inductive branch of Fig. 15-7. This may be seen from the phase angle of I_T in Table 15-3 which is $-60.6°$, the *negative* sign denoting a lagging (or inductive) current.

At $F_{r\text{ parallel}}$ (980 kHz), the circuit is purely resistive as shown by the zero degree I_T in the table. At a frequency just slightly above this, or at the series resonant frequency (F_r = 1000 kHz), the circuit is slightly capacitive since I_T in the table is leading by a small angle of 11.3°.

At a frequency *above* $F_{r\text{ parallel}}$, or 1500 kHz in Table 15-3, X_C now becomes smaller than X_L. More current now flows in the capacitive branch of Fig. 15-7, making the circuit *capacitive*. This is shown by I_T in the table with an angle of $+84.1°$, meaning that the current *leads* E_T (or E_T lags the current).

Figure 15-8*A* shows the graphs of Z_T and I_T of a *parallel resonant circuit*, with Z_T being *maximum* and I_T minimum at $F_{r\text{ parallel}}$. Note that these are just the reverse of Z and I in a series resonant circuit, as shown in the graphs of Fig. 15-2 and 15-3. The parallel resonant circuit is often called an anti-resonant circuit.

Refer to end-of-chapter Problem 15-10

TABLE 15-3 (For Circuit of Fig. 15-7)

Frequency	X_C (Z_2) $(1/2\pi FC)$	I_2 (E_T/X_C) μA	X_L $(2\pi FL)$	Z_1 $(\sqrt{R^2+X_L^2})$ $(\tan\theta_1 = X_L/R)$	I_1 (E_T/Z_1) μA	Z_T $\left(\frac{Z_1 Z_2}{Z_1+Z_2}\right)$	I_T I_{line} (E_T/Z_T)
500 kHz	$0 - j636$ $636\angle -90°$ Ω	$15.7\angle +90°$ μA	$0 + j159$	$63.5 + j159$ $171\angle 68.2°$ Ω	$58.5\angle -68.2°$ μA	$225\angle 60.6°$ Ω	$44.5\angle -60.6°$ μA
[a]980 kHz	$0 - j325$ $325\angle -90°$ Ω	$30.8\angle +90°$ μA	$0 + j312$	$63.5 + j312$ $319\angle 78.5°$ Ω	$31.4\angle -78.5°$ μA	$1630\angle 0°$ Ω	$6.13\angle 0°$ μA
1000 kHz	$0 - j318$ $318\angle -90°$ Ω	$31.4\angle +90°$ μA	$0 + j318$	$63.5 + j318$ $324\angle 78.7°$ Ω	$30.8\angle -78.7°$ μA	$1620\angle -11.3°$ Ω	$6.16\angle +11.3°$ μA
1500 kHz	$0 - j212$ $212\angle -90°$ Ω	$47.1\angle +90°$ μA	$0 + j477$	$63.5 + j477$ $482\angle 82.4°$ Ω	$20.7\angle -82.4°$ μA	$377\angle -84.1°$ Ω	$26.5\angle +84.1°$ μA

[a] F_r parallel.

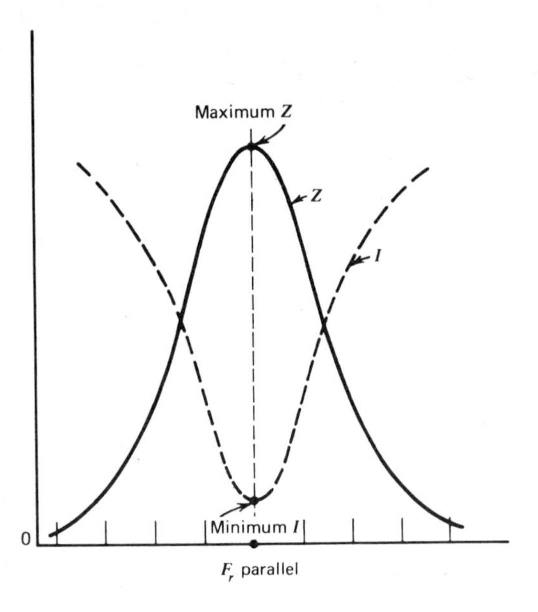

Figure 15-8*a*. *Z* and *I* graphs of parallel resonant circuit.

15-3. Filters

In electronics, a *filter* is a circuit that allows voltages at certain desired frequencies to be passed on to the load while stopping other undesired frequencies. In the next chapter on electron tubes, it will be shown that the voltage at the output of a rectifier tube in the power supply of a radio or television receiver consists of dc and ac components. Since the purpose of the power supply is to produce a dc voltage for the amplifier stages, then a filter permits the dc component to reach the amplifiers, while stopping the alternating current (or ripple). Similarly, as is shown in this next chapter, at the plate (or output point) of an amplifier stage, there exists a dc and an ac voltage. A filter is used here also to permit the ac component (called the *signal*) to be passed on to the grid (the input) of the next amplifier, while stopping the dc voltage. These filters are discussed in the Electron Tube chapter. In this present section, filters that discriminate between ac voltages at different frequencies are discussed.

Resonant Filters. This type of filter consists of either a *series resonant* L and C which has very *low* Z at F_r, or a *parallel resonant* L and C which has a very *high* Z at $F_{r\,\text{parallel}}$, or both circuits. The resonant filter is also called a *wave trap*, since it "traps" or stops an undesired frequency voltage.

Band-Pass Filter. Figure 15-9*a* shows the circuit of a *band-pass filter* using both a series resonant circuit ($L_1 - C_1$) and a parallel resonant circuit ($L_2 - C_2$). Three separate input ac voltages of 10 V each are applied at frequencies much higher than F_r, at F_r and much lower than F_r. L_1 and C_1, the series circuit, is resonant at 1000 kHz. L_2 and C_2, the parallel circuit, is also resonant at 1000 kHz.

As shown in Fig. 15-9*b*, the impedance Z_1 of the series circuit $L_1 - C_1$ is very low (shown as 10 Ω) at the resonant frequency input signal, while Z_2 of the parallel circuit $L_2 - C_2$ is very high (shown as 10 kΩ) at this same input signal. With Z_1 only 10 Ω in series with $Z_2 =$ 10 K in parallel with the 1-K load resistor, practically no ac voltage is developed across the small Z_1, and the entire 10 V ac is developed across the large Z_2 and the load resistor, as shown at the output of part (b).

In Fig. 15-9*c*, the circuit is shown as it appears to an input ac voltage very much *higher* than F_r, and also to an input ac voltage very much *lower* than F_r. The series circuit $L_1 - C_1$ now appears as a high Z_1 (it is low only at F_r), while the parallel circuit $L_2 - C_2$ acts as a low Z_2 (it is high only at F_r). The large Z_1 (shown as 10 K) is in series with the parallel combination of the very small Z_2 (shown as 10 Ω) and the 1-K load resistor. Z_2 (10 Ω) in parallel with R_{load} (1 K) results in a combination of only about 10 Ω. As a result, practically the entire 10 V input ac voltage is developed across the large Z_1 (10 K), and practically zero volts appears across the small $Z_2 - R_{load}$ combination (10 Ω). As shown in Fig. 15-9*c*, the output voltage = 0 at the very high and very low frequency voltages.

Figure 15-9*d* shows the graph of output voltage versus frequency for the band-pass filter circuit of part (a). This graph is usually called the *frequency response* curve. Note that E_{output} is large for a band of frequencies between f_1 and f_2. These are the frequencies passed by the filter. The shaded areas below f_1 and above f_2 denote the low frequencies and the high frequencies not passed by the filter.

Band-Stop Filter. This circuit, shown in Figure 15-10*a*, stops a band of frequencies, but allows lower and higher frequencies to pass on to the load. This is just the reverse of the band-pass filter of Fig. 15-9. Note that the two circuits have the series resonant $L_1 - C_1$ and the parallel resonant $L_2 - C_2$ in opposite positions.

In Fig. 15-10*b*, the band-stop filter is shown as it appears to a 10 V input ac voltage at the *resonant frequency*. The *impedance* Z_2 of the *parallel resonant* $L_2 - C_2$ is *maximum at* F_r, and is shown as 10 kΩ. The *impedance* Z_1 of the *series resonant* $L_1 - C_1$ is *minimum at* F_r and is shown as 10 Ω. With Z_2 so much larger than Z_1, practically the entire 10 V input is developed across Z_2, leaving zero volts across the small Z_1 in parallel with the 1 K R_{load}.

Figure 15-10*c* depicts the band-stop filter as it appears to a 10 V ac input signal at a frequency much *lower* than F_r, and also much *higher* than F_r. The

Figure 15-9. (Opposite) Frequency-response graph of band-pass filter. Band-pass filter circuit theory. (*a*) Band-pass filter. (*b*) Equivalent circuit at F_r. (c) Equivalent circuit at high and low frequencies. (*d*) Frequency-response graph of band-pass filter.

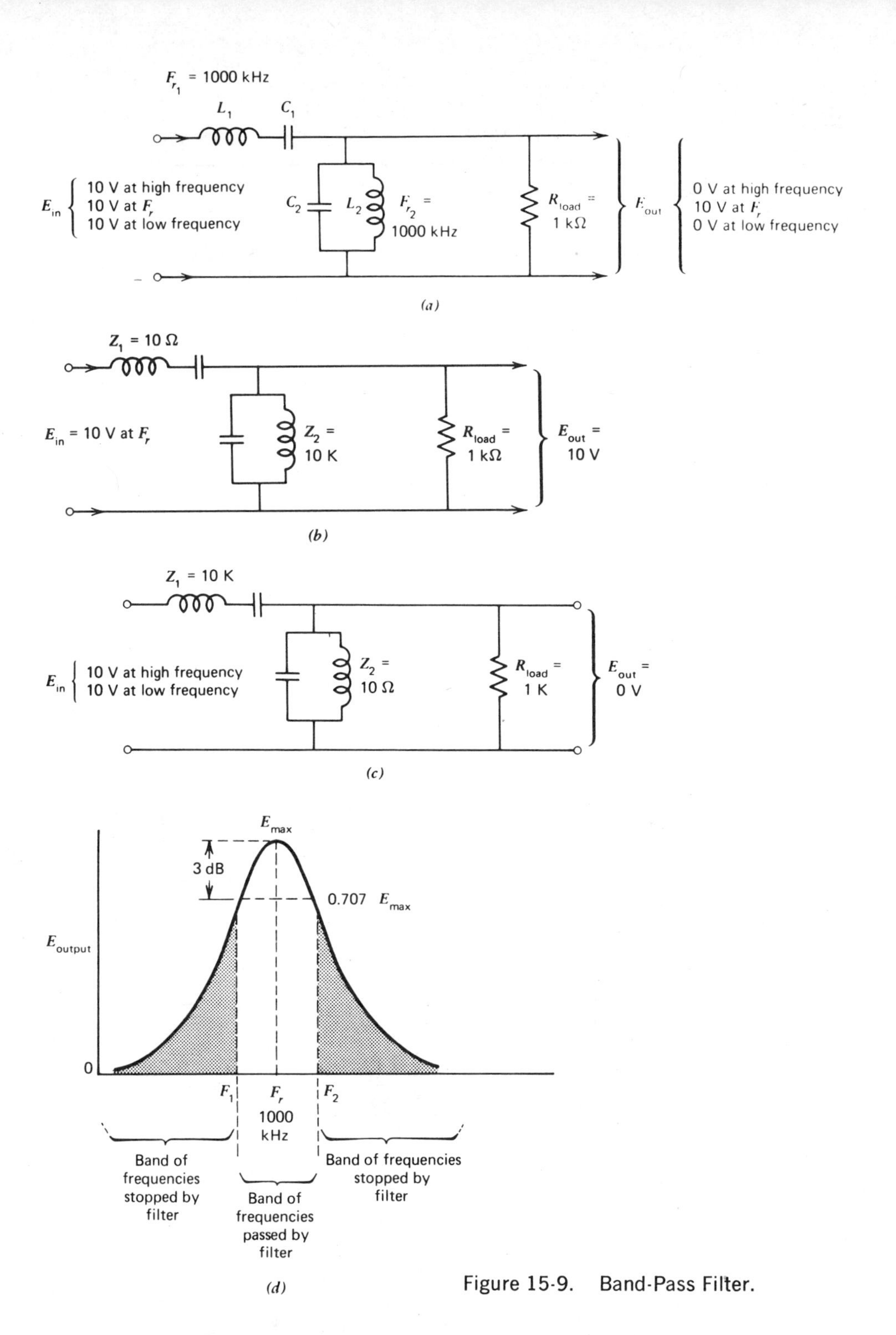

Figure 15-9. Band-Pass Filter.

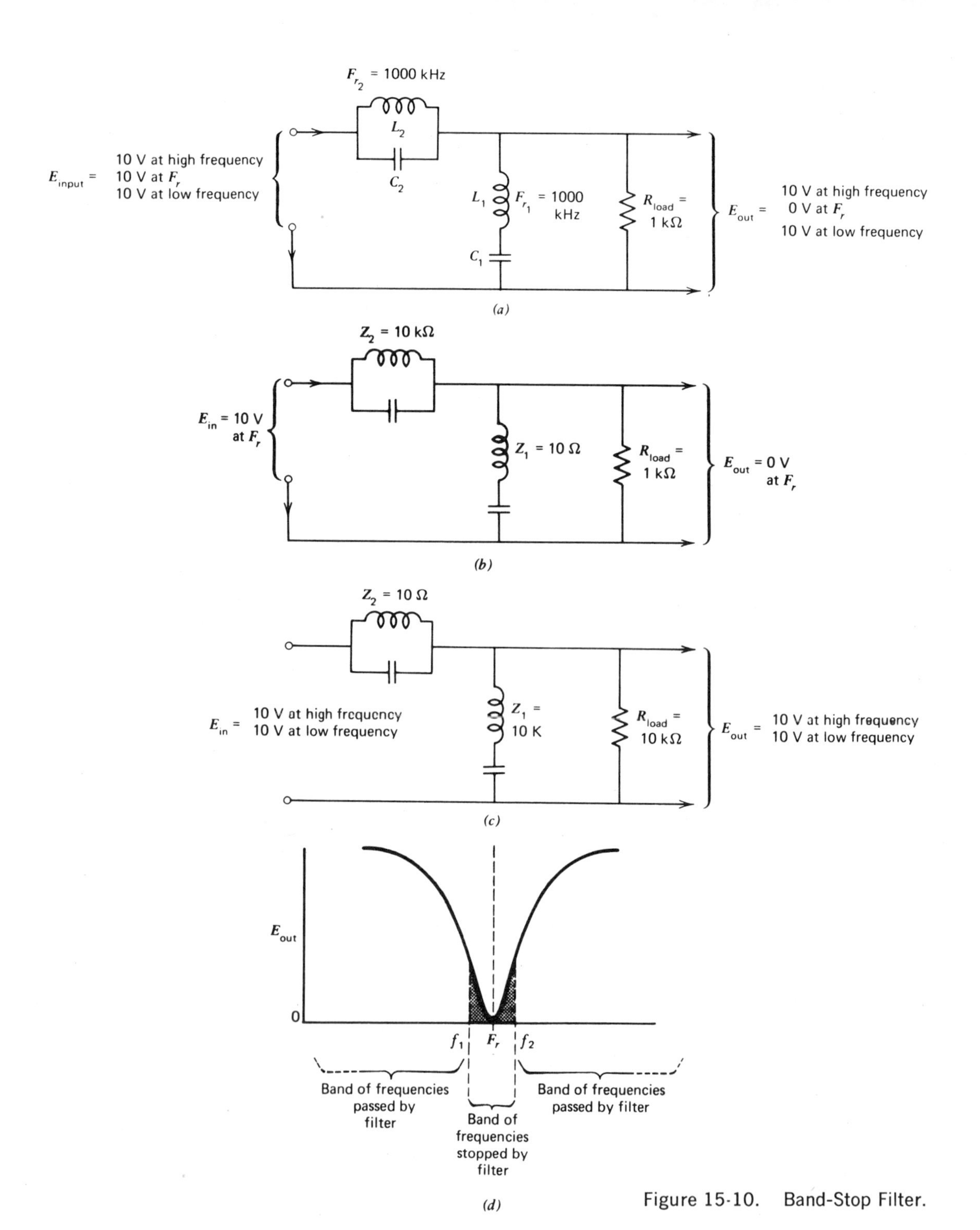

Figure 15-10. Band-Stop Filter.

impedance of the parallel tank Z_2 is very small at frequencies much below and much higher than F_r (*parallel* tank Z_2 is only *high* at F_r). Z_2 is shown as only 10 Ω. Impedance Z_1 of the series circuit is a large value at frequencies much below and much higher than F_r (*series* impedance Z_1 is only *low* at F_r). Z_1 is shown here as 10 K. As a result of Z_1 and R_{load} being very much larger than Z_2, then practically the full 10-V input is developed across Z_1 and R_{load}, and is shown as a 10-V output.

Part (d) of Fig. 15-10 is a graph of E_{output} versus frequency. This is called the *frequency response* curve. Note that at F_r, as shown in part (b), $E_{output} = 0$, and the graph of part (d) is down at zero volts at F_r. The frequency band from f_1 to f_2 is stopped by the filter. E_{output} is large and appears across R_{load} at frequencies below f_1 and also above f_2 as shown in the graph.

Nonresonant Filters. This type employs an L and C, but its resonant frequency is far enough away from the ac signal voltage frequencies being handled that the filter is considered nonresonant.

Low-Pass Filter. Figure 15-11*a* shows the circuit of a *low-pass filter*. At a *low* frequency, X_{L_1} *is very small*, shown as 10 Ω, while X_{c_1} *is very large*, shown as 10 K. As a result, practically the entire 10-V input at the low frequency is developed as E_{output} across the large X_{c_1} and R_{load}.

At a *high* frequency, X_{L_1} *becomes very large*, shown in parentheses as 10 K, while X_{c_1} *becomes very small*, shown in parentheses as 10 Ω. As a result, the 10 V ac input signal produces practically the full 10 V across the large X_{L_1}, leaving zero volts across the small X_{C_1} as E_{output}.

The *frequency response curve* of Fig. 15-11*b* shows that a band of low frequency signals produces a large E_{output} up to the *cutoff frequency*. The high frequency signals above this point produce very small or zero E_{output}.

High-Pass Filter. Figure 15-12*a* shows the circuit of a *high-pass filter*. With a 10-V ac signal input at a *high* frequency, X_{c_2} is very *small*, shown as 10 Ω. At this *high frequency*, X_{L_2} is *large*, shown as 10 K. As a result, the 10-V input produces practically the full 10 V across the large X_{L_2} (10 K) in parallel with the 1 K R_{load}.

At a *low* frequency, the situation reverses. Now, X_{C_2} is *large* in Fig. 15-12*a*, shown in parentheses as 10 K. At this *low* frequency, X_{L_2} *is small*, shown in parentheses as 10 Ω. As a result, the 10 V ac input low frequency signal produces practically the full 10 V across the large X_{C_2}, leaving zero volts across the small X_{L_2}. The output is shown as zero volts at the low frequency.

The *frequency response* curve of a high-pass filter is shown in Fig. 15-12*b*. E_{output} is maximum at the high frequencies. The *cutoff frequency* is defined as that which produces an E_{output} which is 0.707 E_{max}. Frequencies lower than this point are said to be stopped by the filter.

Figure 15-10. (Opposite) Band-stop filter circuit theory. (*a*) Band-stop filter. (*b*) Equivalent circuit at F_r. (*c*) Equivalent circuit at high and low frequencies. (*d*) Frequency-response curve for band-stop filter.

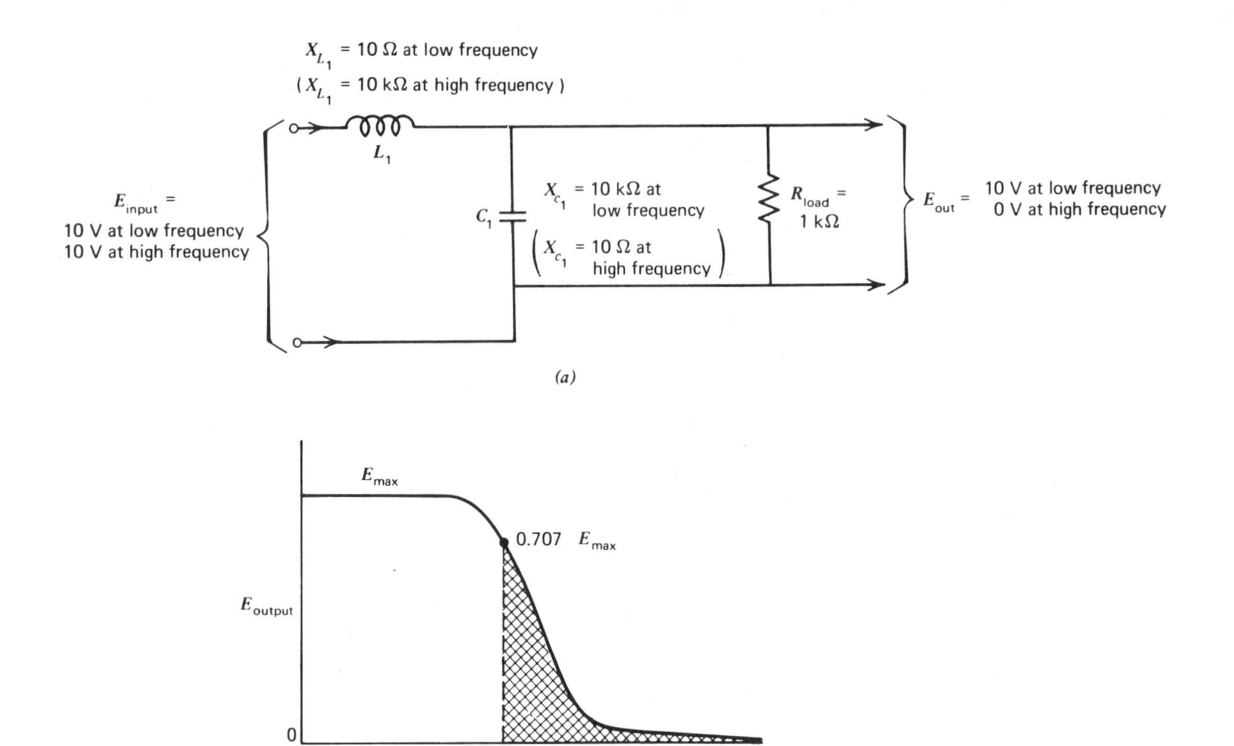

Figure 15-11. Low pass filter circuit theory. (*a*) Low pass filter. (*b*) Frequency response curve for low pass filter.

Example 15-8

A filter has a frequency response curve shown in Fig. 15-13. (a) What type filter is this? (b) Draw the circuit.

Solution

(a) since E_{output} is at zero volts at F_r, then this filter stops F_r and nearby frequencies, but passes all others. The filter is therefore a *band-stop* filter.

(b) The circuit is shown in Fig. 15-10*a*.

For Similar Problems See End-of-Chapter Problems 15-11 to 15-13

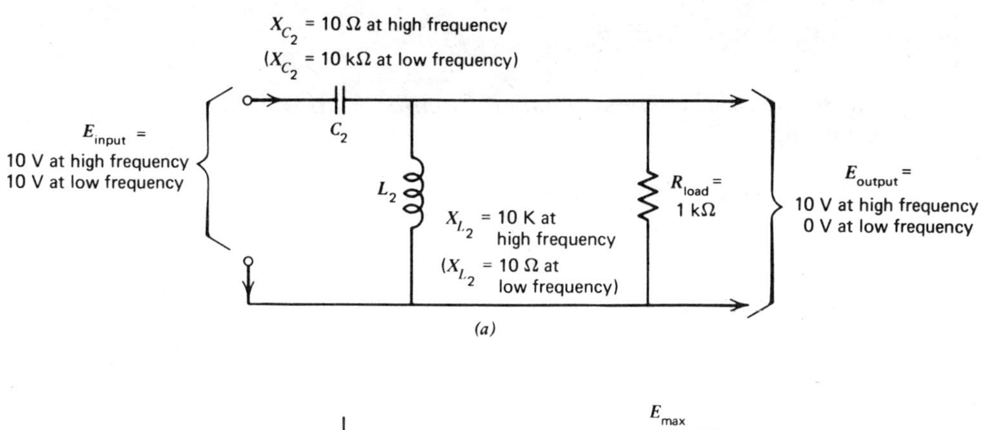

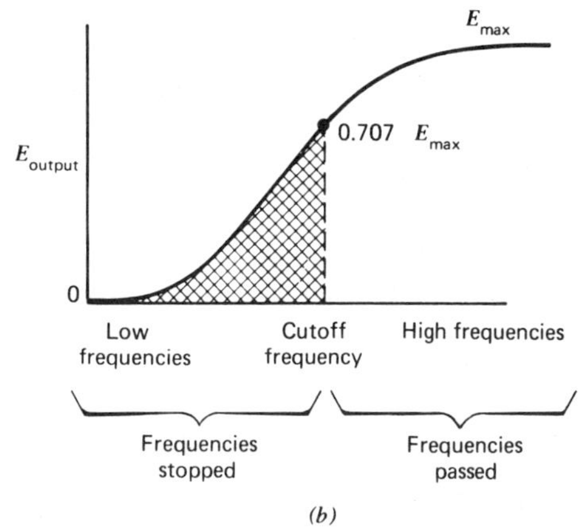

Figure 15-12. High-pass filter circuit theory. (*a*) High-pass filter. (*b*) Frequency-response curve for high-pass filter.

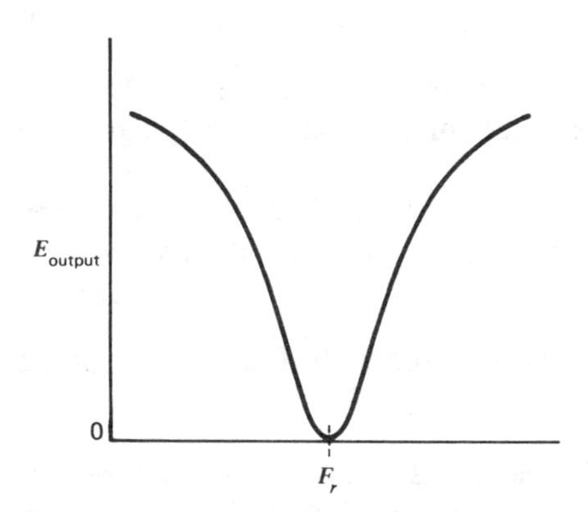

Figure 15-13. Frequency-response curve, Example 15-8.

PROBLEMS

See section 15-1, Series Resonant Circuits, for discussion covered by the following.

15-1. A series circuit consists of L, R, and C. At the resonant frequency $X_L = 2\ k\Omega$, $X_C = 2\ k\Omega$, and $R = 100\ \Omega$. If the ac applied voltage is 50 μV, find: (a) Z, (b) I, (c) E_R, (d) E_{X_L}, and (e) E_{X_C}.

See section 15-1, Quality of Tuned Circuit, for discussion covered by the following.

15-2. A series resonant circuit has a $Q = 50$, and a $F_r = 1500$ kHz. Find: (a) Δf, (b) lowest frequency passed F_1, and (c) highest frequency passed F_2.

See section 15-1, Q Rise of Voltage at F_r, for discussion covered by the following.

15-3. If 40 μV at F_r is applied to a series resonant circuit having a Q of 75, find E_{X_C}.

See section 15-1, Resonant Frequency, for discussion covered by the following.

15-4. What is the resonant frequency F_r of a 10-pF capacitor and a 100-μH coil connected in series?

15-5. What value C will produce resonance at 350 kHz with a 10-mH coil?

See section 15-1, L/C Ratio, for discussion covered by the following.

15-6. A coil and capacitor resonate at 1400 kHz. If the value of the coil is doubled, while the capacitor value is halved, what is the F_r of the new combination?

15-7. In the previous problem, if R remains the same, what happens to (a) the L/C ratio, (b) Q, and (c) the band-width Δf?

See section 15-2, Parallel Resonant Frequency, for discussion covered by the following.

15-8. A 10 pF capacitor is in parallel with a series combination consisting of a 100-μH coil a resistance of 500 Ω. Find: (a) the series resonant frequency F_r, (b) the parallel resonant frequency $F_{r\ parallel}$.

15-9. In the previous problem, if the resistance of the coil were much less, resulting in a Q of 300, find the $F_{r\,parallel}$.

15-10. At $F_{r\,\text{parallel}}$, what are the relative values of (a) Z_T and (b) I_T?

See section 15-3, Filters, for discussion covered by the following.

15-11. A filter has a frequency response curve shown for this problem. (a) What type filter is this? (b) Draw the circuit.

15-12. (a) What type of filter has the frequency response curve shown here? (b) Draw this circuit.

15-13. (a) What type of filter has the frequency response curve shown for this problem? (b) Draw this circuit.

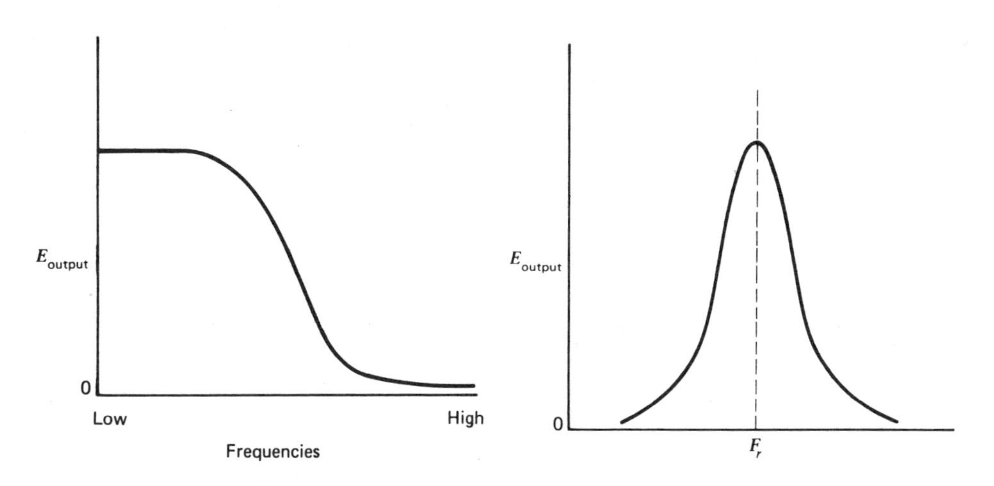

Problem 15-11.

Problem 15-12.

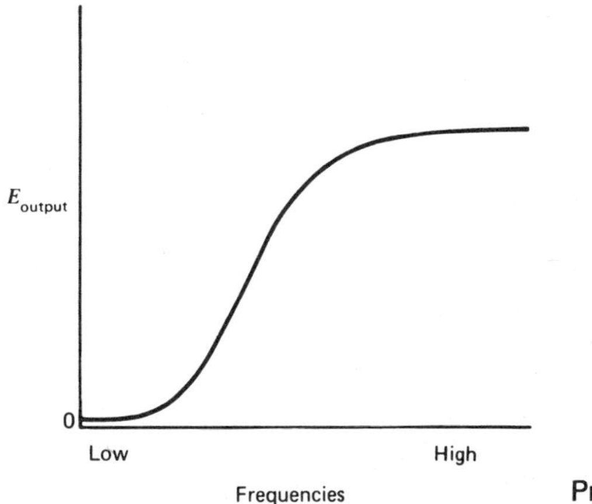

Problem 15-13.

chapter

16

Introduction to Vacuum Tubes

Although the study of electron tubes really belongs in an electronics book rather than in one like this on basic electricity, the simple basic theory of tubes is given here as an introduction only, preceding a course in electron tubes.

16-1. The Diode Vacuum Tube

A *diode* tube is a two-element device, as shown in Figs. 16-1 and 16-2. A *filament* wire is placed inside the hollow, tubular *cathode*. The *plate* or *anode* is a larger tube in which the cathode and filament are located.

When a conductor is heated, the electrons speed up as they travel in their orbits. If the heat is great enough, some electrons are hurled off the metal. This is called *thermionic* or *thermal emission*. The cathode may be the red-hot filament itself, in which case it is called a *directly heated cathode*. Most tubes use the red-hot filament or heater simply to heat up a separate cathode called the *indirectly heated cathode*. To permit a greater number of electrons to be emitted, the cathode is usually coated with an oxide, making the cathode a much greater emitter than a bare metal type.

Edison, in his experimental work with the incandescent light bulb, observed that a very tiny current (a microampere or so) flowed between the heated filament of his light bulb and a metal plate inside the same glass envelope, as shown in Fig. 16-3. This occurrence is called the *Edison effect*. Sometime later,

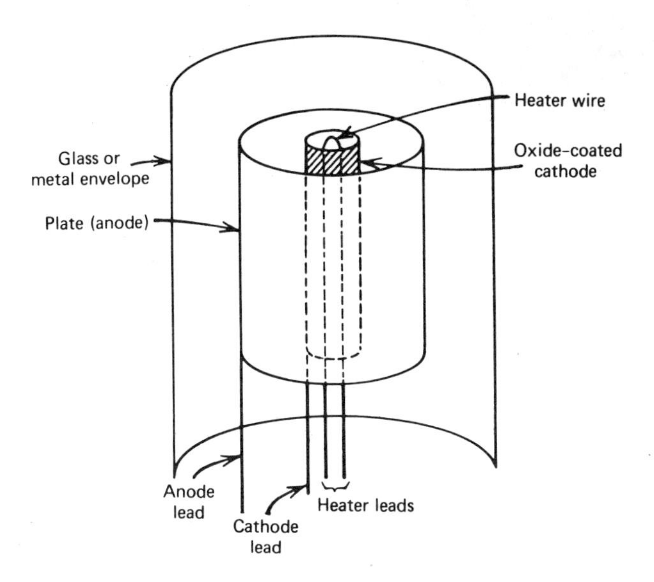

Figure 16-1. Physical construction of diode vacuum tube (indirectly heated cathode).

J. A. Fleming, the Englishman, invented his "valve." This is the diode vacuum tube shown in the symbol diagram of Fig. 16-2. To prevent the heated filament from burning up, as much air as possible is removed from the tube envelope. Air would also impede the movement of the emitted electrons. To improve the vacuum, a magnesium compound called the *getter* is flashed during the tube manufacture to "get" the last traces of air remaining in the tube.

The emitted electrons of Fig. 16-2 leave the hot cathode, but attempt to fall back onto it. Newer emitted electrons repel the previously emitted ones,

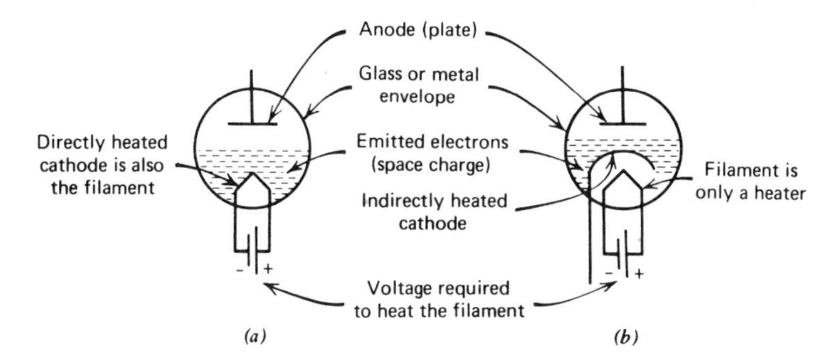

Figure 16-2. The diode vacuum tube. (*a*) Directly heated cathode. (*b*) Indirectly heated cathode.

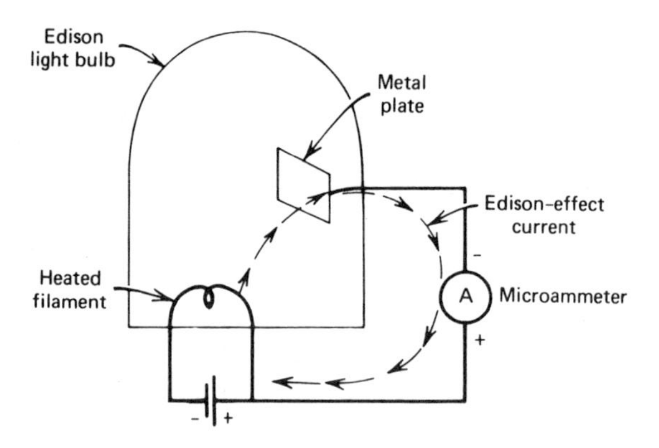

Figure 16-3. The Edison effect.

preventing them from returning to the cathode. A cloud of electrons called the *space charge* then forms, preventing further emission. Fleming found that by connecting a dc voltage source between plate and cathode with the plus end at the plate, as in Fig. 16-4*a*, the plate attracted some of the emitted electrons, and plate current flowed. This voltage is separate from the filament voltage. The electron movement goes from cathode to plate, through the plate supply voltage, and back to the cathode, for a complete path.

In Fig. 16-4*b*, when the *polarity* of the plate supply voltage is reversed, making the plate negative, no current can flow. The negative plate repels the emitted electrons. The diode acts as a valve, permitting a plate current (electron movement) in one direction only, from the cathode to the positive plate.

The amount of plate current, with normal heater voltage, depends on the attraction of electrons by the positive plate. If the plate supply voltage is in-

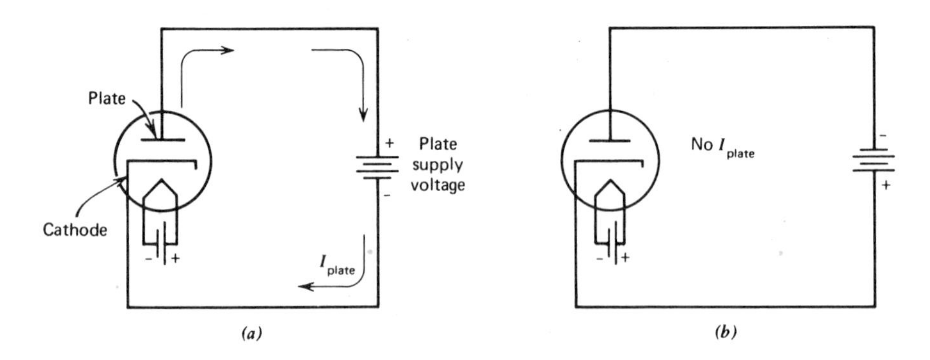

Figure 16-4. The diode (Fleming value) only permits plate current to flow in one direction. (*a*) *I* plate flows when plate is +, attracting emitted electrons. (*b*) No *I* plate flows when plate is −, repelling emitted electrons.

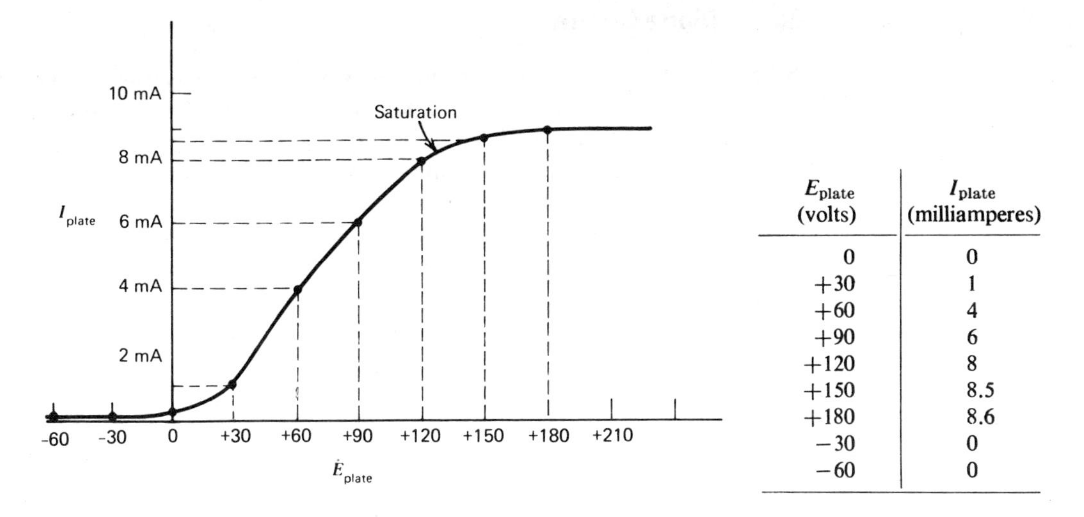

E_{plate} (volts)	I_{plate} (milliamperes)
0	0
+30	1
+60	4
+90	6
+120	8
+150	8.5
+180	8.6
−30	0
−60	0

Figure 16-5. $E_{plate} - I_{plate}$ **diode characteristics.**

creased, making the plate more positive, I_{plate} increases. This is shown in the table and graph of Fig. 16-5. Note that no plate current flows when the plate is either negative or zero. As the plate supply voltage of Fig. 16-4*a* is increased, I_{plate} also increases. However at some point *saturation* is reached. That is, when practically every emitted electron from the cathode gets to the plate, then making the plate more positive results in almost no increase in plate current. This occurs in the $E_{plate} - I_{plate}$ graph as the graph flattens out starting in the +120 to +150 V E_{plate} region, in this example.

Diode Characteristics, Plate Resistance. When any device has voltage across it and a current through it, there is a resistance or opposition in the device. A diode is no exception. Its resistance is called R_{plate}. The diode has an internal *dc resistance* which, using Ohm's law, is: $R_{plate\ dc} = E_{plate}/I_{plate}$ (equation 16-1). At various plate voltages, the internal $R_{plate\ dc}$ is different. The graph of Fig. 16-5 is linear (a straight line) in the E_{plate} area between +60 to +120 V. At the +90 V point, for example, $R_{plate\ dc} = 90\ \text{V}/6\ \text{mA} = 90/(6 \times 10^{-3})\ \Omega$ or 15 kΩ.

There is also an internal *ac resistance* of a diode. This is the resistance of the diode to a *change* or *variation* of plate current. The $R_{plate\ ac}$ = change of E_{plate}/change of I_{plate}, or $\Delta E_{plate}/\Delta I_{plate}$ (equation 16-2). The Δ means a small *change*. Referring to the graph and table of Fig. 16-5, when the plate voltage is changed from +60 to +90 V, $\Delta E_{plate} = 30$ V. This results in I_{plate} going from 4 to 6 mA, or $\Delta I_{plate} = 2$ mA. Therefore, $R_{plate\ ac} = \Delta E_{plate}/\Delta I_{plate} = 30\ \text{V}/2\ \text{mA} = 30/(2 \times 10^{-3}) = 15 \times 10^{3}\ \Omega = 15\ \text{k}\Omega$. Usually $R_{plate\ ac}$ is less than $R_{plate\ dc}$ although in the examples shown here, they are equal.

16-2. Diode Circuit

Since a diode only permits current flow when the plate is positive with respect to the cathode, but no current when the plate is negative, then the diode is called a *rectifier*. That is, if an ac voltage is applied to the diode, it will only conduct when the alternating voltage makes the plate positive. As a result, the *ac applied voltage* produces a *dc output voltage* from the diode rectifier.

Figure 16-6*a* shows a *half-wave rectifier power supply*. The term *half wave* means the diode produces a dc output voltage during only a half cycle of the ac sine-wave input voltage. First consider the circuit without the low-pass π type filter components (C_1, L_1, and C_2). On each positive half cycle of the secondary ac voltage (shown as the shaded area), with the polarity shown across the secondary, I_{plate} flows. This electron movement flows from cathode to plate, down through the secondary, and up through R_{load} back to the cathode. This produces a positive half cycle of voltage across R_{load}. When the secondary ac voltage goes negative, driving the diode plate negative, no current flows. $E_{R\,load}$ now becomes zero. The output voltage as shown (unfiltered) is a series of pulsating dc voltages. This actually consists of a dc component and an ac component (or ripple).

The power supply must furnish a dc voltage with almost no alternating voltage (no ripple) to the amplifier stages. The low-pass π type filter is, therefore, used to allow the dc voltage to appear across R_{load}, while stopping most of the ac component (or ripple). The filter choke or inductance L_1 (about 30 H) acts as a small resistor to the direct current (about 100 Ω), while the filter capacitors C_1 and C_2 (about 50 μF each) act as a huge resistance to the direct current. As a result, practically the *full dc voltage* from the diode cathode appears across R_{load}.

At the 60-Hz ac component, X_{L_1} is very large (about 11 kΩ), while X_{C_1} and X_{C_2} are very small (about 50 Ω) acting as almost a short circuit. As a result, practically *no ac component* appears across R_{load} in the output. The filtered output voltage is then practically only direct with no alternating voltage.

Refer to End-of-Chapter Problems 16-1 to 16-5

16-3. The Triode Vacuum Tube

An American, Lee De Forest, in 1907 added a third element to the vacuum tube. This element, called the *control grid*, consists of widely spaced loops of wire, all connected together, encircling the cathode but not touching it. As shown in Figure 16-7*a*, the grid is between the cathode and plate. The circuit symbol for a triode is shown in part (b) of this figure. The emitted electrons from the heated cathode must pass between the grid wires on the way to the plate or anode. The grid, therefore, has an influence on the number of elec-

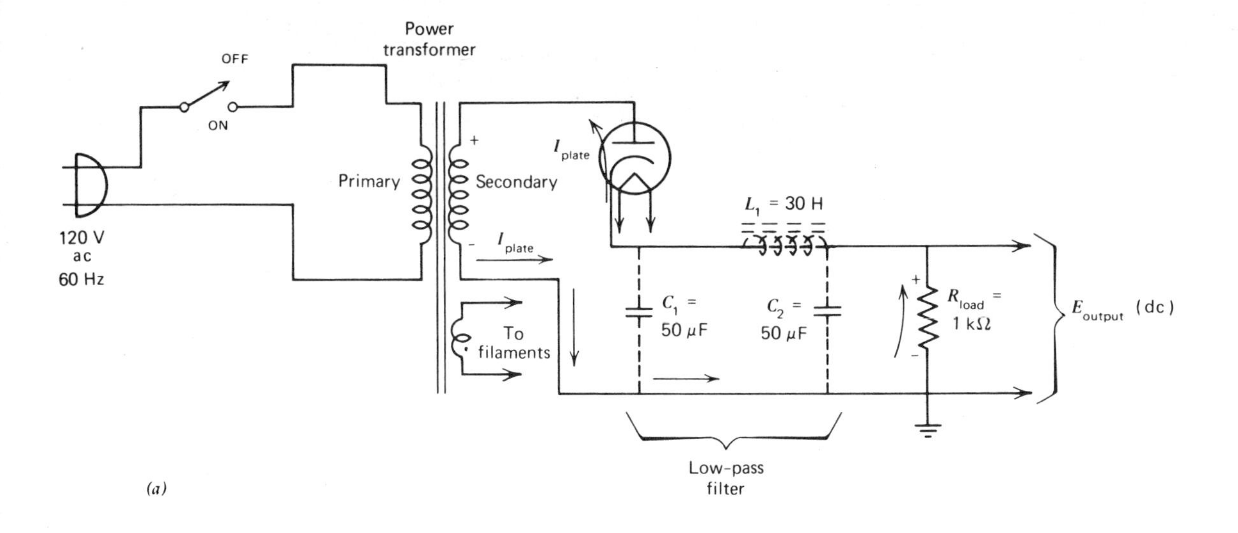

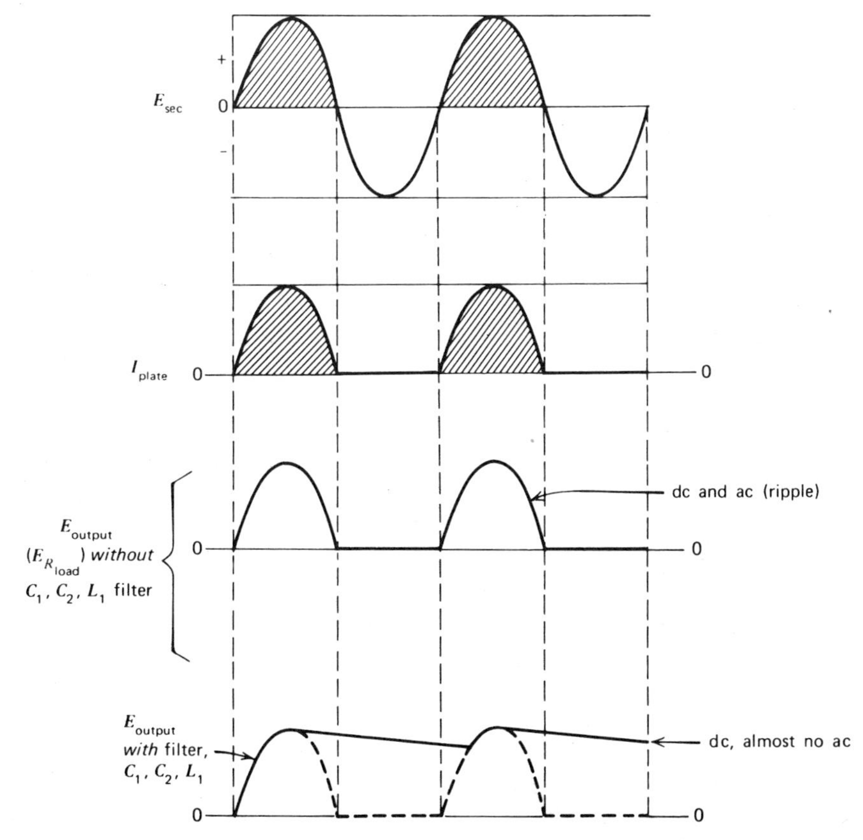

Figure 16-6. Half-wave rectifier power supply. (*a*) Circuit. (*b*) Wave shapes.

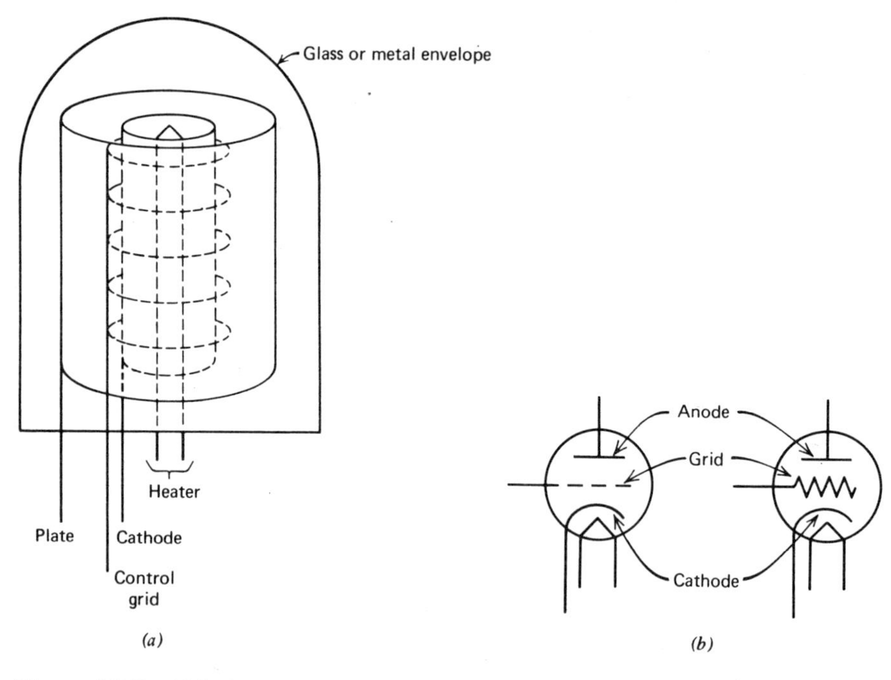

Figure 16-7. Triode vacuum tube. (*a*) Physical construction. (*b*) Circuit symbol.

trons that reach the plate. If the grid is made sufficiently negative, it repels all electrons coming from the cathode, and none get through to the plate. This is called *plate current cutoff*.

As the grid is made less negative, a few electrons get through to the plate. More reach the plate as the grid is made still less negative. The voltage on the grid with *respect to the cathode* is called *grid bias* voltage. The graph of Fig. 16-8 shows the grid voltage-plate current characteristic of a triode. In this *example*, -10 V on the grid, with the cathode at zero volts (grounded), causes I_{plate} to be zero (cutoff). A bias of -6 V, results in 5 mA of I_{plate}. A smaller bias of -2 V produces 15 mA of I_{plate}. Zero bias (grid and cathode at same voltage), results in a still greater I_{plate} of about 18 mA. Note that if the grid is made positive, $+2$ V (with respect to cathode), I_{plate} increases to about 20 mA. This is called *plate current saturation*, since making the grid still more positive produces no appreciable increase in I_{plate}.

16-4. Triode Amplifier Circuit

Figure 16-9 shows the circuit diagram of a *triode voltage amplifier*. The filament or heater voltage is usually called the *A* voltage, while the bias voltage

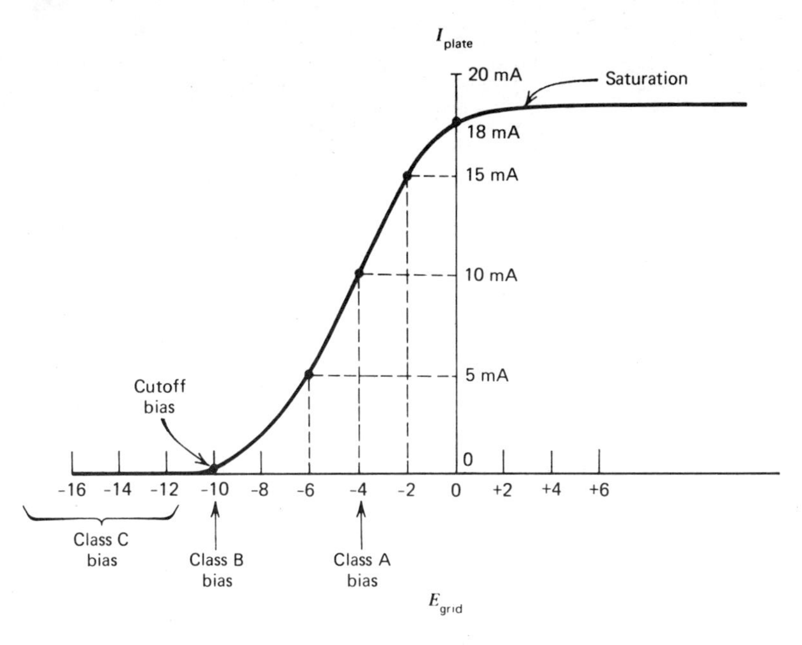

Figure 16-8. $I_{plate} - E_{grid}$ characteristic curve.

of 4 V (which makes the grid negative) is called the *C* voltage. The *B* voltage, 300 V, is the filtered output voltage from the rectifier power supply (Section 16-2) which makes the triode plate positive so that it attracts the emitted electrons from the cathode.

By applying a small ac voltage called the *input signal* to the grid in series with the *C* voltage, as shown in the circuit of Fig. 16-9, and by using the load resistor (14 kΩ), the circuit produces an output signal at the plate, which is larger than the input signal at the grid. This is called *amplification.* The *input signal* always *varies around the bias voltage* as its axis.

To understand how the circuit of Fig. 16-9 produces amplification, we also refer to the characteristic graph of Fig. 16-8. The following discussion is a simplified explanation of how a triode produces amplification. The input signal as shown in the wave shapes diagram of Fig. 16-10 is a 2-V peak ac or 4 V peak-to-peak. When the input signal is zero, as at instants *a*, *c*, and *e*, the only voltage on the grid is the -4 V (bias). From the curve of Fig. 16-8, -4 V on the grid results in 10 mA I_{plate}, as shown in the wave shapes (Fig. 16-10). Voltage across the load resistor is: $E_{R\,load} = I_{plate} R_{load} = (10\text{ mA})(14\text{ K}) = 140$ V, as shown at instants *a*, *c*, and *e* of the wave shapes. With 140 V dropped across R_{load}, then the dc plate voltage is the $+300$ V ($B+$ point) less the 140 V ($E_{R\,load}$). E_{plate} is then $+160$ V as shown.

When E_{input} becomes $+2$ V, at instant *b* of wave shapes (Fig. 16-10), it then subtracts from the -4 V bias, and the grid voltage becomes -2 V. From the

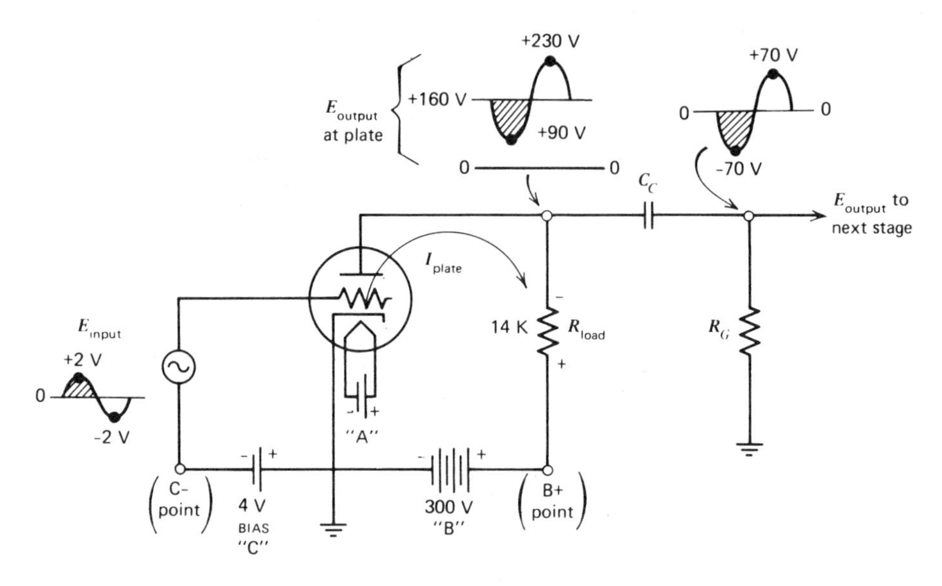

Figure 16-9. Triode amplifier circuit.

graph of Fig. 16-8, the less negative grid results in I_{plate} increasing to 15 mA, shown at instant *b*. $E_{R\,load}$ is now $= I_{plate}\ R_{load}$ = (15 mA) (14 K) = 210 V. E_{plate} is now +300 V (B+) − 210 V ($E_{R\,load}$) = +90 V. Note that when E_{input} at the grid went from zero to + 2 V (positive-going), E_{plate} went from +160 to +90 V (negative-going). This is *amplification* and *phase inversion.*

Finally, when E_{input} at the grid goes to −2 V at instant *d*, it adds to the −4 V bias, and the grid voltage becomes −6 V, as shown. With −6 V on the grid, I_{plate} decreases to 5 mA (from Fig. 16-8 graph). $E_{R\,load}$ is now = (5 mA) (14 K) = 70 V as shown. E_{plate} now becomes +300 V (B+) − 70 V ($E_{R\,load}$) = +230 V, as shown in the wave shapes.

The *input signal* at the grid goes from +2 to −2 V, or 4 *V peak-to-peak*, while the *output signal* at the plate goes from +90 to +230 V, or 140 *V peak-to-peak.* Comparing the two singals, E_{output} is 35 times larger than E_{input}, for a voltage amplification of 35 times. The shaded areas of E_{input} and E_{output} on the wave shapes of Fig. 16-10 show the phase inversion.

Classes of Amplifiers. When the grid is biased about the center of the linear (straight-line) portion of the graph of Fig. 16-8 (−4 V in this example), it is called a *class A* amplifier, as drawn on the graph. I_{plate} flows during the entire input cycle as shown in Fig. 16-11*a*. A *class B* amplifier is biased at the cutoff point, −10 V in the graph of Fig. 16-8. In class B operation, I_{plate} only flows on the positive half of E_{input} sine wave, as shown in Fig. 16-11*b*. *Class C* operation is biasing the grid more negative than cutoff, or −14 V or larger in the graph of Fig. 16-8. Here, I_{plate} only flows for a small portion of the E_{input} signal, on the most positive part of it, as is shown in Fig. 16-11*c*.

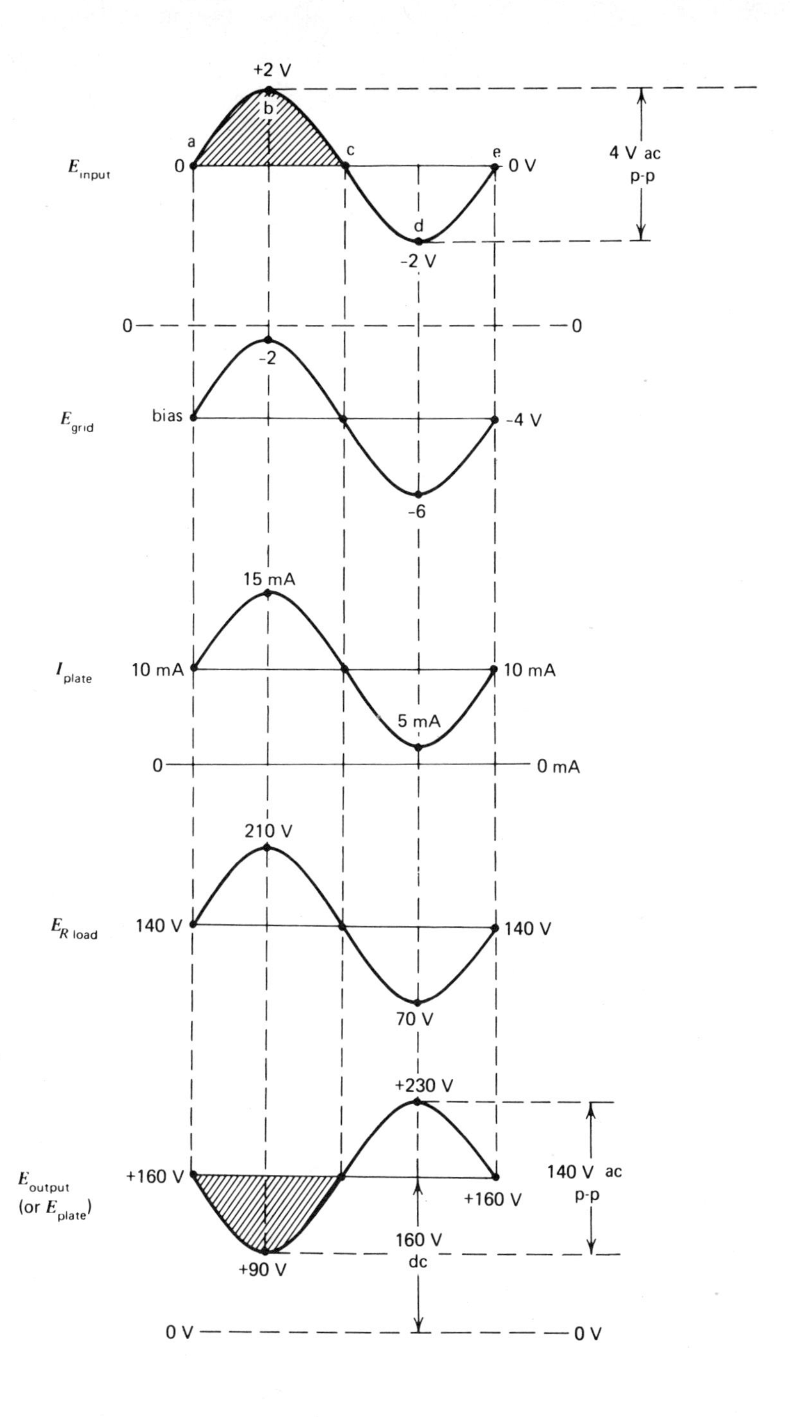

Figure 16-10. Wave shapes of the triode-amplifier circuit of Fig. 16-9.

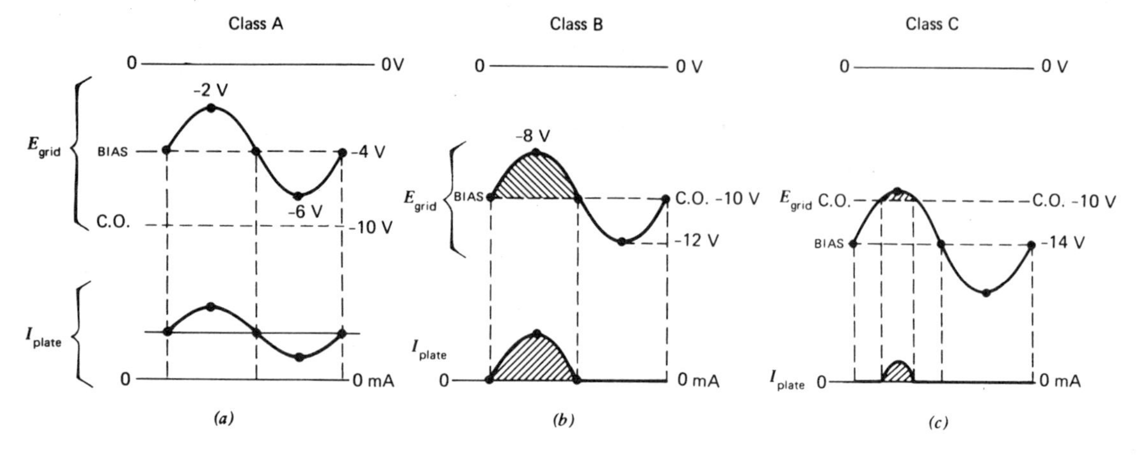

Figure 16-11. Classes of amplifiers, wave shapes. (*a*) Class A, biased less than cutoff; I_{plate} flows during entire cycle. (*b*) Class B, biased at cutoff; I_{plate} flows for half cycle. (*c*) Class C, biased more negative than cutoff; I_{plate} flows for less than half cycle.

Coupling Filter. The output signal voltage at the triode amplifier plate is a varying positive dc voltage, going from +160 V (at instant *a*), to +90 V (at instant *b*), to +230 V (at instant *d*). This is shown in the wave shapes of Fig. 16-10 and in the circuit of Fig. 16-9. This fluctuating dc voltage has a *dc component of* +160 *V*, and an *ac peak-to-peak voltage of* 140 *V*. Only the alternating voltage should be applied to the next stage. The direct voltage must be stopped. This is achieved by capacitor C_c and resistor R_g. C_c is called a *blocking* capacitor and also a *coupling* capacitor. C_c and R_g act as a high-pass filter (Section 15-3).

To the alternating voltage, X_{C_c} is very small compared to R_g, allowing the full alternating voltage to appear across R_g, available to the next stage. To the direct voltage, C_c charges fully to the +160 V, and no direct voltage appears across R_g.

Refer to End-of-Chapter Problems 16-6 to 16-9

16-5. Tetrode and Pentode Vacuum Tube

By adding a second layer of widely spaced loops of wire between the control grid and plate, and connecting this fourth element, called the *screen grid*, to a positive volatge, I_{plate} is greatly increased over that of a triode, resulting in greater amplification. With four elements (cathode, grid, screen, and plate), the tube is called a *tetrode*. The circuit symbol for a tetrode is shown in Fig. 16-12*a*. The positive screen attracts electrons emitted from the cathode, but

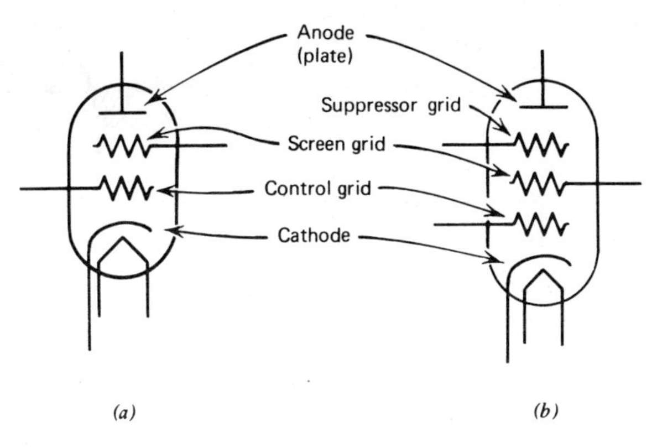

Figure 16-12. (*a*) Tetrode. (*b*) Pentode.

because of the large spaces between the screen grid wires, most of the electrons pass through to the plate. Since the positive screen is much closer to the cathode than is the plate, electrons are accelerated toward the plate with such velocity that they cause other electrons on the plate to be hurled off. This is called *secondary emission*. The positive screen attracts some of these secondary electrons, resulting in less current to the plate than should have been possible.

Pentode. A third grid, called the *suppressor grid* is wound between the screen grid and the plate. With five elements (cathode, grid, screen, suppressor, and plate), the tube is called a *pentode*, and its circuit symbol is shown in Fig. 16-12*b*.

As shown in the pentode amplifier circuit of Fig. 16-13, the suppressor grid is connected to ground, and is therefore at zero volts, the same as the cathode. The suppressor grid *appears negative* with respect to the very positive plate. As a result, secondary electrons that are knocked off the plate are repelled back into the plate by the suppressor grid, thus preventing or suppressing the loss of secondary electrons by the plate. Pentodes have replaced practically all tetrodes.

The circuit of Fig. 16-13 shows a typical operation of a pentode amplifier. Note that $I_{cathode}$ is 15 mA, of which only 2 mA flows to the screen, and the remaining 13 mA goes to the plate. The screen grid is connected to the 300 V B+ through resistor R_s. I_{screen} flowing through R_s produces a voltage drop across the resistor of 200 V. E_{screen} is +100 V (+300 V less the 200 V dropped across R_s). The screen is usually operated at a less positive voltage than the plate. The input signal applied to the control grid causes screen and plate currents to vary, producing voltage changes across R_{load} and R_s. The *plate voltage* changes make up the *output signal*, but screen grid voltage changes should be prevented. If the screen grid voltage *decreases* due to a larger drop across R_s when E_{input} goes positive, the less positive screen would reduce I_{plate} at a time when E_{input} should increase I_{plate}. In other words, variations of E_{screen} would tend to work against the effect of E_{input}. This is called *negative feedback*

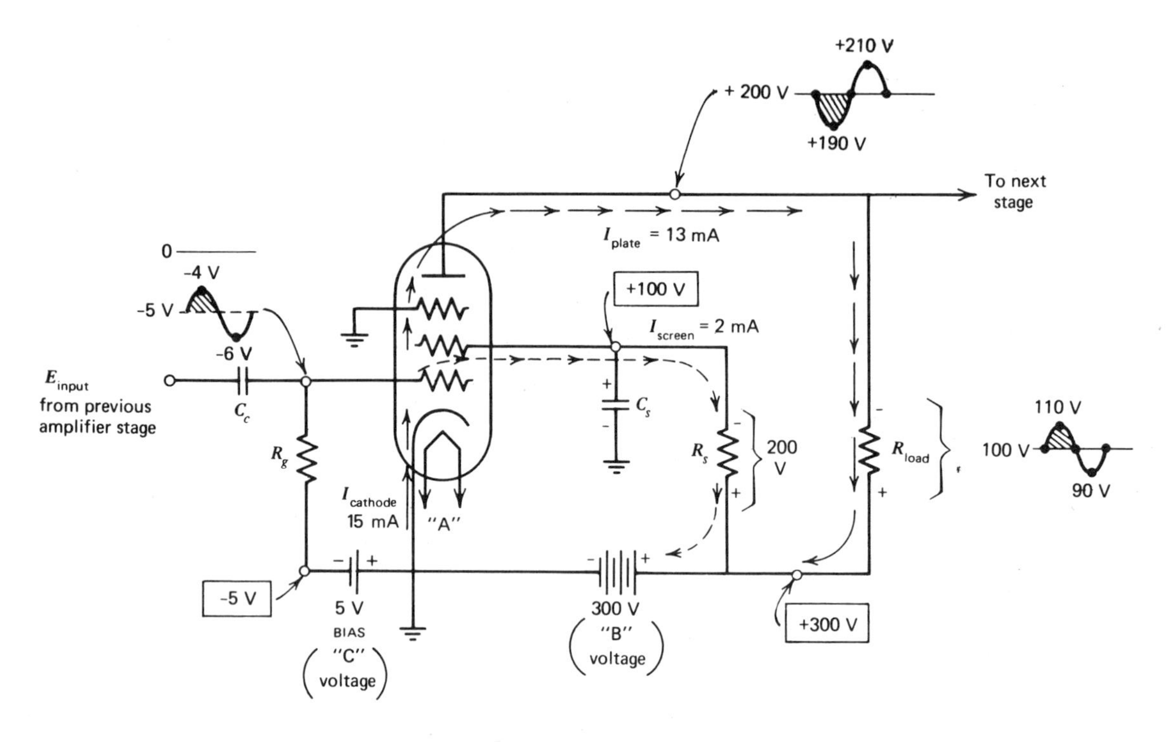

Figure 16-13. Pentode amplifier circuit.

or *degeneration*. To prevent this, capacitor C_s holds the screen voltage constant, since C_s cannot charge and discharge to higher and lower voltages at the fast rate of the input signal. C_s is called a screen *filter* or *bypass* capacitor. Due to C_s, there is no voltage variation or signal present on the screen. The positive screen has a much greater effect on *plate* current than the plate voltage itself has.

Refer to End-of-Chapter Problems 16-10 to 16-14

16-6. Tube Characteristics or Parameters

Three characteristics of vacuum tube triodes and pentodes are: (a) the ac plate resistance $R_{P\,\mathrm{ac}}$, (b) the amplification factor, μ (pronounced "mew"), and (c) the transconductance, G_m.

The *ac plate resistance* is the ratio of a small *change of plate voltage* to the resultant *change of plate current*, with the *bias voltage unchanged*, or

$$R_{\mathrm{plate\ ac}} = \frac{\Delta E_{\mathrm{plate}}}{\Delta I_{\mathrm{plate}}}, \text{ with constant bias} \tag{16-3}$$

Example 16-1

If a triode has $E_{plate} = +100$ V and $I_{plate} = 7$ mA, find $R_{plate\ ac}$ if E_{plate} is increased to +120 V, causing I_{plate} to rise to 9 mA, where grid bias remains fixed.

Solution

$$R_{plate\ ac} = \frac{\Delta E_{plate}}{\Delta I_{plate}} \tag{16-3}$$

$$R_{plate\ ac} = \frac{100 \text{ to } 120 \text{ V}}{7 \text{ to } 9 \text{ mA}}$$

$$R_{plate\ ac} = \frac{20 \text{ V}}{2 \text{ mA}}$$

$$R_{plate\ ac} = \frac{20}{2 \times 10^{-3}}$$

$$R_{plate\ ac} = 10 \text{ k}\Omega$$

Amplification factor describes the required *change of grid voltage* (change of bias) needed to bring I_{plate} back to its *original* value despite making the plate more positive.

$$\text{amplification factor, or } \mu = \frac{\Delta E_{plate}}{\Delta E_{grid}} \text{ to keep } I_{plate} \text{ constant} \tag{16-4}$$

Example 16-2

What is the amplification factor, μ, of a tube if $E_{plate} = +100$ V, $E_{grid\ (bias)} = -5$ V, $I_{plate} = 12$ mA, and then when E_{plate} is made +130 V, E_{grid} must be made −7 V to bring I_{plate} back down to 12 mA?

Solution

$$\mu = \frac{\Delta E_{plate}}{\Delta E_{grid}} \text{ to keep } I_{plate} \text{ constant} \tag{16-4}$$

$$\mu = \frac{100 \text{ to } 130 \text{ V}}{-5 \text{ to } -7 \text{ V}}$$

$$\mu = \frac{30}{2}$$

$$\mu = 15$$

The grid-to-plate *transconductance* or G_m is the ratio of the *plate current change* due to a *grid voltage change*, with E_{plate} remaining fixed, or

$$G_m = \frac{\Delta I_{plate}}{\Delta E_{grid}} \text{ with } E_{plate} \text{ constant} \qquad (16\text{-}5)$$

Transconductance, similar to conductance, is measured in mhos.

Example 16-3

Find the transconductance, G_m, of a pentode if $E_{plate} = +200$ V, $I_{plate} = 20$ mA, $E_{grid\ bias} = -5$ V, and then, making $E_{grid\ bias} = -3$ V causes I_{plate} to increase to 24 mA, with E_{plate} kept unchanged.

Solution

$$G_m = \frac{\Delta I_{plate}}{\Delta E_{grid}} \text{ with } E_{plate} \text{ constant} \qquad (16\text{-}5)$$

$$G_m = \frac{20 \text{ to } 24 \text{ mA}}{-5 \text{ to } -3 \text{ V}}$$

$$G_m = \frac{4 \text{ mA}}{2 \text{ V}}$$

$$G_m = \frac{4 \times 10^{-3}}{2}$$

$$G_m = 2 \times 10^{-3} \text{ mhos (or } \mho)$$

or 2000 micromhos ($\mu\mho$)

For Similar Problems Refer to End-of Chapter Problems 16-15 to 16-17

PROBLEMS

Refer to section 16-1, Diodes, for discussion covered by the following.

16-1. What is the purpose of the filament or heater in a diode vacuum tube using an indirectly heated cathode?
16-2. What is thermionic emission?
16-3. When will electrons flow to the plate of a diode?
16-4. If E_{plate} on a diode at saturation is +150 V and I_{plate} is 10 mA, what is I_{plate} with +200 V on the plate.

16-5. If E_{plate} on a diode is +60 V resulting in I_{plate} of 3 mA, and increasing E_{plate} to +100 V causes I_{plate} to rise to 5 mA, find: (a) $R_{plate\ dc}$, and (b) $R_{plate\ ac}$.

Refer to section 16-4, Triode Amplifier, for discussion covered by the following.

16-6. If the bias voltage on the grid is −10 V, and E_{input} is a 3 V peak ac sine wave, what is the (a) grid voltage at the positive peak of E_{input}, and (b) grid voltage at the negative peak of E_{input}?

16-7. If B+ is +250 V, I_{plate} = 10 mA, and R_{load} = 15 K, find: (a) $E_{R\ load}$, and (b) E_{plate}.

16-8. If E_{input} is a sine wave, draw the plate current wave for (a) class A amplifier, (b) class B amplifier, and (c) class C amplifier.

16-9. What happens to I_{plate} if (a) the grid is made less negative, and (b) the grid is made more negative?

Refer to section 16-5, Tetrode and Pentode, for discussion covered by the following.

16-10. What is the advantage of the screen grid tube over a triode?

16-11. What is the purpose of the suppressor grid?

16-12. In the pentode amplifier circuit of Fig. 16-13, what is the purpose of capacitor C_s in the screen circuit?

16-13. In Fig. 16-13, what is (a) E_{input} peak-to-peak, (b) E_{output} peak-to-peak, and (c) amplification of circuit?

16-14. In a pentode tube, if I_{plate} = 20 mA and I_{screen} = 2 mA, what is $I_{cathode}$?

Refer to section 16-6, Tube Characteristics, for discussion covered by the following.

16-15. Find the ac plate resistance of a triode if E_{plate} = +200 V, I_{plate} = 15 mA, E_{grid} = −10 V, and then making E_{plate} = +170 V, I_{plate} decreases to 10 mA, with grid bias unchanged.

16-16. If E_{plate} = +200 V, I_{plate} = 15 mA, E_{grid} = −10 V, and then decreasing E_{plate} to +190 V, it is necessary to decrease the grid bias to −8 V to bring I_{plate} back to 15 mA. Find the amplification factor.

16-17. A pentode has E_{plate} = +300 V, I_{plate} = 30 mA, E_{grid} = −12 V. Increasing E_{grid} to −15 V decreases I_{plate} to 20 mA, with E_{plate} unchanged. Find the transconductance, G_m.

chapter
17

Introduction to Semiconductors

Semiconductor diode and transistor theory and circuits actually belong in an electronics book instead of one on basic *electricity*. However, as an introduction to a semiconductor course, a brief and simple glimpse of this field is provided here as an incentive, along with some preliminary information.

17-1. Semiconductors

In Chapter 1, the makeup of the atom is discussed. It should be recalled that basically the atom consists of a center portion called the nucleus, having positive charges called protons. Outside the nucleus and rotating around it, as the planets do with the sun, are negative particles called electrons. In an atom, there is an equal number of electrons and protons. Adding an electron to an atom results in a negatively charged ion; removing an electron results in a positively charged ion. Each element (iron, oxygen, copper, hydrogen, etc.) differs from other elements in the number of electrons on its atoms. The number of these planetary electrons determines the number of rings or orbits, with each ring capable of containing a maximum number of electrons. Our discussion here of semiconductors is concerned only with the outermost ring, called the *valence* ring.

The maximum number of electrons on the valence ring of the atom of any element is eight. A good conductor such as silver or copper only has one electron on the valence ring of the atom. A good insulator such as rubber, mica, or some of the plastics has its valence ring filled with eight electrons, or almost filled with six or seven. An atom of the elements *germanium* or *silicon* contains four electrons on the valence ring. These elements are in between the insulator and the conductor, and are called *semiconductors*.

As shown in Chapter 1, Fig. 1-8 the atoms of the elements germanium (Ge) and silicon (Si), in crystalline form, are arranged in a three-dimensional latticework, where one atom (E) is surrounded by other atoms (A, B, C, and D). The atoms of germanium or silicon, as shown in Figs. 1-9 and 1-10, with four valence electrons each, combine together in a method called *covalent bonding*. The electrons on the valence rings of adjacent atoms are shared. As a result, atom E acts as if it has eight valence electrons, its own four, plus one from each of its four surrounding atoms (A, B, C, and D). Each of the other atoms, also surrounded by four atoms, acts as if it also has eight valence electrons. This is the maximum number of electrons possible on a valence ring.

N Material. Figure 17-1 shows several atoms of either germanium or silicon, each having four valence electrons, surrounding a single *pentavalent* atom. This is an atom having five electrons on its valence ring. The elements antimony and arsenic are two examples of a pentavalent substance. *N material* consists of a large quantity of four valence atoms (either germanium or silicon), and a very small number of pentavalent atoms, called the *impurity*. As shown, every atom of the germanium or silicon acts as if it has the maximum eight due to covalent bonding. However, the pentavalent impurity atom acts as if it has nine valence electrons, its own five, plus one from each of the four adjacent atoms. Since eight is the maximum number possible, the ninth electron is an extra one that cannot remain on the valence ring. This extra electron is called a *free electron*, and is free to drift and move about. *N* material is so called because it has negative particles, electrons, as the main or majority carriers. Actually, since the material has equal numbers of protons and electrons, *N* material is not negative but is really neutral or uncharged.

P Material. As we shall see here, it is much more difficult to understand what *P material* is, compared to what constitutes *N* material. Figure 17-2 shows several atoms of either germanium or silicon surrounding a single *trivalent* atom, having only three electrons on its valence ring. Elements such as aluminum, gallium, and indium are trivalent. Note that as a result of covalent bonding, every germanium or silicon atom acts as if it has eight valence electrons, its own four, plus one from each of the four adjacent atoms. However, the lone trivalent atom (called the *impurity*), acts as if it has only seven valence electrons, its own *three*, plus one from each of the four adjacent atoms. Since a valence ring can hold up to eight electrons, with only seven, there is room for one more electron. The "missing" electron area is called a *hole*, as shown in Fig. 17-2, and is regarded as a positive particle. The *hole* has the ability to move about just as the electron does and should be thought of as a positive movable particle.

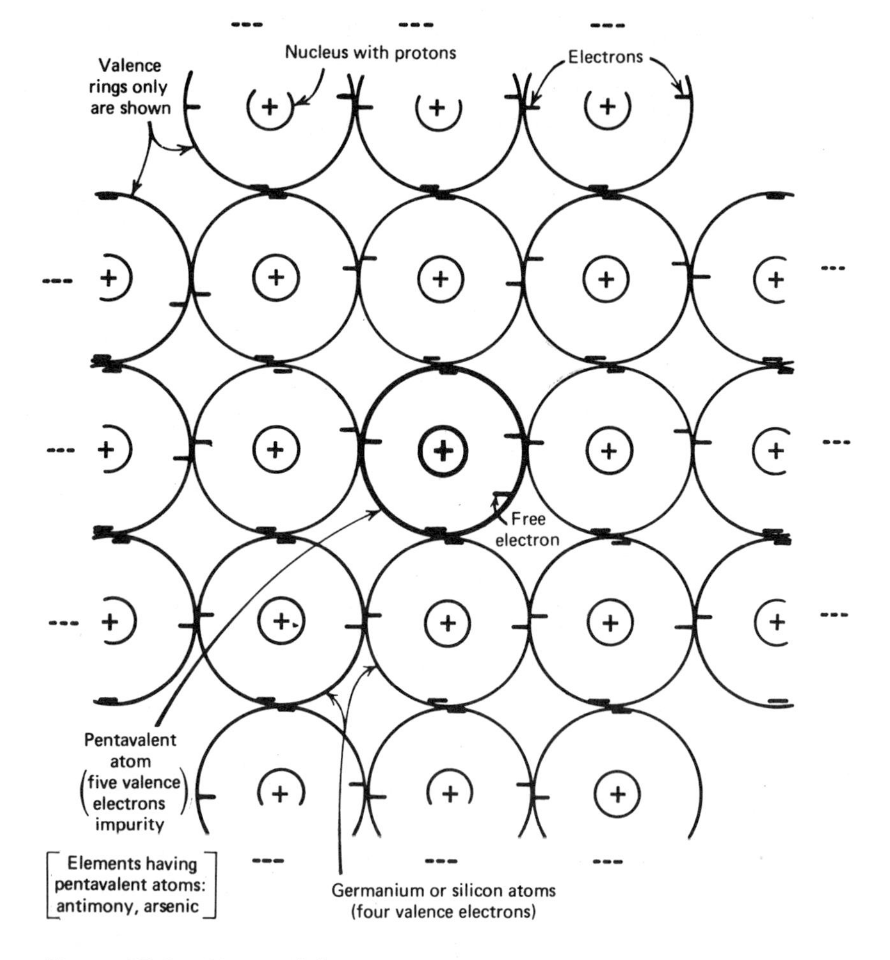

Figure 17-1. *N* material.

The mobility of the positive hole may be understood if one can visualize an electron's leaving an adjacent atom and moving into the hole. The hole is no longer present at its original location, since it is now occupied by an electron. However, the area that has just been vacated by the electron is now a *hole*. In effect then, the hole disappears from its original location, when it becomes occupied by an electron, and the hole reappears at the new location just vacated by the electron.

P material is so called because its main or majority current "carriers" are positive holes. However, *P* material, as *N* material, contains equal numbers of protons and electrons, and is uncharged or neutral. Semiconductor diodes and triodes (transistors) contain *N* and *P* materials.

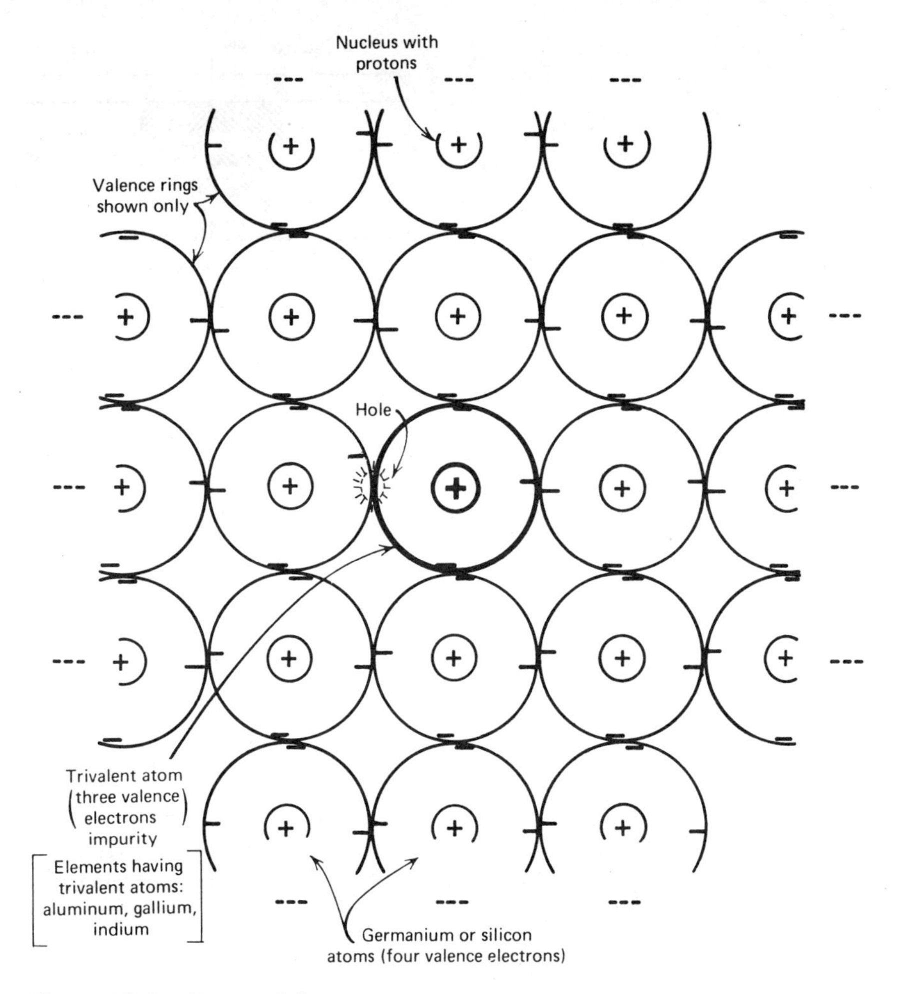

Figure 17-2. *P* material.

17-2. *P-N* Junction

When an *N* and a *P* material are joined together, the boundary between them is called the *P-N junction* as shown in Fig. 17-3. Except for a momentary movement of electrons and holes, no current flows. At first, some electrons from the *N* material cross the junction into the *P* material, filling up some holes. Similarly, holes move from the *P* material, crossing the junction and entering the *N* material.

The electrons, having left that area of the *N* material near the junction, leave behind positive holes. These electrons in moving into the *P* material holes, with the holes disappearing there, result in negative charges in the *P*

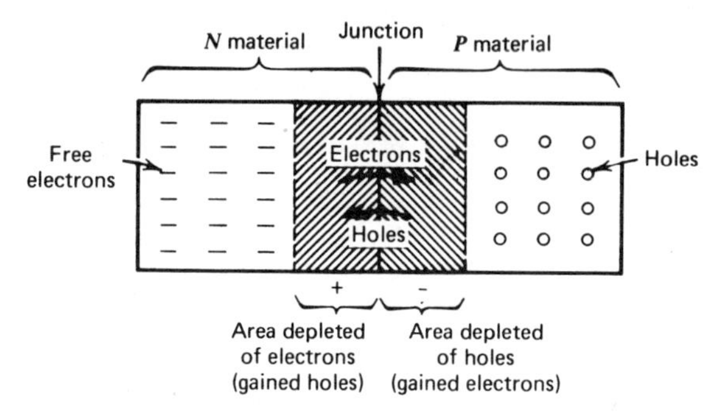

Electrons move across the junction going from the *N* into the *P* material, charging that area negative.

Holes move across the junction going from *P* into the *N* material, charging that area positive.

The charged depletion zones prevent any further movement of electrons toward the *P* material, or holes toward the *N* material.

Figure 17-3. The *P-N* junction.

material, as shown in Fig. 17-3. The shaded area of the *N* material near the junction is called a *depletion zone.* This area has been depleted of electrons, and has a positive charge. The shaded area of the *P* material near the junction is another depletion zone, depleted of holes, and has a negative charge, as shown in Fig. 17-3. Further movement of electrons from the *N* material is prevented by the negatively charged depletion zone in the *P* material. Likewise further movement of holes from the *P* material is prevented by the positively charged depletion zone in the *N* material.

Reverse Bias. When a dc voltage is applied to a *P-N* junction as shown in Fig. 17-4, with the *positive* voltage lead connected to the *N material*, and the *negative* voltage lead connected to the *P material*, the voltage is said to be *reverse bias.*

This voltage prevents current flow, since it adds to the original voltage or difference of charge between the depletion zones. In effect it widens the depletion zones. The positive end of the reverse bias voltage attracts electrons to the left in the *N* material, leaving a new positively charged depletion zone in the *N* material near the junction.

Similarly, the holes in the *P* material are attracted to the right by the negative bias voltage, leaving a new zone depleted of holes in the *P* material near the junction. This zone has become negatively charged. It is important to note that the movement of holes can only exist inside the semiconductor material, and not in the copper wire leading to the applied bias voltage. As in the *P-N* materials joined together in the previous Fig. 17-3, without any applied voltage,

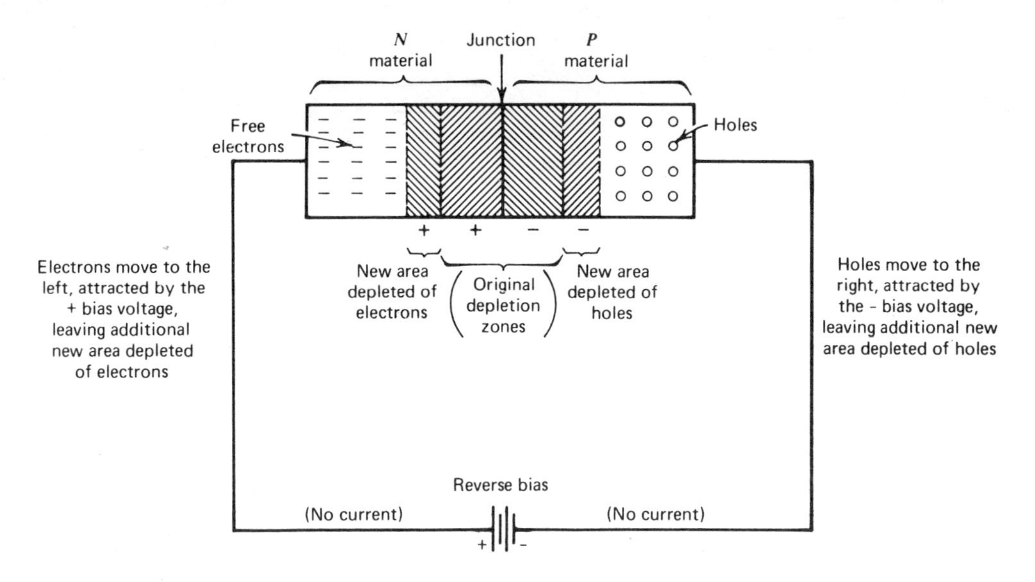

Figure 17-4. *P-N* junction with reverse bias. Reverse bias widens or increases the depletion zones.

no further movement of electrons or holes can take place after the depleted zones have been formed.

Forward Bias. When the applied dc voltage has the polarity shown in Fig. 17-5, that is, the *positive* end is connected to the *P* material and the *negative* end is connected to the *N* material, it is called *forward bias.* Under this condition, current will flow through the semiconductor diode (two-element device).

The negative end of the bias voltage of Fig. 17-5 repels electrons in the *N* material so that they move across the junction and enter the *P* material. There, they are attracted by the positive end of the bias voltage, and the electrons move on through to this voltage source. At the same time, the positive end of the bias voltage repels the holes in the *P* material. The holes move across the junction into the *N* material and are attracted toward the negative end of the bias voltage. However, the holes only move through the *P* and *N* materials, but not through the external circuit, while the electrons move through the entire circuit. Depletion zones cannot form since the applied voltage keeps the electrons and holes moving through the semiconductor materials.

The Diode. A semiconductor *diode* consists of *N* and *P* material joined together. When the applied voltage forward biases the diode, as shown in Fig. 17-6*a*, electrons flow through the circuit. The *N material* is called the *cathode*, and the *P material*, the *anode*. Note that the symbol for the diode has the arrow (the anode) pointing in the reverse direction from electron flow. The diode arrow points in the direction of "hole current" or in the direction

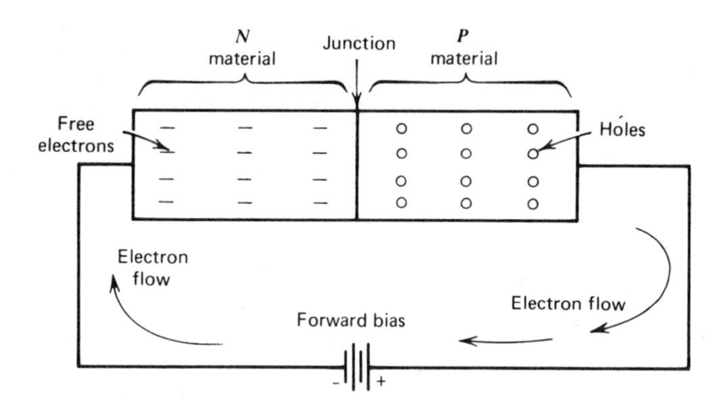

Electrons are repelled by the − bias voltage and move across the junction from N into P material; here they are attracted by the + bias voltage.

Holes are repelled by the + bias voltage and move across the junction from P into N material; here they are attracted by the − bias voltage.

Electrons and holes move in opposite directions through the semiconductor, but only electrons flow in the external circuit.

Figure 17-5. *P-N* junction with forward bias.

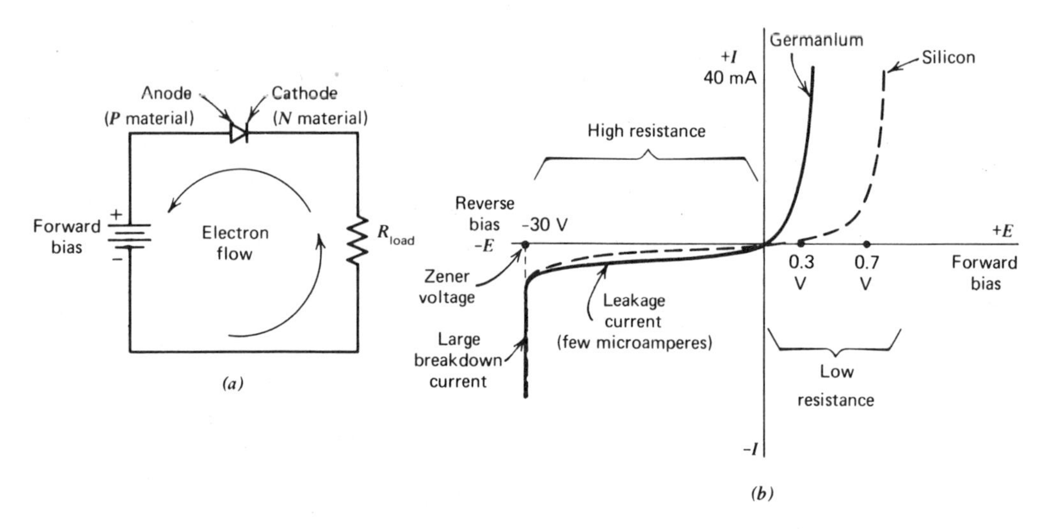

Figure 17-6. Semiconductor diode. (*a*) Semiconductor diode with forward bias. (*b*) Diode voltage-current characteristics.

of "conventional" current from + to −. If the polarity of the applied voltage were reversed so that the negative end connected to the *P* material anode (reverse bias), no current would flow except for a very tiny *leakage* current. This is shown in the voltage-current graph of Fig. 17-6*b*. If the reverse bias voltage is made sufficiently large, called the *Zener* voltage, a reverse direction current flows called the *breakdown* current. A Zener diode operates with this large reverse-bias voltage, producing a constant voltage drop across the diode itself, which is used as a voltage-regulating device.

17-3. The Transistor

A *transistor* consists of two *P-N* junctions formed by sandwiching a very thin slice of *P* material between two *N* materials. This is called an *NPN* transistor such is as shown in the circuit of Fig. 17-7*a*. If the type of materials is reversed, that is, a thin slice of *N* material is placed between two *P* materials, the transistor is called a *PNP* type, as shown in the circuit of Fig. 17-7*b*.

A transistor circuit produces amplification of an input signal in the same manner as the electron tube circuit (Chapter 16). The three elements of the

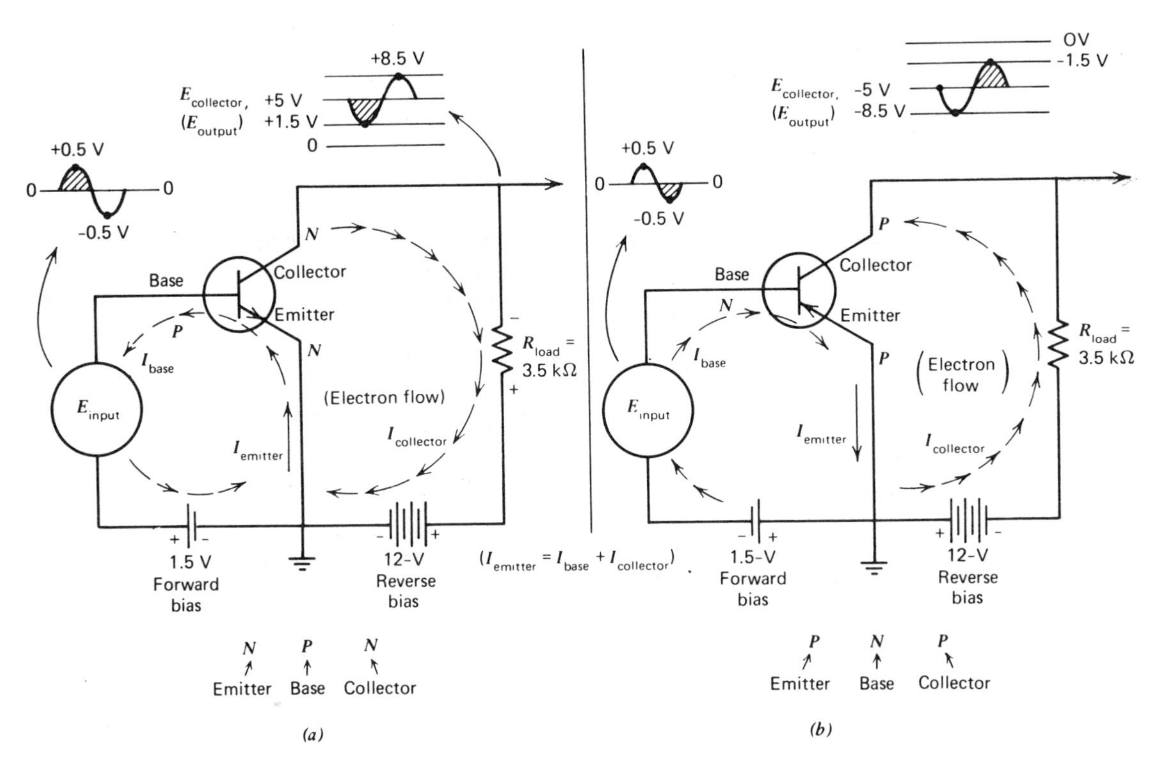

Figure 17-7. Common-emitter (*CE*) transistor amplifier circuit.

transistor are the *emitter*, *base*, and *collector*, as shown in Fig. 17-7. These may be compared to the cathode, grid, and plate (anode) of a triode electron tube. The difference between the circuit diagram symbol for an *NPN* and a *PNP* transistor is simply in the direction that the emitter arrow faces. In the *PNP*, it is *toward* the *base*.

Note that the *P-N* junction between *emitter* and *base* must be *forward biased*, as shown in Fig. 17-7*a*. The *positive* end of the 1.5 V is connected to the *P material base*, while the *negative* end connects to the *N Material emitter*. The *P-N* junction between the *collector* and *base* must be *reverse biased* as shown in Fig. 17-7*a*. The *positive* end of the 12 V is connected to the *N material collector* Note the opposite polarities of the applied dc voltages in the *PNP* circuit of Fig. 17-7*b*.

Transistor Amplifier. In the circuits of Fig. 17-7, the *input* signal voltage is applied between *base* and *emitter*, and the *output* is taken between *collector* and *emitter*. Since the emitter is common to the input and the output, the circuit is called a *common emitter* (*CE*). Two others (not shown here) are called *common base* (*CB*), and *common collector* (*CC*).

In Fig. 17-7*a*, the *NPN* transistor, *emitter current* from the *N*-type emitter consists of electrons leaving the emitter. A small portion of this current flows into the base, attracted by the +1.5 V, as *base current*. Because of the extreme thinness of the base, most fo the emitted electrons penetrate the base and are attracted by the +12 V applied to the collector. This is collector current. Emitter current is the sum of base and collector currents, or $I_{emitter} = I_{base} + I_{collector}$ (equation 17-1). The ratio of *output circuit current* to *input circuit current* is called the *current amplification*, or, $I_{amplification} = I_{output\ circuit}/I_{input\ circuit}$ (*equation* 17-2).

In the *PNP* circuit of Fig. 17-7*b*, the emitter is *P*-type material, emitting not electrons, but holes. *Hole* current flows from the *P* emitter to the *N* base, attracted by the −1.5 V connected to the base. Most of the holes penetrate through the thin base and go to the collector, attracted by the −12 V applied dc voltage. Recalling that when holes move in one direction, electrons have drifted in the opposite direction, then emitter, base, and collector currents are depicted as electron flow in this reverse direction in Fig. 17-7*b*.

Example 17-1

In the circuit of Fig. 17-7*a*, if I_{base} is 0.1 mA, and $I_{collector}$ is 2 mA, what is the value of $I_{emitter}$?

Solution

$I_{emitter} = I_{base} + I_{collector}$ (equation 17-1), then $I_{emitter} = 0.1 + 2 = 2.1$ mA.

For Similar Problems Refer to End-of-Chapter Problems 17-1 to 17-5

In Fig. 17-7, emitter current depends on the amount of forward bias between base and emitter. In Fig. 17-7*a*, the *NPN* circuit, I_{base} increases when the input signal voltage goes *positive*, driving the *P base* more *positive*, *increasing* the *forward bias*. This results in a greater $I_{collector}$. This is shown by the shaded areas of the wave shapes of Fig. 17-8*a*.

In Fig. 17-7*b*, the *PNP* circuit, I_{base} and $I_{collector}$ increase when the input signal goes *negative*, driving the *N base* more *negative*, *increasing* the *forward bias*. This is shown by the shaded areas of the wave shapes of Fig. 17-8*b*.

In Fig. 17-8*a*, the wave shapes of the *NPN C-E* amplifier are shown. The values are all assumed. Actual values are determined from the transistor characteristic curves (not shown). When E_{input} is zero, E_{base} is only the +1.5 V forward bias. Base current brings base voltage down to a fraction of a volt, but for this simplified explanation of transistor amplification it is assumed that base voltage is the forward bias of +1.5 V. As shown, I_{base} is 0.1 mA, and $I_{collector}$ is 2 mA. $E_{R\,load}$ is 7 V (2 mA times 3.5 K). With collector supply voltage +12 V, and 7 V dropped across R_{load}, $E_{collector}$ is +5 V, as shown.

When E_{input} goes from zero to +0.5 V in Fig. 11-8*a*, the *P* base now becomes more positive and E_{base} is +2 V (sum of +1.5 V forward bias and +0.5 V signal). I_{base} rises to 0.15 mA, and $I_{collector}$ rises to 3 mA. $E_{R\,load}$ now becomes 10.5 V (3 mA times 3.5 K). $E_{collector}$ or E_{output} now becomes +1.5 V (+12 V forward bias, less the 10.5 V dropped across R_{load}). Note that when E_{input} goes *positive* 0.5 V (shaded area), E_{output} goes *less-positive* (negative-going) from +5 down to +1.5 V, or 3.5 V decrease. This is *inversion* and *amplification*. Only the common emitter (CE) circuit inverts the signal.

Example 17-2

In the *NPN C-E* amplifier circuit of Fig. 17-7*a*, and from the wave shapes of Fig. 17-8*a*, *solve* for: (a) $E_{R\,load}$ and (b) $E_{collector}$ when E_{input} goes from zero to −0.5 V, (c) $E_{input\ peak\text{-}to\text{-}peak}$, (d) $E_{output\ p\text{-}p}$, (e) voltage amplification, (f) current amplification, using the values shown in the figure.

Solution

The reader should solve for these, and should compare his results with those shown here: (a) $E_{R\,load} = (I_{collector})\ (R_{load} = (1\text{ mA})\ (3.5\text{ K}) = 3.5$ V, (b) $E_{collector} = +12\text{ V} - 3.5\text{ V} = +8.5$ V, (c) $E_{input} = 1$ V p-p, (d) $E_{output} =$ 7 V p-p (from +1.5 to +8.5 V), (e) voltage amplification $= E_{output}/E_{input} = 7/1 = 7$ times, (f) $I_{amplification} = I_{output\ p\text{-}p}/I_{input\ p\text{-}p} = 2\text{ mA}/0.1\text{ mA} = 20$ times.

The wave shapes of Fig. 17-8*b* for the *PNP C-E* amplifier circuit of Fig. 17-7*b* show that when E_{input} goes *negative* (during the shaded area), it drives the *N material base* more *negative*, thus *increasing* the *forward bias*. This increases I_{base} and also $I_{collector}$, producing a larger voltage across R_{load}. $E_{collector}$ (or

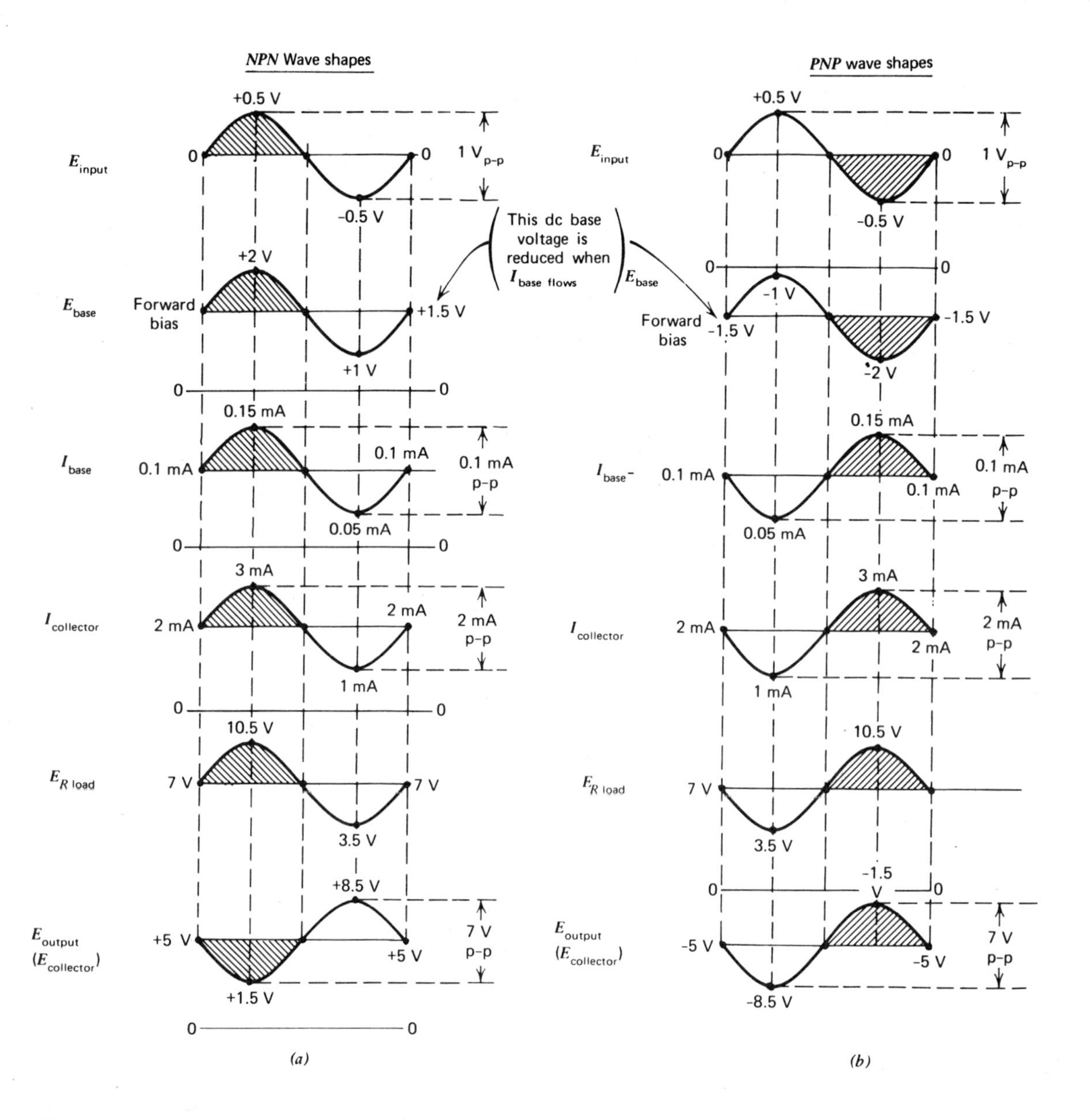

Figure 17-8. Common emitter (*CE*) transistor amplifier wave shapes. (*a*) Wave shapes of *NPN* amplifier of Fig. 17-7*a*. (*b*) Wave shapes of *PNP* amplifier of Fig. 17-7*b*.

E_{output}) becomes *less negative* or is *positive-going*, Since E_{input} is *going negative*, while E_{output} is *positive-going*, as in the shaded portions of the wave shapes, then *phase inversion* is taking place.

When E_{input} is zero in Fig. 17-8*b*, E_{base} is at its -1.5 V forward bias in the circuit of Fig. 17-7*b*. Actually, it is only a fraction of a volt negative due to base-to-emitter current, but for purposes of this simple explanation it will be assumed that the base is at the applied voltage of -1.5 V. As shown, I_{base} is 0.1 mA, and $I_{collector}$ is 2 mA. $E_{R\ load}$ is 7 V (2 mA times 3.5 K), and $E_{collector}$ is the -12 V reverse bias less the 7 V dropped across R_{load}, or $E_{collector}$ is -5 V.

During the shaded portions of the wave shapes, when E_{input} goes to -0.5 V, E_{base} becomes -2 V, increasing the forward bias on the *N*-type base. I_{base} increases to 0.15 mA, and $I_{collector}$ increases to 3 mA. $E_{R\ load}$ is now 10.5 V (3 mA times 3.5 K). $E_{collector}$ now becomes -1.5 V (the -12 V reverse bias less the 10.5 V across R_{load}).

Example 17-3

From the *PNP C-E* amplifier circuit of Fig. 17-7*b*, and the unshaded portions of the wave shapes of Fig. 17-8*b*, when E_{input} goes to $+0.5$ V, determine: (a) $E_{R\ load}$, and (b) $E_{collector}$, using the values shown.

Solution

Solve this example from the information shown in the wave shapes of Fig. 17-8*b*.

For Similar Problems Refer to End-of-Chapter Problems 17-6 to 17-7

PROBLEMS

See section 17-3, Transistors, for discussion covered by the following.

17-1. What is meant by forward bias?

17-2. What is meant by reverse bias?

17-3. Draw the circuit of an *NPN* transistor *C-E* amplifier showing the polarities of the forward and reverse bias voltages, as well as where E_{input} is applied and E_{output} is taken.

17-4. Draw the previous circuit using a *PNP* transistor.

17-5. Find $I_{collector}$ where $I_{emitter}$ is 18.3 mA and I_{base} is 0.4 mA.

See section 17-3, Transitor Amplifiers, for discussion covered by the following.

17-6. In which direction (polarity) should the input signal go when applied to the base to increase the forward bias in (a) a *PNP* transistor, and (b) a *NPN* transistor?

17-7. A *PNP C-E* amplifier uses the same circuit as that of Fig. 17-7*b*, except that the reverse bias voltage is 9 V, and R_{load} is 4 K. When the input signal is zero, $I_{collector}$ is 1.5 mA. When the signal goes to -0.25 V, $I_{collector}$ becomes 2 mA. Find: (a) $E_{collector}$ with no input signal, (b) $E_{collector}$ when E_{input} goes to -0.25 V, and (c) voltage amplification.

APPENDIX

I

Mathematics for Electronics

Simple Algebra. Positive and Negative Numbers

Addition: For like signs, add the numbers, give answer this same sign. For unlike signs, subtract the numbers, give answer the larger number sign.

$$\begin{array}{r} +10 \\ +\ 2 \\ \hline +12 \end{array} \qquad \begin{array}{r} -10 \\ -\ 2 \\ \hline -12 \end{array} \qquad \begin{array}{r} +10 \\ -\ 2 \\ \hline +\ 8 \end{array} \qquad \begin{array}{r} -10 \\ +\ 2 \\ \hline -\ 8 \end{array}$$

Subtraction: First reverse the sign of the subtrahend (the term following the subtraction sign), then follow addition procedure.

$$\begin{array}{lr} & +10 \\ - & +\ 2 \\ \hline & +\ 8 \end{array} \qquad \begin{array}{lr} & -10 \\ - & -\ 2 \\ \hline & -\ 8 \end{array} \qquad \begin{array}{lr} & +10 \\ - & -\ 2 \\ \hline & +12 \end{array} \qquad \begin{array}{lr} & -10 \\ - & +\ 2 \\ \hline & -12 \end{array}$$

Multiplication and Division: Like signs produce a + answer; unlike signs produce a − answer.

Multiplication:

$$(+10)(+2) = +20$$

$$(-10)(-2) = +20$$

$$(+10)(-2) = -20$$

$$(-10)(+2) = -20$$

Division:

$$\frac{+10}{+2} = +5$$

$$\frac{-10}{-2} = +5$$

$$\frac{+10}{-2} = -5$$

$$\frac{-10}{+2} = -5$$

Simple Equations: To solve for the value of the unknown term, this term must be brought to one side of the equation, while all other terms must be brought to the other side of the equation. When transposing a term (moving to the other side of equation) from one side to the other side of the equation, simply move the term over, but *reverse* its *function*. If it is +, it becomes − on the other side; if it is multiplication, it becomes division on the other side.

Solve for x in the following:

Example

$$5x + 10 = 2x - 17$$

$$5x + 10 - 2x = -17$$

$$5x - 2x = -17 - 10$$

Move the $+2x$ to the left side, where it becomes $-2x$. Now move the $+10$ to the right, where it becomes -10.

$$3x = -27$$

$$x = \frac{-27}{3}$$

$3x$ means x times 3. Move the *times* 3 to the right side, where it becomes ÷ 3.

Answer

$$x = -9$$

If the unknown term is given as a squared term, the term may be found by taking the square root of both sides of the equation. This is shown in the following example.

Example

Solve for the value of I, where P is 125 W and R is 5 Ω.

$$P = I^2R$$

$$125 = I^2 5$$

$$\frac{125}{5} = I^2$$

$$25 = I^2$$

$$\sqrt{25} = \sqrt{I^2} \quad \left\{\text{Take the square root of both sides of equation.}\right.$$

$$5 = I$$

Scientific Notation. Powers of Ten

The number 10 raised to a + power (or exponent) is the numeral 1 followed by the number of zeros given by this + power.

Examples

10^0 is 1 followed by no zeros, or 1.

10^3 is 1 followed by 3 zeros, or 1000.

10^4 is 1 followed by 4 zeros, or 10,000.

The number 5000. may be then written as 5×1000, or 5×10^3.

The number 75,000,000. may be written as $75 \times 1{,}000{,}000$ or 75×10^6. Note that this means that the *decimal point* of the 75,000,000. may be placed to the right of the numeral 5. *To get back to where it belongs* (to the right of the sixth zero), it must be moved six places to the *right*, (or 10^6).

The number 620,000. may then be written as 62×10^4, or 6.2×10^5, or even $620. \times 10^3$, and so on.

Negative Powers of Ten

The number 10 raised to a *negative* power is equal to the numeral 1 with the decimal point preceding the 1, or being moved from the *right side* of the 1 to its *left* by the number of places stated by the negative power.

Examples

10^{-2} is 0.01

10^{-3} is 0.001

10^{-5} is 0.00001

{Note that the zero to the left of the decimal point is meaningless. It simply calls the reader's attention to the presence of the decimal point.

The number 0.006 is six times 0.001, or 6×10^{-3}. Note that this means that the *decimal point* may be placed to the right of the 6, but to get back *to where it belongs*, it must be moved three places to the left (or 10^{-3}).

The number 0.0045 may then be written as 45×10^{-4}, or 4.5×10^{-3}, and so on.

Working with Powers of Ten. *Multiplying* numbers using powers of ten requires algebraically adding the powers.

Examples

$$(15{,}000)\ (30{,}000), \text{ is } (1.5 \times 10^4)\ (3 \times 10^4) = 4.5 \times 10^8$$

$$(0.005)\ (400), \text{ is } (5 \times 10^{-3})\ (4 \times 10^2) = 20 \times 10^{-1}, \text{ or } 2.$$

Dividing numbers using powers of ten requires *reversing* the sign of the power of ten of the *divisor*, and then algebraically adding the powers.

Examples

$$\frac{15{,}000}{300} = \frac{15 \times 10^3}{3 \times 10^2} = 5 \times 10^3 \times 10^{-2} = 5 \times 10^1$$

$$\frac{7500}{0.005} = \frac{7.5 \times 10^4}{5 \times 10^{-3}} = 1.5 \times 10^4 \times 10^3 = 1.5 \times 10^7$$

Raising a number to a power, using powers of ten, is performed by algebraically *multiplying* the *powers.*

Examples

$$3000^2 = (3 \times 10^3)^2 = 9 \times 10^6$$

$$0.0005^2 = (5 \times 10^{-4})^2 = 25 \times 10^{-8}$$

The *root* of numbers using powers of ten is performed by algebraically *dividing* the *power* by the *root.*

Examples

$$\sqrt[2]{250{,}000.} = \sqrt[2]{25 \times 10^4} = 5 \times 10^{4/2} = 5 \times 10^2$$

$$\sqrt[2]{.000009} = \sqrt[2]{9 \times 10^{-6}} = 3 \times 10^{-6/2} = 3 \times 10^{-3}$$

Solving two simultaneous equations by determinant method

$$2x + 3y = 8$$

$$-2x - 2y = -6$$

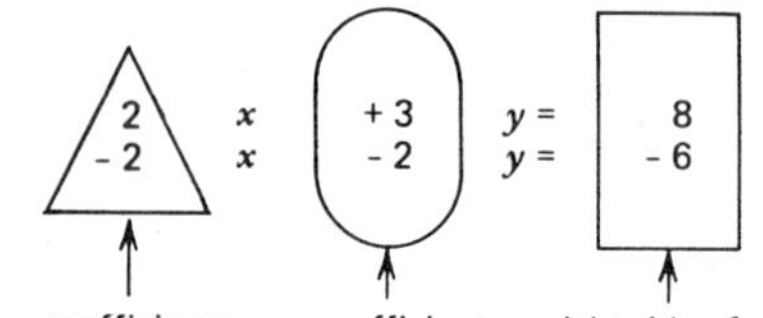

$x = \dfrac{\text{numerator}}{\text{denominator}} = \dfrac{\text{replace the } x \text{ coefficients with the } \boxed{\text{numbers}} \text{ at right side of equation, and list them and the } \textcircled{y} \text{ coefficients in order of appearance.}}{\text{list the } \triangle x \text{ coefficients and the } \textcircled{y} \text{ coefficients in order of appearance.}}$

$$x = \frac{\begin{array}{cc} 8 & 3 \\ -6 & -2 \end{array}}{\begin{array}{cc} 2 & 3 \\ -2 & -2 \end{array}}$$

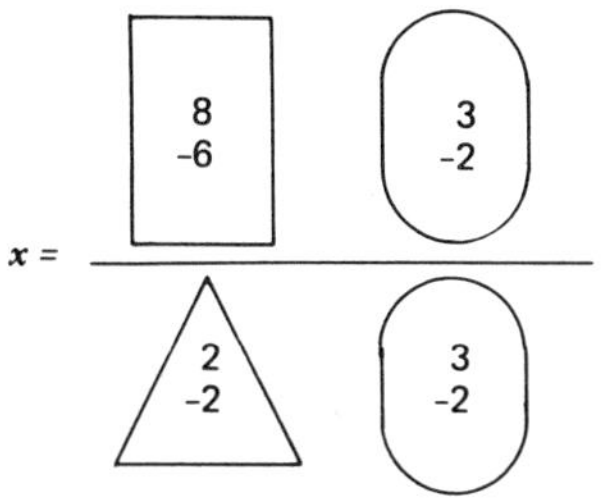

or,

$$x = \frac{\begin{vmatrix} 8 & 3 \\ -6 & -2 \end{vmatrix}}{\begin{vmatrix} 2 & 3 \\ -2 & -2 \end{vmatrix}}$$

In the numerator and also in the denominator, take the *difference of cross products*, starting at the *upper left* corners, as shown in the following.

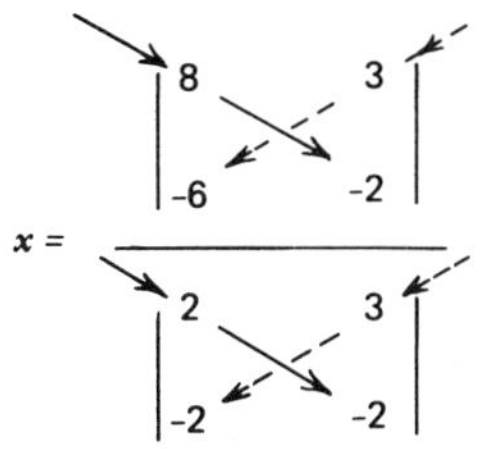

$$x = \frac{\begin{vmatrix} 8 & 3 \\ -6 & -2 \end{vmatrix}}{\begin{vmatrix} 2 & 3 \\ -2 & -2 \end{vmatrix}} = \frac{(8)(-2) - (3)(-6)}{(2)(-2) - (3)(-2)} = \frac{(-16) - (-18)}{(-4) - (-6)} = \frac{-16 + 18}{-4 + 6} = \frac{+2}{+2}$$

$$x = 1$$

Solve for y in the same way:

$y = \dfrac{\text{numerator}}{\text{denominator}} = \dfrac{\text{replace the } y \text{ coefficients with the } \boxed{\text{numbers}} \text{ at right side of equation, and list them and the } \triangle x \text{ coefficients in order of appearance.}}{\text{list the } \triangle x \text{ and } \textcircled{y} \text{ coefficients in order of appearance.}}$

Note that the *denominators* for x and y are identical.

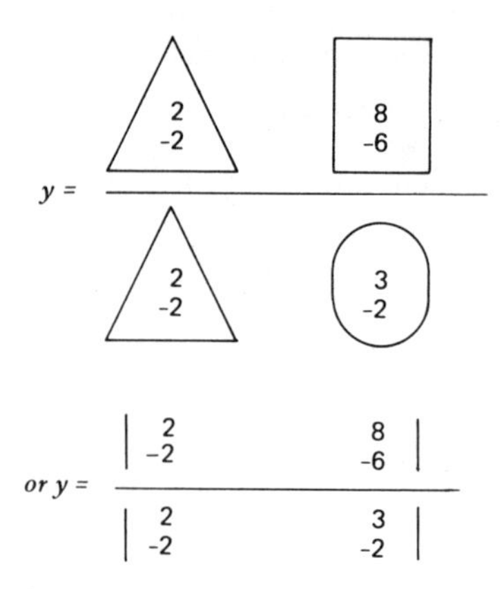

In the numerator and also in the denominator, take the difference of the cross products, starting at the upper left corners, as shown in the following.

$$y = \frac{\begin{vmatrix} 2 & 8 \\ -2 & -6 \end{vmatrix}}{\begin{vmatrix} 2 & 3 \\ -2 & -2 \end{vmatrix}} = \frac{(2)(-6) - (8)(-2)}{(2)(-2) - (3)(-2)} = \frac{(-12) - (-16)}{(-4) - (-6)} = \frac{-12 + 16}{-4 + 6} = \frac{+4}{+2} = +2$$

Solving three simultaneous equations using the determinant method

$$\begin{aligned} 2x + 3y - 2z &= 2 \\ -1x - 2y + 1z &= -2 \\ -2x + 2y - 1z &= -1 \end{aligned}$$

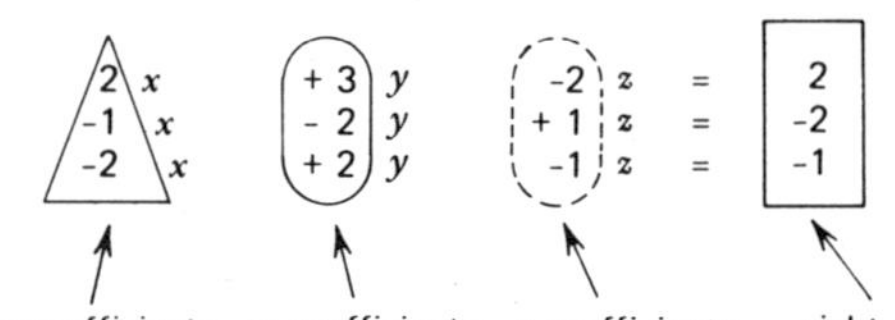

To solve for x:

$x = \dfrac{\text{numerator}}{\text{denominator}} =$ replace the x coefficients with the numbers at right side of equations, and list these numbers, then the y coefficients and then the z coefficients in the order of their appearance.

list the x coefficients, then the y coefficients, and the z coefficients in order of appearance.

$$x = \frac{\boxed{\begin{matrix}2\\-2\\-1\end{matrix}} \quad \begin{matrix}3\\-2\\2\end{matrix} \quad \begin{matrix}-2\\1\\-1\end{matrix}}{\begin{matrix}2\\-1\\-2\end{matrix} \quad \begin{matrix}3\\-2\\2\end{matrix} \quad \begin{matrix}-2\\1\\-1\end{matrix}}$$

$$\text{or, } x = \frac{\begin{vmatrix}2 & 3 & -2\\-2 & -2 & 1\\-1 & 2 & -1\end{vmatrix}}{\begin{vmatrix}2 & 3 & -2\\-1 & -2 & 1\\-2 & 2 & 1\end{vmatrix}}$$

Now, repeat the first two columns, listing them at the right, as shown:

$$x = \frac{\left|\begin{matrix}2 & 3 & -2\\-2 & -2 & 1\\-1 & 2 & -1\end{matrix}\right|\begin{matrix}2 & 3\\-2 & -2\\-1 & 2\end{matrix}}{\left|\begin{matrix}2 & 3 & -2\\-1 & -2 & 1\\-2 & 2 & -1\end{matrix}\right|\begin{matrix}2 & 3\\-1 & -2\\-2 & 2\end{matrix}}$$

Starting at the *upper left number* in the *numerator*, multiply the three digits on the solid-line diagonal arrows sloping down to the right:

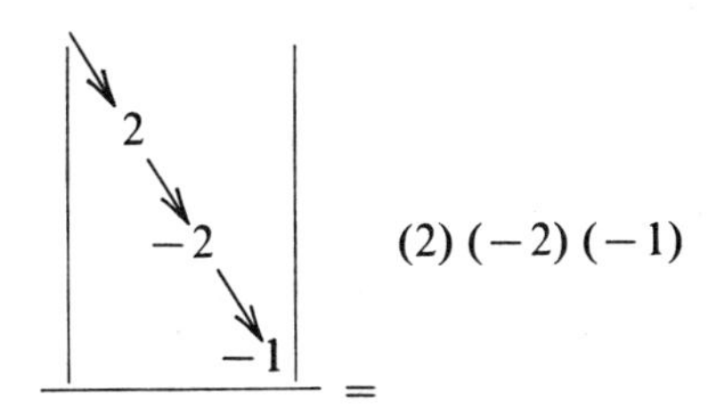

Add to this, the product of the next three numbers on the next solid-line diagonal arrows, moving to the right:

$$\frac{\left|\begin{matrix}2 & 3 & \\ & -2 & 1\\ & & -1\end{matrix}\right| -1}{} = \quad +(3)\,(1)\,(-1)$$

And then, add the product of the next three numbers:

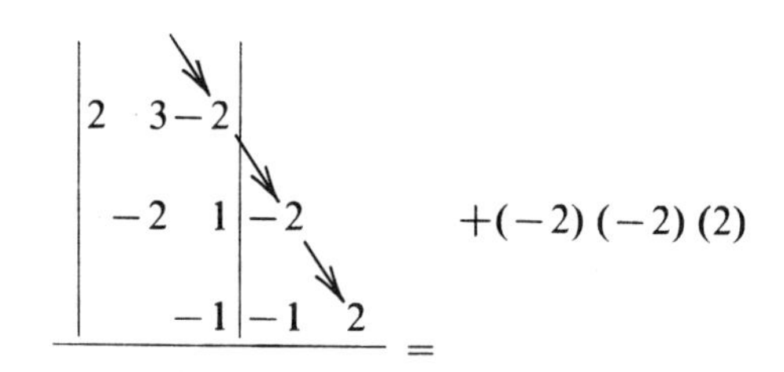

Now, from the previous three products added together, *subtract* the following three products: *First product*: starting at the *upper right number* in the *numerator*, the products of the three digits on the dotted-line arrows sloping down to the left:

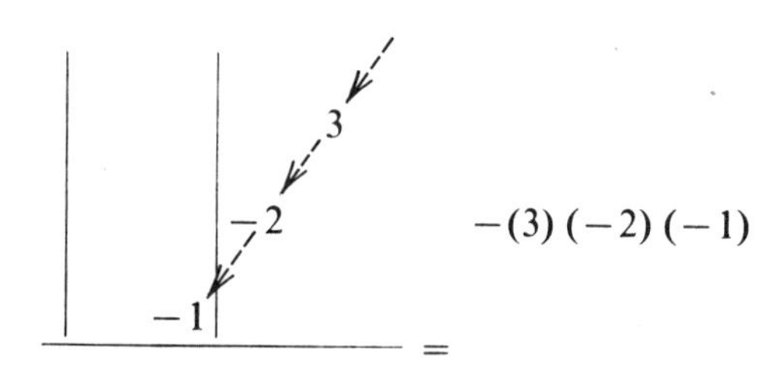

Second Product (moving to the next dotted-line arrows to the left):

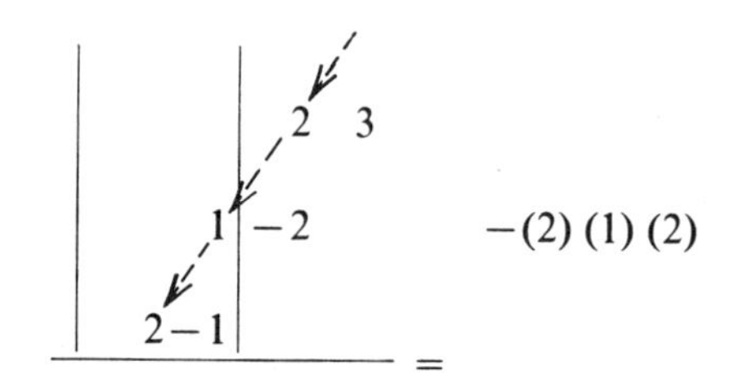

Third product (moving to the next dotted arrows to the left):

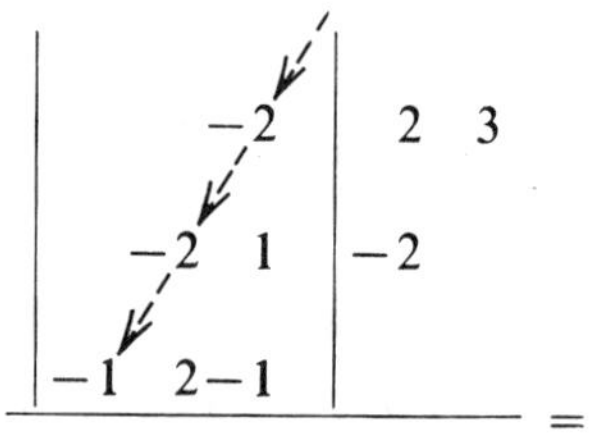

$$-(-2)(-2)(-1)$$

Follow the same procedure for the denominator. The final results are

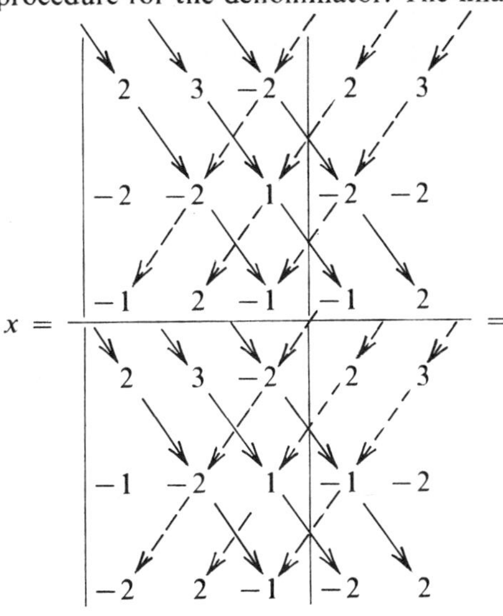

$$x = \frac{(2)(-2)(-1) + (3)(1)(-1) + (-2)(-2)(2) - (3)(-2)(-1)}{(2)(-2)(-1) + (3)(1)(-2) + (-2)(-1)(2) - (3)(-1)(-1)}$$

$$\frac{-(2)(1)(2) - (-2)(-2)(-1)}{-(2)(1)(2) - (-2)(-2)(-2)}$$

$$x = \frac{(4) + (-3) + (8) - (6) - (4) - (-4)}{(4) + (-6) + (4) - (3) - (4) - (-8)}$$

$$x = \frac{4 - 3 + 8 - 6 - 4 + 4}{4 - 6 + 4 - 3 - 4 + 8}$$

$$x = \frac{3}{3}$$

$$x = 1$$

To solve for z:

$z = \dfrac{\text{numerator}}{\text{denominator}} =$ (list the x [in triangle] and y [in circle] coefficients in order, but replace the z coefficient with the [numbers] [in box] at right side of equation) / (same as x denominator)

$$z = \frac{\begin{vmatrix} 2 & 3 & 2 \\ -1 & -2 & -2 \\ -2 & 2 & -1 \end{vmatrix}}{3}$$

Rewrite, repeating the first two columns, listing them at the right.

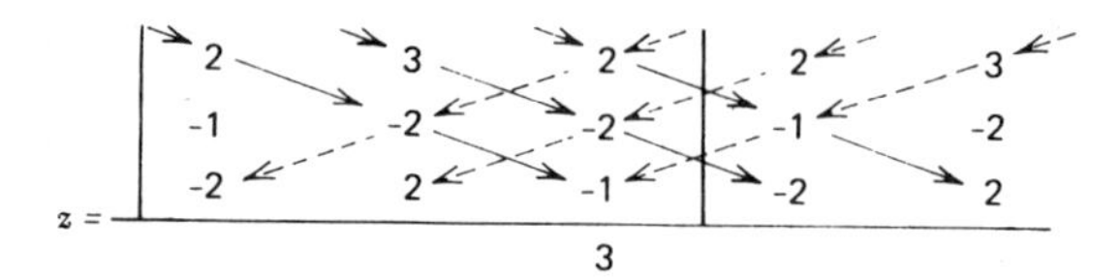

Performing the sums and differences of the cross products above yields

$$z = \frac{(2)(-2)(-1) + (3)(-2)(-2) + (2)(-1)(2) - (3)(-1)(-1)}{3}$$

$$\frac{-(2)(-2)(2) - (2)(-2)(-2)}{3}$$

$$z = \frac{(4) + (12) + (-4) - (3) - (-8) - (8)}{3}$$

$$z = \frac{4 + 12 - 4 - 3 + 8 - 8}{3}$$

$$z = \frac{9}{3}$$

$$z = 3$$

In a similar manner, the value of y could be found to be 2.

Trigonometry of Right Triangles.

Pythagoras' theorem: The hypotenuse squared is equal to the sum of the squares of the sides (see Fig. 1-A).

$$Z^2 = R^2 + X^2$$

or

$$Z = \sqrt{R^2 + X^2}$$

Trigonometric Functions of Angle θ (see Fig. 1-A and also Fig. 12-14, Chapter 12): When angle is in the *first quadrant* (90° or less):

$$\sin\theta = \frac{X}{Z} \quad \text{or} \quad Z = \frac{X}{\sin\theta} \quad \text{or} \quad X = Z(\sin\theta)$$

$$\cos\theta = \frac{R}{Z} \quad \text{or} \quad Z = \frac{R}{\cos\theta} \quad \text{or} \quad R = Z(\cos\theta)$$

$$\tan\theta = \frac{X}{R} \quad \text{or} \quad R = \frac{X}{\tan\theta} \quad \text{or} \quad X = R(\tan\theta)$$

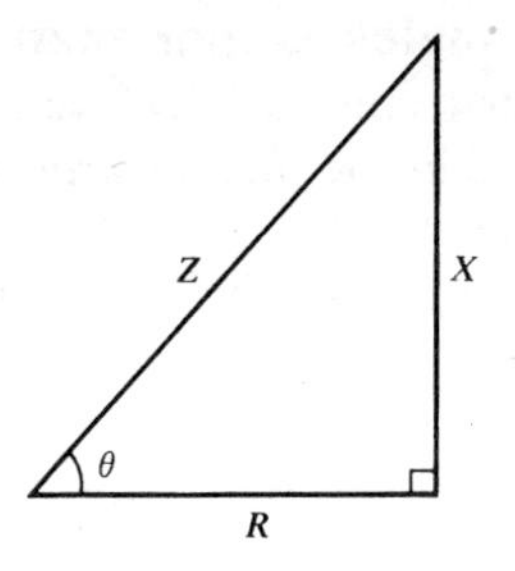

Appendix Figure 1-A.

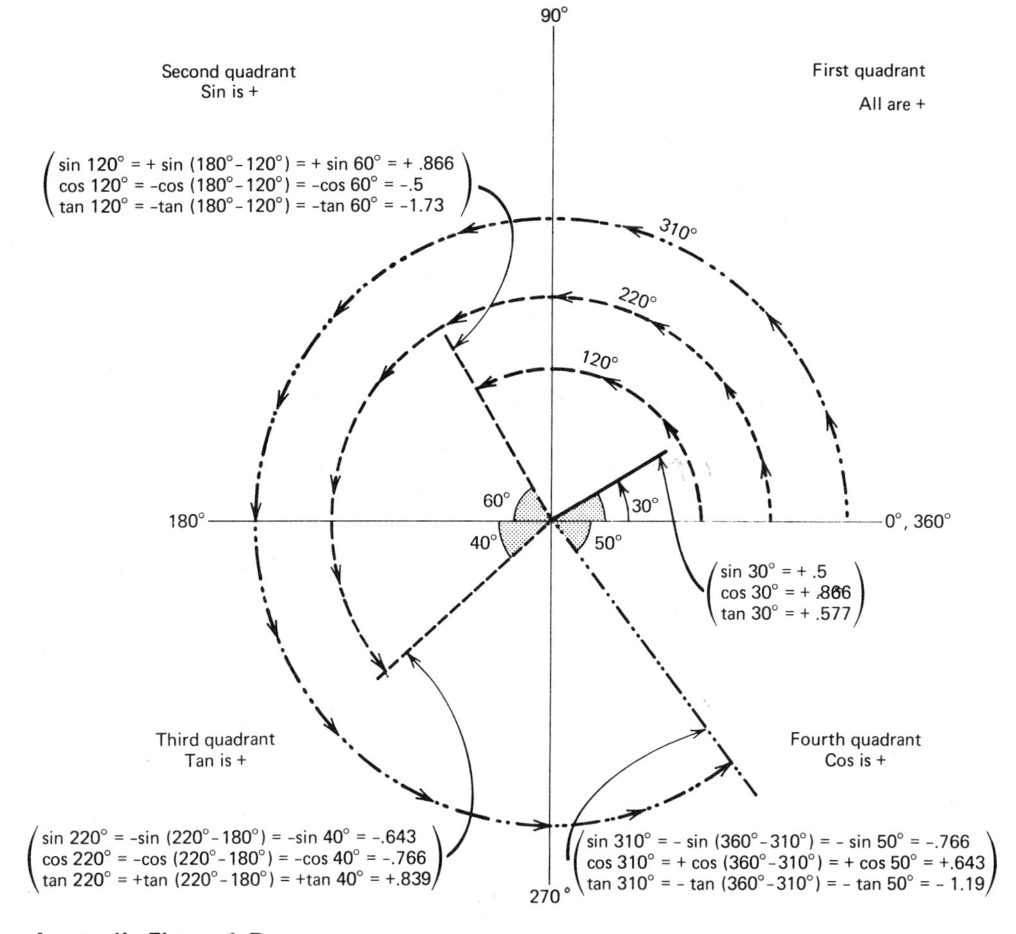

Appendix Figure 1-B.

Trigonometric Functions of Angles Larger than 90° (see Fig. 1-B): When a vector or phasor is rotated counter clockwise (normal direction) going from 0° to 90°, it is said to be in the *first quadrant*; going from 90° to 180°, it is in the *second quadrant*; from 180° to 270°, it is in the *third quadrant*; and from 270° to 360°, it is in the *fourth quadrant*. The trigonometric functions of any of these angles has the same value as the acute angle (less than 90°) formed between the rotated phasor and the *nearest horizontal* axis (either the 0°, 180°, or 360° lines). In the first quadrant, it is simply the angle between the rotated phasor and 0°, with *All* functions having + values.

An angle of 120° falls in the *second* quadrant where only the *S*in is + (cos and tan are −). The acute angle to be used here is the difference between the 120° and the horizontal axis of 180°, or a 60° angle, as shown in Fig. 1-B.

An angle of 220° falls in the *third* quadrant where only the *T*an is + (sin and cos are −). The acute angle to be used here is the difference between the 220° and the horizontal axis of 180° or 40°, as shown in Fig. 1-B.

An angle of 310° falls in the *fourth* quadrant where only the *C*os is + (sin and tan are −). The acute angle to be used here is the difference between the 310° and the horizontal axis of 360°, or 50°, as shown in Fig. 1-B.

The *polarity* of the value of the sin, cos, or tan of these large angles may be easily remembered as shown in Fig. 1-C. The word *CAST* is formed by the first letter of those functions that are *positive*.

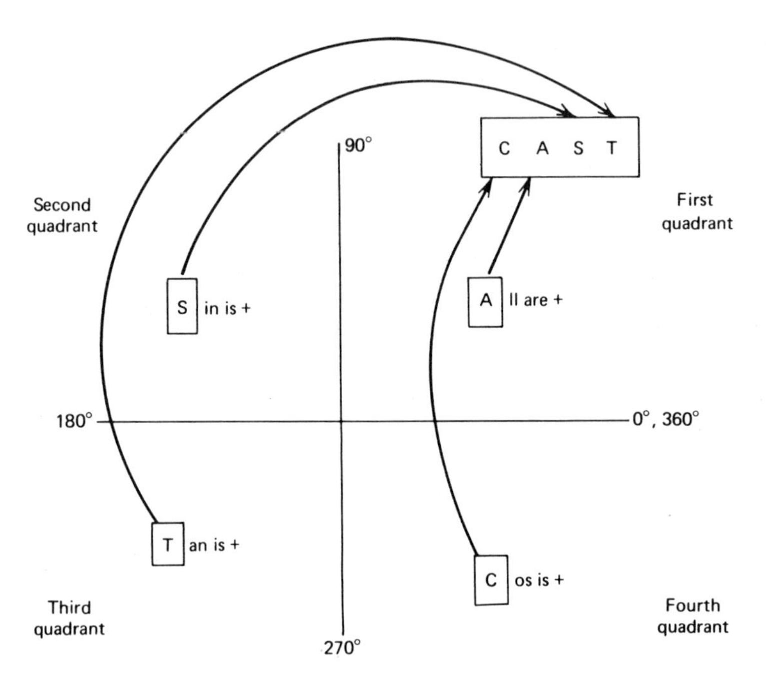

Appendix Figure 1-C.

APPENDIX

II

APPENDIX II More Complete List of Prefixes

Prefix	Symbol	Value	Conversion to Basic Unit (Example)
tera	T	trillion	$\times 10^{12}$ (15 THz = 15 $\times 10^{12}$ Hz)
giga	G	billion	$\times 10^{9}$ (25 GHz = 25 $\times 10^{9}$ Hz)
meg	M	million	$\times 10^{6}$ (2 MΩ = 2 $\times 10^{6}$ Ω)
myria	My	ten thousand	$\times 10^{4}$ (3 myria meter = 3 $\times 10^{4}$ m)
kilo	K	thoudand	$\times 10^{3}$ (5 KΩ = 5 $\times 10^{3}$ Ω = 5000 Ω)
hecto	h	hundred	$\times 10^{2}$ (12 hm = 12 $\times 10^{2}$ m = 1200 m)
deka (also deca)	dk	ten	$\times 10^{1}$ (6 dkm = 6 $\times$ 10 m = 60 m)
deci	d	tenth	$\times 10^{-1}$ (50 dm = 50 $\times 10^{-1}$ m = 5 m)
centi	c	hundredth	$\times 10^{-2}$ (5 cm = 5 $\times 10^{-2}$ m = .05 m)
milli	m	thousandth	$\times 10^{-3}$ (25 mA = 25 $\times 10^{-3}$ A = 0.025 A)
micro	μ	millionth	$\times 10^{-6}$ (15 μsec = 15 $\times 10^{-6}$ sec)
nano	n	thousandth of a millionth (billionth)	$\times 10^{-9}$ (5 nsec = 5 $\times 10^{-9}$ sec)
pico (micro micro)	p ($\mu\mu$)	millionth of a millionth (trillionth)	$\times 10^{-12}$ (250 pF = 250 $\times 10^{-12}$ F)
femto	f	quadrillionth	$\times 10^{-15}$ (2 fm = 2 $\times 10^{-15}$ m)
atto	a	quintillionth	$\times 10^{-18}$ (5 am = 5 $\times 10^{-18}$ m)

APPENDIX III

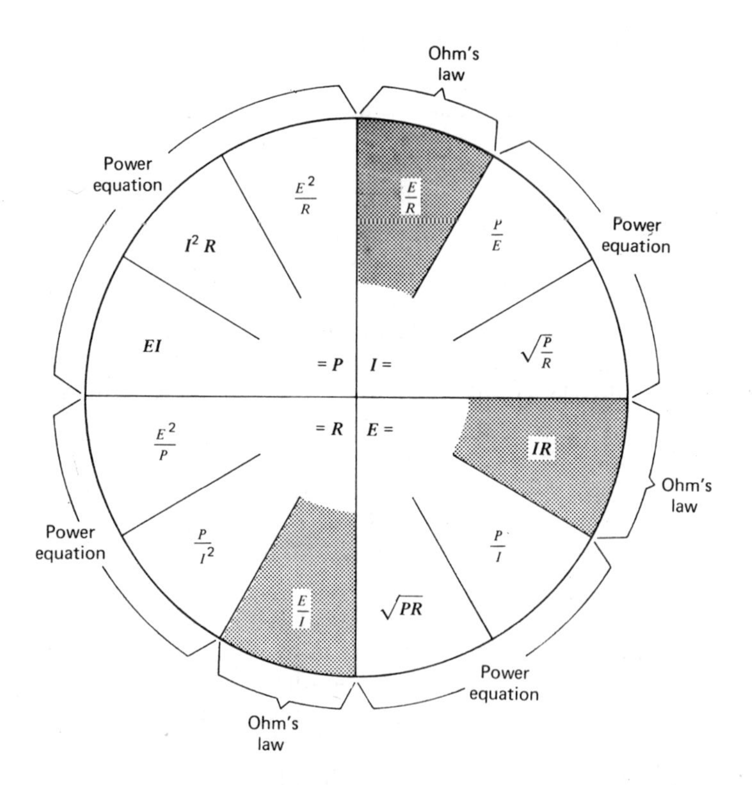

Ohm's law and power equation chart for resistor circuit.

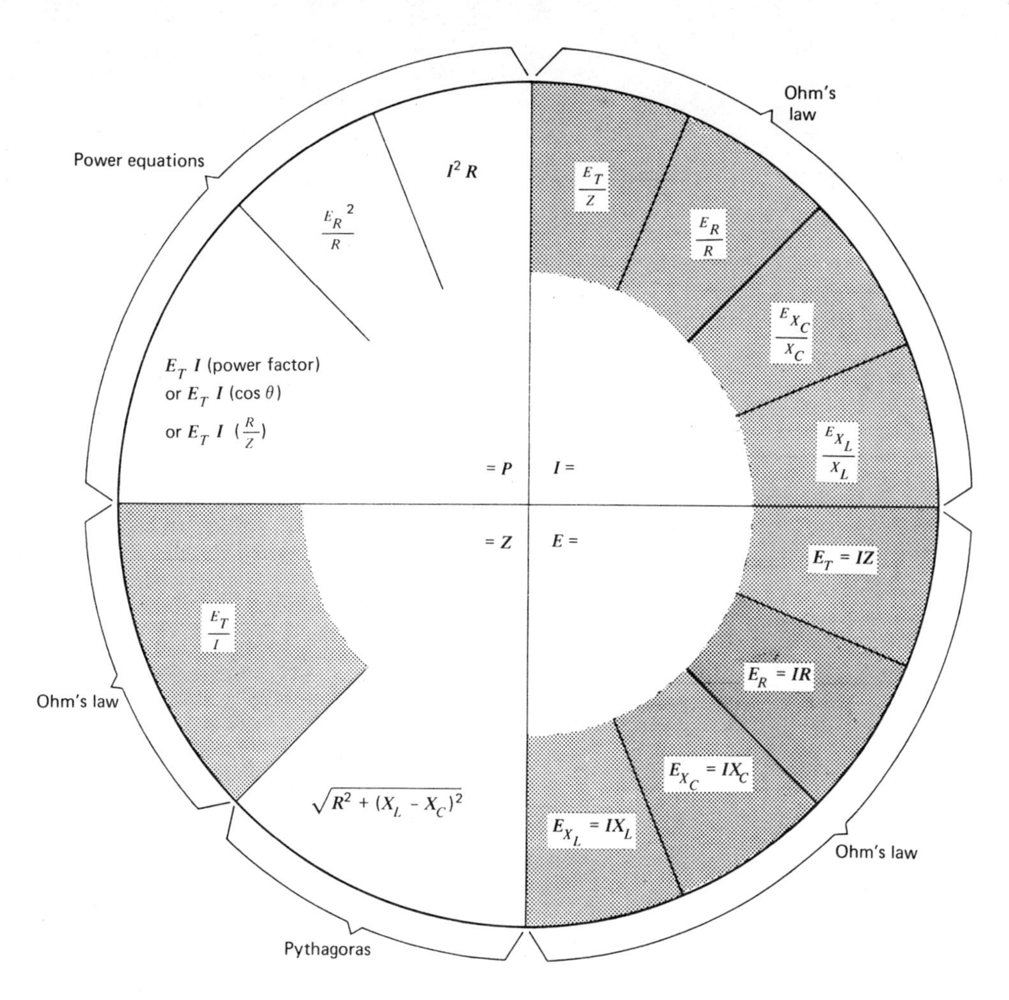

Ohm's law and power equations for ac series circuit of R, X_L, and X_C.

APPENDIX IV

APPENDIX IV Trigonometric Functions

θ (in degrees)	Sin θ	Cos θ	Tan θ	θ (in degrees)	Sin θ	Cos θ	Tan θ
0	.000	1.000	.000	**20**	.342	.940	.361
1	.018	.999	.018	21	.358	.934	.381
2	.035	.999	.035	22	.375	.927	.404
3	.052	.999	.052	23	.394	.921	.424
4	.070	.998	.070	24	.407	.914	.445
5	.087	.996	.088	**25**	.423	.906	.466
6	.105	.995	.105	26	.438	.899	.488
7	.122	.993	.123	27	.454	.891	.510
8	.139	.990	.144	28	.470	.883	.532
9	.156	.988	.158	29	.485	.875	.554
10	.174	.985	.176	**30**	.500	.866	.577
11	.191	.982	.194	31	.515	.857	.601
12	.208	.978	.213	32	.530	.848	.625
13	.225	.974	.231	33	.545	.839	.649
14	.242	.970	.249	34	.559	.829	.675
15	.259	.966	.268	**35**	.571	.819	.700
16	.276	.961	.287	36	.588	.809	.727
17	.292	.956	.306	37	.602	.799	.754
18	.309	.954	.325	38	.646	.788	.781
19	.326	.946	.341	39	.629	.777	.810

θ (in degrees)	Sin θ	Cos θ	Tan θ
40	.643	.766	.839
41	.656	.755	.869
42	.669	.743	.900
43	.682	.731	.933
44	.695	.719	.966
45	.707	.707	1.000
46	.719	.695	1.04
47	.731	.682	1.07
48	.743	.669	1.11
49	.755	.656	1.15
50	.766	.643	1.19
51	.777	.629	1.23
53	.788	.616	1.28
53	.799	.602	1.33
54	.809	.588	1.38
55	.819	.574	1.43
56	.829	.559	1.48
57	.839	.545	1.54
58	.848	.530	1.60
59	.857	.515	1.66
60	.866	.500	1.73
61	.875	.485	1.80
62	.883	.470	1.88
63	.891	.454	1.96
64	.898	.438	2.05
65	.906	.423	2.14
66	.914	.407	2.25
67	.924	.391	2.36
68	.927	.375	2.48
69	.934	.358	2.61
70	.940	.342	2.75
71	.946	.326	2.90
72	.954	.309	3.08
73	.956	.292	3.27
74	.961	.276	3.49
75	.966	.259	3.73
76	.970	.242	4.01
77	.974	.225	4.33
78	.978	.208	4.70
79	.982	.191	5.14
80	.985	.176	5.67
81	.988	.156	6.31
82	.990	.139	7.12
83	.993	.122	8.14
84	.995	.105	9.51
85	.996	.087	11.43
86	.998	.070	14.30
87	.999	.052	19.08
88	.999	.035	28.64
89	.999	.018	57.29
90	1.000	.000	Infinity

APPENDIX V

APPENDIX V Exponential Functions—Values and Common Logarithms

	e^x		e^{-x}		e^x		e^{-x}
x	Value	Log	Value	x	Value	Log	Value
0.00	1.0000	.00000	1.00000	0.20	1.2214	.08686	.81873
0.01	1.0101	.00434	0.99005	0.21	1.2337	.09120	.81058
0.02	1.0202	.00869	.98020	0.22	1.2461	.09554	.80252
0.03	1.0305	.01303	.97045	0.23	1.2586	.09989	.79453
0.04	1.0408	.01737	.96079	0.24	1.2712	.10423	.78663
0.05	1.0513	0.2171	.95123	0.25	1.2840	.10857	.77880
0.06	1.0618	.02606	.94176	0.26	1.2969	.11292	.77105
0.07	1.0725	.03040	.93239	0.27	1.3100	.11726	.76338
0.08	1.0833	.03474	.92312	0.28	1.3231	.12160	.75578
0.09	1.0942	.03909	.91393	0.29	1.3364	.12595	.74826
0.10	1.1052	.04343	.90484	0.30	1.3499	.13029	.74082
0.11	1.1163	.04777	.89583	0.31	1.3634	.13463	.73345
0.12	1.1275	.05212	.88692	0.32	1.3771	.13897	.72615
0.13	1.1388	.05646	.87809	0.33	1.3910	.14332	.71892
0.14	1.1503	.06080	.86936	0.34	1.4049	.14766	.71177
0.15	1.1618	.06514	.86071	0.35	1.4191	.15200	.70469
0.16	1.1735	.06949	.85214	0.36	1.4333	.15635	.69768
0.17	1.1853	.07383	.84366	0.37	1.4477	.16069	.69073
0.18	1.1972	.07817	.85327	0.38	1.4623	.16503	.68386
0.19	1.2092	.08252	.82696	0.39	1.4770	.16937	.67706

APPENDIX V Exponential Functions—Values and Common Logarithms —Cont.

	e^x		e^{-x}		e^x		e^{-x}
x	Value	Log	Value	x	Value	Log	Value
0.40	1.4918	.17372	.67032	0.75	2.1170	.32572	.47237
0.41	1.5068	.17806	.66365	0.76	2.1383	.33006	.46767
0.42	1.5220	.18240	.65705	0.77	2.1598	.33441	.46301
0.43	1.5373	.18675	.65051	0.78	2.1815	.33875	.45841
0.44	1.5527	.19109	.64404	0.79	2.2034	.34309	.45384
0.45	1.5683	.19543	.63763	0.80	2.2255	.34744	.44933
0.46	1.5841	.19978	.63128	0.81	2.2479	.35178	.44486
0.47	1.6000	.20412	.62500	0.82	2.2705	.35612	.44043
0.48	1.6161	.20846	.61878	0.83	2.2933	.36046	.43605
0.49	1.6323	.21280	.61263	0.84	2.3164	.36481	.43171
0.50	1.6487	.21715	.60653	0.85	2.3396	.36915	.42741
0.51	1.6653	.22149	.60050	0.86	2.3632	.37349	.42316
0.52	1.6820	.22583	.59452	0.87	2.3869	.37784	.41895
0.53	1.6989	.23018	.58860	0.88	2.4109	.38218	.41478
0.54	1.7160	.23452	.58275	0.89	2.4351	.38652	.41066
0.55	1.7333	.23886	.57695	0.90	2.4596	.39087	.40657
0.56	1.7507	.24320	.57121	0.91	2.4843	.39521	.40252
0.57	1.7683	.24755	.56553	0.92	2.5093	.39955	.39852
0.58	1.7860	.25189	.55990	0.93	2.5345	.40389	.39455
0.59	1.8040	.25623	.55433	0.94	2.5600	.40824	.39063
0.60	1.8221	.26058	.54881	0.95	2.5857	.41258	.38674
0.61	1.8404	.26492	.54335	0.96	2.6117	.41692	.38289
0.62	1.8589	.26926	.53794	0.97	2.6379	.42127	.37908
0.63	1.8776	.27361	.53259	0.98	2.6645	.42561	.37531
0.64	1.8965	.27795	.52729	0.99	2.6912	.42995	.37158
0.65	1.9155	.28229	.52205	1.00	2.7183	.43429	.36788
0.66	1.9348	.28664	.51685	1.01	2.7456	.43864	.36422
0.67	1.9542	.29098	.51171	1.02	2.7732	.44298	.36060
0.68	1.9739	.29532	.50662	1.03	2.8011	.44732	.35701
0.69	1.9937	.29966	.50158	1.04	2.8292	.45167	.35345
0.70	2.0138	.30401	.49659	1.05	2.8577	.45601	.34994
0.71	2.0340	.30835	.49164	1.06	2.8864	.36035	.34646
0.72	2.0544	.31269	. 8675	1.07	2.9154	.46470	.34301
0.73	2.0751	.31703	.48191	1.08	2.9447	.46904	.33960
0.74	2.0959	.32138	.47711	1.09	2.9743	.47338	.33622

APPENDIX V Exponential Functions—Values and Common Logarithms—Cont.

x	e^x Value	e^x Log	e^{-x} Value	x	e^x Value	e^x Log	e^{-x} Value
1.10	3.0042	.47772	.33287	1.45	4.2631	.62973	.23457
1.11	3.0344	.48207	.32956	1.46	4.3060	.63407	.23224
1.12	3.0649	.48641	.32628	1.47	4.3492	.63841	.22993
1.13	3.0957	.49075	.32303	1.48	4.3929	.64276	.22764
1.14	3.1268	.49510	.31982	1.49	4.4371	.64710	.22537
1.15	3.1582	.49944	.31664	1.50	4.4817	.65144	.22313
1.16	3.1899	.50378	.31349	1.51	4.5267	.65578	.22091
1.17	3.2220	.50812	.31037	1.52	4.5722	.66013	.21871
1.18	3.2544	.51247	.30728	1.53	4.6182	.66447	.21654
1.19	3.2871	.51681	.30422	1.54	4.6646	.66881	.21438
1.20	3.3201	.52115	.30119	1.55	4.7115	.67316	.21225
1.21	3.3535	.52550	.29820	1.56	4.7588	.67750	.21014
1.22	3.3872	.52984	.29523	1.57	4.8066	.68184	.20805
1.23	3.4212	.53418	.29229	1.58	4.8550	.68619	.20598
1.24	3.4556	.53853	.28938	1.59	4.9037	.69053	.20393
1.25	3.4903	.54287	.28650	1.60	4.9530	.69487	.20190
1.26	3.5254	.54721	.28365	1.61	5.0028	.69921	.19989
1.27	3.5609	.55155	.28083	1.62	5.0531	.70356	.19790
1.28	3.5966	.55590	.27804	1.63	5.1039	.70790	.19593
1.29	3.6328	.56024	.27527	1.64	5.1552	.71224	.19398
1.30	3.6693	.56458	.27253	1.65	5.2070	.71659	.19205
1.31	3.7062	.56893	.26982	1.66	5.2593	.72093	.19014
1.32	3.7434	.47327	.26714	1.67	5.3122	.72527	.18825
1.33	3.7810	.57761	.26448	1.68	5.3656	.72961	.18637
1.34	3.8190	.58195	.26185	1.69	5.4195	.73396	.18452
1.35	3.8574	.58630	.25924	1.70	5.4739	.73830	.18268
1.36	3.8962	.59064	.25666	1.71	5.5290	.74264	.18087
1.37	3.9354	.59498	.25411	1.72	5.5845	.74699	.17907
1.38	3.9749	.59933	.25158	1.73	5.6407	.75133	.17728
1.39	4.0149	.60367	.24908	1.74	5.6973	.75567	.17552
1.40	4.0552	.60801	.24660	1.75	5.7546	.76002	.17377
1.41	4.0960	.61236	.24414	1.76	5.8124	.76436	.17204
1.42	4.1371	.61670	.24171	1.77	5.8709	.76870	.17033
1.43	4.1787	.62104	.23931	1.78	5.9299	.77304	.16864
1.44	4.2207	.62538	.23693	1.79	5.9895	.77739	.16696

x	e^x Value	e^x Log	e^{-x} Value	x	e^x Value	e^x Log	e^{-x} Value
1.80	6.0496	.78173	.16530	2.15	8.5849	.93373	.11648
1.81	6.1104	.78607	.16365	2.16	8.6711	.93808	.11533
1.82	6.1719	.79042	.16203	2.17	8.7583	.94242	.11418
1.83	6.2339	.79476	.16041	2.18	8.8463	.94676	.11304
1.84	6.2965	.79910	.15882	2.19	8.9352	.95110	.11192
1.85	6.3598	.80344	.15724	2.20	9.0250	.95545	.11080
1.86	6.4237	.80779	.15567	2.21	9.1157	.95979	.10970
1.87	6.4883	.81213	.15412	2.22	9.2073	.96413	.10861
1.88	6.5535	.81647	.15259	2.23	9.2999	.96848	.10753
1.89	6.6194	.82082	.15107	2.24	9.3933	.97282	.10646
1.90	6.6859	.82516	.14957	2.25	9.4877	.97716	.10540
1.91	6.7531	.82950	.14808	2.26	9.5831	.98151	.10435
1.92	6.8210	.83385	.14661	2.27	9.6794	.98585	.10331
1.93	6.8895	.83819	.14515	2.28	9.7767	.99019	.10228
1.94	6.9588	.84253	.14370	2.29	9.8749	.99453	.10127
1.95	7.0287	.84687	.14227	2.30	9.9742	0.99888	.10026
1.96	7.0993	.85122	.14086	2.31	10.074	1.00322	.09926
1.97	7.1707	.85556	.13946	2.32	10.176	1.00756	.09827
1.98	7.2427	.85990	.13807	2.33	10.278	1.01191	.09730
1.99	7.3155	.86425	.13670	2.34	10.381	1.01625	.09633
2.00	7.3891	.86859	.13534	2.35	10.486	1.02050	.09537
2.01	7.4633	.87293	.13399	2.36	10.591	1.02493	.09442
2.02	7.5383	.87727	.13266	2.37	10.697	1.02928	.09348
2.03	7.6141	.88162	.13134	2.38	10.805	1.03362	.09255
2.04	7.6906	.88596	.13003	2.39	10.913	1.03796	.09163
2.05	7.7679	.89030	.12873	2.40	11.023	1.04231	.09072
2.06	7.8460	.89465	.12745	2.41	11.134	1.04665	.08982
2.07	7.9248	.89899	.12619	2.42	11.246	1.05099	.08892
2.08	8.0045	.90333	.12493	2.43	11.359	1.05534	.08804
2.09	8.0849	.90768	.12369	2.44	11.473	1.05968	.08716
2.10	8.1662	.91202	.12246	2.45	11.588	1.06402	.08629
2.11	8.2482	.91636	.12124	2.46	11.705	1.06836	.08543
2.12	8.3311	.92070	.12003	2.47	11.822	1.07271	.08458
2.13	8.4149	.92505	.11884	2.48	11.941	1.07705	.08374
2.14	8.4994	.92939	.11765	2.49	12.061	1.08139	.08291

APPENDIX V Exponential Functions—Values and Common Logarithms—Cont.

x	e^x Value	e^x Log	e^{-x} Value	x	e^x Value	e^x Log	e^{-x} Value
2.50	12.182	1.08574	.08208	2.85	17.288	1.23774	.05784
2.51	12.305	1.09008	.08127	2.86	17.462	1.24208	.05727
2.52	12.429	1.09442	.08046	2.87	17.637	1.24643	.05670
2.53	12.554	1.09877	.07966	2.88	17.814	1.25077	.05613
2.54	12.680	1.10311	.07887	2.89	17.993	1.25511	.05558
2.55	12.807	1.10745	.07808	2.90	18.174	1.25945	.05502
2.56	12.936	1.11179	.07730	2.91	18.357	1.26380	.05448
2.57	13.066	1.11614	.07654	2.92	18.541	1.26814	.05393
2.58	13.197	1.12048	.07577	2.93	18.728	1.27248	.05340
2.59	13.330	1.12482	.07502	2.94	18.916	1.27683	.05287
2.60	13.464	1.12917	.07427	2.95	19.106	1.28117	.05234
2.61	13.599	1.13351	.07353	2.96	19.298	1.28551	.05182
2.62	13.736	1.13785	.07280	2.97	19.492	1.28985	.05130
2.63	13.874	1.14219	.07208	2.98	19.688	1.29420	.05079
2.64	14.013	1.14654	.07136	2.99	19.886	1.29854	.05029
2.65	14.154	1.15088	.07065	3.00	20.086	1.30288	.04979
2.66	14.296	1.15522	.06995	3.05	21.145	1.32460	.04736
2.67	14.440	1.15957	.06925	3.10	22.198	1.34631	.04505
2.68	14.585	1.16391	.06856	3.15	23.336	1.36803	.04285
2.69	14.732	1.16825	.06788	3.20	24.533	1.38974	.04076
2.70	14.880	1.17260	.06721	3.25	25.790	1.41146	.03877
2.71	15.029	1.17694	.05564	3.30	27.113	1.43317	.03688
2.72	15.180	1.18128	.06587	3.35	28.503	1.45489	.03508
2.73	15.333	1.18562	.06522	3.40	29.964	1.47660	.03337
2.74	15.487	1.18997	.06457	3.45	31.500	1.49832	.03175
2.75	15.643	1.19431	.06393	3.50	33.115	1.52003	.03020
2.76	15.800	1.19865	.06329	3.55	34.813	1.54175	.02872
2.77	15.959	1.20300	.06266	3.60	36.598	1.56346	.02732
2.78	16.119	1.20734	.06204	3.65	38.475	1.58517	.02599
2.79	16.281	1.21168	.06142	3.70	40.447	1.60689	.02472
2.80	16.445	1.21602	.06081	3.75	42.521	1.62860	.02352
2.81	16.610	1.22037	.06020	3.80	44.701	1.65032	.02237
2.82	16.777	1.22471	.05961	3.85	46.993	1.67203	.02128
2.83	16.945	1.22905	.05901	3.90	49.402	1.69375	.02024
2.84	17.116	1.23340	.05843	3.95	51.935	1.71546	.01925

APPENDIX V Exponential Functions—Values and Common Lograithms —Cont.

	e^x		e^{-x}		e^x		e^{-x}
x	Value	Log	Value	x	Value	Log	Value
4.00	54.598	1.73718	.01832	5.60	207.43	2.43205	.00370
4.10	60.340	1.78061	.01657	5.70	298.87	2.47548	.00335
4.20	66.686	1.82404	.01500	5.80	330.30	2.51891	.00303
4.30	73.700	1.86747	.01357	5.90	365.04	2.56234	.00274
4.40	81.451	1.91090	.01227	6.00	403.43	2.60577	.00248
4.50	90.017	1.95433	.01111	6.25	518.01	2.71434	.00193
4.60	99.484	1.99775	.01005	6.50	665.14	2.82291	.00150
4.70	109.95	2.04118	.00910	6.75	854.06	2.93149	.00117
4.80	121.51	2.08461	.00823	7.00	1096.6	3.04006	.00091
4.90	134.29	2.12804	.00745	7.50	1808.0	3.25721	.00055
5.00	148.41	2.17147	.00674	8.00	2981.0	3.47436	.00034
5.10	164.02	2.21490	.00610	8.50	4914.8	3.69150	.00020
5.20	181.27	2.25833	.00552	9.00	8103.1	3.90865	.00012
5.30	200.34	2.30176	.00499	9.50	13360.	4.12580	.00007
5.40	221.41	2.34519	.00452	10.00	22026.	4.34294	.00005
5.50	244.69	2.38862	.00409				

APPENDIX VI

APPENDIX VI Color Code

						Capacitor Tolerance ±%		Ceramic Capacitors		
								Tolerance		
Colors	Numerical Value For Significant Digits of Resistors and Capacitors	Number of Zeros To Be Added	(or Multiply by)	Resistor Tolerance ±%	Cap. Voltage Rating	Molded Paper	Molded Mica	Capacitor 10 pF or Less ±%	Capacitor Larger than 10 pF ±%	Temperature Coefficient in Parts per Million per Degree Centigrade
Black	0	none	(1)		—	20%	20%	2%	20%	0
Brown	1	1 zero	(10)	1%	100		1%		1%	−30
Red	2	2 zeros	(100)	2%	200		2%		2%	−80
Orange	3	3 zeros	(1000)	3%	300	30%	3%		2.5%	−150
Yellow	4	4 zeros	(10,000)	4%	400	40%	4%			−220
Green	5	5 zeros	(100,000)		500	5%	5%(EIA)	0.5%	5%	−330
Blue	6	6 zeros	(1,000,000)		600		6%			−470
Violet	7	7 zeros	(10,000,000)		700		7%			−750
Gray	8	8 zeros	[b](100,000,000)		800		8%	0.25%		+30
White	9	9 zeros	[b](1,000,000,000)		900	10%	9%	1%	10%	+550
Silver	—	—	[a](0.01)	10%	2000	10%	10%			
Gold	—	—	[a](0.1)	5%	1000		5%(MIL)			
				No Color =20%						

[a] For *resistors* only, 3rd band.

[b] For *ceramic* capacitors: gray is × 0.01; white is × 0.1.

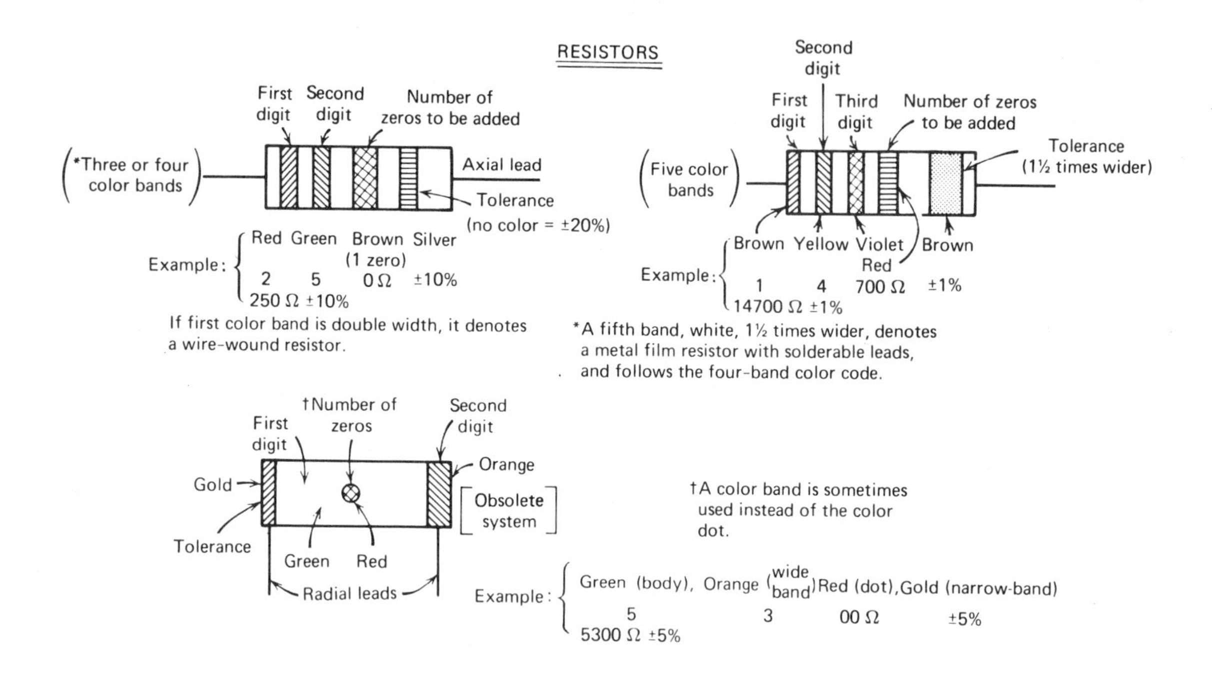

CAPACITORS

Molded mica, (flat capacitor)

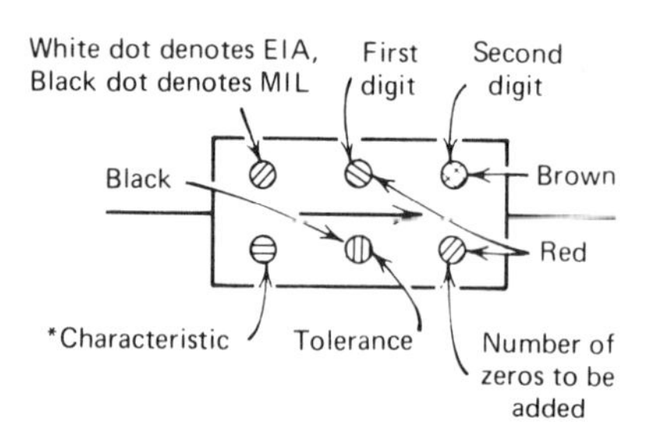

*Characteristic refers to temperature, coefficient and capacitance drift

Example:
Red Brown Red Black
2 1 00 pF ±20%
2100 pF ±20%

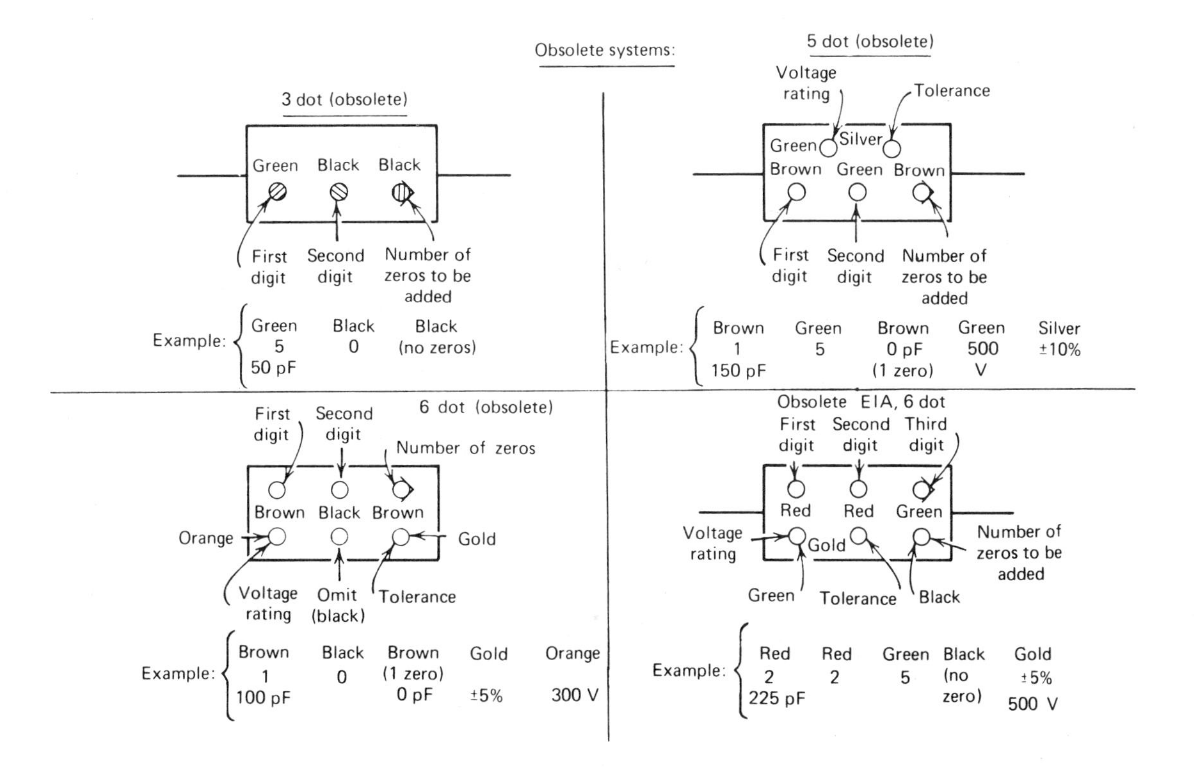
Obsolete systems:
3 dot (obsolete)
Green
Black
Black
First digit
Second digit
Number of zeros to be added
Example:
Green
5
50 pF
Black
0
Black
(no zeros)
5 dot (obsolete)
Voltage rating
Tolerance
Green
Silver
Brown
Green
Brown
First digit
Second digit
Number of zeros to be added
Example:
Brown
1
150 pF
Green
5
Brown
0 pF
(1 zero)
Green
500
V
Silver
±10%
6 dot (obsolete)
First digit
Second digit
Number of zeros
Brown
Black
Brown
Orange
Gold
Voltage rating
Omit (black)
Tolerance
Example:
Brown
1
100 pF
Black
0
Brown
(1 zero)
0 pF
Gold
±5%
Orange
300 V
Obsolete EIA, 6 dot
First digit
Second digit
Third digit
Red
Red
Green
Voltage rating
Gold
Number of zeros to be added
Green
Tolerance
Black
Example:
Red
2
225 pF
Red
2
Green
5
Black
(no zero)
Gold
±5%
500 V

CAPACITORS

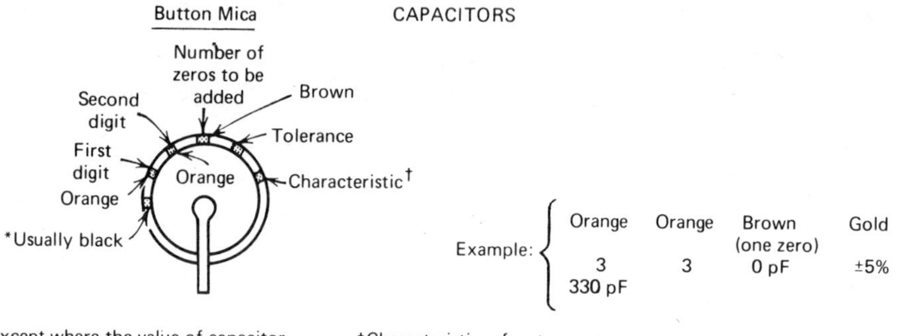

Example:

Orange	Orange	Brown (one zero)	Gold
3	3	0 pF	±5%
330 pF			

*Except where the value of capacitor is given in three digits; In such a case, this color becomes the first digit.

†Characteristic refers to temperature coefficient and capacitance drift

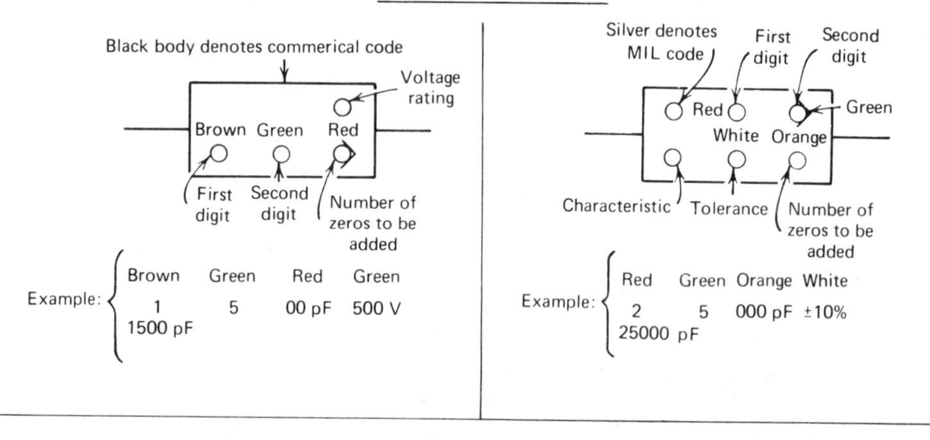

Example:

Brown	Green	Red	Green
1	5	00 pF	500 V
1500 pF			

Example:

Red	Green	Orange	White
2	5	000 pF	±10%
25000 pF			

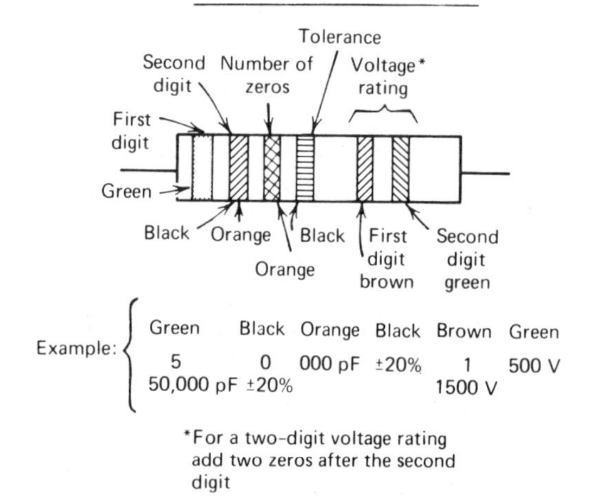

Example:

Green	Black	Orange	Black	Brown	Green
5	0	000 pF	±20%	1	500 V
50,000 pF ±20%				1500 V	

*For a two-digit voltage rating add two zeros after the second digit

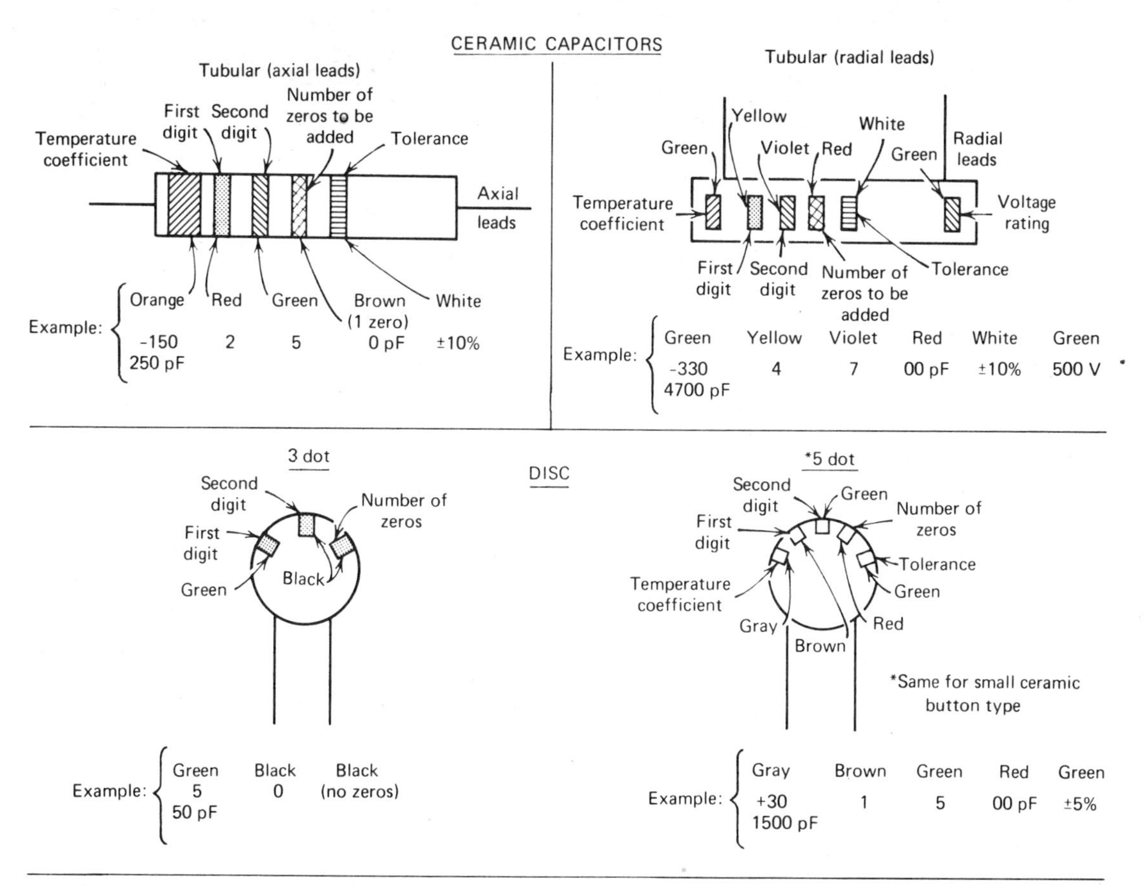

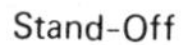

Example: Green 5 50 pF | Black 0 | Black (no zeros)

Example: Gray +30 1500 pF | Brown 1 | Green 5 | Red 00 pF | Green ±5%

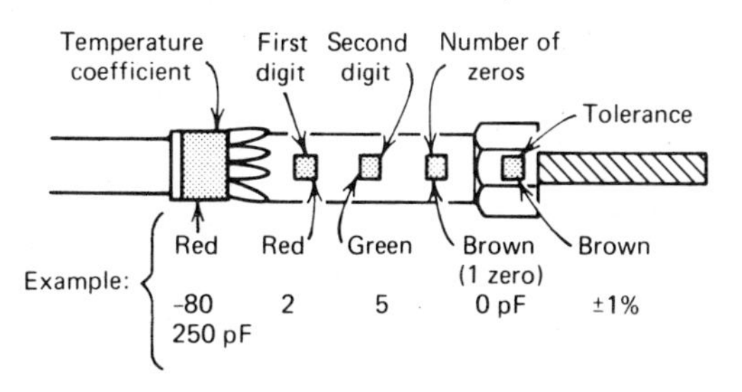

Answers to Odd-Numbered Problems

CHAPTER 1

1-1 Anything occupying space and having mass. **1-3** Air, gas. **1-5** Substance made of two or more elements. **1-7** A group of two or more atoms comprising an element or a compound. **1-9** 103, see Table 1-2. **1-11** (a) −, (b) +. **1-13** $2(n^2)$. **1-15** Valence ring. **1-17** Positive. **1-19** Negative. **1-21** Loosely. **1-23** Repel. **1-25** Insulator. **1-27** Silver, copper, gold, aluminum, nickel (Table 1-3). **1-29** Tightly bound. **1-31** When sufficient electrical force is applied. **1-33** When a sufficiently high voltage is applied, an insulator allows some current to flow. **1-35** Silicon, germanium. **1-37** Proton. **1-39** $1 \text{ C} = 6.24 \times 10^{18}$ electrons. **1-41** Q represents the *quantity* of electrons or protons in coulombs. **1-43** The second ring actually consists of two rings or subshells; the third orbit has three subshells, etc. **1-45** Two on the first, and progressively four more on each of any following subshells. (Fig. 1-4). **1-47** When atoms of one or more elements combine. **1-49** One or more electrons leave an atom of one element to join an atom of another element. The former atom is now a + ion, while the latter is now a − ion. The ions attract and the atoms adhere. **1-51** The atoms become oppositely charged ions. **1-53** +. **1-55** Four-sided, see Fig. 1-8. **1-57** Eight. **1-59** Ability to do work. **1-61** Its motion traveling in its orbit. **1-63** Greater speed or velocity. **1-65** Increases. **1-67** The electron gives up energy in the form of light or heat. **1-69** Conductor.

CHAPTER 2

2-1 Voltage or difference of potential. **2-3** The volt. **2-5** (a) E_T or V_T, (b) V_R or E_R. **2-7** Voltage and a path. **2-9** Amperes. **2-11** I. **2-13** Ohms. **2-15** Conductor. **2-17** Mho.

2-19 ℧. **2-21** (a) $0.05 \times 10^{-6} f$ or $5 \times 10^{-8} f$, (b) $0.05 \times 10^{6}\ \Omega$ or $5 \times 10^{4}\ \Omega$, (c) $75. \times 10^{-3}$ A or 7.5×10^{-2} A, (d) $350. \times 10^{-6}$ V or 3.5×10^{-4} V. **2-23** 6×10^{-4} A or 0.6 mA. **2-25** 120 V. **2-27** 168 kV. **2-29** 200 Ω . **2-31** 0.4 Ω. **2-33** Newton. **2-35** 120 N. **2-37** 60 J. **2-39** 746 J/sec. **2-41** 0.0675 W. **2-43** 2 A. **2-45** 1200 W. **2-47** 605 mW. **2-49** 242 Ω.

CHAPTER 3

3-1 (a) 1000 Ω. (b) 9 mA. **3-3** (a) 2 V, (b) $1\frac{1}{2}$ V, (c) $\frac{1}{2}$ V, (d) 4 V. **3-5** (a) 2 mA, (b) 2 V, (c) 4 V, (d) 8 V, (e) 10 V, (f) 30 V. **3-7** (a) 3 mA, (b) 6 V, (c) 30 V, (d) 18 mW, (e) 90 mW, (f) 108 mW. **3-9** (a) 2 mA, (b) 4 V, (c) 32 mW, (d) 20 V, (e) 40 mW, **3-11** 10 V. **3-13** 150 V. **3-15** (a) 90 V, (b) 60 V, (c) 30 V, (d) 120 V. **3-17** (a) +90 V, (b) 0 V, (c) −60 V, (d) −90 V, (e) −210 V. **3-19** −39 V. **3-21** (a) 60 V, (b) 40 V, (c) +10 V, (d) +60 V. **3-23** Too much I; too much E across each filament; and probably a filament would burn out. **3-25** (a) 0 V, (b) +100 V. **3-27** A = 0 V, B = 0 V, C = + 300 V, D = +300 V, E = +300 V. **3-29** A = 0 V, B = 0 V, C = 0 V, D = 0 V, E = +300 V. **3-31** B = −50 V, C = −20 V. **3-33** X = −50 V, Y = −50 V. **3-35** X = +100 V, Y = +100 V. **3-37** (a) 30 V, (b) $C - W$, (c) 18 V, + at top, (d) 12 V, − at top, (e) −72 V.

CHAPTER 4

4-1 60 Ω. **4-3** (a) 12 V, (b) 12 V. **4-5** 60 Ω. **4-7** (a) 9 V, (b) 9 V. **4-9** (a) 1.5 A. (b) 1 A, (c) 2.5 A. **4-11** (a) 0.18 V, (b) 9 mA, (c) 20 Ω, (d) 18 Ω. **4-13** (a) 3 V, (b) 3 V, (c) 0.2 A. **4-15** (a) 5 mA, (b) 15 mA. **4-17** 8 Ω. **4-19** 54.1 Ω. **4-21** 50 Ω. **4-23** 150 Ω. **4-25** (a) 10 Ω, (b) 15 Ω. **4-27** 180 and 90 in parallel, or 60 Ω. **4-29** Disconnect one end of R_3. **4-31** (a) 12 K, (b) 432 V, (c) 432 V, (d) 432 V, (e) 432 V, (f) 432 V, (g) 3.6 mA, (h) 10.8 mA, (i) 7.2 mA, (j) 14.4 mA. **4-33** (a) 3 mA, (b) 2.5 mA. **4-35** 900 Ω. **4-37** 18 K.

CHAPTER 5

5-1 (a) 60 K, (b) 120 K, (c) 40 K, (d) 0.4 mA, (e) 0.2 mA, (f) 0.6 mA, (g) 6 V, (h) 18 V, (i) 14 V, (j) 10 V. **5-3** (a) 12 K, (b) 20 K, (c) 5 mA, (d) 40 V, (e) 60 V, (f) 3 mA, (g) 2 mA, (h) 5 mA. **5-5** (a) 6 K, (b) 6 K, (c) 40 K, (d) 10 mA, (e) 60 V, (f) 6 mA, (g) 4 mA, (h) 30 V, (i) 60 V, (j) 8.57 mA, (k) 1.43 mA, (l) 60 V, (m) 28.6 V, (n) 31.4 V, (o) 250 V. **5-7** (a) 20 mA, (b) 0.4 V, (c) 2 V, (d) 13.33 mA, (e) 33.33 mA, (f) 0.333 V, (g) 1.67 V, (h) 4 V. **5-9** 29 K. **5-11** 67.4 μA. **5-13** (a) + 20 V, (b) +20 V, (c) 0 V. **5-15** 2400 Ω. **5-17** 2 K. **5-19** (a) 18 V, (b) +6 V, (c) +6 V. **5-21** (a) 10 mA, (b) 10 mA, (c) 20 mA, (d) 40 mA, (e) 15 K, (f) 5 K, (g) 6250 Ω. **5-23** (a) 100 mA, (b) 20 mA, (c) 100 mA, (d) 20 mA, (e) 30 mA, (f) 50 mA, (g) 50 V, (h) 200 V, (i) 100 V, (j) 150 V, (k) 500 V, (l) 500 Ω, (m) 10 K, (n) 3333 Ω, (o) 3 K. **5-25** (a) 0 A, (b) 0.2 A, (c) 0.2 A, (d) 36 V, (e) 24 V, (f) 0 V, (g) 60 V. **5-27** (a) 0 V, (b) +15 V. **5-29** (a) All currents and voltages are zero, except V_3 = 18 V, (b) same as (a) except V_7 = 18 V. **5-31** (a) 0 V, (b) 0 V, (c) 110 V, (d) 0 V, (e) 0 V, (f) 0 mA. **5-33** (a) R_{AB} = R_2 in parallel with the series combination of R_3, R_4, and R_5, (b) increases. **5-35** R_{AB} = R_2 in parallel with the series combination of R_3, R_6, R_7, R_8, and R_5. **5-37** (a) 300 V, (b) 0 V, (c) 0 V, (d) 0 V, (e) 0 V, (f) 0 mA. **5-39** (a) Decrease, (b) increase.

CHAPTER 6

6-1 18 Ω. **6-3** 15 Ω. **6-5** 24 V (D is +) 4 Ω. **6-7** 11.35 V (X is −) 15.15 Ω. **6-9** 2 V (X is −) 4 Ω. **6-11** 9.32 V (Y is −) 13.8 Ω. **6-13** 66.7 mA 2 V (X is −). **6-15** −2 V. **6-17** 0.4 A flowing down. **6-19** (a) $-7\frac{1}{2}$ V, (b) + 5 V.

6-21 (a) +6 V, (b) +3.5 V, (c) +9.5 V. **6-23** (a) −6.67 V, (b) +11.67 V, (c) +5 V. **6-25** (a) −10 V, (b) +4 V, (c) −6 V. **6-27** (a) 6 V, (b) 6 V, (c) 5.15 V, (d) 17.15 V all + at the upper end. **6-29** 4 Ω, 68 V (point *F* is +). **6-31** 20 Ω, −6 V. **6-33** 1.5 Ω, 9.15 V (*X* is +). **6-35** 1.5 V (*A* is −). **6-37** 30.8 V, + at top. **6-39** 3 V − at top. **6-41** 17.15 V + at top. **6-43** 14.5 V − at top. **6-45** 46.5 V + at top. **6-47** 0.2 A left to right. **6-49** (a) 0.95 A down, (b) 2.86 A up, (c) 2.67 A down, (d) 0.763 A up. **6-51** (a) 0.44 A up, (b) 0.44 A left to right, (c) 0.24 A down, (d) 0.2 A left to right, (e) 0.2 A down, (f) 0.2 A right to left. **6-53** (a) 0.95 A down, (b) 2.86 A up, (c) 2.67 A down, (d) 0.763 A up. **6-55** −6 V. **6-57** 10.64 V + at *K*. **6-59** 14.5 V − at *A*. **6-61** 1.8 V − at *D*. **6-63** 3.61 Ω. **6-65** R_f = 10 Ω. R_e = 7.5 Ω. R_d = 4.28 Ω. **6-67** 33.75 Ω. **6-69** 3.61 Ω. **6-71** 12 Ω, 18 Ω, 6 Ω.

CHAPTER 7

7-1 Two or more cells. **7-3** One that is not designed to be recharged. **7-5** Carbon-zinc or Leclanche cell, mercury cell, alkaline cell, silver cell. **7-7** See Fig. 7-1. **7-9** See Fig. 7-2. **7-11** (a) Voltage and current are unchanged, (b) current capacity or ampere-hours are tripled. **7-13** (a) 9 V, (b) 8.6 V, (c) 0.5 Ω.

CHAPTER 8

8-1 Yes. **8-3** (a) Value cannot normally be changed, (b) value can be changed by turning a shaft. **8-5** A resistor that automatically varies its resistance inversely with temperature. **8-7** Its E-I graph is a straight line. **8-9** 5 W will run cooler. **8-11** It would get very hot and could cause damage. **8-13** The band nearest one end of the resistor. **8-15** Zero. **8-17** 64 cmil. **8-19** Ohms per circular-mil-foot. **8-21** 0.038 Ω. **8-23** 312 Ω.

CHAPTER 9

9-1 Magnetism. **9-3** A piece of metal that has been magnetized. **9-5** Magnetic field. **9-7** 10^8 lines. **9-9** At the ends or poles. **9-11** North to south. **9-13** Attraction. **9-15** Practically all electrons rotate in the same direction. **9-17** The lines travel along the steel and do not go through it to the air beyond. **9-19** About 100,000. **9-21** It becomes magnetized. **9-23** It becomes strongly magnetized when placed in a magnetic field; iron. **9-25** Becomes weakly magnetized when placed in a magnetic field, but is repelled by the magnet; gold, silver, and copper. **9-27** Electrons moving in the same direction around the nucleus of each atom. **9-29** Magnetic lines encircle the wire. **9-31** A compass needle near the wire will have its north pole pointing in the direction of the wire's magnetic lines. **9-33** Increase the current. **9-35** (a) N at bottom, (b) N at bottom, (c) N at top. **9-37** Weaker field. **9-39** Number of turns, and current, (ampere-turns), and core material. **9-41** When current flows in the coil. **9-43** Contacts are made and broken in each position of the armature. **9-45** Pivoted coil, permanent horseshoe magnet, spring, pointer, and calibrated dial scale. **9-47** Coil would rotate counterclockwise, and dial pointer reads off-scale to the left. **9-49** The ac audio signal magnetizes the coil with a reversing polarity. The coil vibrates as it is attracted and repelled by the permanent magnet. The moving coil makes the large cone move. **9-51** How often it reverses. **9-53** A coil mounted on an axle, and placed between the poles of a permanent magnet. **9-55** To the left where the magnetic field is weaker. **9-57** To the right. **9-59** South. **9-61** Orbital motion and spin. **9-63** Prevents magnetic saturation. **9-65** Magnetic fields or lines of force. Symbol is ϕ. **9-67** Magnetic field strength or the number of flux lines per unit area. Symbol is B. **9-69** Permeance ($\mathcal{P}$) is the ability of a material to permit the production of a magnetic field. It is also the reciprocal of reluctance. Permeability (μ) is permeance per unit length and unit area. **9-71** Webers. **9-73** One Mxw = 1 line in the cgs system. **9-75** 50 At. **9-77** 12,000 turns. **9-79** 278.5 At. **9-81** The mmf = $\phi\mathcal{R}$. **9-83** B = ϕ/A. **9-85** Tesla is the mks unit for flux density. It is a weber per m². **9-87** Gauss is the cgs unit for flux density. It is a maxwell per cm². **9-89** (a) 0.015 Wb. (b) 1.5×10^6 lines. **9-91** (a) 12,000 G. (b) 0.6 T. **9-93** H. **9-95** Oersted. **9-97** 795.8 At/m. **9-99** 0.2512 Oe. **9-101** μ. **9-103** $\mathcal{P}$. **9-105** Reluctance is the opposition to the production of a magnetic field. **9-107** Eq. 9-15, $\mathcal{P} = 1/\mathcal{R}$. **9-109** (a) 750 At, (b) 1.27 At/Wb. (c) 75,000 T. **9-111** *B* also doubles. **9-113** (a) 150 At. (b) 300 At/m. (c) 376.8 μT. **9-115** Saturation is being approached, and μ is decreasing. **9-117** Not a straight line. **9-119** μ is smaller but constant. **9-121** The magnetism that resides in the iron in the absence of a magnetizing force. **9-123** *H* must be reversed and increased sufficiently. **9-125** The force required to demagnetize the iron, and reduce flux density to zero. **9-127** Hysteresis loss.

CHAPTER 10

10-1 A coil, placed between the poles of a permanent magnet, is on an axle so that it can rotate. A spiral spring holds it in its normal position. A pointer is attached to the coil. **10-3** The coil would rotate in an opposite direction to its normal movement. **10-5** Galvanometer. **10-7** Add a resistor of the correct value in parallel with the meter coil. **10-9** 9.9 mA or 9900 μA. **10-11** 1 Ω. **10-13** The rotor of this switch is never disconnected from all switch contacts. When changing switch positions, the rotor makes the next contact before breaking the previous one. **10-15** Current (electron drift) should enter the negative end of the meter, and would leave from the positive one. **10-17** Negative lead to point X, and positive lead to R_5. **10-19** The low resistance of the ammeter circuit (coil and *R* shunt) is placed in series with the much larger resistance of the circuit being measured, adding negligible resistance. **10-21** Highest range at first, then reduce as needed. **10-23** 250 mA. **10-25** (a) 10 V, (b) 500 V. **10-27** (a) R_1 = 99 KΩ, (b) R_2 = 999 KΩ, (c) R_3 = 1.999 MΩ, (d) R_4 = 9.999 MΩ, (e) R_5 = 19.999 MΩ. **10-29** 5000 Ω/V. **10-31** (a) 5 μA, (b) 6 Meg Ω. **10-33** Negative to point A, positive to point B. **10-35** Negative to point D, positive to E. **10-37** 1.67 V. **10-39** High as possible. **10-41** (a) 2.98 V. (b) 0.667%. **10-43** (a) Infinity, (b) zero. **10-45** R_T = 900 K. **10-47** (a) 45 μA, (b) 22.5 μA, (c) 15 μA. **10-49** With the test leads shorted together, vary the zero adjustment control until pointer indicates 0 Ω. Then with leads separated, vary the ohms adjustment control until pointer indicates infinity. **10-51** There must be no voltage in the circuit to be measured. **10-53** The capacitor must be completely discharged first so that no voltage exists to damage the meter. **10-55** A large movement of pointer toward zero, and then it moves slowly back toward infinity. After about 10 or 20 sec, the pointer should read at least a few hundred thousand ohms or larger. **10-57** No pointer movement at all. Pointer remains on infinity. **10-59** They are sliding contacts, making contact through the rotating split rings to the rotating armature. **10-61** To keep the armature rotating. At one time the armature magnetic pole is attracted by the field coil pole. An instant later they must repel. **10-63** D'Arsonval uses a permanent magnet; dynamometer uses field coils as electromagnets. **10-65** L_1, L_2, and L_3 are in series with a multiplier resistor, and all are connected across the circuit being measured. **10-67** (a) Read normally, (b) pointer moves off-scale to the left. **10-69** $R_1 = 10\ \Omega$, $R_2 = 0.9\ \Omega$, $R_3 = 0.2\ \Omega$. **10-71** $R_1 = 900\ \Omega$, $R_2 = 90\ \Omega$, $R_3 = 8.08\ \Omega$, $R_4 = 1.92\ \Omega$. **10-73** $R_b R_x = R_a R_c$. **10-75** No. **10-77** $R_c = 21\ \text{K}\Omega$. **10-79** A bank of 10 resistors. **10-81** 9296 Ω. **10-83** 1000's on 10 is 10 KΩ. 100's on 10 is 1 KΩ. 10's on 10 is 100 Ω. 1's on 10 is 10 Ω. Ratio on 10 K, giving a maximum value of 111.1 MΩ. **10-85** $R_3 = 1\ \Omega$. **10-87** (a) 1.25 μA, (b) 0.238 μA.

CHAPTER 11

11-1 No voltage and no current. **11-3** Voltage polarity and current direction are reversed. **11-5** Smaller induced voltage. **11-7** Voltage increases. **11-9** Electron movement would reverse, flowing into the wire. **11-11** Having a voltage induced in it when cut by a magnetic field. **11-13** Magnetic lines build up, cutting the coil wires, inducing a voltage in the coil. **11-15** 1 H. **11-17** The field of one cuts across the wires of a second coil. **11-19** Bring the coils closer together; place them parallel to each other; wind one over the other; wind each coil on the same iron core. **11-21** 600 V. **11-23** The magnetic field is stationary and does not cut across the secondary wires. **11-25** 840 V. **11-27** 141 W. **11-29** Eddy currents, hysteresis, and copper losses. **11-31** (a) Losses due to the reluctance of the iron to being magnetized and demagnetized. (b) Use a low reluctance soft iron. **11-33** 141.4 V. **11-35** The rms or effective value. **11-37** One half of 90 V, or 45 V. **11-39** 56.4 V. **11-41** 169.5 H. **11-43** 1.1 H. **11-45** 250 mV. **11-47** 5.4 H. **11-49** 1 H. **11-51** 0.943. **11-53** 510 V. **11-55** 114 W. **11-57** (a) 40 W, (b) 50 W, (c) 416 mA. **11-59** (a) 60 V, (b) 120 mA, (c) 7.2 W, (d) 8 W, (e) 267 mA. **11-61** 10 KΩ. **11-63** A split ring. **11-65** The commutator split ring changes the ac to dc. **11-67** (a) 5 KHz or 5×10^3 Hz, (b) 660 KHz or 660×10^3 Hz, (c) 105 MHz or 105×10^6 Hz, (d) 150 GHz or 150×10^9 Hz. **11-69** 0.5×10^{-9} sec or 0.5 nsec. **11-71** 0.004×10^9 Hz or 4 MHz. **11-73** 200 MHz. **11-75** 51.3 V. **11-77** (a) 21.6°, (b) 14.72 V. **11-79** 20 V. **11-81** (a) +200 V. (b) 5 volt-peak ac. **11-83** (a) dc, (b) dc.

CHAPTER 12

12-1 0.0005 sec or 500 μsec. **12-3** (a) 6 V, (b) 0, (c) 0. **12-5** (a) 0 V, (b) 20 mA. (c) 6 V. **12-7** 628. KΩ. **12-9** (a) 20 V, (b) 40 V, (c) 44.7 V. **12-11** $Z = 762\ \Omega$. **12-13** (a) 628 Ω, (b) 695 Ω, (c) 0.144 A, (d) 43.2 V, (e) 90.5 V. **12-15** 1. **12-17** (a) 12.5 W, (b) 7.5 W, (c) 0.6. **12-19** (a) 19 Ω, (b) 1.58 A, (c) 45 W. **12-21** (a) 0.6 A, (b) 3 A, (c) 0.9 A, (d) 3.6 A, (e) 3.6 A, (f) 4.5 A, (g) 5.76 A, (h) 31.2 Ω. **12-23** (a) 20 μsec, (b) 60.6 V, (c) 39.4 V, (d) 3.94 A. **12-25** (a) 2.5 μsec, (b) 4.02 μsec. **12-27** (a) 1.5 μsec, (b) 0.132 A. **12-29** $\theta = 30°$. **12-31** $\theta = 76°$. **12-33** $1000 \angle 30°$. **12-35** *Trig. form*: $(50 \cos 30° + j \sin 30°)$, *rect. form*: $43.3 + j\,25$. **12-37** $720 \angle 70°$. **12-39** $75\,000 \angle 65°$. **12-41** $2 \angle 20°$. **12-43** (a) $31.6 \angle 71.5°\ \Omega$, (b) $44.7 \angle 63.5°\ \Omega$, (c) $18.55 \angle 68°\ \Omega$, (d) $2.7 \angle -68°$ A. **12-45** (a) $55.9 \angle 63.5°\ \Omega$, (b) $13.9 \angle 67°\ \Omega$, (c) $3.6 \angle -67°$ A. **12-47** (a) 1.174 W, (b) 1.175 W. **12-49** (a) 0.676 W, (b) 0.675 W. **12-51** 0.2 J. **12-53** $Q = 50$. **12-55** 1.5 Ω. **12-57** 0.05 μsec. **12-59** (a) 173.2 Ω, (b) 100 Ω.

CHAPTER 13

13-1 Two conducting plates with insulation between them. **13-3** Applying a dc voltage to a capacitor causes electrons to move onto one plate, making that plate negative. This repels electrons off the other plate, making that plate positive. **13-5** Area of plates adjacent to each other; spacing between plates; dielectric material. **13-7** When plates are closest together. **13-9** 300 multiplied by 5, or 1500 pF. **13-11** It gives the maximum voltage that can safely be applied to a capacitor without causing the dielectric to become damaged. **13-13** Turn the plates in or out of mesh and vary the spacing between plates **13-15** 410. **13-17** 3000. **13-19** This is correct and the circuit would operate normally. **13-21** Pointer immediately swings toward zero, then slowly moves toward infinity. After about 10 or 20 sec, the pointer reads about 200 kΩ or larger. **13-23** Pointer momentarily moves slightly in the direction of zero, but returns to its infinity reading, practically staying at infinity. **13-25** Pointer immediately moves to and remains at 5 K. **13-27** 500 pF. **13-29** (a) 14,700 pF, (b) 0.0147 μF. **13-31** 4 μF. **13-33** 2×10^{-6} C or 2 μC. **13-35** The same current flows in all parts of a series circuit. **13-37** (a) 4 pF, (b) 40×10^{-12} C, (c) 40×10^{-12} C, (d) 40×10^{-12} C, (e) 40×10^{-12} C, (f) 0.4 V, (g) 1.6 V, (h) 8 V. **13-39** Equal value capacitors would have equal charging currents. **13-41** (a) 300×10^{-6} C, (b) 1020×10^{-6} C. **13-43** (a) 120 μF, (b) 160 μF, (c) 32 μF, (d) 0.01536 C, (e) 0.01536 C, (f) 0.01536 C, (g) 0.01536 C, (h) 256 V, (i) 128 V, (j) 96 V, (K) 0.00896 C, (l) 0.0064 C, (m) 0.00384 C, (n) 0.01152. **13-45** 15 μsec. **13-47** (a) 25.8 V, (b) 4.2 V, (c) 0.21×10^{-3} A. **13-49** (a) 74 V, (b) 74 V, (c) 1.48 mA. **13-51** 1000 V. **13-53** After C has completely charged, I becomes zero. **13-55** Capacitive Reactance (X_C). **13-57** 106 Ω. **13-59** 1.59 μF. **13-61** 291 V. **13-63** 194 V. **13-65** 500 Ω. **13-67** (a) 1060 Ω, (b) 2260 Ω, (c) 22.1 mA, (d) 44.2 V, (e) 23.4 V. **13-69** (a) 1060 Ω, (b)7 10 Ω, (c) 6 pF, (d) 1770 Ω, (e) 1000 Ω, (f) 2030 Ω, (g) 14.8 mA, (h) 8.85 V, (i) 5.9 V, (j) 15.65 V, (k) 10.5 V, (l) 30 V. **13-71** (a) 2 W, (b) 1.2 W. **13-73** 0.3. **13-75** (a) 30 W, (b) 24 W. **13-77** (a) 5 K, (b) 6 K, (c) 3.84 K. **13-79** (a) 5 μF, (b) 3 V, (c) 15 V. **13-81** (a) 24 μF, (b) 120 V, (c) 48 V, (d) 73 V, (e) 72 V, (f) 72 V. **13-83** (a) 12 μF, (b) 60 V, (c) clockwise, (d) 24 V, (e) 36 V, (f) 24 V, (g) 24 V. **13-85** (a) 5 μF, (b) 30 V, (c) clockwise, (d) 25 V, (e) 5 V, (f) 5 V, negative at the lower plate, or reversed polarity from its starting voltage, (g) 5 V, negative at the lower plate, or no polarity reversal. **13-87** (a) 1 msec. (b) 55.03 V, (c) 4.97 V, (d) 0.249 mA. **13-89** 2.49 msec. **13-91** (a) 5 msec, (b) 10.05 V, (c) 10.05 V, (d) 0.402 mA. **13-93** -63.5°. **13-95** (a) $300 - j\,400$, (b) $500 \angle -53^\circ$ Ω, (c) $0.1 \angle 53^\circ$ A, (d) $200 - j\,600$, (e) $633 \angle -71.5^\circ$ Ω, (f) $0.079 \angle 71.5^\circ$ A, (g) $283 \angle -61^\circ$ Ω, (h) $0.177 \angle 61^\circ$ A. **13-97** (a) .019 VA, (b) 0.53, (c) 10.1 mW, (d) 10.1 mW. **13-99** 0.09 J. **13-101** 0.0208 μF. **13-103** (a) 20 mA, (b) 0, (c) 10 mA, (d) 150 mA. **13-105** (a) 0 V, (b) +20 V, (c) +20 V, (d) +20 V, (e) 0 V, (f) −20 V.

CHAPTER 14

14-1 41.3 Ω. **14-3** (a) 82.5 V, (b) 165 VA, (c) 40 W. **14-5** (a) $510 \angle 78.7^\circ$ Ω, (b) $200 \angle -90^\circ$ Ω, (c) $323 \angle -82.9^\circ$ Ω, (d) $0.589 \angle -78.7^\circ$ A, (e) $1.5 \angle 90^\circ$ A, (f) $0.929 \angle 82.85^\circ$ A, (g) 39.9 Ω, (h) 320 Ω X_C.

CHAPTER 15

15-1 (a) 100 Ω, (b) 0.5 μA, (c) 50 μV, (d) 1 mV, (e) 1 mV. **15-3** 3000 μV. **15-5** 20.8 pF. **15-7** (a) Ratio increased 4 times, (b) Q doubles since X_L has doubled, (c) Δf has decreased or narrowed. **15-9** 5.03 MHz. **15-11** (a) Low-pass, (b) see Fig. 15-11*a*. **15-13** (a) High-pass, (b) see Fig. 15-12*a*.

CHAPTER 16

16-1 To heat the cathode. **16-3** When plate is +. **16-5** (a) 20 K, (b) 20 K. **16-7** (a) 150 V, (b) +100 V. **16-9** (a) I plate increases, (b) I plate decreases. **16-11** Prevents plate secondary emission. **16-13** (a) 2 V p-p, (b) 20 V p-p, (c) 10. **16-15** 6 K. **16-17** 3330 μ mho.

CHAPTER 17

17-1 Negative voltage to N material, and + voltage to P material. **17-3** See Fig. 17-7*a*. **17-5** I collector = 17.9 mA. **17-7** (a) −3 V, (b) −1 V, (c) 8.

Index

A

Admittance (Y), 628
Algebra, Appendix I, 687–688
Alternating Current and Voltage
  ac generator, 382–387, 419–420
  frequency, 397, 421–424
  Hertz, 397
  Peak, 397
  R.M.S., 398
  Sine-wave, 397–400, 425–432
  three-finger rule, 383
  wavelength, 424
Ampere, 51
Anti-Resonance, *see* Resonant Frequency
Atom 7–10, 16
Atomic number, 8

B

Batteries
  alkaline cell, 244
  carbon-zinc cell, 244
  Edison, 245
  fuel cell, 245
  internal resistance of, 249–251
  lead-acid cell, 245
  mercury cell, 244
  nickel-cadmium, 245
  nickel-iron, 245
  parallel, 246–247
  photovoltaic, 245
  primary cell, 243–245
  secondary (rechargeable) cells, 245
  series, 246
  silver-oxide cell, 245
  solar cell, 245
  testing, 252
Bonding
  atomic, 20
  covalent, 23–25
  ionic, 20–21
  metallic, 21–23
Bridge Circuits
  resistors, 146–148
  Wheatstone, 354–356

C

Capacitive Reactance (X_C), 530–532
Capacitors
  ac voltage, 528, 532
  color code, Appendis VI, 711–715
  dielectric of, 504–506
  electrolytic, 509–510
  inductors, Resistors, and, in parallel, 619–629
    in series, 616–619
  Ohmmeter check of, 510
  plate area of, 502

plate distance of, 504
resistor and, 532, 547–548, 580–585
series and parallel, 510–515
types of, 506–510
voltage divider, charged and uncharged, 556–571
Complex numbers, 472–477
Compounds, 2, 6
Conductance (G), 38, 628
Conductors, 12–14, 260
table of, 14
Coulomb, 17
Current, 33–37
ampere, 51
parallel resistor, 103–113, 116–119
series resistor, 61–66
see also Ohm's law

D

D'Arsonval movement, *see* Meters
Delta-to-Wye, *see* Networks
Determinants, Appendix I, 690–696
Dielectric, *see* Capacitors
Differentiating circuit, 489–490, 600–602
Diode, *see* Vacuum Tubes

E

Electromagnetism
ampere-turns, 283–284
coil, 280–284
C.R.T. deflection, 289–291
electric motor, 288–289
left-hand generator rules, 279, 281
loudspeaker, 287–288
meter, 286
right-hand motor rule, 291
wire, 278–280
Electron, 7, 16
energy, 26–27
spin, 18
Electron Tubes, *see* Vacuum Tubes
Elements, 2
table of, 2–6
Energy, 26–27, 51
forbidden gap, 26
in capacitor, 588
in coil, 484
kinetic, 26
valence, 27
Equations
simple, Appendix I, 688
simultaneous, Appendix I, 690–696
Exponential Functions
table of, Appendix V, 704–709

F

Faraday's Law, 382
Filaments
series, 79–81
Filters
band-pass, 650–653
band-stop, 650–653
coupling, 668
high-pass, 653–655
low-pass, 653–655
resonant, 650

G

Generators
ac, 382–387, 419–420
dc, 420–421
Ground, 70–74

I

Impedance, (Z), 452, 453–455, 539–545, 549–555
Induced Voltage, 380–381, 405–407
Inductance, 387, 401–404
capacitance, Resistance, and
in parallel, 619–629
in series, 616–619
coefficient of coupling in, 388–389
Henry, 401
internal resistance in, 490–493
mutual, 388, 407–411

self, 387
see also Inductors
Inductive Reactance (XL), 446
resistance and, in parallel, 458–461
in series, 453–455
Inductors
in ac, 445
resistors and, in dc, 441–444, 462–469
in ac parallel, 458–461, 477–480
in ac series, 447–448, 453–458, 469
see also Inductance
Insulators, 13–15
table of, 15
Integrator, 600
Ions, 11, 12

K

Kirchhoff
current law, 98
voltage law, 65, 66

L

L/C Ratio, 640–643
Lenz's Law, 381–382
Loop currents, see Networks

M

Magnetism, 269–316
diamagnetic, 276
domain, 293
ferromagnetic, 275
field, 270–273
flux, 294
flux density, 270, 298–299
Gauss, 270
hysteresis, 310–316
keeper, 277
lines, 270
materials, 273
Maxwell, 270
paramagnetic, 275
permeability, 274, 302
reluctance, 274, 303
saturation, 273
table of terms, 296–297, 305
Weber, 294
Matter, 2
Maxwell's Cyclic Currents, see Networks
Meters, dc
Ammeter conversion, 327–329
Ayrton shunt, 356–360
electrodynamometer, 354–356
loading effect of, 342–345
Ohmmeter, 345–350, 366–372
Ohms-per-Volt, 338–340
voltmeter conversion, 334–337
Wheatstone Bridge, 361–366
Millman's Theorem, see Networks
MKS System, 48–50
table of systems, 49
Molecule, 7
Motor, dc, 351–354
Armature coil, 351
commutator, 351
field coil, 351
split-ring, 351

N

Networks, 161–235
Delta-to-Wye, 222–229
Maxwells cyclic currents, 168–173, 201–204
Millman's Theorem, 217–221
Nodal Analysis, 173–178, 204–217
Norton's Theorem, 166–168, 193–201
superposition, 180–187
Thevenin, 161–166, 188–193
Wye-to-Delta, 229–235
Nodal Analysis, see Networks
Norton's Theorem, see Networks
Nucleus, 7

O

Ohm's Law, 41–48
  charts, Appendix III, 700–701
  current, 41–44
  resistance, 46–48
  voltage, 44–46
Orbit, 7–11

P

Parallel Resistors, *see* Resistors
Pentode, *see* Vacuum Tubes
Phasors and Vectors, 448–451
Phonon, 26
Photon, 26
Potential
  difference of, 32–33
  see also Ohm's Law, voltage
Power, 51–57, 456–458, 586
  apparent, 456, 480–484, 545–546
  factor, 457, 480–484, 546–547, 586
  in resistor circuits, 66–68
  in X_C and R, 545–547
  in X_L and R, 456–458
  True, 456
  VAR, 587–588
Power equations
  chart of, Appendix III, 701
Powers of Ten, Appendix I, 689–690
Prefixed, 38–41
  table of, 39, 699
Proton, 7, 16
Pythagoras Theorem, Appendix I, 696

Q

Quality (Q)
  in coil, 485–487
  of tuned circuit, 635–643
Quantity (Q), 17
  charge on capacitor, 515–521

R

Resistance, 37
  see also Ohm's Law, resistors
Resistors
  capacitors and, 532–539
  capacitors, Inductors, and,
    in parallel, 619–629
    in series, 616–619
  color code of, 258–260, Appendix VI, 711–712
  inductors and, in dc, 441–444, 462–469
    in ac parallel, 458–461, 477–480
    in ac series, 447–448, 453–458, 469
  in Networks, *see* Networks
  in parallel, 96–121
  in parallel defective, 113–116
  in series, 62–66, 68–89
  in series-parallel, 126–155
  open, Series, 81–84
  types of, 254–257
  voltage dividers, 74–78, 148–153
  Wattage rating of, 257
Resonant Circuits
  parallel, 643–649
  series, 631–643
Resonant Frequency, 639–640, 644–649
Ring, 7–11

S

Scientific Notation, Appendix I, 680–690
Semiconductors, 16
  bias, forward, 678
    reverse, 679
  Diode, 679–681
  N material, 675–676
  P material, 675–676
  P–N Junction, 677–678
  transistor, 681–685
Series-Parallel Resistors, *see* Resistors
Series Resistors, *see* Resistors
Shell, 7–11

sub-, 18–19
Square Wave
voltage, in L R, 487
voltage, in R C, 596–604
Superposition, *see* Networks
Susceptance (B), 628

T

Tetrode, *see* Vacuum Tubes
Thevenin's Theorem, *see* Networks
Time Constant
capacitor and resistor, 521–528, 571–580
inductor and resistor, 443–444, 462–469
square waves in, 487–490
Transformers, 389–396, 411–418
currents in, 393, 395–396, 413–415
efficiency in, 412–413
impedance matching, 416–418
power losses in, 394
tuned, 412
turns ratio, 389–392
voltage in, 389–392
Transistors, *see* Semiconductors
Trigonometry
Cosine, Appendix I, 696–698
functions in, 471–477
right triangles, Appendix I, 696–698
Sine, Appendix I, 696–698
table of, Appendix IV, 702–703
tangent, Appendix I, 696–698
Triode, *see* Vacuum Tubes

V

Vacuum Tubes
characteristics of, 670–672
classes of amplifiers, 666–668
Diode, 658–663
Pentode, 669–670
Tetrode, 668
Triode, 662–666
Valence ring, 11
Vectors and Phasors, 448–451
Voltage
dividers no load, 74–78
dividers with load, 148–153
in Parallel resistors, 96–121
in Series resistors, 63–66, 68–70
more than one, 84–89
negative, 70–74
positive, 70–74
see also Ohm's Law, networks
Voltmeters ac
hot-wire, 433
iron-vane, 433
rectifier, 432
thermocouple, 433
Voltmeter dc, *see* Meters

W

Watts, 52–57
Law, 52
see also Power
Weston movement, *see* Meters
Wheatstone Bridge, 361–366
Wires
sizes and resistance, 261–265
temperature coefficient, 265
Work, 51
Wye-to-Delta, *see* Networks